The Coronavirus Pandemic and the Future Volume 1

Virology, Epidemiology, Translational Toxicology and Therapeutics

Edited by

Michael D. Waters
Consultant, Integrated Laboratory Systems (ILS) Inc., USA
Email: mdwaters@centurylink.net

Alok Dhawan
Centre of Biomedical Research, Lucknow, India
Email: alokdhawan@iitr.res.in

Tim Marrs
Edentox Associates, UK
Email: timothymarrs05@aol.com

Diana Anderson
University of Bradford, UK
Email: d.anderson1@bradford.ac.uk

Stafford Warren
Anne Arundel Medical Center, Annapolis, USA
Email: staffwarren@yahoo.com

and

Claude L. Hughes
IQVIA, USA
Email: claude.hughes@IQVIA.com

ROYAL SOCIETY OF CHEMISTRY

Print ISBN: 978-1-83916-306-7
PDF ISBN: 978-1-83916-364-7
EPUB ISBN: 978-1-83916-365-4
Two-volume set print ISBN: 978-1-83916-676-1

A catalogue record for this book is available from the British Library

The Royal Society of Chemistry is a charity, registered in England and Wales, Number 207890, and a company incorporated in England by Royal Charter (Registered No. RC000524), registered office: Burlington House, Piccadilly, London W1J 0BA, UK, Telephone: +44 (0) 20 7437 8656.

Visit our website at www.rsc.org/books

Printed in the United Kingdom by CPI Group (UK) Ltd, Croydon, CR0 4YY, UK

Dedication

To Professor Diana Anderson, with our sincerest thanks and congratulations. During the completion of this project Professor Anderson was elected Member of the British Empire in the 2022 New Year Honours for services to Genetic and Reproductive Toxicology.

The Coronavirus Pandemic and the Future Volume 1: Virology, Epidemiology, Translational Toxicology and Therapeutics
Edited by Michael D. Waters, Alok Dhawan, Tim Marrs, Diana Anderson, Stafford Warren and Claude L. Hughes

Published by the Royal Society of Chemistry, www.rsc.org

Foreword

Plagues: An Inevitable Consequence of Civilization

We humans, particularly in America, perseverate on diseases of individuals: heart disease, breast cancer, prostate cancer, *etc*. All of us have had family members and friends who have died from these afflictions. But these diseases, as devastating as they can be when they involve family and friends, take away only individuals. Infectious diseases – plagues – can take out civilizations. Yet, ironically, plagues are an inevitable consequence of civilization. The word civilization is derived from the Latin *civitas*, meaning city, and it is this characteristic of civilization that marks the process whereby humans came to live close together in significant numbers. Since plagues require enough vulnerable people collected in one location for the germ to spread and cause its attendant devastation, civilization has provided the substrate of vulnerable people needed to nourish a contagion.

The first well-described plague was the Plague of Athens. Although Greek cities in the 5th century BCE did not ordinarily provide the crowding needed for a plague to sustain itself, at the start of the Peloponnesian War the infantry of Sparta was able to force the citizens and allies of the naval power, Athens, into close quarters, resulting in the first of humanity's "great mortalities". Its ravages were essentially confined to Athens and the adjacent other most populous towns. It killed between 80 and 100 000 people, including Pericles, and was described in vivid detail by Thucydides, who had the disease himself but recovered.[1] Recent scholarly work with DNA points to typhoid as the likely cause.[2]

The Coronavirus Pandemic and the Future Volume 1: Virology, Epidemiology, Translational Toxicology and Therapeutics
Edited by Michael D. Waters, Alok Dhawan, Tim Marrs, Diana Anderson, Stafford Warren and Claude L. Hughes

Published by the Royal Society of Chemistry, www.rsc.org

While large enough collections of humans to sustain plagues were not common in the 5th century BCE, for the past 2000 years, civilization has provided many cities with sufficient crowding to sustain many plagues, and the list of these recurrent contagions that have changed history is not a short one. To name two, the Plague of Justinian[3] was a bubonic plague that so devastated the "civilized" resurgent Roman Empire and the "civilized" Persian Empire, that in the next century the relatively unaffected "uncivilized" Arabs were able to dominate that part of the world under the banner of Islam; in the New World discovered in 1492, there were between 50 million and 125 million Native Americans, but the "plagues" of the Old World, particularly measles and smallpox, killed 90% of the Indians, permitting the Europeans to occupy and totally control the Americas.[4] It is extremely unlikely that the eruption of Islam out of the Arabian Peninsula in the 6th century would have been possible without the Justinianic Plague first decimating the Roman and Persian Empires, and it is certain the takeover of the Americas by Europeans in the 15th, 16th, and 17th centuries would not have been possible without European disease "plagues" with their "great mortalities" first paving the way.

Many of man's most virulent sources of infection, including those causing plagues, have as a common denominator that originally they were infections of the animals with which man lived in close contact; many of the animals were domesticated like cattle, but man has lived cheek by jowl with rats that have never been domesticated. However, the infective agents of these animals often mutated enough to make the "jump" from host animal to man. The plagues of the Black Death are caused by *Yersinia pestis*, a disease of rats transmitted by the rats' fleas, but when there is an insufficient supply of rats the fleas go to "plan B", jump to man, and *Yersinia pestis* carried by the fleas then infects man. Tuberculosis made the jump from cattle. Malaria jumped from birds. Pertussis – whooping cough – jumped from pigs and/or dogs.[5]

And then there are the viruses. Measles is probably a mutated variant of rinderpest which infects cattle. Smallpox, a descriptive term that distinguished it from the larger poxes of syphilis, likely jumped from cowpox in cattle. Flu makes its annual mutation to jump from birds, and also perhaps pigs. Ebola likely originated in primates and bats, as did the human immunodeficiency virus. Today's version of this recurrent zoonotic transmission theme is the coronavirus which normally inhabits a number of animals, particularly bats, but has shown ready adaptability in jumping from other animals to man and then demonstrating its plague capability in being transmissible from human to human. This large family of viruses first showed its truly malignant potential with severe acute respiratory syndrome coronavirus (SARS-CoV) in 2002, then with Middle East respiratory syndrome in 2012. The viral structure of the coronavirus makes mutation relatively easy, and this decade's version, SARS-CoV-2, is causing our current coronavirus disease 2019 pandemic.

It is human nature, almost a visceral need, to concoct conspiracy theories and to blame others when plagues hit. Even in Athens there was a conspiracy theory that someone was poisoning the water supply. Conspiracy theories in Christian Western Europe usually accused Jews of poisoning the water, and Jews were also often scapegoated as being the cause of God's displeasure in sending the current plague; and such anti-Semitic conspiracy fervour did, through the centuries, result in the murder, dare we say sacrifice, of thousands of Jews. Today's coronavirus also has the traditional conspiracy theory that God has sent this plague, although at least this time I haven't seen these fundamentalists blaming the Jews. And this time, the usual poisoning-the-water conspiracy theory has been replaced by an up-to-date conspiracy theory that blames fifth-generation wireless transmissions, convincing some that those transmission towers must be burned down. Today's conspiracy theory, trying to blame China for America's being hit by this pandemic has no basis in fact, as genomic evidence gives no credence to this politically driven idea that the virus was hatched in a lab in China.

But surely we can blame someone for our plague problems. The Native Americans would certainly "blame" the Europeans for infecting them with their diseases. But as the world got "smaller", as exploration proceeded, it was the inevitable consequence of living on the same planet. It was going to happen; it was only a matter of when. When a new influenza virus first emerged in Kansas, causing the "Great Influenza" during and after World War I, it spread throughout the world killing millions;[6] and when a novel coronavirus first emerged in China 100 years later, it too spread throughout the world. Trying to throw "blame" on either Kansas and America or Wuhan and China where these mutated viruses emerged only indicates a total lack of understanding of the process. The never-ending mutations of viruses in all of the world's animals means that it is also inevitable that events such as the coronavirus pandemic now being experienced are inevitable. No conspiracy theory is needed to explain that this is the "nature of things", and the only people who should appropriately be "blamed" are those who fail to understand this fact but who fail to make plans prospectively to manage plagues as well as possible.

The exact details of each plague, such as the mortality rate, will change, and how readily human-to-human transmission occurs will determine how bad each infectious event will be. But the interconnectivity of our world today on land, air, and sea also means that John Donne's phrase, "no man is an island entire of itself; every man is...a part of the main"; applies not only to individuals, but to each and every country on our planet. Understanding that we humans are all in this together should focus our attention on this fact; only by working cooperatively with all of the world's countries involved can the damage from these worldwide plagues be minimized.

It is worthwhile to paraphrase the narrator Rieux's thoughts from the end of Albert Camus' Nobel Prize awarded book *The Plague*:[7] in times of pestilence we learn that there are more things to admire in men than to despise.

But we understand that plagues never die or disappear for good; they may lie dormant, but will come again for the bane of men; perhaps this bane may also serve to enlighten men as to what really matters in life, but when that inevitable contagion comes the rats will again rouse up and good people will die in previously happy cities.

Charles T. Rowe, MD
Asheville, NC, USA

References

1. Thucydides, The History of the Peloponnesian War, *The Encyclopedia of Britannica*, published with editorial advice of The University of Chicago, 1952, pp. 399–401.
2. D. Biello, *Scientific American*, 25 January 2006.
3. W. Rosen, *Justinian's Flea: The First Great Plague and the End of the Roman Empire*, Penguin Books, London, England, 2007.
4. C. C. Mann, *1491: New Revelations of the Americas before Columbus*, Knopf, 2005.
5. J. Diamond, *Guns, Germs and Steel: The Fates of Human Societies*, W. W. Norton & Co., New York, 1999, p. 207.
6. J. M. Barry, *The Great Influenza: The Story of the Deadliest Pandemic in History*, Penguin Books, London, England, 2005, p. 92.
7. A. Camus, *The Plague*, Vintage Books, A Division of Random House, New York, 1991, p. 308.

Preface

Our hope and our goal in producing this book is that scientific knowledge will enable us to survive this pandemic and to be better prepared for future viral epidemics and pandemics. These articles are dedicated to those who lost their lives to coronavirus disease 2019 (COVID-19), and to the health-care workers who valiantly tried to save them, as well as to the scientists who worked in the field to develop new test methods, pharmaceuticals, and vaccines.

Our perception is that few Western scientists or non-scientists took much note of either the 2003 outbreak of severe acute respiratory syndrome coronavirus (SARS-CoV) in China, or the 2012 outbreak of Middle East respiratory syndrome coronavirus (MERS-CoV) in Saudi Arabia. Both viruses spread to approximately 30 countries, and each killed between 800 and 900 people, with mortality rates of 10% and 35%, respectively. By contrast, with a lower mortality rate, estimated at 3–5%, SARS-CoV-2 spread around the globe in just 4 months, and in fewer than 8 months killed more than a million people, and we're still counting. Those countries that immediately began lockdown, testing, and contact tracing were able to contain the virus; those that did not continue to experience large numbers of hospitalizations and deaths.

Here are the facts: the first case of SARS emerged in Guangdong Province, China, in November 2002. SARS-CoV was identified in 2003 and spread to 32 countries or regions, primarily in mainland China, but also Hong Kong, Taiwan, Singapore, Vietnam, and Canada. There were 8437 reported cases of SARS and 813 deaths reported to the World Health Organization (WHO). The first case of MERS was reported in 2012 in Saudi Arabia. Since then,

The Coronavirus Pandemic and the Future Volume 1: Virology, Epidemiology, Translational Toxicology and Therapeutics
Edited by Michael D. Waters, Alok Dhawan, Tim Marrs, Diana Anderson, Stafford Warren and Claude L. Hughes

Published by the Royal Society of Chemistry, www.rsc.org

27 countries have reported cases, although 80% of cases were in Saudi Arabia. From September 2012 until the end of November 2019 (the time of writing), the WHO has been notified of 2494 confirmed cases of MERS and 858 deaths.

Transmission of SARS-CoV is person-to-person through unprotected contact, while MERS is transmitted *via* dromedary camels or infected people.[1] Although there are gross similarities in the clinical spectra of SARS, MERS and COVID-19, the new viral disease is far more contagious and dangerous than its predecessors. In part, because SARS and MERS were more deadly than COVID-19, they have all but disappeared. However, the morbidities suffered from COVID-19 infections are more pervasive and more persistent than in the case of SARS or MERS, and people afflicted with moderate to severe cases of COVID-19 are more likely to suffer chronic debilitation.

Like MERS-CoV, the clinical spectrum of SARS-CoV-2 infection ranges from no symptoms (asymptomatic) or mild respiratory symptoms to severe acute respiratory disease and death. A typical presentation of MERS involves fever, cough, and shortness of breath. Pneumonia is a common finding, but is not always present. Gastrointestinal symptoms, including diarrhoea, have also been reported. In both MERS and COVID-19 severe illness can cause respiratory failure that requires mechanical ventilation and life support in an intensive care unit. Both viruses appear to cause more severe disease in older people, people with weakened immune systems, and those with chronic diseases such as renal disease, cancer, chronic lung disease, and diabetes.

Like SARS-CoV, transmission of SARS-CoV-2 is primarily from person to person. However, according to the WHO,[2] SARS-CoV transmission appears to occur mainly during the second week of illness, which corresponds to the peak of virus excretion in respiratory secretions and stool, and when hospitalized patients with severe disease start to deteriorate clinically. Symptoms are readily apparent, including fever, cough, and shortness of breath, which facilitate detection and hospitalization. Many cases of human-to-human transmission of SARS-CoV have occurred in a healthcare setting in the absence of adequate infection control precautions.

Transmission of SARS-CoV-2 is altogether different in that pre-symptomatic or asymptomatic individuals can shed and transmit the virus within days of infection, and well before the onset of symptoms. In fact, viral infection, reproduction, and shedding occur entirely without symptoms in a large percentage of patients who test positive for SARS-CoV-2. This alteration in the pattern of disease transmission may relate to the acquisition of the furin cleavage site in the spike protein and/or increased binding affinity for angiotensin-converting enzyme 2 in the case of SARS-CoV-2 as compared with SARS-CoV.

COVID-19 can progress in severity from mild to moderate to severe, sometimes with a precipitous drop in oxygen saturation in severe cases such that mechanical ventilatory life support is required. This is can be because the virus is equipped to circumvent the immune system such that a cytokine storm, in which cells release massive quantities of small proteins called cytokines, results in extensive tissue damage with dysfunctional coagulation and multi-organ failure. Cytokine storm is a component of the respiratory diseases caused by SARS and MERS as well as by COVID-19, and is particularly detrimental to patients with comorbidities such as hypertension, diabetes, and cardiovascular disease.

The main threat of SARS-CoV-2 is its propensity to be spread by unknowing asymptomatic carriers in exhaled aerosols containing their respiratory droplets. In fact, this attribute presents the greatest challenge to effective public and global health management, since the virus can survive for hours as indoor aerosols and on non-porous surfaces. Social distancing, mask wearing, hand washing, and disinfection are effective tools in slowing the spread of the virus, but gatherings of large numbers of people are to be avoided until community spread is contained.

The initial series of 11 *Chemistry World* reports chronicled the outbreak of the novel coronavirus (SARS-CoV-2), beginning with the first known disease case in China on 1 December 2019; it described what had been learnt and practised during the first 6 months after the declaration of the COVID-19 pandemic on 11 March 2020. While the world awaited a safe and efficacious vaccine, many efforts were undertaken to find suitable pharmaceutical therapies and appropriate patient management interventions.

The second release, consisting of 19 reports, continued to document our progress in fighting COVID-19 during the second 6 months of the pandemic. The two volumes of this work include all of the articles originally published in the first two releases to *Chemistry World* and additional updates and new information that became available, as well as recommendations for the future. As of 15 February 2021, according to the WHO there have been 108 153 741 confirmed cases of COVID-19 and 2 381 295 deaths (an overall 2.20% case fatality ratio). We noted earlier that SARS-CoV-2 had killed more than a million people in just under 8 months. Here we are after four more months, and well over twice as many people are dead. The regional breakdown of confirmed cases is illustrated in Figure 1.

As may be seen in Figure 2, histograms for 15 February 2021, the number of cases in every region was on the decline; however, cumulative death numbers (Figure 3) are staggering (2 381 295 total deaths). Colours on the histogram plots correspond to WHO regions, as indicated on Figure 1.

"No one is protected until everyone is protected", said Dr Tedros Adhanom Ghebreyesus, the Director General of the WHO, early in the pandemic. All WHO regions, except for South-East Asia where the pandemic began, had faced increasing infections and deaths from COVID-19 from 11 March until the end of 2020. The United States led the world in infections and hospitalizations in part because of failure to implement appropriate control measures and take actions to test and trace quickly enough to mitigate the spread of the virus. Early in the pandemic, it became clear that the longer any country waited to do so the greater would be the difficulty, economic

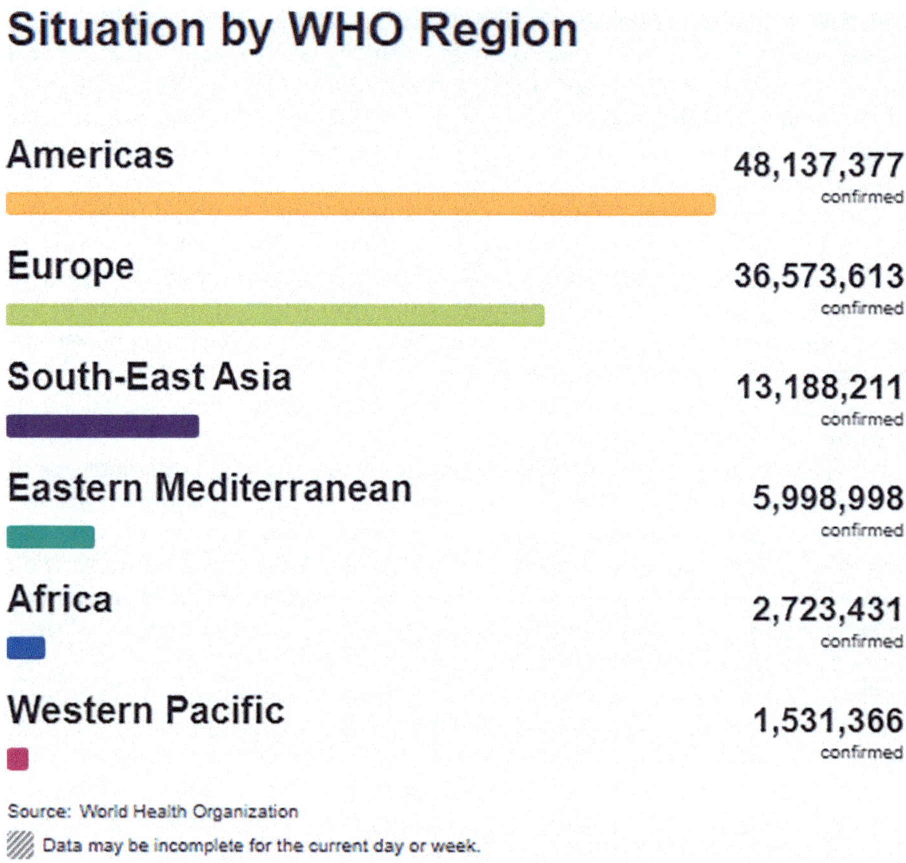

Figure 1 Situation by World Health Organization (WHO) region. WHO Coronavirus (COVID-19) Dashboard https://covid19.who.int/.

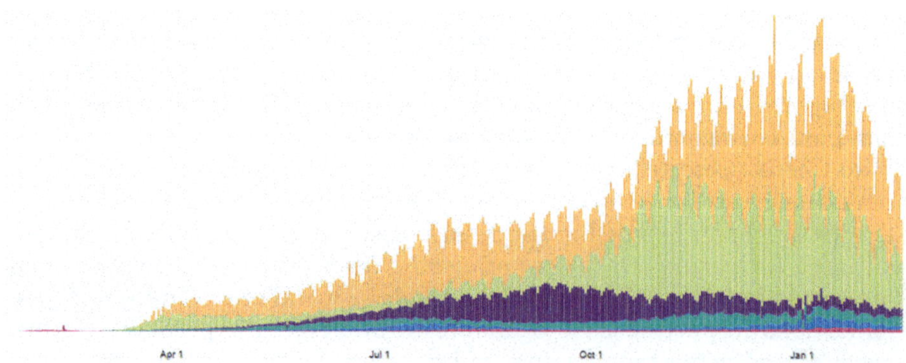

Figure 2 The number of cases in every region. Colours correspond to World Health Organization (WHO) regions, as indicated on Figure 1. WHO Coronavirus (COVID-19) Dashboard https://covid19.who.int/.

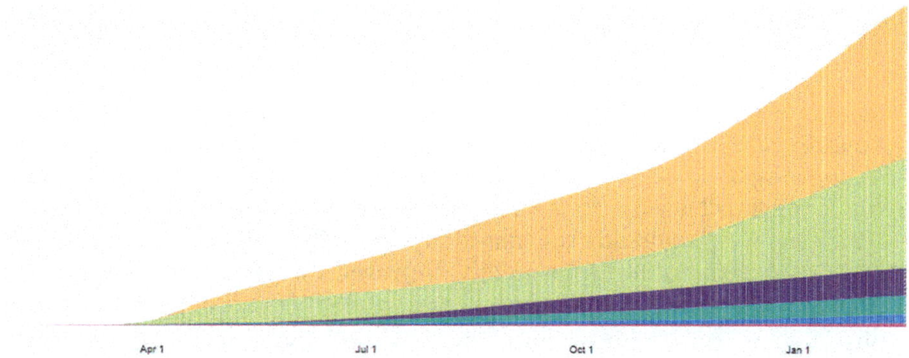

Figure 3 Cumulative death numbers. Colours correspond to World Health Organization (WHO) regions, as indicated on Figure 1. WHO Coronavirus (COVID-19) Dashboard https://covid19.who.int/.

cost, and number of lives lost. We include with this release "country reports" from New Zealand, Turkey, and Sweden, each of which has dealt with the pandemic differently.

After a year of fighting to slow the global spread of SARS-CoV-2, the scientific and public health communities have learned a great deal. As of 15 February 2021, lifesaving vaccines have been delivered to vaccination sites in many parts of the world. While the development of these vaccines has taken place in record time, the virus has continued to win the race to infect as many people as possible. Furthermore, our early concern that viral mutation would be a major problem has been realized. When specific mutations, or sets of mutations, are selected through several rounds of viral replication, new variants can emerge. Since the beginning of 2021, dangerous variants have been detected in the United Kingdom, South Africa, Brazil, and California. These variants are more easily transmitted, such that they can easily become dominant elsewhere.

From the beginning of the pandemic, vaccine development, testing, manufacture, and distribution has been an international priority, and this has worked well in much of the developed world. Indeed, it is encouraging that more than 240 COVID-19 vaccines are in various stages of development globally. In this *Chemistry World* release we feature a detailed report on vaccination and vaccines that describes eight vaccines that are currently being distributed and administered in various countries. Very few individuals that have been vaccinated thus far in the United States and United Kingdom have had serious side-effects. However, vaccination has not happened rapidly enough in most countries to mitigate the spread of SARS-CoV-2. And, unfortunately, poorer countries are not able to afford the costs of mass vaccination on their own.

It should be noted that because of inadequacies in GIS-based sequencing, it has not been possible to monitor the emergence and translocation of SARS-CoV-2 variants in real time. Furthermore, if we do not completely understand the import of mutational changes in the variant viral genome, and how it interacts with the host genome, we cannot know precisely how to mitigate its spread by vaccination. It now appears that vaccines will need to be modified relatively frequently, and "boosters" will be required to maintain immunity against SARS-CoV-2 variants. These topics will be dealt with in the e-book and hardback editions, as our knowledge is currently incomplete.

However, we can already say that international surveillance and testing of SARS-CoV-2 variants could be greatly improved by the sharing of validated, variant-specific assays. Similarly, a central repository of sera and cells from individuals with past infection, with or without past COVID-19 immunization, could facilitate testing of seroneutralization and cellular immunity against newly discovered variants. Importantly, global access, availability, and affordability of COVID-19 vaccines that work is necessary to ensure that no country or population is left behind.

There is significant concern whether vaccines will be available timely, and whether they alone will be adequate to save endangered lives around the world. Vaccines induce a natural immune response by exposing the body to a weakened, dead, or replica virus. Another class of therapeutics, monoclonal antibody preparations, can provide immediate protection when infused directly into the body. With increasing production and availability, these preparations could be administered at infusion centres to high-risk patients in the early stages of COVID-19 infection.

Pharmaceuticals can offer an addition line of defence against the virus. Again, there are economic disparities in the ability of countries, and certain populations within countries, to afford newly developed drugs or costly monoclonal antibodies. For this reason, and because of the protracted time course necessary for development and testing of new drugs and other therapies, repurposed drugs have been investigated by scientists internationally to provide ready treatments for those who become infected with SARS-CoV-2.

At the beginning of the pandemic there were few drugs specific to the coronavirus in the pipeline. Drug repurposing efforts have not been well funded, but many candidates have been found and pre-clinical investigations as well as clinical trials have been initiated. Some candidate repurposed drugs that have proven records of safety and indications of efficacy have been widely used in countries where there is a critical need for antiviral agents. Several reports in this release bear on the issue, including detailed coverage of candidate pharmaceuticals in *in silico* screening and animal-alternative model systems.

Although several repurposed drugs are promising and available globally, none should be used without a medical prescription, nor at dosages above those recommended. In the meantime, public health policies regarding prevention, biosafety measures, large-scale testing, and adequate epidemiological surveillance must be followed assiduously. The bottom line is that all available approved vaccines, pharmaceutical interventions, public health methods, clinical, and research approaches are needed to treat COVID-19 and to slow the global spread of SARS-CoV-2. Every effort should be made to do what is necessary to save human lives.

Michael D. Waters, Alok Dhawan, Diana Anderson, Timothy Marrs, Stafford Warren, and Claude L. Hughes.

References

1. World Health Organization (WHO), 2019.
2. World Health Organization (WHO), 2020.

Contents

Volume 1 The Coronavirus Pandemic and the Future: Virology, Epidemiology, Translational Toxicology and Therapeutics

Chapter 1 What We Know About the Life-threatening Novel Human Coronavirus 1
Claude Hughes, Arul Narayanaswamy, Balachandar Vellingiri, Dhivya Venkatesan, Stafford Warren and Michael Waters

1.1 Introduction 1
1.2 The Chronology of the Outbreak 2
1.3 Initial Transmission 8
1.3.1 Worldwide Spread of the Virus 10
1.4 SARS-CoV-2 and Its Etiology 11
1.5 Classification and Grouping 13
1.5.1 Alpha-CoVs 14
1.5.2 Beta-CoVs 14
1.5.3 Gamma-CoVs 15
1.5.4 Delta-CoVs 15
1.6 Epidemiological History of Human Coronaviruses 15
1.6.1 Severe Acute Respiratory Syndrome 16
1.6.2 Middle East Respiratory Syndrome 17
1.6.3 Detailed Documentation of SARS-CoV-2 and COVID-19 17
1.7 Introductory Virology and Pathophysiology of SARS-CoV-2 20

The Coronavirus Pandemic and the Future Volume 1: Virology, Epidemiology, Translational Toxicology and Therapeutics
Edited by Michael D. Waters, Alok Dhawan, Tim Marrs, Diana Anderson, Stafford Warren and Claude L. Hughes

Published by the Royal Society of Chemistry, www.rsc.org

1.7.1 Viral Genome 20
1.7.2 Interaction of the Spike and Receptor for Entry 20
1.7.3 Genomic Differences 21
1.7.4 Pathophysiology 22
1.8 Executive Summary and Conclusions 25
List of Abbreviations 26
Acknowledgements 26
References 26

Chapter 2 Severe Acute Respiratory Syndrome Coronavirus-2 (SARS-CoV-2): Structure, Lifecycle, and Replication Machinery 36
Arghadip Samaddar and Vijaya Lakshmi Nag

2.1 Introduction 36
2.2 Classification of Coronaviruses 37
2.3 Genomic Structure of SARS-CoV-2 38
2.4 Life Cycle of SARS-CoV-2 41
2.4.1 Attachment and Entry 41
2.4.2 Replicase Protein Expression 42
2.4.3 Replication and Transcription 42
2.4.4 Assembly and Release 42
2.5 SARS-CoV-2 RNA Replication Machinery 43
References 45

Chapter 3 Accumulated Epidemiological Lessons and China's COVID-19 Response 50
Chao Xie and Liangmin Gao

3.1 Introduction 50
3.2 Accumulated Lessons from China's Response to SARS in 2003 51
3.3 Epidemiologicial Perspective on China's COVID-19 Response 53
3.4 China's Control and Prevention Measures in COVID-19 54
3.4.1 Mask-wearing and its Comforting Effect 54
3.4.2 Epidemiological Source-tracking and Slow-down of Community Transmission 55
3.5 Comparing China's Response Against SARS and COVID-19 56
3.6 Concluding Remarks 58
3.7 Executive Summary 59
References 59

Chapter 4 Clinical Presentation, Pathophysiology and Histopathology **63**
Claude Hughes, Iyer Mahalaxmi, Nachimuthu Senthil Kumar, Suresh Selvapuram Sudalaimuthu Raja, Raviminickam Thangarasu, Stafford Warren and Michael Waters

4.1 Introduction 63
4.2 The Distinct Clinical Trajectory and Disease Spectrum of COVID-19 65
4.3 A Novel Application of Smartphone Technology 70
4.4 COVID-19 and Comorbidities in Severe Cases 71
4.5 ARDS and its Sequelae in COVID-19 73
4.5.1 ARDS in COVID-19 73
4.5.2 Sepsis 73
4.5.3 Septic Shock 74
4.6 Pathophysiology 74
4.6.1 Mechanism of SARS-Cov-2 Invasion into Host Cells 78
4.7 Histopathology and Disease Manifestations: Introduction 83
4.7.1 Histopathology 83
4.7.2 Pulmonary Manifestations 83
4.7.3 Cardiovascular Manifestations 85
4.7.4 Gastrointestinal Manifestations 86
4.7.5 Renal Manifestations 86
4.7.6 Neurological Manifestations 87
4.7.7 Musculocutaneous Manifestations 88
4.7.8 Hematological Manifestations 88
4.8 Executive Summary 89
Acknowledgements 89
References 90

Chapter 5 Evaluation of the Disease, Sample Collection and Diagnostics **99**
Claude Hughes, Kaavya Jayaramayya, Balachandar Vellingiri, Stafford Warren and Michael Waters

5.1 Introduction 99
5.1.1 Symptoms of COVID-19 101
5.1.2 Detection of Disease Based on Symptoms 102
5.1.3 Evaluation Criteria for COVID-19 Based on ICMR and WHO Guidelines 103

5.2 Sample Collection 107
5.2.1 Overview of Sample Collection Guidelines from the WHO 107
5.3 Recent Diagnostics 109
5.3.1 Molecular Diagnostics 111
5.3.2 Laboratory Examinations (Modified from Cascella *et al.*) 112
5.3.3 Imaging 112
5.3.4 Blood Parameter Levels (Laboratory Diagnostics) 115
5.3.5 Genetic Susceptibility to COVID-19 117
5.3.6 Importance of Real-time RT-PCR in the Detection of COVID-19 119
5.3.7 Serological Assays 120
5.3.8 Enzyme-linked Immunosorbent Assay 120
5.3.9 Magnetic Chemiluminescence Enzyme Immunoassay 121
5.3.10 Colloidal Gold Immunochromatographic Assay 121
5.3.11 Lateral Flow Immunoassay 122
5.3.12 Rapid Antigen Test 122
5.3.13 Neutralization Antibody Assay 122
5.3.14 Other Useful Diagnostics for COVID-19 123
5.3.15 False Negatives and Positives 123
5.4 Advantages and Disadvantages of Different Diagnostic Approaches 124
5.4.1 Detection Based on Patient History 124
5.4.2 Detection of Disease Based on Symptoms 125
5.4.3 Scanning Techniques 125
5.4.4 Laboratory Serum Diagnostics (*e.g.* WBC, Inflammatory Markers) 127
5.4.5 Molecular Diagnosis 127
5.4.6 CRISPR-based Diagnosis 128
5.4.7 Serological Assays 128
5.5 Executive Summary 130
Acknowledgements 131
References 131

Chapter 6 Therapeutic Options Initially Available for COVID-19 Patients and Initial Clinical Trials 136
Ssang-Goo Cho, Claude Hughes, Iyer Mahalaxmi, Mohana Devi Subramaniam, Balachandar Vellingiri and Stafford Warren

6.1 Introduction to Therapeutic Options Initially Available for COVID-19 Patients 136
6.1.1 General Guidelines for the Treatments Provided to COVID-19 Patients 138

6.1.2 Common Treatment Regimens Provided to COVID-19 Patients 138
6.1.3 Medicinal Plants as Therapy for COVID-19 144
6.1.4 Convalescent Plasma Therapy 146
6.1.5 Mesenchymal Stem Cell-Derived Exosomes as Therapy for COVID-19 146
6.1.6 Other Treatment Options 147
6.2 Introduction to Ongoing Clinical Trials 148
6.2.1 Ongoing Trials on Various Antiviral Agents 148
6.2.2 Immunosuppressants/ Immunomodulators 148
6.2.3 Current Trials on MSCs for COVID-19 149
6.3 Executive Summary 159
Acknowledgements 159
References 159

Chapter 7 Clinical Epidemiology of Coronavirus Disease 2019: Infectivity, Clinical Spectrum and Presentation, and Population Distribution 165
Fengyu Zhang and Claude Hughes

7.1 Introduction 165
7.2 Infection and Tests 168
7.2.1 Viral Tests 169
7.2.2 Population-based Infection Rate 170
7.2.3 Antibody Tests 172
7.3 Clinical Spectrum of COVID-19 173
7.3.1 Classification by Clinical Type 173
7.3.2 Asymptomatic Cases 174
7.3.3 Severe Cases and Comorbidity 175
7.3.4 Clinical and Immunological Response 176
7.4 Regional Variation in Clinical Presentation 177
7.4.1 Clinical Manifestation 177
7.4.2 Chemosensory Dysfunction 178
7.4.3 Common Symptoms 179
7.5 Population Distribution 183
7.5.1 Pregnancy and Pregnant Women 183
7.5.2 Infants and Pediatric Patients 184
7.5.3 Asymptomatic Cases and Clinical Severity in Children 186
7.5.4 Older Adults 187
7.5.5 Sex Differences 187
7.5.6 Occupation 188
7.6 Case-fatality Rate 190
7.7 Executive Summary and Conclusions 193
References 195

Chapter 8 SARS-CoV-2 Genomics and Host Cellular Susceptibility Factors of COVID-19 **203**
Fengyu Zhang and Michael D. Waters

8.1 Introduction 203
8.2 Coronavirus and Human Diseases 204
8.2.1 Classification of Coronaviruses 204
8.2.2 Human Coronaviruses and Diseases 206
8.2.3 The Natural Origin of HCoVs 206
8.3 Genomic Structure and Genetics 207
8.3.1 Viral Genomic Structure 207
8.3.2 The Replicase Locus and Proteins 208
8.3.3 Genomic Mutations 214
8.4 Viral Replication 215
8.4.1 Attachment and Cell Entry 215
8.4.2 Translation and RNA Replication 216
8.4.3 Assembly and Release of Virion 217
8.5 Host Cellular Factors 217
8.5.1 ACE2 218
8.5.2 TMPRSS2 220
8.5.3 Furin 221
8.5.4 DPP4 221
8.5.5 CD147 222
8.5.6 GRP78 223
8.5.7 Cathepsins 223
8.5.8 Ly6E 223
8.5.9 L-SIGN and DC-SIGN 224
8.5.10 Sialic Acid 224
8.5.11 Plasmin and Other Proteases 225
8.6 Executive Summary 226
References 227

Chapter 9 Infection and Pathogenesis of SARS-CoV-2: an Immunological Perspective **242**
Sneha Mohanty and Prabhanshu Tripathi

9.1 Introduction 242
9.2 SARS-CoV-2 Virology 244
9.2.1 Structure 244
9.2.2 Genome 244
9.2.3 Proteome 244
9.2.4 Mutation 245
9.2.5 Transmission 245
9.3 Pathogenesis 246
9.3.1 Attachment to the Host Cell 246
9.3.2 Penetration of SARS-CoV-2 Inside the Host Cell 246

9.3.3 Assembly of the RTC 247
9.3.4 Genome Replication, Transcription, and Translation 247
9.3.5 Virion Assembly and Release 248
9.3.6 Incubation Period 248
9.4 Disease Progression 248
9.5 Immunology 249
9.5.1 Innate Immune Response 249
9.5.2 Adaptive Immune Response 257
9.5.3 Cytokine Storm 262
9.6 Factors Affecting COVID-19 Severity 264
9.6.1 Human Genetic Variation 264
9.6.2 Blood Group 266
9.6.3 Gender 266
9.6.4 Age 267
9.6.5 Comorbidities 268
9.6.6 Diet and Microbiota 270
9.6.7 BCG Vaccine 271
9.7 Therapeutics 272
9.7.1 Chloroquine and Hydroxychloroquine 272
9.7.2 Corticosteroids 274
9.7.3 Cytokine Inhibition 275
9.7.4 Convalescent Plasma Therapy 278
9.7.5 Cell Therapy 278
9.8 Conclusion 280
9.9 Executive Summary 280
References 281

Chapter 10 Neurological Manifestations in COVID-19 **311**
A. Samhita Panda, B. Saksham Jain, C. Sanjiv Sharma and D. Sarbesh Tiwari

10.1 Introduction 311
10.2 Pathophysiology of Neurotropism of SARS-CoV-2 312
10.2.1 Risk Factors Potentiating Neurotropism 313
10.2.2 Mechanisms for Neurological Complications 313
10.2.3 Pathogenesis of Neuronal Autoimmunity in COVID-19 314
10.3 Neurological Manifestations of COVID-19 315
10.4 CNS Involvement in COVID-19 316
10.4.1 Vascular Manifestations 316
10.4.2 Seizures 318
10.4.3 Encephalitis 319
10.4.4 Encephalopathy 320
10.4.5 CNS Demyelination 320

10.4.6 Myelitis 321
10.4.7 Movement Disorders 321
10.5 PNS Involvement in COVID-19 322
10.5.1 Loss of Smell and Taste Sensation 322
10.5.2 Other Cranial Nerves 322
10.5.3 Guillain-Barré Syndrome and its Variants 322
10.5.4 Neuromuscular Junction Disorders 323
10.5.5 Myalgia and Myositis 323
10.6 Neuroradiological Findings in COVID-19 324
10.6.1 Acute Ischemic Stroke 324
10.6.2 Acute Hemorrhagic Stroke 324
10.6.3 Acute Inflammatory/Infectious CNS Syndromes 326
10.6.4 Guillain-Barré Syndrome 326
10.6.5 Olfactory Bulb Involvement 326
10.6.6 Global Hypoxia and Diffuse Post-hypoxic Leukoencephalopathy 328
10.7 Treatment of Neurological Aspects of COVID-19 328
10.8 Outcomes 329
10.9 Future Perspectives 329
List of Abbreviations 330
Acknowledgements 330
References 330

Chapter 11 COVID-19 and Coagulopathy **336**
Shruti Vaswani, Deepak Kumar, Gopal Krishana Bohra and Abhishek Purohit

11.1 Introduction 336
11.1.1 SARS-CoV-2 337
11.2 Pathogenesis 337
11.2.1 Virchow's Triad and COVID-19 337
11.2.2 Disseminated Intravascular Coagulation and COVID-19 339
11.2.3 Viraemia and Disease Severity 340
11.2.4 Neutrophilic Extracellular Traps 340
11.2.5 Role of Cytokines in COVID-19 341
11.3 Clinical Autopsy Findings 341
11.4 Clinical Features 343
11.4.1 Features of Venous Thromboembolism 343
11.4.2 Features of Arterial Thromboembolism 344
11.5 Investigations 344
11.6 Management 345
11.7 Conclusion 347
References 348

Chapter 12 COVID-19 Endocrinopathies: Implications for Care **353**
Maya Gopalakrishnan, Bharat Kumar, Prakrati Yadav and M. K. Garg

12.1 Introduction 353
12.2 Hypothalamo-pituitary Axis 354
12.3 Adrenals 355
12.4 Pancreas 356
12.5 Thyroid Gland 357
12.6 Gonads 359
12.7 Conclusion 361
References 361

Chapter 13 Potential Risks of COVID-19 in Reproductive Health **368**
Khaled Habas, Claude Hughes and Diana Anderson

Highlights 368
13.1 Introduction 369
13.2 COVID-19 and the Female Reproductive System 369
13.3 Early Pregnancy 370
13.4 Clinical Presentation of COVID-19 Infection During Pregnancy 372
13.5 COVID-19 and the Male Reproductive System 374
13.6 Conclusion 375
References 375

Chapter 14 Clinical Autopsy in COVID-19 **380**
Deepika Gupta, Vikarn Vishwajeet, Aasma Nalwa, Abhishek Purohit and Tanuj Kanchan

14.1 Introduction 380
14.1.1 Viral Transmission 381
14.1.2 Pathogenesis 381
14.1.3 Clinical Features 382
14.2 Importance of Autopsy 382
14.3 Autopsy Protocols in Routine Practice 382
14.4 Autopsy Protocols in COVID-19 383
14.4.1 Handling of the Dead Body in the Ward 383
14.4.2 Transportation to the Mortuary 383
14.4.3 Mortuary Handling 384
14.4.4 Prosector Precautions 384
14.4.5 Laboratory Biosafety Guidelines 385
14.4.6 Requirement of Dedicated Autopsy Facility, Including Engineering and Environmental Controls 386

14.5 Different Types of Autopsy 386
14.5.1 Complete Autopsy 386
14.5.2 Partial Autopsy 388
14.5.3 Minimally Invasive Autopsy 388
14.6 Sample Acquisition 389
14.6.1 Histopathology Examination 389
14.6.2 Molecular Testing (RT-PCR) 389
14.6.3 For Electron Microscopy 391
14.7 Pathological Examination 391
14.7.1 Gross Examination of Organs 391
14.7.2 Light Microscopic Examination of Organs 393
14.7.3 Immunohistochemistry 399
14.7.4 *In Situ* Hybridization 399
14.7.5 Electron Microscopy 400
14.8 Molecular Testing in Autopsies 401
References 402

Chapter 15 COVID-19 Pharmacotherapy – Present and Future 412
Deepak Kumar, Ravi Mangalia, Mahadev Meena, Varatharajan Sakthivadivel, Gopal Krishna Bohra and Stafford Warren

15.1 Introduction 412
15.2 Antivirals 413
15.2.1 Remdesivir 413
15.2.2 Favipiravir 416
15.2.3 Molnupiravir (MK-4482) 417
15.2.4 Umifenovir 418
15.2.5 Lopinavir/Ritonavir 418
15.3 Drugs with Antiviral Activity 418
15.3.1 Ivermectin 418
15.3.2 Nitazoxanide 423
15.3.3 Merimepodib 423
15.3.4 Niclosamide 423
15.3.5 Rintatolimod 424
15.3.6 β-D-N4-Hydroxycytidine 424
15.3.7 Bemcentinib 424
15.3.8 Plitidepsin 424
15.3.9 Emetine Hydrochloride 425
15.3.10 Trabedersen 425
15.3.11 Stannous Protoporphyrin 425
15.3.12 Antroquinonol 425
15.3.13 Apilimod Dimesylate 425

15.3.14 VIR-2703 426
15.3.15 Colchicine 426
15.4 Immunomodulators 426
15.4.1 Corticosteroids 427
15.4.2 Interleukin Inhibitors 428
15.4.3 NAK and JAK Inhibitors 430
15.4.4 Convalescent Plasma 431
15.5 Monoclonal Antibodies 434
15.6 Ancillary Nutritional Therapies 435
15.6.1 Vitamin D 435
15.6.2 Vitamin C 437
15.6.3 Zinc 438
15.7 Anticoagulants 439
15.8 Executive Summary 440
References 443

Chapter 16 Knowledge from Human Relevant Cell, Tissue and Mathematics-based Methods as Key Tools for Understanding COVID-19 Dynamics, Kinetics, Symptoms, Risk Factors and Non-conventional Treatments **455**
Sandra Coecke, Amalia Munoz, Vito D'Alessandro, Francesca De Bernardi, Pietro Romeo, Felipe Torres, Georgina Harris and Surat Parvatam

16.1 Cell, Tissue and Mathematical-based Methods Used for COVID-19 Research 456
16.2 Human Body Barriers and Cell and Mathematics-based Methods as Tools for Understanding COVID-19 Infection Kinetics 462
16.2.1 Respiratory Barrier 464
16.2.2 Intestinal Barrier 464
16.2.3 Anterior Segment Barrier 465
16.2.4 Blood–Retinal Barrier 465
16.2.5 Skin Barrier 466
16.2.6 Blood–Brain Barrier 466
16.2.7 Placental Barrier 467
16.2.8 Blood–Testis Barrier 467
16.3 Tissue-specific Distribution Patterns 468
16.4 Human Pulmonary and Extra-pulmonary Target Organ Systems Cell and Mathematics-based Methods as Tools for Understanding COVID-19 Infection Dynamics 469

16.4.1 Potential SARS-CoV-2 Target Organ Systems 471
16.4.2 SARS-CoV-2 Target Organ Systems Relevant Cell Types, Molecular and Cellular Events, Methods and Symptoms 483
16.4.3 SARS-CoV-2 Cellular Mechanisms 493
16.4.4 Immunity Dynamics Following SARS-CoV-2 Infection 494
16.5 Risk Factors 495
16.5.1 Genetic Factors 496
16.5.2 Cellular Factors 497
16.5.3 Tissue/Organ Level 498
16.5.4 Organism Level 498
16.6 Vulnerabilities 498
16.6.1 Malnutrition 500
16.6.2 Immunocompromised Conditions 500
16.6.3 Presence of Comorbidities 500
16.7 Non-conventional or Supportive Treatments or Approaches Based on Underlying Cellular Mechanisms 501
16.7.1 *In Silico* Strategies 503
16.7.2 Fit-for-purpose Cellular Assays 504
16.7.3 Cell Lines 505
16.7.4 Mesenchymal Stem Cells 505
16.7.5 3D Tissue/Organoids 506
16.7.6 Organ-on-a-chip 506
16.7.7 Diamagnetism as an Alternative or Integrating Cellular Therapy for COVID-19 507
16.8 Concluding Remarks 510
Acknowledgements 511
References 511

Subject Index **526**

Volume 2 The Coronavirus Pandemic and the Future: Virology, Epidemiology, Translational Toxicology and Therapeutics

Chapter 1 ***In Silico* Approaches for Drug Repurposing for SARS-CoV-2 Infection** 1
Sudhir A. Kulkarni and Kundan Ingale

1.1 Introduction 1
1.1.1 What is Drug Repurposing? 2
1.1.2 Why Drug Repurposing? 2
1.1.3 Why Drug Repurposing for COVID-19? 3
1.1.4 Why *In Silico* Approaches for Drug Repurposing? 3
1.2 Understanding SARS-CoV-2 from an *In Silico* Perspective 5
1.2.1 Structural Proteins of SARS-CoV-2 7
1.2.2 Non-structural and Accessory Proteins of SARS-CoV-2 8
1.2.3 Structure of SARS-CoV-2 Proteins 10
1.3 Structure-based Approaches for Drug Repurposing 18
1.3.1 Docking Studies in the Main Protease (Mpro/3CLpro) 19
1.3.2 Docking Studies in RNA-dependent RNA Polymerase 30
1.3.3 Docking Studies in Papain-like Protease 32
1.3.4 Docking Studies in the Nucleocapsid Protein (N-protein) 33
1.3.5 Docking Studies in the Spike Glycoprotein (S-protein) 34
1.3.6 Docking Studies in NSP1 35
1.3.7 Docking Studies in NSP13/Helicase 36
1.3.8 Docking Studies in NSP15/Endonucleases 36
1.3.9 Docking Studies in NSP16 36
1.3.10 Docking in Main Protease and Spike Glycoprotein 37
1.3.11 Docking in Multiple Structural Proteins 39
1.3.12 Docking in Proteases 40
1.3.13 Docking in Multiple Targets 41
1.3.14 Discussion and Consensus Screening Protocol from the Reviewed Literature 49
1.4 Ligand-based Approaches for Drug Repurposing 51
1.4.1 QSAR-based Approaches 51
1.4.2 Pharmacophore-based Approaches 52

1.5 Other Approaches for Drug Repurposing 52
1.5.1 Machine Learning-based Methods 52
1.5.2 Pharmacology-based Network Analysis Methods 54
1.6 Understanding Human Targets in COVID-19 From an *In Silico* Perspective 55
1.6.1 Host Proteins Involved in the SARS-CoV-2 Life Cycle 55
1.6.2 Host Response to SARS-CoV-2 Infection 56
1.6.3 Structural Information of Human Proteins in COVID-19 56
1.7 Structure-based Approaches for Drug Repurposing Using Human Proteins 61
1.7.1 Docking Studies in Angiotensin Converting Enzyme-2 61
1.7.2 Docking Studies in Transmembrane Protease, Serine 2 (TMPRSS2) 62
1.7.3 Docking Studies in Glucose-Regulated Protein 78 (GRP78) 64
1.7.4 Docking Studies in Furin 64
1.7.5 Docking Studies in ARDS Targets 64
1.8 Summary of Hits from Reviewed Literature 64
1.9 Concluding Remarks 68
1.10 Executive Summary 69
Author Contributions 69
Acknowledgements 69
References 69

Chapter 2 Vaccination and Vaccines for COVID-19 81
Ashoka Jahnavi Prasad

2.1 Introduction 81
2.2 Vaccination in the Context of a Pandemic Outbreak 85
2.3 COVID-19 89
2.3.1 Specific Strategies for COVID-19 90
2.4 Mental Health Aspects of Immunization and Vaccination 92
2.5 Vaccines for COVID-19 98
2.5.1 SARS-CoV-2 98
2.5.2 Immune Response to SARS-CoV-2 and Previous Coronavirus Infections 100
2.5.3 Platforms for COVID-19 Vaccine Development 100
2.6 Executive Summary 109
References 110

Chapter 3 Understanding the Emergence of SARS-CoV-2 Viral Variants From a Genomic Perspective **116**
Vineeta Sharma, Samiya Baby, Sukhveer Singh and Vikas Srivastava

3.1 Introduction 116
3.2 A History of Pandemics 117
3.3 Origin and Classification of SARS-CoV-2 119
3.4 Genomic Architecture of SARS-CoV-2 120
3.4.1 Overview 120
3.4.2 Structural Proteins 120
3.4.3 Non-structural Proteins 121
3.5 The Emergence of SARS-CoV-2 Variants 122
3.6 Antigenic Variations and Immune Escape Mutations 127
3.7 Leading Vaccines Across the Globe 128
3.8 Conclusion 130
References 130

Chapter 4 Susceptibility and Spread of SARS-CoV-2 in Animals **137**
Nicola Pusterla and Emir Hodzic

4.1 Introduction 137
4.2 Wild Animals 138
4.3 Laboratory Animals 140
4.4 Companion Animals 141
4.5 Farmed Animals 142
4.6 Conclusions 143
References 144

Chapter 5 Profiling Some Plant-based Immunomodulatory Bioactive Compounds for COVID-19 Prophylaxis and Treatment Based on Indian Traditional Medicine **151**
Prakrity Singh, Shraddha Pandit and Ramakrishnan Parthasarathi

5.1 Introduction 151
5.2 Prevention of SARS-CoV-19 Infection: Stage 1 164
5.3 Incubation to Non-severe Symptomatic Phase – Stage 2 166
5.4 Severe Respiratory Symptomatic Phase with Hyperinflammation – Stage 3 167
5.5 Multi-organ Response to COVID-19 Infection – Stage 4 169

5.6 Summary and Outlook 170
5.7 Executive Summary 170
List of Abbreviations 171
Authors' Contributions 171
Declaration of Competing Interest 171
Acknowledgements 171
References 171

Chapter 6 The Potential Therapeutic Effects of Natural Products, Herbs, and Mushrooms Against COVID-19 **177**
Mojgan Najafzadeh, Fanila Shahzad and Diana Anderson

6.1 Origins and Pathogenesis of COVID-19 177
6.2 The Immune Profile in COVID-19 Infection 180
6.3 Natural Marine Compounds as Potent Inhibitors Against SARS-CoV-2 182
6.4 Potential Use of Mushrooms and Herbs Against COVID-19 Infection 183
6.5 Conclusion 189
References 190

Chapter 7 Application of Chinese Herbal Medicine in COVID-19 **197**
Yehong Tian, Xiaowei Qiu, Xin Jiang, Jinchang Huang and Fengyu Zhang

7.1 Introduction 197
7.2 Representative Prescriptions 200
7.2.1 *Qingfei Paidu* Decoction (QFPDD, 清肺排毒汤) 200
7.2.2 *Huashi Baidu* Formula (HSBDF, 化湿败毒方) 202
7.2.3 *XuanFei Baidu* Formula (XFBDF, 宣肺败毒方) 203
7.3 Finished Patent Medicines 205
7.3.1 *Lianhua Qingwen* Capsules (LH-C, 莲花清瘟胶囊) 205
7.3.2 *Jinhua Qingan* Granules (JH-G, 金花清感颗粒) 205
7.3.3 *Xuebijing* Injection (XBJ, 血必净注射液) 206
7.4 Other Chinese Herbs 207
7.4.1 *Ephedra Herba* (Ma Huang, 麻黄) 207
7.4.2 *Honeysuckle* (Jin Yin Hua, 金银花) 208
7.4.3 *Scutellariae* (Huang Qin, 黄芩) 209
7.4.4 *Glycyrrhizae Radix* (Gan Cao, 甘草) 209
7.4.5 *Armeniacae Semen* (Ku Xing Ren, 苦杏仁) 210
7.4.6 *Sophorae flavescentis Radix* (Ku Shen, 苦参) 210
7.4.7 *Curcuma longa* (Jiang Huang, 姜黄) 211
7.5 Conclusion 211
List of Abbreviations 212
References 213

Chapter 8 COVID-19 Interventional and Therapeutic Clinical Trials: Small Molecules, Interactions, Outcomes and Opportunities **221**
Edward Croom, Fengyu Zhang and Claude Hughes

8.1 Introduction 221
8.2 Removal-of-treatment Preventative Clinical Trials 223
8.2.1 Trial Aims 223
8.2.2 Trial Populations 224
8.2.3 Potential Therapeutics 224
8.2.4 Trial Designs 226
8.2.5 Primary Outcomes 226
8.2.6 Secondary Outcomes 226
8.2.7 Toxicities 227
8.2.8 Opportunities 227
8.3 Addition-of-treatment Preventative and/or Curative Trials 228
8.3.1 Trial Aims 229
8.3.2 Trial Populations 230
8.3.3 Potential Therapeutics 230
8.3.4 Trial Designs 231
8.3.5 Primary Outcomes 232
8.3.6 Secondary Outcomes 232
8.3.7 Toxicity 232
8.3.8 Opportunities 234
8.4 Supportive Clinical Trials 235
8.4.1 Trial Aims 235
8.4.2 Trial Populations 235
8.4.3 Potential Therapeutics 236
8.4.4 Trial Designs 236
8.4.5 Primary Outcomes 237
8.4.6 Secondary Outcomes 237
8.4.7 Toxicity 237
8.4.8 Opportunities 238
8.5 Extended Investigation for More Iterative Interventional Trials 239
8.6 Executive Summary 241
List of Abbreviations 241
References 241

Chapter 9 A Trend Analysis of the COVID-19 Pandemic in Turkey and in the Indian State of Kerala **243**
H. Seval Akgun and K. Rajasekharan Nayar

9.1 Introduction 243
9.2 COVID-19 in Turkey 245

9.2.1 Deaths 249
9.2.2 Hospital Care 251
9.2.3 Seroprevalence Study 251
9.2.4 How did Turkey Control the Virus? 252
9.3 Conclusions 253
References 254

Chapter 10 The Swedish Approach to COVID-19 **255**
Sebastian Rushworth

10.1 Introduction 255
10.2 Herd Immunity 256
10.3 Early Cases 256
10.4 Advice and Restrictions 256
10.5 Face Masks 257
10.6 Temporary Hospitals 258
10.7 Removal of Restrictions and Advice 258
10.8 Local Measures 259
10.9 A Change of Tack 260
10.10 Conclusion 264
References 264

Chapter 11 New Zealand's Response to the First Wave of COVID-19 **267**
Sarah Jefferies, Anne French, Petra Muellner, Virginia Hope, Nigel French, Caroline McElnay and Patricia Priest

11.1 Introduction 267
11.2 Key Features of Surveillance and Response to the First Wave 268
11.2.1 The Border 268
11.2.2 Surveillance 270
11.2.3 National Alert Level NPIs for COVID-19 Elimination 271
11.2.4 Further Aspects of National Governance 273
11.3 Impacts of the National Response 274
11.3.1 COVID-19 Epidemiology 274
11.3.2 New Zealand's Pacific Island Neighbours 275
11.3.3 New Zealand-specific Risks 276
11.4 Conclusion 276
References 276

Chapter 12 Significant SARS-CoV-2 Variants Circulating Around the World: Origin, Spread and Phylodynamics **279**
Zainy Zehra, Naseem A. Gaur and Asimul Islam

12.1 Introduction 279
12.2 General Features 280
12.3 Nomenclature for SARS-CoV-2 Variants 281
12.3.1 GISAID 281
12.3.2 PANGOLIN 282
12.3.3 Nextstrain 282
12.4 Assessment of Reported Variants 283
12.5 Significant Identified Variants 284
12.5.1 Cluster 5 Variant 284
12.5.2 B.1.1.7/20I/501Y.V1 285
12.5.3 B.1.351/501Y.V2 286
12.5.4 P.1/501Y.V3 286
12.6 Significant Mutations Found in Reported Variants 287
12.6.1 D614G Mutation 287
12.6.2 N501Y Mutation 287
12.6.3 E484K Mutation 288
12.6.4 A930V Mutation 289
12.6.5 Y453F Mutation 289
12.7 Evidence of Evolution Subject to Exogenous Factors 290
12.7.1 Climate 290
12.7.2 Pressure 290
12.7.3 Health Index 291
12.7.4 Blood Type 291
12.8 Exceptions are Everywhere 291
12.9 Conclusion 292
12.10 Executive Summary 292
Acknowledgements 293
References 293

Chapter 13 Survival of SARS-CoV-2 Outside the Body and Methods of Disinfection **299**
By Timothy C. Marrs and Robert L. Maynard

13.1 Introduction 299
13.2 The Problem of Disinfection as Applied to the SARS-CoV-2 Virus 301
13.3 Survival of the SARS-CoV-2 Virus Outside the Body 302
13.4 Use of Disinfectants 303
13.4.1 Ethanol and Other Alcohols 304
13.4.2 Bleach 305

13.4.3 Hydrogen Peroxide 305
13.4.4 Soaps and Detergents 306
13.4.5 Other Preparations 306
13.4.6 Natural Products 307
13.4.7 Physical Methods of Disinfection 307
13.5 Conclusions 307
References 308

Chapter 14 US Regulatory Approaches to Disinfectants for Coronavirus 313
Sol Bobst

14.1 List N 313
14.2 Safe Use of Disinfectants 314
14.3 Regulation of Hand Sanitizers Under the FDA Temporary Policy 314
References 441

Chapter 15 Diagnostic Strategies for Early and Point-of-Care Detection of COVID-19 in Clinical and Personalized Settings 442
Supratim Mahapatra, Buddhadev Purohit, Mahesh M. Shanbhag, Nagaraj P. Shetti, Sanjeev Kumar Mahto, Rajiv Prakash and Pranjal Chandra

15.1 Introduction 443
15.2 Molecular Assessments and Detection of SARS-CoV-2 Nucleic Acid 444
15.2.1 Reverse Transcription-PCR 444
15.2.2 Isothermal Amplification of Nucleic Acid 454
15.2.3 Microarray-based Assessment 457
15.2.4 Amplicon-based MinION Sequencing 458
15.2.5 Molecular Genetic Assays for SARS-CoV-2 Detection 458
15.3 Immunological Assays 459
15.3.1 Enzyme-linked Immunosorbent Assay 460
15.3.2 Lateral Flow Immunoassay 460
15.3.3 High-throughput Neutralizing Antibody Assay 466
15.3.4 Chemiluminescent Immunological Assay 467
15.3.5 Antigenic Assessment 467
15.4 Biosensor Tests 468
15.4.1 Plasmonic Transducer-based Biosensor 469
15.4.2 Field-effect Transistor-based Biosensor 470
15.4.3 A Graphene-based Multiplexed Immunosensing Platform 470

15.5 Computed Tomography 471
15.6 Existing Serological Assessments 474
15.7 Future Diagnostic Approaches 475
15.8 The Importance of Identifying Novel Biomarkers 476
15.9 Conclusion 477
15.10 Executive Summary 477
Declaration of Competing Interest 478
Acknowledgements 478
References 478

Chapter 16 Human Resource Management, Preparedness, and Outcomes 486
Satyendra Khichar, Naresh Midha, M. K. Garg, Akhilesh Kumar and Kamlesh Ahari

16.1 Introduction 486
16.2 Objectives and Plan of Action 487
16.2.1 Command and Control Centre: System 488
16.2.2 Management of Healthcare Resources 488
16.2.3 Environment, Capacity Building and Supply Management 492
16.2.4 Quarantine Facilities and Social Care 493
16.2.5 Outpatient Department Management 494
16.3 Conclusion 494
List of Abbreviations 495
References 495

Chapter 17 The Role of Telehealth During the COVID-19 Pandemic 497
Binit Sureka, Puneet Pareek, Aliza Mittal and Stafford Warren

17.1 Introduction 497
17.2 Types of Telehealth Services 499
17.2.1 Social Media Platforms, Including Messaging Software, WhatsApp 499
17.2.2 Appointments and Consultancy Through Email 499
17.2.3 Live Video Conferencing 499
17.2.4 Electronic Health Records (Remote Patient Monitoring) 499
17.2.5 Patient Portal Messaging 500
17.2.6 Digital Photography 500
17.2.7 Appointments Through Hospital Websites 500
17.2.8 Mobile Health 500

17.3 Key Outputs Expected from Telehealth Services in COVID-19 500
17.4 Effects of Telehealth Services 501
17.5 Limitations of Telehealth and Telemedicine 501
17.6 AIIMS Jodhpur Telehealth Services in India During COVID-19 as One Example of Implementation Logistics 502
17.7 Future of Telemedicine/Future Directions 503
17.8 Conclusions 504
References 505

Chapter 18 Medical Education During the COVID Pandemic: Challenges and Solutions 507
Kuldeep Singh, Shilpi Gupta Dixit, Pradeep Dwivedi, Shival Srivastav, Abhinav Dixit, Abhishek Bhardwaj, Mukesh Swami and Sanjeev Misra

18.1 Introduction 507
18.2 Online Teaching Tools 509
18.3 Challenges Posed by Pandemics 510
18.4 COVID Pandemic – Global Experience 510
18.4.1 Experience of Developed Countries and Latin America 510
18.4.2 Response in Asia 511
18.4.3 The Indian Perspective 511
18.4.4 AIIMS Jodhpur Experience 512
18.5 COVID-19: a Learning Opportunity for Both Teachers and Students 525
18.5.1 Opportunity for Teachers 525
18.5.2 Opportunity for Students 525
18.6 Hybrid Model 526
References 527

Chapter 19 Containment Strategies to Prevent Spread in Different Societies and Cultures 532
Nitin Joshi, Akhil Goel and Pankaj Bhardwaj

References 539

Chapter 20 Effective Health Education, Patient Education and Health Behaviour Change 542
Suman Saurabh, Manoj Kumar Gupta and Pankaj Bhardwaj

20.1 Introduction 542
20.2 Brief Overview of the Areas of Education Related to COVID-19 543

20.3 What Constitutes COVID-appropriate Behaviour for Different Population Groups? 544
20.4 Steps in Planning Health Education and Patient Education: COVID-19 Considerations 545
20.5 Translating Health Education Into Behaviour Change 545
20.6 Examples of Application of Behaviour Change Models in COVID-19 548
20.6.1 The Transtheoretical Model in Understanding COVID-19-related Behaviour Change in India 548
20.6.2 A Case Study: the "Badlao Programme" for COVID-19 Prevention in Schools 549
20.7 Modalities of Health Education in COVID-19 549
20.8 Conclusion 552
References 552

Chapter 21 Strengthening the Biological and Toxin Weapons Convention After COVID-19: Reaching Agreement on a Code of Conduct and Biological Security Education at the 2021 9th Review Conference 554
Lijun Shang, Simon Whitby and Malcolm Dando

21.1 Introduction 554
21.2 The Gap in the Web of Prevention 555
21.3 The Hague Ethical Guidelines and the Work of the Advisory Board for Education and Outreach 556
21.4 Codes of Conduct, Awareness-Raising and Education for Life Scientists 558
21.4.1 Introduction: the Biological and Toxin Weapons Convention 558
21.4.2 The BTWC and the World Health Organization (WHO) 559
21.4.3 The BTWC and Dual Use 560
21.4.4 Codes of Conduct and Education 560
21.5 Conclusions 563
References 564

Chapter 22 Covid-19 Prevention 568
Chris Ringwalt

22.1 Introduction 568
22.2 Potential Preventive Measures 569
22.3 The Haddon Matrix 570
22.4 Preventive Measures: Evidence of Effectiveness 570
22.4.1 Air Disinfection 574

22.4.2 Personal Protective Measures: Cloth Masks, Social Distancing, and Eye Protection 574
22.4.3 Surveillance and Health Promotion Measures 575
22.5 Executive Summary and Conclusions 576
22.6 Postscript 576
List of Abbreviations 577
Acknowledgements 577
References 577

Chapter 23 Human Responses to Pandemics in History and Medico-anthropological Correlates 580
Ashoka Jahnavi Prasad

23.1 Introduction 580
23.2 The Athenian Plague of 430 BC 582
23.3 The Antonine Plague 582
23.4 The Justinian Plague 583
23.5 The Black Death 584
23.5.1 The Plague Doctor 586
23.5.2 Quarantine 586
23.6 "Spanish Flu" Pandemic 1918–1920 587
23.7 HIV Pandemic 588
23.8 Smallpox Outbreak in Former Yugoslavia (1972) 589
23.9 SARS 590
23.10 "Swine Flu" or H1N1/09 Pandemic 590
23.11 Ebola Outbreak (2014–2016) 591
23.12 Zika (2015–2016) 591
23.13 Medico-anthropological Correlates 592
23.14 A Brief Literature Review 593
23.15 Addressing the Challenges of Religious and Cultural Practices 595
23.16 Lessons Learned for Future Prevention Strategies 596
23.17 Conclusion 597
References 597

Chapter 24 The COVID-19 Pandemic – Global Lessons for the Future 604
Alok Dhawan, Claude Hughes, Timothy Marrs, Stafford Warren, Michael Waters and Diana Anderson

24.1 Introduction 604
24.2 Coronaviruses and SARS-CoV-2 605

24.3 Global Efforts to Halt or Slow the Spread of COVID-19 606
24.4 Vaccine Development 606
24.5 Antiviral Drug Development 609
24.6 Beyond Vaccines and Pharmacotherapies 611
24.7 Managing the Pandemic of COVID-19 612
24.8 Clinical Epidemiology 616
24.9 Recommendations for the Future 617
24.9.1 Databases to Inform, Track and Share Progress 618
24.9.2 Strategic Advice and Control of Outbreaks and Community Spread 618
24.9.3 Pharmacotherapies 619
24.10 Ideas for the Future 619
24.10.1 Testing – Rapid Futuristic Diagnostic Methods 619
24.10.2 Serum Repository 620
24.10.3 Drugs and Vaccines of the Future 620
24.10.4 Ventilators and Protective Equipment 621
24.11 Executive Summary 622
References 623

Subject Index **625**

CHAPTER 1

What We Know About the Life-threatening Novel Human Coronavirus

CLAUDE HUGHES[†*a,b], ARUL NARAYANASWAMY[†c], BALACHANDAR VELLINGIRI[†*d], DHIVYA VENKATESAN[†d], STAFFORD WARREN[†e] AND MICHAEL WATERS[†f]

[a]IQVIA, Durham, NC, USA; [b]Duke University Medical Center, Durham, NC, USA; [c]Department of Zoology, Bharathiar University, India; [d]Department of Human Genetics and Molecular Biology, Bharathiar University, India; [e]Anne Arundel Medical Center, Annapolis MD, USA; [f]Michael Waters Consulting USA, Hillsborough, NC, USA

*E-mail: clhughes@duke.edu, geneticbala@buc.edu.in

1.1 Introduction

Coronaviruses are a large family of zoonotic single-stranded RNA viruses that occur in numerous wild and domesticated animal species and are known to occasionally cross species barriers. Human coronaviruses were first identified by Dr June Almeida in 1964 at her laboratory in St Thomas' Hospital in London, UK. The new discovery from strain B814 was described in the *British Medical Journal* in 1965.[1] The severe acute respiratory syndrome coronavirus 2 (SARS-CoV-2) is the seventh member of enveloped RNA coronaviruses. The emergence of severe life-threatening human coronavirus-related diseases

[†]All authors contributed equally and are listed alphabetically by surname.

The Coronavirus Pandemic and the Future Volume 1: Virology, Epidemiology, Translational Toxicology and Therapeutics
Edited by Michael D. Waters, Alok Dhawan, Tim Marrs, Diana Anderson, Stafford Warren and Claude L. Hughes

Published by the Royal Society of Chemistry, www.rsc.org

began intensely during the 21st century with the severe acute respiratory syndrome coronavirus (SARS-CoV) and the Middle East Respiratory Syndrome coronavirus (MERS-CoV). Human coronaviruses cause illness in humans ranging from the common cold to more severe diseases such as MERS and SARS. SARS-CoV and MERS-CoV likely originated in bats and then infected other mammalian hosts – the Himalayan palm civet for SARS-CoV, and the dromedary camel for MERS-CoV – before being transmitted to humans. The dynamics of the novel coronavirus (SARS-CoV-2) are currently unknown, but it is likely that the new virus also originated in the same manner, as its overall genome sequence is 96.2% identical to that of bat coronavirus (BatCoV RaTG13) and its overall genetic architecture is very similar to SARS-CoV.

On 11 January 2020 Yong-Zhen Zhang and his colleagues at Fudan University in Shanghai, China deposited into a public database the genome sequence of a virus isolated from a 41 year-old man who had worked at the Huanan seafood wholesale market.[2] This seafood market, since closed, also sold live animals of many species, including bats and pangolins. In so doing, Zhang *et al.* alerted the world to the existence of a new coronavirus that was obviously related to SARS-CoV.[2] Their findings were subsequently published in *Nature* (Wu, Zhao *et al.* 2020).[3]

Although Zhang's team had sequenced the virus from only a single patient, simultaneous work by other groups identified the same virus from other people with pneumonia. Together, these researchers firmly implicated this new coronavirus as the cause of the disease. One of the teams, led by Shi Zhengli at the Wuhan Institute of Virology, also determined that the closest known relative of the new virus was a bat coronavirus.[2] In another study, full-length genome sequences were obtained from five patients at an early stage of the outbreak. These sequences are almost identical and share 79.6% sequence identity with SARS-CoV.[4] Although not definitive, this suggested that the new virus was potentially closely linked to the zoonotic virus source and that bats were the primary reservoir for SARS-CoV-2.[4]

1.2 The Chronology of the Outbreak

Beginning in the month of December, during a festive season of celebrations, the SARS-CoV-2 outbreak in Wuhan rapidly engulfed an unprepared world. A red-letterhead document bearing the official seal of the Wuhan Municipal Health Commission had circulated on social media in China and had issued an emergency notice: "Cases of a pneumonia of unknown cause have intermittently surfaced at Wuhan's Huanan Seafood Market". On 31 December 2019, the first of many cases of "pneumonia of unknown etiology" were reported to the World Health Organization (WHO) country office in China.

The timeline of initial events in China over the first 10 weeks of the pandemic as documented by Yan *et al.*[5] is illustrated in Figure 1.1. Chapter 3 of this volume provides a unique Chinese perspective on that period and its significance. The symptom onset date of the first patient identified could be traced to 1 December 2019.

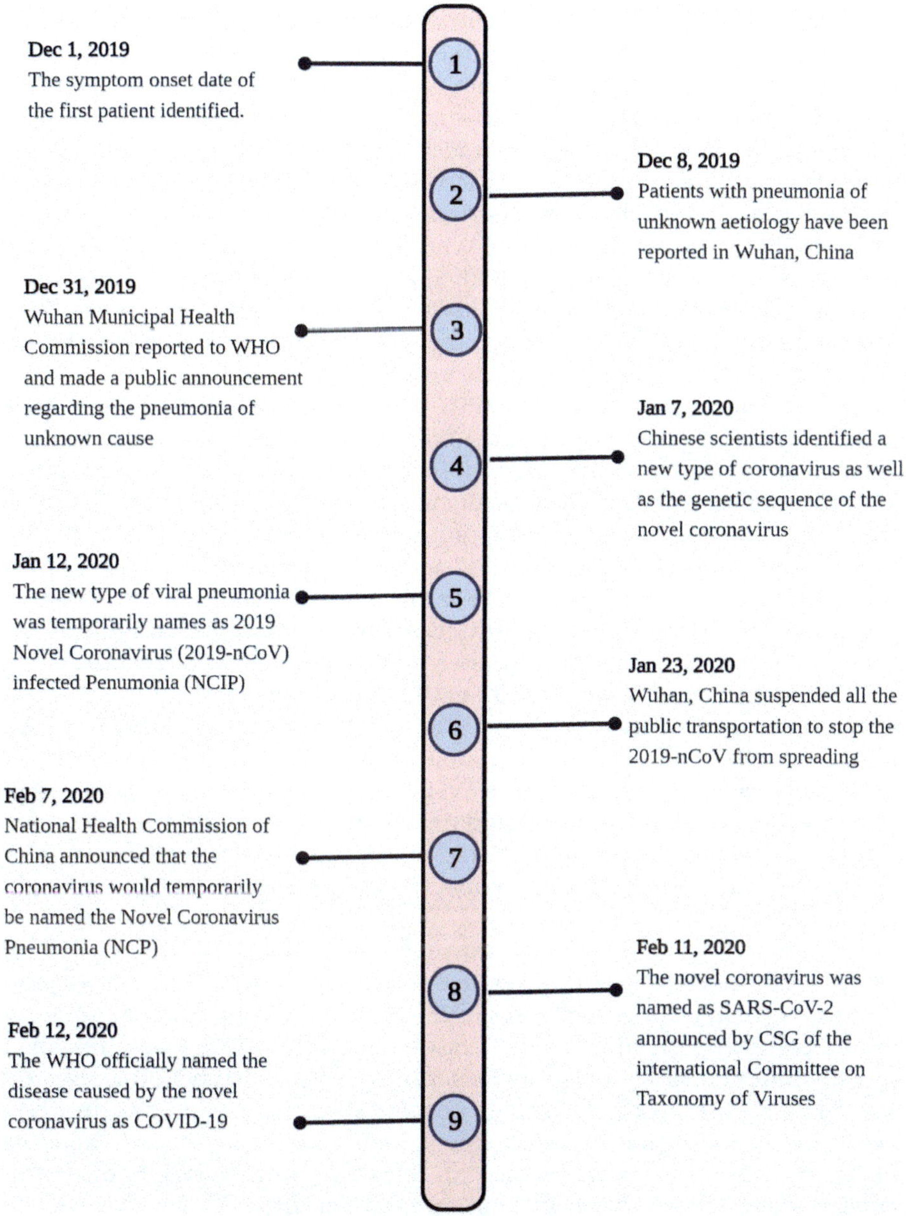

Figure 1.1 The timeline of the development of SARS-CoV-2. CSG: Coronaviridae Study Group. Reproduced from ref. 5, https://doi.org/10.3390/ijerph17072323, under the terms of the CC BY 4.0 license, https://creativecommons.org/licenses/by/4.0/.

In response, the Chinese Center for Disease Control and Prevention (China CDC) and local health groups organized an intensive outbreak investigation program, and the etiology of this illness was soon attributed by the Chinese CDC to a novel virus belonging to the coronavirus family. An epidemic of unexplained lower respiratory tract infections with distinctive bilateral "ground-glass" radiographic findings on chest X-ray and computed tomography was occurring in Wuhan, the largest metropolitan area in China's Hubei province (estimated population: 10 607 700). From 3 January 2020, information concerning newly identified coronavirus disease 2019 (COVID-19) cases was reported daily by China to the World Health Organization (WHO). The virology of SARS-CoV-2 was first studied from a bronchoalveolar lavage fluid sample revealing its genome size to be 29.9 kb, which includes genomic RNA, phosphorylated nucleocapsid (N) protein, spike (S) protein, hemagglutinin esterase, membrane (M) protein, and envelope (E) protein.[6]

The common symptoms of COVID-19 reported in infected individuals included fever, shortness of breath, tiredness, dry cough, and abnormalities in smell and/or taste. These symptoms were often accompanied by a number of clinical observations that were documented in the literature.[7,8] Clinical laboratory and imaging findings demonstrated that in more advanced cases the disease could result in pneumonia and even profound multisystem clinical pathologies.[9] However, it was observed very early that a patient could test positive for the virus and yet be entirely without symptoms, and this is now thought to be the case in approximately 40% of persons who test positive.[10]

Full genomic sequencing of the new virus was shared with the WHO and the international community soon after the pathogen was identified on 7 January. On 10 January, an expert group involving Hong Kong, Macao and Taiwanese technical experts and a WHO team was invited to visit Wuhan. The full disease outbreak news update, which appeared on the WHO website on 12 January 2020, can be found at https://www.who.int/docs/default-source/coronaviruse/situation-reports/20200121-sitrep-1-2019-ncov.pdf?sfvrsn=20a99c10_4.

A set of nucleic acid primers and probes for polymerase chain reaction (PCR) detection of COVID-19 was released on 21 January 2020. As of 23 January 2020, China had reported cases in 25 provinces (autonomous regions and municipalities). 25% of confirmed cases reported by China were classified by the Chinese health authorities as seriously ill (from Hubei Province: 16% severely ill, 5% critically ill, and 4% having died). Cases infected in China had been exported to the United States, Thailand, Japan, and the Republic of Korea, with the expectation that more cases would be exported to other countries and that further transmission might occur.

Although the initial source of the 2019 novel coronavirus (2019-nCoV) remained unknown, it was clear that the growing outbreak was no longer due to ongoing exposures at the Huanan seafood market, as <15% of new cases reported having visited the Huanan market. There was more evidence that 2019-nCoV spreads from human to human and across generations of cases. Moreover, family clusters involving persons with no reported travel to Wuhan had been reported from Guangdong Province.

The WHO initially assessed the risk of this event to be "moderate" in situation reports published on 23, 24, and 25 January, and later revised them to "very high" in China, "high" at the regional level, and "high" at the global level[11] (Figure 1.2). The distribution of 2019-nCoV cases in China (and other countries, territories, or areas with reported known cases) is shown in Figure 1.2.

By the time of the total lockdown of the city of Wuhan on 23 January, it was reported[12] that 5 million people had left the city for other destinations and an outbreak of pandemic scale had begun worldwide.

A harbinger of future revelations, Ma Xiaowei, the Director of China's National Health Commission, was reported on 27 January 2020 to have said[12] "From observations, the virus is capable of transmission even during incubation period".

And as a fifth case of the coronavirus was confirmed in the United States on that same date, the mayor of the city of Wuhan warned that the world that it should expect COVID-19 infections to keep climbing.[12] The complete report, as published by MarketWatch, can be found at: https://www.marketwatch.com/story/mayor-of-wuhan-epicenter-of-coronavirus-outbreak-says-5-million-people-left-the-city-before-travel-restrictions-were-imposed-2020-01-26.

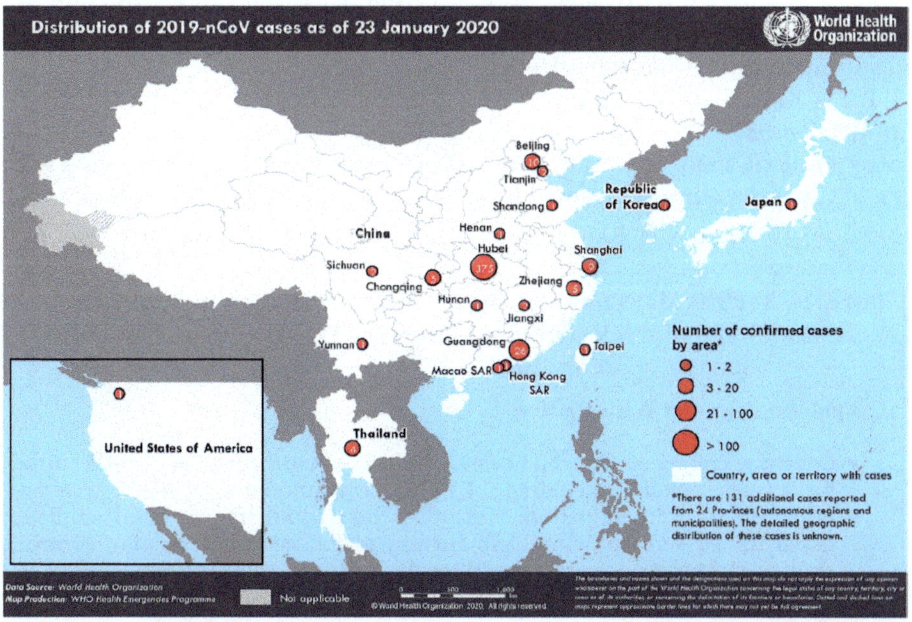

Figure 1.2 Countries, territories, or areas with reported confirmed cases of 2019-nCoV, 23 January 2020. Reproduced from the World Health Organization https://www.who.int/docs/default-source/coronaviruse/situation-reports/20200209-sitrep-20-ncov.pdf?sfvrsn=6f80d1b9_4.

The world had been warned as to what to expect with the pandemic of COVID-19. Some countries heeded the warning and suffered relatively few fatalities at the hand of SARS-CoV-2, but others dismissed it and continue to suffer the consequences.

The chronology of the ensuing events in the outbreak is as follows:

On **30 January 2020**, the WHO declared the novel coronavirus outbreak a Public Health Emergency of International Concern as it had spread to 18 countries, with four countries reporting human-to-human transmission (per the International Health Regulations, 2005[13]).

On **11 February 2020**, the WHO Director General, Dr Tedros Adhanom Ghebreyesus, announced that the symptomatic disease caused by this new coronavirus was "COVID-19", which is the acronym of "coronavirus disease 2019".

By **13 February 2020**, the disease had spread to 25 different countries and there had been a total of 60 412 confirmed cases with 1370 fatalities.[5]

Between **16 and 24 February 2020**, the WHO–China Joint Mission on Coronavirus Disease 2019 (COVID-19) met in Wuhan, China.

On **28 February 2020**, the WHO raised the threat of a SARS-CoV-2 epidemic to the "very high" level and the comprehensive report of the WHO–China Joint Mission on Coronavirus Disease 2019 (COVID-19) was issued.[14] See the concluding technical recommendations from the report in Box 1.1.

On **11 March 2020**, the WHO declared COVID-19 a global pandemic, as the number of COVID-19 cases outside China increased 13-fold, the number of countries involved tripled, with more than 118 000 cases in 114 countries resulting in more than 4000 deaths.[15]

By **31 March 2020**, COVID-19 had infected 750 890 people globally with 36 405 deaths.[16]

By **14 April 2020**, approximately 1 844 683 cases had been confirmed with 117 021 deaths reported from at least 213 countries, areas, or territories.[17]

Box 1.1 Technical Recommendations from the Report of the WHO–China Joint Mission on Coronavirus Disease 2019 (COVID-19) issued 28 February 2020

Epidemiology and Transmission

- Continue enhanced surveillance across the country through existing respiratory disease systems, including ILI, SARI or pneumonia surveillance systems
- Prioritize early investigations, including household transmission studies, age-stratified sero-epidemiologic surveys including children, case–control studies, cluster investigations, and serologic studies in healthcare workers

Severity

- Continue to share information on patient management, disease progression and factors leading to severe disease and favorable outcomes

Box 1.1 (*continued*)

- Review and analyze the possible factors associated with the disease severity, which may include:
  - natural history studies to better understand disease progression in mild, severe, and fatal patients
  - medical chart reviews about disease severity among vulnerable groups (*e.g.* those with underlying conditions, older age groups, pregnant women, and children) to develop appropriate standards of care
  - evaluation of factors leading to favorable outcomes (*e.g.* early identification and care)

Clinical Care and Infection Prevention and Control

- Suspect patients who have not yet been tested should be isolated in single normal-pressure rooms; cohorting of positive cases is acceptable
- Physicians and all healthcare workers need to maintain a high level of clinical alert for COVID-19
- For affected countries, standardize training for clinical care and IPC and scale with the development of local (*e.g.* district-level) experts
- Ensure concurrent testing for other viral pathogens to support a negative COVID-19 test
- Ensure maintenance of usual and essential services during the outbreak
- Ensure processes are in place for infection prevention among the most vulnerable, including the elderly
- Ensure readiness to provide clinical care and to meet IPC needs, including:

  a. anticipated respiratory support requirements (*e.g.* pulse oximeters, oxygen, and invasive support where appropriate)
  b. national guidelines for clinical care and IPC, revised for COVID-19
  c. nationally standardized trainings for disease understanding and PPE use for healthcare workers
  d. community engagement
  e. PPE and medication stockpiles
  f. early identification protocols; triage, temperature screening, holding bays (triage, including pulse oximetry)
  g. treatment protocols including designated facilities, patient transportation
  h. enhanced uptake of influenza and pneumococcal vaccine according to national guidelines
  i. laboratory testing
  j. rapid response teams

Laboratory and Virology

- Continue to perform whole-genome analysis of COVID-19 viruses isolated from different times and places, to evaluate virus evolution
- Conduct pathogenesis studies using biopsy/post-mortem specimens of COVID-19 patients or infected animal models
- Evaluate available nucleic acid PCR diagnostics
- Rapidly develop and evaluate rapid/point-of-care diagnostics and serologic assays
- Conduct further study to interpret the result of positive COVID-19 RNA detection in feces in patients recovering from COVID-19

(*continued*)

Box 1.1 (*continued*)

- Enhance international cooperation, especially in terms of biosafety and information sharing for increased understanding of the COVID-19 virus and traceability of the virus
- Consider monitoring proinflammatory cytokines *via* multiplex assays to predict the development of "cytokine storm"

Research and Development

- Additional effort should be made to find the animal source, including the natural reservoir and any intermediate amplification host, to prevent any new epidemic foci or resurgence of similar epidemics
- Efforts should be made to consistently evaluate existing and future diagnostic tests for detection of COVID-19 using a harmonized set of standards for laboratory tests and a biorepository that can be used for evaluating these tests
- Consider the establishment of a centralized research program in China to oversee that portfolio and ensure the most promising research (vaccines, treatments, pathogenesis) are adequately supported and studied first; program staff dedicated to the clinical research would work at the clinical research site(s) to decrease the research workload of the clinicians at the site
- Consider including one or more sites within China in the ongoing and future multicenter, international trials; Chinese investigators should be actively engaged in international trials
- Continue to develop additional animal models, making every effort to ensure these mimic human infection and virus transmission as closely as possible
- Conduct studies to determine which of the commonly used forms of PPE are most effective in controlling the spread of COVID-19

ILI: influenza-like illness; IPC: infection prevention and control; PPE: personal protective equipment; SARI: severe acute respiratory infection.

1.3 Initial Transmission

As described in WHO reports, the first cases of the novel coronavirus disease were linked to direct exposure at the Huanan seafood wholesale market of Wuhan. For this reason, animal-to-human transmission was presumed to be the main mechanism of transfer. However, since subsequent cases were not associated with this location, it was determined that the virus could be directly transmitted from one human to another. The key epidemiological study in this regard was reported by Li and associates in the *New England Journal of Medicine* on 26 March 2020.[18]

Li *et al.*[18] analyzed data on the first 425 confirmed cases in Wuhan to determine the epidemiological characteristics of the disease. They collected information on demographic characteristics, exposure history, and illness timelines of laboratory-confirmed cases that had been reported by 22 January 2020. The median age was 59 years and 56% of patients were male. The majority of cases (55%) had a disease onset before 1 January 2020, and were linked to the Huanan seafood wholesale market, as compared with 8.6% of subsequent cases. The mean incubation period was determined to be 5.2 days (95% CI 4.1–7.0 days), with the 95th percentile of the distribution at 12.5 days. In its early stages, the epidemic doubled in size every 7.4 days. With a mean serial interval of 7.5 days (95% CI 5.3–19 days), the basic reproductive number (R_0) was estimated to be 2.2 (95% CI 1.4–3.9). In general, an epidemic will increase as long as R_0 is >1, and control measures aim to reduce the reproductive number to <1. The longest time frame from infection to symptoms was 12.5 days (95% CI 9.2–18 days).[18] In other words, these data showed that the infection rate for this novel virus doubled approximately every 7 days. For comparison, the R_0 of the SARS-CoV epidemic in 2002–2003 was ~3 (that is, one person infects approximately three other people; if this occurred within approximately 10 days, the prevalence of the epidemic would double within 1 week).[19] On the basis of this information, Li *et al.* concluded that evidence suggested that human-to-human transmission had occurred among close contacts since the middle of December 2019 and that considerable efforts to reduce transmission would be required to control outbreaks if similar dynamics applied elsewhere. Apart from China, human-to-human transmission has been reported in other countries such as Germany, Japan, Thailand, the United States and elsewhere.[20]

As with other respiratory pathogens, including flu and rhinovirus, transmission was believed to occur through respiratory droplets (particles >5–10 μm in diameter) from coughing and sneezing. Aerosol transmission was also believed possible in cases of protracted exposure to elevated aerosol (≤5 μm) concentrations in closed spaces. Analysis of data related to the spread of SARS-CoV-2 in China indicated that close contact (6 feet, 1.8 m) between individuals is necessary for transmission, most of which occurs between family members, healthcare professionals, and other close contacts. Surprisingly, it was speculated that pre- and asymptomatic individuals account for up to 80% of viral transmission.[21] All of the above suggested that the virus could easily spread among totally asymptomatic individuals in close contact with one another and remain undetected for days.

Transmission of SARS-CoV-2 can occur *via* contact with contaminated surfaces. The virus can live for 2–3 days on plastic and stainless steel, up to 1 day on cardboard, and up to 4 hours on copper.[22] However, because data so far available have been generated under experimental conditions, these estimates must be interpreted with caution. Additionally, assessing risk from exposures to contaminated surfaces must also consider that the presence of viral RNA does not necessarily indicate that the virus is viable and potentially infectious. Higher contamination rates have been reported in intensive care units than general wards, as SARS-CoV-2 has been reported on floors, computers, trash cans, and sickbed handrails, as well as 4 m in the air above patient beds.[23]

Based on these data from the first cases in Wuhan and investigations conducted by the China CDC and local CDCs, the incubation time is generally estimated to be 3–7 days (median 5.1 days, similar to SARS).[24] This early observation combined with the likelihood that approximately 40–45% of those infected with SARS-CoV-2 will remain asymptomatic suggested that the virus might have greater potential than previously estimated to spread silently and extensively through human populations.[25] He *et al.* estimated that 44% (95% CI 30–57%) of secondary cases were infected during the index cases' presymptomatic stage and argued that control measures should be adjusted to account for probable substantial presymptomatic transmission.[26] Indeed, the long duration of viral shedding and high proportion of asymptomatic patients make asymptomatic transmission a critical concern for effective public health management.

It must be emphasized that much of the information presented thus far was the result of the early reports. Thus, further studies are needed to understand mechanisms of transmission, incubation times, clinical symptoms and course, and duration of infectivity. Differing symptoms experienced by different people might be dictated by a combination of factors including: (1) which cells and tissues are infected; (2) the direct damage the virus causes to these cells; (3) interference with the normal function of host cells' angiotensin-converting enzyme (ACE)2 receptors, to which the virus binds; and (4) individual variations in the immune response to the virus.[2] For example, severely ill patients show hyperactivation of their immune response, which damages the lungs and other organs. What is generally known to date regarding these topics is covered in subsequent sections.

1.3.1 Worldwide Spread of the Virus

At the time of this writing the data provided by the WHO Coronavirus Disease (COVID-19) Dashboard is "Globally, as of 12:44 pm CEST, 15 August 2020, there have been 20 995 433 confirmed cases of COVID-19, including 760 774 deaths, reported to WHO". The most up-to-date information on the emerging pandemic can be found at the following sources:

The WHO Coronavirus (COVID-19) Dashboard Situation Board: https://covid19.who.int/, and the Johns Hopkins Center for Systems Science and Engineering site for Coronavirus Global Cases COVID-19 has reported the Coronavirus COVID-19 Global Cases since the beginning of the pandemic: https://coronavirus.jhu.edu/map.html

World governments are currently at work to establish countermeasures to stem the devastating effects of the virus. Health organizations coordinate information flow and issue directives and guidelines to best mitigate the impact of the threat. At the same time, scientists around the world are working tirelessly to discern information about transmission mechanisms, the clinical spectrum of disease, new diagnostics, prevention, and therapeutic strategies. Many uncertainties remain regarding the virus, virus–host interactions, and the evolution of the pandemic, with specific reference to the times when it will reach its peaks in various regions and globally. At this moment, therapeutic strategies to treat COVID-19 are only supportive, and prevention aimed at reducing transmission in the community remains our best tactic. Aggressive isolation measures in China have led to a progressive reduction of cases.

In northern Italy, where the coronavirus ravaged cities from late February through April, doctors credited the turnaround to the country's strict nationwide lockdown, widespread testing, robust contact tracing.

When Italy's "father of the swabs", Andrea Crisanti, learned of the first Italian COVID-19 fatality in Vo, Italy, he saw it as an ideal place to perform an epidemiological experiment: a small population, universally tested, whose progress could be monitored closely. He tested everyone in the village and then retested them after 9 days. In the first round, 73 residents were positive for the virus and more than 40% of them had no symptoms, yet they had levels of the virus similar to those who were visibly ill. The Vo study confirmed asymptomatic transmission and showed that isolating people helps stop transmission, since everyone who tested positive was confined to their home, regardless of whether they had symptoms. By the second round of testing, a week and a half later, the number of positives had dropped to 29, and they too were isolated. A third round of tests 2 months later found no positive cases.[27]

In Germany, mass testing was not performed at the highest rates – as was seen in South Korea with 10 000 tests per day – but Germany was meticulous about the process. Once an individual tested positive, officials traced each one of their contacts and proceeded to test and quarantine those individuals, again essentially breaking "infection chains".[28]

In the United States, before 24 August 2020, the CDC website stated that testing was recommended "for all close contacts of persons with SARS-CoV-2 infection". On 24 August, that was changed to say that someone who was in close contact (within 6 feet) of a person with COVID-19 for at least 15 minutes, but who didn't have symptoms does not "necessarily need a test".[29] As of this writing, that change has caused significant concern among public health and medical professionals, especially since most states have been unable to perform contact tracing due to the high numbers of individuals testing positive.[30]

Details on the public management of COVID-19 in various countries are provided in Chapters 9–11 of Volume 2 of this book.

1.4 SARS-CoV-2 and Its Etiology

As mentioned previously, coronaviruses are positive-stranded RNA viruses with a crown-like (*coronam* is Latin for crown) appearance under an electron microscope due to the presence of spike glycoproteins on the outer envelope. SARS-CoV-2 is a *Betacoronavirus* enveloped positive-sense RNA virus (subgenus *Sarbecovirus*, Orthocoronavirinae subfamily).[7] As be discussed in greater detail later, the subfamily Orthocoronavirinae of the Coronaviridae family (order Nidovirales) classifies into four genera of coronaviruses:

Alphacoronavirus (alphaCoV), *Betacoronavirus* (betaCoV), *Deltacoronavirus* (deltaCoV), and *Gammacoronavirus* (gammaCoV). The *Betacoronavirus* genus further divides into five subgenera.[31] The coronaviruses found in various groups of bat and bird species serve as natural reservoirs. Genomic characterization has shown that bats and rodents are probably the sources of alphaCoVs and betaCoVs. Avian species are the likely sources of deltaCoVs and gammaCoVs. How coronaviruses in each species originated is elusive, and associated with varied evolution models. The first evolutionary model on coronaviruses proposed bat species as the gene source for all coronaviruses.[32] However, confirmation studies over the following 2 years (after 2007) did not support this hypothesis.[33] Figure 1.3 illustrates the animal origins of coronaviruses.

Initially, these viruses had not been considered harmful to humans, but investigations gained momentum after SARS spread in 2002 and 2003 in China.[34–36] Coronaviruses of each genus are found in many species including cows, cats, pigs, birds, horses, dogs, and ferrets, which result in enteric hepatitis and respiratory, neurologic, renal and other disease.

To date, seven human coronaviruses (HCoVs) capable of infecting humans have been identified. Gastrointestinal and pulmonary infections are caused in both humans and animals by coronaviruses.[37] In general, estimates suggest that 2% of the population are healthy carriers of a coronavirus and that these viruses are responsible for approximately 10–20% of common

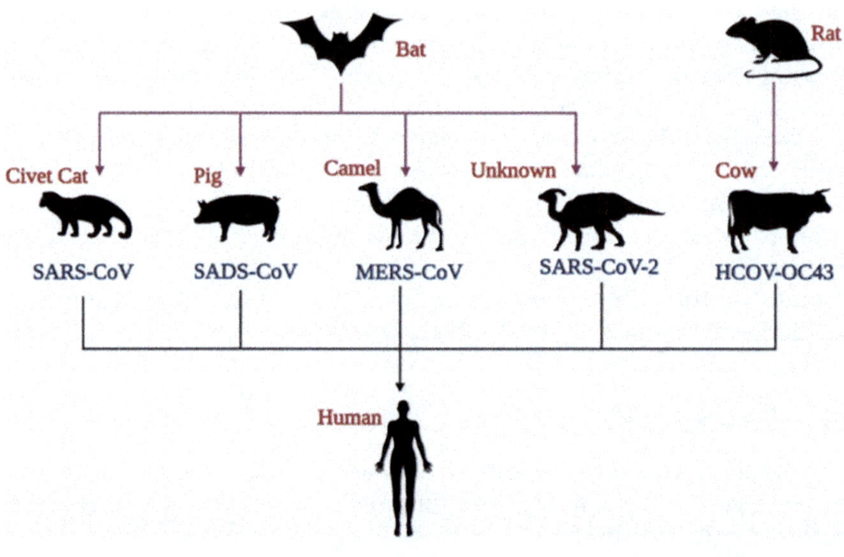

Figure 1.3 Animal origin of human coronaviruses. The transmission chain of severe acute respiratory syndrome coronavirus (SARS-CoV) and Middle East respiratory syndrome coronavirus (MERS-CoV) began with bats, who passed it on to civet cats and camels, followed by humans. SARS-CoV-2 has possibly come from pangolins.[86]

colds.[38–40] Common human alphaCoVs (HCoV-229E and HCoV-NL63) and betaCoVs of the A lineage (HCoV-OC43 and HCoV-HKU1) typically cause common colds and self-limiting upper respiratory infections in immunocompetent individuals; they can cause lower respiratory infections in immunocompromised subjects and the elderly. Human betaCoVs of the B and C lineage (SARS-CoV, SARS-CoV-2, and MERS-CoV) have resulted in epidemics with variable clinical severity featuring respiratory and extra-respiratory manifestations.

Mortality rates reported for SARS-CoV were ~10% and for MERS-CoV ~35%. The mortality rate for SARS-CoV-2 is considerably lower, but many survivors display enhanced morbidities.

The genome of SARS-CoV-2, isolated from a cluster of patients displaying atypical pneumonia after visiting Wuhan, had 89% nucleotide identity with bat SARS-like-CoVZXC21 and 82% with human SARS-CoV.[41] For this reason, the new virus was called SARS-CoV-2 by the International Committee on Taxonomy of Viruses, and it was thought that several SARS-CoV-2 variants probably exist. Its single-stranded RNA genome contains 29 891 nucleotides, encoding for 9860 amino acids.

Although the origins of SARS-CoV-2 are not entirely understood, genomic analyses suggest that SARS-CoV-2 probably evolved from a strain found in bats. It is still not definitively known if there is an intermediate mammalian host between bats and humans, or if a mutation in the original strain directly enabled virulence towards humans.

SARS-CoV-2 is ~60–140 nM in diameter and has a round or elliptic and often pleomorphic form. Like other coronaviruses, SARS-CoV-2 is sensitive to ultraviolet rays and heat. As high temperature decreases the replication of any species of virus, SARS-CoV-2 is estimated to be inactivated at ~27 °C (80–81 °F), but may resist the cold even below 0 °C (32 °F). This class of viruses can be effectively inactivated by lipid solvents including ether (75%), ethanol (60%), chlorine-containing disinfectant, peroxyacetic acid, and chloroform (except for chlorhexidine).

Some 6 months into the pandemic, the coronavirus has continued to spread and has ravaged the world. Scientists have now generated more than 80 000 viral sequences. This new genetic information has allowed transmission chains to be traced – revealing, for example, cryptic community transmission in the United States – as well as showing that a variant that seems to be particularly infectious to cultured cells has now become dominant around the world.[42,43] What this altered infectivity means for transmission and disease is not yet clear.[2]

1.5 Classification and Grouping

Phylogenetic trees have been projected for different regions of the coronavirus genome, in which four coronavirus clusters are distinguished. Three have been classified as genera alpha-CoV, beta-CoV and gamma-CoV. The fourth type includes recently detected coronaviruses of birds and seems to

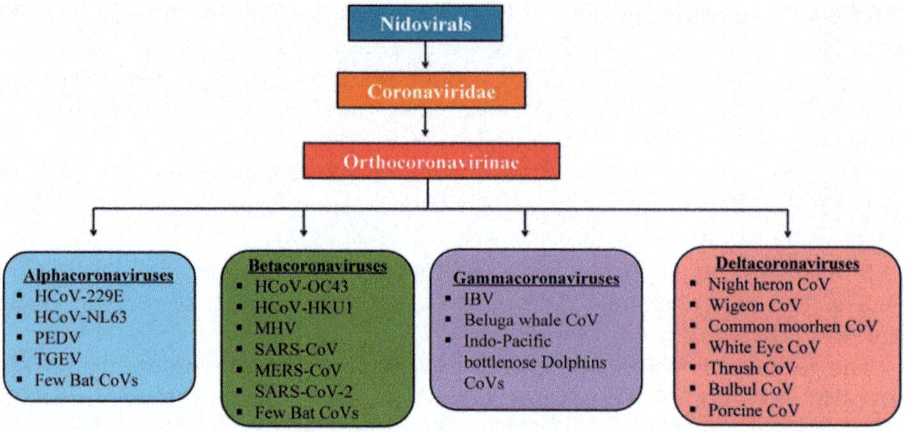

Figure 1.4 Classification of coronaviruses. Coronaviruses come under the order Nidovirals, suborder Coronaviridae and subfamily Orthocoronavirinae. The four coronavirus clusters are distinguished in this figure, in which they have been classified as the genera alpha-CoV, beta-CoV, gamma-CoV and Delta-CoV.

characterize a novel genus named delta-CoV. In the beta-CoV, four separate lineages were distinguished, and they are designated as A through D, in accordance with previous subgroups 2A through D correspondingly. Figure 1.4 depicts the classification of coronaviruses.

1.5.1 Alpha-CoVs

The viruses in this genus are a distinct monophyletic group within the Coronavirinae subfamily. The general characteristics that are different from other coronaviruses are (1) a distinctive type of non-structural protein (nsp) 1, different in size and sequence from beta-CoV nsp1 and without a counterpart in the gamma-CoV and (2) the presence of a shared gene for an unnecessary multispanning alpha-CoV membrane protein (αmp). As mentioned previously, while for some alpha-CoVs, αmp is the only additional protein, others may include up to six additional genes.

1.5.2 Beta-CoVs

Beta-CoVs fall under murine coronaviruses, which form a diverse monophyletic group in the Coronavirinae subfamily. Excepting for their comparatively close phylogenetic association, the only known general characteristics that would be different from other coronaviruses is their nsp1, distinctive in size and sequence from alpha-CoV nsp1 and without equivalent in the gamma-CoV genus.

1.5.3 Gamma-CoVs

Gamma-CoVs are avian coronaviruses which form a distinct monophyletic group in the Coronavirinae subfamily. Besides their moderately close phylogenetic relationship, there are no known mutual characteristics of virion morphology, genome organization and gene replication, composition or biology that would differ from other coronaviruses. This group lacks an nsp1 moiety.

1.5.4 Delta-CoVs

Delta-CoVs fall under the type Bulbul coronavirus. Recombination is common in delta-CoV and occurs commonly in the viral genome region that encrypts the receptor binding protein. Recombination between different viral roots may contribute to the advent of novel viruses capable of interspecies transmission and adaptation to novel animal hosts.

The main targets of the alpha-CoVs and beta-CoVs are mammals, while gamma-CoVs and delta-CoVs are found in birds, along with mammals.[44] Members of this large family of viruses can cause respiratory, enteric, hepatic, and neurological diseases in many species, including camels, cattle, cats, and bats. The dangerous viruses that have spread among humans, the SARS-CoV and MERS-CoV, cause severe respiratory illness in humans, whereas the other four human coronaviruses (HCoV-NL63, HCoV-229E, HCoV-OC43, and HKU1) only cause mild upper respiratory infections in humans, often with a compromised host immune system.[45] Alpha-CoVs and beta-CoVs cause various diseases in domestic animals; they include porcine transmissible gastroenteritis virus,[46] porcine enteric diarrhea virus,[47] and swine acute diarrhea syndrome-CoV.[48] From recent studies, it has been established that the origin of coronaviruses in general can be attributed to animals. SARS-CoV, MERS-CoV, HCoV-NL63, and HCoV-229E are thought to have originated from bats; HCoV-OC43 and HKU1 supposedly from rodents.[45,49] These animals are associated with the spreading of the virus from their normal hosts to humans. Moreover, these animals also succumb to certain diseases that have originated in bats.[50]

1.6 Epidemiological History of Human Coronaviruses

Serological techniques applied in epidemiological studies have demonstrated that human respiratory coronaviruses occur mostly during the winter and spring seasons. The first human coronavirus was detected in 1965, when Tyrrell and Bynoe discovered a virus designated B814.[51] It was found in adult human embryonic tracheal organ cultures from people affected with a common cold. The infectious agent was studied by inoculating the cultures in the medium intranasally in human subjects; cold symptoms were noticed in these subjects, but the virus could not be grown in tissue culture. Simultaneously, Hamre and Procknow cultured the virus (229E) in tissue culture,[52] and B814 and 229E were found to be ether-sensitive and to possess a lipid-coated membrane to facilitate entry into host cells. However, B814 was not similar to either OC43 or 229E.[53] OC43 is the virus that is primarily involved with the common cold. Later, Almeida and Tyrrell viewed the organ cultures infected with B814 using electron microscopy and found particles resembling the bronchitis virus.[54] The morphologies of 229E and other isolated viruses by McIntosh *et al.* were similar.[52,55] In the late 1960s, Tyrrell, along with other virologists working with human and animal viruses confirmed that these viruses had similar morphology.[55,56] This group of viruses were then named coronaviruses (corona meaning crown-like appearance), which was later officially accepted as a new virus genus.[57]

The coronaviruses are associated with many respiratory illnesses that are considered to have pathogenicity.[52,53,58,59] The major illness that was associated was pneumonia, which was predominantly found in infants and young adults.[60,61] These viruses caused asthma in children and chronic bronchitis in older people and adults.[62–64] Because many researchers explored the pathogenicity and epidemiology of human coronaviruses, the importance of animal coronaviruses also grew rapidly. In the family of coronaviruses with various strain differences, McIntosh *et al.* noticed that a few strains were distantly related to 229E or OC43.[65] Human and animal coronaviruses were divided into three groups based on their genomic structures. Group I includes the 229E viruses and group II includes the OC43 virus as well as the avian infectious bronchitis and related viruses.[57]

1.6.1 Severe Acute Respiratory Syndrome

SARS is a disease that has transmitted among populations, causing a significant number of deaths and other health problems. From 2002 to 2003, a total of 8098 of SARS cases were identified in 29 countries with 774 fatalities.[66] The first patient was from Guangdong in southern China. This patient had infected three family members at the initial stages of the disease.[67] This initial infection resulted in the surge of infections in various areas from November 2002 to January 2003. International transmission began when an infected doctor who had treated people with the condition travelled to Hong Kong and stayed at a hotel.[68] He eventually began to exhibit SARS symptoms, which progressed to respiratory failure and death.[69] During his stay, the virus had infected other individuals who were concurrently staying at the hotel. When these people flew back to their respective countries, the rise of the epidemic began in North America, Europe, and Asia.[70,71] Another patient from the same hotel spread the illness to many healthcare workers and medical students, who eventually spread it to people they were in contact with, thereby starting the transmission chain.[72] In addition, another patient reportedly having diarrhea had the infection and further increased community spread. The probable route of infection was through defective sewage pipes permitting infectious aerosol fecal materials to travel between buildings in expanding air currents. Despite the efforts to control the spread, by 23 June 2003, Hong Kong reported 1755 cases with 299 deaths in a population of 6.9 million.[66] SARS was defined by Dr Carlo Urbani from Hanoi in Vietnam, a pioneer in the field of infectious disease who worked for the WHO, and was the president of Medecins sans Frontières (Italy).[73] Later it was determined that SARS is disseminated through droplet transmission.[35,74,75]

1.6.2 Middle East Respiratory Syndrome

A novel coronavirus was initially found in a patient with SARS in Saudi Arabia in June 2012. This virus was then termed the Middle East Respiratory Syndrome coronavirus (MERS-CoV).[76] From there, several outbreaks occurred in the Arabian Peninsula and up to April 2016, Saudi Arabia reported 1386 cases, of whom 587 individuals died.[77] With an elevated death rate, a dearth of specific antiviral medication, and lack of readily available vaccines, there was tremendous panic among the people of the region at that time. It was later indicated by a large study that MERS-CoV had originated from bats.[78] Interestingly, a bat-derived coronavirus was also reported in South Africa with a very similar in structure to MERS-CoV.[79] In Saudi Arabia, where bat feces was tested for MERS-CoV nucleic acids, it was reported that virus samples from the Egyptian tomb bat *Taphozous perforatus* shared 100% similarity to MERS-CoV.[80] However, the viral nucleic acid was not present in other bats from Egypt and Lebanon.[81] An additional investigation of bats as potential carriers of MERS-CoV was reported in an experiment on the Jamaican fruit bat (*Artibeus jamaicensis*). When these bats were infected with the virus, they did not exhibit any clinical symptoms, but viral shedding was observed in their respiratory and intestinal tracts.[82] This report suggested that MERS-CoV can exist and survive in bats without causing any specific symptoms. This pointed to the hypothesis that these bats probably carried the ancestor of this virus, and that they could carry the virus for further transmission into subsequent hosts.

1.6.3 Detailed Documentation of SARS-CoV-2 and COVID-19

As mentioned earlier, recent studies have stated that SARS-CoV-2 probably originated in bats and then adapted to become the virus as we know it.[7,83] The infection of humans by this virus can be ascribed to zoonotic transmission.[7,83,84] A coronavirus from the Malayan pangolin shared 99% homology with SARS-CoV-2. In addition, the receptor-binding domain (RBD) of the pangolin coronavirus had only a single amino acid change when aligned with SARS-CoV-2. Moreover, it was also observed that the pangolins that had the disease exhibited symptoms similar to those of patients infected with COVID-19, and the antibodies in the pangolin also showed specific binding to the spike (S) protein of SARS-CoV-2.[83,85,86] Although a coronavirus specific to bats had a 96% similarity with SARS-CoV-2, the RBDs were distinctive, which resulted in a low binding capability to the ACE2 receptor in humans.[87] Interestingly, it was also discovered that six residues of the RBD of SARS-CoV-2 were analogous to SARS and MERS.[83,86] From these data it can be inferred that SARS-CoV-2 is an amalgamation of pangolin-CoV and the bat-CoV-RaTG13-like virus.[86,88,89] Hence, pangolins can be considered an intermediary between the bats and humans. However, other possible intermediate hosts such as snakes, minks, and turtles are currently being studied to enhance the understanding of the virus.[86,90,91] Interestingly, scientists observed a five (out of six) amino acid residue difference within the RBD of the S protein between SARS-CoV and SARS-CoV-2, which discredited speculation that the virus was constructed in a laboratory by genetic manipulation.[90]

The rapid spread of this virus has made it a public health concern worldwide. As described earlier, COVID-19 was declared a pandemic on 11 March 2020, once the number of cases outside China had increased 13 times and 114 countries were affected with more than 118 000 cases and more than 4000 deaths.

Governmental organizations around the world had tried to slow the spread of the disease by delivering "social-distancing" guidelines and directives. Concurrently, scientists worked tirelessly to develop novel diagnostics, prevention methodologies, and therapies. Unfortunately, clarity on virus–host interaction and the disease outbreak is still lacking. Currently, therapeutic strategies are being developed and preventive measures have been implemented to reduce community transmission. In China, severe isolation strategies accomplished the reduction in progressive situations. Less severe isolation strategies have met with success in early outbreaks where infected individuals were identified through testing and contact tracing was employed. The number of COVID-19 cases continues to rise internationally. Data for 15 countries as of 31 May 2020 are depicted in Figure 1.5.

Five simple personal health measures (masking, social distancing, avoiding crowded places, convening outside in preference to indoors, and hand washing) and five straightforward public health measures (testing, contact tracing, quarantining of suspected and confirmed COVID-19 cases, travel restrictions, and lockdown of some or all places where the public gathers) would probably suffice to contain the virus in most instances (Staff Warren, personal observation).

Table 1.1 presents a comparison of the respiratory viruses SARS-CoV, MERS-CoV, and SARS-CoV-2 in terms of their infectivity and morbidity. These characteristics tend to explain the relative pathogenicity and probable persistence of SARS, MERS, and COVID-19, even in the absence of the discovery in real time, especially given the apparent possible asymptomatic transmission of COVID-19 and the large number of deaths already observed globally.

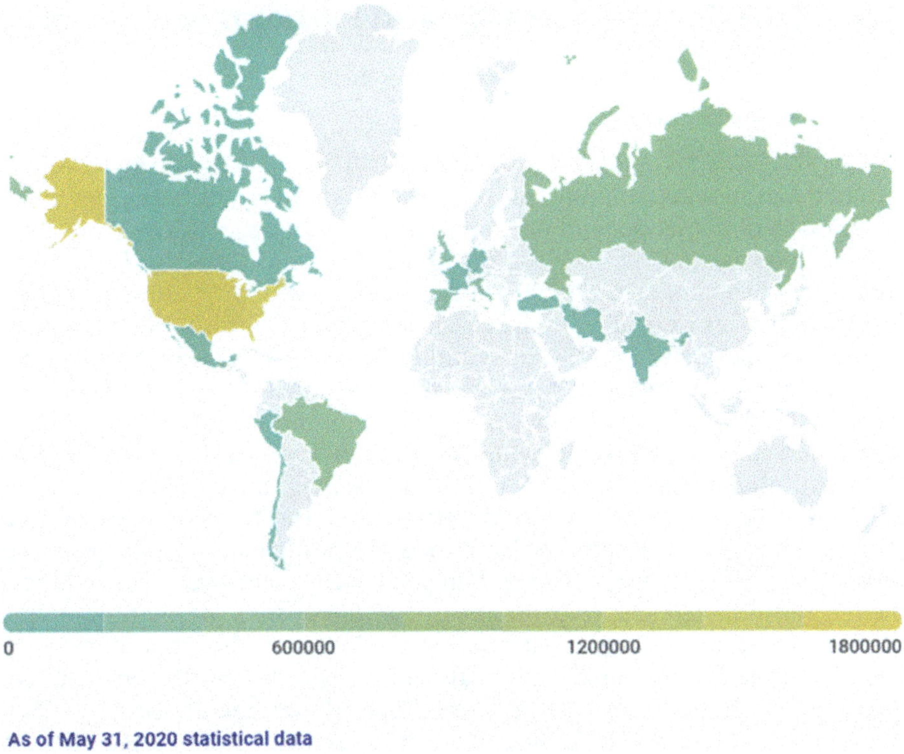

Figure 1.5 Numbers of COVID-19 cases in 15 affected countries as of 31 May 2020.

Table 1.1 Comparison of SARS-CoV, MERS-CoV, and SARS-CoV-2.

	SARS-CoV	MERS-CoV	SARS-CoV-2
Disease-causing pathogen	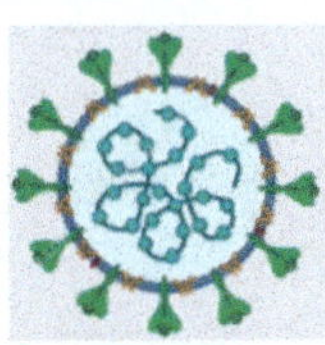	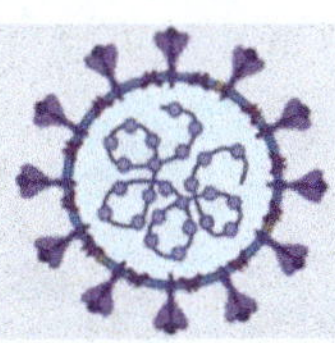	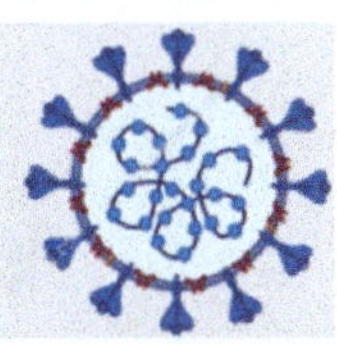
Incubation time	2–7 days	6 days	4–14 days
Community attack rate	10–60%	4–13%	30–40%
Global annual infection	8098 (in 2003)	2442	Ongoing
Global annual deaths	774 (in 2003)	842	Ongoing

1.7 Introductory Virology and Pathophysiology of SARS-CoV-2

1.7.1 Viral Genome

SARS-CoV-2 was first isolated from three COVID-19 patients using broncho-alveolar lavage fluid from Wuhan Jinyintan Hospital on 30 December 2019.[7] After molecular analysis and phylogenetic tree study, SARS-CoV-2 was determined to be a beta-CoV.[4] The reports noted that SARS-CoV-2 had 79.5% and 50% sequence similarity with SARS-CoV and MERS-CoV.[92] However, 94.6% of the sequences were alike in the seven replicase domains in the open reading frame (ORF)1ab of SARS-CoV-2 and SARS-CoV;[4] SARS-CoV-2 also shared likeness with other beta-CoVs,[7] suggesting that SARS-CoV-2 belonged to the lineage B (*Sarbecovirus*) of beta-CoVs.[93] The genome size of SARS-CoV-2 was calculated to be approximately 29.9 kb, which was inclusive of the genomic RNA and phosphorylated nucleocapsid (N) protein surrounded by a nucleocapsid.[93] The nucleocapsid is present within the phospholipid bilayers and covered with two different kinds of spike proteins: the spike glycoprotein trimmer (S) that is conserved in most of the coronaviruses, and the hemagglutinin-esterase, uncommon in coronaviruses. The membrane (M) protein and the envelope (E) protein are in between the S proteins in the viral envelope.[93] The 5′ and 3′ sequences of SARS-CoV-2 is like those of beta-CoVs, with a 5′-replicase ORF1ab-S-envelope (E)-membrane (M)-N-3′. Similar to SARS-CoV, SARS-CoV-2 has an ORF8 gene which is 366 nucleotides in length between the M and N ORF genes.[93] The basic structure of SARS-CoV-2 is illustrated in Figure 1.6.

1.7.2 Interaction of the Spike and Receptor for Entry

SARS-CoV and SARS-CoV-2 both utilize the same receptor, ACE2, for entry into the host cells.[4,94] ACE2 is a type I membrane protein found in the kidney, lung, heart, and intestines.[95] ACE2 consists of an N-terminal peptidase

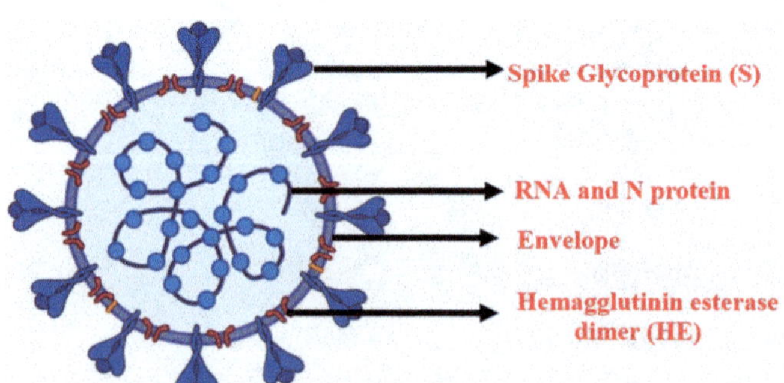

Figure 1.6 SARS-CoV-2 virion structure.

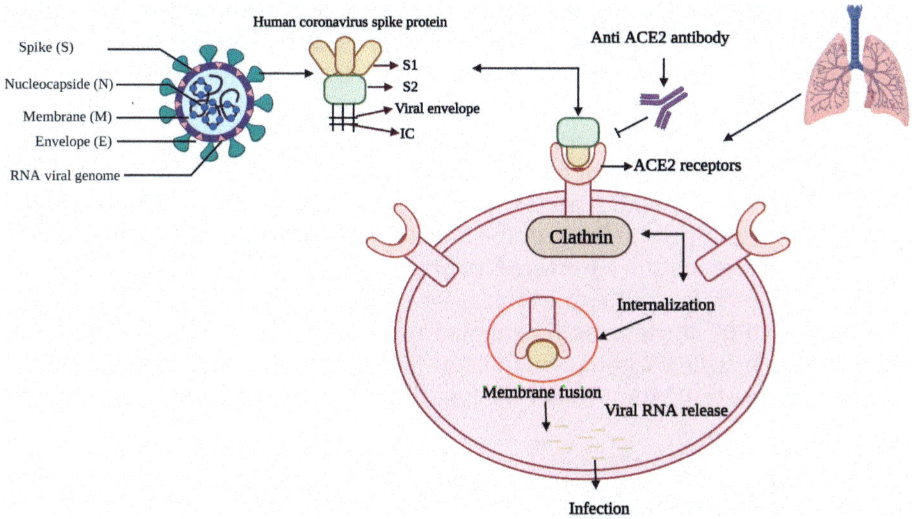

Figure 1.7 Structure and binding of the SARS-CoV-2 virus to the ACE2 receptor: The S protein has two subunits, the S1 receptor-binding subunit and S2 the membrane fusion subunit; the former attaches itself to the ACE2 receptor of the human host cell and the latter internalizes and creates the membrane fusion of the viral subunit with the ACE2 receptor. This leads to the release of the viral RNA into the host cell and results in respiratory infection. Reproduced from ref. 125 with permission from Elsevier, Copyright 2020.

domain and a C-terminal collectrin-like domain.[95] ACE2 is principally involved in the conversion of angiotensin (Ang) I to Ang-(1–7).[95] The S protein of the virus undergoes a change in its confirmation to promote the interaction between the cell membrane of the host and the viral membrane.[96] This process occurs after the receptor binds to the virus. This causes the cleavage of the S protein into its subunits S1 and S2. This results in the shedding of the S1 subunit that promotes a change in the S2 domain.[96] Then, the RBD of the S1 subunit experiences changes that can either display or conceal the factors associated with receptor binding.[97] Moreover, it has also been established that the that SARS-CoV-2 S binds to human ACE2 with an increased affinity compared to its SARS-CoV counterpart.[98] Furthermore, it has been unearthed that the ACE2-BOAT1 has the capability to bind two S proteins concurrently[4] (Figure 1.7).

1.7.3 Genomic Differences

Tang *et al.* reported that there are two genotypes of SARS-CoV-2 designated L and S, depending on an amino acid site 84 (S84L) of ORF8 gene.[99] Genomic sequences of closely related coronaviruses such as bat-CoV RaTG13 and pangolin-CoV suggested that the ancestral genotype was S-type.[99] However, the L-type emerged with the SARS-CoV-2 outbreak and is the major type spreading worldwide (https://nextstrain.org/ncov/global). Zhang *et al.* analyzed clinical and immunological data from 326 patients with confirmed COVID-19 and compared genetic variation among them, including the S84L mutation, but could not find any association.[100] Korber *et al.* reported a mutation at an amino acid site 614 (D614G) of the S protein, which is currently dominant in Europe.[101] Since the S protein is essential in infecting cells and is a primary target for neutralizing antibodies, the mutations in the S protein could be related to the virulence; however, this hypothesis must be evaluated experimentally using reverse genetics. Although more than 5000 mutations have accumulated in the infected population CoV-GLUE (http://cov-glue.cvr.gla.ac.uk), there is no evidence to supporting the contention that SARS-CoV-2 genomes are separating into distinct genotypes.[102]

1.7.4 Pathophysiology

In addressing the pathogenetic mechanisms of SARS-CoV-2, its viral structure and genome must be considered. Coronaviruses are the largest known RNA viruses, approximately 30 kb in length, with a 5′-cap and 3′-poly-A tail. Starting from the viral RNA, the synthesis of polyprotein 1a/1ab (pp1a/pp1ab) occurs within the host cell. The transcription works through the replication–transcription complex organized in double-membrane vesicles *via* the synthesis of subgenomic mRNA sequences. Of note, transcription termination occurs at transcription regulatory sequences, located between the ORFs that work as templates to produce subgenomic mRNAs. In the typical coronavirus genome, at least six ORFs can be present. Among these, a frameshift between ORF1a and ORF1b guides the production of both pp1a and pp1ab polypeptides that are processed by virally encoded chymotrypsin-like protease (3CLpro) or main protease (Mpro), as well as one or two papain-like proteases for producing 16 nsps. Apart from ORF1a and ORF1b, other ORFs encode for structural proteins, including spike, membrane, envelope, and nucleocapsid proteins, and accessory protein chains.[103] Different coronaviruses present special structural and accessory proteins translated by dedicated single-stranded guided RNAs.

Pathophysiology and virulence mechanisms of coronaviruses, and therefore also of SARS-CoV-2, have links to the function of the nsps and structural proteins. For instance, research underlined that nsp is able to block the host innate immune response.[104] Among functions of structural proteins, the envelope plays a crucial role in virus pathogenicity, as it promotes viral assembly and release. However, many of these features (*e.g.* those of nsp 2, and 11) have not yet been described. Among the structural elements of human coronaviruses are the spike glycoproteins composed of two subunits (S1 and S2). Homotrimers of S proteins comprise the spikes on the viral surface, guiding the link to host receptors.[105]

Of note, in SARS-CoV-2, the S2 subunit – containing a fusion peptide, a transmembrane domain, and cytoplasmic domain – is highly conserved. Thus, it could be a target for antiviral (anti-S2) compounds. Indeed, the spike RBD presents only a 40% amino acid identity with other SARS-CoVs. Other structural elements on which research must necessarily focus are the ORF3b that has no homology with that of SARS-CoVs and a secreted protein (encoded by ORF8) that is structurally different from those of SARS-CoV.

As mentioned earlier, both SARS-CoV and SARS-CoV-2 use the same cell receptor, ACE2,[4] and the same protease, TMPRSS2,[106] to enter cells. However, SARS-CoV-2 can also infect cell lines that don't express TMPRSS2, which could stymie drug development.[2] Researchers used Vero cells – which do not express TMPRSS2 – in an early study that suggested that the drug chloroquine might work as a treatment for COVID-19.[107] Chloroquine did not prove effective in clinical trials, and scientists discovered that it does not inhibit the virus in lung cells that express TMPRSS2.[108] Despite the overall structural similarity between the spike proteins of the two SARS coronaviruses, scientists found that the SARS-CoV-2 spike binds the ACE2 receptor at least 10 times more tightly than SARS-CoV-1 does.[98] This might explain some of the differences between how the two viruses infect people and cause disease.

The SARS-CoV-2 spike also has a feature that SARS-CoV-1 lacks: a sequence of amino acids that allows it to be recognized and cleaved by an enzyme called furin.[109] How this sequence contributes to the virulence of SARS-CoV-2 is not yet known. But similar sequences are also found in the receptor-binding protein of some influenza viruses and contribute to their virulence.

In international gene banks such as GenBank, researchers have published numerous SARS-CoV-2 gene sequences. This gene mapping is of fundamental importance allowing researchers to trace the phylogenetic tree of the virus and, above all, the recognition of strains that differ according to the mutations. According to recent research, a spike mutation, which probably occurred in late November 2019, triggered the jump to humans. Angeletti *et al.* compared the SARS-CoV-2 gene sequence with that of SARS-CoV. They analyzed the transmembrane helical segments in the ORF1ab encoded 2 (nsp2) and nsp3 and found that position 723 presents a serine instead of a glycine residue, while the position 1010 is occupied by proline instead of isoleucine.[110] The matter of viral mutations is key for explaining potential disease relapses. Further research will be needed to determine the structural characteristics of SARS-CoV-2 that underlie its pathogenetic mechanisms.

Initial reports of the disease, named COVID-19 on 11 February, described a severe respiratory illness similar to that caused by SARS-CoV. Chest scans showed patchy shadows – known as "ground-glass opacities" – in the lungs of many patients, according to early studies from hospitals in Wuhan.[111] Moreover, older people, men, and those with other diseases were more likely to be admitted to intensive care, whereas children seemed to have milder disease.[112]

The pathogenic mechanism that produces pneumonia seems to be particularly complex. But it quickly became apparent that SARS-CoV-2 is not just a respiratory virus. It also affects blood vessels, causing thrombosis[113] and strokes.[114] In rare cases, children can develop what is called a multisystem inflammatory syndrome, reminiscent of Kawasaki disease.[115] Clinical and preclinical research will have to explain many aspects that underlie the clinical presentations of the disease.

The data so far available seem to indicate that the viral infection can produce an excessive immune reaction in the host. In some cases, a reaction takes place called a "cytokine storm", resulting in extensive tissue damage with dysfunctional coagulation and damage to multiple organs. A new diagnostic term called MicroCLOTS (microvascular COVID-19 lung vessels obstructive thrombo-inflammatory syndrome) has been initiated to identify the underlying lung viral injury associated with the inflammatory reaction and with microvascular pulmonary thrombosis.[116] While several cytokines such as tumor necrosis factor-α, interleukin (IL)-1β, IL-8, IL-12, interferon-gamma inducible protein (IP10), macrophage inflammatory protein 1A (MIP1A), and monocyte chemoattractant protein 1 (MCP1) are implicated in the pathogenic cascade of the disease, the protagonist of this storm is IL-6. IL-6 is produced primarily by activated leukocytes and acts on many cells and tissues. While it can promote the differentiation of B lymphocytes and growth of some categories of cells, it inhibits the growth of others. It also stimulates the production of acute-phase proteins and plays an important role in thermoregulation, bone maintenance, and the functionality of the central nervous system. Although the main role played by IL-6 is pro-inflammatory, it can also have anti-inflammatory effects. In turn, IL-6 increases during inflammatory diseases, infections, autoimmune disorders, cardiovascular diseases, and some types of cancer. It is implicated in the pathogenesis of the cytokine release syndrome that is an acute systemic inflammatory syndrome characterized by fever and multiple organ dysfunction. In addition to the IL-6 pathway, SARS-CoV-2 binds to the Toll-like receptor to induce the release of pro-IL-1β, which is cleaved into the active mature IL-1β, mediating lung inflammation, and later fibrosis.[117]

Tian *et al.*[118] and others reported histopathological data obtained from the lungs of two patients who underwent lung lobectomies for adenocarcinoma and were found to have had SARS-CoV-2 infection at the time of surgery. Apart from the tumors, the lungs of both "accidental" cases showed edema and large proteinaceous exudative globules. The authors also reported vascular congestion combined with inflammatory clusters of fibrinoid material and multinucleated giant cells and hyperplasia of pneumocytes.

More recently, Zhang *et al.*[119] performed a post-mortem transthoracic needle lung biopsy in a patient who died of COVID-19. Immunostaining showed diffuse alveolar injury and an important alveolar expression of viral antigens. The pathogenic mechanism that produces pneumonia seems to be particularly complex. However, it became apparent very early on that

SARS-CoV-2 is not just a respiratory virus. It also affects blood vessels, causing thrombosis[113] and strokes.[114] In autopsies on COVID-19 cases, the authors[120] offered a detailed picture of the histological patterns in lung and extrapulmonary tissues. This picture was characterized by capillary congestion, necrosis of pneumocytes, hyaline membranes, interstitial edema, pneumocyte hyperplasia, and reactive atypia. Platelet–fibrin thrombi in small arterial vessels were the expression of intravascular coagulopathy. Moreover, in the lung they found infiltrates expressed as macrophages in alveoli and lymphocytes in the interstitium. Like SARS and MERS, severe COVID-19 lung damage was manifested in terms of diffuse alveolar disease with severe capillary congestion. Again, many findings were suggestive of vascular dysfunction in lung and other tissues. Other autopsies also have found the virus in organs other than the lungs, including the kidneys, liver, heart, and brain, as well as in the blood.[121] It is now known that symptoms of COVID-19 can include gastrointestinal, neurological, renal, cardiovascular, and other complications.[9]

Many existing and approved drugs have been considered for use in the treatment of COVID-19.[122] The steroid drug dexamethasone, which calms the overactive immune response, has been shown to reduce mortality from severe COVID-19.[123] In May 2020, the United States Food and Drug Administration (FDA) issued an emergency use authorization (EUA) that authorized remdesivir (Veklury) for the treatment of hospitalized adult and pediatric patients with severe COVID-19. The EUA was limited to those patients with severe disease, which was defined as patients with low blood oxygen levels or needing oxygen therapy or more intensive breathing support such as a mechanical ventilator. On 28 August 2020, based ongoing review of the totality of scientific information available, the FDA expanded its authorization to include suspected or laboratory-confirmed COVID-19 in all hospitalized adult and pediatric patients. The Administration's review concluded that benefits inherent in the use of remdesivir in these patients outweigh the known and potential risks.[124] Other antiviral drugs, convalescent serum, and synthetic monoclonal antibodies are being studied in clinical trials as discussed in subsequent chapters. See Chapters 6 and 15 in Volume 1 and Chapter 8 in Volume 2.

1.8 Executive Summary and Conclusions

- The spread of this novel coronavirus began in China and the numbers of persons now infected have drastically increased globally, such that the virus is uncontrollable except by the most extreme public health measures.
- The transmission of SARS-CoV-2 is yet to be fully determined, but it appears to transmit quite readily among both symptomatic and asymptomatic persons, especially *via* respiratory droplets and aerosols.

- Although SARS-CoV-2 is similar to SARS-CoV, the likelihood of dying from SARS-CoV-2 is less, but the likelihood of becoming infected is greater than for SARS-CoV. Part of the explanation for this may be that that SARS-CoV-2 can be frequently transmitted from an asymptomatic person, whereas SARS-CoV is typically transmitted only from a symptomatic person.
- Further research will be needed to fully determine the structural and biological characteristics of SARS-CoV-2 that underlie its ability to cause severe disease in some persons with derangement of their immune system and multi-organ failure, but cause no apparent disease in others who nevertheless test positive for the virus.
- For current pharmacological therapy, the steroid dexamethasone has been shown to reduce mortality in those with severe COVID-19, and the antiviral drug remdesivir has been shown to abbreviate the course of the disease and may also reduce mortality. Other antiviral drugs, convalescent serum, and synthetic monoclonal antibodies are being studied in clinical trials.
- There are at least a million zoonotic viruses that could potentially jump species to infect humans, *via* mammals or birds, and more careful attention will need to be given to safeguards in "wet markets" like the one in Wuhan, as well as avoiding the capture and ingestion of wild animals.
- The global spread of this virus and the economic and social disruption it has caused should serve as a warning to all countries that international collaboration and 10 simple personal and public health measures need to be priorities for all countries so that no future outbreak can again trigger a pandemic.

List of Abbreviations

CDC	Centers for Disease Control and Prevention
CoVs	Coronaviruses, positive-sense, single-stranded RNA enveloped viruses belonging to the subfamily coronavirinae
MERS-CoV	Middle East respiratory syndrome coronavirus
SARS-CoV	Severe acute respiratory syndrome coronavirus
SARS-CoV-2	Severe acute respiratory syndrome coronavirus 2
+ssRNA	Single-stranded RNA (viruses)

Acknowledgements

The authors and editors thank Melanie K. Shadoan and Chris L. Ringwalt for their excellent editing of this chapter.

References

1. D. Tyrrell and M. Bynoe, Cultivation of a novel type of common-cold virus in organ cultures, *Br. Med. J.*, 1965, **1**(5448), 1467.
2. Progress report on the coronavirus pandemic, *Nature*, 2020, **584**(325), DOI: 10.1038/d41586-020-02414-1.

3. F. Wu, S. Zhao, B. Yu, Y. M. Chen, W. Wang, Z. G. Song, Y. Hu, Z. W. Tao, J. H. Tian, Y. Y. Pei, M. L. Yuan, Y. L. Zhang, F. H. Dai, Y. Liu, Q. M. Wang, J. J. Zheng, L. Xu, E. C. Holmes and Y. Z. Zhang, A new coronavirus associated with human respiratory disease in China, *Nature*, 2020, **579**, 265–269.
4. P. Zhou, X. L. Yang, X. G. Wang, B. Hu, L. Zhang and W. Zhang, *et al.*, A pneumonia outbreak associated with a new coronavirus of probable bat origin, *Nature*, 2020, **579**(7798), 270–273.
5. Y. Yan, W. I. Shin, Y. X. Pang, Y. Meng, J. Lai and C. You, *et al.*, The First 75 Days of Novel Coronavirus (SARS-CoV-2) Outbreak: Recent Advances, Prevention, and Treatment, *Int. J. Environ. Res. Public Health*, 2020, **17**(7), 2323.
6. Y. Jin, H. Yang, W. Ji, W. Wu, S. Chen and W. Zhang, *et al.*, Virology, Epidemiology, Pathogenesis, and Control of COVID-19, *Viruses*, 2020, **12**(4), 372.
7. N. Zhu, D. Zhang, W. Wang, X. Li, B. Yang and J. Song, *et al.*, A Novel Coronavirus from Patients with Pneumonia in China, 2019, *N. Engl. J. Med.*, 2020, **382**(8), 727–733.
8. D. Wu, T. Wu, Q. Liu and Z. Yang, The SARS-CoV-2 outbreak: what we know, *Int. J. Infect. Dis.*, 2020, 44–48.
9. A. Gupta, M. V. Madhavan, K. Sehgal, N. Nair, S. Mahajan and T. S. Sehrawat, *et al.*, Extrapulmonary manifestations of COVID-19, *Nat. Med.*, 2020, **26**(7), 1017–1032.
10. CDC CfDCaP, *COVID-19 Pandemic Planning Scenarios – Updated 10 July, 2020*, 2020.
11. WHO, *Novel Coronavirus(2019-nCoV) Situation Report – 3*, 2020. Available from: https://www.who.int/docs/default-source/coronaviruse/situation-reports/20200123-sitrep-3-2019-ncov.pdf?sfvrsn=d6d23643_8.
12. Q. Fottrell, Mayor of Wuhan, epicenter of coronavirus outbreak, says 5 million people left the city before travel restrictions were imposed, MarketWatch, 27 January, 2020.
13. WHO, *IHR Procedures Concerning Public Health Emergencies of International Concern*, PHEIC, 2005.
14. WHO, *WHO China Joint Mission on COVID-19 Final Report*, World Health Organization, Geneva, Switzerland, 2020.
15. WHO, *Coronavirus Disease 2019 (COVID-19): Situation Report. Updated 11 March, 2020*, World Health Organization, Geneva, 2020. Available from: https://apps.who.int/iris/handle/10665/331475.
16. WHO, *Coronavirus Disease 2019 (COVID-19) Situation Report – 71*, World Health Organization, 2020. Data as Reported by National Authorities by 10:00 CET 31 March, 2020. Available from: https://www.who.int/docs/default-source/coronaviruse/situation-reports/20200331-sitrep-71-covid-19.pdf?sfvrsn=4360e92b_8.
17. WHO, *Coronavirus Disease 2019 (COVID-19) Situation Report – 85*, 2020. Available from: https://www.who.int/docs/default-source/coronaviruse/situation-reports/20200414-sitrep-85-covid-19.pdf?sfvrsn=7b8629bb_4.

18. Q. Li, X. Guan, P. Wu, X. Wang, L. Zhou and Y. Tong, *et al.*, Early Transmission Dynamics in Wuhan, China, of Novel Coronavirus-Infected Pneumonia, *N. Engl. J. Med.*, 2020, **382**(13), 1199–1207.
19. C. T. Bauch, J. O. Lloyd-Smith, M. P. Coffee and A. P. Galvani, Dynamically modeling SARS and other newly emerging respiratory illnesses: past, present, and future, *Epidemiology*, 2005, **16**(6), 791–801.
20. K. Kupferschmidt, Outbreak of Virus from China Declared Global Emergency, *Science*, 2020.
21. M. Cascella, M. Rajnik, A. Cuomo, S. C. Dulebohn and R. Di Napoli, *Features, Evaluation and Treatment Coronavirus (COVID-19)*, StatPearls, Treasure Island (FL), 2020, available at https://www.ncbi.nlm.nih.gov/books/NBK.
22. N. Van Doremalen, T. Bushmaker, D. H. Morris, M. G. Holbrook, A. Gamble and B. N. Williamson, *et al.*, Aerosol and surface stability of SARS-CoV-2 as compared with SARS-CoV-1, *N. Engl. J. Med.*, 2020, **382**(16), 1564–1567.
23. Z. D. Guo, Z. Y. Wang, S. F. Zhang, X. Li, L. Li and C. Li, *et al.*, Aerosol and Surface Distribution of Severe Acute Respiratory Syndrome Coronavirus 2 in Hospital Wards, Wuhan, China, 2020, *Emerging Infect. Dis.*, 2020, **26**(7), 1583–1591.
24. S. A. Lauer, K. H. Grantz, Q. Bi, F. K. Jones, Q. Zheng and H. R. Meredith, *et al.*, The incubation period of coronavirus disease 2019 (COVID-19) from publicly reported confirmed cases: estimation and application, *Ann. Intern. Med.*, 2020, **172**(9), 577–582.
25. D. P. Oran and E. J. Topol, Prevalence of Asymptomatic SARS-CoV-2 Infection: A Narrative Review, *Ann. Intern. Med.*, 2020, M20-3012.
26. X. He, E. H. Y. Lau, P. Wu, X. Deng, J. Wang and X. Hao, *et al.*, Temporal dynamics in viral shedding and transmissibility of COVID-19, *Nat. Med.*, 2020, **26**(5), 672–675.
27. D. Starr, How Italy's 'father of the swabs' fought the coronavirus, *Science*, 2020.
28. L. Morris, Why Germany's coronavirus death rate is so much lower than other countries' rates, *The Washington Post*. 24 March, 2020.
29. CDC CfDCaP, *Overview of Testing for SARS-CoV-2 (COVID-19)*, 2020. Available from: https://www.cdc.gov/coronavirus/2019-ncov/hcp/testing-overview.html.
30. E. R. Weise, Adrianna, CDC clarifies surprise guidelines that people without COVID-19 symptoms don't need testing, *USA Today*, 2020. Published 3:36 pm EDT 27 August 2020. Updated 11:54 am EDT 28 August 2020.
31. J. F. Chan, K. K. To, H. Tse, D. Y. Jin and K. Y. Yuen, Interspecies transmission and emergence of novel viruses: lessons from bats and birds, *Trends Microbiol.*, 2013, **21**(10), 544–555.
32. D. Vijaykrishna, G. J. D. Smith, J. X. Zhang, J. S. M. Peiris, H. Chen and Y. Guan, Evolutionary insights into the ecology of coronaviruses, *J. Virol.*, 2007, **81**(8), 4012–4020.

33. P. C. Woo, S. K. Lau, Y. Huang and K. Y. Yuen, Coronavirus diversity, phylogeny and interspecies jumping, *Exp. Biol. Med.*, 2009, **234**(10), 1117–1127.
34. R. A. Fouchier, T. Kuiken, M. Schutten, G. Van Amerongen, G. J. Van Doornum and B. G. Van Den Hoogen, *et al.*, Koch's postulates fulfilled for SARS virus, *Nature*, 2003, **423**(6937), 240.
35. T. G. Ksiazek, D. Erdman, C. S. Goldsmith, S. R. Zaki, T. Peret and S. Emery, *et al.*, A novel coronavirus associated with severe acute respiratory syndrome, *N. Engl. J. Med.*, 2003, **348**(20), 1953–1966.
36. N. S. Zhong, B. J. Zheng, Y. M. Li, L. L. Poon, Z. H. Xie and K. H. Chan, *et al.*, Epidemiology and cause of severe acute respiratory syndrome (SARS) in Guangdong, People's Republic of China, in February, 2003, *Lancet*, 2003, **362**(9393), 1353–1358.
37. P. Masters and S. Perlman, *Fields Virology*, Coronaviridae, Wolters Kluwer Health/Lippincott Williams & Wilkins, Philadelphia, 6th edn, 2013, pp. 825–858.
38. S. B. Greenberg, Update on Human Rhinovirus and Coronavirus Infections, *Semin. Respir. Crit. Care Med.*, 2016, **37**(4), 555–571.
39. M. J. Mäkelä, T. Puhakka, O. Ruuskanen, M. Leinonen, P. Saikku and M. Kimpimäki, *et al.*, Viruses and bacteria in the etiology of the common cold, *J. Clin. Microbiol.*, 1998, **36**(2), 539–542.
40. M. Singh, Heated, humidified air for the common cold, *Cochrane Database Syst. Rev.*, 2006, **3**, Cd001728.
41. J. F. Chan, S. Yuan, K. H. Kok, K. K. To, H. Chu and J. Yang, *et al.*, A familial cluster of pneumonia associated with the 2019 novel coronavirus indicating person-to-person transmission: a study of a family cluster, *Lancet*, 2020, **395**(10223), 514–523.
42. T. Bedford, A. L. Greninger, P. Roychoudhury, L. M. Starita, M. Famulare and M.-L. Huang, *et al.*, Cryptic transmission of SARS-CoV-2 in Washington State, *Science*, 2020, **370**, 571–575.
43. B. Korber, W. M. Fischer, S. Gnanakaran, H. Yoon, J. Theiler and W. Abfalterer, *et al.*, Tracking changes in SARS-CoV-2 Spike: evidence that D614G increases infectivity of the COVID-19 virus, *Cell*, 2020, 812–827. e19.
44. P. C. Y. Woo, S. K. P. Lau, C. S. F. Lam, C. C. Y. Lau, A. K. L. Tsang and J. H. N. Lau, *et al.*, Discovery of seven novel Mammalian and avian coronaviruses in the genus deltacoronavirus supports bat coronaviruses as the gene source of alphacoronavirus and betacoronavirus and avian coronaviruses as the gene source of gammacoronavirus and deltacoronavirus, *J. Virol.*, 2012, **86**(7), 3995–4008.
45. D. Forni, R. Cagliani, M. Clerici and M. Sironi, Molecular evolution of human coronavirus genomes, *Trends Microbiol.*, 2017, **25**(1), 35–48.
46. D. A. Brian and R. S. Baric, *Coronavirus Genome Structure and Replication. Coronavirus Replication and Reverse Genetics*, Springer, 2005, pp. 1–30.

47. C. M. Lin, L. J. Saif, D. Marthaler and Q. Wang, Evolution, antigenicity and pathogenicity of global porcine epidemic diarrhea virus strains, *Virus Res.*, 2016, **226**, 20–39.
48. P. Zhou, H. Fan, T. Lan, X. L. Yang, W. F. Shi and W. Zhang, *et al.*, Fatal swine acute diarrhoea syndrome caused by an HKU2-related coronavirus of bat origin, *Nature*, 2018, **556**(7700), 255–258.
49. S. Su, G. Wong, W. Shi, J. Liu, A. C. K. Lai and J. Zhou, *et al.*, Epidemiology, Genetic Recombination, and Pathogenesis of Coronaviruses, *Trends Microbiol.*, 2016, **24**(6), 490–502.
50. Y. W. Huang, A. W. Dickerman, P. Pineyro, L. Li, L. Fang and R. Kiehne, *et al.*, Origin, evolution, and genotyping of emergent porcine epidemic diarrhea virus strains in the United States, *mBio*, 2013, **4**(5), e00737-13.
51. D. A. Tyrrell and M. L. Bynoe, Cultivation of viruses from a high proportion of patients with colds, *Lancet*, 1966, **1**(7428), 76–77.
52. D. Hamre and J. J. Procknow, A new virus isolated from the human respiratory tract, *Proc. Soc. Exp. Biol. Med.*, 1966, **121**(1), 190–193.
53. A. Bradburne, M. Bynoe and D. Tyrrell, Effects of a "new" human respiratory virus in volunteers, *Br. Med. J.*, 1967, **3**(5568), 767.
54. J. D. Almeida and D. A. J. Tyrrell, The morphology of three previously uncharacterized human respiratory viruses that grow in organ culture, *J. Gen. Virol.*, 1967, **1**(2), 175–178.
55. K. McIntosh, W. B. Becker and R. M. Chanock, Growth in suckling-mouse brain of "IBV-like" viruses from patients with upper respiratory tract disease, *Proc. Natl. Acad. Sci. U. S. A.*, 1967, **58**(6), 2268.
56. K. H. Witte, M. Tajima and B. C. Easterday, Morphologic characteristics and nucleic acid type of transmissible gastroenteritis virus of pigs, *Arch. Gesamte Virusforsch.*, 1968, **23**(1), 53–70.
57. M. M. Lai, *Coronaviridae: the Viruses and Their Replication*, 2001.
58. A. F. Bradburne and B. A. Somerset, Coronavirus antibody titres in sera of healthy adults and experimentally infected volunteers, *Epidemiol. Infect.*, 1972, **70**(2), 235–244.
59. K. McIntosh, A. Z. Kapikian, H. C. Turner, J. W. Hartley, R. H. Parrott and R. M. Chanock, Seroepidemiologic studies of coronavirus infection in adults and children, *Am. J. Epidemiol.*, 1970, **91**(6), 585–592.
60. K. McIntosh, R. K. Chao, H. E. Krause, R. Wasil, H. E. Mocega and M. A. Mufson, Coronavirus infection in acute lower respiratory tract disease of infants, *J. Infect. Dis.*, 1974, **130**(5), 502–507.
61. R. P. Wenzel, J. O. Hendley, J. A. Davies and J. M. Gwaltney Jr, Coronavirus infections in military recruits. Three-year study with coronavirus strains OC43 and 229E, *Am. Rev. Respir. Dis.*, 1974, **109**(6), 621–624.
62. A. R. Falsey, R. M. McCann, W. J. Hall, M. M. Criddle, M. A. Formica and D. Wycoff, *et al.*, The "common cold" in frail older persons: impact of rhinovirus and coronavirus in a senior daycare center, *J. Am. Geriatr. Soc.*, 1997, **45**(6), 706–711.

63. A. R. Falsey, E. E. Walsh and F. G. Hayden, Rhinovirus and coronavirus infection-associated hospitalizations among older adults, *J. Infect. Dis.*, 2002, **185**(9), 1338–1341.
64. K. McIntosh, E. F. Ellis, L. S. Hoffman, T. G. Lybass, J. J. Eller and V. A. Fulginiti, Association of viral and bacterial respiratory infection with exacerbations of wheezing in young asthmatic children, *Chest*, 1973, **63**(Suppl. 4), 43S.
65. K. McIntosh, A. Z. Kapikian, K. A. Hardison, J. W. Hartley and R. M. Chanock, Antigenic relationships among the coronaviruses of man and between human and animal coronaviruses, *J. Immunol.*, 1969, **102**(5), 1109–1118.
66. C. W. Lam, M. H. Chan and C. K. Wong, Severe acute respiratory syndrome: clinical and laboratory manifestations, *Clin. Biochem. Rev.*, 2004, **25**(2), 121–132.
67. Z. Zhao, F. Zhang, M. Xu, K. Huang, W. Zhong and W. Cai, *et al.*, Description and clinical treatment of an early outbreak of severe acute respiratory syndrome (SARS) in Guangzhou, PR China, *J. Med. Microbiol.*, 2003, **52**(Pt 8), 715–720.
68. M. Chan-Yeung and W. C. Yu, Outbreak of severe acute respiratory syndrome in Hong Kong Special Administrative Region: case report, *BMJ*, 2003, **326**(7394), 850–852.
69. G. W. Wong and D. S. Hui, Severe acute respiratory syndrome (SARS): epidemiology, diagnosis and management, *Thorax*, 2003, **58**(7), 558–560.
70. L. Y. Hsu, C. C. Lee, J. A. Green, B. Ang, N. I. Paton and L. Lee, *et al.*, Severe acute respiratory syndrome (SARS) in Singapore: clinical features of index patient and initial contacts, *Emerging Infect. Dis.*, 2003, **9**(6), 713–717.
71. S. M. Poutanen, D. E. Low, B. Henry, S. Finkelstein, D. Rose and K. Green, *et al.*, Identification of severe acute respiratory syndrome in Canada, *N. Engl. J. Med.*, 2003, **348**(20), 1995–2005.
72. B. Tomlinson and C. Cockram, SARS: Experience at Prince of Wales Hospital, Hong Kong, *Lancet*, 2003, **361**(9368), 1486–1487.
73. J. M. Drazen, Case clusters of the severe acute respiratory syndrome, *N. Engl. J. Med.*, 2003, **348**(20), e6–e7.
74. J. S. M. Peiris, S. T. Lai, L. L. M. Poon, Y. Guan, L. Y. C. Yam and W. Lim, *et al.*, Coronavirus as a possible cause of severe acute respiratory syndrome, *Lancet*, 2003, **361**(9366), 1319–1325.
75. C. Drosten, S. Gunther, W. Preiser, S. van der Werf, H. R. Brodt and S. Becker, *et al.*, Identification of a novel coronavirus in patients with severe acute respiratory syndrome, *N. Engl. J. Med.*, 2003, **348**(20), 1967–1976.
76. A. M. Zaki, S. van Boheemen, T. M. Bestebroer, A. D. Osterhaus and R. A. Fouchier, Isolation of a novel coronavirus from a man with pneumonia in Saudi Arabia, *N. Engl. J. Med.*, 2012, **367**(19), 1814–1820.

77. H. A. Mohd, J. A. Al-Tawfiq and Z. A. Memish, Middle East Respiratory Syndrome Coronavirus (MERS-CoV) origin and animal reservoir, *Virol. J.*, 2016, **13**(1), 87.
78. A. Annan, H. J. Baldwin, V. M. Corman, S. M. Klose, M. Owusu and E. E. Nkrumah, *et al.*, Human betacoronavirus 2c EMC/2012-related viruses in bats, Ghana and Europe, *Emerging Infect. Dis.*, 2013, **19**(3), 456–459.
79. N. L. Ithete, S. Stoffberg, V. M. Corman, V. M. Cottontail, L. R. Richards and M. C. Schoeman, *et al.*, Close relative of human Middle East respiratory syndrome coronavirus in bat, South Africa, *Emerging Infect. Dis.*, 2013, **19**, 1697–1699.
80. Z. A. Memish, N. Mishra, K. J. Olival, S. F. Fagbo, V. Kapoor and J. H. Epstein, *et al.*, Middle East respiratory syndrome coronavirus in bats, Saudi Arabia, *Emerging Infect. Dis.*, 2013, **19**(11), 1819–1823.
81. M. M. Shehata, D. K. Chu, M. R. Gomaa, M. AbiSaid, R. El Shesheny and A. Kandeil, *et al.*, Surveillance for Coronaviruses in Bats, Lebanon and Egypt, 2013–2015, *Emerging Infect. Dis.*, 2016, **22**(1), 148–150.
82. V. J. Munster, D. R. Adney, N. van Doremalen, V. R. Brown, K. L. Miazgowicz and S. Milne-Price, *et al.*, Replication and shedding of MERS-CoV in Jamaican fruit bats (Artibeus jamaicensis), *Sci. Rep.*, 2016, **6**(1), 21878.
83. L. Zhang, F. M. Shen, F. Chen and Z. Lin, Origin and evolution of the 2019 novel coronavirus, *Clin. Infect. Dis.*, 2020, **71**(15), 882–883.
84. R. Lu, X. Zhao, J. Li, P. Niu, B. Yang and H. Wu, *et al.*, Genomic characterisation and epidemiology of 2019 novel coronavirus: implications for virus origins and receptor binding, *Lancet*, 2020, **395**(10224), 565–574.
85. K. Xiao, J. Zhai, Y. Feng, N. Zhou, X. Zhang and J.-J. Zou, *et al.*, Isolation and characterization of 2019-nCoV-like coronavirus from Malayan pangolins, *Nature*, 2020, **583**(7815), 286–289; Erratum in: *Nature*, 2021, **600**(7887), E8–E10.
86. K. Xiao, J. Zhai, Y. Feng, N. Zhou, X. Zhang and J. J. Zou, *et al.*, Isolation of SARS-CoV-2-related coronavirus from Malayan pangolins, *Nature*, 2020, 286–289.
87. M. Paniagua, D. Cartmel and C. Dominguez, *Receptor Recognition by the Novel Coronavirus from Wuhan: an Analysis Based on Decade-long Structural Studies of SARS*, 2020.
88. J. Cui, F. Li and Z. L. Shi, Origin and evolution of pathogenic coronaviruses, *Nat. Rev. Microbiol.*, 2019, **17**(3), 181–192.
89. P. Liu, W. Chen and J. P. Chen, Viral Metagenomics Revealed Sendai Virus and Coronavirus Infection of Malayan Pangolins (Manis javanica), *Viruses*, 2019, **11**(11), 979.
90. K. G. Andersen, A. Rambaut, W. I. Lipkin, E. C. Holmes and R. F. Garry, The proximal origin of SARS-CoV-2, *Nat. Med.*, 2020, **26**(4), 450–452.
91. Z. Liu, X. Xiao, X. Wei, J. Li, J. Yang and H. Tan, *et al.*, Composition and divergence of coronavirus spike proteins and host ACE2 receptors predict potential intermediate hosts of SARS-CoV-2, *J. Med. Virol.*, 2020, **92**(6), 595–601.

92. R. Lu, X. Zhao, J. Li, P. Niu, B. Yang and H. Wu, *et al.*, Genomic characterisation and epidemiology of 2019 novel coronavirus: implications for virus origins and receptor binding, *Lancet*, 2020, **395**(10224), 565–574.
93. A. Wu, Y. Peng, B. Huang, X. Ding, X. Wang and P. Niu, *et al.*, Genome Composition and Divergence of the Novel Coronavirus (2019-nCoV) Originating in China, *Cell Host Microbe*, 2020, **27**(3), 325–328.
94. W. Li, M. J. Moore, N. Vasilieva, J. Sui, S. K. Wong and M. A. Berne, *et al.*, Angiotensin-converting enzyme 2 is a functional receptor for the SARS coronavirus, *Nature*, 2003, **426**(6965), 450–454.
95. M. Donoghue, F. Hsieh, E. Baronas, K. Godbout, M. Gosselin and N. Stagliano, *et al.*, A novel angiotensin-converting enzyme-related carboxypeptidase (ACE2) converts angiotensin I to angiotensin 1–9, *Circ. Res.*, 2000, **87**(5), E1–E9.
96. A. H. de Wilde, E. J. Snijder, M. Kikkert and M. J. van Hemert, Host factors in coronavirus replication, *Roles of Host Gene and Non-coding RNA Expression in Virus Infection*, Springer, 2017, pp. 1–42.
97. F. Li, Structure, Function, and Evolution of Coronavirus Spike Proteins, *Annu. Rev. Virol.*, 2016, **3**(1), 237–261.
98. D. Wrapp, N. Wang, K. S. Corbett, J. A. Goldsmith, C. L. Hsieh and O. Abiona, *et al.*, Cryo-EM structure of the 2019-nCoV spike in the prefusion conformation, *Science*, 2020, **367**(6483), 1260–1263.
99. X. Tang, C. Wu, X. Li, Y. Song, X. Yao and X. Wu, *et al.*, On the origin and continuing evolution of SARS-CoV-2, *Natl. Sci. Rev.*, 2020, **7**(6), 1012–1023.
100. X. Zhang, Y. Tan, Y. Ling, G. Lu, F. Liu and Z. Yi, *et al.*, Viral and host factors related to the clinical outcome of COVID-19, *Nature*, 2020, **583**(7816), 437–440.
101. B. Korber, W. M. Fischer, S. Gnanakaran, H. Yoon, J. Theiler and W. Abfalterer, *et al.*, Spike mutation pipeline reveals the emergence of a more transmissible form of SARS-CoV-2, *Cell*, 2020, **182**, 812–827.
102. O. A. MacLean, R. J. Orton, J. B. Singer and D. L. Robertson, No evidence for distinct types in the evolution of SARS-CoV-2, *Virus Evol.*, 2020, **6**(1), veaa034.
103. S. Perlman and J. Netland, Coronaviruses post-SARS: update on replication and pathogenesis, *Nat. Rev. Microbiol.*, 2009, **7**(6), 439–450.
104. J. Lei, Y. Kusov and R. Hilgenfeld, Nsp3 of coronaviruses: Structures and functions of a large multi-domain protein, *Antiviral Res.*, 2018, **149**, 58–74.
105. W. Song, M. Gui, X. Wang and Y. Xiang, Cryo-EM structure of the SARS coronavirus spike glycoprotein in complex with its host cell receptor ACE2, *PLoS Pathog.*, 2018, **14**(8), e1007236.
106. M. Hoffmann, H. Kleine-Weber, S. Schroeder, N. Kruger, T. Herrler and S. Erichsen, *et al.*, SARS-CoV-2 Cell Entry Depends on ACE2 and TMPRSS2 and Is Blocked by a Clinically Proven Protease Inhibitor, *Cell*, 2020, **181**(2), 271–280.e8.

107. M. Wang, R. Cao, L. Zhang, X. Yang, J. Liu and M. Xu, *et al.*, Remdesivir and chloroquine effectively inhibit the recently emerged novel coronavirus (2019-nCoV) *in vitro*, *Cell Res.*, 2020, **30**(3), 269–271.
108. M. Hoffmann, K. Mösbauer, H. Hofmann-Winkler, A. Kaul, H. Kleine-Weber and N. Krüger, *et al.*, Chloroquine does not inhibit infection of human lung cells with SARS-CoV-2, *Nature*, 2020, **585**(7826), 588–590.
109. M. Hoffmann, H. Kleine-Weber and S. Pöhlmann, A Multibasic Cleavage Site in the Spike Protein of SARS-CoV-2 Is Essential for Infection of Human Lung Cells, *Mol. Cell*, 2020, **78**(4), 779–784.e5.
110. S. Angeletti, D. Benvenuto, M. Bianchi, M. Giovanetti, S. Pascarella and M. Ciccozzi, COVID-2019: The role of the nsp2 and nsp3 in its pathogenesis, *J. Med. Virol.*, 2020, 584–588.
111. X. Yang, Y. Yu, J. Xu, H. Shu, J. Xia and H. Liu, *et al.*, Clinical course and outcomes of critically ill patients with SARS-CoV-2 pneumonia in Wuhan, China: a single-centered, retrospective, observational study, *Lancet Respir. Med.*, 2020, **8**(5), 475–481.
112. X. Lu, L. Zhang, H. Du, J. Zhang, Y. Y. Li and J. Qu, *et al.*, SARS-CoV-2 Infection in Children, *N. Engl. J. Med.*, 2020, **382**(17), 1663–1665.
113. M. Ackermann, S. E. Verleden, M. Kuehnel, A. Haverich, T. Welte and F. Laenger, *et al.*, Pulmonary Vascular Endothelialitis, Thrombosis, and Angiogenesis in Covid-19, *N. Engl. J. Med.*, 2020, **383**(2), 120–128.
114. T. J. Oxley, J. Mocco, S. Majidi, C. P. Kellner, H. Shoirah and I. P. Singh, *et al.*, Large-vessel stroke as a presenting feature of Covid-19 in the young, *N. Engl. J. Med.*, 2020, **382**(20), e60.
115. L. R. Feldstein, E. B. Rose, S. M. Horwitz, J. P. Collins, M. M. Newhams and M. B. F. Son, *et al.*, Multisystem Inflammatory Syndrome in U.S. Children and Adolescents, *N. Engl. J. Med.*, 2020, **383**(4), 334–346.
116. F. Ciceri, L. Beretta, A. M. Scandroglio, S. Colombo, G. Landoni and A. Ruggeri, *et al.*, Microvascular COVID-19 lung vessels obstructive thromboinflammatory syndrome (MicroCLOTS): an atypical acute respiratory distress syndrome working hypothesis, *Crit. Care, Resusc.*, 2020, 95–97.
117. P. Conti, G. Ronconi, A. Caraffa, C. E. Gallenga, R. Ross and I. Frydas, *et al.*, Induction of pro-inflammatory cytokines (IL-1 and IL-6) and lung inflammation by Coronavirus-19 (COVI-19 or SARS-CoV-2): anti-inflammatory strategies, *J. Biol. Regul. Homeostatic Agents*, 2020, **34**(2), 327–331.
118. S. Tian, W. Hu, L. Niu, H. Liu, H. Xu and S. Y. Xiao, Pulmonary Pathology of Early-Phase 2019 Novel Coronavirus (COVID-19) Pneumonia in Two Patients With Lung Cancer, *J. Thorac. Oncol.*, 2020, **15**(5), 700–704.
119. H. Zhang, P. Zhou, Y. Wei, H. Yue, Y. Wang and M. Hu, *et al.*, Histopathologic Changes and SARS-CoV-2 Immunostaining in the Lung of a Patient With COVID-19, *Ann. Intern. Med.*, 2020, **172**(9), 629–632.
120. C. Deshpande, Thromboembolic Findings in COVID-19 Autopsies: Pulmonary Thrombosis or Embolism? *Ann. Intern. Med.*, 2020, M20-3255.

121. V. G. Puelles, M. Lutgehetmann, M. T. Lindenmeyer, J. P. Sperhake, M. N. Wong and L. Allweiss, *et al.*, Multiorgan and Renal Tropism of SARS-CoV-2, *N. Engl. J. Med.*, 2020, 590–592.
122. P. I. Andersen, A. Ianevski, H. Lysvand, A. Vitkauskiene, V. Oksenych and M. Bjoras, *et al.*, Discovery and development of safe-in-man broad-spectrum antiviral agents, *Int. J. Infect. Dis.*, 2020, **93**, 268–276.
123. RECOVERY Collaborative Group, P. Horby, W. S. Lim, J. R. Emberson, M. Mafham, J. L. Bell, L. Linsell, N. Staplin, C. Brightling, A. Ustianowski, E. Elmahi, B. Prudon, C. Green, T. Felton, D. Chadwick, K. Rege, C. Fegan, L. C. Chappell, S. N. Faust, T. Jaki, K. Jeffery, A. Montgomery, K. Rowan, E. Juszczak, J. K. Baillie, R. Haynes and M. J. Landray, Dexamethasone in Hospitalized Patients with Covid-19, *N. Engl. J. Med.*, 2021, **384**(8), 693–704.
124. FDA, *COVID-19 Update: FDA Broadens Emergency Use Authorization for Veklury (Remdesivir) to Include All Hospitalized Patients for Treatment of COVID-19 – Updated 28 August, 2020*, 2020. Available from: https://www.fda.gov/news-events/press-announcements/covid-19-update-fda-broadens-emergency-use-authorization-veklury-remdesivir-include-all-hospitalized.
125. B. Vellingiri, *et al.*, COVID-19: A promising cure for the global panic, *Sci. Total Environ.*, 2020, **725**, 138277.

CHAPTER 2

Severe Acute Respiratory Syndrome Coronavirus-2 (SARS-CoV-2): Structure, Lifecycle, and Replication Machinery

ARGHADIP SAMADDAR[a] AND VIJAYA LAKSHMI NAG*[a]

[a]Department of Microbiology, All India Institute of Medical Sciences, Jodhpur, Rajasthan, India
*E-mail: vijayalakshmi005@gmail.com

2.1 Introduction

Throughout history, infectious diseases have been the major cause of morbidity and mortality worldwide. Emerging and re-emerging infectious diseases have been a major concern within the recent past. According to the World Health Organization (WHO), more than 30 new infectious agents have emerged worldwide in the past three decades, out of which 60% are of zoonotic origin and have accounted for 26% of annual deaths worldwide.[1] Several viral epidemics, such as the severe acute respiratory syndrome coronavirus (SARS-CoV) in 2003, H1N1 influenza in 2009 and Middle East respiratory syndrome coronavirus (MERS-CoV) in 2012 took a heavy toll of life

The Coronavirus Pandemic and the Future Volume 1: Virology, Epidemiology, Translational Toxicology and Therapeutics
Edited by Michael D. Waters, Alok Dhawan, Tim Marrs, Diana Anderson, Stafford Warren and Claude L. Hughes

Published by the Royal Society of Chemistry, www.rsc.org

and had a considerable impact on social, economic, and public health systems worldwide.[2] In December 2019, a cluster of cases of unexplained acute pneumonia was reported from Wuhan City in China's Hubei province. These initial cases were classified as "pneumonia of unknown etiology", as the causative agent could not be ascertained. Later on, the etiology of this illness was attributed to a novel virus belonging to the *Coronaviridae* family and was designated as 2019-nCoV by the WHO. On 30 January 2020, as per the International Health Regulations (2005), the outbreak was declared a Public Health Emergency of International Concern, as it had spread to 18 countries with four countries reporting human-to-human transmission.[3] On 11 February 2020, the WHO renamed the disease as coronavirus disease 2019 (COVID-19). The same day, the Coronavirus Study Group of the International Committee on Taxonomy of Viruses designated 2019-nCoV as severe acute respiratory syndrome coronavirus 2 (SARS-CoV-2) on the basis of a phylogenetic analysis of related coronaviruses.[4] Considering its contagiousness and potential to evolve into a pandemic, the WHO upgraded the threat alarm to the "very high" level on 28 February 2020. With the dismal increase in the number of COVID-19 cases outside China, affecting more than 118 000 people across 114 countries and more than 4000 deaths, the WHO declared COVID-19 a pandemic on 11 March 2020.[5] COVID-19 has emerged has one of the worst pandemics in human history, with a staggering number of >100 million affected people across 216 countries and territories, with >2 million deaths as of January 2021.[6] COVID-19 is a respiratory illness with a clinical spectrum comprising mild to moderate disease (80%), severe disease (15%), and critical illness (5%) with an overall case fatality rate of 0.5–2.8%.[7] The severe and critical illness categories (approximately 20% of all infections) have overwhelmed healthcare systems worldwide.

2.2 Classification of Coronaviruses

Coronaviruses (CoVs) are a group of enveloped viruses with positive sense single-stranded RNA genome and having a crown-like appearance under an electron microscope (corona is the Latin term for crown) due to the presence of spike glycoproteins on the envelope. Coronaviruses are classified under the order *Nidovirales*, family *Coronaviridae*, and subfamily *Orthocoronavirinae*.[8] Coronaviruses have the largest genome among RNA viruses, with genome sizes ranging from 26 to 32 kilobases (kb) in length. Based on genetic and antigenic criteria, coronaviruses are classified into four genera: *Alphacoronavirus* (α-CoV), *Betacoronavirus* (β-CoV), *Deltacoronavirus* (δ-CoV), and *Gammacoronavirus* (γ-CoV).[9] The β-CoV genus is further divided into five subgenera or lineages. Members of this large family of viruses can cause respiratory, enteric, hepatic, and neurological diseases in different animals, including camels, cattle, cats, and bats. To date, seven coronaviruses capable of infecting humans (HCoVs) have been identified.[10] According to an estimate, 2% of the population are healthy carriers of a coronavirus and these viruses account for 5–10% of acute respiratory infections.[11]

The common HCoVs – HCoV-OC43 and HCoV-HKU1 (β-CoVs of the A lineage) and HCoV-229E and HCoV-NL63 (α-CoVs) – cause common colds and self-limiting upper respiratory infections in immunocompetent individuals. In immunocompromised subjects and the elderly, lower respiratory tract infections can occur. Other human coronaviruses – SARS-CoV, SARS-CoV-2, and MERS-CoV (β-CoVs of the B and C lineages, respectively) cause epidemics with variable clinical severity featuring respiratory and extra-respiratory manifestations.[5] SARS-CoV and MERS-CoV infection possess pandemic potential and can cause life-threatening disease with mortality rates up to 10% and 35%, respectively.[5] The γ-CoVs infect avian species, while δ-CoVs tend to infect both mammals and birds. Phylogenetic analysis has placed SARS-CoV-2 under the subgenus *Sarbecovirus* of the genus *Betacoronavirus*. According to next-generation sequencing data, SARS-CoV-2 possesses 79% homology to SARS-CoV and 50% to MERS-CoV. Besides, it also shows 88% homology to two bat-derived CoV-like coronaviruses, bat-SL-CoVZC45 and bat-SL-CoVZXC21, which suggests a zoonotic origin.[12]

2.3 Genomic Structure of SARS-CoV-2

SARS-CoV-2 has a positive-sense single-stranded RNA genome, approximately 30 kb in length.[13] The genomic RNA (gRNA) has a 5′-cap and a 3′-poly (A) tail and can act as an mRNA for immediate translation of the viral polyproteins. In addition, both 5′ and 3′ ends of the gRNA present a highly structured untranslated region that plays an important role in the regulation of RNA replication and transcription.[14] The SARS-CoV-2 genome consists of 14 open reading frames (ORFs) which are preceded by transcription regulatory sequences (TRSs). The two main transcriptional units, ORF1a and ORF1ab, comprise 67% of the viral genome and encode two major polyproteins, PP1a (~500 kDa) and PP1ab (~800 kDa), respectively.[15] The synthesis of PP1ab involves programmed ribosomal frame shifting during translation of ORF1a.[16] These polypeptides are processed by virally encoded proteases like chymotrypsin-like protease (3CLpro), main protease (Mpro) and papain-like protease (PLpro) into 16 non-structural proteins (Nsp1-16) which assemble to form the replication–transcription complex (RTC) involved in genome transcription and replication.[15] Nsp1 to Nsp10 are encoded by PP1a, and Nsp12 to Nsp16 are encoded by PP1ab (Figure 2.1).[17] The putative functions of non-structural proteins are summarized in Table 2.1.

The other ORFs at the 3′ end of the viral genome encode four structural proteins: the spike surface glycoprotein (S), membrane (M), envelope (E) and the nucleocapsid (N) proteins, which play a crucial role in structural integrity and pathogenesis (Figure 2.2).[18–20] Besides the structural proteins, the 3′ end also contains eight putative ORFs for accessory proteins: 3a, 3b, p6, 7a, 7b, 8b, 9b, and ORF14. The structural and accessory proteins are translated from a set of nested subgenomic RNAs (sgRNAs).[18]

The spike or S glycoprotein is a homotrimeric viral surface protein with a molecular weight of approximately 150 kDa that mediates attachment of the virus to the host cell receptors, resulting in fusion and subsequent

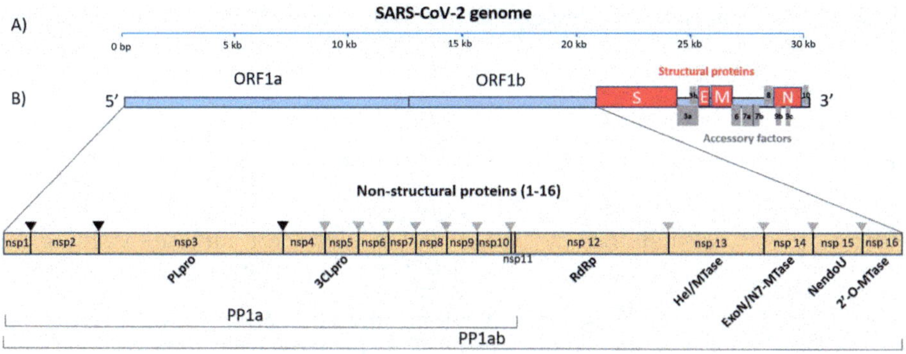

Figure 2.1 (A) Genome of SARS-COV-2 showing individual ORFs. (B) Polyprotein 1ab (PP1ab) is cleaved into 16 non-structural proteins (Nsps) by viral proteases; the black and grey triangles denote the cleavage sites of PLpro and 3CLpro, respectively. Reproduced from ref. 17, https://doi.org/10.3390/cells9051267, under the terms of the CC BY 4.0 license, https://creativecommons.org/licenses/by/4.0/.

viral entry. It consists of an extracellular N-terminus, a transmembrane domain anchored to the viral membrane, and a short intracellular C-terminal domain. In the native state, the S protein exists as an inactive precursor. During infection, target cell proteases like transmembrane protease serine 2 (TMPRSS2) and cathepsin *L* activate the S protein by cleaving it into S1 and S2 subunits.[21] S1 comprises signal peptide, N-terminal domain, and receptor-binding domain (RBD) and determines the host range and cellular tropism and facilitates viral attachment to the target cells.[22] Like SARS-CoV, SARS-CoV-2 uses the angiotensin-converting enzyme 2 (ACE2) receptor for internalization. The RBD binds to the cell ACE2 receptor in the region of aminopeptidase N. S2 mediates the fusion of the viral envelope protein with the host cell membrane, ensuring viral entry through endocytosis.[22,23] The S protein of SARS-CoV-2 has been found to exhibit a novel furin-like cleavage site (PRRAR↓SV) at the interface between S1 and S2 subunits that is processed during viral egress for S-protein priming.[23]

The membrane or M protein (~25–30 kDa) has three transmembrane domains and is the most abundant structural protein, which defines the shape of the viral envelope.[24] It has a small N-terminal glycosylated ectodomain and a large C-terminal endodomain that extends into the viral particle.[25] Studies have shown that the M protein exists as a heterodimer and may adopt two different conformations, thus allowing it to endorse membrane curvature as well as bind to the nucleocapsid.[24] Interaction of S with M protein is necessary for the retention of S protein in the Golgi complex and its incorporation into new virions. Binding of M to N protein stabilises the nucleocapsid (N protein–RNA complex) as well as the internal core of virions, which eventually helps to complete the viral assembly.[26] Together, M and E proteins constitute the viral envelope, and their interaction is necessary for the production and release of virus-like particles (VLPs).[27,28]

Table 2.1 Non-structural proteins of SARS-CoV-2 and their roles in replication machinery[a].[18,19]

Non-structural proteins	Amino acids	Proposed functions
Nsp1	180 amino acids	Leader protein host translation inhibitor, degrades cellular mRNA and inhibits IFN signalling
Nsp2	638 amino acids	Modulation of host cell survival signalling pathway by interacting with host proteins: PHB1 and PHB2
Nsp3	1945 amino acids	PLpro involved in polypeptides cleaving, blocking host innate immune response and promoting cytokine expression
Nsp4	500 amino acids	DMV formation, viral replication
Nsp5	306 amino acids	3CLpro involved in polypeptides cleaving, inhibiting IFN signalling
Nsp6	290 amino acids	Restricting autophagosome expansion, promoting DMV formation
Nsp7	83 amino acids	Cofactor with Nsp8 and Nsp12, promotes RNA polymerase activity of Nsp8
Nsp8	198 amino acids	Cofactor with Nsp7 and Nsp12, primase
Nsp9	113 amino acids	Dimerization and RNA binding
Nsp10	139 amino acids	Scaffold protein for Nsp14 and Nsp16
Nsp11	13 amino acids	Unknown
Nsp12	932 amino acids	RNA-dependent RNA polymerase
Nsp13	601 amino acids	RNA helicase, 5′ triphosphatase involved in transcription and replication
Nsp14	527 amino acids	5′-cap RNA (3′ to 5′ exonuclease, guanine N7-methyltransferase)
Nsp15	346 amino acids	Endoribonuclease, degrades viral polyuridine sequences to evade host dsRNA sensors
Nsp16	298 amino acids	5′-cap RNA (2′-*O*-ribose-methyltransferase—potential antiviral drug target) methylation (adenine)

[a]IFN: interferon; PHB: prohibitin; PLpro: papain-like protease; DMV: double-membrane vesicles; 3CLpro: 3-chymotrypsin-like cysteine protease; dsRNA: double-stranded RNA.

The envelope or E protein (~8–12 kDa) is the smallest of the major structural proteins and plays a significant role in viral morphogenesis,[29,30] assembly,[30] release,[31] and pathogenesis.[32] This transmembrane protein has a short hydrophilic N-terminal ectodomain, a hydrophobic transmembrane domain and a C-terminal endodomain with ion channel activity.[33] During the replication cycle, E is abundantly expressed inside the infected cell, but only a small portion is incorporated into the virus envelope.[30] The E protein is being investigated for vaccine development and as a therapeutic target against SARS-CoV-2.[33]

The nucleocapsid or N protein is the only structural protein to bind to the RNA genome.[34] It binds to viral RNA in a "bead on a string" manner and plays an important role in the packaging of viral RNA into the ribonucleocapsid.[35] N protein of SARS-CoV-2 is highly conserved across coronaviruses, sharing

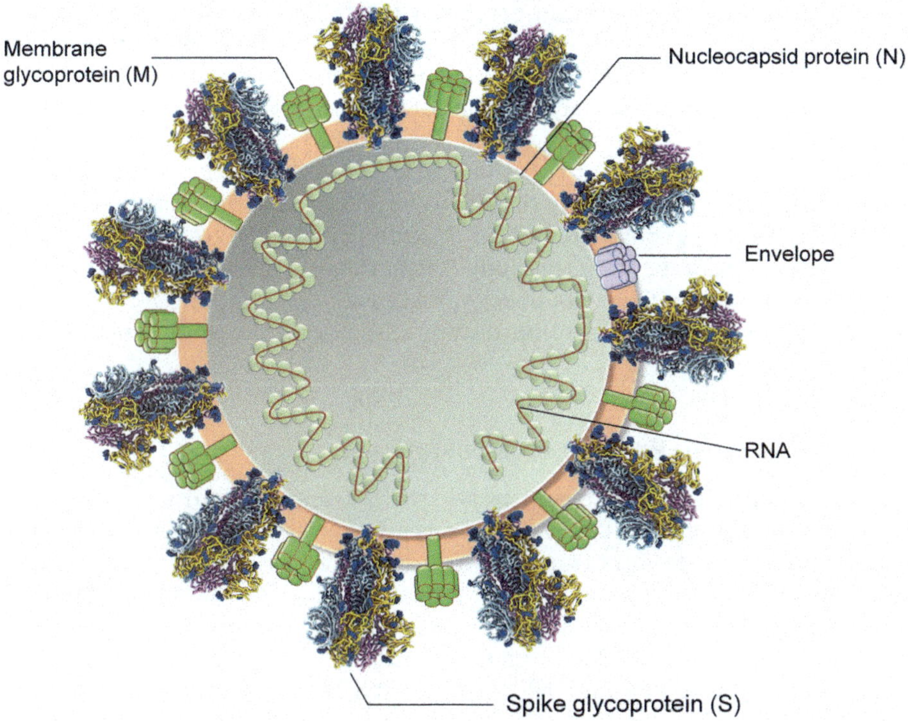

Figure 2.2 Structure of SARS-CoV-2 showing surface viral proteins, such as the spike (S) glycoprotein which mediates interaction with cell surface receptor ACE2, the membrane (M) glycoprotein and envelope (E) protein embedded in a host membrane-derived lipid bilayer enclosing the helical nucleocapsid (N) comprising viral RNA. Reproduced from ref. 20 with permission from Springer Nature, Copyright © 2020.

~90% sequence identity with that of SARS-CoV.[36] The protein is composed of two separate domains, an N-terminal domain and a C-terminal domain. It facilitates viral assembly by interacting with the viral genome and M protein, which are essential for viral RNA transcription and replication.[36]

2.4 Life Cycle of SARS-CoV-2

The life cycle of SARS-CoV-2 consists of attachment and entry, replication and transcription, and assembly and release of progeny virions.

2.4.1 Attachment and Entry

Attachment of the virus to the host cell is initiated by interactions between the S protein and ACE2 receptors which are found in various organs, such as heart, lungs, kidneys, and gastrointestinal tract, thus facilitating viral entry into target cells.[37,38] The S protein binds to ACE2 through the RBD region of the S1 subunit. The S-protein/receptor interaction is the primary determinant in infecting a host species and also controls viral tissue tropism.[38] Following receptor binding, the virus enters the host cell cytosol *via* acid-dependent proteolytic cleavage of S protein by host proteases like TMPRRS2 and cathepsin *L*, followed by fusion of the viral and cellular membranes.[39–41] S protein cleavage occurs at two sites within the S2 portion of the protein: the first cleavage for separating the RBD and fusion domains of the S protein, and the second to expose the fusion peptide (cleavage at S2′).[42] Cleavage at S2′ exposes a fusion peptide that lodges into the membrane, followed by the joining of two heptad repeats to form a six-helix bundle. The formation of this bundle results in fusion and ultimate release of the viral genome into the cytoplasm.[40,43]

2.4.2 Replicase Protein Expression

The next step is the translation of the replicase gene from viral genomic RNA. Following internalization, there is uncoating of genomic RNA into cytoplasm that undergoes translation and generates replicase polyproteins PP1a and PP1ab, which are further cleaved by virus-encoded proteinases into 16 Nsps.[44] Many of these Nsps assemble to form the RTC in double-membrane vesicles (DMVs) which are mainly a congregation of RNA-dependent RNA polymerase (RdRp)- and helicase-containing subunits.[45] RTC is responsible for RNA replication and transcription of the subgenomic RNAs.[44,46] The Nsps also harbor other enzyme domains and functions, as listed in Table 2.1.

2.4.3 Replication and Transcription

Viral RNA synthesis follows the translation and assembly of the viral replicase complexes. The replication of SARS-CoV-2 involves ribosomal frame shifting during the translation process and generates both genomic and multiple copies of subgenomic RNA by discontinuous transcription required for relevant viral proteins.[20,44] sgRNAs serve as mRNAs for the structural and accessory genes which reside downstream of the replicase polyproteins. These subgenomic messenger RNAs (mRNAs) possess common 5′ leader and 3′ terminal sequences.[47] Transcription termination and subsequent acquisition of a leader RNA occurs at the TRS, located between ORFs.[17] SARS-CoV-2, like other coronaviruses, is able to recombine using both homologous and non-homologous recombination, a phenomenon attributed to the strand-switching ability of the RdRp.[48,49]

2.4.4 Assembly and Release

After replication and synthesis of subgenomic RNA, the structural proteins S, E, and M are translated and inserted into the endoplasmic reticulum. These proteins move along the secretory pathway into the endoplasmic reticulum–Golgi intermediate compartment (ERGIC).[35,50] In the compartment, the viral genomes that are encapsidated by the N protein bud into the membrane, are enveloped, and form the mature virus. The M protein directs most of the protein–protein interactions required for virus assembly.[51] However, VLPs can only be formed when M protein is co-expressed with E protein, suggesting the need for these two proteins to produce viral envelope.[52,53] Additional roles of the E protein include promoting membrane curvature and preventing M protein aggregation.[54] Following assembly, the mature virions are transported to the cell surface in vesicles and released by exocytosis.

2.5 SARS-CoV-2 RNA Replication Machinery

The viral RNA synthesis is initiated by the RTC, a complex network of several non-structural proteins.[55] It comprises convoluted membranes and DMVs derived from the endoplasmic reticulum. These membranous elements represent a precise spatial distribution of different units of the RTC. The viral replication machinery is anchored to the convoluted membranes by means of transmembrane proteins Nsp3, Nsp4, and Nsp6, while the double-stranded RNA (dsRNA) originating from the replication-transcription process is mainly housed within the DMVs. Therefore, these structures confer a protective environment to evade the host innate immune system and subsequent degradation.[56] The replication machinery of SARS-CoV-2 involves an array of functional proteins from the N- to C-termini of the polyprotein, PP1ab (Figure 2.1). These include RdRp (Nsp12),[57] the zinc-binding helicase (HEL, Nsp13)[58] and a number of other enzymatic functions related to post-translational modification of viral RNA, such as mRNA capping (Nsp14, Nsp16), RNA proofreading (Nsp14),[59,60] and uridylate-specific endoribonuclease activity (NendoU, Nsp15) which counteracts dsRNA sensing and detection.[61] The activity of these enzymes is further regulated by other non-structural proteins (Nsp7–Nsp10) which play a crucial role in replication and transcription processes.[62,63] These protein subunits assemble to form a replication–transcription enzyme complex which is anchored to the membranes derived from the host cell endoplasmic reticulum and drives the synthesis of new genomic and subgenomic mRNAs.[64] The main non-structural proteins involved in SARS-CoV-2 RNA replication along with their putative functions are summarised in Table 2.1. In addition to these functions, several other activities are essential to boost the efficiency of the replication machinery. Suppression of host gene expression and inhibition of host innate immune responses are attributed to Nsp1, which is considered as a major virulence factor.[65,66] The primary role of N protein is to protect the viral genome by enclosing it into a helical ribonucleocapsid.[50,67] In addition, the N protein interacts with the viral envelope protein E and facilitates genome condensation and packaging inside the viral particle.[68,69] Nsp12 and Nsp8 play a vital role in the assembly of the entire RNA polymerase replicative machinery (Figure 2.3). Nsp12, an RdRp, is the key enzyme regulating the synthesis of all viral RNA molecules. It interacts with and enhances the activity of Nsp13 helicase through a direct

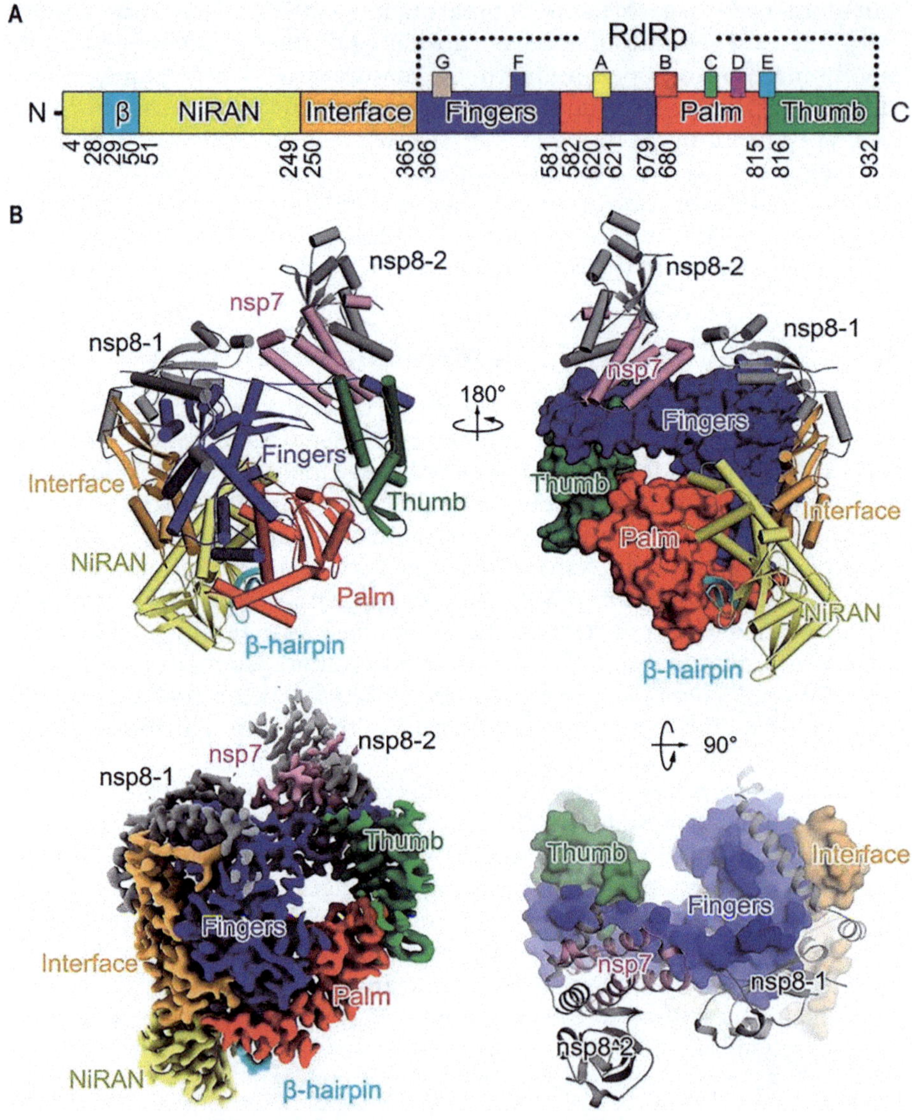

Figure 2.3 Structure of SARS-CoV-2 replication machinery. (A) Domain organization of SARS-CoV-2 Nsp12. The interdomain borders are labeled with residue numbers. The polymerase motifs are colored as follows. Motif A: yellow; motif B: red; motif C: green; motif D: violet; motif E: cyan; motif F: blue; and motif G: light brown. (B) Ribbon diagram of SARS-CoV-2 Nsp12 polypeptide chain in three perpendicular views. Domains are colored the same as in (A). The individual Nsp8 (Nsp8-1) bound to Nsp12 and that in the Nsp7–Nsp8 pair (Nsp8-2) are in grey; the Nsp7 is in pink. The bottom left panel shows electron microscopic reconstruction of the Nsp12–Nsp8 complex. Reproduced from ref. 48, https://doi.org/10.1126/science.abb7498, under the terms of the CC BY 4.0 license, https://creativecommons.org/licenses/by/4.0/.

protein–protein interaction.[70] The positive regulation of Nsp13 by Nsp12 is an important event in viral replication. Like SARS-CoV, Nsp12 can associate with Nsp14, Nsp5, and Nsp9.[71] Thus, Nsp12–Nsp8 complex represents the central hub for viral replication machinery.

References

1. World Health Organization, Disease Outbreaks by Year, Accessed 13 August, 2020. Available from: https://www.who.int/csr/don/archive/year/en/.
2. A. A. Balkhair, COVID-19 Pandemic: A New Chapter in the History of Infectious Diseases, *Oman Med. J.*, 2020, **35**, e123.
3. T. K. Burki, Coronavirus in China, *Lancet Respir. Med.*, 2020, **8**, 238.
4. Coronaviridae Study Group of the International Committee on Taxonomy of Viruses, The species Severe acute respiratory syndrome-related coronavirus: classifying 2019-nCoV and naming it SARS-CoV-2, *Nat. Microbiol.*, 2020, **5**, 536–544.
5. M. Cascella, M. Rajnik and A. Cuomo, *et al.*, *Features, Evaluation, and Treatment of Coronavirus*, Updated 4 October, 2020. In: StatPearls [Internet], StatPearls Publishing, Treasure Island (FL), January 2020. Available from: https://www.ncbi.nlm.nih.gov/books/NBK554776/.
6. World Health Organization (WHO), *WHO Coronavirus Disease (COVID-19) Dashboard*, WHO, Geneva, 2021. Available from: https://covid19.who.int/. Accessed 31 January, 2021.
7. Z. Wu and J. M. McGoogan, Characteristics of and Important Lessons From the Coronavirus Disease 2019 (COVID-19) Outbreak in China: Summary of a Report of 72 314 Cases From the Chinese Center for Disease Control and Prevention, *JAMA*, 2020, **323**, 1239–1242.
8. C. I. Paules, H. D. Marston and A. S. Fauci, Coronavirus Infections-More Than Just the Common Cold, *JAMA*, 2020, **323**, 707–708.
9. A. R. Sahin, A. Erdogan and P. M. Agaoglu, *et al.*, 2019 novel coronavirus (COVID-19) outbreak: a review of the current literature, *Eurasian J. Med. Oncol.*, 2020, **4**, 1–7.
10. J. F. Chan, K. K. To, H. Tse, D. Y. Jin and K. Y. Yuen, Interspecies transmission and emergence of novel viruses: lessons from bats and birds, *Trends Microbiol.*, 2013, **21**, 544–555.
11. Y. Chen, Q. Liu and D. Guo, Emerging coronaviruses: Genome structure, replication, and pathogenesis, *J. Med. Virol.*, 2020, **92**, 418–423.
12. Z. W. Ye, S. Yuan, K. S. Yuen, S. Y. Fung, C. P. Chan and D. Y. Jin, Zoonotic origins of human coronaviruses, *Int. J. Biol. Sci.*, 2020, **16**, 1686–1697.
13. A. Wu, Y. Peng and B. Huang, *et al.*, Genome Composition and Divergence of the Novel Coronavirus (2019-nCoV) Originating in China, *Cell Host Microbe*, 2020, **27**, 325–328.
14. I. Sola, F. Almazán, S. Zúñiga and L. Enjuanes, Continuous and Discontinuous RNA Synthesis in Coronaviruses, *Annu. Rev. Virol.*, 2015, **2**, 265–288.

15. A. E. Gorbalenya, L. Enjuanes, J. Ziebuhr and E. J. Snijder, Nidovirales: evolving the largest RNA virus genome, *Virus Res.*, 2006, **117**, 17–37.
16. S. G. Sawicki, D. L. Sawicki and S. G. Siddell, A contemporary view of coronavirus transcription, *J. Virol.*, 2007, **81**, 20–29.
17. M. Romano, A. Ruggiero, F. Squeglia, G. Maga and R. Berisio, A Structural View of SARS-CoV-2 RNA Replication Machinery: RNA Synthesis, Proofreading and Final Capping, *Cells*, 2020, **9**, 1267.
18. A. Wu, Y. Peng and B. Huang, *et al.*, Genome Composition and Divergence of the Novel Coronavirus (2019-nCoV) Originating in China, *Cell Host Microbe*, 2020, **27**, 325–328.
19. F. K. Yoshimoto, The Proteins of Severe Acute Respiratory Syndrome Coronavirus-2 (SARS CoV-2 or n-COV19), the Cause of COVID-19, *Protein J.*, 2020, **39**, 198–216.
20. S. Kumar, R. Nyodu, V. K. Maurya and S. K. Saxena, Morphology, Genome Organization, Replication, and Pathogenesis of Severe Acute Respiratory Syndrome Coronavirus 2 (SARS-CoV-2), in *Coronavirus Disease 2019 (COVID-19). Medical Virology: From Pathogenesis to Disease Control*, ed. Saxena S., Springer, Singapore, 2020, pp. 23–31.
21. J. Lan, J. Ge and J. Yu, *et al.*, Structure of the SARS-CoV-2 spike receptor-binding domain bound to the ACE2 receptor, *Nature*, 2020, **581**, 215–220.
22. J. Shang, G. Ye and K. Shi, *et al.*, Structural basis of receptor recognition by SARS-CoV-2, *Nature*, 2020, **581**, 221–224.
23. A. C. Walls, Y. J. Park, M. A. Tortorici, A. Wall, A. T. McGuire and D. Veesler, Structure, Function, and Antigenicity of the SARS-CoV-2 Spike Glycoprotein, *Cell*, 2020, **181**, 281–292.e6.
24. B. W. Neuman, G. Kiss and A. H. Kunding, *et al.*, A structural analysis of M protein in coronavirus assembly and morphology, *J. Struct. Biol.*, 2011, **174**, 11–22.
25. Y. L. Siu, K. T. Teoh and J. Lo, *et al.*, The M, E, and N structural proteins of the severe acute respiratory syndrome coronavirus are required for efficient assembly, trafficking, and release of virus-like particles, *J. Virol.*, 2008, **82**, 11318–11330.
26. K. Narayanan, A. Maeda, J. Maeda and S. Makino, Characterization of the coronavirus M protein and nucleocapsid interaction in infected cells, *J. Virol.*, 2000, **74**, 8127–8134.
27. E. Mortola and P. Roy, Efficient assembly and release of SARS coronavirus-like particles by a heterologous expression system, *FEBS Lett.*, 2004, **576**, 174–178.
28. H. Vennema, G.-J. Godeke, J. Rossen, W. Voorhout, M. Horzinek and D. Opstelten, *et al.*, Nucleocapsid-independent assembly of coronavirus-like particles by co-expression of viral envelope protein genes, *EMBO J.*, 1996, **15**, 2020–2028.
29. P. Venkatagopalan, S. M. Daskalova, L. A. Lopez, K. A. Dolezal and B. G. Hogue, Coronavirus envelope (E) protein remains at the site of assembly, *Virology*, 2015, **478**, 75–85.

30. J. L. Nieto-Torres, M. L. DeDiego and E. Álvarez, *et al.*, Subcellular location and topology of severe acute respiratory syndrome coronavirus envelope protein, *Virology*, 2011, **415**, 69–82.
31. T. R. Ruch and C. E. Machamer, The coronavirus E protein: assembly and beyond, *Viruses*, 2012, **4**, 363–382.
32. S. R. Weiss and S. Navas-Martin, Coronavirus pathogenesis and the emerging pathogen severe acute respiratory syndrome coronavirus, *Microbiol. Mol. Biol. Rev.*, 2005, **69**, 635–664.
33. M. Sarkar and S. Saha, Structural insight into the role of novel SARS-CoV-2 E protein: A potential target for vaccine development and other therapeutic strategies, *PLoS One*, 2020, **15**, e0237300.
34. M. E. Grunewald, A. R. Fehr, J. Athmer and S. Perlman, The coronavirus nucleocapsid protein is ADP-ribosylated, *Virology*, 2018, **517**, 62–68.
35. A. R. Fehr and S. Perlman, Coronaviruses: an overview of their replication and pathogenesis, *Methods Mol. Biol.*, 2015, **1282**, 1–23.
36. Y. Cong, M. Ulasli and H. Schepers, *et al.*, Nucleocapsid Protein Recruitment to Replication-Transcription Complexes Plays a Crucial Role in Coronaviral Life Cycle, *J. Virol.*, 2020, **94**, e01925-19.
37. W. Ni, X. Yang and D. Yang, *et al.*, Role of angiotensin-converting enzyme 2 (ACE2) in COVID-19, *Crit. Care*, 2020, **24**, 422.
38. M. A. Tortorici and D. Veesler, Structural insights into coronavirus entry, *Adv. Virus Res.*, 2019, **105**, 93–116.
39. M. Letko, A. Marzi and V. Munster, Functional assessment of cell entry and receptor usage for SARS-CoV-2 and other lineage B betacoronaviruses, *Nat. Microbiol.*, 2020, **5**, 562–569.
40. J. Shang, G. Ye and K. Shi, *et al.*, Structural basis of receptor recognition by SARS-CoV-2, *Nature*, 2020 May, **581**(7807), 221–224.
41. M. Hoffmann, H. Kleine-Weber and S. Schroeder, *et al.*, SARS-CoV-2 Cell Entry Depends on ACE2 and TMPRSS2 and Is Blocked by a Clinically Proven Protease Inhibitor, *Cell*, 2020, **181**, 271–280.e8.
42. B. Coutard, C. Valle, X. de Lamballerie, B. Canard, N. G. Seidah and E. Decroly, The spike glycoprotein of the new coronavirus 2019-nCoV contains a furin-like cleavage site absent in CoV of the same clade, *Antiviral Res.*, 2020, **176**, 104742.
43. J. Shang, Y. Wan and C. Luo, *et al.*, Cell entry mechanisms of SARS-CoV-2, *Proc. Natl. Acad. Sci. U. S. A.*, 2020, **117**, 11727–11734.
44. S. Perlman and J. Netland, Coronaviruses post-SARS: update on replication and pathogenesis, *Nat. Rev. Microbiol.*, 2009, **7**, 439–450.
45. E. J. Snijder, E. Decroly and J. Ziebuhr, The Nonstructural Proteins Directing Coronavirus RNA Synthesis and Processing, *Adv. Virus Res.*, 2016, **96**, 59–126.
46. Y. Finkel, O. Mizrahi and A. Nachshon, *et al.*, The coding capacity of SARS-CoV-2, *Nature*, 2021, **589**, 125–130.
47. D. Kim, J. Y. Lee, J. S. Yang, J. W. Kim, V. N. Kim and H. Chang, The Architecture of SARS-CoV-2 Transcriptome, *Cell*, 2020, **181**, 914–921.e10.

48. Y. Gao, L. Yan and Y. Huang, *et al.*, Structure of the RNA-dependent RNA polymerase from COVID-19 virus, *Science*, 2020, **368**, 779–782.
49. H. S. Hillen, G. Kokic, L. Farnung, C. Dienemann, D. Tegunov and P. Cramer, Structure of replicating SARS-CoV-2 polymerase, *Nature*, 2020, **584**, 154–156.
50. C. A. de Haan and P. J. Rottier, Molecular interactions in the assembly of coronaviruses, *Adv. Virus Res.*, 2005, **64**, 165–230.
51. M. Ujike and F. Taguchi, Incorporation of spike and membrane glycoproteins into coronavirus virions, *Viruses*, 2015, **7**, 1700–1725.
52. H. Vennema, G. J. Godeke and J. W. Rossen, *et al.*, Nucleocapsid-independent assembly of coronavirus-like particles by co-expression of viral envelope protein genes, *EMBO J.*, 1996, **15**, 2020–2028.
53. Y. L. Siu, K. T. Teoh and J. Lo, *et al.*, The M, E, and N structural proteins of the severe acute respiratory syndrome coronavirus are required for efficient assembly, trafficking, and release of virus-like particles, *J. Virol.*, 2008, **82**, 11318–11330.
54. D. Schoeman and B. C. Fielding, Coronavirus envelope protein: current knowledge, *Virol. J.*, 2019, **16**, 69.
55. I. Sola, F. Almazán, S. Zúñiga and L. Enjuanes, Continuous and Discontinuous RNA Synthesis in Coronaviruses, *Annu. Rev. Virol.*, 2015, **2**, 265–288.
56. K. Knoops, M. Kikkert and S. H. Worm, *et al.*, SARS-coronavirus replication is supported by a reticulovesicular network of modified endoplasmic reticulum, *PLoS Biol.*, 2008, **6**, e226.
57. A. Cheng, W. Zhang and Y. Xie, *et al.*, Expression, purification, and characterization of SARS coronavirus RNA polymerase, *Virology*, 2005, **335**, 165–176.
58. K. A. Ivanov, V. Thiel, J. C. Dobbe, Y. van der Meer, E. J. Snijder and J. Ziebuhr, Multiple enzymatic activities associated with severe acute respiratory syndrome coronavirus helicase, *J. Virol.*, 2004, **78**, 5619–5632.
59. E. J. Snijder, P. J. Bredenbeek and J. C. Dobbe, *et al.*, Unique and conserved features of genome and proteome of SARS-coronavirus, an early split-off from the coronavirus group 2 lineage, *J. Mol. Biol.*, 2003, **331**, 991–1004.
60. M. Sevajol, L. Subissi, E. Decroly, B. Canard and I. Imbert, Insights into RNA synthesis, capping, and proofreading mechanisms of SARS-coronavirus, *Virus Res.*, 2014, **194**, 90–99.
61. K. A. Ivanov, T. Hertzig and M. Rozanov, *et al.*, Major genetic marker of nidoviruses encodes a replicative endoribonuclease, *Proc. Natl. Acad. Sci. U. S. A.*, 2004, **101**, 12694–12699.
62. I. Imbert, J. C. Guillemot and J. M. Bourhis, *et al.*, A second, non-canonical RNA-dependent RNA polymerase in SARS coronavirus, *EMBO J.*, 2006, **25**, 4933–4942.
63. M. Bouvet, A. Lugari and C. C. Posthuma, *et al.*, Coronavirus Nsp10, a critical co-factor for activation of multiple replicative enzymes, *J. Biol. Chem.*, 2014, **289**, 25783–25796.

64. S. G. Sawicki, D. L. Sawicki and S. G. Siddell, A contemporary view of coronavirus transcription, *J. Virol.*, 2007, **81**, 20–29.
65. W. Kamitani, C. Huang, K. Narayanan, K. G. Lokugamage and S. Makino, A two-pronged strategy to suppress host protein synthesis by SARS coronavirus Nsp1 protein, *Nat. Struct. Mol. Biol.*, 2009, **16**, 1134–1140.
66. K. Narayanan, C. Huang and K. Lokugamage, *et al.*, Severe acute respiratory syndrome coronavirus nsp1 suppresses host gene expression, including that of type I interferon, in infected cells, *J. Virol.*, 2008, **82**, 4471–4479.
67. R. McBride, M. van Zyl and B. C. Fielding, The coronavirus nucleocapsid is a multifunctional protein, *Viruses*, 2014, **6**, 2991–3018.
68. L. Kuo and P. S. Masters, Genetic evidence for a structural interaction between the carboxy termini of the membrane and nucleocapsid proteins of mouse hepatitis virus, *J. Virol.*, 2002, **76**, 4987–4999.
69. C. A. De Haan, L. Kuo, P. S. Masters, H. Vennema and P. J. Rottier, Coronavirus particle assembly: Primary structure requirements of the membrane protein, *J. Virol.*, 1998, **72**, 6838–6850.
70. Z. Jia, L. Yan and Z. Ren, *et al.*, Delicate structural coordination of the Severe Acute Respiratory Syndrome coronavirus Nsp13 upon ATP hydrolysis, *Nucleic Acids Res.*, 2019, **47**, 6538–6550.
71. S. M. Brockway, C. T. Clay, X. T. Lu and M. R. Denison, Characterization of the expression, intracellular localization, and replication complex association of the putative mouse hepatitis virus RNA-dependent RNA polymerase, *J. Virol.*, 2003, 77, 10515–10527.

CHAPTER 3

Accumulated Epidemiological Lessons and China's COVID-19 Response

CHAO XIE*[a] AND LIANGMIN GAO[b]

[a]Institute of International Studies, Fudan University, Shanghai, China;
[b]Institute for International and Area Studies, Tsinghua University, Beijing, China
*E-mail: xiechaocn@aliyun.com

3.1 Introduction

There is a long history of human fear of infectious disease, and that fear does not necessarily subside with advancements in medical science. The primal fear of infectious diseases plays a due role and the emergence of any new virus tends to awaken fearful recollections of past epidemics and pandemics. Collateral damage from human greed and government corruption in these situations will paradoxically lead to more infection and higher death rates as well as ethical decline.[23] Hence, it should not seem strange to consider that increased development of drugs and vaccines under the threat of a rapidly growing infectious disease benefits from increased public expenditure.[28] Nonetheless, the coronavirus disease 2019 (COVID-19) outbreak reminds the world of the very real threat and substantial lethality resulting from infectious disease.

The Coronavirus Pandemic and the Future Volume 1: Virology, Epidemiology, Translational Toxicology and Therapeutics
Edited by Michael D. Waters, Alok Dhawan, Tim Marrs, Diana Anderson, Stafford Warren and Claude L. Hughes

Published by the Royal Society of Chemistry, www.rsc.org

This chapter attempts to examine the logic of China's response to COVID-19, and thus provide a critical analysis from the perspective of fear and trust, with a particular focus on its choice of mitigation measures with high political costs involving massive quarantine and lockdowns, as quoted in a report by *The Beijing News*.[6] An opposing argument is that such drastic measures themselves would create panic. However, its marginal effect decreases to a minimum when there is a high level of extant fear due to the contagion. As noted in a *Time* report,[14] with a spreading viral infection and panic in sight, the countermeasures from the Chinese perspective should be sweeping and a physical isolation strategy could serve to stabilize public reaction. Such drastic measures carried the message that as long as people stayed at home, they would be safe. With initial resentment, the public was quick to accept that it would be in their best interest to suspend social activities and thus cut short the transmission chains. Again, it helps to comfort the public in a time of high uncertainty.

There are different logical arguments on how to mitigate public fear and regain trust among the people and with the government. The discursive or conversational approach stresses that the government should carefully design narratives to calm people down; such a functional approach consists only of concrete measures that serve to control the disease. Be that as it may, the biggest challenge is how to strike a delicate balance between considering public opinion and initiating exclusionary protective measures. Despite all the experience and lessons from its fight against severe acute respiratory syndrome (SARS) in 2003, China's response to COVID-19 witnessed setbacks with a heavy human loss at the initial stage and recognition of that fact when it had finally controlled the situation. Discussions in summarizing useful experience and lessons learned become extremely relevant in the likely event that a second or even third wave of infection might occur before a working vaccine could be found.

3.2 Accumulated Lessons from China's Response to SARS in 2003

In 2003, the SARS epidemic spread to 32 countries or regions, primarily in mainland China, Hong Kong, Taiwan, Singapore, Vietnam, and Canada. In China, the first case emerged in Guangdong Province on 16 November 2002. 3 months later, the SARS pandemic spread rapidly to Beijing and other provinces due to person-to-person transmission. Beijing was hit the hardest and experienced the largest outbreak of SARS. Beginning on 5 March 2003, cases from Shanxi province and Hong Kong were imported to Beijing, and then the epidemic reached its peak within 6 weeks on 25 April 2003, when 173 newly suspected SARS cases were hospitalized daily.[19] At that time, Beijing became the epicenter of the SARS outbreak and its position as the capital city of China had put many other cities in grave danger of infection, since it is a central traffic hub.

It was during this fight against SARS that China accumulated knowledge of a series of effective control measures. The measures included reduction of the rate of population contact, promotion of personal and environmental hygiene, such as hand-washing, mask-wearing, and disinfection of living quarters and shared facilities such as elevators, and contact tracing to ensure detection and isolation of SARS cases. Other lessons also included increasing the speed to react to an outbreak; it was demonstrated that a timely response could prevent overwhelming the capacity of the Chinese health system and thus play a critical role in addressing public fear of a spreading virus.

There were many important lessons in the details of implementation of preventive measures. First, surveillance is now a key component of the public health system after SARS, and a project for sharing scientific data was funded which includes an international monitoring system in China.[9] Second, coordination and collaboration are also keys to infectious disease control. An important lesson from SARS was that coordination and collaboration is required at every level of the Center for Disease Control and Prevention, health departments, hospitals, and institutes nationwide and globally.[36] Since SARS, reforming the mechanisms for managing infectious diseases in China is ongoing, even to the present day. Third, training of medical staff and education of the public is routine, with more resources going to the public health sector.

In 2003, these lessons were learned at heavy cost due to the general lack of preparation for emergencies, including ineffective surveillance, poor communication among health departments, and delay in reporting.[2] The World Health Organization (WHO) (2003)[30] noted that on 20 June 2003, the last group of 18 SARS patients recovered and were discharged from Beijing's hospital, signaling the end of the pandemic in China. By 3 June 2003, 5328 cases had been reported in China, and 349 deaths had occurred. Success in the fight against SARS was partly based on the epidemiological and biological characterization of SARS and the coronavirus, SARS-CoV. But at the beginning of the outbreak, very little was known about SARS-CoV and it took approximately 3 months to complete the sequencing of the virus. The understanding of transmission patterns and the ability to diagnose and determine which patients were infected ultimately led to effective containment measures and slowing of the spread of the disease. But due to the lack of medications or vaccines, early interventions were limited to public health measures.

In addition, China recognized that it is necessary to seek international cooperation in response to public health emergencies and create a good international environment for combating the epidemic. For example, China actively established contact and communication mechanisms with global organizations such as the WHO, and shared new disease knowledge and experience with other countries and international organizations. China also sought assistance from other countries and sent aid to other countries in the form of medical personnel and supplies. The experience and lessons from SARS had a tremendous impact on China's response plan for epidemic control and prevention, which was reflected in its efforts and measures to control the COVID-19 epidemic in 2020.

3.3 Epidemiologicial Perspective on China's COVID-19 Response

In late December 2019, Wuhan's health authorities detected several pneumonia cases of unknown origin. With the Chinese New Year approaching, millions of Chinese people would travel back to their homes for the celebration, and any loopholes in control and prevention measures would mean that the virus would rapidly spread from one city to the whole country.[32] On 23 January, when Wuhan announced the lockdown, approximately 5 million people had left Wuhan. By the end of January, the number of infections had surged to over 10 000. On 12 February, thousands of more cases were confirmed in Wuhan after an improvement in the method of diagnosis, when China's National Health Commission (NHC) recommended not only laboratory diagnosis, but also comprehensive consideration of epidemiological history, clinical manifestations, and imaging examinations,[34] resulting in a sudden apparent surge in confirmed cases. On 31 March 2020, the NHC announced that it would begin reporting the infection cases of symptom-free individuals who tested positive with coronavirus.

Later on, several other cities announced lockdowns, and subsequently nation-wide social distancing measures were implemented to prevent the epidemic spreading. Eventually, a quasi-national lockdown remained in place for approximately 2 months. On every count, it was a risky political decision to make. First, it might work to flatten the curve, but public stress would become more and more intense with the duration of lockdown and the rising numbers of confirmed cases. The sooner the curve would decline, and no one knew when, the lower political costs there would be. Second, Chinese New Year would mean a lot of domestic and international travel, traditional celebrations and processions, as well as extensive visits of relatives and friends. As a reference point from the Xinhua News Agency,[8] the Chinese New Year travel rush is known as "Spring movement", and in 2019 there were 3.0 billion trips in 40 days. The power of tradition should be properly understood here; no one could be sure the public would cooperate. It was as if people were asked to stay at home during the Holi in India, or Christmas in the West. It would be a huge task to obtain public understanding and cooperation on such dramatic countermeasures.

Quarantine and lockdowns are ancient tools to conquer epidemic disease; modern technology only transforms these tools into more powerful public health armaments. In China, people got used to online shopping and communication during the SARS quarantines, and COVID-19 would only further drive the public subscription to an online lifestyle in the largely cash-less society that developed after SARS.[24] Despite these capabilities, it is not easy to ensure a basic supply of daily commodities, and the sheer size of the population affected is enough to demonstrate the scale of the challenge. After initial chaos, big-data technology and intelligent supply chains took charge, and features of door-to-door and zero-contact delivery regained public trust in their environment.

After quarantine measures took effect, the country reported no new local COVID-19 transmissions for the first time on 18 March 2020.[18] Since that time, there have been small-scale epidemics in some cities, but most have been imported cases from abroad, and have been dealt with expeditiously. Targeted quarantine and lockdowns have been instituted quickly in one or two regions, such as the small outbreaks in Beijing and Xinjiang. As noted by Worldometer,[31] as of 30 August 2020, the novel coronavirus SARS-CoV-2 has infected 85 048 people, and killed 4634 in China. It is safe to conclude that the Chinese government has employed the lessons from 2003 and that through the fight against COVID more lessons will be derived to help in future epidemic control and prevention.

3.4 China's Control and Prevention Measures in COVID-19

Concrete measures must be taken to address and mitigate the stress caused to human beings by the COVID-19 pandemic. In the case of Zika, studies found that individuals who suppressed their fears would have higher levels of fear later.[12] The first thing a spreading virus destroys is people's trust in their environment. Any working public health strategy should aim at maximizing that trust and minimizing the stress thus caused.[10] As mentioned earlier, for centuries, measures such as quarantine, lockdowns, and contact tracing have been central in responding infectious disease outbreaks.[27] The same applies today, albeit with much larger mega-cities involved. That said, the root causes of fear is always the same, that is uncertainty due to the fatality rate and the lack of effective vaccines and therapies for a cure.

3.4.1 Mask-wearing and its Comforting Effect

There were debates in China about whether people should wear masks in confronting COVID-19 and the Chinese government's positive prompting generated a discussion on whether face masks would help to control a highly contagious disease like COVID-19 and about when and how to wear them.[29] Similar discussions ensued when other governments made the same suggestion to their citizens.[3] With all the valid points on the measure, it is problematic to construct a positive logical link between mask-wearing and the eventual control of the pandemic. The discussion does not have a firm conclusion from a medical science perspective, yet its psychological effects in addressing public fear and rebuilding trust among people and with the government should be properly recognized.

Wearing a mask does not change the fact that the invisible spreading virus is making the environment dangerous. But a psychological suggestion it carries is that others in the environment share a sense of insecurity and no one would intentionally or unintentionally spread the disease. It is not only about the effectiveness of mask-wearing. Subconscious measures like this reduce individual stress and thus mitigate fear caused by environmental uncertainty. Mask-wearing serves as an indicator to others that their chance of infection is reduced when people in their immediate environment wear masks.

When the Chinese government suggested it, and the public was willing to wear masks, the next question was how the supply could meet the demand. In China, the National Medical Products Administration (NMPA) is responsible for ensuring the availability of medical supplies, and on 28 August 2009 it released standard procedures for emergency medical supplies for public health needs.[17] On 7 February 2020, NMPA issued a further administrative notice to expedite registration and approval procedures on personal protective equipment (PPE) production.[16] Then, provincial and local levels would take over to make detailed plans, and local NMPA offices were drafted guidelines and temporary measures on emergency and temporary approvals.[4] But the situation became chaotic, as fear caused people to store masks at all costs, and the supply chain was further severed since most manufacturers were on holiday leave for the Chinese lunar new year. Only when major players like Sinopec, a state-owned enterprise, and BYD, a private company, launched mask manufacturing in massive quantities was the supply–demand gap filled.

It proved an equal challenge to transport PPE to affected areas, since major highways were closed due to holiday leave and forthcoming lockdowns. The lack of intra-provincial coordination only made the situation worse when local governments were storing and competing for available PPE products. Public opinion backfired when reports showed that some local governments expropriated pandemic prevention materials for their districts.[26] These obstacles could only be dealt with when the State Council stepped in to press provincial governments and local governments to ensure smooth transportation of anti-epidemic supplies and materials.

3.4.2 Epidemiological Source-tracking and Slow-down of Community Transmission

Besides wearing masks and lockdowns, tracking close contacts is "probably the most important thing", as noted by Dr Zhong Nanshan on 21 January 2020.[15] Dr Zhong is the leading medical scientist to contain the spread of COVID-19 in China and a public health veteran who led China's campaign against SARS. Modern technology again intervened to make these prior outbreak toolkits more effective. In China, popular apps such as WeChat and Alipay were used to monitor and track travel routes, and possible contacts if needed. As reported by the *South China Morning Post*,[11] colored health codes of green, yellow, and red were generated after people provided information about their temperature and whereabouts in the past 14 days in a quick health survey. These tools and information permitted them to visit other cities or enter residential areas or official buildings. Some apps were developed to enable people to check nearby infection and transmission status so that, depending on that information, they could feel relieved or keep high alert. Despite all the controversies over invasion of privacy, these massive surveillance measures proved useful in tracking close contacts and places that had been visited by confirmed COVID-19 patients.

Combined with source- or exposure-tracking, testing all close contacts and admitting all suspected and confirmed cases for treatment will be the ultimate tool to control the disease. Without proper testing, affected persons might have to delay seeking medical help and remain undetected in the community.[21] In case of a pandemic, the number of tests being performed in a particular situation could have easily discouraged any attempt at source-tracking or contact-tracing. There is no available data of how many tests China has done in the fight against COVID-19, but as a reference, as of 25 April 2020, Guangdong province alone accomplished 4.54 million tests.[35] High demand for testing early in the outbreak caused a shortage of test kits, and a huge influx of confirmed patients overwhelmed hospitals and public health workers in pandemic epicenters such as Wuhan and Huangshi. Before the outbreak of COVID-19, a city in China would have dozens of beds for infectious disease, or hundreds in the case of a mega-city, but during the COVID-19 outbreak in China it was tens of thousands of confirmed patients, and the number would be multiplied if all suspect cases had been admitted.

The situation was only partially addressed when differentiated admission and a phased treatment plan was adopted, such that the situation was stabilized. Suspect cases were quarantined in independent houses and hotel rooms; confirmed cases with light and modest syndromes went to temporary hospitals that were specially established to monitor and provide them basic medical care; and those with severe syndromes were sent to designated hospitals with epidemiological experts and intensive care units.[20] At a later stage, health workers from cities and regions that had largely controlled the situation came to help, with each province assisting one city, or region of Wuhan, the city hit first and hardest by the disease.

3.5 Comparing China's Response Against SARS and COVID-19

China's SARS cases showed the importance of research and public health capacity in China.[13] After SARS in 2003, huge infrastructure investments made China much better prepared for COVID-19 than it had been for SARS. When COVID-19 emerged in December 2019, Chinese scientists and doctors were quickly able to identify the virus and they shared genomic sequencing data internationally on 11 January 2020. They published many research papers in international journals sharing the clinical features of patients, information on person-to-person transmission, genomic characteristics, and epidemiology with other countries, and they warned the world about the threat of COVID-19. And since 2003, the Chinese government has improved its epidemic response capacity. In the 2002–2003 SARS outbreak, 300 cases and five deaths had occurred by the time China reported the outbreak to the WHO, whereas in the COVID-19 outbreak, only 27 cases and zero deaths had occurred when the WHO was notified on 3 January 2020.[33] From the time of WHO notification, 2 months elapsed before SARS-CoV was identified, compared with only 1 week until COVID-19 was identified. In short, China's experience shows the importance of investing in national health and research systems to enhance laboratory capacity as well as the workforce. These things are fundamental to a quick and effective national response to health emergencies and global health security.

Another requirement is strong top-down political action and engagement to tackle the outbreak decisively, to make sure to guarantee effective control. At the beginning of the outbreak, China's NHC sent three groups of top infectious disease experts to Wuhan to investigate COVID-19, and their recommendations informed the decision to lock down Wuhan on 23 January 2020. In addition, according to their advice, local governments were quick to respond to build multiple *Fangcang* shelter hospitals in Wuhan and other cities.[7] Community engagement was key for achieving effective implementation of control measures in China. Control measures such as mandatory mask-wearing in public areas were accepted readily by the public, and millions of China's community workers have built a first line of defense against COVID-19 in their work to provide essential healthcare and support for people with fever, severe diseases, pregnant women, the elderly, and those quarantined at home.[22] Thus, China's experiences show the importance of community solidarity and its social significance in the fight against COVID-19.

Much remains to be done and many questions remain unanswered for SARS-CoV-2. China is grateful for the help it is receiving from the international scientific, health, and public health communities in fighting COVID-19. At the same time, China is also actively engaged in foreign aid, especially to Africa and other developing countries. Since March 2020, China has sent medical teams to assist with medical supplies and shared knowledge of prevention and its control experience with the international community through multiple channels. In summary, China's experiences show that proactive investment in public health infrastructure and capacity is crucial to effectively respond to pandemics like COVID-19, and it is critical to continue to improve global surveillance, cooperation, coordination, and communication about this pandemic to be even better prepared to respond to new public health threat in the future.

The psychological perspective is one lens through which we can understand the logic of China's response plan against COVID-19, and other sociological factors should get fair credit as well. For instance, a high public saving rate in China ensures that the public is more likely to get through the hardships caused by a viral pandemic. A public memory from the effective fight against SARS in 2003, when rural areas were cordoned off and kept untouched, even though they were most vulnerable to the epidemic disease,[1] assured quick public acceptance of similar measures in urban areas. And the necessity of drastic measures was obvious in that the week-long holiday and month-long vacation period, coinciding with the outbreak in China meant that extensive human movement, interactions and gatherings, had already occurred. People had already traveled from Wuhan by the millions, and only a total lockdown would slow down, if not prevent community transmission, in any tangible way.

Chinese people tend to prioritize rights of development and survival before they can enjoy freedom and liberty in social life, and most of them tend to accept that there will be limits for the latter rights. In the West, individual freedom and liberty supersede other rights such that the public will defend and protest any attempt to sabotage their individual rights, and even a pandemic will not easily change that belief. A fair note is that this doesn't mean that in China people don't have freedom and liberty and in the West people don't have development and survival rights, but differences in rights priorities seem to influence their acceptance of certain drastic measures to fight pandemics like COVID-19. Therefore, mass surveillance tools for purposes of source-tracking were accepted as a temporary measure in China, and in the West, people are more critical of such measures and focus more on the potential abuse and misuse of their personal information.

3.6 Concluding Remarks

It is understood from a historical perspective that governments of different polities have records in succeeding and failing in epidemic prevention and control. We recognize that a people with limited government tend to rely more on themselves rather than expecting assistance and protection from the government. Public resentment is obvious, for example, in the United States against wearing masks and lockdowns. In Michigan, a supermarket security guard was shot dead when he asked a customer to wear a mask.[25] It seems that cultural and historical factors make major differences in public acceptance. Chinese culture is full of stories about how ancient kings like King Shennong tasted a hundred herbs and Dayu tamed the flood to conquer diseases and natural disasters. Such a cultural background makes the public more cooperative in tolerating costs and in observing government countermeasures to epidemics.

Human history evolves by overcoming natural disasters and conquering diseases, and COVID-19 will not be the last pandemic. The danger of a second or even a third wave of the pandemic is real until a safe and effective vaccine is produced, and hence discussions about public health strategies that complement countries of different national characteristics, and in their unique situations, are extremely relevant. In the discussion, it is important for countries to mitigate public fear and hence restore public trust among themselves and with the government. With a high cost, China has largely controlled the situation, but this doesn't mean that its public health strategy can be simply copied by others. Besides domestic debates on the controversial decisions, China's response to COVID-19 came under international criticism that it was still using medieval tools for containment in the 21st century.[5] Even now a haunting question for China is whether measures other than massive lockdowns could have the same effect.

3.7 Executive Summary

For centuries, measures such as quarantine, lockdowns and contact tracing have been central in responding infectious disease outbreaks.

There is a long history of human fear of infectious disease and that fear does not necessarily subside with advancements in medical science.

The biggest challenge for any public health strategy is to strike a delicate balance between concern for public opinion and the initiation of exclusionary protective measures.

After SARS in 2003, huge infrastructure investments made China much better prepared for COVID-19 than it had been for SARS.

Human history evolves by overcoming natural disasters and conquering diseases, and COVID-19 will not be the last pandemic.

References

1. A. Ahmad, R. Krumkamp and R. Reintjes, Controlling SARS: a review on China's response compared with other SARS-affected countries, *Trop. Med. Int. Health*, 2009, **14**(suppl.1), 36–45.
2. X. J. An, D. M. Li, L. Wang, H. D. Zhu and H. Y. Wang, Reflection on the problems and countermeasures of the institution of public health from the SARS outbreak, *Chin. J. Public Health Manage.*, 2004, **20**(1), 10–11.
3. H. Bastian, The Face Mask Debate Reveals a Scientific Double Standard, *The Wired*, 8 April 2020, available at: https://www.wired.com/story/the-face-mask-debate-reveals-a-scientific-double-standard/ (accessed 25 August 2020).
4. China Regulatory Consulting for Medical Device, *Collection of Emergency Approval Policies on Medical Masks and Protective Gears* (医用口罩及防护服等医疗器械应急审批政策汇总), 2 March 2020, available at: http://www.cirs-group.com/md/news/news/12721.html (accessed 25 August 2020).
5. F. Kretschmer, Wuhan lockdown: China takes extreme measures to stop virus spread, *Deutsche Welle*, 23 January 2020, available at: https://www.dw.com/en/wuhan-lockdown-china-takes-extreme-measures-to-stop-virus-spread/a-52120126 (accessed 25 August 2020).
6. X. Y. Die, Wuhan Lockdowns Aims at Calming Down Public (武汉"封城"，重要的是让人心安), *The Beijing News*, 23 January 2020, available at: http://www.bjnews.com.cn/opinion/2020/01/23/678263.html (accessed 25 August 2020).
7. D. Fang, S. Pan, Z. Li, T. Yuan, B. Jiang, D. Gan, B. Sheng, J. Han, T. Wang and Z. Liu, Large-scale public venues as medical emergency sites in disasters: lessons from COVID-19 and the use of Fangcang shelter hospitals in Wuhan, China, *BMJ Global Health*, 2020, **5**(6), e002815.

8. X. Fang, Z. X. Qi and W. J. Zhao, Nearly 3 billion trips will be completed in the Spring Festival in 2019 (2019 年春运收官 近30亿人次出行), *Xinhua News Agency*, 1 March 2019, available at: http://www.xinhuanet.com/politics/2019-03/01/c_1124181637.htm (accessed 25 August 2020).
9. D. Feng, S. J. De Vlas, L. Q. Fang, X. N. Han, W. J. Zhao, S. Sheng, H. Yang, Z. W. Jia, J. H. Richardus and W. C. Cao, The SARS epidemic in mainland China: bringing together all epidemiological data, *Trop. Med. Int. Health*, 2009, **14**(suppl. 1), 4–13.
10. P. D. Harms, M. Credé, M. Tynan, M. Leon and W. Jeung, Leadership and stress: A meta-analytic review, *Leadership Q.*, 2017, **28**(1), 178–194.
11. M. H. Hu, Beijing rolls out colour-coded QR system for coronavirus tracking despite concerns over privacy, inaccurate ratings, *South China Morning Post*, 2 March 2020, available at: https://www.scmp.com/tech/apps-social/article/3064574/beijing-rolls-out-colour-coded-qr-system-coronavirus-tracking (accessed 25 August 2020).
12. E. Keet, New Study Examines Public Fears During Disease Outbreaks, *ContagionLive: Infectious Diseases Today*, 11 July 2018, available at: https://www.contagionlive.com/news/new-study-examines-public-fears-during-disease-outbreaks (accessed 25 August 2020).
13. T. Lancet, COVID-19 and China: lessons and the way forward, *Lancet*, 2020, **396**(10246), 213.
14. B. Luscomb, Why Overreacting to the Threat of the Coronavirus May Be Rational, *Time*, 11 March 2020, available at: https://time.com/5801010/coronavirus-overreaction/ (accessed 25 August 2020).
15. NHC, Timeline of China releasing information on COVID-19 and advancing international cooperation on epidemic response, 6 April 2020, available at: http://en.nhc.gov.cn/2020-04/06/c_78840.htm (accessed 25 August 2020).
16. NMPA, *Notice on Accelerating Registration and Approval of Manufacturing Medical Protective Gears* (国家药监局综合司关于加快医用防护服注册审批和生产许可的通知), 7 February 2020, available at: http://www.nmpa.gov.cn/WS04/CL2078/374630.html (accessed 25 August 2020).
17. NMPA, Notice on Emergency Approval Procedures on Medical Apparatus and Instruments (关于印发医疗器械应急审批程序的通知), 28 August 2009, available at: http://www.nmpa.gov.cn/WS04/CL2197/324626.html (accessed 25 August 2020].
18. A. Pan, L. Liu, C. Wang, H. Guo, X. Hao, Q. Wang, J. Huang, N. He, H. Yu, X. Lin and S. Wei, Association of public health interventions with the epidemiology of the COVID-19 outbreak in Wuhan, China, *JAMA*, 2020, **323**(19), 1915–1923.
19. X. Pang, Z. Zhu, F. Xu, J. Guo, X. Gong, D. Liu, Z. Liu, D. P. Chin and D. R. Feikin, Evaluation of control measures implemented in the severe acute respiratory syndrome outbreak in Beijing, 2003, *JAMA*, 2003, **290**(24), 3215–3221.
20. F. Peng, L. Tu, Y. Yang, P. Hu, R. Wang, Q. Hu, F. Cao, T. Jiang, J. Sun, G. Xu and C. Chang, Management and treatment of COVID-19: the Chinese experience, *Can. J. Cardiol.*, 2020, **36**(6), 915–930.

21. B. Person, F. Sy, K. Holton, B. Govert and A. Liang, Fear and stigma: the epidemic within the SARS outbreak, *Emerging Infect. Dis.*, 2004, **10**(2), 358.
22. K. Prem, Y. Liu, T. W. Russell, A. J. Kucharski, R. M. Eggo, N. Davies, S. Flasche, S. Clifford, C. A. Pearson, J. D. Munday and S. Abbott, The effect of control strategies to reduce social mixing on outcomes of the COVID-19 epidemic in Wuhan, China: a modelling study, *Lancet Public Health*, 2020, **5**(5), e261–e270.
23. M. A. Riva, M. Benedetti and G. Cesana, Pandemic fear and literature: observations from Jack London's The Scarlet Plague, *Emerging Infect. Dis.*, 2014, **20**(10), 1753–1757.
24. C. Shuyang, S. Meng and L. Haotong, The comparative study between China and South Korea on online shopping, in 2014 International Conference on Management of e-Commerce and e-Government, 2014, pp. 16–19, available at: https://www.computer.org/csdl/proceedings-article/icmecg/2014/6543a016/12OmNx4Q6H0/ (accessed 25 August 2020).
25. M. Spelbring, Security guard, father of 9, shot and killed in Michigan after telling customer to put on face mask, *USA Today*, 4 May 2020, available at: https://www.usatoday.com/story/news/nation/2020/05/04/family-dollar-security-guard-killed-after-telling-shopper-wear-mask/3081674001/ (accessed 25 August 2020).
26. Y. Z. Tan, Dali apologizes for requisition of Chongqing masks, *China Daily*, 6 February 2020, available at: https://www.chinadaily.com.cn/a/202002/07/WS5e3c9eb2a3101282172759a6.html (accessed 25 August 2020).
27. E. Tognotti, Lessons from the history of quarantine, from plague to influenza A, *Emerging Infect. Dis.*, 2013, **19**(2), 254.
28. M. A. Vance, Disease mongering and the fear of pandemic influenza, *Int. J. Health Serv.*, 2011, **41**(1), 95–115.
29. M. J. Wei, *Under Coronavirus, Do We Need to Wear Masks?* (疫情下，我们需要一直佩戴口罩吗), Xinhua News Agency, 10 February 2020, available at: http://www.xinhuanet.com/politics/2020-02/10/c_1125554775.htm, (accessed 25 August 2020).
30. WHO, Summary table of SARS cases by country, 7 August 2003, available at: http://www.who.int/csr/sars/country/2003_08_15/en/ (accessed 25 August 2020).
31. Worldometers, *COVID-19 Coronavirus Pandemic*, 31 August 2020, available at: https://www.worldometers.info/coronavirus/ (accessed 25 August 2020).
32. J. T. Wu, K. Leung and G. M. Leung, Nowcasting and forecasting the potential domestic and international spread of the 2019-nCoV outbreak originating in Wuhan, China: a modelling study, *Lancet*, 2020, **395**(10225), 689–697.
33. Z. Wu and J. M. McGoogan, Characteristics of and important lessons from the coronavirus disease 2019 (COVID-19) outbreak in China: summary of a report of 72 314 cases from the Chinese Center for Disease Control and Prevention, *JAMA*, 2020, **323**(13), 1239–1242.

34. Y. Yang, F. Peng, R. Wang, K. Guan, T. Jiang, G. Xu, J. Sun and C. Chang, The deadly coronaviruses: The 2003 SARS pandemic and the 2020 novel coronavirus epidemic in China, *J. Autoimmun.*, 2020, 102434.
35. L. L. Zhang, As of 25 April, Guangdong Province Completed 4.54 million tests (截至4月25日 广东共完成新冠病毒核酸检测454.10万人份), *Sina.com*, 28 April 2020, available at: http://news.sina.com.cn/c/2020-04-28/doc-iirczymi8877014.shtml (accessed 25 August 2020).
36. N. S. Zhong, B. J. Zheng, Y. M. Li, L. L. M. Poon, Z. H. Xie, K. H. Chan, P. H. Li, S. Y. Tan, Q. Chang, J. P. Xie and X. Q. Liu, Epidemiology and cause of severe acute respiratory syndrome (SARS) in Guangdong, People's Republic of China, in February, 2003, *Lancet*, 2003, **362**(9393), 1353–1358.

CHAPTER 4

Clinical Presentation, Pathophysiology and Histopathology

CLAUDE HUGHES[†*a,b], IYER MAHALAXMI[†c], NACHIMUTHU SENTHIL KUMAR[†d], SURESH SELVAPURAM SUDALAIMUTHU RAJA[†e], RAVIMINICKAM THANGARASU[†f], STAFFORD WARREN[†g] AND MICHAEL WATERS[†h]

[a]IQVIA, Durham, NC, USA; [b]Duke University Medical Center, Durham, NC, USA; [c]Department of Zoology, Avinashilingam Institute of Home Science and Higher Education for Women, India; [d]Department of Biotechnology, Mizoram University, India; [e]Department of Microbiology, Government Arts and Science College (affiliated to Bharathidasan University, Trichy), India; [f]Department of Zoology, Tamil Nadu Open University, India; [g]Anne Arundel Medical Center, Annapolis, MD, USA; [h]Michael Waters Consulting USA, Hillsborough, NC, USA
*E-mail: clhughes@duke.edu

4.1 Introduction

In December 2019, severe acute respiratory syndrome coronavirus 2 (SARS-CoV-2) caused what was initially characterized as a pneumonia outbreak in Wuhan City, China; this was followed by global spread of the virus seeded mainly by airline travel worldwide. Chinese clinicians described the early

[†]All authors contributed equally and are listed alphabetically by surname.

The Coronavirus Pandemic and the Future Volume 1: Virology, Epidemiology, Translational Toxicology and Therapeutics
Edited by Michael D. Waters, Alok Dhawan, Tim Marrs, Diana Anderson, Stafford Warren and Claude L. Hughes

Published by the Royal Society of Chemistry, www.rsc.org

symptoms of coronavirus disease 2019 (COVID-19) and the progression of the disease in terms of severity, as discussed in detail later.[1] Based on an initial assessment made by the Chinese government, the World Health Organization (WHO) indicated that approximately 80% of affected individuals had symptoms similar to those of mild seasonal influenza. However, the remaining 20% developed viral interstitial pneumonia, requiring hospitalization. In approximately 5% of these patients, the pneumonia was critical and required intensive care. Overall mortality was seen to range between 0.5% and 5%, with a clear positive correlation with age and pre-existing conditions.[2] The main mode of direct transmission of SARS-CoV-2 was initially thought to be through fomites and respiratory droplets. With an emerging viral infection, the WHO provided guidelines about how to protect oneself from and prevent infection using simple public health measures such as social distancing and frequent hand washing with soap and warm water. On 11 March 2020, the WHO declared COVID-19 a pandemic, citing over 118 000 cases of the disease in more than 110 countries and territories around the world, with the sustained risk of further global spread.

What was not immediately appreciated was the extent to which the virus can be spread by asymptomatic individuals. Among individuals infected with SARS-CoV-2, the duration of viral shedding was found to be quite similar between asymptomatic and symptomatic patients.[3] Likewise, rates of transmission were found to be similar for asymptomatic and symptomatic individuals. A review of 16 reports indicates that asymptomatic patients could account for 40–96% of patients identified in the outbreaks of COVID-19.[4] The long duration of viral shedding and high proportion of asymptomatic patients make asymptomatic transmission a critical concern for effective public health management. 6 months into the outbreak, millions of confirmed cases of SARS-CoV-2 infection (COVID-19) have been reported, and the global death toll has reached hundreds of thousands of victims.

Advice for personal protection around the world went from social distancing, to that plus masks, to that plus face shields or goggles. The mantra voiced by WHO Director-General Dr Tedros Adhanom Ghebreyesus was, "No one is protected until everyone is protected". Simple face coverings were proved effective in reducing the community spread of the virus, while medical professionals were advised to wear high-efficiency N95 masks and face shields from the beginning. The literature concerning the transmission of virus-laden droplets and aerosols in various environmental settings has been reviewed.[5] In a laboratory investigation, SARS-CoV-2 remained viable in aerosols for at least 3 hours. It was detected for up to 3 days on plastic or stainless steel, but for less than 24 hours on cardboard and for less than 4 hours on copper.[6] Data released by the European Centre for Disease Prevention and Control confirmed these numbers, stating that the environmental sustainability of active SARS-CoV-2 is up to 3 hours in the air after aerosolization, approximately 24 hours on cardboard and almost 3 days on stainless steel or plastic. Environmental factors such as low humidity, reduced temperature and reduced sunlight serve to increase transmission.

As suggested by its name, SARS-CoV-2 is closely related to severe acute respiratory syndrome coronavirus (SARS-CoV).[7] Full-genome sequencing analysis was performed early on and indicated that SARS-CoV-2 shares a high sequence similarity with SARS-CoV.[8] It was generally known that the spike protein (S-protein) of coronaviruses interacts with cell receptors to mediate viral entry into target cells.[9] Upon entering the host, SARS-CoV-2 has been reported to remain in the respiratory tract for several days. This is the asymptomatic viral incubation period. Once symptoms appear, the virus typically persists for 8 days in non-severe cases and up to 14 days in severe cases. In contrast to SARS-CoV, infectivity of patients with SARS-CoV-2 peaks before symptom onset.[10] The viral load profile is found to peak at the time of onset of symptoms and a high viral load is found in severe cases and thus can be potentially used as a marker of case severity and prognosis. That said, in individual cases the disease may progress in severity without symptoms. Viral RNA particles are found in the feces as well as in whole blood, urine, serum and saliva for 5 days after the onset of symptoms and this persists for up to 4–5 weeks.[11]

Recent communications from the United States Centers for Disease Control and Prevention (CDC) have stated that symptoms may appear 2–14 days after exposure to the virus and that people with the following symptoms (updated 13 May 2020) may have COVID-19: fever or chills, cough, shortness of breath or difficulty breathing, fatigue, muscle or body aches headache, new loss of taste or smell, sore throat, congestion or runny nose, nausea or vomiting or diarrhea. Symptoms vary greatly from mild to moderate and severe, depending especially upon age and whether comorbidities are present. Thus, in some such patients with mild disease, but with comorbidities such as hypertension, cardiovascular disease, obesity and diabetes, the symptoms may quickly worsen after approximately 1 week following onset. Furthermore, if a person had more than one comorbidity, it was found that they were frequently in the severe category. These individuals were typically older and exhibited more symptoms than those with a single comorbid condition. Understanding the course of the virus under different comorbid and living conditions will help determine the best treatment and prevention strategies to limit spread and mortality.

4.2 The Distinct Clinical Trajectory and Disease Spectrum of COVID-19

It was reported early on that symptoms of COVID-19 arise on average approximately 11.5 days post-incubation period.[12] Fever and dry cough are initial milder symptoms with fewer indications of difficulty in breathing, muscle and/or joint pain, headache/dizziness, diarrhea, nausea, and the coughing up of blood.[12] Upper respiratory tract symptoms are common early in the disease; nasopharyngeal and throat swabs testing positive at presentation.[13] The viral infection peaks 5–6 days after the commencement of symptoms.[14] Severe

COVID-19 cases frequently progress to acute respiratory distress syndrome (ARDS), on average approximately 8–9 days after symptom onset.[15] Symptoms are described in Table 4.1. The clinical spectrum of COVID-19 cases vary considerably, from persons who exhibit either no or mild symptoms to those with respiratory failure who require mechanical ventilation in an intensive care unit (ICU), or those who experience multi-organ and systemic manifestations that are manifested in sepsis, septic shock, and multiple organ dysfunction syndromes (MODS). In one early study of the disease, Huang and colleagues reported that patients ($n = 41$) suffered from fever, malaise, dry cough, and dyspnea. Their chest computed tomography (CT) scans uniformly showed pneumonia with "ground-glass" opacities, which proved characteristic of COVID-19. About a third of their small sample required ICU care, and there were six (15%) fatalities.[15] A larger study from the New York metropolitan area found that in 2634 patients requiring hospitalization who had been discharged or died at the study endpoint (4 April 2020), the mortality rate was 24.5%.[16]

In some patients, the clinical onset of COVID-19 involves fever which is not very responsive to antipyretics, as well as a state of malaise and a dry cough. In more fragile patients, dyspnea may appear at the onset of symptoms. After 5–7 days, typically older patients, who may already have impaired lung function, often begin to experience shortness of breath and an increased respiratory rate. In younger subjects and in those who do not have any underlying respiratory impairments or other comorbidities, dyspnea may appear somewhat later. In otherwise healthy COVID-19 patients

Table 4.1 Mild, moderate and severe symptoms in COVID-19.

Mild	Moderate	Severe
Tiredness	Tiredness	Tiredness
Fever	Fever	Fever (moderate, high or absent)
Dry cough	Persistent cough	Persistent cough
Slight breathlessness	Breathlessness	Extreme breathlessness
Sore throat	Sore throat	Sometimes
Muscle pain	Sometimes	Sometimes
Diarrhea	Sometimes	Sometimes
Headache	Headache	Sometimes
X	Lethargy	Extreme lethargy
X	Dry mouth	Sometimes
X	Dyspnea	Dyspnea
X	X	Multi-organ damage
X	X	Hypoxia
X	X	Lung infiltrates
X	X	Pain and tightness of chest
X	X	Anorexia
X	X	Bluish lips and face
X	X	ARDS
Mild or no pneumonia	Mild or no pneumonia	Pneumonia
X	X	Septic shock

who experience a deteriorating inflammatory-induced lung injury there is a decrease in oxygen saturation (<94%). This seems to be a crucial phase of the disease, since from this point onwards, respiratory functions may deteriorate rapidly. The scenario is surprising, because, for those patients who are minimally symptomatic and slightly hypoxic, the first therapeutic approach is oxygen therapy. Although this strategy is typically effective, respiratory failure may be exacerbated in some patients. The drive for oxygen movement from blood to tissue is the oxygen tension difference that exists across the vascular wall; percentage oxygen saturation of the hemoglobin is not the driving force. With the drive preserved, the next step is typically non-invasive ventilation (NIV). This therapy can rapidly succeed by increasing the ratio between the arterial partial pressure of oxygen ($P\text{aO}_2$) and the percentage of oxygen supplied (fraction of inspired oxygen, $F\text{iO}_2$), *i.e.* $P\text{aO}_2/F\text{iO}_2$ or P/F. However, in some patients, there is a sudden, unexpected deterioration of clinical conditions. Patients may collapse under the operator's eyes and require rapid intubation and invasive mechanical ventilation. However, after 24–48 hours of mechanical ventilation, the patient can rapidly improve with a demonstrable increase in the P/F ratio. Operators are therefore tempted to proceed with weaning the patient from the ventilator. Often, however, after an initial apparent success, the patient's respiratory condition may deteriorate once again and require renewed invasive therapy. Therefore, maintaining mechanical ventilation, once initiated, for 1–2 weeks has been suggested; see Alía and Esteban for general guidance.[17] Some patients experiencing mild respiratory distress who do not need a ventilator, or even those who progress to severe respiratory distress, show improved oxygenation from proning. Proning is the process of turning a patient with precise, safe motions from their back onto their abdomen (stomach) so the individual is lying face down. The face-down position allows for better expansion of the dorsal lung regions, improved body movement and enhanced removal of secretions, which may ultimately lead to advances in oxygenation.

The heterogenous symptoms and manifestations observed clinically in patients are the result of two main factors: (1) the multiplication and spread of the virus within the body and (2) aberrant, inappropriate, excessive, and unregulated host immune response. Uncontrolled overproduction of inflammatory cytokines from innate and adaptive immune responses contributes to acute lung injury and ARDS. The secretion of more than 150 inflammatory cytokines, also termed cytokine release syndrome or cytokine storm is closely related to the development of clinical symptoms in severe cases, which results in damage to the lung microvascular and alveolar epithelial cell barrier and leads to vascular leakage, alveolar edema and hypoxia. Structural and functional damage to pulmonary architecture brings about multi-organ failure involving the kidneys, liver, heart and/or a disseminated intravascular coagulopathy picture with thromboembolic sequelae. Thus, immune-mediated inflammation plays an important role in the pathogenesis of COVID-19. Understanding the pathogenesis and associated abnormalities is a critical and essential step in the management and treatment of the disease.[18]

Case studies reported by Li and colleagues in late January 2020 include the first 425 cases recorded in Wuhan.[19] Their data indicate that the patients' median age was 59 years, with a range of 15–89 years. They found no significant sex-based differences in the sample. However, in other, later, reports, females have manifested a lower infection prevalence. Gemmati *et al.* have confirmed that males are more susceptible to SARS-CoV-2 than females, with males 65% more likely to die from the infection than females, citing data from the WHO and Chinese scientists that show that of all cases approximately 1.7% of women who contract the virus will die, compared with 2.8% of men. Data from Hong Kong hospitals stated that 32% of male and 15% of female COVID-19 patients required intensive care or died.[20]

Data from the Chinese CDC including 72 314 records that involved confirmed, suspected, diagnosed and asymptomatic cases provided the first important illustration of the epidemiologic trajectory of the Chinese outbreak.[21] There were 62% confirmed cases, which included 1% that were asymptomatic but laboratory-positive, as determined by a viral nucleic acid test. The overall case-fatality rate among confirmed cases was 2.3%. Of note was the fact that the highest number of fatal cases were in elderly patients aged 80 years and above (15% of the sample) and between 70 and 79 years (8.0%). Approximately half (49.0%) of the critically ill patients died, as did those affected by pre-existing conditions such as cardiovascular disease, diabetes, chronic respiratory disease and oncological diseases. There were no fatalities among the 1% of patients who were aged 9 years or younger. The overall fatality rate for COVID-19 is estimated to be approximately 3.4% (WHO press conference, 3 March 2020). Counting asymptomatic persons, it may be as low as 0.6% or 0.7%, or in vulnerable populations such as nursing homes, over 25%. By comparison, seasonal influenza has a mortality rate of 0.1%.

The authors of the Chinese CDC report categorized in hierarchical fashion the clinical manifestations of the disease in their sample, in ascending order of severity, as follows:

- Mild (81%): non-pneumonia and mild pneumonia.
- Severe (14%): dyspnea, respiratory rate ≥30 breaths per minute, blood oxygen saturation (SpO_2) ≤93%, PaO_2/FiO_2 ratio or P/F (the ratio between the blood pressure of the oxygen (PaO_2) and the percentage of oxygen supplied (FiO_2)) <300 mmHg, and/or lung infiltrates >50% within 24–48 hours.
- Critical (5%): respiratory failure, septic shock and/or multiple organ dysfunction or failure.[21,22]

The clinical course of the disease suggests a favorable trend in a majority of patients affected. However, as a point of reference, the criteria for the severity of respiratory insufficiency and the diagnostic criteria for sepsis and septic shock also can be used.[23] A general hierarchy of infection severity, therefore, may be specified as follows:

- Uncomplicated (mild) illness: these patients usually present with symptoms of an upper respiratory tract viral infection. Symptoms may include a mild fever, a dry cough, sore throat, nasal congestion, malaise, headache and muscle pain. Other symptoms may include a recent loss of taste and/or smell, diarrhea and vomiting. There are no signs and symptoms of a more serious condition like dyspnea.
- Moderate pneumonia: respiratory symptoms such as cough and shortness of breath (or tachypnea in children) are present with radiographic evidence of pneumonia, but without symptoms of severe pneumonia.
- Severe pneumonia: fever is associated with severe dyspnea, respiratory distress, tachypnea (>30 breaths per minute) and hypoxia (SpO_2 <90% on room air). That said, fever should be interpreted carefully, as when fever is present, the underlying disease can be moderate or occasionally even asymptomatic. Children can manifest cyanosis. Radiological imaging is necessary to confirm the diagnosis and extent of pneumonia. Multi-organ involvement often accompanies severe pneumonia.

ARDS is a distinct clinical diagnosis[24] apart from COVID-19. The diagnosis of this condition requires both clinical and ventilatory criteria. This syndrome is suggestive of a serious new-onset respiratory failure or of an exacerbation of an already identified respiratory picture. Different forms of ARDS are distinguished based on the degree of hypoxia. The reference parameter is the PaO_2/FiO_2 or P/F ratio:

- Mild ARDS: 200 mmHg < PaO_2/FiO_2 ≤ 300 mmHg. In unventilated patients or in those managed with NIV by using positive end-expiratory pressure or a continuous positive airway pressure ≥5 cm H_2O.
- Moderate ARDS: 100 mmHg < PaO_2/FiO_2 ≤ 200 mmHg.
- Severe ARDS: PaO_2/FiO_2 ≤ 100 mmHg.

When PaO_2 is not available, a SpO_2/FiO_2 ratio ≤315 mmHg is suggestive of ARDS.

It is important to differentiate COVID-19-induced pneumonia from other causes of pulmonary infiltrates such as congestive heart failure and fluid overload. Therefore, chest imaging should include a chest radiograph, CT scan or lung ultrasound that can reveal the extent of any bilateral opacities typical of COVID-19 (lung infiltrates > 50%) that are not fully explained by effusions or lobar or lung collapse. While in some cases the clinical scenario and ventilator data could suggest pulmonary edema, the primary respiratory origin of the edema is demonstrated after excluding cardiac failure or other causes such as fluid overload. Echocardiography can help to assess the presence or absence of left ventricular dysfunction that could cause or contribute to pulmonary edema.

4.3 A Novel Application of Smartphone Technology

In a novel application of smartphone application (app) technology, data from the COVID Symptom Study app, analyzed by researchers from King's College London and the health technology company ZOE, revealed that there are six distinct "types" of COVID-19, each distinguished by a particular cluster of symptoms.[25] These disease types differed in severity and requirement for respiratory support during hospitalization. The findings have major implications for clinical management of COVID-19 and may help doctors predict those most at risk and likely to need hospital care in coronavirus infections. The COVID Symptom Study was launched in March in the UK and subsequently extended to the United States and Sweden. The app asks participants to log their health and any new potential symptoms of COVID-19 on a daily basis. With more than 4 million users, this study is the largest study of its kind in the world.

Although continuous cough, fever and loss of smell (anosmia) are typically highlighted as the three key symptoms of COVID-19, data gathered from app users shows that people experience a wide range of symptoms including headaches, muscle pains, fatigue, diarrhea, confusion, loss of appetite, shortness of breath, *etc.* The disease progression and outcomes vary significantly, ranging from mild flu-like symptoms or a simple rash to severe or fatal disease. To learn whether particular symptoms tended to appear together and how they related to the progression of the disease, the research team used machine learning to analyze data from a subset of approximately 1600 users in the UK and US having confirmed COVID-19 who had regularly logged their symptoms using the app in March and April.

The analysis revealed six groupings of symptoms emerging at characteristic timepoints in the progression of the illness, thus representing six distinct types of COVID-19. The machine-learning algorithm was then tested on a second independent dataset of 1000 users in the UK, US and Sweden, who had logged their symptoms during May. All people reporting symptoms experienced headache and loss of smell, with varying combinations of additional symptoms at various times. Some of these, such as confusion, abdominal pain and shortness of breath, are not widely known as COVID-19 symptoms, yet are hallmarks of the most severe forms of the disease. The six clusters are as follows:

(1) "Flu-like" with no fever: headache, loss of smell, muscle pains, cough, sore throat, chest pain, no fever.
(2) "Flu-like" with fever: headache, loss of smell, cough, sore throat, hoarseness, fever, loss of appetite.
(3) Gastrointestinal: headache, loss of smell, loss of appetite, diarrhea, sore throat, chest pain, no cough.
(4) Severe level one, fatigue: headache, loss of smell, cough, fever, hoarseness, chest pain, fatigue.

(5) Severe level two, confusion: headache, loss of smell, loss of appetite, cough, fever, hoarseness, sore throat, chest pain, fatigue, confusion, muscle pain.
(6) Severe level three, abdominal and respiratory: headache, loss of smell, loss of appetite, cough, fever, hoarseness, sore throat, chest pain, fatigue, confusion, muscle pain, shortness of breath, diarrhea, abdominal pain.

Next, the team investigated whether people experiencing particular symptom clusters were more likely to require breathing support in the form of additional oxygen or mechanical ventilation. They discovered that only 1.5% of people with cluster 1, 4.4% of people with cluster 2 and 3.3% of people with cluster 3 COVID-19 required breathing support. These figures were 8.6%, 9.9% and 19.8% for clusters 4, 5 and 6, respectively. Nearly half of the patients in cluster 6 ended up in a hospital, compared with just 16% of those in cluster 1. Broadly, people with cluster 4, 5 or 6 COVID-19 symptoms tended to be older and frailer, and were more likely to be overweight and have pre-existing conditions such as diabetes or lung disease than those with type 1, 2 or 3. The researchers then developed a model combining information about age, sex, body mass index (BMI) and pre-existing conditions together with symptoms gathered over just 5 days from the onset of the illness. This model was able to predict which cluster a patient falls into and their risk of requiring hospitalization and breathing support with a higher likelihood of being correct than an existing risk model based purely on age, sex, BMI and pre-existing conditions alone. Given that most people who require breathing support come to hospital approximately 13 days after their first symptoms, this extra 8 days represents an important "early warning" as to who is most likely to need more intensive care (https://covid.joinzoe.com/post/covid-clusters).[25]

4.4 COVID-19 and Comorbidities in Severe Cases

As noted earlier, symptomatic cases can be classified into mild, moderate, severe or critically ill. Some patients may first present for care with severe symptoms, leading to immediate admission to an ICU, with a subsequent diagnosis of COVID-19 through viral testing. This pattern of presentation was seen in the early epidemic in many regions where the viral testing capacity was limited, and testing was only performed for individuals with severe symptoms who were suspected of having the disease. In this circumstance, most COVID-19-positive patients were already in a severe or critical condition. In other circumstances, infected individuals identified based on viral testing could have been mostly mild or moderate cases previously in close contact with persons confirmed to have the disease.[26] China's national statistics estimated that 13.8% of early cases were severe and 4.7% were critical, while mild or moderate cases accounted for 80% of patients overall.[27] These estimates would not have fully considered all asymptomatic patients, which were unknown in the community since population-based widespread testing was not performed. Individuals who developed severe and critical cases of COVID-19 predominantly had underlying chronic health conditions.

In a clinical study of 1099 patients from a national sample in which severe cases were defined at hospital admission by the American Thoracic Society guideline for community-acquired pneumonia,[28] severe cases had an approximately two- or three-fold higher likelihood of presenting with diabetes, hypertension or coronary heart disease, and were five-fold more likely to have chronic obstructive pulmonary disease (COPD), but 75% less likely to have hepatitis B viral infection, compared with non-severe patients.[28] Since these comorbidities of cardiovascular disease and hypertension are prevalent and associated with older age,[29] the elderly who are infected with SARS-CoV-2 are likely to be severely or critically ill and may have a high case-fatality rate.[30] As pointed out by Zhang and Hughes, one must be aware that all hospital-based patients are highly selected.[31] Therefore, findings from a hospital-based observational study are likely confounded[32] and subject to selection bias. Obviously, underreporting of existing comorbidities also can confound the prediction of the course of an illness.

Many patients diagnosed with COVID-19 present with comorbidities that appear to influence the disease outcome. These individuals should be constantly monitored using a pulse oximeter to detect further progression of COVID-19. Early reports about the role of comorbidities as risk factors for severe or critical COVID-19 show some inconsistencies, but also a substantial amount of agreement. While many comorbid conditions have been assessed, several appear to be common and important, such as COPD, obesity, diabetes, hypertension and cardiovascular and cerebrovascular diseases.[28] Patients with pre-existing illnesses tended to have a more severe form of the disease and were also older than those without comorbidities. Shortness of breath, vomiting, nausea and abnormal CT findings were also found in these individuals and were more frequent in patients with other conditions.[28] In a meta-analysis undertaken in a total of 1558 patients, it was found that hypertension, diabetes, COPD, cardiovascular disease and cerebrovascular disease were risk factors for COVID-19. However, no correlation was found between liver disease, malignancy and renal disease and COVID-19.[33] In a study performed in New York, it was found that hypertension, obesity and diabetes were the most common comorbidities.[16] Likewise, in Italy, individuals with cardiometabolic diseases were found to be more at risk of mortality if infected by COVID-19.[34] Furthermore, it was also found that children with underlying conditions like hydronephrosis, leukemia and intussusception, and lacrimal sac dredge were more likely to require intensive care support and ventilation, while children without these conditions recovered much more readily from COVID-19.[35] Hence, it is very important to stay vigilant for comorbidities while treating patients with COVID-19 as well as to conduct regular follow-up for recovered patients.[36] Figure 4.1 depicts the comorbidities of COVID-19.

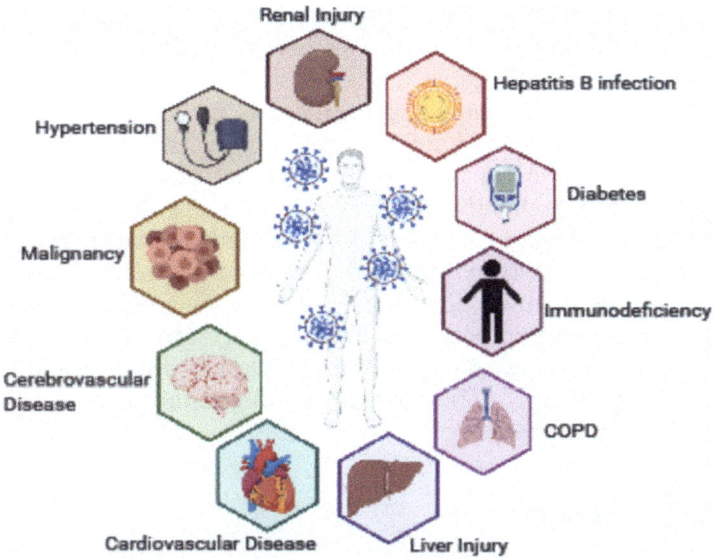

Figure 4.1 Comorbidities associated with COVID-19.

4.5 ARDS and its Sequelae in COVID-19

4.5.1 ARDS in COVID-19

ARDS occurs in patients with COVID-19 typically 1 week after symptom onset with deterioration of respiratory status. When ARDS occurs, various CT abnormalities are found; these include bilateral ground-glass opacities, lobar or lung collapse, lung volume overload and presence of nodules.[37] Patients may have some of these radiographic signs, which qualifies them as "moderate" rather than "mild", and they may not yet be in the ARDS group. It is important to consider echocardiography to exclude the possibility of hydrostatic issues causing pulmonary edema. When a patient fails to improve with non-invasive oxygen therapy, mechanical ventilation is necessary. In adults with severe ARDS, at least 12–16 hours of ventilation per day is recommended (many patients require 24 hours per day).[38]

4.5.2 Sepsis

Sepsis is a potentially life-threatening condition whereby the body's response to infection is dysfunctional, triggering changes that can damage multiple organ systems.[39] The clinical condition of patients with both COVID-19 and sepsis is particularly serious, characterized by a wide range of signs and symptoms of multi-organ involvement. In adults, the signs of organ dysfunction include rapid and difficult breathing, varying mental status, reduction in urine quantity, increased heart rate, weakening of the pulse, low blood pressure, cool extremities, skin mottling, coagulopathy, thrombocytopenia, acidosis and increased lactate.

In children, sepsis is usually accompanied by fever and elevated or depressed white blood cell count. The evaluation of MOD and its prognostic significance is conducted by means of the Sequential Organ Failure Assessment (SOFA) score, which indicates the risk of ICU mortality based on a combination of lab results and clinical data.[40] A validated pediatric version of the score is also available.[41] Even if the patient with sepsis is considered to have COVID-19, antimicrobial medication(s) are typically administered immediately after signs of sepsis appear, as bacterial superinfection as a contributing cause of sepsis cannot be excluded. The medication given should be based on the clinical symptoms, laboratory and radiographic findings and treatment guidelines of the country.

4.5.3 Septic Shock

Septic shock is sepsis with circulatory, cellular and metabolic dysfunction with a dangerously low blood pressure and a high risk of mortality. Cellular/metabolic abnormalities such as serum lactate level >2 mmol L^{-1} (18 mg dL^{-1}) are typically present. Because patients with septic shock usually suffer from persisting hypotension despite volume resuscitation, the prompt administration of vasopressors is typically required to maintain an adequate blood pressure, *e.g.* a mean arterial pressure ≥65 mmHg when septic shock is strongly suspected or identified. Prompt treatment of septic shock is essential if mortality is to be avoided. In children, indicators of septic shock include hypotension, change in mental status, tachycardia, weak pulse, tachypnea, skin mottling, rash, increased lactate, cold extremities, fluctuation in temperature and prolonged capillary refill.[38] In addition, as soon as septic shock is identified, antimicrobial therapy, vasopressors and fluid bolus initiation are recommended. Central venous lines are preferred for vasopressor administration but if unavailable, vasopressors can also be given through a peripheral intravenous line.

4.6 Pathophysiology

The expression and distribution of entry receptors regulates tropism, determining the tissues that are infected and therefore the pathogenesis of the disease.[42] SARS-CoV-2 is the third (after SARS-CoV and NL63) human coronavirus known to co-opt the peptidase angiotensin-converting enzyme 2 (ACE2) to gain cell entry.[43] ACE2, a metallopeptidase, is the functional receptor for SARS-CoV-2. In addition to acting as SARS-CoV-2 receptor, ACE2 plays an essential role in the pathogenesis of SARS-CoV-2. A member of the ACE family of dipeptidyl-carboxydipeptidases, ACE2 is a transmembrane protein best characterized for its homeostatic role in counterbalancing the effects of

ACE on the cardiovascular system.[44] ACE2 is the receptor for both the SARS-CoV and NL63, but not the Middle East respiratory syndrome (MERS) coronavirus. NL63 is the related human respiratory coronavirus first identified in the Netherlands. ACE2 is expressed in human airway epithelia as well as lung parenchyma.[45]

Earlier studies in mice had shown that ACE2 is generated mainly in club cells and type II alveolar epithelial cells, and it was known that the production of ACE2 is severely impaired after epithelial injury in the development of ARDS.[46] ACE2 is not only a functional receptor for coronaviruses, but also acts as an important negative regulator of the renin–angiotensin system (RAS) through conversion of the vasoconstrictor angiotensin II (Ang II) to its metabolite angiotensin-(1–7) (Ang 1–7) and angiotensin I (Ang I) to angiotensin-(1–9) (Ang 1–9).[47–49] The ACE2/Ang 1–7 axis plays a series of roles in the improvement of endothelial dysfunction, anti-inflammation, anti-hypertension, anti-thrombus and antifibrosis activity, as well as cardiovascular protection.[50–54] The protective effect of ACE2 is associated with attenuating Ang II levels and increasing Ang 1–7 levels in lung pathophysiology.[50] Prior evidence indicated that RAS signaling and ACE2 play crucial roles in SARS-CoV-induced ARDS and lethal avian influenza A (H5N1, H7N9)-induced acute lung injury.[54,55] Additional evidence developed during the COVID-19 pandemic suggested that both SARS-CoV[56] and SARS-CoV-2 employ ACE2 as the entry receptor and that the receptor-binding domain (RBD) of the S-protein binds directly to ACE2, triggering endocytosis of virus particles.[47,57,58] Research has indicated that the binding affinity between ACE2 and the RBD of SARS-CoV-2 is 10–20 times stronger than that with the RBD of SARS-CoV,[57] which probably explains the increased infectivity of SARS-CoV-2 that was immediately observed. The COVID-19 infection in the respiratory system of the host peaks 5–6 days after the commencement of symptoms.[14] It is at this time that severe COVID-19 cases progress to ARDS, on average, approximately 8–9 days after symptom onset.[15] When ARDS occurs, various CT abnormalities are found; these include bilateral ground-glass opacities, lobar or lung collapse, lung volume overload and the presence of nodules.[37] ARDS seen in severe COVID-19 is characterized by dyspnea and hypoxemia.[26] ARDS caused respiratory dysfunction resulting in death in approximately 70% of the COVID-19 individuals who died.[26] In addition to this, sepsis and cytokine storm were implicated in the death of 28% of the fatal COVID-19 cases.[59] An extended innate immune response subverting adaptive response is cited as the prime reason for pulmonary pathology. Following is an explanation of the likely sequence of events beginning at the alveolar level.

The alveolus is the basic structural unit of lungs. Alveolar architecture is made of two types of epithelial cells or pneumocytes. Type I pneumocytes are concerned with gaseous exchange and type II pneumocytes are sites of surfactant production. Under normal circumstances, these surfactants decrease surface tension within the alveoli at a liquid–air interface and thereby reduce its collapsing pressure, ensuring the well-being of the lungs. But type II pneumocytes are the cells initially targeted and lysed by SARS-CoV-2. Pyroptosis, a highly inflammatory form of programmed cell death usually occurring in response to an infectious pathogen, is likely to trigger subsequent inflammatory responses. Interleukin (IL)-1β, an important cytokine released during pyroptosis, is elevated during SARS-CoV-2 infection.[15] A primary inflammation ensues, with increased secretion of the pro-inflammatory cytokines and chemokines into the blood of afflicted patients.[60] Interferon (IFN) and other cytokines from virus-disintegrated pneumocytes recruit alveolar macrophages and dendritic cells, which perform crucial antigen processing and presentation, which then lead to a secondary wave of inflammation.

Dendritic cells reside beneath the epithelium and macrophages are located at the apical side of the epithelium. T-cell responses are initiated by antigen presentation through phagocytosed antigen-enclosed major histocompatibility complex (MHC)-bearing dendritic cells and macrophages. These antigen-presenting cells (APCs) move to the draining lymph nodes to present viral antigens to T-cells. CD4+ and CD8+ T-cells play a critical role in cell-mediated immunity. CD4 T-cells produce cytokines to activate immune cells including B-cells, CD8 T-cells and macrophages, while CD8 T-cells destroy virus-infected cells and tumor cells. Helper T-cells become activated when they are presented with peptide antigens by MHC class II molecules, which are expressed on the surface of APCs. Secretion of cytokines and chemokines at the primary or secondary level attracts immune cells.[61] Recruitment of immune cells leads to lymphopenia.[62] This response, if appropriate, neutralizes the infection. However, a dysfunctional immune response triggers a cytokine storm (Figure 4.2), that mediates widespread lung inflammation.

Histopathology from cynomolgus monkeys, 4 days after inoculation with SARS-CoV-2, shows a viral pneumonitis with alveolar edema, capillary leakage, inflammatory cell infiltration, interstitial thickening and cell fusion (a feature of coronavirus infection), with viral spike expression on alveolar epithelial cells.[63] Approximately 80% of patients with COVID-19 pneumonitis recover without specific treatment.[21] However, ~20% of patients deteriorate, often rapidly, ~7–10 days after symptom onset. This is when patients are most frequently admitted to hospital, with worsening fever, hypoxia, lymphopenia, rising inflammatory markers (C-reactive protein, IL-1 and IL-6), coagulopathy, and cardiovascular involvement. Approximately 25% of these patients will require mechanical ventilation, which is associated with high mortality (40–80%). The demographic of this "at-risk" group is reproduced across many countries: older men with hypertension, diabetes and obesity, as well as a less well-defined contribution of ethnicity.[21] Similar factors regulate ACE2 expression, which may therefore contribute to disease severity. Nonetheless, the amount and distribution of ACE2 expression cannot be the only factor affecting disease progression, because the three human coronaviruses that use ACE2 for cell entry exhibit markedly different pathogenicity (Figure 4.3).

In severe COVID-19, patients exhibited higher blood plasma levels of IL-2, IL-7, IL- 6, IL-10, granulocyte colony-stimulating factor, IP-10, monocyte

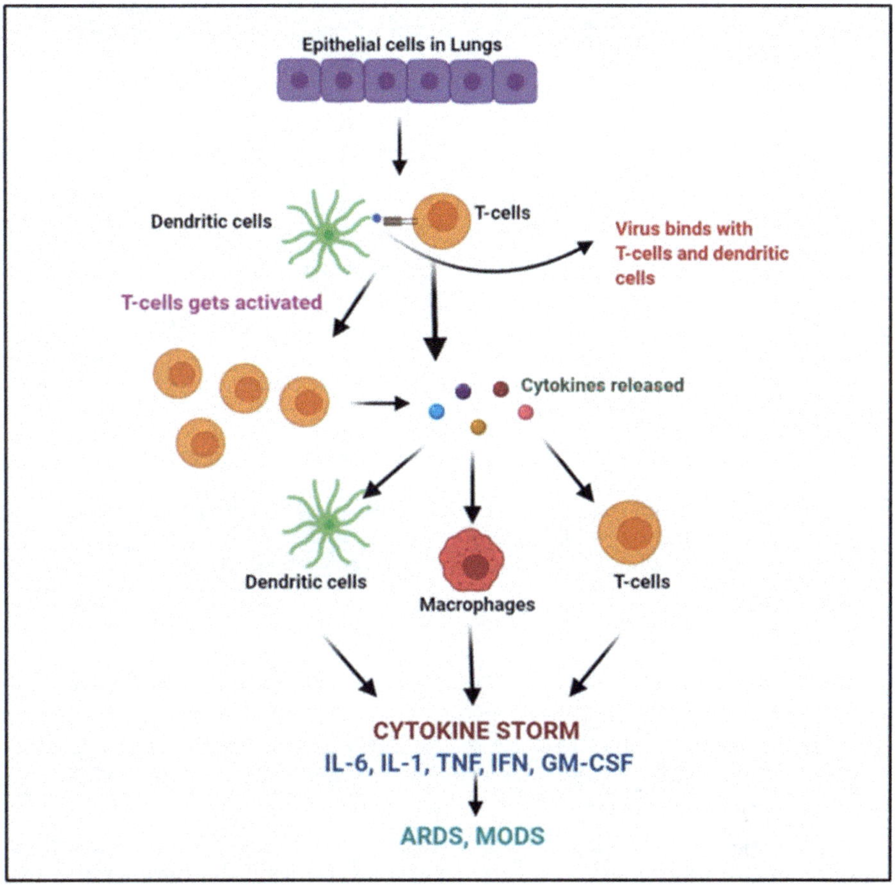

Figure 4.2 Cytokine storm. Once the virus binds to the dendritic cells, the T-cells are activated, producing cytokines. This results in further production of more cytokines. The multiple cytokines produced cause a cytokine storm.

chemoattractant protein 1, macrophage inflammatory protein 1α (MIP1α) and tumor necrosis factor (TNF).[15] Cytokines are intercellular communication mediators synthesized by all the participating immune cells, chiefly APCs and CD4+ T-cells. High expression of IL-6 in particular continues to increase over time, accelerating the progression of the systemic inflammatory response.[64] In severe patients, pathological cytotoxic T-cells from CD4+ T-cells[65] kill the virus, while contributing to lung injury.[66] Granulocyte–macrophage colony-stimulating factor (GM-CSF) is synthesized by aberrant CD4+ T cells.[67] Virus infects lung epithelial cells which produce IL-8 and IL-6,[68] and pulmonary neutrophils act to intercept the virus but also add to the local inflammatory response.[69] Hyaluronan is often associated with ARDS in the lungs.[70] The cytokines are strong inducers of HA-synthase-2 (HAS2) in the endothelium, lung alveolar epithelial cells and fibroblasts.[71]

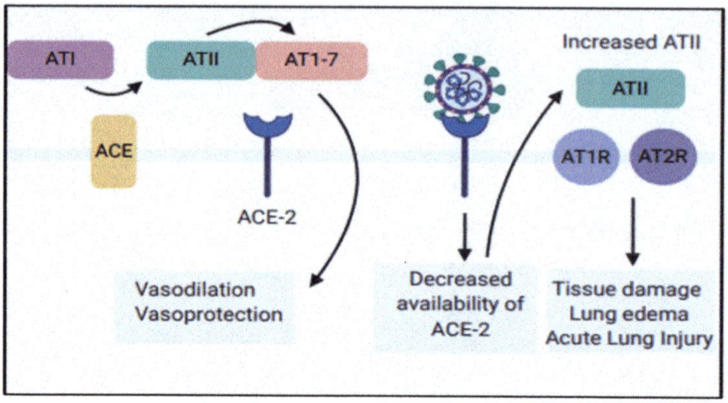

Figure 4.3 ACE2 and the renin–angiotensin system (RAS). ACE2 is involved with the conversion of angiotensin (AT)II to AT 1–7. Increased levels of ATII can cause lung edema and tissue damage. ACE in the cell is involved with the conversion of ATI to ATII. When SARS-CoV-2 binds to ACE2, the availability of ACE2 receptor decreases, thereby increasing the levels of ATII, causing the symptoms associated with COVID-19.

Antagonism occurs at certain stages of the IFN signaling pathway, which prevents pattern recognition receptor recognition of viral RNA.[72] Inhibition of the IFN response helps in the replication of virus, resulting in increased release of pyroptosis-related products, which leads to altered pathological inflammatory responses. In addition to the direct damage, secretion of proteases and reactive oxygen species due to uncontrolled inflammatory cell infiltration can mediate damage by the virus.[61] Overall, the pathophysiology of SARS-CoV-2 infection is a result of aggressive inflammatory responses in the respiratory pathway, causing damage to the cellular architecture of airways.[73] Hence, it is the interplay of viral infection and host immune response which determines disease severity.

4.6.1 Mechanism of SARS-Cov-2 Invasion into Host Cells

ACE2 is highly homologous to ACE1 in structure and function. ACE1 converts Ang I to Ang II, a highly active octapeptide that exhibits both vasopressor (vascular contraction to increase blood pressure) and pro-inflammatory activities. The carboxypeptidase activity of ACE2 converts Ang II to the heptapeptide Ang,[1–7] a functional antagonist of Ang II. Thus, ACE2 counteracts the effects of ACE1 in the RAS and down-regulates the production of Ang II (Figure 4.3). Because ACE1 is highly expressed in vascular endothelial cells of the lungs, Ang II is also likely to be high in the pulmonary vasculature. Indeed, ACE2 deletion in mouse models of acute lung injury results in more severe disease, suggesting a protective role for ACE2 in lung tissue.[44]

SARS-CoV-2 binds to host cells by a transmembrane S-glycoprotein that forms homotrimers arising from the viral surface. Composite S-protein is made of two subunits, S1 and S2; the former helps in binding to the host cell, while the latter helps it to fuse to the host cell. After this, the S-protein is cleaved at the S1/S2 junction. The S1 subunit maintains the integrity of the pre-fusion state of the S2 subunit. This subunit consists of the factors necessary for fusion of the membranes.

Three different host proteases are involved in cleavage and activation of the S-0 protein with a significant cleavage at the S2 site. This results in the change in conformation of the proteins which activates the protein enabling fusion of the membranes.[74] This conformational modification in S-glycoprotein trimers is initiated by the intact binding of virus and host cell receptor ACE2 described earlier. The S2 subunit consists of a fusion peptide region and two heptad repeat regions: HR1 and HR2.[72] The S1 subunit is cleaved in the endosome and with the help of fusion peptide it fuses with the host cell. This leads to fusion of the membrane and further releasing the viral particles into the cytoplasm of the host cell[72] (Figure 4.4).

After fusing with the host cell, the SARS-CoV-2 releases its genome into the cell; this in turn marks the beginning of replication of the viral genes. The virus releases a single-stranded positive-sense RNA. The size of the SARS-CoV-2 genome ranges from 26 to 32 kb and comprises of 6–11 open reading frames (ORFs) encoding 9680 amino acid polyproteins. The first ORF comprises 67% of the genome that encodes for 16 non-structural RNA-binding proteins (nsps), and the other ORFs encode accessory and structural proteins. The four major structural proteins, S, N, E and M, and accessory proteins are encoded by ORFs. The nsps include two viral cysteine proteases including

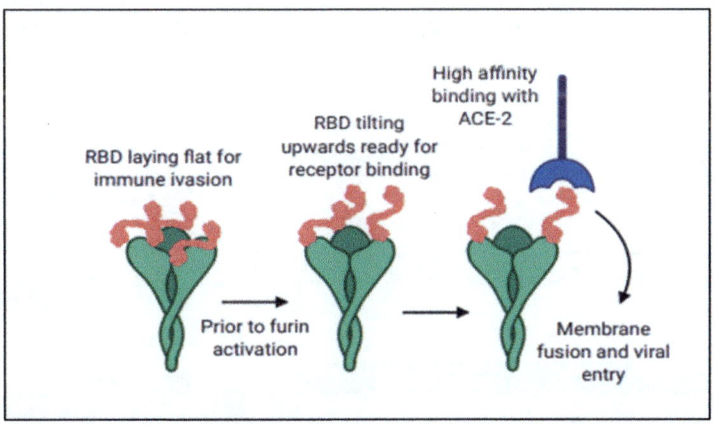

Figure 4.4 S-protein and ACE2 binding dimensions. The RBD of the spike protein is often present in a recumbent position to evade the immune system. When the receptor is available, the RBD tilts upwards and binds with the receptor with high affinity. This facilitates membrane fusion and viral entry into the cell.

papain-like protease (nsp3, PLpro), chymotrypsin-like, 3C-like or main protease (nsp5, Mpro or 3CLpro), RNA-dependent RNA polymerase (nsp12, RdRp), helicase (nsp13) and other proteins that are probably involved in transcription and replication of SARS-CoV-2 genes[75] (Figure 4.5).

The gene segments ORF1a and ORF1b that encode for non-structural polyproteins are the genes first to be translated into pp1a and pp1ab, which are large overlapping polyproteins, by executing ribosomal frameshifting event.[76] Polyprotein synthesis is facilitated by PLpro and a serine-type Mpro (3CLpro) encoded by nsp3 and nsp5, respectively. Cleavage of pp1a and pp1ab forms the replicase–transcriptase complex, assembled by RdRp- and helicase-containing subunits. This complex transcribes an endogenous genome template into intermediate negative-sense genomic and subgenomic RNAs which finally gets replicated and discontinuously transcribed into positive-sense genomic RNA and nine subgenomic mRNA respectively, mediated chiefly by RdRp.[77] The subgenomic mRNAs are translated into structural M-, S- and E-proteins and other accessory polypeptides. The polypeptides are structurally modified into functional nucleoproteins in the endoplasmic reticulum–Golgi intermediate compartment (ERGIC). Genomic RNA is packed into the assembling nucleoprotein to form a functional nucleocapsid in the ERGIC, which releases virions containing small wallet vesicles out of the cell through exocytosis[78] (Figure 4.6).

In many host–virus interactions, expression of the viral receptor is down-regulated in infected cells, and expression of ACE2 in the lungs of mice was reduced by SARS-CoV infection.[42] When SARS-CoV-2 binds to ACE2, there is probably reduced availability of the receptor for its normal protective functions. This, in turn, may result in acute respiratory failure and other organ-associated damage.[44,79] Depletion of ACE2 may thus play a causative role in the lung injury caused by SARS-CoV and SARS-CoV-2, and high plasma Ang II is reported in patients with COVID-19. However, MERS-CoV causes a similar

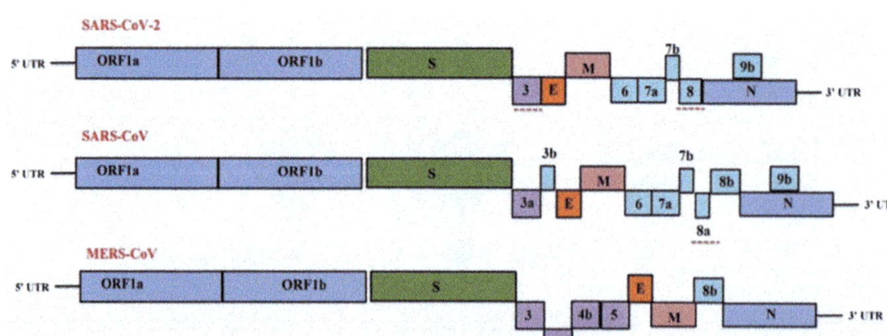

Figure 4.5 Genomic organization of SARS-CoV-2. The SARS-CoV-2 genome consists of the genes for the structural and accessory proteins. The genomic areas of SARS-CoV-2 differs in various domains from SARS-CoV and MERS-CoV, while maintaining a few conserved sequences present in coronaviruses.

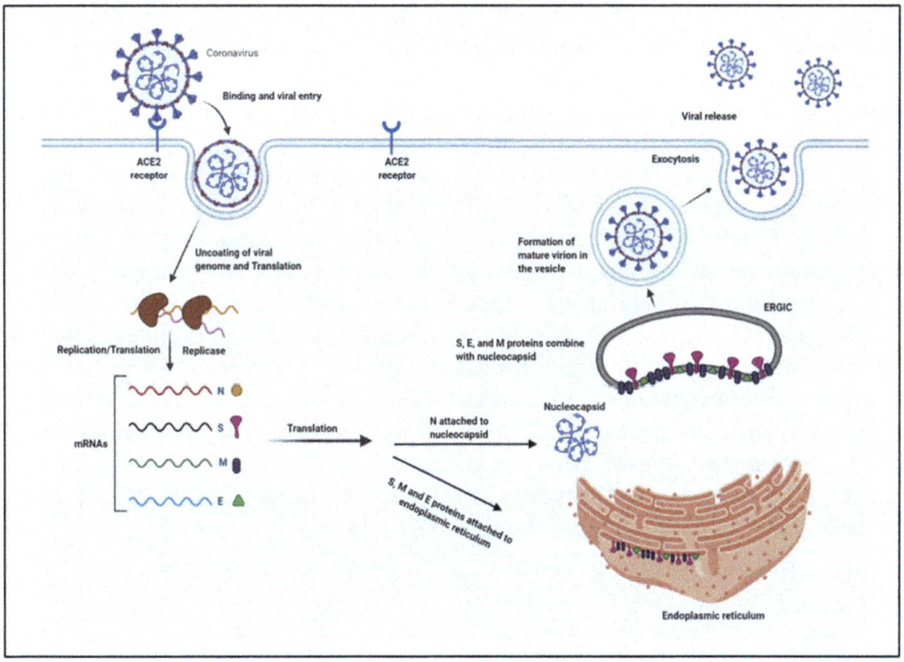

Figure 4.6 Replication of SARS-CoV-2. Upon binding of SARS-CoV-2 to the receptor, it fuses and enters the cell. The viral proteins are then translated within the cell. The proteins are then processed by the ERGIC system and capsulated into vesicles and viral particles are sent outside the cell by exocytosis.

lung disease without targeting ACE2, so other factors must also be important. These studies suggest that SARS-CoV-2 not only affects the respiratory system, but also other major functioning organ systems in the human body, and this is borne out by multi-organ histopathological manifestations.

To reiterate and amplify the mechanistic detail, SARS-CoV-2 makes its entry and exit through the apical surface of the well-differentiated epithelium of the respiratory tract, where ACE2 is expressed more abundantly than the basolateral area.[80] ACE2 is widely expressed in various organs, and attachment of the virus to these extrapulmonary ACE2 receptors contributes to the added MOD that has been observed in COVID-19. Once the virus reaches the circulatory system, it spreads to the other organs *via* the vascular system.[81] The SARS-CoV-2 RBD has a significantly higher ACE2 binding affinity as compared to the SARS-CoV RBD.[82] Despite RBD exhibiting high affinity to ACE2, the entire SARS-CoV-2 S-protein does not bind to ACE2 any more strongly than SARS-CoV S-protein. It is less accessible to binding due to its recumbent state (Figure 4.4), and hence has a lower ACE2 binding affinity for SARS-CoV-2. To maintain its high infectivity while keeping its RBD less accessible, SARS-CoV-2 relies on a second strategy – host protease activation. The RBD of the SARS-CoV-2 S-protein is mostly in the partially opened recumbent state (Figure 4.4),[57] and is activated through cleavage by the host cell-surface proteases TMPRSS2 (transmembrane serine protease), lysosomal proteases (cathepsin L) and the proprotein convertase (PPC) furin. Furin is a ubiquitously expressed serine protease which cleaves a plethora of proteins at polybasic recognition motifs in many parts of the body.[83]

The unique feature of SARS-CoV-2 among other coronaviruses is the existence of the furin cleavage site ("RPPA" sequence) at the S1/S2 site, conferring the higher degree of pathogenicity of the virus. The hidden RBD can evade immune surveillance as one of the conformational masking strategies, potentially leading to insufficient immune responses and prolonged recovery time.[82] As discussed earlier, unlike SARS-CoV, infectivity of patients with SARS-CoV-2 peaks before symptom onset.[10] To be sure, massive presymptomatic transmission makes SARS-CoV-2 impossible to contain through case isolation alone and is a key driver of the pandemic.[10] This alteration in the pattern of disease may relate to the acquisition of the furin cleavage site in the spike or increased binding affinity for ACE2 in SARS-CoV-2 as compared with SARS-CoV[57] (Figure 4.7).

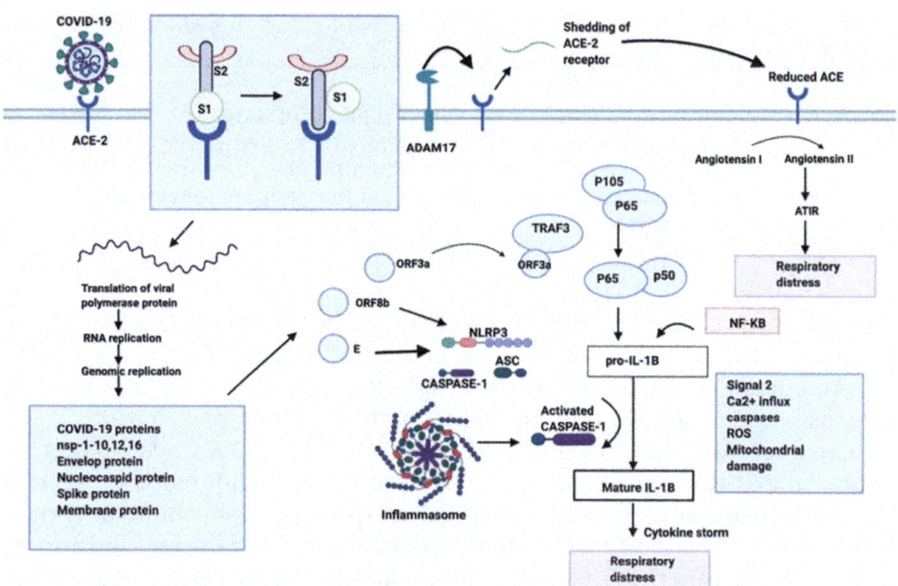

Figure 4.7 Possible mechanism of action of SARS-COV-2. The figure illustrates the binding of SARS-COV-2 to its receptor ACE2. The S1 and S2 subunits are subsequently cleaved followed by the shedding of ACE2 by ADAM 17. This resulting in an increased amount of angiotensin II, leading to respiratory distress. Upon binding, the virus fuses with the membrane and enters the cell, followed by translation and replication of the proteins. ORF3a, ORF8b, E-proteins and the nuclear factor κB (NF-κB) pathway activates the inflammasome pathway through various means, leading to the activation of cytokines. Reproduced from ref. 116 with permission from Elsevier, Copyright 2020.

4.7 Histopathology and Disease Manifestations: Introduction

Many of the more severe clinical signs and symptoms of COVID-19 result from cytokine release syndrome, an acute systemic inflammatory syndrome. IFN-γ causes fever, chills, headaches, dizziness and fatigue and TNF-α causes flu-like symptoms.[84] IL-6 can lead to vascular leakage, activation of complement and the coagulation cascade, leading to a diffuse intravascular coagulation scenario,[85] and cardiomyopathy.[86] Activation of endothelial cells may lead to hypotension and coagulopathy[84] (Figure 4.8).

4.7.1 Histopathology

Tian *et al.*[87] and others have reported findings from histopathological data obtained from the lungs of two patients who had undergone lung lobectomies for adenocarcinoma and who were discovered retrospectively to have been infected at the time of surgery. Apart from the tumors, the lungs of both cases provided evidence of edema and important proteinaceous exudates such as large protein globules. The authors also reported vascular congestion combined with inflammatory clusters of fibrinoid material and multinucleated giant cells, as well as hyperplasia of pneumocytes.

More recently, Zhang *et al.*[88] performed a post-mortem transthoracic needle lung biopsy on a patient who died of COVID-19. Immunostaining revealed diffuse alveolar injury and an important alveolar expression of viral antigens. In autopsies on COVID-19 cases, the authors[89] offered a detailed picture of the histological patterns in the patient's lung and extrapulmonary tissues. This picture was characterized by capillary congestion, necrosis of pneumocytes, hyaline membrane, interstitial edema, pneumocyte hyperplasia and reactive atypia. Platelet–fibrin thrombi in small arterial vessels were the expression of intravascular coagulopathy. Moreover, an examination of the patient's lung revealed infiltrates expressed as macrophages in alveolar lumens and lymphocytes in the interstitium. In summary, and as in the case of SARS and MERS, severe COVID-19 lung damage was manifested in terms of diffuse alveolar disease with severe capillary congestion. Again, many findings were suggestive of vascular dysfunction, in lung as well as other tissues.

4.7.2 Pulmonary Manifestations

In COVID-19 infection, the common symptom is the dry cough,[90] and an uncommon manifestation is sore throat and sputum production ($\leq$5%).[28] The lungs show diffuse alveolar damage and display the following features:[91] extensive edema; lung basement zone thickening due to hyaline membrane formation; alveolar edema diluting surfactant and hence contributing to the collapse of alveoli;[64] and abnormal chest CT[28] with ground-glass opacities and/or consolidation.[90] ARDS complications are observed in 5% of COVID-19-positive patients. There is evidence of a higher incidence of

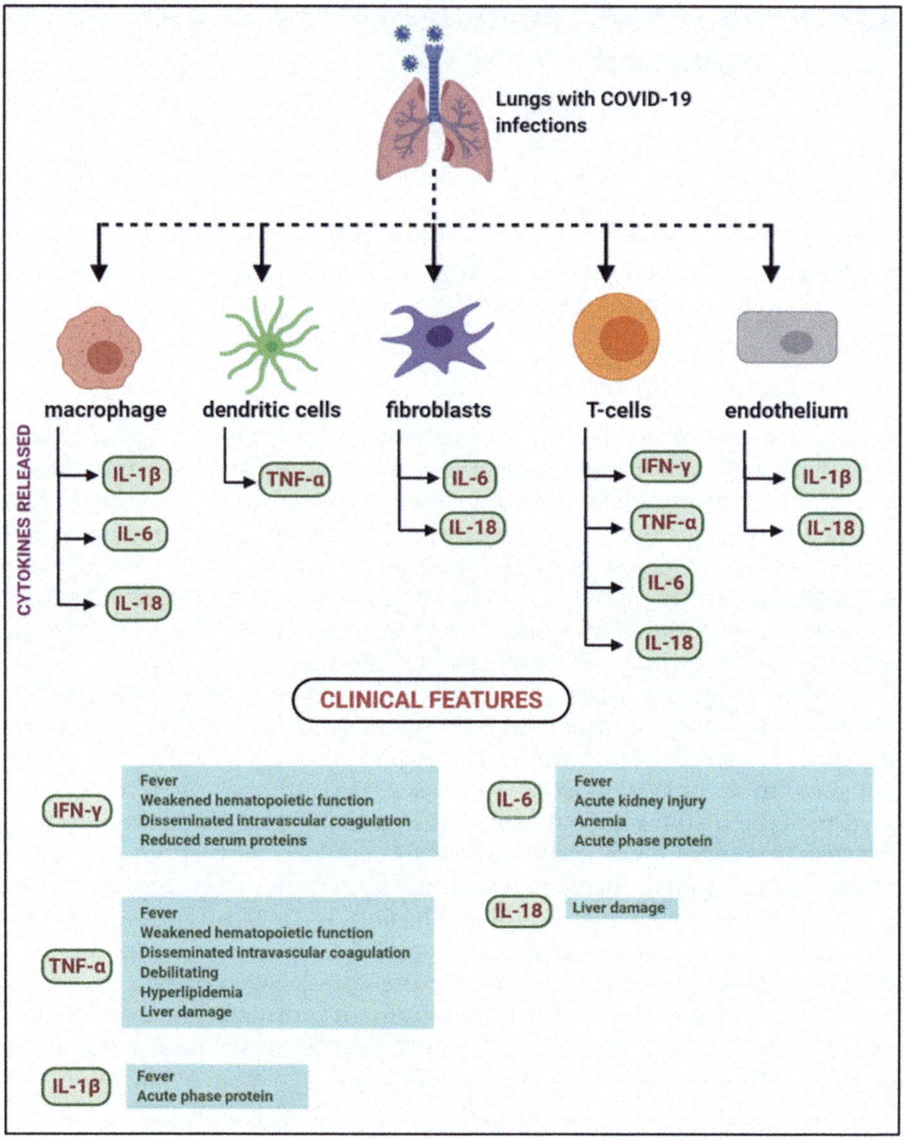

Figure 4.8 Cytokines and associated manifestations. The figure depicts the association of clinical manifestations linked with each of the cytokines released by macrophages, dendritic cells, fibroblasts, T-cells and endothelium.

thromboembolism in COVID-19 patients than in patients with ARDS caused by other pathogens.[64] The interstitial and alveolar edema resulting from COVID-19 can be compounded by acute and/or chronic congestive heart failure (CHF) due to pre-existing disease and/or a SARS-COV-2 direct effect on the myocardium.

4.7.3 Cardiovascular Manifestations

ACE2, the functional receptor of SARS-CoV-2 is expressed in the myocardium.[92] The incidence of troponin release, consistent with acute myocardial injury, has been reported to be between 8% and 28% in hospitalized patients with COVID-19.[93–95] The likelihood of myocardial injury increases with the severity of disease. Clerkin *et al.* reported a 7.2% incidence of elevated troponin in hospitalized patients with COVID-19, but a 22% incidence in those who were critically ill.[96] The report from the National Health Commission of China revealed that nearly 12% of patients without known cardiovascular disease had either elevated troponin levels or cardiac arrest during hospitalization, and among the non-survivors, 46% had elevated troponin levels ante-mortem.[97] Gao *et al.* in a retrospective study of 188 patients found elevated troponin in 27.8% and of these, 59.6% died *versus* 8.9% in the group without elevated troponin.[98] The rise in serum troponin levels trends with inflammatory markers, such as IL-6, C-reactive protein and D-dimer, raising the possibility that it may reflect cytokine storm more often than viral myocarditis.[64] However, COVID-19 patients may have acute myocarditis, which can be fulminant with resulting CHF. In a small series of 21 patients in Washington State, Arentz *et al.* found that one-third appeared to have developed cardiomyopathy in the setting of COVID-19.[99] Inciardi *et al.* published a case report with cardiac magnetic resonance evidence of acute myocarditis in a person who presented with an electrocardiogram (ECG) picture consistent with acute myocardial infarction.[100] Similarly, there have been anecdotal reports in on-line forums of patients presenting with ST elevations on ECG, but found to have no significant coronary artery blockages, yet evidence of new-onset CHF. It is not yet clear at present whether most cases of new-onset CHF are due to direct viral myocyte invasion, cytokine storm or as a form of reversible stress cardiomyopathy. A report from the Cleveland Clinic found a doubling in the incidence of stress cardiomyopathy or "broken heart syndrome" during the COVID pandemic.[101]

Arrhythmias are also common. Up to 50% of COVID-19 patients admitted to the ICU may develop atrial fibrillation; inflammation and myocardial injury both being substrates for atrial arrhythmias. Ventricular arrhythmias have also been noted and associated with cardiac arrest.[102–104] Pre-existing cardiovascular disease and hypertension are important comorbidities in severe disease requiring critical care.[103] The proposed mechanisms of cardiac injury in COVID-19 patients includes overexpression of ACE2 in patients with chronic cardiovascular disease, hypoxemia and cytokine storm triggered by an imbalanced response of type 1 and type 2 helper cells, which may trigger myocarditis, plaque rupture, coronary vasospasm, arrhythmias and/or direct vascular injury.[15] A significant number of patients with cardiovascular disease such as CHF or hypertension are on ACE inhibitors (ACEIs) or angiotensin receptor blockers (ARBs). Therefore, it is conceivable that the up-regulation of ACE2 by ACEIs/ARBs might increase the virulence of SARS-CoV-2. However, to date this has not been proven clinically and in a recent report of 1128 patients with COVID-19, 188 patients who were on ACEs or ARBs had a lower mortality, even after adjusting for imbalanced variables in a propensity score matched analysis.[105] Given the absence of evidence for harm, it has been recommended that patients remain compliant taking these medications, although taking them specifically to treat COVID-19 is currently not recommended.[106]

An unknown number of survivors of severe COVID-19 may be left with impaired left ventricular function, either subclinical or with chronic CHF. Long-term follow-up studies will need to assess these potential outcomes.

4.7.4 Gastrointestinal Manifestations

Gastrointestinal manifestations are observed in 50% of COVID-19 patients.[14] Anorexia and diarrhea are the most common symptoms, and the symptoms that occur less often are vomiting and abdominal pain.[14] In children, vomiting is more common.[107] Patients with delayed diagnosis were more likely to suffer from hepatic dysfunction.[107] Obesity is a predisposing factor and such individuals are at higher risk of severe disease. The virus is detectable in stool in up to half of COVID-19 patients.[108] ACE2 and viral protein have been detected in gastrointestinal epithelial cells.[109] Fecal polymerase chain reaction (PCR) testing has been shown to be as accurate as PCR detection from sputum samples, and in some cases, fecal PCR is positive even before sputum PCR becomes positive.[109]

Abnormal liver function and elevated liver enzymes, bilirubin and lactate dehydrogenase levels are common in approximately half of COVID-19-positive patients.[110] Patients with abnormal liver function tests usually exhibit a high degree of fever.[110] Reports suggest that liver dysfunction in patients with COVID-19 may be related to damage to the cholangiocytes.[111] Pre-existing metabolic fatty liver disease and co-existing obesity were implicated in the severity of disease.[15]

4.7.5 Renal Manifestations

Albuminuria or proteinuria, hematuria, elevated urea, creatinine and low estimated glomerular filtration rate have been observed. Acute renal dysfunction is known to be common in individuals with COVID-19.[112] The incidence of acute kidney injury is as high as 15% and carries a high mortality rate in the 60–90% range.[60] There may also be imaging evidence of active renal edema and inflammation.[112] Since renal dysfunction is early, an immunopathology response or direct viral injury may be contributing to it along with other systemic factors.[113] The COVID-19 virus has been detected in renal tissue and in the urine.[90] Additionally, the male reproductive tract appears to be directly subject to COVID-19 injury since Leydig cells and seminiferous tubules have ACE2 receptors.[110]

4.7.6 Neurological Manifestations

Neurological symptoms of COVID-19 infection include sudden onset of partial or complete sense of smell (anosmia), dysgeusia or ageusia,[114] stroke, encephalitis, delirium, psychosis and dementia-like symptoms. The association between SARS-CoV-2 infection and anosmia (loss of smell) may reflect ACE2 and TMPRSS2 expression in sustentacular cells, which maintain the integrity of olfactory sensory neurons. Olfactory epithelial cells also express NRP1 and could provide a direct route to the brain.[115] Vellingiri *et al.* have speculated that entry of SARS-CoV-2 into the central nervous system (CNS) is through the olfactory bulb upon nasal infection, which causes inflammation and demyelination. In addition, the virus reaches the brain *via* the blood–brain barrier and the cerebrospinal fluid (CSF) *via* the blood–CSF barrier in less than 7 days. The possible entry of SARS-CoV-2 into the brain and CNS is important for the design of effective antiviral drugs. Effective drugs that may cross the blood–brain barrier and blood–CSF barrier may be taken into consideration while designing and this could be a promising treatment strategy[116] (Figure 4.9).

Of 43 patients admitted to the National Hospital for Neurology and Neurosurgery in London (29 with definite, eight with probable, and six with possible COVID-19 by WHO criteria), 10 had transient encephalopathy with

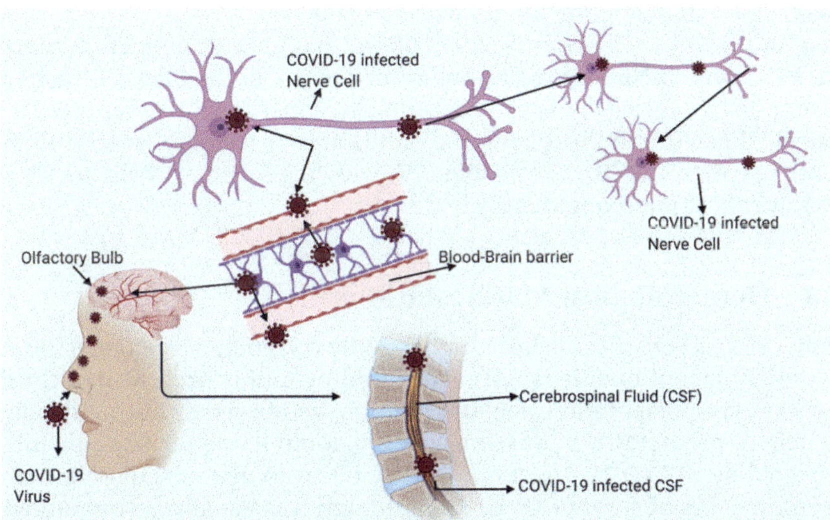

Figure 4.9 Potential routes of SARS-CoV-2 entry into the CNS. See text for explanation. Reproduced from ref. 116 with permission from Elsevier, Copyright 2020.

delirium, 12 had brain inflammation, nine of whom had acute disseminated encephalomyelitis (a rare disease typically seen in children and triggered by viral infection), eight had strokes and another eight had nerve damage, all but one of which were due to Guillain–Barré, the other a brachial plexopathy.[117] Of particular interest is that the virus was not detected in any patient's CSF, suggesting that the virus may not have directly attacked the brain, but that the immune response may have been the culprit. CSF analysis in Filatov *et al.*'s series revealed oligoclonal bands or elevated immunoglobulin G levels.[114] A larger series of 125 patients with severe COVID-19 (114 confirmed, six probable and five possible, again by WHO criteria) compiled from several UK hospitals reported that 62% had strokes and 31% had altered mental states which included encephalitis, psychosis and dementia-like symptoms.[118] In Paterson *et al.*'s patient series, the neurologic involvement did not always correlate with the severity of respiratory symptoms.[117] While animal studies have shown that ACE2 and TMPRSS2 are expressed in the conjunctiva,[119] any likelihood that this could be a possible route to the brain is speculative.

4.7.7 Musculocutaneous Manifestations

Myalgia seems to be a common symptom of COVID-19 infection.[120] Rhabdomyolysis was reported in patients with COVID-19 with myoglobin levels $>12\,000$ μg L^{-1} and creatine kinase levels $>11\,000$ U L^{-1}.[15]

In a study with 88 patients, 20% of the patients developed cutaneous manifestations including widespread urticaria, erythematous rash and chickenpox-like vesicles.[59] Several recent case series have also reported a red or purple discoloration involving the hands and/or feet similar to chilblains in patients with COVID-19,[121] called "Covid fingers" or "Covid toes", that probably reflect microangiopathy.[122,123]

4.7.8 Hematological Manifestations

Patients with severe disease have higher white cell counts (median 6100 cells per mm^3).[28] Lymphopenia is a frequent finding and is present in a third of patients.[28] On presentation, leukopenia may be observed, but neutrophilia may help to predict ICU admissions. Hemoglobin levels seems to be mostly unaffected by COVID-19 infection. Diffuse intravascular coagulation may be present in cases of severe COVID-19, although it is usually accompanied by microangiopathy with thrombosis rather than with a bleeding diathesis.[28] In general, mild thrombocytopenia is present in one-third of patients.[28] Patients who require ICU admission are seen to have higher levels of D-dimer.[124] Increased levels of other inflammatory molecules such as ferritin, circulatory cytokines, C-reactive protein as well as procalcitonin also seem to correlate with the severity of the disease.[125]

4.8 Executive Summary

- The pandemic caused by COVID-19 is a serious health issue with rising cases and fatalities affecting people the world over.
- COVID-19 symptoms range from mild to moderate to severe to critical, and most deaths occur in patients with severe or critical disease, usually with multi-organ failure involving the lungs, and frequently the heart and kidneys.
- In persons with COVID-19, those who have these comorbid conditions are far more likely to progress to severe or critical status and are at far higher risk of fatality.
- The most common comorbidities associated with this disease are cardiac issues, renal problems, immune dysfunction, diabetes, hypertension and obesity.
- Elderly COVID-19 patients and those who have these comorbid conditions are far more likely to progress to severe or critical status and are at greater risk of dying.
- The common complications requiring intensive care are ARDS, septic shock and multi-organ failure. Myocardial injury and congestive heart failure as well as renal failure may accompany severe or critical disease.
- Extensive reports about the pathology and pathogenesis of COVID-19 present a complex picture of the etiological factors involved and interplay of intricate host–virus interactions.
- Although the major clinical manifestations of the disease involve the host respiratory system, the major reason for systemic pathogenesis is linked to the immune system.
- Hypercytokinemia or "cellular storm", insufficient interferon reaction, and an inadequate cellular immune response caused by direct infection or indirect injury of immune cells may all contribute to pathological changes.
- An aberrant host immune response leads to aggravation of SARS-CoV-2-induced lung injury and probably injury in other organs as well.
- Controlling the inflammatory response may be as important as targeting the virus.
- A multi-pronged approach of therapies targeting both viral infection and regulation of inappropriate immune responses will block pathologies at multiple steps.
- At the same time, the association between immune dysfunction and outcome of disease severity in patients with COVID-19 should serve as a note of caution in vaccine development and evaluation.

Acknowledgements

The authors and editors thank Melanie K. Shadoan and Chris L Ringwalt for their excellent editing of this chapter.

References

1. N. Zhu, D. Zhang, W. Wang, X. Li, B. Yang and J. Song, *et al.*, A Novel Coronavirus from Patients with Pneumonia in China, 2019, *N. Engl. J. Med.*, 2020, **382**(8), 727–733.
2. Epidemiology Working Group for NCIP Epidemic Response CCfDCaP, The epidemiological characteristics of an outbreak of 2019 novel coronavirus diseases (COVID-19) in China, *Zhonghua Liu Xing Bing Xue Za Zhonghua Liuxingbingxue Zazhi*, 2020, **41**(2), 145–151.
3. Q. X. Long, X. J. Tang, Q. L. Shi, Q. Li, H. J. Deng and J. Yuan, *et al.*, Clinical and immunological assessment of asymptomatic SARS-CoV-2 infections, *Nat. Med.*, 2020, **26**(8), 1200–1204.
4. D. P. Oran and E. J. Topol, Prevalence of Asymptomatic SARS-CoV-2 Infection: A Narrative Review, *Ann. Intern. Med.*, 2020, DOI: 10.7326/M20-3012.
5. M. Jayaweera, H. Perera, B. Gunawardana and J. Manatunge, Transmission of COVID-19 virus by droplets and aerosols: A critical review on the unresolved dichotomy, *Environ. Res.*, 2020, **188**, 109819.
6. N. Van Doremalen, T. Bushmaker, D. H. Morris, M. G. Holbrook, A. Gamble and B. N. Williamson, *et al.*, Aerosol and surface stability of SARS-CoV-2 as compared with SARS-CoV-1, *N. Engl. J. Med.*, 2020, **382**(16), 1564–1567.
7. Coronaviridae Study Group of the International Committee on Taxonomy of V, The species Severe acute respiratory syndrome-related coronavirus: classifying 2019-nCoV and naming it SARS-CoV-2, *Nat. Microbiol.*, 2020, **5**(4), 536–544.
8. P. Zhou, X.-L. Yang, X.-G. Wang, B. Hu, L. Zhang and W. Zhang, *et al.*, A pneumonia outbreak associated with a new coronavirus of probable bat origin, *Nature*, 2020, **579**(7798), 270–273.
9. D. P. Han, A. Penn-Nicholson and M. W. Cho, Identification of critical determinants on ACE2 for SARS-CoV entry and development of a potent entry inhibitor, *Virology*, 2006, **350**(1), 15–25.
10. X. He, E. H. Y. Lau, P. Wu, X. Deng, J. Wang and X. Hao, *et al.*, Temporal dynamics in viral shedding and transmissibility of COVID-19, *Nat. Med.*, 2020, **26**(5), 672–675.
11. FDA UFaDA, *C.I. COVID, Convalescent Plasma-emergency INDs*, Administration UFaD, 2020.
12. W.-j. Guan, Z.-y. Ni, Y. Hu, W.-h. Liang, C.-q. Ou and J.-x. He, *et al.*, Clinical characteristics of 2019 novel coronavirus infection in China, *N. Engl. J. Med.*, 2020, **382**, 1708–1720.
13. R. Wölfel, V. M. Corman, W. Guggemos, M. Seilmaier, S. Zange and M. A. Müller, *et al.*, Virological assessment of hospitalized patients with COVID-2019, *Nature*, 2020, **581**(7809), 465–469.
14. Y. Pan, D. Zhang, P. Yang, L. L. M. Poon and Q. Wang, Viral load of SARS-CoV-2 in clinical samples, *Lancet Infect. Dis.*, 2020, **20**(4), 411–412.
15. C. Huang, Y. Wang, X. Li, L. Ren, J. Zhao and Y. Hu, *et al.*, Clinical features of patients infected with 2019 novel coronavirus in Wuhan, China, *Lancet*, 2020, **395**(10223), 497–506.

16. S. Richardson, J. S. Hirsch, M. Narasimhan, J. M. Crawford, T. McGinn and K. W. Davidson, *et al.*, Presenting Characteristics, Comorbidities, and Outcomes Among 5700 Patients Hospitalized With COVID-19 in the New York City Area, *J. Am. Med. Assoc.*, 2020, **323**(20), 2052–2059.
17. I. Alía and A. Esteban, Weaning from mechanical ventilation, *Crit. Care*, 2000, **4**(2), 72–80.
18. J. C. Marshall, D. J. Cook, N. V. Christou, G. R. Bernard, C. L. Sprung and W. J. Sibbald, Multiple organ dysfunction score: a reliable descriptor of a complex clinical outcome, *Crit. Care Med.*, 1995, **23**(10), 1638–1652.
19. Q. Li, X. Guan, P. Wu, X. Wang, L. Zhou and Y. Tong, *et al.*, Early Transmission Dynamics in Wuhan, China, of Novel Coronavirus–Infected Pneumonia, *N. Engl. J. Med.*, 2020, **382**(13), 1199–1207.
20. D. Gemmati, B. Bramanti, M. L. Serino, P. Secchiero, G. Zauli and V. Tisato, COVID-19 and Individual Genetic Susceptibility/Receptivity: Role of ACE1/ACE2 Genes, Immunity, Inflammation and Coagulation. Might the Double X-chromosome in Females Be Protective against SARS-CoV-2 Compared to the Single X-Chromosome in Males? *Int. J. Mol. Sci.*, 2020, **21**(10), 3474.
21. Z. Wu and J. M. McGoogan, Characteristics of and Important Lessons From the Coronavirus Disease 2019 (COVID-19) Outbreak in China: Summary of a Report of 72314 Cases From the Chinese Center for Disease Control and Prevention, *J. Am. Med. Assoc.*, 2020.
22. M. Yang, Y.-X. Shang, Z.-Y. Tian, M. Xiong, C.-L. Lu and Y. Jiang, *et al.*, Characteristics of registered studies for Coronavirus disease 2019 (COVID-19): A systematic review, *Integr. Med. Res.*, 2020, **9**(3), 100426.
23. A. Kogan, M. J. Segel, E. Ram, E. Raanani, Y. Peled-Potashnik and S. Levin, *et al.*, Acute Respiratory Distress Syndrome following Cardiac Surgery: Comparison of the American-European Consensus Conference Definition *versus* the Berlin Definition, *Respiration*, 2019, **97**(6), 518–524.
24. A. Saguil and M. Fargo, Acute respiratory distress syndrome: diagnosis and management, *Am. Fam. Physician*, 2012, **85**(4), 352–358.
25. C. H. Sudre, K. Lee, M. Ni Lochlainn, T. Varsavsky, B. Murray and M. S. Graham, *et al.*, Symptom clusters in Covid19: A potential clinical prediction tool from the COVID Symptom study app, *Sci. Adv.*, 2021, **7**(12), eabd4177.
26. L. Zhang, F. M. Shen, F. Chen and Z. Lin, Origin and evolution of the 2019 novel coronavirus, *Clin. Infect. Dis.*, 2020, 882–883.
27. China CDC The Novel Coronavirus Pneumonia Emergency Response Epidemiology Team, The Epidemiological Characteristics of an Outbreak of 2019 Novel Coronavirus Diseases (COVID-19) – China, *China CDC Wkly.*, 2020, **2**(8), 113–122.
28. W.-J. Guan, Z.-Y. Ni, Y. Hu, W.-H. Liang, C.-Q. Ou and J.-X. He, *et al.*, Clinical characteristics of coronavirus disease 2019 in China, *N. Engl. J. Med.*, 2020, **382**(18), 1708–1720.

29. J. R. Lechien, C. M. Estomba, S. Place, Y. Van Laethem, P. Cabaraux and Q. Mat, *et al.*, Clinical and Epidemiological Characteristics of 1420 European Patients with mild-to-moderate Coronavirus Disease 2019, *J. Intern. Med.*, **2020**, 335–344.
30. G. Onder, G. Rezza and S. Brusaferro, Case-Fatality Rate and Characteristics of Patients Dying in Relation to COVID-19 in Italy, *J. Am. Med. Assoc.*, 2020, **323**(18), 1775–1776.
31. F. Zhang and C. Hughes, Clinical Epidemiology of Coronavirus Disease 2019: Defined on Current Research, *Global Clinical and Translational Global Clin. Transl. Res.*, 2020, **2**(3), 54–72.
32. F. Zhang and C. Hughes, A Caution in Association of ABO Blood Group with COVID-19, *Global Clinical and Translational Global Clin. Transl. Res.*, 2020, **2**(3), 54–72.
33. D. Wang, B. Hu, C. Hu, F. Zhu, X. Liu and J. Zhang, *et al.*, Clinical Characteristics of 138 Hospitalized Patients With 2019 Novel Coronavirus-Infected Pneumonia in Wuhan, China, *J. Am. Med. Assoc.*, 2020, **323**(11), 1061–1069.
34. T. M. C. de Lucena, A. F. da Silva Santos, B. R. de Lima, M. E. de Albuquerque Borborema and J. de Azevedo Silva, Mechanism of inflammatory response in associated comorbidities in COVID-19, *Diabetes Metab. Syndr.*, 2020, **14**(4), 597–600.
35. D. Dayal, We urgently need guidelines for managing COVID-19 in children with comorbidities, *Acta Paediatr.*, 2020, 1497–1498.
36. V. Balachandar, I. Mahalaxmi, M. Subramaniam, J. Kaavya, N. Senthil Kumar and G. Laldinmawii, *et al.*, Follow-up studies in COVID-19 recovered patients - is it mandatory? *Sci. Total Environ.*, 2020, **729**, 139021.
37. W. T. Gibson, D. M. Evans, J. A. Jones and S. J. Jones, ACE2 Coding Variants: A Potential X-linked Risk Factor for COVID-19 Disease, bioRxiv, 2020, DOI: 10.1101/2020.04.05.026633.
38. WHO, *Novel Coronavirus (2019-nCoV) Situation Report – 3*, 2020. Available from: https://www.who.int/docs/default-source/coronaviruse/situation-reports/20200123-sitrep-3-2019-ncov.pdf?sfvrsn=d6d23643_8.
39. M. Singer, C. S. Deutschman, C. W. Seymour, M. Shankar-Hari, D. Annane and M. Bauer, *et al.*, The Third International Consensus Definitions for Sepsis and Septic Shock (Sepsis-3), *J. Am. Med. Assoc.*, 2016, **315**(8), 801–810.
40. C. W. Seymour, J. N. Kennedy, S. Wang, C. H. Chang, C. F. Elliott and Z. Xu, *et al.*, Derivation, Validation, and Potential Treatment Implications of Novel Clinical Phenotypes for Sepsis, *J. Am. Med. Assoc.*, 2019, **321**(20), 2003–2017.
41. T. J. Matics and L. N. Sanchez-Pinto, Adaptation and Validation of a Pediatric Sequential Organ Failure Assessment Score and Evaluation of the Sepsis-3 Definitions in Critically Ill Children, *JAMA Pediatr.*, 2017, **171**(10), e172352.
42. N. J. Matheson and P. J. Lehner, How does SARS-CoV-2 cause COVID-19? *Science*, 2020, **369**(6503), 510–511.

43. K. G. Andersen, A. Rambaut, W. I. Lipkin, E. C. Holmes and R. F. Garry, The proximal origin of SARS-CoV-2, *Nat. Med.*, 2020, **26**(4), 450–452.
44. K. Kuba, Y. Imai, S. Rao, H. Gao, F. Guo and B. Guan, *et al.*, A crucial role of angiotensin converting enzyme 2 (ACE2) in SARS coronavirus-induced lung injury, *Nat. Med.*, 2005, **11**(8), 875–879.
45. H. P. Jia, D. C. Look, L. Shi, M. Hickey, L. Pewe and J. Netland, *et al.*, ACE2 receptor expression and severe acute respiratory syndrome coronavirus infection depend on differentiation of human airway epithelia, *J. Virol.*, 2005, **79**(23), 14614–14621.
46. R. S. Wiener, Y. X. Cao, A. Hinds, M. I. Ramirez and M. C. Williams, Angiotensin converting enzyme 2 is primarily epithelial and is developmentally regulated in the mouse lung, *J. Cell. Biochem.*, 2007, **101**(5), 1278–1291.
47. M. Hoffmann, H. Kleine-Weber, S. Schroeder, N. Kruger, T. Herrler and S. Erichsen, *et al.*, SARS-CoV-2 Cell Entry Depends on ACE2 and TMPRSS2 and Is Blocked by a Clinically Proven Protease Inhibitor, *Cell*, 2020, **181**(2), 271–280.e8.
48. K. R. Acharya, E. D. Sturrock, J. F. Riordan and M. R. Ehlers, Ace revisited: a new target for structure-based drug design, *Nat. Rev. Drug Discovery*, 2003, **2**(11), 891–902.
49. T. C. O. Unger, T. Csikos, J. Culman, S. Gallinat and P. Gohlke, *et al.*, Angiotensin receptors, *J. Hypertens. Suppl.*, 1996, **14**, S95–S103.
50. V. Shenoy, Y. Qi, M. J. Katovich and M. K. Raizada, ACE2, a promising therapeutic target for pulmonary hypertension, *Curr. Opin. Pharmacol.*, 2011, **11**(2), 150–155.
51. Y. Zhu, Y.-C. Zhu, J. Li, H. Schäfer, W. Schmidt and T. Yao, *et al.*, Effects of losartan on haemodynamic parameters and angiotensin receptor mRNA levels in rat heart after myocardial infarction, *J. Renin-Angiotensin-Aldosterone Syst.*, 2000, **1**, 257–262.
52. Y. Z. Zhu, Y. C. Zhu, Z. J. Wang, Q. Lu, H. S. Lee and T. Unger, Time-dependent apoptotic development and pro-apoptotic genes expression in rat heart after myocardial infarction, *Jpn. J. Pharmacol.*, 2001, **86**(3), 355–358.
53. A. J. Turner and N. M. Hooper, The angiotensin-converting enzyme gene family: genomics and pharmacology, *Trends Pharmacol. Sci.*, 2002, **23**(4), 177–183.
54. Z. Zou, Y. Yan, Y. Shu, R. Gao, Y. Sun and X. Li, *et al.*, Angiotensin-converting enzyme 2 protects from lethal avian influenza A H5N1 infections, *Nat. Commun.*, 2014, **5**(1), 3594.
55. P. Yang, H. Gu, Z. Zhao, W. Wang, B. Cao and C. Lai, *et al.*, Angiotensin-converting enzyme 2 (ACE2) mediates influenza H7N9 virus-induced acute lung injury, *Sci. Rep.*, 2014, **4**(1), 7027.
56. W. Li, M. J. Moore, N. Vasilieva, J. Sui, S. K. Wong and M. A. Berne, *et al.*, Angiotensin-converting enzyme 2 is a functional receptor for the SARS coronavirus, *Nature*, 2003, **426**(6965), 450–454.

57. D. Wrapp, N. Wang, K. S. Corbett, J. A. Goldsmith, C. L. Hsieh and O. Abiona, *et al.*, Cryo-EM structure of the 2019-nCoV spike in the prefusion conformation, *Science*, 2020, **367**(6483), 1260–1263.
58. R. Yan, Y. Zhang, Y. Li, L. Xia and Q. Zhou, Structural basis for the recognition of SARS-CoV-2 by full-length human ACE2, *Science*, 2020, **367**, 1444–1448.
59. X. Zhang, H. Cai, J. Hu, J. Lian, J. Gu and S. Zhang, *et al.*, Epidemiological, clinical characteristics of cases of SARS-CoV-2 infection with abnormal imaging findings, *Int. J. Infect. Dis.*, 2020, **94**, 81–87.
60. G. Chen, D. Wu, W. Guo, Y. Cao, D. Huang and H. Wang, *et al.*, Clinical and immunological features of severe and moderate coronavirus disease 2019, *J. Clin. Invest.*, 2020, **130**(5), 2620–2629.
61. J. Xu, S. Zhao, T. Teng, A. E. Abdalla, W. Zhu and L. Xie, *et al.*, Systematic Comparison of Two Animal-to-Human Transmitted Human Coronaviruses: SARS-CoV-2 and SARS-CoV, *Viruses*, 2020, **12**(2), 244.
62. C. Qin, L. Zhou, Z. Hu, S. Zhang, S. Yang and Y. Tao, *et al.*, Dysregulation of immune response in patients with COVID-19 in Wuhan, China, *Clin. Infect. Dis.*, 2020, ciaa248.
63. B. Rockx, T. Kuiken, S. Herfst, T. Bestebroer, M. M. Lamers and B. B. O. Munnink, *et al.*, Comparative pathogenesis of COVID-19, MERS, and SARS in a nonhuman primate model, *Science*, 2020, **368**(6494), 1012–1015.
64. F. Zhou, T. Yu, R. Du, G. Fan, Y. Liu and Z. Liu, *et al.*, Clinical course and risk factors for mortality of adult inpatients with COVID-19 in Wuhan, China: a retrospective cohort study, *Lancet*, 2020, **395**(10229), 1054–1062.
65. M. Fang, N. A. Siciliano, A. R. Hersperger, F. Roscoe, A. Hu and X. Ma, *et al.*, Perforin-dependent CD4+ T-cell cytotoxicity contributes to control a murine poxvirus infection, *Proc. Natl. Acad. Sci. U. S. A.*, 2012, **109**(25), 9983–9988.
66. B. A. Small, S. A. Dressel, C. W. Lawrence, D. R. Drake 3rd, M. H. Stoler and R. I. Enelow, *et al.*, CD8(+) T cell-mediated injury *in vivo* progresses in the absence of effector T cells, *J. Exp. Med.*, 2001, **194**(12), 1835–1846.
67. J.-M. Sallenave and L. Guillot. Host signaling and proteolytic pathways in the resolution or the exacerbation of coronavirus (CoV-2) infection in COVID-19 disease: what therapeutic targets? 2020.
68. J. Zhong, J. Tang, C. Ye and L. Dong, The immunology of COVID-19: is immune modulation an option for treatment? *Lancet Rheumatol.*, 2020, e428–e436.
69. R. E. Young, R. D. Thompson, K. Y. Larbi, M. La, C. E. Roberts and S. D. Shapiro, *et al.*, Neutrophil elastase (NE)-deficient mice demonstrate a nonredundant role for NE in neutrophil migration, generation of proinflammatory mediators, and phagocytosis in response to zymosan particles in vivo, *J. Immunol.*, 2004, **172**(7), 4493–4502.
70. R. Hallgren, T. Samuelsson, T. C. Laurent and J. Modig, Accumulation of hyaluronan (hyaluronic acid) in the lung in adult respiratory distress syndrome, *Am. Rev. Respir. Dis.*, 1989, **139**(3), 682–687.

71. T. J. Bell, O. J. Brand, D. J. Morgan, S. Salek-Ardakani, C. Jagger and T. Fujimori, *et al.*, Defective lung function following influenza virus is due to prolonged, reversible hyaluronan synthesis, *Matrix Biol.*, 2019, **80**, 14–28.
72. M. Z. Tay, C. M. Poh, L. Renia, P. A. MacAry and L. F. P. Ng, The trinity of COVID-19: immunity, inflammation and intervention, *Nat. Rev. Immunol.*, 2020, **20**(6), 363–374.
73. R. J. Jose and A. Manuel, COVID-19 cytokine storm: the interplay between inflammation and coagulation, *Lancet Respir. Med.*, 2020, e46–e47.
74. A. C. Walls, Y. J. Park, M. A. Tortorici, A. Wall, A. T. McGuire and D. Veesler, Structure, Function, and Antigenicity of the SARS-CoV-2 Spike Glycoprotein, *Cell*, 2020, **181**(2), 281–292.e6.
75. S. Kumar, R. Nyodu, V. K. Maurya and S. K. Saxena, Morphology, Genome Organization, Replication, and Pathogenesis of Severe Acute Respiratory Syndrome Coronavirus 2 (SARS-CoV-2), *Coronavirus Disease 2019 (COVID-19)*, Springer, 2020, pp. 23–31.
76. A. Llanes, C. M. Restrepo, Z. Caballero, S. Rajeev, M. A. Kennedy and R. Lleonart. *Betacoronavirus* genomes: How genomic information has been used to deal with past outbreaks and the COVID-19 pandemic. Int. J. Mol. Sci., 2020 Jun; 21(12): 4546.
77. B. Kim, J.-Y. Park, D.-Y. Cho, H. M. Ko, S.-H. Yoon and D.-K. Choi, 2-(5-(4-Chlorophenyl)-1-(2, 4-dichlorophenyl)-4-methyl-1H-pyrazol-3-yl)-*N*-(2-hydroxyethyl)-2-oxoacetamide (CDMPO) has anti-inflammatory properties in microglial cells and prevents neuronal and behavioral deficits in MPTP mouse model of Parkinson's disease, *Neuropharmacology*, 2020, **166**, 107928.
78. I. Astuti and Ysrafil, Severe Acute Respiratory Syndrome Coronavirus 2 (SARS-CoV-2): An overview of viral structure and host response, *Diabetes Metab. Syndr.*, 2020, **14**(4), 407–412.
79. H. Qiu, J. Wu, L. Hong, Y. Luo, Q. Song and D. Chen, Clinical and epidemiological features of 36 children with coronavirus disease 2019 (COVID-19) in Zhejiang, China: an observational cohort study, *Lancet Infect. Dis.*, 2020, **20**(6), 689–696.
80. Z. Liu, X. Xiao, X. Wei, J. Li, J. Yang and H. Tan, *et al.*, Composition and divergence of coronavirus spike proteins and host ACE2 receptors predict potential intermediate hosts of SARS-CoV-2, *J. Med. Virol.*, 2020, **92**(6), 595–601.
81. Y. Cao, *Suggestion Using Alcohol Vaporization or Nebulization Inhalation Therapy for Pneumonitis Caused by Coronavirus.* Available at SSRN 3545744. 2020.
82. J. Shang, G. Ye, K. Shi, Y. Wan, C. Luo and H. Aihara, *et al.*, Structural basis of receptor recognition by SARS-CoV-2, *Nature*, 2020, **581**(7807), 221–224.
83. E. Braun and D. Sauter, Furin-mediated protein processing in infectious diseases and cancer, *Clin. Transl. Immunol.*, 2019, **8**(8), e1073.

84. K. A. Hay, L. A. Hanafi, D. Li, J. Gust, W. C. Liles and M. M. Wurfel, *et al.*, Kinetics and biomarkers of severe cytokine release syndrome after CD19 chimeric antigen receptor-modified T-cell therapy, *Blood*, 2017, **130**(21), 2295–2306.
85. T. Tanaka, M. Narazaki and T. Kishimoto, Immunotherapeutic implications of IL-6 blockade for cytokine storm, *Immunotherapy*, 2016, **8**(8), 959–970.
86. N. Pathan, C. A. Hemingway, A. A. Alizadeh, A. C. Stephens, J. C. Boldrick and E. E. Oragui, *et al.*, Role of interleukin 6 in myocardial dysfunction of meningococcal septic shock, *Lancet*, 2004, **363**(9404), 203–209.
87. S. Tian, W. Hu, L. Niu, H. Liu, H. Xu and S. Y. Xiao, Pulmonary Pathology of Early-Phase 2019 Novel Coronavirus (COVID-19) Pneumonia in Two Patients With Lung Cancer, *J. Thorac. Oncol.*, 2020, **15**(5), 700–704.
88. H. Zhang, P. Zhou, Y. Wei, H. Yue, Y. Wang and M. Hu, *et al.*, Histopathologic Changes and SARS-CoV-2 Immunostaining in the Lung of a Patient With COVID-19, *Ann. Intern. Med.*, 2020, **172**(9), 629–632.
89. T. Menter, J. D. Haslbauer, R. Nienhold, S. Savic, H. Deigendesch and S. Frank, *et al.*, Postmortem examination of COVID-19 patients reveals diffuse alveolar damage with severe capillary congestion and variegated findings in lungs and other organs suggesting vascular dysfunction, *Histopathology*, 2020, **77**(2), 198–209.
90. W. Wang, Y. Xu, R. Gao, R. Lu, K. Han and G. Wu, *et al.*, Detection of SARS-CoV-2 in Different Types of Clinical Specimens, *J. Am. Med. Assoc.*, 2020, **323**(18), 1843–1844.
91. Y. Ding, H. Wang, H. Shen, Z. Li, J. Geng and H. Han, *et al.*, The clinical pathology of severe acute respiratory syndrome (SARS): a report from China, *J. Pathol.*, 2003, **200**(3), 282–289.
92. Q. Ruan, K. Yang, W. Wang, L. Jiang and J. Song, Clinical predictors of mortality due to COVID-19 based on an analysis of data of 150 patients from Wuhan, China, *Intensive Care Med.*, 2020, 1–3.
93. A. R. Chapman, A. Bularga and N. L. Mills, High-Sensitivity Cardiac Troponin Can Be an Ally in the Fight Against COVID-19, *Circulation*, 2020, **141**(22), 1733–1735.
94. G. L. Lippi, J. Carl and F. Sanchis-Gomar, Cardiac troponin I in patients with coronavirus disease 2019 (COVID-19): Evidence from a meta-analysis, *Prog. Cardiovasc. Dis.*, 2020, **63**, 390–391.
95. S. Shi, M. Qin, B. Shen, Y. Cai, T. Liu and F. Yang, *et al.*, Association of Cardiac Injury With Mortality in Hospitalized Patients With COVID-19 in Wuhan, China, *JAMA Cardiol.*, 2020, **5**(7), 802–810.
96. K. J. Clerkin, J. A. Fried, J. Raikhelkar, G. Sayer, J. M. Griffin and A. Masoumi, *et al.*, COVID-19 and Cardiovascular Disease, *Circulation*, 2020, **141**(20), 1648–1655.
97. J. Yang, Y. Zheng, X. Gou, K. Pu, Z. Chen and Q. Guo, *et al.*, Prevalence of comorbidities and its effects in patients infected with SARS-CoV-2: a systematic review and meta-analysis, *Int. J. Infect. Dis.*, 2020, **94**, 91–95.

98. C. Gao, Y. Cai, K. Zhang, L. Zhou, Y. Zhang and X. Zhang, *et al.*, Association of hypertension and antihypertensive treatment with COVID-19 mortality: a retrospective observational study, *Eur. Heart J.*, 2020, **41**(22), 2058–2066.
99. M. Arentz, E. Yim, L. Klaff, S. Lokhandwala, F. X. Riedo and M. Chong, *et al.*, Characteristics and outcomes of 21 critically ill patients with COVID-19 in Washington State, *J. Am. Med. Assoc.*, 2020, **323**(16), 1612–1614.
100. R. M. Inciardi, L. Lupi, G. Zaccone, L. Italia, M. Raffo and D. Tomasoni, *et al.*, Cardiac Involvement in a Patient With Coronavirus Disease 2019 (COVID-19), *JAMA Cardiol.*, 2020, **5**(7), 1–6.
101. A. Jabri, A. Kalra, A. Kumar, A. Alameh, S. Adroja and H. Bashir, *et al.*, Incidence of Stress Cardiomyopathy During the Coronavirus Disease 2019 Pandemic, *JAMA Netw. Open*, 2020, **3**(7), e2014780.
102. P. P. Liu, A. Blet, D. Smyth and H. Li, The Science Underlying COVID-19, *Circulation*, 2020, **142**(1), 68–78.
103. B. Long, W. J. Brady, A. Koyfman and M. Gottlieb, Cardiovascular complications in COVID-19, *Am. J. Emerg. Med.*, 2020, **38**(7), 1504–1507.
104. E. Y. Wang, O. L. Hulme, S. Khurshid, L.-C. Weng, S. H. Choi and A. J. Walkey, *et al.*, Initial Precipitants and Recurrence of Atrial Fibrillation, *Circ.: Arrhythmia Electrophysiol.*, 2020, **13**(3), e007716.
105. P. Zhang, L. Zhu, J. Cai, F. Lei, J.-J. Qin and J. Xie, *et al.*, Association of Inpatient Use of Angiotensin-Converting Enzyme Inhibitors and Angiotensin II Receptor Blockers With Mortality Among Patients With Hypertension Hospitalized With COVID-19, *Circ. Res.*, 2020, **126**(12), 1671–1681.
106. B. K. Bozkurt, R. Kovacs and B. Harrington, *HFSA/ACC/AHA Statement Addresses Concerns Re: Using RAAS Antagonists in COVID-19. Updated 17 March, 2020*, 2020. Available from: https://www.acc.org/latest-in-cardiology/articles/2020/03/17/08/59/hfsa-acc-aha-statement-addresses-concerns-re-using-raas-antagonists-in-covid-19.
107. Q. Huang, X. Deng, Y. Li, X. Sun, Q. Chen and M. Xie, *et al.*, Clinical characteristics and drug therapies in patients with the common-type coronavirus disease 2019 in Hunan, China, *Int. J. Clin. Pharm.*, 2020, **42**(3), 837–845.
108. C. Xie, L. Jiang, G. Huang, H. Pu, B. Gong and H. Lin, *et al.*, Comparison of different samples for 2019 novel coronavirus detection by nucleic acid amplification tests, *Int. J. Infect. Dis.*, 2020, **93**, 264–267.
109. W. Tian, W. Jiang, J. Yao, C. J. Nicholson, R. H. Li and H. H. Sigurslid, *et al.*, Predictors of mortality in hospitalized COVID-19 patients: A systematic review and meta-analysis, *J. Med. Virol.*, 2020, **92**(10), 1875–1883.
110. Z. Fan, L. Chen, J. Li, X. Cheng, J. Yang and C. Tian, *et al.*, Clinical Features of COVID-19-Related Liver Functional Abnormality, *Clin. Gastroenterol. Hepatol.*, 2020, **18**(7), 1561–1566.

111. X. Chai, L. Hu, Y. Zhang, W. Han, Z. Lu and A. Ke, *et al.*, Specific ACE2 expression in cholangiocytes may cause liver damage after 2019-nCoV infection, bioRxiv, 2020, DOI: 10.1101/2020.02.03.931766.
112. Z. Xu, L. Shi, Y. Wang, J. Zhang, L. Huang and C. Zhang, *et al.*, Pathological findings of COVID-19 associated with acute respiratory distress syndrome, *Lancet Respir. Med.*, 2020, **8**(4), 420–422.
113. S. Naicker, C. W. Yang, S. J. Hwang, B. C. Liu, J. H. Chen and V. Jha, The Novel Coronavirus 2019 epidemic and kidneys, *Kidney Int.*, 2020, **97**(5), 824–828.
114. A. Filatov, P. Sharma, F. Hindi and P. S. Espinosa, Neurological Complications of Coronavirus Disease (COVID-19): Encephalopathy, *Cureus*, 2020, **12**(3), e7352.
115. L. Cantuti-Castelvetri, R. Ojha, L. Pedro, M. Djannatian, J. Franz and S. Kuivanen, *et al.*, Neuropilin-1 facilitates SARS-CoV-2 cell entry and infectivity, *Science*, 2020, **370**(6518), 856–860.
116. B. Vellingiri, K. Jayaramayya, M. Iyer, A. Narayanasamy, V. Govindasamy and B. Giridharan, *et al.*, COVID-19: A promising cure for the global panic, *Sci. Total Environ.*, 2020, **725**, 138277.
117. R. W. Paterson, R. L. Brown, L. Benjamin, R. Nortley, S. Wiethoff and T. Bharucha, *et al.*, The emerging spectrum of COVID-19 neurology: clinical, radiological and laboratory findings, *Brain*, 2020, **143**(10), 3104–3120.
118. A. Varatharaj, N. Thomas, M. A. Ellul, N. W. S. Davies, T. A. Pollak and E. L. Tenorio, *et al.*, Neurological and neuropsychiatric complications of COVID-19 in 153 patients: a UK-wide surveillance study, *Lancet Psychiatry*, 2020, **7**(10), 875–882.
119. P. K. Bhatraju, B. J. Ghassemieh, M. Nichols, R. Kim, K. R. Jerome and A. K. Nalla, *et al.*, Covid-19 in Critically Ill Patients in the Seattle Region - Case Series, *N. Engl. J. Med.*, 2020, **382**(21), 2012–2022.
120. B. Li, J. Yang, F. Zhao, L. Zhi, X. Wang and L. Liu, *et al.*, Prevalence and impact of cardiovascular metabolic diseases on COVID-19 in China, *Clin. Res. Cardiol.*, 2020, **109**(5), 531–538.
121. A. Alramthan and W. Aldaraji, A case of COVID-19 presenting in clinical picture resembling chilblains disease. First report from the Middle East, *Clin. Exp. Dermatol.*, 2020, 746–748.
122. V. Piccolo, I. Neri, F. Manunza, C. Mazzatenta and A. Bassi, Chilblain-like lesions during the COVID-19 pandemic: should we really worry? *Int. J. Dermatol.*, 2020, **59**(8), 1026–1027.
123. S. Recalcati, Cutaneous manifestations in COVID-19: a first perspective, *J. Eur. Acad. Dermatol. Venereol.*, 2020, **34**(5), e212–e213.
124. G. Lippi and E. J. Favaloro, D-dimer is associated with severity of coronavirus disease 2019: a pooled analysis, *Thromb. Haemostasis*, 2020, **120**(05), 876–878.
125. E. Terpos, I. Ntanasis-Stathopoulos, I. Elalamy, E. Kastritis, T. N. Sergentanis and M. Politou, *et al.*, Hematological findings and complications of COVID-19, *Am. J. Hematol.*, 2020, **95**(7), 834–847.

CHAPTER 5

Evaluation of the Disease, Sample Collection and Diagnostics

CLAUDE HUGHES[†*a], KAAVYA JAYARAMAYYA[†b], BALACHANDAR VELLINGIRI[†*c], STAFFORD WARREN[†d] AND MICHAEL WATERS[†e]

[a]IQVIA, and Duke University Medical Center, Durham, NC, USA; [b]Department of Zoology, Avinashilingam Institute for Home Science and Higher Education for Women, India; [c]Department of Human Genetics and Molecular Biology, Bharathiar University, India; [d]Anne Arundel Medical Center, Annapolis MD, USA; [e]Michael Waters Consulting USA, Hillsborough, NC, USA
*E-mail: clhughes@duke.edu, geneticbala@buc.edu.in

5.1 Introduction

The rapid propagation of coronavirus disease 2019 (COVID-19) across various populations has allowed researchers to pinpoint symptoms that are specific to COVID-19. However, the interpretation of symptoms is somewhat situation-dependent, as described in detail in Chapter 4 (Clinical Presentation, Pathology and Histopathology) and further elaborated in Chapter 7 (Clinical Epidemiology of Coronavirus Disease 2019: Infection, Clinical Spectrum and Distribution, and Case-fatality). Briefly stated, patients with COVID-19 can be asymptomatic or have mild, moderate or severe illness.

[†]All authors contributed equally and are listed alphabetically by surname.

The Coronavirus Pandemic and the Future Volume 1: Virology, Epidemiology, Translational Toxicology and Therapeutics
Edited by Michael D. Waters, Alok Dhawan, Tim Marrs, Diana Anderson, Stafford Warren and Claude L. Hughes

Published by the Royal Society of Chemistry, www.rsc.org

The predominant manifestation of moderate or severe COVID-19 appears to be dyspnea, together with fever. During this manifestation, and particularly when a reduced concentration of oxygen in the blood is encountered, patients exhibiting these symptoms must be watched closely. Government agencies around the world have recommended that persons with symptoms of COVID-19 as well as persons with a recent history of exposure should be tested for the presence of the virus, as they may be carriers. Moreover, a variety of guidelines have been issued for the appropriate collection of samples to avoid faulty results. Extreme vigilance is required while identifying those patients who need to have nucleic acid-based testing to detect the presence of COVID-19. This is the case because of the large number of asymptomatic individuals who test positive for COVID-19. Most jurisdictions now utilize some type of clinical and epidemiological information, in conjunction with the availability of tests and the personnel required to administer them, to determine who should be tested and with what frequency. In the United States, criteria have been developed to identify persons under investigation (PUI) for COVID-19. According to the United States Centers for Disease Control and Prevention (CDC), most patients with confirmed COVID-19 have developed a fever and/or symptom of acute respiratory illness, such as cough and difficulty breathing. If a person is a PUI, the CDC recommends that practitioners immediately impose appropriate infection control and prevention measures. Initially, the CDC recommends testing for all other sources of respiratory infection. In addition, the CDC recommends assessing epidemiological factors to assist in decision-making concerning which other persons should be tested. This includes anyone who has been in close contact with a patient who has a laboratory-confirmed case of COVID-19 within 14 days of symptom onset, or a history of travel within 14 days of symptom onset from geographic areas where the virus is prevalent.

The inability of governments and private-sector companies to rapidly field appropriate reliable tests and supplies has hindered widespread public testing as well as clinical diagnosis. Due to the limitations in the supply of test kits and reagents, with few exceptions it has not been possible to test the public in an aggressive manner. Furthermore, clear and consistent rules need to be developed by which to identify and test vulnerable populations. Perhaps the greatest impediment in the application of tests is the large proportion of asymptomatic individuals (variously estimated at 25–80% of the exposed population). In any case, the use of test results must be reported in a timely manner to diagnose cases and to implement tracing of contacts and isolation to suppress disease transmission.

Recently, governments have considered multiplex reverse transcriptase polymerase chain reaction (RT-PCR) assays to "get ahead of the curve" in virus testing. Thus, the International Reagent Resource at https://www.internationalreagentresource.org/ is distributing the Flu SC2 Multiplex Assay and supplies to registered public health laboratories so that they can perform this assay to test simultaneously for influenza and COVID-19. To begin to decipher the genetic epidemiology of severe acute respiratory syndrome coronavirus 2 (SARS-CoV-2), India has implemented the COVIDSeq protocol, which involves multiplex-PCR, barcoding and sequencing of samples for high-throughput detection of the virus.[1]

5.1.1 Symptoms of COVID-19

People worldwide have been exposed to SARS-CoV-2 in some manner and many have developed symptoms of COVID-19 that range from asymptomatic to mild, moderate and severe. Patients with a mild form of the disease do not show any radiographic abnormalities. Those patients with moderate disease exhibit respiratory symptoms, fever and radiographic features such as ground-glass opacities (GGOs). Those with severe forms of the condition may have dyspnea, a respiratory rate >30 breaths min^{-1}, oxygen saturation <94% in ambient air, and arterial oxygen tension/inspiratory oxygen fraction ratio <300 mmHg. The first report of patients with COVID-19 described bilateral lung involvement on initial chest computed tomography (CT) in 40 out of 41 patients, with a consolidative pattern seen in patients in the intensive care unit (ICU) and a predominantly ground-glass pattern in patients who were not in the ICU.[2,3] Critically ill patients may present with respiratory failure, septic shock or multiple organ failure.[4]

In a retrospective study involving 8697 patients with COVID-19, fever was found in 78.4% of cases, and 58.3% of them exhibited cough.[5] Importantly, other investigators found that 5.4% of the patients were asymptomatic[6] and subsequent studies employing more widespread testing have revealed higher percentages of asymptomatic individuals. Some COVID-19 patients have also exhibited neurological symptoms such as anosmia, hyposmia and dysgeusia, and these symptoms were present prior to the other clinical abnormalities, suggesting that these symptoms may be the first ones that can be observed in individuals with COVID-19.[7] Moreover, in pediatric patients, the most commonly occurring abnormalities were fever (59%) and cough (46%). 12% of the children exhibited gastrointestinal signs and/or symptoms and 26% were asymptomatic.[8]

When considering all these symptoms, it is important to note that other organ involvement may also constitute an important tool by which to differentiate between the common flu and COVID-19. Regardless of travel history, patients exhibiting fever and cough must be tested for COVID-19 during the pandemic. In this situation, individuals most dangerous to others are those who are asymptomatic and may not reach out to the medical care centers to be tested. This points to the importance of contact tracing, which is likely to identify some of these individuals. Furthermore, the period between initial infection and a positive nucleic acid test may be as long as 8 to 24 days, which provides the virus with plenty of time to infect neighboring individuals.[9] It is essential to understand the differences between the symptoms of COVID-19 and other illnesses like the common cold and flu. Clinicians and the public must be trained to be vigilant for the symptoms of COVID-19, and the public should also be advised to report even the slightest occurrences of fever and cough, leading to testing, in order to minimize the further spread of the pandemic.

5.1.2 Detection of Disease Based on Symptoms

The detection of COVID-19 based on symptoms is extremely challenging, as they overlap with other coronaviruses and common ailments such as the flu and the common cold. It is thus essential to identify individuals exhibiting these characteristics. Many signs of respiratory ailments are present in COVID-19, which can serve as identifiers that distinguish COVID-19 from other illnesses.[10,11] In addition, many individuals may report common symptoms of COVID-19 such as fever and dyspnea, but these may result from prior conditions such as urinary tract infections, chronic obstructive pulmonary disease and others that can cause an inflammatory state in the body. Furthermore, it has been observed that fever is typically persistent throughout a COVID-19 infection; thus, if the fever subsides after an antimicrobial medication is administered, it is highly probable that COVID-19 is not the causative factor.[7] Thus, taking a patient history is extremely important to confirm a COVID-19 diagnosis. Moreover, the timeline of COVID-19 infection is truly relevant in identifying the condition. When patients exhibit any sign of acute or chronic respiratory failure, such as central cyanosis, tachypnea and use of accessory muscles, these parameters should be carefully assessed with transfer to an ICU with consideration for proning, non-invasive ventilation or mechanical ventilation (when other measures fail). In typical COVID-19 patients, the onset of respiratory symptoms usually occurs from day 5 to day 7 (Figure 5.1).

Regarding non-specific symptoms, care should be taken to ensure that these symptoms are not attributable to other pre-existing conditions or the common cold or seasonal flu. This will help stem unnecessary panic that is associated with fear of contracting COVID-19. Thus, the most common early sign of the virus is alteration of smell and/or taste followed by cough

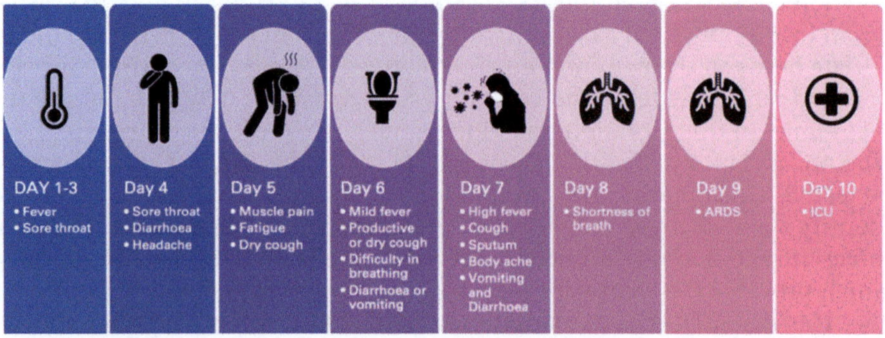

Figure 5.1 Progression of COVID-19 and its symptoms over the course of 10 days.

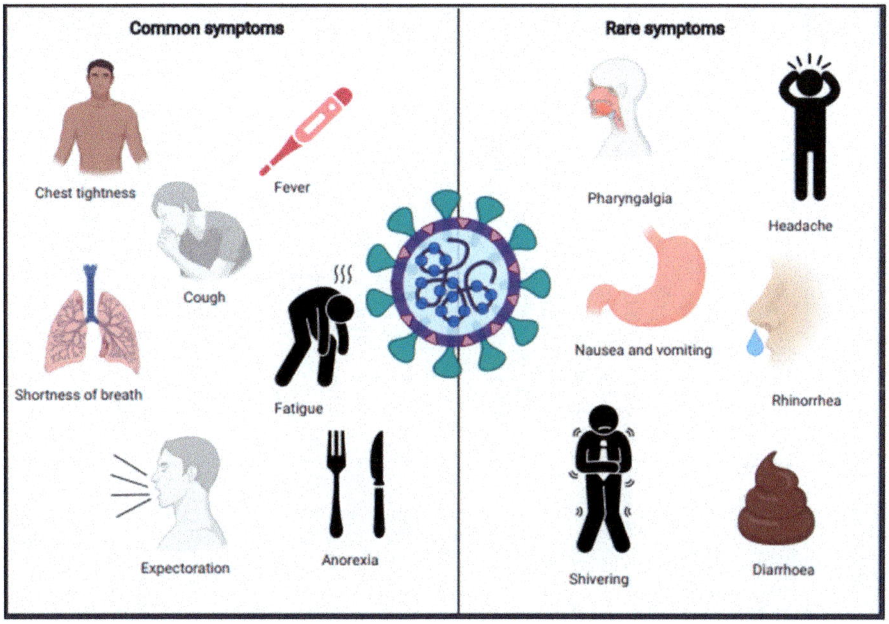

Figure 5.2 Symptoms of COVID-19. These are the initial common and rare symptoms observed in COVID-19 patients.

and dyspnea (Figure 5.2). This sequence is not usually seen in individuals infected with the flu or common cold. Furthermore, sneezing is a common symptom of a cold, but is not typically present in COVID-19 nor with the flu. When compared to the symptoms of SARS-CoV and Middle East respiratory syndrome (MERS), the clinical symptoms of COVID-19 are similar to SARS-CoV and overlap with those of MERS[12] (Table 5.1). Testing is helpful to identify that SARS-CoV-2 is the pathogen, but clinical confusion with SARS-CoV and MERS-CoV is unlikely, as the former has not been reported since 2004 and the latter has caused infection only infrequently and sporadically since 2012–2013.

5.1.3 Evaluation Criteria for COVID-19 Based on ICMR and WHO Guidelines

As the importance of molecular detection of the virus continues to rise, the World Health Organization (WHO) has disseminated criteria concerning who should be tested first in countries with limited resources. The WHO has prioritized testing in people who are vulnerable and who may require intensive care if they contract the virus, as well as healthcare workers and non-clinical healthcare staff, even if they have not come into contact with an infected individual. Also considered a priority for testing are symptomatic

Table 5.1 Comparative symptoms between COVID-19, MERS, SARS, flu and the common cold. This table represents the parallel investigation of SARS, MERS, COVID-19 and common flu with their symptoms, onset of disease, incubation period, recovery, transmission of disease, complications and available treatments.

Diseases	Symptoms	Onset of disease	Incubation period	Recovery	Transmission of disease	Complications, if any	Treatments, if available
Novel coronavirus (COVID-19)	Fever Cough Shortness of breath Fatigue	Sudden	2–14 days after exposure	2–8 weeks	Human to human	Acute pneumonia, septic shock, respiratory failure in adverse condition	No vaccines available, only symptoms can be treated
Severe acute respiratory syndrome (SARS)	Fever Dry cough Headache Difficulty in breathing Muscle aches Loss of appetite Diarrhea	Sudden	2–7 days after exposure	5– 6 weeks	Human to human	Heart, liver and respiratory failure in adverse condition	Breathing ventilator to deliver oxygen Pneumonia-treating antibiotics Antiviral medicines Steroids to reduce lung swelling
Middle East respiratory syndrome (MERS)	Fever Chills Diarrhea Nausea Vomiting Congestion Sneezing Sore throat	Sudden	5–6 days after exposure	6–7 weeks	Human to human	Acute pneumonia, kidney failure in adverse condition	Treatment only for symptoms, such as fluid replacement Oxygen therapy
Common flu	Runny or stuffy nose Sneezing Sore throat Mild headache Low-grade fever	Gradual	2–3 days after exposure	7–10 days	Human to human	Extremely rare or none	Symptoms can be treated by medication

people in enclosed living quarters or those who live or otherwise spend time in institutional settings like schools, hospitals, prisons and long-term care facilities. Individuals with symptoms who are close to these areas can be considered as probable cases and should be tested based on the availability of standard tests. As countries can have geographic areas with no cases as well as those with a high level of community transmission, the WHO has recognized that different areas may require different strategies. Countries with no known transmission are recommended to test suspected individuals in new, unevaluated populations to monitor and contain the virus. In addition, communities where the virus is prevalent must increase their testing for early detection, contact tracing and quarantine when positive, as well as medical intervention when needed. These recommendations are also designed to prevent the progression of the disease in individuals with pre-existing conditions. The WHO global influenza surveillance standards define the surveillance case definitions for influenza-like illness (ILI) and severe acute respiratory infections (SARIs). These standards have served as a starting point for in-country recommendations.

Key messages when using the case definitions (as of January 2014):

- Influenza infection causes a clinical syndrome not easily distinguished from other respiratory infections.
- The case definitions for ILI and SARI are not necessarily intended to capture all cases, but to describe trends over time.
- Using one common case definition globally will allow national health authorities to interpret their data in an international context.

ILI case definition: an acute respiratory infection with a measured fever of ≥38 °C and cough, with onset within the past 10 days.

SARI case definition: an acute respiratory infection with a history of fever or measured fever of ≥38 °C and cough, with onset within the past 10 days and requiring hospitalization.

Based on the WHO's recommendations, the CDC has also issued some guidelines concerning testing, as follows. Medical practitioners are encouraged to use their judgement while testing for COVID-19. As many individuals are asymptomatic, clinicians are encouraged to adhere to local testing guidelines to ensure that community transmission is limited. Testing is particularly encouraged in long-term care facilities, especially where one or more residents have already contracted the virus. Testing is also encouraged for healthcare professionals who are in close contact with patients (Figure 5.3). Priority for testing should be given to individuals presenting with the symptoms discussed herein and to others who are considered as potential carriers by clinicians.

Various countries have conditions and guidelines appropriate for their context. For example, the Indian Council of Medical Research (ICMR) has established rules to be followed by hospitals in that country. Patients are considered as a SARI case if they have an acute respiratory infection with fever

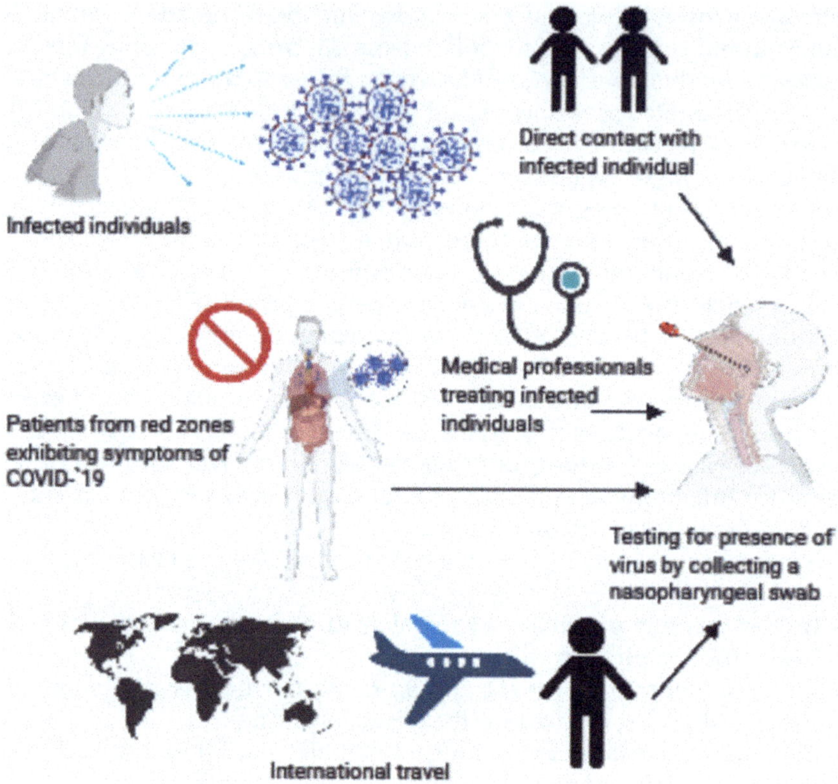

Figure 5.3 Spread of the SARS-CoV-2 virus. The figure depicts the various modes of SARS-CoV-2 transmission with testing procedures.

≥38 °C, cough and require hospitalization and medical intervention. When a patient shows ILI symptoms and has traveled internationally within the previous 2 weeks, molecular diagnostic tests using real time RT-PCR (described later) are performed to identify the presence of viral proteins. Individuals who exhibit all ILI symptoms and who have been in physical contact with other patients who have tested positive through the RT-PCR method are also eligible for other molecular diagnostics such as rapid antigen tests.

In addition, all healthcare workers and frontline medical professionals who exhibit ILI symptoms are tested for viral proteins. All individuals who then progress from ILI to SARI are again tested for these proteins to reconfirm the presence of COVID-19. Asymptomatic people who have previously been in contact with affected individuals must be tested 5–10 days following contact. Moreover, patients presenting with ILI symptoms in containment or hotspot zones should immediately be tested for the presence of the virus. Likewise, patients who have been admitted to the hospital due to prior illnesses and who then develop ILI symptoms are also tested for the COVID-19. In addition, people returning to their homes from elsewhere and migrant populations are also tested within 7 days of displaying ILI symptoms. These rigorous testing strategies are essential to identify carriers of COVID-19 and to ensure that it is contained.

The WHO recommends collecting specimens from both the upper respiratory tract (*i.e.* nasopharyngeal and oropharyngeal samples) and the lower respiratory tract when possible (*e.g.* expectorated sputum, endotracheal aspirate or bronchoalveolar lavage). The collection of bronchoalveolar lavage samples should only be performed in mechanically ventilated patients or patients undergoing bronchoscopy, but lower respiratory tract samples are particularly useful, as they seem to remain positive for a more extended period. These samples require storage at 4 °C (39.2 °F).

The viral genomic material for SARS-CoV-2 is ribonucleic acid (RNA), which remains in the body only while the virus is still replicating. In the laboratory, amplification of the genetic material extracted from the saliva or mucus sample can be performed by means of a RT-PCR, which involves the synthesis of a double-stranded DNA molecule from a sample containing viral RNA. If sufficient viral genetic material is contained in the sample, portions of the conserved coronavirus sequence may be examined and confirmed. The probes used for the amplification process are based on the initial gene sequence released by the Shanghai Public Health Clinical Center and School of Public Health, Fudan University, Shanghai, China on https://virological.org/ with subsequent confirmatory evaluation by additional laboratories. Ideally, positive test results should be repeated for verification. In patients with a confirmed COVID-19 diagnosis, the laboratory evaluation should be repeated to assess viral clearance before the patient is released from observation. While the availability of testing will vary greatly depending on the country in which a person lives, tests are increasingly available in most countries.

5.2 Sample Collection

5.2.1 Overview of Sample Collection Guidelines from the WHO

As samples collected from COVID-19 patients are hazardous, the WHO and CDC[13,14] have developed standard operating procedures to ensure the careful collection, storage, packaging and transport of specimens. The specimens are generally handled by trained and qualified staff. The COVID-19 samples collected are upper respiratory specimens including nasopharyngeal and oropharyngeal swabs. Specimens obtained from washes have also been recommended in ambulatory patients. As the aerosolization (aerosols are usually thought of as droplets ≤5 μm in size) of the virus has been observed, stringent measures must be undertaken to prevent infection. As viral shedding has also been detected in blood, urine and stool, these samples have also been considered to test for COVID-19. In patients who have died as a consequence of COVID-19, lung tissue has been collected after autopsy to determine infection.[15]

Both the WHO and CDC have recommended that the specimens must reach the testing laboratory soon after collection. Appropriate handling of the samples during transport is necessary to prevent infection. The specimens are predominantly transported in a viral medium that contains an antifungal and antibiotic supplement if a delay in reaching the testing center is anticipated. International transport of samples is required to adhere to United Nations model regulations. Moreover, it is essential to ensure proper labeling of the samples while ensuring their timely processing. Key details concerning the samples must be communicated to the laboratory that is processing them. Various types of samples and their related specifications are as follows.[16]

5.2.1.1 *Nasopharyngeal or Oropharyngeal Swabs*

It has been recommended that dacron- or polyester-flocked swabs should be used to collect specimens. Calcium-alginate containing swabs are to be avoided to prevent inactivation of the virus, which will result in faulty PCR results. If both nasopharyngeal and oropharyngeal swabs are collected, they are combined in a single tube to increase the amount of virus present and reduce the waste of resources. The swab must be rubbed and rolled against the mucosa, and the swab is left for some time so that the virus can be absorbed. The swab is then removed while rotating it and the same process is repeated on the other side if necessary. For the oropharyngeal specimen, the swab is inserted into the pharynx and tonsillar area. The swab must be rubbed over the tonsillar pillars and care should be taken to avoid teeth, tongue and gums.

5.2.1.2 *Nasal Mid-turbinate Swab*

A flocked tapered swab is used for this specimen. The patient's head is tilted back to 70°. While rotating the swab, it is inserted less than one inch into the nostril. The process is then repeated to ensure adequate collection of the sample.

5.2.1.3 *Anterior Nares Specimen*

A flocked or polyester swab is inserted 1 cm into the nostril and rotated and left in place for 10–15 seconds.

5.2.1.4 *Nasopharyngeal Wash/Aspirate or Nasal Wash/Aspirate*

A catheter is first attached to the suction apparatus, and the tube is then inserted into the nostril parallel to the palate. Once the catheter enters a desired depth, suction/aspiration is begun, and the catheter is removed while rotating gently. The specimen is then placed into the transport medium.

5.2.1.5 Bronchoalveolar Lavage, Tracheal Aspirate, Pleural Fluid, Lung Biopsy

A ~2–3 mL screw cap sputum collection cup or sterile container is used for collection.

5.2.1.6 Sputum (Phlegm)

Patients are asked to wash their mouth using water and then to expectorate sputum. This is collected in a sterile dry container.

5.2.1.7 Saliva

Patients or individuals at home simply spit into a sterile dry tube and mail it to a lab for processing. Some tests can be processed at home. The United States Food and Drug Administration (FDA) has issued an emergency use authorization (EUA) to Yale School of Public Health for its SalivaDirect COVID-19 diagnostic (www.fda.gov/media/141194/download), which uses a new method of processing saliva samples when testing for COVID-19 infection.

5.2.1.8 Other

Stool is collected in an appropriate container. Urine samples may also be collected, as may autopsy tissues, using a sterile container or viral transport medium (VTM). For serum, 3–5 mL of whole blood is collected in serum separator tubes with gel separator.

During collection, appropriate biosafety precautions must be observed to prevent the spread of the virus to the involved healthcare workers. Home collection of samples is allowed for private laboratories to prevent local travel to the testing centers and potential contact with infected individuals. Medical waste must be disposed of in accordance with national standards. Additionally, professionals can open the samples only in a class II A2 biosafety cabinet. During specimen disposal, the VTM must be discarded in a bag containing 2% Lysol or 5% hypochlorite solution. Care should be taken to prevent contamination from the specimens.

5.3 Recent Diagnostics

Substantial efforts have been made worldwide to develop accurate tools by which to diagnose COVID-19 (Table 5.2). These various tools have specific advantages and disadvantages as described in Section 5.4. The diagnostic testing field for COVID-19 is evolving rapidly and improving in quality every day, with many tests focused on diagnosing patients with active viral infections. Diagnostics able to detect current, active infections are typically molecular-based and inform investigators of the presence of the virus, either

Table 5.2 Properties of different diagnostic methods.

		Efficiency			
Diagnostic method	Detection	Specificity	Sensitivity	Accuracy	Place of testing
Clinical examination	Symptoms				Hospital
Chest CT	Ground-glass opacity, irregular interlobular septal thickening, air bronchogram	25%	97%	68%	Hospital
Laboratory diagnosis	Increased cytokines, CRP, blood cell count, abnormal levels of myocardial, renal and liver enzymes, elevated D-dimer, ESR, procalcitonin, LDH and low platelets and albumin				Hospital
RT-PCR	Viral nucleic acid	Considered as gold standard for detection			Biosafety laboratory 3
ELISA	Host antibody	100%	44.4%	64.3%	Biosafety laboratory 2
MCLIA	Host antibody	High specificity, sensitivity and accuracy.			Biosafety laboratory 2
GICA	Host antibody	100%	57.1%	69%	Biosafety laboratory 2
Lateral flow immunoassay	Host antibody	90.63%	88.66%		Biosafety laboratory 2
Rapid antigen test	Host antibody	100%	68%	72%	Biosafety laboratory 2
Neutralization assay	Ability of antibodies to prevent infection *in vitro*				Biosafety laboratory 3

by identifying its genetic material or identifying unique markers of the pathogen. There are also rapid antigen tests for the presence of virus under development that act by detecting specific surface markers on the outside of the virus.

5.3.1 Molecular Diagnostics

Molecular diagnostics usually require samples from the patient that are likely to contain virus, such as nasopharyngeal swabs or sputum samples. For COVID-19, nasopharyngeal swabs have been the most reliable, as they sample an area of the respiratory tract where the virus appears to first infect an individual. This site is relatively easily accessed, compared to the final site of viral infection: the lower respiratory tract. Consequently, the nasopharyngeal tract likely has active virus replication and enough virus to be detected by kits. Many tests now available or in development can use saliva or nasal swabs that facilitate easier sampling procedures for healthcare providers and patients. A number of these tests have received FDA EUA and are available now. This was not the case at the beginning of the outbreak. See the FDA's list of *in vitro* diagnostics EUAs at www.fda.gov/medical-devices/coronavirus-disease-2019-covid-19-emergency-use-authorizations-medical-devices/vitro-diagnostics-euas#individual-antigen/.

In vitro diagnostic (IVD) devices are tests performed on samples taken from the human body, such as swabs of mucus from inside the nose or back of the throat, or blood taken from a vein or fingerstick. IVDs can detect diseases or other conditions and can be used to monitor a person's overall health to help cure, treat or prevent diseases. There are several types of SARS-CoV-2- and COVID-19-related IVDs:

- Diagnostic tests: tests that detect parts of the SARS-CoV-2 virus and can be used to diagnose infection with the SARS-CoV-2 virus. These include molecular tests and antigen tests. Currently EUA has 141 molecular diagnostic test centers for the detection of SARS-CoV-2. There are only three individual EUAs for antigen diagnostic tests for SARS-CoV-2.
- Serology/antibody tests: tests that detect antibodies (*e.g.* immunoglobulin (Ig)M, IgG) to the SARS-CoV-2 virus. Serology/antibody tests cannot be used to diagnose a current infection. There are currently 39 individual EUAs for serology tests for SARS-CoV-2.
- Tests for management of COVID-19 patients: beyond tests that diagnose or detect SARS-CoV-2 virus or antibodies, there are also tests that are authorized for use in the management of patients with COVID-19, such as to detect biomarkers related to inflammation. Once patients are diagnosed with COVID-19 disease, these additional tests can be used to inform patient management decisions. There is currently only one individual EUA for an IVD for management of COVID-19 patients. It is

an *in vitro* diagnostic test for the quantitative measurement of interleukin (IL)-6 in human serum and plasma. This assay is used to assist in identifying severe inflammatory response in patients with confirmed COVID-19 illness to aid in determining the risk of intubation with mechanical ventilation, in conjunction with clinical findings and the results of other laboratory testing.

5.3.2 Laboratory Examinations (Modified from Cascella *et al.*[17])

In the early stage of COVID-19, a normal or decreased total white blood cell count (WBC) and a decreased lymphocyte count can be demonstrated. Interestingly, lymphopenia appears to be a negative prognostic factor (see Chapter 5, Section 5.3.4 for discussion). Increased values of liver enzymes, lactate dehydrogenase (LDH), muscle enzymes and C-reactive protein (CRP) can be detected. Unless there is a bacterial infection overlap, a normal procalcitonin value is found. The elevated neutrophil-to-lymphocyte ratio (NLR), derived NLR ratio (d-NLR) (neutrophil count divided by the result of WBC count minus neutrophil count) and platelet-to-lymphocyte ratio can be the expression of the inflammatory storm.[18] The correction of these indices is an expression of a favorable trend. In critical patients, D-dimer value is increased, blood lymphocytes decrease persistently and laboratory alterations of multiorgan imbalance (high amylase, coagulation disorders, *etc.*) are found.

5.3.3 Imaging

5.3.3.1 Chest X-ray

Since the disease manifests itself as pneumonia, radiological imaging has a fundamental role in the diagnostic process, management and follow-up. Standard radiographic examination (X-ray) of the chest has a low sensitivity in identifying early lung changes in the initial stages of the disease. At this stage, it can be completely negative. In the more advanced stages of infection, the chest X-ray examination generally shows bilateral multifocal alveolar opacities, which tend to conflate up to the complete opacity of the lung. Pleural effusion can be associated.

5.3.3.2 Chest CT

Given the high sensitivity of the method, chest CT, high-resolution CT (HRCT), is the method of choice in the study of COVID-19 pneumonia, even in the initial stages. Several non-specific HRCT findings and patterns can be found. Most of these findings may also be observed in other lung infections, such as influenza A (H1N1), cytomegalovirus, SARS, MERS, *Streptococcus*, *Chlamydia* and *Mycoplasma*. The most common findings are multifocal bilateral ground areas, which may be associated with patchy areas of consolidation, mainly peripheral/subpleural, and with greater involvement of the posterior regions and lower lobes. The "crazy paving" pattern can be also observed. This latter finding is characterized by the presence of ground-glass areas with superimposed interlobular septal thickening and intralobular septal thickening. It is a non-specific finding that can be detected in different conditions. Other findings are the "reversed halo sign" which is a focal area of ground glass delimited by a peripheral ring with consolidation, and the finding of cavitations, calcifications, lymphadenopathies and pleural effusion.

5.3.3.3 Lung Ultrasound

The ultrasound approach can allow evaluation of the evolution of the disease, from a focal interstitial pattern up to "white lung" with evidence often of subpleural consolidations. When ultrasound is utilized It should be performed within the first 24 hours and it may be useful every 24/48 hours to follow the progression of disease in the lungs. The main sonographic features are:

- Pleural lines, often thickened and irregular. Subpleural lesions can be seen as small patchy areas of consolidation or nodules.
- B lines: they are often motionless, coalescent and may "flow into" adjacent areas of consolidation.
- Thickenings: they are most evident in the posterior and bilateral fields, especially in the lower fields; the dynamic air bronchogram within the consolidation is a manifestation of disease evolution.
- Perilesional pleural effusion.

In summary, during the course of the disease, it is possible to identify the first phase with focal areas of fixed B lines, a phase of numerical increase of the number of B lines up to the white lung with small subpleural thickenings, and further progression usually to evidence of posterior lung consolidation.

As most individuals with COVID-19 have lung abnormalities, scanning techniques can be considered to detect the presence of the virus. However, it would be erroneous to rely on these methods alone as the representation of the disease can vary between individuals. This is also the case for blood parameter-based diagnoses as well. Despite the numerous antibody tests already designed, to date, serological diagnoses are limited regarding both specificity and sensitivity. Again, results from different tests may vary. A CDC study of a test developed by the Vaccine Research Center at the US National Institutes of Health is now ongoing. Of note, this test seems to have extremely high specificity (>99%) with a sensitivity of 96%.[9] Nevertheless, further research is needed to elucidate several attributes related to these tests, including whether IgG antibodies will provide immunity from future SARS-CoV-2 infection and, if so, for how long.

As of now, RT-PCR appears to be the gold standard for COVID-19 testing, as it can be used to identify individuals who have the virus present in their system. Moreover, antibody-based testing can be used to identify asymptomatic individuals and those who have been exposed to the virus. Hence, a combination of various diagnostic tools can be employed to detect COVID-19. However, precautions must be taken into consideration to prevent spreading of the virus.

5.3.3.4 *Scanning Techniques*

A chest CT scan is a widely used conventional method that provides additional information beyond a standard chest X-ray. It combines a series of X-rays taken from different angles and uses artificial intelligence to integrate them into a comprehensive picture. This procedure is simple to perform and can provide diagnostic results rapidly; it is especially beneficial in this pandemic situation where rapid and accurate testing measures can aid in the identification of individuals who may be contagious.[19] When chest CT scans have been performed on COVID-19 patients, the most consistent abnormalities observed have been ground glass opacities in 65% of patients; smooth or irregular interlobular septal thickening observed in 35% of the patients; air bronchogram exhibited by 47%; and thickening of the adjacent pleura in 32%. Rarer CT patterns have included nodules, cystic changes, bronchiectasis, pleural effusion and lymphadenopathy. The major anomalies were often located in the right lower lobe of the lung. Asymptomatic individuals had a comparatively low number of lung segments that were aberrant relative to symptomatic individuals. Patients exhibited greater numbers of abnormalities in more segments for approximately 1–2 weeks after onset of symptoms. Observed CT manifestations correlated with the progression of disease. Asymptomatic individuals sometimes exhibited CT abnormalities such as unilateral or multifocal ground-glass opacification. At approximately 1 week from the onset of symptoms, the lesions may rapidly advance to become bilateral and diffuse, but the appearance is most commonly that of GGOs. At this stage, some patients may also exhibit pleural effusion and lymphadenopathy. Investigators have also found that COVID-19 CT manifestations can appear similar to those of other non-infectious conditions.[20] However, CT scans can be utilized as a technique to identify individuals who are likely to have COVID-19 in conjunction with other signs and symptoms and can also be used as a tool to monitor progression of the disease in patients after treatment.

To determine the accuracy of CT scans, 888 patients with positive CT finding were tested for SARS-CoV-2 using nucleic acid testing (NAT). The sensitivity of the chest CT was 97%, and it had a specificity rate of 25% and accuracy of 68%. CT scans were more precise in individuals who were older (≥60 years) relative to their younger counterparts. Moreover, some patients whose scans exhibited abnormalities did not test positive for the virus. As CT scans can yield false-positive results, it is essential to use it in combination with other diagnostic strategies to yield accurate and highly specific results.[19]

For example, lung ultrasounds have also been conducted in COVID-19 patients, and the abnormalities found in the ultrasounds have been congruent with progression of disease. This diagnostic method has also been useful in the diagnosis of other respiratory conditions such as pneumonia and is worthwhile considering as a diagnostic method for the detection of COVID-19.[21,22] As another example, when combined with a CT scan, 2-fluoro-2-deoxy-D-glucose positron emission tomography (^{18}F-FDG PET) scans can provide additional details on the localization of lesions. When GGOs in COVID-19 patients have been observed using ^{18}F-FDG PET scans, investigators have found that the lesions utilize a substantial amount of the glucose analogue.[22] In comparison, despite the presence of bilateral lesions, tumors that exhibit GGOs do not normally have an increased uptake of FDG. Thus, the increased uptake of FDG in COVID-19 patients could be due to the body's increased inflammatory response because of the viral infection.

Recently, the lung lesions of COVID-19 were shown to have increased ^{18}F-FDG uptake on PET scanning in a limited number of patients and it was suggested that PET/CT might be helpful in patients at an early stage of disease when clinical symptoms are not specific and differential diagnosis is challenging.[23] However, the findings on PET/CT are not specific for COVID-19 and there are currently no data suggesting an added value of ^{18}F-FDG PET/CT compared to chest CT in the management or outcome of patients with COVID-19 infection.

These tools can be used prior to the occurrence of symptoms as a means by which to identify high-risk individuals. It is important to understand that there may be practical difficulties in conducting CT scans. These adversities are primarily due to the contamination of the rooms in which the scans are conducted, and in addition many infected individuals may lack access to CT facilities. The safe transfer of patients to CT rooms may also pose a hazard in terms of infections. In this situation, portable chest radiography devices have gained momentum. Portable chest X-ray devices can be made available in various countries that lack space in hospitals. Although chest radiography is not as accurate as CT scanning, it is exposes the patient to significantly less radiation and often helps in preventing the need for CT scanning.[24] Moreover, various artificial intelligence-associated technologies are under development that will increase the automation of imaging techniques. These innovative methods may provide additional protection to technicians during this pandemic.[20] Despite advances in the utilization of various radiology-based scanning techniques regarding COVID-19, nucleic acid testing still appears to be more reliable than the scanning tools. These techniques can be used concomitantly with nucleic acid tests to help establish the presence of the virus. They can also be used to monitor disease progression in infected individuals and to monitor recovery in follow-up studies in the post-COVID-19 era.

5.3.4 Blood Parameter Levels (Laboratory Diagnostics)

Changes in the blood parameters of individuals with COVID-19 can be compared to their healthy counterparts and, in general, the more abnormal the values, the more severe the disease. Findings from a retrospective study performed with 8697 patients with COVID-19[5] suggested that the major laboratory abnormalities included higher CRP levels, abnormal myocardial enzymes, an elevated erythrocyte sedimentation rate (ESR) and elevated procalcitonin. A variety of other inflammatory markers such as interleukins and D-dimer values also may be elevated, depending on the severity of the disease. CRP is commonly synthesized by hepatocytes and is increased when there is a proinflammatory state in the body. It is normally involved with the activation of the complement pathway on binding with foreign pathogens. CRP is commonly an indicator of inflammation and is sometimes used to monitor treatment of inflammatory conditions in patients. In addition, it can be utilized as a predictor of prognosis in COVID-19 individuals.[10]

In a meta-analysis of 28 studies that included 4189 COVID-19 patients, higher troponin levels consistent with myocardial damage were observed in individuals with the severe form of the disease. Likewise, acute cardiac injury was correlated with disease severity and worse prognosis; it was also observed that the levels of troponin and N-terminal pro-brain natriuretic peptide (consistent with heart failure) subsequently increased during the hospital stay in people who died.[25] These reports suggest that cardiac biomarkers may serve as diagnostic markers for severity of infection. In another meta-analysis of 14 studies documenting the outcomes of 4659 patients infected with SARS-CoV-2, cardiac involvement, in addition to liver injury with elevated aspartate aminotransferase (AST) and decreased levels of albumin has been reported.[26]

Additionally, hypercoagulation of the blood has been seen in COVID-19 patients. Reports show that COVID-19 patients may be predisposed to coagulopathies with microangiopathy and thrombosis that can be life-threatening.[27] During infection, elevated serum fibrinogen levels can make the erythrocytes more likely to adhere to one another, resulting in an elevated ESR secondary to infection. The relative rise or fall in the ESR can be used as a simple, albeit imprecise, guide as to whether the infection is lessening or worsening if clinical parameters are insufficient.

A recent meta-analysis was performed on levels of procalcitonin in SARS-CoV-2 infection and the levels were found to reflect the severity of the disease.[28] Normally, when there is systemic infection, procalcitonin is released into the bloodstream from extrathyroidal sources and amplified. Procalcitonin levels are also sustained by the cytokines released during infection.[28] Abnormal renal function is often seen in patients infected with SARS-CoV-2. Higher levels of serum urea and creatinine have been consistent in patients with severe COVID-19.[29] Results of a meta-analysis of 1799 patients revealed that the majority had low platelet counts, which was correlated with mortality.[29] In addition to these factors, LDH was also predominantly increased relative to the control population. Serum LDH can be a sign of pyroptosis and cell death.[30]

Similarly, lymphopenia (47.6%), leukopenia (23.5%) and leukocytosis (9.9%) have all been found in patients with COVID-19.[6] Lymphopenia refers to a reduction in natural killer and B- and T-cells, which can either be attributed to the cytokine surge in COVID-19 or to direct injury of these cells by the virus.[10,69] Leukopenia, a reduction in the WBC count, can be seen as well. However, it is important to note that leukocytosis has been seen in ~10% of COVID-19 patients. A WBC level $<4 \times 10^9$ cells L^{-1} was associated with better clinical outcome.[31] A meta-analysis of 21 studies and 3377 patients found that severe leukopenia was associated with higher mortality.[32] Hence, it may be beneficial to closely monitor WBC levels as prognostic indicators in SARS-CoV-2 infection.[32] Patients exhibiting a rapid increase in the plasma levels of inflammatory molecules, including IL-1β, IL-1RA, IL-7, IL-8, IL-10, interferon-γ, monocyte chemoattractant protein-1, macrophage inflammatory protein (MIP)1α, MIP-1B, granulocyte colony-stimulating factor and tumor necrosis factor-α relative to controls, are also more likely to require ICU treatment and develop severe disease.[33] Very high levels of neutrophils, AST, LDH and creatinine, as well as depressed levels of platelets and albumin also carry a worse prognosis in COVID-19 patients. Considered in aggregate, these laboratory findings measured serially can aid in the management and prognosis of hospitalized COVID-19 patients and may suggest the diagnosis before more specific testing measures are available.[32]

5.3.5 Genetic Susceptibility to COVID-19

Non-genetic studies reported as preprints[34,35] have previously implicated the involvement of ABO blood groups in COVID-19 susceptibility, with blood type A being more susceptible and blood type O being less susceptible. During the SARS epidemic, ABO blood groups also were implicated in susceptibility to SARS-CoV infection.[36] (See also Chapter 5, Section 5.3.4.[37]).

Using simplified inclusion criteria and teams of clinicians at the European COVID-19 epicenters in Italy and Spain as well as scientists in the less-affected countries of Germany and Norway, Ellinghaus *et al.*[38] performed a genome-wide association study that included *de novo* genotyping for COVID-19 with respiratory failure in approximately 2 months. They detected a novel susceptibility locus at a chromosome 3p21.31 gene cluster and confirmed a potential involvement of the ABO blood group system in COVID-19.

On chromosome 3p21.31, the peak association signal covered a cluster of six genes (SLC6A20, LZTFL1, CCR9, FYCO1, CXCR6 and XCR1), several of which have functions that are potentially relevant to COVID-19. A causative gene cannot be reliably implicated, but one candidate is SLC6A20, which encodes the sodium–imino acid (proline) transporter 1 (SIT1) and functionally interacts with angiotensin-converting enzyme 2, the SARS-CoV-2 cell-surface receptor.[39,40] However, the locus also contains genes encoding chemokine receptors, including the CC motif chemokine receptor 9 (CCR9) and the C-X-C motif chemokine receptor 6 (CXCR6), the latter of which regulates the specific location of lung-resident memory CD8 T-cells throughout the sustained immune response to airway pathogens, including influenza viruses.[41] Flanking genes (*e.g.*, CCR1 and CCR2) also have relevant functions.[42] Further studies are required to delineate the functional consequences of detected associations.

Preliminary results from the COVID-19 Host Genetics Initiative[43] included suggestive associations within the same locus at chromosome 3p21.31, which lend substantial support to the Ellinghaus *et al.*[38] findings. The Host Genetics Initiative analysis also used population-based controls, but the patients included persons with both mild COVID-19 and severe COVID-19. The parallel findings in the two studies suggest an important point about the ascertainment of patients and controls in genetic studies of COVID-19. Because most patients with SARS-CoV-2 infection are asymptomatic, any sample involving patients with a positive nasopharyngeal RNA test is likely to hold a bias toward some degree of symptomatic burden. Two of the identifiers for inclusion in the Ellinghaus *et al.*[38] study were a positive result for the presence of SARS-CoV-2 according to RT-PCR testing and provision of respiratory support (an extreme COVID-19 phenotype). As such, it seems reasonable to conclude that the chromosome 3p21.31 locus is involved in COVID-19 susceptibility *per se*, with a possible enrichment in patients with severe disease. This latter interpretation is supported by the significantly higher frequency of the risk allele among patients who received mechanical ventilation than among those who received supplemental oxygen only, as well as by the finding of younger age among patients who were homozygous for the risk allele compared to patients who were heterozygous or homozygous for the non-risk allele.[38]

Genetic data from the Ellinghaus *et al.*[38] study confirm the results of non-genetic studies cited earlier[34,35] that blood group O is associated with a risk of acquiring COVID-19 that was lower than that in non-O blood groups, whereas blood group A was associated with a higher risk than non-A blood groups.[34,35] The biological mechanisms undergirding these findings may have to do with the ABO group *per se* (*e.g.* with the development of neutralizing antibodies against protein-linked N-glycans)[44] or with other biological effects of the identified variant,[45–47] including the stabilization of von Willebrand factor.[48,49] The ABO locus holds considerable risk for population stratification,[50] which is increased by the inclusion of randomly selected blood donors in the current study (for which there is an inherent risk of blood group O enrichment). Alignment of the allele frequencies at the ABO locus in our control population with those in several non-blood-donor control populations would suggest that blood type does not play a major role in susceptibility to COVID-19. Pragmatic aspects leading to the feasibility of this massive undertaking in a very short period of time during the extreme clinical circumstances of the pandemic imposed limitations that will be important to explore in follow-up studies. For example, to enable the recruitment of study participants, a bare minimum of clinical metadata was requested. For this reason, extensive genotype–phenotype elaboration of current findings could not be conducted, and adjustments for all potential sources of bias (*e.g.*, underlying cardiovascular and metabolic factors relevant to COVID-19) could not be performed. Furthermore, we have limited information about the SARS-CoV-2 infection status in the control participants; this concern is mitigated by the fact that the presence of susceptible persons in the control group would only bias the tests toward the null. In addition, few restrictions were imposed during inclusion, which led to genotyped samples having to be excluded owing to differing ethnic groups (population outliers). Further exploration of current findings, both as to their usefulness in clinical risk profiling of patients with COVID-19 and toward a mechanistic understanding of the underlying pathophysiology, is warranted.

Another gene variant was described among four young men, average age 26 years, with severe COVID-19 associated with loss-of-function variants of X-chromosome TLR.[51]

5.3.6 Importance of Real-time RT-PCR in the Detection of COVID-19

5.3.6.1 Viral Nucleic Acid Detection

Nucleic acid-based testing strategies have been considered as the gold standards for COVID-19 diagnosis. RT-PCR is a tool that is widely used to study gene expression in various disease conditions and has been commonly used in testing for COVID-19. RT-PCR can help determine if a gene of interest is present in a sample, and the quantification of the gene present can also be determined based on the intensity of the bands during amplification. During RT-PCR testing, specific primers or probes are used to bind to the gene of interest for its amplification. The viral genes E, N and RdRp are primarily targeted in the SARS-CoV-2 genome for detection purposes.[52] The RdRp-P2 has demonstrated cross reactivity with SARS-CoV.[53] In addition, the N gene has been found to be more sensitive in these tests.[54] Another protocol developed in Paris probes for two targets in RdRp genes including IP2 and IP4, and the confirmatory gene in this assay was E. The US CDC has also developed a test in which there are three targets in the N gene. An assay developed in Japan consists of multiple targets and the S-protein of the virus. An assay developed in Germany targets the RdRp, E and N genes. Likewise, open reading frame (ORF)1b-nsp14 and N are targeted by testing kits developed in Hong Kong. Similarly, researchers in Thailand have developed tests, approved by the WHO, that specifically target the N genes. Among these, the probes for E (Paris), ORF-1 and N from Hong Kong and the N gene from the US CDC have been considered the most sensitive as compared to the other probes.[55] When RT-quantitative PCR was used to identify a single positive sample among an array of negative samples, it detected the viruses with a 10% false-negative rate. This technique can be utilized to test a large number of suspected individuals.[56] Due to the high costs, delays in obtaining results and requirements for trained professionals, RT-PCR may not be feasible in some settings. Hence, scientists have developed rapid, easy to use, point-of-care assays for detection of the virus. These tests use isothermal amplification for detection. As RT-PCR requires various temperature settings, it requires thermal cycling equipment. This method allows for the detection of the virus at approximately 480 RNA copies. Transcription-mediated amplification is another type of isothermal amplification that uses a single tube to amplify the genes. The amplified product is detected by torches that bind specifically to the amplicon. The torch is then conjugated to a fluorophore which emits an appropriate signal allowing for the detection of the virus. This process targets the sequences of the E and N genes. A reporter RNA is cleaved and made to bind specifically to the gene of interest, and then the product is amplified and detected. These tests do not require sophisticated instruments, are quick, cost relatively little and can be used in point-of-care settings.[57] These tests can be done in a closed tube with fluorescence- or calorimetric-based detection. They have the potential to reduce false negatives while being relatively easy to use without extensive training.[58] Utilization of this technique may be fruitful in the long run for COVID-19 diagnosis.[57]

In summary, these novel nucleic acid-based diagnostic approaches have paved the way for more rapid testing methods that are highly expedient in this pandemic situation. The false-positive and false-negative rates for these different point-of-care assays await the results of larger studies.

5.3.7 Serological Assays

Although RT-PCR is the most useful test by which to detect the virus, it cannot be utilized to determine prior infection, the body's immune response and/or disease progression. In serological assays the immune response of the body against the virus can be identified by the presence of antibodies. The common antibodies present include IgM and IgG. In addition to IgM and IgG, IgA has been described as an early humoral immune responder to SARS-CoV-2[59,60] and some laboratories (*e.g.* the Johns Hopkins immunology laboratory) utilize IgA instead of IgM in their assays (personal communication, Stafford Warren, Anne Arundel Medical Center, Annapolis, MD, USA). IgA antibody detection is of great value in developing vaccines against the virus. It will aid in the determination of dosage of an eventual vaccine and its appropriate administration. The body initially produces IgM as a response to the virus, and it is produced in the first few days after the body encounters the virus and lasts for a few weeks. After this, starting at approximately 3–4 weeks, IgG is produced in response to the virus. Hence, the detection of IgM and/or IgA may indicate the early stage of infection and the presence of IgG indicates a later stage of the disease or prior infection. Due to these limitations, it may be undesirable to rely on serological assays alone for the detection of COVID-19.[61]

5.3.8 Enzyme-linked Immunosorbent Assay

Enzyme-linked immunosorbent assay (ELISA) is a technique that is used for the detection of peptides, proteins, antibodies and hormones. In this method, the microwell is coated with a specific virus protein. If the patient's serum contains antibodies specific to the virus, it will bind to the viral protein in the well. The bound antibody is then made to bind to another antibody that is conjugated to a molecule that can produce a colorimetric or fluorescence signal. ELISA is a relatively rapid test and can be handled in point-of-care settings.[57] After the human body encounters the virus, it produces IgM in low concentration and affinity. This antibody disappears within a short span of time. IgG antibody affinity tests can be utilized for early detection.[62] When serum was collected from 63 COVID-19 patients and tested using the ELISA method, it was found that IgM was present in 28 patients and the accuracy was 64.3%, while the test was 44.4% sensitive and 100% specific. IgG was found in 52 patients and had a higher range of accuracy, sensitivity and specificity than IgM. When a combination of IgM and IgG were used, the results yielded improved sensitivity.[63] This indicates that antibody detection will be beneficial in identifying individuals after the nucleic acid has waned from their body. This could be important for contact tracing and led to further research on the time course of IgG and IgM persistence in the serum of recovered patients who had experienced either mild, moderate or severe disease. It has been reported that the sensitivity of the antibody detection against the S-protein is higher than the N-protein.[64] Testing for IgG and IgM and/or IgA antibody levels can be useful in a pandemic setting as a means to increase the identification of infected individuals and potentially to use convalescent serum from recovered patients as a therapeutic option.

5.3.9 Magnetic Chemiluminescence Enzyme Immunoassay

Detection of the virus using chemiluminescence is a novel approach based on the principle of a photon being released because of a chemical reaction. When antigens bind to antibodies in the serum of patients, this will produce a signal depicting the presence of viral-specific antibodies. The peptides are labeled with biotin and bound to streptavidin-coated magnetic beads. The sera of the patients are then mixed with the beads and allowed to react. The results are then detected by a luminescence reader.[65] One study that used a magnetic chemiluminescence enzyme immunoassay (MCLIA) kit to detect antibodies in patients' plasma found that IgG antibody was present in all patients infected after approximately 22 days and IgM was seen in 94.1% of the patients. The concentration of the antibodies increased gradually within 3 weeks after the onset of symptoms. Hence, this method is useful to detect the virus in patients with COVID-19.[66]

5.3.10 Colloidal Gold Immunochromatographic Assay

Colloidal gold immunochromatographic assay (GICA) is a rapid diagnostic method used to detect pathogens. The sample serum will flow in the nitrocellulose membrane due to capillary action. If the sample is positive for the antibodies, it will conjugate with the antigen and this will generate a red line displaying the presence of antibodies. When GICA was utilized to test for antibodies in COVID-19 patients' serum, IgM and IgG detection were sensitive and capable of identifying the antibodies.[63] A subsequent study found that levels of these antibodies increased in parallel with the progression of the condition. GICA has also shown improved sensitivity in cases where the nucleic acid test rendered a negative result. This method can be used as an auxiliary test to detect COVID-19.[67]

5.3.11 Lateral Flow Immunoassay

This type of rapid detection test can reveal results in 30 minutes. The sample flows over a material that contains the immobilized viral antigens; if the antibodies are present, they bind to the antigens and produce a colored result. In the detection of COVID-19, these tests have demonstrated significant sensitivity and specificity. This method can be used in remote areas and small laboratories as the methodology is simple and results are available quickly,[25] but, unfortunately, due to their false-positive and false-negative rates, it may not be wise to rely on this test alone for detection.

5.3.12 Rapid Antigen Test

As the viral antigens appear before the production of antibodies against the virus, these tests can be used to test for the virus in the early stages of infection. This test has demonstrated 68% sensitivity while detecting the N-protein. Additionally, the test was 100% accurate in predicting a negative viral infection.[68] This test can serve as a simple and rapid diagnostic measure during the pandemic.

5.3.13 Neutralization Antibody Assay

When the body detects a virus, its immune system produces neutralizing antibodies. The presence of these antibodies can indicate exposure to the virus. Specimens collected from the patients are added to the viral cultures and measured based on the inactivation of the virus in the culture. These antibodies have been associated with CRP in patients with COVID-19.[69] This method takes a few days and requires biosafety labs and trained professionals. This technique may be especially beneficial in developing plasma therapy for the treatment of COVID-19.

Numerous antibody tests have already designed, but to date serologic diagnoses are limited regarding both specificity and sensitivity. Results from different tests may vary. A CDC study of a serology or antibody test developed by the Vaccine Research Center at the US National Institutes of Health is now ongoing. Of note, this test seems to have very high specificity (>99%) with a sensitivity of 96%.[70] Nevertheless, further research is needed to elucidate several attributes related to these tests, including whether IgG antibodies will provide immunity from future SARS-CoV-2 infection, and if so, for how long.

5.3.14 Other Useful Diagnostics for COVID-19

Although various new diagnostic approaches are on the market, few have been approved by the FDA. It is essential to continue to discover and develop approaches that are cheap, easy to use, portable and reliable.

A CDC study of a serology or antibody test developed by the Vaccine Research Center at the US National Institutes of Health is now ongoing. Of note, this test seems to have very high specificity (>99%) with a sensitivity of 96%.[70] Nevertheless, further research is needed to elucidate several attributes related to these tests, including whether IgG antibodies will provide immunity from future SARS-CoV-2 infection, and if so, for how long. An example of a promising novel approach was reported in *ACS Nano*[71] wherein researchers wanted to develop a nanomaterial-based sensor that could detect COVID-19 in exhaled breath, similar to a breathalyzer test for alcohol intoxication.[71]

5.3.15 False Negatives and Positives

When testing for any illness it is essential to secure a correct and appropriate result. In this case, false positives and false negatives can be detrimental to combatting the COVID-19 pandemic. In a study conducted at Johns Hopkins University, 1330 patient samples were analyzed and subjected to PCR testing. On the first day of infection, the false-negative rate was 100%, and on the first day of symptoms, the test produced a 38% false-negative rate. This number dropped to 20% 3 days after the occurrence of symptoms, but after this the rates tended to deteriorate. This report suggests that there is a very small window of time in which the accuracy is higher. If the false-negative rates were 20% it would mean that 20 out of 100 tested patients could be given a false result. In this situation it would be highly dangerous if these 20 people were to infect more people with whom they came into contact. People who wrongly test positive for the virus may be given medication and advised to self-quarantine or may be held in government-approved quarantine areas. Any person with a false positive might then end up contracting the virus in the facility itself. In addition to this, the stigma of COVID-19 and fear of disease progression can result in high levels of anxiety. Moreover, these people may also be given medication that could have adverse effects if they are uninfected. If a person is in fact positive while the test indicates that they are negative, they will feel free to interact with others, at the risk of infecting them. Moreover, healthcare professionals may not wear appropriate personal protective equipment to treat patients whom they believe are negative, which could result in a vicious circle of infection. Normally, in other situations a false-negative patient can be retested easily, but given limited testing capability, it may be impractical to test patients multiple times. A faulty test could be a result of a variety of factors, including the following.[72]

- Patients are tested too early and are in the initial stage of infection, so they may not have enough of a viral load to be detected by the test.
- If patients have no respiratory tract symptoms such as the production of sputum or cough, there might not be enough virus to be detected by standard tests.
- Sometimes the improper collection of the sample can yield a negative result. To avoid faulty results, it is important to ensure that the professionals who collect the specimens are properly trained.
- A faulty test result can also occur from incorrect handling and transport of the samples. The virus may be inactivated in such conditions, yielding an inaccurate result.
- Many different tests have been developed in a short span of time without going through the usual vetting process, which could result in uncertainty in the test results and in the reagents used.

Based on these considerations, it is essential to develop strategies to minimize faulty results. The tests can be repeated, and samples can be obtained from different sites in the patient's body such as the respiratory tract, urine, stool and blood. Specimens should optimally be taken from the lower respiratory tract, as the accuracy is increased in these samples, but this is often difficult to carry out. Stringent quality control should be maintained in testing labs so that the consistency of the testing procedure is maintained in laboratories around the world.[72] More research is required to develop accurate diagnostic techniques for the detection of COVID-19 and to more conclusively delineate the likelihood of false-positive and false-negative results with different assays and under different ambient circumstances (*e.g.* time from collection of specimen to test run, ambient temperature, *etc.*).

5.4 Advantages and Disadvantages of Different Diagnostic Approaches

5.4.1 Detection Based on Patient History

Advantages

- Contact tracing based on patient history is important for early detection of the virus.
- This method will allow for the testing of asymptomatic individuals who may be potential carriers of the virus.
- This will help identify high-risk individuals who may contract the virus.
- Potential detection of patients prior to the onset of symptoms.
- Can help in quarantining the potential carrier to contain the virus.

Disadvantages

- Individuals who had no prior contact with infected individuals will be untested. These individuals can be carriers of the virus.
- Patients exhibiting symptoms may not be tested as they do not have any history of contact with infected individuals.

5.4.2 Detection of Disease Based on Symptoms

Advantages

- Allows testing of individuals who display symptoms.
- Prevents spread of disease through early detection.

Disadvantages

- Many individuals are asymptomatic, but may be potential carriers of the virus.
- Symptoms can be quite variable, which could lead to some symptoms from COVID-19 being overlooked as due to another health problem.
- The symptoms could be a consequence of underlying disease or a different acute illness.
- People with the common cold or seasonal flu may be accidently considered as high-risk individuals leading to quarantine, possibly alongside COVID-19 patients.

5.4.3 Scanning Techniques

5.4.3.1 CT Scan

Advantages

- Reliable technique that has been used for detection of pneumonia, and the pattern is often suggestive or diagnostic for COVID-19.
- It is a relatively rapid method.
- Medical professionals are already well trained in the procedure.
- Highly sensitive for COVID-19 detection.
- Procedure is easy to perform.
- A majority of the infected individuals exhibit lung lesions and CT abnormalities.

Disadvantages

- Similar CT manifestations can be seen with other pathogens and other non-infectious diseases.
- CT rooms and suites may not be available during the pandemic.
- Possible infection transmission to other individuals during transport to and from the CT room.
- Infections can also arise from the room in which the scan is performed.

- Can yield false-positive results.
- Exposes the patient to radiation and several-fold more radiation than from a chest X-ray.

5.4.3.2 Chest X-ray

Advantages

- Take minimal time to complete.
- If chest X-ray is normal, the tested individuals can be sent home until RT-PCR tests are ready.
- Can be utilized to monitor disease progression.
- Portable options are available so that the individuals can be tested easily at appropriate settings.
- Can be made available in areas with resource constraints.

Disadvantages

- Similar manifestations can be seen in other pathogens and other non-infectious diseases.
- Can yield false-positive results.
- Normal results cannot be used to rule out COVID-19.
- Less sensitive than other scanning techniques.

5.4.3.3 ^{18}F-FDG PET/CT

Advantages

- Can detect localizations of various lesions.
- Imaging of the whole body can be obtained from this scan; hence, organ damage may be identified.
- High resolution.
- Increased patient tolerance.
- Highly accurate images.
- The test is highly sensitive.

Disadvantages

- Can yield false-positive results as the manifestations can be due to other factors.
- The patient is exposed to several-fold more radiation than from a chest X-ray.
- Infections can arise during transport of the patient and in the scan rooms.
- The scan rooms may not be available, and many infected individuals may not have access to the rooms.
- Infections can arise from contact with other patients in the immediate surroundings of the scan rooms.

5.4.4 Laboratory Serum Diagnostics (*e.g.* WBC, Inflammatory Markers)

Advantages

- Provides prognostic markers for disease progression.
- Can be utilized as tools to help determine false-positive and false-negative nucleic acid studies.
- Mass screening can be done using this method in countries that may not have immediate access to RT-PCR techniques.
- Appropriate software and large data sets can further increase the accuracy of these tests.
- Relatively inexpensive.

Disadvantages

- The abnormalities in the blood parameters could be due to other underlying illnesses.
- May not accurately distinguish between people with and without COVID-19.
- Gender discrepancies have been observed in these parameters.
- All infected individuals may not exhibit abnormalities in these blood tests.

5.4.5 Molecular Diagnosis

5.4.5.1 RT-PCR

Advantages

- RT-PCR is a commonly used method by scientists and healthcare workers.
- RT-PCR is a commonly used technique in research. Hence the process is well established and can be accurately replicated in testing for COVID-19.

Disadvantages

- RT-PCR for COVID-19 is only positive if a person is currently or very recently infected with SARS-CoV-2.
- The procedure takes time to complete and by its delay may not be beneficial in patient management.
- A biosafety lab is required for the process, so many areas that do not have this facility will have to safely transport the specimens. Contamination can occur during this time.

5.4.5.2 RT-LAMP

Advantages

- Highly specific.
- Reaction can be followed in real time.

- Results can be obtained rapidly.
- Available in point-of-care settings.
- Relatively inexpensive.
- Can detect the genes in low-sample specimens.
- Results can be read visually.

Disadvantages

- Primer design is relatively complicated.
- Restricted availability of reagents.

5.4.5.3 Transcription-mediated Amplification

Advantages

- Capable of screening for other respiratory pathogens simultaneously.
- Rapid technique and can result in amplification quickly.
- Can be available in point-of-care settings.
- Highly sensitive assay.

Disadvantages

- The nucleic acid is susceptible to enzymatic and chemical degradation.

5.4.6 CRISPR-based Diagnosis

Advantages

- Highly sensitive.
- Inexpensive.
- Can be detected with paper strips without compromising sensitivity.
- Does not require complex instruments.

Disadvantages

- RNA extraction is time consuming.
- Off-target effects can be observed.

5.4.7 Serological Assays

5.4.7.1 ELISA

Advantages

- Simple and cheap.
- Is a well-known procedure that is used in many laboratory settings.
- Results can be obtained within 3 hours of specimen collection.
- Multiple samples can be tested at one time.

Disadvantages

- The tests are not fully established for COVID-19.

5.4.7.2 GICA

Advantages

- Extremely quick, taking only ~15 minutes.
- Non-invasive procedure, but blood samples can also be used for the test.
- Neither a laboratory nor special equipment are required to perform the test.
- Can be performed with minimal training.

Disadvantages

- More expensive than ELISA when testing in bulk.
- Cannot confirm whether the person is still infected with the virus.
- Infection will not be detected in early stages of the disease.
- It is a new technology and the accuracy is still being tested.

5.4.7.3 MCLIA

Advantages

- Sensitive technique.
- No requirement for extraction of nucleic acid.
- Rapid analysis time.
- Can detect trace amounts of analytes.

Disadvantages

- Infection will not be detected in early stages of the disease.
- Antibodies will appear only 3–6 days after infection.

5.4.7.4 Lateral Flow Immunoassay

Advantages

- Easy to use.
- Can be used at a large scale.
- Additional equipment is not required.
- It takes only 15 minutes to obtain results.
- Highly sensitive and specific.
- Can identify asymptomatic individuals.
- Identification of recovered patients.

Disadvantages

- Infection will not be detected in the early stages of the disease.
- Antibodies will appear only 3–6 days after infection.

5.4.7.5 Neutralization Bioassay

Advantages

- Beneficial for vaccine development.
- Important for convalescent plasma therapy.

Disadvantages

- Relatively long time to obtain results (3–5 days).
- Requires a level 3 biosafety laboratory.
- Requires SARS-CoV-2 culture.

5.4.7.6 Rapid Antigen Test

Advantages

- Results can be obtained within 30 minutes.
- Utilized for acute or initial infection.
- Early diagnosis.

Disadvantages

- Sensitivity of tests may vary.
- Can result in false positives due to cross-reactivity.

5.5 Executive Summary

- As COVID-19 has an unpredictable and potentially fatal outcome, the public needs to become aware of the various symptoms of the disease so that prompt testing can be carried out with quarantine and contact tracing when appropriate. But since asymptomatic persons can and do also transmit the disease, public health measures need to be in place for those populations at risk and testing available for asymptomatic persons as well as those with symptoms.
- A variety of tests are available for SARS-CoV-2 detection. None of these are 100% sensitive or specific and different assay methods, *e.g.* for PCR, may give divergent results. Some of the advantages and disadvantages of these different methods are listed in this chapter.
- Considering all the advantages and limitations associated with the various diagnostic tests, a combination of them can often be useful to accurately diagnose COVID-19.
- The major risk in all these tests concerns the generation of false positives or negatives. A false negative may lead the individual to infect a significant number of people, while thinking that he or she is free of disease. Conversely, a false positive may result in the individual's contracting the virus in a hospital setting and/or being inappropriately quarantined.
- Given the wide dissemination of this viral pandemic to more than 200 countries, diagnostic methods that are reliable, rapid, inexpensive, require little training and yield few false-positive and -negative results

are sought after and will become the "gold standard" in the future, but further research is needed to determine which method or methods best fit this description and for what stage of the disease they can be best utilized.

Acknowledgements

The authors wish to thank Chris Ringwalt for skillful editing and referencing of the manuscript.

References

1. R. C. Bhoyar, A. Jain, P. Sehgal, M. K. Divakar, D. Sharma and M. Imran, *et al.*, High throughput detection and genetic epidemiology of SARS-CoV-2 using COVIDSeq next generation sequencing, *PLoS One*, 2021, **16**(2), e0247115.
2. M. Hosseiny, S. Kooraki, A. Gholamrezanezhad, S. Reddy and L. Myers, Radiology Perspective of Coronavirus Disease 2019 (COVID-19): Lessons From Severe Acute Respiratory Syndrome and Middle East Respiratory Syndrome, *Am. J. Roentgenol.*, 2020, **214**(5), 1078–1082.
3. W. H. Huang, L. C. Teng, T. K. Yeh, Y. J. Chen, W. J. Lo and M. J. Wu, *et al.*, 2019 novel coronavirus disease (COVID-19) in Taiwan: Reports of two cases from Wuhan, China, *J. Microbiol., Immunol. Infect.*, 2020, **53**(3), 481–484.
4. J. Wang, Fast Identification of Possible Drug Treatment of Coronavirus Disease-19 (COVID-19) through Computational Drug Repurposing Study, *J. Chem. Inf. Model.*, 2020, **60**(6), 3277–3286.
5. J. Zhu, Z. Zhong, P. Ji, H. Li, B. Li and J. Pang, *et al.*, Clinicopathological characteristics of 8697 patients with COVID-19 in China: a meta-analysis, *Fam. Med. Community Health*, 2020, **8**(2), e000406.
6. N. Zhu, D. Zhang, W. Wang, X. Li, B. Yang and J. Song, *et al.*, A Novel Coronavirus from Patients with Pneumonia in China, 2019, *N. Engl. J. Med.*, 2020, **382**(8), 727–733.
7. L. Bertolino, M. Vitrone and E. Durante-Mangoni, Does this patient have COVID-19? A practical guide for the internist, *Intern. Emerg. Med.*, 2020, 791–800.
8. T. H. Chang, J. L. Wu and L. Y. Chang, Clinical characteristics and diagnostic challenges of pediatric COVID-19: A systematic review and meta-analysis, *J. Formosan Med. Assoc.*, 2020, **119**(5), 982–989.
9. J. Gao, Z. Tian and X. Yang, Breakthrough: Chloroquine phosphate has shown apparent efficacy in treatment of COVID-19 associated pneumonia in clinical studies, *BioSci. Trends*, 2020, 72–73.
10. G. Chen, D. Wu, W. Guo, Y. Cao, D. Huang and H. Wang, *et al.*, Clinical and immunological features of severe and moderate coronavirus disease 2019, *J. Clin. Invest.*, 2020, **130**(5), 2620–2629.
11. W.-J. Guan, Z.-Y. Ni, Y. Hu, W.-H. Liang, C.-Q. Ou and J.-X. He, *et al.*, Clinical characteristics of 2019 novel coronavirus infection in China, *N. Engl. J. Med.*, 2020, **382**(18), 1708–1720.

12. A. A. Rabaan, S. H. Al-Ahmed, S. Haque, R. Sah, R. Tiwari and Y. S. Malik, *et al.*, SARS-CoV-2, SARS-CoV, and MERS-COV: A comparative overview, *Infez. Med.*, 2020, **28**(2), 174–184.
13. CDC CfDCaP, *Information for Clinicians on Investigational Therapeutics for Patients with COVID-19*, CfDCaP, 2020.
14. WHO, *Novel Coronavirus (2019-nCoV) Situation Report – 3*, 2020. Available from: https://www.who.int/docs/default-source/coronaviruse/situation-reports/20200123-sitrep-3-2019-ncov.pdf?sfvrsn=d6d23643_8.
15. CDC CfDCaP, *Interim Guidance for Collection and Submission of Postmortem Specimens from Deceased Persons with Known or Suspected COVID-19*, CfDCaP, 2020.
16. M. Iyer, K. Jayaramayya, M. D. Subramaniam, S. B. Lee, A. A. Dayem and S. G. Cho, *et al.*, COVID-19: an update on diagnostic and therapeutic approaches, *BMB Rep.*, 2020, **53**(4), 191–205.
17. M. Cascella, M. Rajnik, A. Cuomo, S. C. Dulebohn and R. Di Napoli, *Features, Evaluation and Treatment Coronavirus (COVID-19)*, StatPearls, Treasure Island (FL), 2020.
18. A. P. Yang, J. P. Liu, W. Q. Tao and H. M. Li, The diagnostic and predictive role of NLR, d-NLR and PLR in COVID-19 patients, *Int. Immunopharmacol.*, 2020, **84**, 106504.
19. T. Ai, Z. Yang, H. Hou, C. Zhan, C. Chen and W. Lv, *et al.*, Correlation of Chest CT and RT-PCR Testing in Coronavirus Disease 2019 (COVID-19) in China: A Report of 1014 Cases, *Radiology*, 2020, 200642.
20. H. Shi, X. Han, N. Jiang, Y. Cao, O. Alwalid and J. Gu, *et al.*, Radiological findings from 81 patients with COVID-19 pneumonia in Wuhan, China: a descriptive study, *Lancet Infect. Dis.*, 2020, **20**(4), 425–434.
21. M. Paniagua, D. Cartmel and C. Dominguez, *Receptor Recognition by the Novel Coronavirus from Wuhan: an Analysis Based on Decade-Long Structural Studies of SARS*, 2020.
22. C. Qin, L. Zhou, Z. Hu, S. Zhang, S. Yang and Y. Tao, *et al.*, Dysregulation of immune response in patients with COVID-19 in Wuhan, China, *Clin. Infect. Dis.*, 2020, 762–768.
23. C. Qin, F. Liu, T.-C. Yen and X. Lan, 18F-FDG PET/CT findings of COVID-19: a series of four highly suspected cases, *Eur. J. Nucl. Med. Mol. Imaging*, 2020, **47**(5), 1281–1286.
24. A. Jacobi, M. Chung, A. Bernheim and C. Eber, Portable chest X-ray in coronavirus disease-19 (COVID-19): A pictorial review, *Clin. Imaging*, 2020, **64**, 35–42.
25. L.-Q. Li, T. Huang, Y.-Q. Wang, Z.-P. Wang, Y. Liang and T.-B. Huang, *et al.*, COVID-19 patients' clinical characteristics, discharge rate, and fatality rate of meta-analysis, *J. Med. Virol.*, 2020, **92**(6), 577–583.
26. W. Tian, W. Jiang, J. Yao, C. J. Nicholson, R. H. Li and H. H. Sigurslid, *et al.*, Predictors of mortality in hospitalized COVID-19 patients: A systematic review and meta-analysis, *J. Med. Virol.*, 2020, 1875–1883.
27. E. Terpos, I. Ntanasis-Stathopoulos, I. Elalamy, E. Kastritis, T. N. Sergentanis and M. Politou, *et al.*, Hematological findings and complications of COVID-19, *Am. J. Hematol.*, 2020, **95**(7), 834–847.

28. G. Lippi, M. Plebani and B. M. Henry, Thrombocytopenia is associated with severe coronavirus disease 2019 (COVID-19) infections: a meta-analysis, *Clin. Chim. Acta*, 2020, 145–148.
29. M. Kermali, R. K. Khalsa, K. Pillai, Z. Ismail and A. Harky, The role of biomarkers in diagnosis of COVID-19 – A systematic review, *Life Sci.*, 2020, **254**, 117788.
30. P. Mo, Y. Xing, Y. Xiao, L. Deng, Q. Zhao and H. Wang, *et al.*, Clinical characteristics of refractory COVID-19 pneumonia in Wuhan, China, *Clin. Infect. Dis.*, 2020, ciaa270.
31. Z. Zheng, F. Peng, B. Xu, J. Zhao, H. Liu and J. Peng, *et al.*, Risk factors of critical & mortal COVID-19 cases: A systematic literature review and meta-analysis, *J. Infect.*, 2020, e16–e25.
32. B. M. Henry, M. H. S. de Oliveira, S. Benoit, M. Plebani and G. Lippi, Hematologic, biochemical and immune biomarker abnormalities associated with severe illness and mortality in coronavirus disease 2019 (COVID-19): a meta-analysis, *Clin. Chem. Lab. Med.*, 2020, **58**(7), 1021–1028.
33. M. Merad and J. C. Martin, Pathological inflammation in patients with COVID-19: a key role for monocytes and macrophages, *Nat. Rev. Immunol.*, 2020, 1–8.
34. J. Zhao, Y. Yang, H. Huang, D. Li, D. Gu and X. Lu, *et al.*, Relationship between the ABO Blood Group and the COVID-19 Susceptibility, *Clin. Infect. Dis.*, 2021, **73**(2), 328–331.
35. M. Zietz, J. Zucker and N. P. Tatonetti, Testing the association between blood type and COVID-19 infection, intubation, and death, *Nat. Commun.*, 2020, **11**(1), 5761.
36. Y. Cheng, G. Cheng, C. H. Chui, F. Y. Lau, P. K. Chan and M. H. Ng, *et al.*, ABO blood group and susceptibility to severe acute respiratory syndrome, *JAMA*, 2005, **293**(12), 1450–1451.
37. F. Zhang and C. Hughes, A Caution in Association of ABO Blood Group with COVID-19, *Global Clin. Transl. Res.*, 2020, **2**(3), 54–72.
38. D. Ellinghaus, F. Degenhardt, L. Bujanda, M. Buti, A. Albillos and P. Invernizzi, *et al.*, Genomewide Association Study of Severe Covid-19 with Respiratory Failure, *N. Engl. J. Med.*, 2020, **383**, 1522–1534.
39. K. Kuba, Y. Imai, T. Ohto-Nakanishi and J. M. Penninger, Trilogy of ACE2: a peptidase in the renin-angiotensin system, a SARS receptor, and a partner for amino acid transporters, *Pharmacol. Ther.*, 2010, **128**(1), 119–128.
40. R. N. Vuille-dit-Bille, S. M. Camargo, L. Emmenegger, T. Sasse, E. Kummer and J. Jando, *et al.*, Human intestine luminal ACE2 and amino acid transporter expression increased by ACE-inhibitors, *Amino Acids*, 2015, **47**(4), 693–705.
41. A. N. Wein, S. R. McMaster, S. Takamura, P. R. Dunbar, E. K. Cartwright and S. L. Hayward, *et al.*, CXCR6 regulates localization of tissue-resident memory CD8 T cells to the airways, *J. Exp. Med.*, 2019, **216**(12), 2748–2762.
42. M. J. Hickey, K. S. Held, E. Baum, J. L. Gao, P. M. Murphy and T. E. Lane, CCR1 deficiency increases susceptibility to fatal coronavirus infection of the central nervous system, *Viral Immunol.*, 2007, **20**(4), 599–608.

43. The COVID-19 Host Genetics Initiative, The COVID-19 Host Genetics Initiative, a global initiative to elucidate the role of host genetic factors in susceptibility and severity of the SARS-CoV-2 virus pandemic, *Eur. J. Hum. Genet.*, 2020, **28**(6), 715–718.
44. A. Breiman, N. Ruvën-Clouet and J. Le Pendu, Harnessing the natural anti-glycan immune response to limit the transmission of enveloped viruses such as SARS-CoV-2, *PLoS Pathog.*, 2020, **16**(5), e1008556.
45. M. Aziz, R. Fatima and R. Assaly, Elevated interleukin-6 and severe COVID-19: A meta-analysis, *J. Med. Virol.*, 2020, 2283–2285.
46. A. G. Comuzzie, S. A. Cole, S. L. Laston, V. S. Voruganti, K. Haack and R. A. Gibbs, *et al.*, Novel genetic loci identified for the pathophysiology of childhood obesity in the Hispanic population, *PLoS One*, 2012, **7**(12), e51954.
47. S. Naitza, E. Porcu, M. Steri, D. D. Taub, A. Mulas and X. Xiao, *et al.*, A genome-wide association scan on the levels of markers of inflammation in Sardinians reveals associations that underpin its complex regulation, *PLoS Genet.*, 2012, **8**(1), e1002480.
48. M. Franchini, S. Crestani, F. Frattini, C. Sissa and C. Bonfanti, ABO blood group and von Willebrand factor: biological implications, *Clin. Chem. Lab. Med.*, 2014, **52**(9), 1273–1276.
49. G. P. Murray, S. R. Post and G. R. Post, ABO blood group is a determinant of von Willebrand factor protein levels in human pulmonary endothelial cells, *J. Clin. Pathol.*, 2020, **73**(6), 347–349.
50. G. Thomson and W. F. Bodmer, Letter: Population stratification as an explanation of IQ and ABO association, *Nature*, 1975, **254**(5498), 363–364.
51. C. I. van der Made, A. Simons, J. Schuurs-Hoeijmakers, G. van den Heuvel, T. Mantere and S. Kersten, *et al.*, Presence of Genetic Variants Among Young Men With Severe COVID-19, *JAMA*, 2020, **324**(7), 663–673.
52. V. M. Corman, O. Landt, M. Kaiser, R. Molenkamp, A. Meijer and D. K. Chu, *et al.*, Detection of 2019 novel coronavirus (2019-nCoV) by real-time RT-PCR, *Eurosurveillance*, 2020, **25**(3), 2000045.
53. J. F. Chan, K. H. Kok, Z. Zhu, H. Chu, K. K. To and S. Yuan, *et al.*, Genomic characterization of the 2019 novel human-pathogenic coronavirus isolated from a patient with atypical pneumonia after visiting Wuhan, *Emerging Microbes Infect.*, 2020, **9**(1), 221–236.
54. D. K. W. Chu, Y. Pan, S. M. S. Cheng, K. P. Y. Hui, P. Krishnan and Y. Liu, *et al.*, Molecular Diagnosis of a Novel Coronavirus (2019-nCoV) Causing an Outbreak of Pneumonia, *Clin. Chem.*, 2020, **66**(4), 549–555.
55. M. Venter and K. Richter, Towards effective diagnostic assays for COVID-19: a review, *J. Clin. Pathol.*, 2020, **73**(7), 370–377.
56. I. Yelin, N. Aharony, E. Shaer Tamar, A. Argoetti, E. Messer and D. Berenbaum, *et al.*, Evaluation of COVID-19 RT-qPCR test in multi-sample pools, *Clin. Infect. Dis.*, 2020, 2073–2078.
57. L. J. Carter, L. V. Garner, J. W. Smoot, Y. Li, Q. Zhou and C. J. Saveson, *et al.*, Assay Techniques and Test Development for COVID-19 Diagnosis, *ACS Cent. Sci.*, 2020, **6**(5), 591–605.

58. J. Song, M. El-Tholoth, Y. Li, J. Graham-Wooten, Y. Liang and J. Li, *et al.*, Single and Two-Stage, Closed-Tube, Point-of-Care, Molecular Detection of SARS-CoV-2, *Anal. Chem.*, 2021, **93**(38), 13063–13071.
59. D. Yan, X.-Y. Liu, Y.-N. Zhu, L. Huang, B.-T. Dan and G.-J. Zhang, *et al.*, Factors associated with prolonged viral shedding and impact of lopinavir/ritonavir treatment in hospitalised non-critically ill patients with SARS-CoV-2 infection, *Eur. Respir. J.*, 2020, **56**(1), 2000799.
60. H.-Q. Yu, B.-Q. Sun, Z.-F. Fang, J.-C. Zhao, X.-Y. Liu and Y.-M. Li, *et al.*, Distinct features of SARS-CoV-2-specific IgA response in COVID-19 patients, *Eur. Respir. J.*, 2020, 2001526.
61. L. Qin, Q. Sun, Y. Wang, K. F. Wu, M. Chen and B. C. Shia, *et al.*, Prediction of Number of Cases of 2019 Novel Coronavirus (COVID-19) Using Social Media Search Index, *Int. J. Environ. Res. Public Health*, 2020, **17**(7), 2365.
62. X. Zhang, H. Cai, J. Hu, J. Lian, J. Gu and S. Zhang, *et al.*, Epidemiological, clinical characteristics of cases of SARS-CoV-2 infection with abnormal imaging findings, *Int. J. Infect. Dis.*, 2020, **94**, 81–87.
63. J. Xiang, M. Yan, H. Li, T. Liu, C. Lin and S. Huang, *et al.*, Evaluation of Enzyme-Linked Immunoassay and Colloidal Gold-Immunochromatographic Assay Kit for Detection of Novel Coronavirus (SARS-Cov-2) Causing an Outbreak of Pneumonia (COVID-19), medRxiv, 2020, DOI: 10.1101/2020.02.27.20028787.65.
64. W. Liu, L. Liu, G. Kou, Y. Zheng, Y. Ding and W. Ni, *et al.*, Evaluation of Nucleocapsid and Spike Protein-Based Enzyme-Linked Immunosorbent Assays for Detecting Antibodies against SARS-CoV-2, *J. Clin. Microbiol.*, 2020, **58**(6), e00461-20.
65. X. Cai, J. Chen, J. Hu, Q. Long, H. Deng and K. Fan, *et al.*, A Peptide-based Magnetic Chemiluminescence Enzyme Immunoassay for Serological Diagnosis of Corona Virus Disease 2019, *J. Infect. Dis.*, 2020, **222**(2), 189–193.
66. Q.-X. Long, H.-J. Deng, J. Chen, J. Hu, B.-Z. Liu and P. Liao, *et al.*, Antibody responses to SARS-CoV-2 in COVID-19 patients: the perspective application of serological tests in clinical practice, *Nat. Med.*, 2020, **26**(6), 845–848.
67. Y. Pan, D. Zhang, P. Yang, L. L. M. Poon and Q. Wang, Viral load of SARS-CoV-2 in clinical samples, *Lancet Infect. Dis.*, 2020, **20**(4), 411–412.
68. B. Diao, C. Wang, Y. Tan, X. Chen, Y. Liu and L. Ning, *et al.*, Reduction and Functional Exhaustion of T Cells in Patients With Coronavirus Disease 2019 (COVID-19), *Front. Immunol.*, 2020, **11**, 827.
69. F. Wu, S. Zhao, B. Yu, Y.-M. Chen, W. Wang and Y. Hu, *et al.*, A new coronavirus associated with severe human respiratory disease in China, *Nature*, 2020, **579**, 265–269.
70. CDC CfDCaP, *Serology Testing for COVID-19 at CDC – Updated to 23 May, 2020*, 2020.
71. ACS, *Toward a Coronavirus Breathalyzer Test Online*, 2020. Available from: https://www.sciencedaily.com/releases/2020/08/200819120714.htm.
72. H. Wu, Q. Hao and Q. Wang, *Difficulties in False Negative Diagnosis of Coronavirus Disease 2019: A Case Report*. 2020.

CHAPTER 6

Therapeutic Options Initially Available for COVID-19 Patients and Initial Clinical Trials

SSANG-GOO CHO[†a], CLAUDE HUGHES[†*b,c], IYER MAHALAXMI[†d], MOHANA DEVI SUBRAMANIAM[†e], BALACHANDAR VELLINGIRI[†*f] AND STAFFORD WARREN[†g]

[a]Department of Stem Cell and Regenerative Biotechnology, Konkuk University, South Korea; [b]IQVIA, Durham, NC, USA; [c]Duke University Medical Center, Durham, NC, USA; [d]Department of Zoology, Avinashilingam Institute of Home Science and Higher Education for Women, India; [e]Department of Genetics and Molecular Biology, Vision Research Foundation, Sankara Nethralaya Campus, India; [f]Department of Human Genetics and Molecular Biology, Bharathiar University, India; [g]Anne Arundel Medical Center, Annapolis, MD, USA
*E-mail: clhughes@duke.edu, geneticbala@buc.edu.in

6.1 Introduction to Therapeutic Options Initially Available for COVID-19 Patients

The severe acute respiratory syndrome coronavirus 2 (SARS-CoV-2), which appeared in China in December 2019, can cause a very dangerous viral infection, coronavirus disease 2019 (COVID-19), when compared with the four

[†]All authors contributed equally and are listed alphabetically by surname.

The Coronavirus Pandemic and the Future Volume 1: Virology, Epidemiology, Translational Toxicology and Therapeutics
Edited by Michael D. Waters, Alok Dhawan, Tim Marrs, Diana Anderson, Stafford Warren and Claude L. Hughes

Published by the Royal Society of Chemistry, www.rsc.org

coronaviruses that result only in the common cold or other common respiratory viral infections, such as the rhinovirus. The rapid global spread of COVID-19 was facilitated by increased globalization, particularly air transportation and industrialization. A complicating factor in the spread of the virus is that asymptomatic healthy individuals may unknowingly reproduce and shed the virus, especially during the early days of their infection.[1,2] As a result, public health officials, researchers and healthcare workers have been compelled to create new strategies in terms of mitigation and therapy to fight this deadly viral infection. Hence, investigations of a wide range of various drugs and therapeutic approaches have been initiated to test potential efficacy and safety for use against COVID-19 (Figure 6.1). The current biomedical research imperative is to determine the essential therapeutic performance characteristics of efficiency and safety/tolerability for the several repurposed antiviral drugs, plasma and stem cell therapies and vaccines as well as traditional medicines (*e.g.* ethnobotany, *etc.*). Details on such alternatives are provided in later chapters.

The only specific antiviral agent, remdesivir, was given temporary emergency use authorization on 1 May 2020 by the United States Food and Drug Administration (FDA) for use in patients with severe COVID-19.[3] Currently, no vaccine is commercially available, although phase 2 and phase 3 vaccine trials are underway in several countries. Treatment is based on symptoms, and oxygen therapy represents the first step in addressing respiratory impairment. In cases where respiratory compromise is refractory to oxygen therapy, non-invasive (NIV) and invasive mechanical ventilation (IMV) may

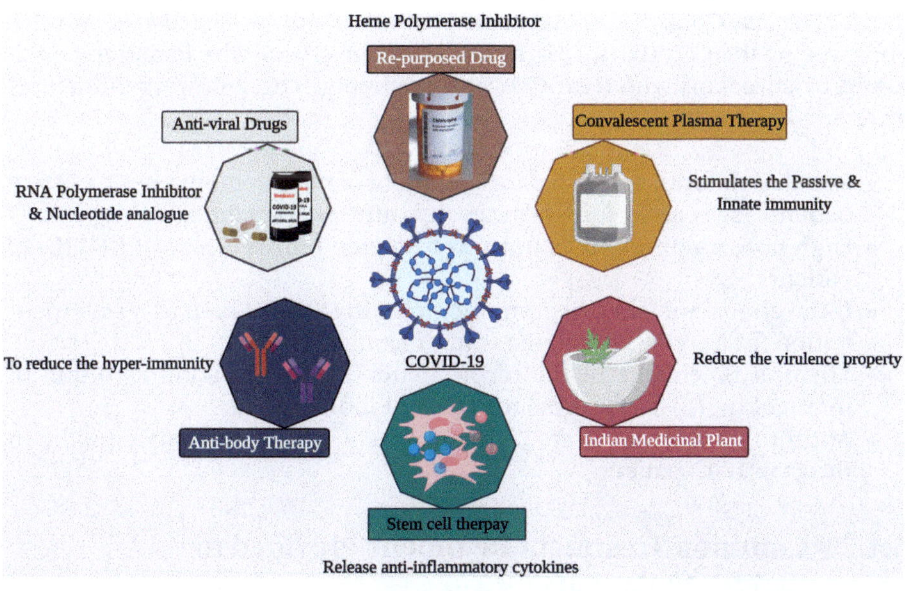

Figure 6.1 Treatment options for COVID-19 disease. The figure depicts the various possible therapeutic options available to treat COVID-19 and their modes of action.

be needed. Intensive care is often required to deal with low oxygen saturation levels and with other complicated forms of the disease. Knowledge has accumulated on acute respiratory distress syndrome (ARDS), requiring clinicians to review strategies for dealing with respiratory failure. Gattinoni *et al.*[4] have asserted that COVID-19-induced ARDS (CARDS) is not a "typical" ARDS. According to Cascella *et al.*:

> This aspect of the disease is of fundamental importance and has probably negatively affected the therapeutic approach in the early stages of the pandemic. While at the beginning of the pandemic, early IMV was postulated as the better strategy for addressing CARDS, in COVID-19 pneumonia the typical ARDS respiratory mechanics featuring reduced lung compliance (*i.e.* ability to stretch and expand lungs) cannot be found. On the contrary, in CARDS, good pulmonary compliance can be demonstrated. As a consequence, and in contrast to what was initially believed, NIV can have a key initial role in CARDS therapy.[5]

6.1.1 General Guidelines for the Treatments Provided to COVID-19 Patients

With the number of individuals affected with COVID-19 increasing at a rapid pace and as yet limited information available about outcomes of randomized clinical trials to treat SARS-CoV-2, the World Health Organization (WHO) suggested that unlicensed therapies which are administered within ethically approved clinical trials can be provided to COVID-19 affected cases. The Centers for Disease Control and Prevention (CDC, 2020d)[6] provided a list of rules approved by the COVID-19 Treatment Guidelines Panel (the Panel) to be followed by clinicians who treat COVID-19 patients. The following guidelines were provided by the CDC:

- Remdesivir is not suggested for treatment of mild or moderate COVID-19 patients (see earlier for use in severely affected patients).
- High-dose chloroquine is not recommended for the cure of COVID-19 patients.
- Intravenous non-specific immunoglobulin initially used as a treatment option for COVID-19 patients is not recommended.
- There is no change in the recommendation of the administration of interleukin (IL)-6 inhibitors for affected individuals.
- Antithrombotic agents are used to prevent venous thromboembolic in the COVID-19 patients.

6.1.2 Common Treatment Regimens Provided to COVID-19 Patients

While no antiviral treatments have been approved excepting remdesivir as noted above, at least 25 drugs are under investigation, including 10 in clinical trials[7] (Table 6.1). The first published major randomized controlled

Table 6.1 Some commercially available drugs used for the treatment of various coronavirus infections.[a]

Name of drug	Illnesses treated
α-Interferon	Spectrum of respiratory infections, RSV and SARS
Ritonavir and lopinavir	SARS, MERS
Ribavirin	RSV and RSV pneumonia
Reverse transcriptase inhibitors: zidovudine, didanosine, zalcitabine, stavudine, lamivudine, abacavir and emtricitabine	SARS
Nucleotide reverse transcriptase inhibitor: tenofovir disoproxil fumarate	SARS
Non-nucleoside reverse transcriptase inhibitors: nevirapine, delavirdine and efavirenz	SARS
Protease inhibitors: saquinavir, ritonavir, indinavir, nelfinavir, amprenavir, lopinavir, atazanavir and fosamprenavir	SARS
Fusion inhibitors: enfuvirtide, lamivudine and adefovir dipivoxil	SARS
Umifenovir	ARVI, influenza, rhinovirus, adenovirus, parainfluenza, RSV, coronavirus, including the causative agent of atypical pneumonia Used in the phase III trials of COVID-19 virus, SARS, MERS
3-Chymotrypsin-like protease	SARS, MERS
Papain-like protease	SARS, MERS and human coronavirus NL63
RNA-dependent RNA polymerase	SARS, murine coronavirus
Capsid spike glycoprotein (hCoV-EMC)	SARS, human coronavirus
Guanosine-analog RNA-synthesis inhibitors	Coronavirus
Nitazoxanide	SARS, MERS and influenza
Influenza drugs	MERS
Remdesivir	COVID-19, SARS, MERS
Favipiravir	COVID-19
Darunavir	COVID-19
Lopinavir	COVID-19, SARS, MERS
Alcohol vaporization or nebulization inhalation	COVID-19
Chloroquine	SARS, human coronavirus OC43
ASC09	ARDS, respiratory distress syndrome, SARS, MERS
TMPRSS2 inhibitor: camostat mesylate	SARS, MERS, coronavirus 229E and COVID-19
Baricitinib	COVID-19
Ruxolitinib	COVID-19
Saquinavir	SARS and feline coronavirus
Indinavir	SARS and COVID-19
Carfilzomib	COVID-19
Oseltamivir	COVID-19
Azvudine	COVID-19

(*continued*)

Table 6.1 (*continued*)

Name of drug	Illnesses treated
Baloxavir marboxil	COVID-19
Thymosin α1	MERS
Methylprednisolone	SARS, MERS
Tocilizumab	COVID-19
Interferon subtypes β-1b, α-n1, α-n3 and human leukocyte interferon α	SARS
Acyclovir	SARS, MERS, coronavirus 229E and COVID-19
Cathespin L	SARS

[a]RSV: respiratory syncytial virus, SARS: severe acute respiratory syndrome, MERS: Middle East respiratory syndrome, ARVI: acute respiratory viral infection; HCoV-EMC: human coronavirus EMC.

open-label clinical trial of an antiviral drug combination, lopinavir/ritonavir for severe COVID-19 (400/100 mg orally every 12 hours) began enrolling patients in China just a week after the virus had been identified,[8] with negative results in 199 patients compared to standard care. Preclinical studies suggested that remdesivir (GS5734) – an inhibitor of RNA polymerase with *in vitro* activity against multiple RNA viruses, including Ebola – could be effective for both prophylaxis and therapy of human coronavirus infections.[9] This drug gave positive results in a rhesus macaque model of Middle East respiratory syndrome coronavirus (MERS-CoV) infection[10] in conjunction with α-interferon (5 million units by aerosol inhalation twice per day). There have now been three randomized phase 3 trials in humans demonstrating benefit from remdesivir. The Adaptive Covid-19 Treatment Trial (ACTT), sponsored by the National Institute of Allergy and Infectious Diseases, a division of the National Institutes of Health, enrolled 1063 patients in 68 sites in 21 countries including the United States, Europe and Asia. Preliminary results reported in April 2020 showed a 31% faster time to recovery (11 days *versus* 15 days, $p < 0.001$) and a non-statistically significant mortality rate benefit (8% *versus* 11.6%, $p = 0.059$). A second randomized double-blind multicenter trial was done in 10 hospitals in Hubei Province, China, in 237 patients with severe disease. It was stopped early, but revealed a non-statistically significant improvement in time to clinical improvement and in survival rate.[11] The third trial, the Simple Trial, was sponsored by Gilead, and was a randomized (1 : 1 : 1 5 days : 10 days : standard care) open-label multicenter trial in more than 180 sites in 15 countries. It divided patients into the Simple Severe Trial (397 patients) and the Simple Moderate Trial (600 patients). The preliminary results reported by Gilead on 1 June 2020 revealed clinical improvement and an apparent mortality benefit with remdesivir, but there was no additional benefit from the 10 day arms *versus* the 5 day arms of the two trials. Both trials have now been expanded: the Simple Severe Trial up to 5600 patients, including mechanically ventilated patients, and the Simple Moderate Trial includes up to 1000 patients.

As mentioned earlier, chloroquine and hydroxychloroquine were proposed as immunomodulatory therapy. In an initial small 20-patient non-randomized trial, 600 mg daily of hydroxychloroquine[12] was associated with a reduced viral presence on day 6 (the study end-point) and this effect was felt to be enhanced by the macrolide azithromycin. Studies *in vitro* and *in vivo* have demonstrated that macrolides may mitigate inflammation and modulate the immune response by inducing the downregulation of adhesion molecules on the cell surface, reducing pro-inflammatory cytokines, and stimulating phagocytosis by alveolar macrophages, as well as by inhibiting the activation and mobilization of neutrophils.[13] However, a study of 1438 hospitalized patients with laboratory-confirmed COVID-19 in 25 hospitals in the New York metropolitan region (representing 88% of the region's hospitalized patients) found no significant survival benefit for patients receiving either hydroxychloroquine+azithromycin, hydroxychloroquine alone or azithromycin alone.[14] Disturbingly, in logistic models, cardiac arrest was significantly more likely in patients receiving hydroxychloroquine+azithromycin compared with patients receiving only hydroxychloroquine or only azithromycin or neither drug, When the subset of patients not on ventilators was analyzed, the risk of cardiac arrest remained statistically significantly elevated for hydroxychloroquine only compared to azithromycin only, suggesting that the risk was a drug effect and not mediated by mechanical ventilation. Cavalcanti *et al.*[15] confirmed the lack of benefit of hydroxychloroquine with or without azithromycin in a 660-patient multicenter randomized open-label controlled trial in hospitalized patients with mild or moderate disease in a 1 : 1 : 1 ratio of standard care : standard care plus hydroxychloroquine 400 mg twice daily : hydroxychloroquine 400 mg twice daily plus azithromycin 500 mg daily for 7 days, with assessment at 15 days. There was no clinical improvement with either drug regimen.[16] It is known that the concomitant use of hydroxychloroquine or chloroquine with azithromycin increases the risk of QT interval prolongation and potentially lethal cardiac arrhythmias;[17] QT prolongation was noted in the Cavalcanti study.[15] On 15 June 2020, the FDA revoked the emergency use authorization (which had been given on 28 March 2020 to treat COVID-19 patients with chloroquine or hydroxychloroquine from the Strategic National Stockpile) to use chloroquine or hydroxychloroquine outside a clinical trial. On 4 July 2020, in a news release, the WHO accepted the recommendation from the Solidarity trial's international steering committee to discontinue the trial's hydroxychloroquine arm as well as its lopinavir/ritanovir arm. The Solidarity trial was established by the WHO to find an effective drug treatment for hospitalized patients with COVID-19. The international steering committee made these recommendations from the Solidarity trial interim results and from a review of all trials presented at the 1–2 July summit on COVID-19 research and innovation.

It has been shown that COVID-19 patients have a higher incidence of venous thromboembolism and that anticoagulant therapy is associated with reduced intensive care unit (ICU) mortality. The COVID-19 treatment panel has recommended antithrombotic therapy.[18] Severe COVID-19 can mimic disseminated intravascular coagulopathy (DIC), with the main clinical difference being thrombosis with COVID-19 *versus* bleeding with DIC.[19] For this reason, it was initially suggested that patients should receive thromboprophylaxis and in cases of known thrombophilia or thrombosis, full-intensity therapeutic anticoagulation.[20] However, a more recent multicenter retrospective review of 400 hospitalized patients with COVID-19 (144 of them with critical disease), all of whom were on thromboprophylaxis,[21] revealed a nearly 10% overall thrombotic complication rate and a nearly 5% incidence of venous thromboembolism, raising the as-yet unanswered question as to whether all patients with severe COVID-19 should receive full-dose anticoagulation. In addition, evidence for "hyperactive platelets" was reported by researchers at the University of Utah who found increased platelet activation and aggregation as well as altered platelet gene expression in patients with COVID-19.[19] They also postulated that platelets may be affected by SARS-CoV-2 independently of angiotensin-converting enzyme (ACE)-2 receptor attachment. There are no current reports of using antiplatelet therapy or combination anticoagulant and antiplatelet therapy for COVID-19, although the University of Utah researchers noted inhibition of platelet hyperactivity by aspirin *in vitro*. Hyperactive platelets could lead to cardiovascular complications such as stroke and heart attack, especially in patients with diabetes and those with underlying atherosclerotic disease who develop COVID-19.

The deterioration of lung function with COVID-19 has been attributed to a maladaptive immune response rather than increased viral loads. The complement system represents the first response of the host immune system to SARS-CoV-2 and there is increasing evidence that unrestrained activation of complement induced by the virus in the lungs and other organs plays a major role in contributing to the release of pro-inflammatory cytokines, resulting in acute and chronic inflammation, endothelial cell dysfunction, thrombus formation, and in the most severe cases, multiorgan failure and death.[22–24] There are ongoing clinical trials with IL-6 and IL-1 blockade with initial promising results.[23] An investigation led by the Istituto Nazionale Tumori, Fondazione Pascale di Napoli in Italy focused on the use of tocilizumab in addition to standard therapies. Tocilizumab is a humanized IgG1 monoclonal antibody, which is directed against the IL-6 receptor, commonly used in the treatment of rheumatoid arthritis, juvenile arthritis, giant cell arteritis, Castleman syndrome, and for managing toxicity due to immune checkpoint inhibitors. In 544 patients with severe COVID-19, the use of tocilizumab was associated with a reduced risk of mechanical ventilation or death (adjusted hazard ratio 0.61, $p = 0.02$), and the risk of death was independently significantly reduced ($p < 0.015$).[25] Another anti-IL-6R antibody, sarilumab, is being studied in the United States in a phase 2/3, randomized, double-blind, placebo-controlled clinical trial.[26] Drugs with other mechanisms of action may soon be in drug trials, alone or in combination with other drugs, such as eculizumab which blocks complement protein C5, baricitinib, a Janus kinase inhibitor, affecting the pathway that communicates information from chemical signals outside a cell to the cell nucleus resulting in the activation of genes through transcription; ribavirin, which is a synthetic guanine analogue that inhibits RNA-dependent RNA polymerase and delivered promising results in a Middle East respiratory syndrome coronavirus (MERS-CoV) rhesus monkey model.[27,28] Type 1 interferons may provide benefit for patients with COVID-19 in combination with other drug therapies. An open-label phase 2 trial of β1 interferon in a small triple drug combination study showed a significant benefit to the arm with β1 interferon in terms of reduced duration of viral shedding and a shortened time to symptom alleviation.[29]

In the search for repurposed drugs based on mechanism of action it may be useful to distinguish viral- *versus* host-directed agents (HDA). Ivermectin is a good example of the latter. An HDA can reduce viral load by inhibiting a key cellular process that the virus hijacks to enhance infection by suppressing the host antiviral response. Reducing viral load by even a modest amount by using an HDA at low dose early in infection can be the key to enabling the body's immune system to begin to mount the full antiviral response before the infection takes control. Caly *et al.*[30] hypothesized that an anti-helminthic drug, ivermectin, inhibits IMPα/β1-mediated nuclear import of viral proteins, as shown for other RNA viruses, and showed that with a single addition to Vero-hSLAM cells 2 hours post-infection with SARS-CoV-2 it effected a ~5000-fold reduction in viral RNA at 48 hours. They concluded that ivermectin therefore warrants further investigation for possible benefits in humans. Their report provoked widespread interest, an FDA warning, and two letters to the editor.[31] Rayner *et al.* pointed out that published studies show that blood levels of ivermectin achieved during standard therapy are much lower than the concentrations that were inhibitory for SARS-CoV-2 in cell culture.[31] Noël and Lima[32] voiced concern that if high concentrations of ivermectin could be achieved, they would likely be toxic. Caly *et al.*[30] agreed, but asserted that the points made do not address the reported mechanism of action of ivermectin;[33] its key direct target in mammalian cells is not a viral component, but a host protein important in intracellular transport,[33] which is almost certainly the basis of its broad-spectrum activity against a number of different RNA viruses *in vitro*.[33,34]

When COVID-19 results in complex multiorgan dysfunction, organ function support is critical in addition to respiratory support. Extracorporeal membrane oxygenation should be considered on a case-by-case basis for patients with refractory hypoxemia despite lung-protective ventilation. The prone position has been used successfully for many years to improve oxygenation in patients who require mechanical ventilatory support and has been found to be particularly helpful in treating COVID-19 pneumonia with hypoxemia.[5]

Currently, treatments for COVID-19 are administered based on symptoms and severely affected patients are provided with organ support.[35]

6.1.3 Medicinal Plants as Therapy for COVID-19

Indian traditional medicinal practices such as Ayurveda, Siddha, Unani, Yoga, naturopathy and homoeopathy are considered some of the oldest therapeutic measures in the history of humankind. They have had a significant impact in combatting health issues globally.[36–38] In total, there are almost 25 000 plant-based formulations in which approximately 8000 different species of plants have been used by traditional practitioners to treat various diseases.[39] It has been reported that a single herb or species of plant contains multiple phytochemical constituents which can function in combination or alone to produce desired pharmacological effects.[39] For many years these traditional medicinal plant constituents have been formulated into drugs to treat various diseases.[40,41] Indian medicinal plant-derived pharmacological content has a major role in treating various viral infections.[42–44] Although studies related to treating coronaviruses using Indian medicinal plants are minimal, Vimalanathan *et al.* demonstrated efficacy in a mouse model with SARS-CoV infection by using a composition of Indian medicinal plants such as *Indigofera tinctoria*, *Vitex trifolia*, *Gymnema sylvestre*, *Abutilon indicum*, *Leucas aspera*, *Cassia alata*, *Sphaeranthus indicus*, *Clitoria ternatea* and others from Tamil Nadu, India (Vimalanathan, Ignacimuthu, and Hudson).[45]

It has also been reported that the active compounds extracted from Indian medicinal plants such as ajwain (*Trachyspermum ammi*), neem (*Azadirachta indica*), athimaduram (*Glycyrrhiza glabra*), turmeric (*Curcuma longa*), tulsi (*Ocimum tenuiflorum*) and southern blue gum (*Eucalyptus globulus*) could be used as a potential antiviral agents against COVID-19 (Table 6.2). Furthermore, authors have elaborately explained the importance of each active ingredient, such as the fact that the combination of neem (*Azadirachta indica*) and tulsi (*Ocimum tenuiflorum*) could block viral attachment, whereas the combination of turmeric (*Curcuma longa*) and eucalyptus (*Eucalyptus globulus*) have been used to treat various infections such as common cold, flu and other respiratory illnesses for many decades in India. Furthermore, the combination of athimaduram (*Glycyrrhiza glabra*) and ajwain (*Trachyspermum ammi*) has been beneficial against pulmonary diseases.[46] In an earlier study, it was found that the compounds from *Vitex trifolia* and *Sphaeranthus indicus* could decrease the effects of cytokine storm caused by SARS-CoV by modifying the nuclear factor-κB pathway.[47] Maity *et al.* identified *Clitoria ternatea* as a metalloproteinase inhibitor.[48] A metalloproteinase inhibitor might be used to mitigate the ACE-2 receptor, which is the entry route of SARS-CoV as well as SARS-CoV-2. Plant-based compounds such as *Glycyrrhiza glabra* and *Allium sativum* can be a promising target to inhibit viral replication.[47] Tsai *et al.* have reported that a compound isolated from *Strobilanthes cusia* could block the viral genome and induce papain-like protease in human coronaviruses.[49]

Table 6.2 Indian medicinal herbs which might inhibit human coronaviruses and other viruses.

Plant source	Mechanism of action	Target	Virus
Acacia nilotica	Inhibition		HIV-protease
Allium sativum	Proteolytic and hemagglutinating activity and viral replication		SARS
Andrographis paniculata	Suppression	NLRP3, capase-1 and IL-1β	SARS-CoV and probably SARS-CoV-2
Boerhaavia diffusa	Inhibition	ACE	
Cassia occidentalis	Inhibition	ACE	
Clerodendrum inerme Gaertn.	Inactivation	Ribosome	SARS-CoV-2
Clitoria ternatea	Metalloproteinase inhibitor	ADAM17	
Coriandrum sativum	Inhibition	ACE	
Coscinium fenestratum	Inhibition	ACE	
Cynara scolymus	Inhibition	ACE	
Embelia ribes	Inhibition	ACE	
Eugenia jambolana	Inhibition	Protease	
Euphorbia granulata	Inhibition		HIV-1 protease
Glycyrrhiza glabra	Inhibition of viral replication; modulation of membrane fluidity		SARS, HIV-1
Gymnema sylvestre	Inhibition of viral DNA synthesis		
Hyoscyamus niger	Inhibition and bronchodilation	Ca^{2+}	
Ocimum kilimandscharicum	Inhibition		HIV-1
Ocimum sanctum	Inhibition		HIV-1
Punica granatum	Inhibition	ACE	
Salacia oblonga	Suppression	Angiotensin II, AT1 signal	
Sambucus ebulus	Inhibition		Enveloped virus
Solanum nigrum			HIV-1
Sphaeranthus indicus	Inhibition		Mouse coronavirus and herpesvirus
Strobilanthes callosa	Blocking		HCoV-NL63
Strobilanthes cusia	Blocking		HCoV-NL63
Vitex negundo	Inhibition		HIV-1
Vitex trifolia	Reduction		SARS-CoV

6.1.4 Convalescent Plasma Therapy

Convalescent plasma therapy is passive immunotherapy which has been used practically in various infectious disease treatment situations for the past several decades. During the SARS and MERS outbreaks, this therapy achieved successful and satisfactory outcomes.[50] A meta-analysis from 32 studies conducted on SARS-CoV infection showed a statistically significant decrease in the pooled odds ratio of mortality rate when compared to placebo or no therapy.[51] Thus, these studies and the sequence homology of SARS-CoV-2 has suggested that convalescent plasma therapy may be also appropriate for the treatment of COVID-19 cases.[52,53] Convalescent plasma therapy used soon after early signs and symptoms of SARS-CoV reduced the mortality rate when compared to placebo or no treatment. In one test, fluid was isolated from a COVID-19 patient and could be neutralized by sera from many SARS-CoV-2 patients.[54] Because the antibody content in a recovered patient will only last for few weeks, it should be collected early in the convalescent period. Clinical trials are essential to evaluate efficiency and safety while using convalescent plasma therapy in COVID-19 patients. It is believed by clinicians that convalescent plasma therapy was effective during past viral epidemics, and in this current scenario it could be a potential treatment with a low risk of side-effects for acutely ill COVID-19 patients. In the convalescent plasma procedure, approximately 1 L of plasma is collected, of which approximately only 250 mL of plasma is required for one patient.[55] A study in China revealed that the convalescent plasma therapy could rescue COVID-19 patients from some of the severe effects of the disease. After the introduction of convalescent plasma therapy, patients tested negative for SARS-CoV-2, displayed increased oxygen saturation levels and lymphocyte counts and improved organ function.[56]

6.1.5 Mesenchymal Stem Cell-Derived Exosomes as Therapy for COVID-19

Mesenchymal stem cells (MSCs) are multipotent stromal cells which are derived from many sources including human dental pulp, bone marrow, umbilical cord and olfactory mucosa. MSCs have demonstrated value in many clinical therapeutic applications.[57] For example, in asthmatic mouse models, treatment with MSC-derived exosomes led to a remarkable recovery including reduced chronic inflammation and induced rebuilding of damaged pulmonary tissues. When peripheral blood mononuclear cells from asthmatic patients were co-cultured with MSC exosomes, increased expansion of regulatory T-cells (Tregs) and induced immunosuppression was observed together with upregulation of suppressive cytokines including IL-10 and transforming growth factor-β1.[58] MSC-derived exosomes have multilineage differentiation potential, aid in tissue repair, have neuroprotective properties and also may mediate immunomodulation.[59] These properties of MSC-derived exosomes may avert the multiorgan damage crisis that is associated with COVID-19 by their anti-inflammatory, immunomodulatory and antifibrotic potential.[60,61] In the current pandemic, MSC-derived exosomes may be a beneficial treatment option to alleviate the effect of SARS-CoV-2 infection on multiple organ systems. While many stem cell-based therapies are being studied for treatment of COVID-19, a widely accepted treatment is not yet available. Ground-breaking research is required to bring novel therapies of this kind to fruition.

6.1.6 Other Treatment Options

6.1.6.1 Corticosteroids

Initially, the use of corticosteroids against SARS-CoV-2 was not recommended by the WHO. However, a multicenter randomized controlled trial of dexamethasone in the pre-COVID-19 era had shown benefit for the treatment of ARDS, conducted between March 2013 and December 2018 in 17 ICU' in teaching hospitals in Spain, randomizing 277 patients with a diagnosis of ARDS to receive either 20 mg dexamethasone for 5 days then 10 mg for 5 days with a 28 day primary outcome of days on a ventilator and a 60 day secondary outcome of mortality. That trial revealed 4.8 fewer days on mechanical ventilators for the dexamethasone-treated patients as well as a significantly lower mortality at 60 days (21% *versus* 36%, $p = 0.0047$) One small trial with methylprednisolone reported improved clinical outcomes in patients with COVID-19, but the breakthrough came with the United Kingdom (UK) RECOVERY trial which enlisted approximately 15% of all hospitalized patients with COVID-19 in the UK and randomized in an open-label controlled fashion 2104 patients treated with up to 10 days of dexamethasone 6 mg day^{-1} (orally or intravenously) *versus* 4321 who received usual care. The outcome was determined at 28 days. Up until this time no therapy had been shown to significantly reduce mortality from COVID-19. The use of dexamethasone resulted in lower 28 day mortality than usual care in patients who were receiving invasive mechanical ventilation at randomization (by 12.3 age-adjusted percentage points, a proportional reduction of approximately one-third) and those who were receiving oxygen without invasive mechanical ventilation (by 4.1 age-adjusted percentage points, a proportional reduction of approximately one-fifth). However, there was no evidence that dexamethasone provided any benefit among patients who were not receiving respiratory support at randomization, and the results were consistent with possible harm in this subgroup. The benefit was also clear in patients who were being treated >7 days after symptom onset, when inflammatory lung damage is likely to have been more common. These findings are consistent with the thesis that mortality from COVID-19 is at least in part related to an excessive and harmful inflammatory host response. The mortality for those on mechanical ventilation of 29.3% was lower than the mortality in the usual-care group (41.4%), but the usual-care group's mortality was similar to the mortality in the UK at large during that time (37%).[62]

6.2 Introduction to Ongoing Clinical Trials

In response to the COVID-19 emergency, clinical trial research assessing the efficacy and safety of clinical candidate interventions to treat COVID-19 is emerging at an unprecedented rate. COVID-19 rapidly developed into a global pandemic with a significant health and economic impact; several national and international scientific groups are working collaboratively on a variety of preventative and therapeutic interventions. Focusing on the viral target, the therapies can be split into two groups which either prevent the virus from accessing human cells or constrain key viral enzymes from replication. Focusing on the host, one approach would be to enhance the host's immune response and another approach would be to inhibit or modulate the production of pro-inflammatory cytokines. Some of the drugs which are undergoing multiple trials for SARS-CoV-2 targeting vaccines and antibodies are listed here.

6.2.1 Ongoing Trials on Various Antiviral Agents

Follow-up data will be available in the next few months for remdesivir from the expanded SIMPLE Trial, which is expected to show mortality benefit. In cell-based techniques, remdesivir delayed RNA chain termination, which clarified the high potency of the drug towards RNA viruses. Combination therapy of remdesivir and cytokine and/or complement inhibitors or remdesivir in combination with other antiviral drugs or with convalescent serum may result in further benefit. Ritonavir was investigated for effectiveness against SARS-CoV and confirmed to be successful.[63] A number of other antiviral drugs including favipiravir, umifenovir, triazavirin and baloxavir marboxil are undergoing clinical trials. Furthermore, at least 26 studies are investigating the impact of antivirals with interferon-based modes of action. Hydroxychloroquine is being studied in several trials outside the United States as a preventive for COVID-19, given some experimental data that it has benefit *in vitro*, but as it has not been shown to be effective in the treatment of COVID-19 and in light of its pro-arrhythmic potential, it holds little promise, even as a preventive option.

6.2.2 Immunosuppressants/Immunomodulators

Dexamethasone, having been shown to reduce mortality for COVID-19 patients with severe disease, will now probably be used on a widespread basis for patients requiring oxygen therapy alone or by mechanical ventilation. Trials using inhaled steroids for this group of patients are likely in the future to try to better control the host's hyperinflammatory pulmonary response. Immunosuppressants such as tocilizumab, adalimumab, eculizumab, sarilumab, ixekizumab and fingolimod are registered for evaluation. Meplazumab functions by inhibition of T-cell chemotaxis and viral entry.[64] Various studies are also examining the immune-modulation properties of antibodies and recombinant human granulocyte colony-stimulating factor. Thalidomide has been shown to reduce the synthesis of tumor necrosis factor (TNF)-α, which has encouraged its use in various immune diseases.[65] However, preclinical studies showed that thalidomide had positive effects in mice infected with H1N1 infection by secreting various pro-inflammatory cytokines,[66] but whether it would be of benefit for "cytokine storm" in SARS-CoV-2 infection still remains to be answered. Another immunomodulating drug is fingolimod, which primarily treats refractory multiple sclerosis as it is an antagonist for S1P1 receptors and might help recover T-cell function and potentially boost host immunity.[67]

6.2.3 Current Trials on MSCs for COVID-19

Numerous stem cell-based therapies have been under consideration for the possible treatment of COVID-19. Wharton's jelly MSCs have been used in some studies as therapy. In addition, MSCs derived from human dental pulp, bone marrow, umbilical cord and olfactory mucosa are undergoing trials for COVID-19. The stem cells are predominantly given through the intravenous route and the dosage is based on the bodyweight of the patient. One particular trial consists of delivering MSC-derived exosomes in an inhalable form. This study is still in phase 1 clinical trials. In a stem cell educator therapy trial, the patients' blood is taken, which then goes through a cell separator; the immune cells of the person are then co-cultured with cord blood stem cells and both are then returned to the patient. The stem cells are either given at different day intervals or given at one dosage. These trials exist as promising therapies for the treatment of COVID-19.[68] In a study of seven patients with COVID-19, five of whom had either severe or critical disease, Leng reported decreased viral titers in 2–4 days after only one stem cell injection and laboratory evidence of a robust anti-inflammatory effect with a 10-fold drop in C-reactive protein, decreased white blood count, elevation of the anti-inflammatory IL-10, decreased levels of the pro-inflammatory TNF-α, and other laboratory indicators of reduced inflammation.[69] At present, at least 30 cell-based trials for treating COVID-19 are registered, 22 of which are based on the use of MSC from different human cell lines (umbilical cord, dental pulp, *etc.*) and two are based on MSC-derived exosomes (Table 6.3; www.clinicaltrials.gov).

Table 6.3 Early clinical trials addressing COVID-19.

Study	Drug	Status	Organization/location
Ongoing clinical trials for the management of the COVID-19 pandemic PMC7144665/[70]	Umifenovir, triazavirin, baloxavir marboxil, danoprevir/ritonavir, azvudine, sofosbuvir/ledipasvir, sofosbuvir/daclatasvir, darunavir/cobicistat and emtricitabine/tenofovir, dexamethasone, methylprednisolone, ASC09/oseltamivir, ixekizumab, bevacizumab	Recruiting	China, Iran, Spain, UK
Study to evaluate the safety and antiviral activity of remdesivir (GS-5734™) in participants with moderate coronavirus disease (COVID-19) compared to standard of care treatment NCT04292730	Remdesivir	Completed	Hoag Memorial Hospital Presbyterian, Newport Beach, CA, USA The New Stanford Hospital, Stanford, CA, USA Providence Regional Medical Center Everett, Everett, WA, USA
Fingolimod in COVID-19 NCT04280588	Fingolimod 0.5 mg	Recruiting	Wan-Jin Chen Fuzhou, China
The clinical study of carrimycin on treatment patients with COVID-19 NCT04286503	1. Carrimycin 2. Lopinavir/ritonavir tablets or Arbidol (umifevonir) or chloroquine phosphate	Not yet recruiting	

Efficacy and safety of corticosteroids in COVID-19 NCT04273321	Methylprednisolone	Completed	Hubei Province Hospital of Integrated Chinese and Western Medicine, Wuhan, Hubei, China Yichang First People's Hospital, Yichang, Hubei, China RenMin Hospital of Wuhan University, Wuhan, China
Evaluation of the efficacy and safety of sarilumab in hospitalized patients with COVID-19 NCT04315298	Sarilumab	Active, not recruiting	Regeneron Study Site, New York, NY, USA
A Trial of Remdesivir in Adults With Mild and Moderate COVID-19 NCT04252664	Remdesivir	Recruiting	Jin Yin-tan Hospital, Wu Han, Hubei, China
Adaptive COVID-19 Treatment Trial (ACTT) NCT04280705	Remdesivir	Completed	National Institutes of Health – Clinical Center, National Institute of Allergy and Infectious Diseases Laboratory of Immunoregulation, Clinical Research Section, Bethesda, MD, USA University of Nebraska Medical Center – Infectious Diseases, Omaha, NE, USA University of Texas Medical Branch – Division of Infectious Disease, Galveston, TX, USA Providence Sacred Heart Medical Center, Spokane, WA, USA
Ongoing clinical trials for the management of the COVID-19 pandemic	Thymosin, suramin; conventional treatment + adalimumab	Not yet recruiting	China
A Trial of Remdesivir in Adults with Severe COVID-19 NCT04257656	Remdesivir	Terminated	Bin Cao; Beijing, China

(continued)

Table 6.3 (*continued*)

Study	Drug	Status	Organization/location
Nitric Oxide Gas Inhalation for Severe Acute Respiratory Syndrome in COVID-19 (NOSARSCOVID) NCT04290871	Nitric oxide gas	Withdrawn	
Efficacy and safety of IFN-α2β in the treatment of novel coronavirus patients NCT04293887	Recombinant human interferon α1β	Not yet recruiting	
Evaluating and comparing the safety and efficiency of ASC09/ritonavir and lopinavir/ritonavir for novel coronavirus infection NCT04261907	1. ASC09/ritonavir group 2. Lopinavir/ritonavir group	Not yet recruiting	
Safety and Immunogenicity study of 2019-nCoV vaccine (mRNA-1273) to prevent SARS-CoV-2 infection (COVID-19) NCT04283461	mRNA-1273	Active, not recruiting	Kaiser Permanente Washington Health Research Institute – Vaccines and Infectious Diseases, Seattle, WA, USA
Study to evaluate the safety and antiviral activity of remdesivir (GS-5734™) in participants with severe coronavirus disease (COVID-19) NCT04292899	Remdesivir	Completed	Hoag Memorial Hospital Presbyterian, Newport Beach, CA, USA The New Stanford Hospital, Stanford, CA, USA Providence Regional Medical Center Everett, Everett, WA, USA
Lopinavir/ritonavir, ribavirin and IFN-beta combination for nCoV treatment NCT04276688	1. Lopinavir/ritonavir 2. Ribavirin 3. Interferon β-1B	Completed	University of Hong Kong, Queen Mary Hospital, Hong Kong
Efficacy of chloroquine and lopinavir/ritonavir in mild/general novel coronavirus (CoVID-19) infections: a prospective, open-label, multicenter randomized controlled clinical study ChiCTR2000029741	1. Chloroquine 2. Lopinavir/ritonavir		The Fifth Affiliated Hospital Sun Yat-Sen University, Zhuhai, China

A prospective, randomized, open-label, parallel controlled trial for the preventive effect of Hydroxychloroquine on medical personnel after exposure to COVID-19 ChiCTR2000029803	Hydroxychloroquine		RenMin Hospital of Wuhan University, Wuhan, China
Glucocorticoid therapy for novel coronavirus Critically ill patients with severe acute respiratory failure NCT04244591	Methylprednisolone	Completed	Medical ICU, Peking Union Medical College Hospital, Beijing, China
The efficacy and safety of carrimycin treatment in patients with novel coronavirus infectious disease (COVID-19): a multicenter, randomized, open-label controlled trial ChiCTR2000029867	Carrimycin		Beijing You'an Hospital, Capital Medical University, Beijing, China
A prospective clinical study for recombinant human interferon alpha 1b spray in preventing novel coronavirus (COVID-19) infection in highly exposed medical staffs ChiCTR2000030013	Recombinant human interferon alpha 1b		Chinese PLA General Hospital, Beijing, China
A pilot study of sildenafil in COVID-19 NCT04304313	Sildenafil citrate	Recruiting	Department and Institute of Infectious Disease, Wuhan, Hubei, China
A study for the efficacy of hydroxychloroquine for mild and moderate COVID-19 infectious diseases ChiCTR2000029760	Hydroxychloroquine	Terminated	The Second Affiliated Hospital of Chongqing Medical University, Chongqing, China
The efficacy and safety of thalidomide combined with low-dose hormones in the treatment of severe COVID-19 NCT04273581	Thalidomide	Not yet recruiting	
Various combination of protease inhibitors, oseltamivir, favipiravir, and chloroquine for treatment of COVID19: a randomized control trial NCT04303299	Oral	Not yet recruiting	Rajavithi Hospital, Bangkok, Thailand

(*continued*)

Table 6.3 (*continued*)

Study	Drug	Status	Organization/location
Randomized controlled trial of losartan for patients with COVID-19 not requiring hospitalization NCT04311177	Losartan	Not yet recruiting	Hennepin County Medical Center, Minneapolis, MN, USA M Health Fairview University of Minnesota Medical Center, Minneapolis, MN, USA University of Minnesota, Minneapolis, MN, USA
Chloroquine/hydroxychloroquine prevention of coronavirus disease (COVID-19) in the healthcare setting (COPCOV) NCT04303507	Chloroquine	Recruiting	
Favipiravir combined with tocilizumab in the treatment of corona virus disease 2019 NCT04310228	Favipiravir combined with tocilizumab	Recruiting	Anhui Medical University Affiliated First Hospital, Hefei, Anhui, China Guiqiang Wang, Beijing, Beijing, China Peking University First Hospital, Beijing, Beijing, China
Trial of treatments for COVID-19 in hospitalized adults (DisCoVeRy) NCT04315948	1. Remdesivir 2. Lopinavir/ritonavir 3. Interferon β-1A	Recruiting	
Randomized controlled trial of losartan for patients with COVID-19 requiring hospitalization NCT04312009	Losartan	Recruiting	Hennepin County Medical Center, Minneapolis, MN, USA M Health Fairview University of Minnesota Medical Center, Minneapolis, MN, USA University of Minnesota, Minneapolis, MN, USA

Comparison of lopinavir/ritonavir or hydroxychloroquine in patients with mild coronavirus disease (COVID-19) NCT04307693	1. Lopinavir/ritonavir 2. Hydroxychloroquine sulfate	Terminated	Asan Medical Center, University of Ulsan College of Medicine, Seoul, Republic of Korea
Evaluation of ganovo (danoprevir) combined with ritonavir in the treatment of SARS-CoV-2 infection NCT04291729	Ganovo with ritonavir ± interferon	Completed	The Ninth Hospital of Nanchang Nanchang, Jiangxi, China
Eculizumab (soliris) in covid-19 infected patients (SOLID-C19) NCT04288713	Eculizumab	Available	
Norwegian coronavirus disease 2019 study (NO COVID-19) NCT04316377	Hydroxychloroquine sulfate	Active, not recruiting	
Post-exposure prophylaxis for SARS-coronavirus-2 (COVID-19 PEP) NCT04308668	Hydroxychloroquine	Completed	University of Minnesota, Minneapolis, MN, USA
The efficacy and safety of pirfenidone capsules in the treatment of severe new coronavirus pneumonia (COVID-19) ChiCTR2000031138	Pirfenidone		Third Xiangya Hospital of Central South University, Changsha, China
Clinical characteristics of 138 hospitalized patients with 2019 novel coronavirus-infected pneumonia in Wuhan, China PMC7042881/[71]	Oseltamivir	Recruiting	Zhongnan Hospital of Wuhan University, Wuhan, China
Expanded access remdesivir (RDV; GS-5734™) NCT04302766	Remdesivir	Available	
Duvelisib to combat COVID-19 NCT04372602	Duvelisib	Not yet recruiting	Washington University School of Medicine, Saint Louis, MO, USA
Application of iron chelator (desferal) to reduce the severity of COVID-19 manifestations NCT04333550	Deferoxamine	Recruiting	Kermanshah University of Medical Sciences, Kermanshah, Iran

(*continued*)

Table 6.3 (*continued*)

Study	Drug	Status	Organization/location
A multi-center, randomized, double-blind, placebo-controlled, phase III clinical study evaluating the efficacy and safety of favipiravir in the treatment of patients with COVID-19-moderate type NCT04336904	Favipiravir	Active, not recruiting	ASST Fatebenefratelli Sacco, Milan, Italy
An open-label, randomized, cross-over interventional study to evaluate the efficacy and safety of tocilizumab *versus* corticosteroids in hospitalised COVID-19 patients with high risk of progression NCT04345445	Tocilizumab and methylprednisolone	Not yet recruiting	University of Malaya, Kuala Lumpur, Malaysia
Isotretinoin in treatment of COVID-19 (randomized) NCT04361422	Isotretinoin product in oral dose form	Not yet recruiting	Tanta University, Tanta, Egypt
Prevention of COVID-19 progression through early administration of inhaled nitric oxide NCT04388683	Nitric oxide	Recruiting	Tufts Medical Center, Boston, MA, USA
The randomized, open, phase 2 study to evaluate the safety and efficacy of clevudine in patients with moderate COVID-19 NCT04347915	Clevudine	Recruiting	Korea University Guro Hospital, Seoul, South Korea
Randomized controlled trial of angiotensin 1–7/TXA127 for the treatment of Severe COVID-19 NCT04401423	TXA127	Not yet recruiting	Columbia University, New York, NY, USA
Preventing cardiac complication of COVID-19 disease with early acute coronary syndrome therapy: a randomised controlled trial NCT04333407	1. Aspirin 2. Clopidogrel 3. Rivaroxaban 4. Atorvastatin 5. Omeprazole	Recruiting	Imperial College London, London, UK

InterMediateProphylACtic *versus* therapeutic dose anticoagulation in critically ill patients with COVID-19: a prospective randomized study (The IMPACT Trial) NCT04406389	1. Enoxaparin sodium 2. Unfractionated heparin 3. Fondapariniux 4. Argatroban	Recruiting	Weill Cornell Medicine, New York, NY, USA
The efficacy and safety of Huaier in the adjuvant treatment of COVID-19: a prospective, multicenter, randomized, parallel controlled clinical study NCT04291053	Huaier granule	Not yet recruiting	Tongji Hospital, China
ABX464 in treating inflammation and preventing acute respiratory failure in patients with COVID-19 (Mir-age) NCT04393038	ABX464	Recruiting	Universitaire de Nice, Nice, France
Study of FT516 safety and feasibility for the treatment of coronavirus disease 2019 (COVID-19) in hospitalized patients with hypoxia NCT04363346	FT516	Recruiting	University of Minnesota, Minneapolis, MN, USA
DAS181 for severe COVID-19: compassionate use NCT04324489	DAS181	Completed	Renmin Hospital of Wuhan University, Wuhan, China
Dapagliflozin in respiratory failure in patients with COVID-19 (DARE-19) NCT04350593	Dapagliflozin	Recruiting	Saint Luke's Mid America Heart Institute, Kansas City, MO, USA
Efficacy and safety of ozanimod for the treatment of COVID-19 patients requiring oxygen support NCT04405102	Ozanimod	Not yet recruiting	Laval University, Laval, Canada
Efficacy of addition of naproxen in the treatment of critically ill patients hospitalized for COVID-19 infection (ENACOVID) NCT04325633	Naproxen	Not yet recruiting	Hôpitaux de Paris, Paris, France

(*continued*)

Table 6.3 (*continued*)

Study	Drug	Status	Organization/location
Randomized, double-blind, placebo-controlled multi-center study to assess the efficacy and safety of ruxolitinib in patients with COVID-19 associated cytokine storm (RUXCOVID) NCT04362137	Ruxolitinib	Recruiting	Novartis Pharmaceuticals, USA
Safety and efficacy of baricitinib for COVID-19 NCT04340232	Baricitinib	Not yet recruiting	University of Colorado, Denver, CO, USA
Open label, randomized study of the efficacy and safety of acalabrutinib with best supportive care *versus* best supportive care in subjects hospitalized with COVID-19 NCT04380688	Acalabrutinib	Recruiting	1. Washington, DC, USA 2. Louisville, KY, USA 3. Baltimore, MD, USA
Alpha-1 adrenergic receptor antagonism to prevent COVID-19 cytokine storm syndrome and acute respiratory distress syndrome: a randomized study comparing the efficacy of prazosin *vs.* standard of care for SARS-CoV-2 infection NCT04365257	Prazosin	Recruiting	Johns Hopkins University, Baltimore, MD, USA
Multicentre, open label, randomised, adaptative clinical trial of efficacy and safety of treatment regimens in adult COVID-19 patients in Senegal NCT04390594	Hydroxychloroquine + azithromycin	Not yet recruiting	Fann Hospital, Dakar, Senegal

6.3 Executive Summary

- Currently, remdesivir, an antiviral drug, has been approved by the regulatory authorities to treat COVID-19 in Japan and approved by emergency use authorization to treat severe COVID-19 disease in the United States by the FDA. It shortens the duration of the disease and there are encouraging data regarding mortality reduction.
- Dexamethasone, a systemic steroid, has been shown to reduce mortality for oxygen-dependent COVID-19 patients, but it may enhance the disease process when given to patients with milder forms of the disease.
- The use of mesenchymal stem cells and interleukin-6 inhibitors as well as other cytokine blockers/modulators have a promising role in controlling the excessive inflammatory host response to COVID-19.
- Convalescent plasma therapy has been useful in SARS and MERS and is under investigation in treating COVID-19. Immuneglobulin may be helpful as adjunctive therapy.
- Combinations of various Indian medicinal plants may be beneficial in the treatment of COVID-19.
- Chloroquine and hydroxychloroquine have not been shown to be effective in the treatment of COVID-19 and may cause harm in a small percentage of patients due to cardiac side-effects.
- A number of antiviral drugs including favipiravir, umifenovir, triazavirin, and baloxavir marboxil are undergoing clinical trials.

Acknowledgements

The authors and editors thank Melanie K. Shadoan and Chris L Ringwalt for their excellent editing of this chapter.

References

1. Q. X. Long, X. J. Tang, Q. L. Shi, Q. Li, H. J. Deng and J. Yuan, *et al.*, Clinical and immunological assessment of asymptomatic SARS-CoV-2 infections, *Nat. Med.*, 2020, **26**, 1200–1204.
2. Y. Bai, L. Yao, T. Wei, F. Tian, D. Y. Jin and L. Chen, *et al.*, Presumed Asymptomatic Carrier Transmission of COVID-19, *JAMA*, 2020, **323**(14), 1406–1407.
3. S. Mulangu, L. E. Dodd, R. T. Davey Jr., O. Tshiani Mbaya, M. Proschan and D. Mukadi, *et al.*, A Randomized, Controlled Trial of Ebola Virus Disease Therapeutics, *N. Engl. J. Med.*, 2019, **381**(24), 2293–2303.
4. L. Gattinoni, S. Coppola, M. Cressoni, M. Busana, S. Rossi and D. Chiumello, COVID-19 Does Not Lead to a "Typical" Acute Respiratory Distress Syndrome, *Am. J. Respir. Crit. Care Med.*, 2020, **201**(10), 1299–1300.
5. M. Cascella, M. Rajnik, A. Cuomo, S. C. Dulebohn, R. Di Napoli, *Features, Evaluation and Treatment Coronavirus (COVID-19)*, StatPearls, Treasure Island (FL), 2020.

6. Information for Clinicians on Investigational Therapeutics for Patients with COVID-19, https://www.cdc.gov/coronavirus/2019-ncov/hcp/therapeutic-options.html.
7. B. N. Rome and J. Avorn, Drug Evaluation during the Covid-19 Pandemic, *N. Engl. J. Med.*, 2020, **382**(24), 2282–2284.
8. B. Cao, Y. Wang, D. Wen, W. Liu, J. Wang and G. Fan, *et al.*, A Trial of Lopinavir-Ritonavir in Adults Hospitalized with Severe Covid-19, *N. Engl. J. Med.*, 2020, **382**(19), 1787–1799.
9. C. J. Gordon, E. P. Tchesnokov, J. Y. Feng, D. P. Porter and M. Gotte, The antiviral compound remdesivir potently inhibits RNA-dependent RNA polymerase from Middle East respiratory syndrome coronavirus, *J. Biol. Chem.*, 2020, **295**(15), 4773–4779.
10. E. de Wit, F. Feldmann, J. Cronin, R. Jordan, A. Okumura and T. Thomas, *et al.*, Prophylactic and therapeutic remdesivir (GS-5734) treatment in the rhesus macaque model of MERS-CoV infection, *Proc. Natl. Acad. Sci. U. S. A.*, 2020, **117**(12), 6771–6776.
11. Y. Wang, D. Zhang, G. Du, R. Du, J. Zhao and Y. Jin, *et al.*, Remdesivir in adults with severe COVID-19: a randomised, double-blind, placebo-controlled, multicentre trial, *Lancet*, 2020, **395**(10236), 1569–1578.
12. P. Gautret, J. C. Lagier, P. Parola, V. T. Hoang, L. Meddeb and M. Mailhe, *et al.*, Hydroxychloroquine and azithromycin as a treatment of COVID-19: results of an open-label non-randomized clinical trial, *Int. J. Antimicrob. Agents*, 2020, 105949.
13. P. Zarogoulidis, N. Papanas, I. Kioumis, E. Chatzaki, E. Maltezos and K. Zarogoulidis, Macrolides: from in vitro anti-inflammatory and immunomodulatory properties to clinical practice in respiratory diseases, *Eur. J. Clin. Pharmacol.*, 2012, **68**(5), 479–503.
14. E. S. Rosenberg, E. M. Dufort, T. Udo, L. A. Wilberschied, J. Kumar and J. Tesoriero, *et al.*, Association of Treatment With Hydroxychloroquine or Azithromycin With In-Hospital Mortality in Patients With COVID-19 in New York State, *JAMA*, 2020, **323**(24), 2493–2502.
15. A. B. Cavalcanti, F. G. Zampieri, R. G. Rosa, L. C. P. Azevedo, V. C. Veiga, A. Avezum, L. P. Damiani, A. Marcadenti, L. Kawano-Dourado, T. Lisboa, D. L. M. Junqueira, P. G. M. de Barros E Silva, L. Tramujas, E. O. Abreu-Silva, L. N. Laranjeira, A. T. Soares, L. S. Echenique, A. J. Pereira, F. G. R. Freitas, O. C. E. Gebara, V. C. S. Dantas, R. H. M. Furtado, E. P. Milan, N. A. Golin, F. F. Cardoso, I. S. Maia, C. R. Hoffmann Filho, A. P. M. Kormann, B. B. Amazonas, M. F. Bocchi de Oliveira, A. Serpa-Neto, M. Falavigna, R. D. Lopes, F. R. Machado and O. Berwanger, Coalition Covid-19 Brazil I Investigators, *N. Engl. J. Med.*, 2020, **383**(21), 2041–2052. Erratum in: *N. Engl. J. Med.*, **383**(21), e119.
16. E. Callaway and D. Cyranoski, China coronavirus: Six questions scientists are asking, *Nature*, 2020, **577**(7792), 605–607.
17. N. J. Mercuro, C. F. Yen, D. J. Shim, T. R. Maher, C. M. McCoy and P. J. Zimetbaum, *et al.*, Risk of QT Interval Prolongation Associated With Use of Hydroxychloroquine With or Without Concomitant Azithromycin

Among Hospitalized Patients Testing Positive for Coronavirus Disease 2019 (COVID-19), *JAMA Cardiol.*, 2020, **5**(9), 1036–1041.
18. B. Bikdeli, M. V. Madhavan, A. Gupta, D. Jimenez, J. R. Burton and C. Der Nigoghossian, *et al.*, Pharmacological Agents Targeting Thromboinflammation in COVID-19: Review and Implications for Future Research, *Thromb. Haemostasis*, 2020, **120**(07), 1004–1024.
19. B. K. Manne, F. Denorme, E. A. Middleton, I. Portier, J. W. Rowley and C. J. Stubben, *et al.*, Platelet Gene Expression and Function in COVID-19 Patients, *Blood*, 2020, **136**, 1317–1329.
20. A. Kollias, K. G. Kyriakoulis, E. Dimakakos, G. Poulakou, G. S. Stergiou and K. Syrigos, Thromboembolic risk and anticoagulant therapy in COVID-19 patients: emerging evidence and call for action, *Br. J. Haematol.*, 2020, **189**(5), 846–847.
21. H. Al-Samkari, R. S. Karp Leaf, W. H. Dzik, J. C. T. Carlson, A. E. Fogerty and A. Waheed, *et al.*, COVID-19 and coagulation: bleeding and thrombotic manifestations of SARS-CoV-2 infection, *Blood*, 2020, **136**(4), 489–500.
22. M. Noris, A. Benigni and G. Remuzzi, The case of complement activation in COVID-19 multiorgan impact, *Kidney Int.*, 2020, **98**(2), 314–322.
23. A. M. Risitano, D. C. Mastellos, M. Huber-Lang, D. Yancopoulou, C. Garlanda and F. Ciceri, *et al.*, Complement as a target in COVID-19? *Nat. Rev. Immunol.*, 2020, **20**(6), 343–344.
24. C. Magro, J. J. Mulvey, D. Berlin, G. Nuovo, S. Salvatore and J. Harp, *et al.*, Complement associated microvascular injury and thrombosis in the pathogenesis of severe COVID-19 infection: A report of five cases, *Transl. Res*, 2020, **220**, 1–13.
25. G. Guaraldi, M. Meschiari, A. Cozzi-Lepri, J. Milic, R. Tonelli and M. Menozzi, *et al.*, Tocilizumab in patients with severe COVID-19: a retrospective cohort study, *Lancet Rheumatol.*, 2020, **2**(8), e474–e484.
26. F. M. Buonaguro, I. Puzanov and P. A. Ascierto, Anti-IL6R role in treatment of COVID-19-related ARDS, *J. Transl. Med.*, 2020, **18**(1), 165.
27. A. A. Elfiky, Anti-HCV, nucleotide inhibitors, repurposing against COVID-19, *Life Sci.*, 2020, **248**, 117477.
28. D. Falzarano, E. de Wit, A. L. Rasmussen, F. Feldmann, A. Okumura and D. P. Scott, *et al.*, Treatment with interferon-alpha2b and ribavirin improves outcome in MERS-CoV-infected rhesus macaques, *Nat. Med.*, 2013, **19**(10), 1313–1317.
29. I. F. Hung, K. C. Lung, E. Y. Tso, R. Liu, T. W. Chung and M. Y. Chu, *et al.*, Triple combination of interferon beta-1b, lopinavir-ritonavir, and ribavirin in the treatment of patients admitted to hospital with COVID-19: an open-label, randomised, phase 2 trial, *Lancet*, 2020, **395**(10238), 1695–1704.
30. L. Caly, J. D. Druce, M. G. Catton, D. A. Jans and K. M. Wagstaff, The FDA-approved drug ivermectin inhibits the replication of SARS-CoV-2 in vitro, *Antiviral Res.*, 2020, **178**, 104787.

31. M. Bray, C. Rayner, F. Noël, D. Jans and K. Wagstaff, Ivermectin and COVID-19: A report in Antiviral Research, widespread interest, an FDA warning, two letters to the editor and the authors' responses, *Antiviral Res.*, 2020, **178**, 104805.
32. F. Noël and J. Lima, Pharmacological aspects and clues for the rational use of Chloroquine/Hydroxychloroquine facing the therapeutic challenges of COVID-19 pandemic, *Lat. Am. J. Clin. Sci. Med. Technol.*, 2020, **2**, 28–34.
33. S. N. Y. Yang, S. C. Atkinson, C. Wang, A. Lee, M. A. Bogoyevitch and N. A. Borg, *et al.*, The broad spectrum antiviral ivermectin targets the host nuclear transport importin α/β1 heterodimer, *Antiviral Res.*, 2020, **177**, 104760.
34. M. Z. Tay, C. M. Poh, L. Renia, P. A. MacAry and L. F. P. Ng, The trinity of COVID-19: immunity, inflammation and intervention, *Nat. Rev. Immunol.*, 2020, **20**(6), 363–374.
35. A. Zumla, J. F. Chan, E. I. Azhar, D. S. Hui and K. Y. Yuen, Coronaviruses - drug discovery and therapeutic options, *Nat. Rev. Drug Discovery*, 2016, **15**(5), 327–347.
36. M. Gomathi, S. Padmapriya and V. Balachandar, Drug Studies on Rett Syndrome: From Bench to Bedside, *J. Autism Dev. Disord.*, 2020, 1–25.
37. B. Ravishankar and V. J. Shukla, Indian systems of medicine: a brief profile, *Afr. J. Tradit., Complementary Altern. Med.*, 2007, **4**(3), 319–337.
38. B. Vellingiri, K. Jayaramayya, M. Iyer, A. Narayanasamy, V. Govindasamy and B. Giridharan, *et al.*, COVID-19: A promising cure for the global panic, *Sci. Total Environ.*, 2020, **725**, 138277.
39. S. Parasuraman, G. S. Thing and S. A. Dhanaraj, Polyherbal formulation: Concept of ayurveda, *Pharmacogn. Rev.*, 2014, **8**(16), 73–80.
40. D. S. Fabricant and N. R. Farnsworth, The value of plants used in traditional medicine for drug discovery, *Environ. Health Perspect.*, 2001, **109**(suppl 1), 69–75.
41. M. Li-Weber, New therapeutic aspects of flavones: the anticancer properties of Scutellaria and its main active constituents Wogonin, Baicalein and Baicalin, *Cancer Treat. Rev.*, 2009, **35**(1), 57–68.
42. M. Akram, I. M. Tahir, S. M. A. Shah, Z. Mahmood, A. Altaf and K. Ahmad, *et al.*, Antiviral potential of medicinal plants against HIV, HSV, influenza, hepatitis, and coxsackievirus: A systematic review, *Phytother. Res.*, 2018, **32**(5), 811–822.
43. W. R. Sandsborg, Natural products in drug discovery and development, *Bioassay Methods in Natural Product Research and Drug Development*, Springer, 1999, pp. 143–149.
44. T. Scior, A. Bender, G. Tresadern, J. L. Medina-Franco, K. Martinez-Mayorga and T. Langer, *et al.*, Recognizing pitfalls in virtual screening: a critical review, *J. Chem. Inf. Model.*, 2012, **52**(4), 867–881.
45. S. Vimalanathan, S. Ignacimuthu and J. Hudson, *Pharm. Biol.*, 2009, **47**, 422–429.

46. V. Balachandar, I. Mahalaxmi, J. Kaavya, G. Vivekanandhan, S. Ajithkumar and N. Arul, *et al.*, COVID-19: emerging protective measures, *Eur. Rev. Med. Pharmacol. Sci.*, 2020, **24**(6), 3422–3425.
47. S. Pal, A. Banerjee and A. Maity, Traditional Indian Medicinal Plants in the Treatment of Patients Infected With Human Coronavirus 2019 (SARS-CoV-2): A. Review, *IOSR Journal Of Pharmacy And Biological Sciences*, 2020, **15**, 46–56.
48. N. Maity, N. K. Nema, B. K. Sarkar and P. K. Mukherjee, Standardized Clitoria ternatea leaf extract as hyaluronidase, elastase and matrix-metalloproteinase-1 inhibitor, *Indian J. Pharmacol.*, 2012, **44**(5), 584.
49. Y. C. Tsai, C. L. Lee, H. R. Yen, Y. S. Chang, Y. P. Lin and S. H. Huang, *et al.*, Antiviral Action of Tryptanthrin Isolated from Strobilanthes cusia Leaf against Human Coronavirus NL63, *Biomolecules*, 2020, **10**(3), 366.
50. Y. Cheng, R. Wong, Y. O. Y. Soo, W. S. Wong, C. K. Lee and M. H. L. Ng, *et al.*, Use of convalescent plasma therapy in SARS patients in Hong Kong, *Eur. J. Clin. Microbiol. Infect. Dis.*, 2005, **24**(1), 44–46.
51. J. Mair-Jenkins, M. Saavedra-Campos, J. K. Baillie, P. Cleary, F. M. Khaw and W. S. Lim, *et al.*, The effectiveness of convalescent plasma and hyper-immune immunoglobulin for the treatment of severe acute respiratory infections of viral etiology: a systematic review and exploratory meta-analysis, *J. Infect. Dis.*, 2015, **211**(1), 80–90.
52. G. Chen, D. Wu, W. Guo, Y. Cao, D. Huang and H. Wang, *et al.*, Clinical and immunological features of severe and moderate coronavirus disease 2019, *J. Clin. Invest.*, 2020, **130**(5), 2620–2629.
53. P. I. Lee and P. R. Hsueh, Emerging threats from zoonotic coronaviruses-from SARS and MERS to 2019-nCoV, *J. Microbiol., Immunol. Infect.*, 2020, **53**(3), 365–367.
54. A. Casadevall and L. A. Pirofski, The convalescent sera option for containing COVID-19, *J. Clin. Invest.*, 2020, **130**(4), 1545–1548.
55. S. S. Bobati and K. R. Naik, Therapeutic plasma exchange-An emerging treatment modality in patients with neurologic and non-neurologic diseases, *J. Clin. Diagn. Res.*, 2017, **11**(8), EC35.
56. K. Duan, B. Liu, C. Li, H. Zhang, T. Yu and J. Qu, *et al.*, Effectiveness of convalescent plasma therapy in severe COVID-19 patients, *Proc. Natl. Acad. Sci. U. S. A.*, 2020, **117**(17), 9490–9496.
57. G. Lazennec and C. Jorgensen, Concise Review: Adult Multipotent Stromal Cells and Cancer: Risk or Benefit? *Stem Cells*, 2008, **26**(6), 1387–1394.
58. Y.-m. Du, Y.-x. Zhuansun, R. Chen, L. Lin, Y. Lin and J.-g. Li, Mesenchymal stem cell exosomes promote immunosuppression of regulatory T cells in asthma, *Exp. Cell Res.*, 2018, **363**(1), 114–120.
59. V. Sengupta, S. Sengupta, J. A. Lazo, P. Woods, A. Nolan and N. Bremer, Exosomes Derived from Bone Marrow Mesenchymal Stem Cells as Treatment for Severe COVID-19, *Stem Cells Dev.*, 2020, **29**(12), 747–754.

60. E. Bari, I. Ferrarotti, L. Saracino, S. Perteghella, M. L. Torre and A. G. Corsico, Mesenchymal stromal cell secretome for severe COVID-19 infections: premises for the therapeutic use, *Cells*, 2020, **9**(4), 924.
61. S. Liu, D. Peng, H. Qiu, K. Yang, Z. Fu and L. Zou, Mesenchymal stem cells as a potential therapy for COVID-19, *Stem Cell Res. Ther.*, 2020, **11**(1), 169.
62. Group RC, Dexamethasone in Hospitalized Patients with Covid-19 — Preliminary Report, *N. Engl. J. Med.*, 2021, **384**, 693–704.
63. C. M. Chu, V. C. C. Cheng, I. F. N. Hung, M. M. L. Wong, K. H. Chan and K. S. Chan, *et al.*, Role of lopinavir/ritonavir in the treatment of SARS: initial virological and clinical findings, *Thorax*, 2004, **59**(3), 252–256.
64. H. Bian, Z.-H. Zheng, D. Wei, Z. Zhang, W.-Z. Kang and C.-Q. Hao, *et al.*, Meplazumab treats COVID-19 pneumonia: an open-labelled, concurrent controlled add-on clinical trial, medRxiv, 2020, DOI: 10.1101/2020.03.21.20040691.
65. N. Vargesson, Thalidomide-induced teratogenesis: history and mechanisms, *Birth Defects Res., Part C*, 2015, **105**(2), 140–156.
66. A. X. Zhu, D. R. Borger, Y. Kim, D. Cosgrove, A. Ejaz and S. Alexandrescu, *et al.*, Genomic profiling of intrahepatic cholangiocarcinoma: refining prognosis and identifying therapeutic targets, *Ann. Surg. Oncol.*, 2014, **21**(12), 3827–3834.
67. D. Pelletier and D. A. Hafler, Fingolimod for multiple sclerosis, *N. Engl. J. Med.*, 2012, **366**(4), 339–347.
68. A. Golchin, E. Seyedjafari and A. Ardeshirylajimi, Mesenchymal Stem Cell Therapy for COVID-19: Present or Future, *Stem Cell Rev. Rep.*, 2020, **16**(3), 427–433.
69. Z. Leng, R. Zhu, W. Hou, Y. Feng, Y. Yang and Q. Han, *et al.*, Transplantation of ACE2-mesenchymal stem cells improves the outcome of patients with COVID-19 pneumonia, *Aging Dis.*, 2020, **11**(2), 216.
70. M. P. Lythgoe and P. Middleton, Ongoing clinical trials for the management of the COVID-19 pandemic, *Trends Pharmacol. Sci.*, 2020, **41**, 353–382.
71. D. Wang and B. Hu, *et al.*, Clinical Characteristics of 138 Hospitalized Patients With 2019 Novel Coronavirus–Infected Pneumonia in Wuhan, China, *JAMA*, 2020, **323**, 1061–1069.

CHAPTER 7

Clinical Epidemiology of Coronavirus Disease 2019: Infectivity, Clinical Spectrum and Presentation, and Population Distribution

FENGYU ZHANG*[a,b] AND CLAUDE HUGHES*[c,d]

[a]Global Clinical and Translational Research Institute, Bethesda, MD, USA; [b]Beijing Huilongguan Hospital and Peking University Huilongguan Clinical Institute, Beijing, China; [c]IQVIA, Durham, NC, USA; [d]Duke University Medical Center, Durham, NC, USA
*E-mail: zhangfy@gcatresearch.org, clhughes@duke.edu

7.1 Introduction

Coronavirus disease 2019 (COVID-19) is a novel respiratory disease caused by severe acute respiratory syndrome coronavirus 2 (SARS-CoV-2), discovered in 2019. SARS-CoV-2 is a positive-strand RNA virus and belongs to the beta coronavirus group, has a genome that is highly similar (96%) to that of bat coronavirus, and is 80% identical to the genome of SARS-CoV.[1,2] The new virus originated from a natural host and then transformed to gain an ability to support human-to-human transmission.[3] While the exact source of the virus

The Coronavirus Pandemic and the Future Volume 1: Virology, Epidemiology, Translational Toxicology and Therapeutics
Edited by Michael D. Waters, Alok Dhawan, Tim Marrs, Diana Anderson, Stafford Warren and Claude L. Hughes

Published by the Royal Society of Chemistry, www.rsc.org

and its transmission mode from its natural host to humans have not been determined, the human-to-human transmission of SARS-CoV-2 has caused a worldwide public health emergency. Compared with the coronaviruses that cause severe acute respiratory syndrome (SARS) and Middle East respiratory syndrome (MERS), the SARS-CoV-2 virus exhibits not only higher infectivity and more rapid transmission (Box 7.1) than SARS-CoV and MERS-CoV, but also more severe clinical symptoms than the coronaviruses that cause seasonal influenza or the common cold.[4] The high infectivity and rapid transmission mode from person to person have posed a substantial challenge in managing the initial outbreak and subsequent pandemic.

Epidemiology is the study of the distribution of the incidence and occurrence of a disease or an event in three dimensions (*i.e.* where, who, and when) and various factors associated with the disease's risk, so that preventive measures can be taken to reduce the risk for exposure. Understanding the epidemiology of the pandemic is fundamental for developing effective prevention, control, and treatment strategies. However, a considerable amount of time and formal population-based studies[5] may be required to systematically define the epidemiology of COVID-19, which should consider various factors of both the viruses (including variant strains) themselves and their human hosts. As an acute infectious respiratory disease characterized by direct person-to-person transmission, the development of COVID-19 begins with an individual's exposure to the virus pathogen. Factors associated with an increased likelihood of exposure and an enhanced interaction between the pathogen and host genomes may facilitate the virus's entry into the host, and then increase the risk of developing the disease. An individual's susceptibility to the SARS-CoV-2 virus is also determined by the host's innate immune system and adaptive immune response. The specific protective immunity or capacity for defending against the pathogen is likely to vary. It may be associated with whether an individual has underlying chronic health conditions that may enhance or reduce the individual's ability to mount an effective immune response to the pathogen.

Clinical epidemiology is the study of the distribution of clinical presentations and factors associated with clinical subtypes or outcomes in patient populations that enhance the understanding of clinical heterogeneity. Patients across the clinical spectrum of COVID-19 (*e.g.* asymptomatic, mild to moderate, severe, and critical illness) serve as numerators for estimating epidemiological parameters, but they constitute entire study populations for clinical epidemiology. In other words, patient populations are the important component for epidemiology and clinical epidemiology, in which objectively characterizing patient populations is critical for designing clinical studies or trials to test new drug applications for COVID-19.

When the virus enters a host, an individual may experience one of several possible outcomes. Some individuals in a population tested positive with the viral infection may not develop symptoms, but can spread the virus within their family or community and cause others to be infected. The asymptomatic patients are critical in describing the disease's clinical spectrum,[5] assessing its infectivity and transmissibility,[6] selecting appropriate public health

Box 7.1 Related terminology for epidemiology of infectious diseases

Term	Definition
Infectivity	Infectivity is defined as a pathogen's ability to establish infection at a horizontal level, *i.e.* from host to host in humans, excluding maternal–fetal vertical transmission and transmission from humans to other hosts. Infectivity is determined by the pathogen's ability to enter, survive and replicate in the host, and is measured at a population level by the rate of a new infection after exposure to a source of confirmed infection. An observed infectivity is equivalent to the incidence rate of a new infection. When a recurrent infection is rare within a short time for a pathogen (*e.g.* SARS-CoV-2), the observed infectivity is equivalent to the prevalence of infection, determined by the rate of viral replication and the chance of exposure to the pathogen. The infection rate of SARS-CoV-2 is estimated at 15–20% in the general population when personal protective equipment is not used, according to the statistics from the outbreaks, such as the one that occurred on the *Diamond Princess* cruise ship.
Transmission	Transmission is the ability to pass on a pathogen that causes an infectious disease from an infected individual to other individuals or a group. It usually occurs by the following means: a. Airborne transmission of tiny dry and wet particles <5 μm b. Droplet transmission of small and usually wet particles >5 μm c. Direct physical contact (*e.g.* handshaking) d. Indirect physical contact (*e.g.* contaminated surfaces) e. Fecal–oral transmission Because of its multiple means of transmission (a–e above), the SARS-CoV-2 virus has a rapid transmission mode. So far, there has been evidence for maternal–fetal vertical transmission (but see Egloff *et al.* 2020).[7]
Transmissibility	Transmissibility is defined as the number of cases directly caused by a given infected case in the absence of any deliberate intervention in disease transmission. It can be expressed as $R_0 = b \times t$; where b represents an infected individual, who makes an average of b infection-producing contacts per unit time within a certain infectious period t. When $R_0 > 1$, the outbreak will expand; when $R_0 < 1$, it will die out.

strategies,[8] and describing its epidemiology and clinical epidemiology. Because asymptomatic patients are less likely to be noticed and identified promptly unless an active testing strategy is implemented, they may pose a dangerous threat to public health. Therefore, asymptomatic patients should be quarantined or, whenever possible, treated with existing medications including traditional medicine that can reduce their viral load and prevent further transmission.

Characterizing the clinical manifestations of patients who require healthcare would help depict clinical course with COVID-19 in patients. Since there are currently no vaccines or effective pharmacological treatment for the disease, it is urgent to assess the need for healthcare and define its clinical epidemiology. This section describes the clinical epidemiology of infection with SARS-CoV-2, based on a review of clinical reports, mostly hospital-based case series, of the clinical types and features seen in patients with COVID-19, in order to understand distinct patient subpopulations and identify potential clinical outcomes. In addition to asymptomatic individuals, those infected with SARS-CoV-2 may have various clinical manifestations ranging from mild to moderate to severe cases; some may become critically ill or die. These outcomes vary with patients' underlying health conditions, immunity, and demographic characteristics, including their race, ethnicity, socioeconomic status, and accessibility to treatment. Information on the classification of clinical types and presentation would help plan healthcare services for key subgroups of patient populations and encourage etiological and clinical studies or trials that aim to test the efficacy and effectiveness of medical treatments and develop novel therapeutic applications.

7.2 Infection and Tests

Infection is defined as the invasion and multiplication of micro-organisms in a host's body tissues after exposure to a source of confirmed infection. When a foreign pathogen enters the host and invades the body's tissues, it colonizes and begins to replicate. Susceptible individuals may develop biological evidence of infection, which could be detected through related assessments such as pathogen-related nucleic acid testing and genomic sequencing, as well as examinations of host-related responses such as the production of antibodies or physiological response (*e.g.* inflammation). Some individuals infected with a pathogen may develop clinical illness and may not exhibit positive viral tests. Additionally, it is now established that some individuals who test positive for SARS-CoV-2 nucleic acids do not develop symptoms (*i.e.* they are asymptomatic), or they may show only minimal symptoms, but can still spread the virus and infect others. An early case report[9] revealed that an individual who traveled to an epicenter might have become infected with SARS-CoV-2 but developed no symptoms or fever; a computed tomography (CT) scan of the individual's chest and laboratory tests were normal, and the initial nucleic acid testing was negative. However, that asymptomatic person might have infected five family members. A recent study found that among individuals infected with SARS-CoV-2, the duration of viral shedding of asymptomatic and symptomatic patients was similar.[10] Furthermore, a review of 16 reports indicates that asymptomatic patients could account for 40–96% of patients identified in the outbreaks of COVID-19.[11] The similar long duration of viral shedding and a high proportion of asymptomatic patients suggest that asymptomatic transmission is a critical concern for effective public health management.

7.2.1 Viral Tests

Viral testing is an objective assessment used to confirm infection and make a diagnosis of COVID-19. Reverse transcriptase polymerase chain reaction (RT-PCR) is a method to detect the genetic material of a pathogen, including a virus, and it is the most widely used method for testing the presence of viral particles of SARS-CoV-2 in human biospecimens. All coronaviruses can only replicate and survive in living cells; therefore, the SARS-CoV-2 virus enters the cells of the mucosal membrane of the mouth and nose of the host and replicates there before moving down the respiratory and gastrointestinal systems. In general, viral testing requires samples of nasopharyngeal mucosa.

To use the RT-PCR technique, RNA must be converted into DNA (called reverse transcription), so that DNA can be copied and amplified. The amplification is only on a pre-selected part or parts of transcribed viral DNA. In some guidelines,[12,13] the open reading frame 1ab (ORF1ab) or nucleocapsid protein (N) gene regions are included as testing targets. Other tests also target the RNA-dependent RNA polymerase (RdRP), N and E gene, in which the E gene is used for screening to eliminate the bat SARS-related CoV, while RdRP and N gene were used to confirm the infection of SARS-CoV-2.[14] A positive viral test is defined as the RT-PCR cycle threshold value of less than 37 or 40, which means that higher viral loading may require fewer RT-PCR cycles to detect the components of the viral genome in a human biospecimen. Biotechnology companies have also developed a rapid qualitative test, which involves testing for the presence of specific SARS-CoV-2 antigen and can be used at the point of care.

Sequencing-based testing is another tool for viral testing. On 10 June 2020, Illumina announced that it had received a United States Food and Drug Administration (FDA) emergency use authorization (EUA) for its COVIDSeq test designed to target and amplify 98 segments of the full genome of SARS-CoV-2, instead of the few segments on which the RT-PCR approach has usually focused. Therefore, the COVIDSeq test may increase the accuracy of tests for the virus, because it targets the whole genome, leading to increased testing sensitivity. More importantly, with the genome-wide sequencing test, much viral genomic data will be produced to facilitate population-based genomic research on the pathogen.

Viral nucleic acid testing can determine if an individual is infected with SARS-CoV-2 and can quickly confirm cases of COVID-19. Due to limited capacity or great demands for testing, viral tests are sometimes performed only in people with suspected exposure to a source of confirmed infection or who have already developed fever or respiratory symptoms. Testing based on this strategy may not reflect the actual infection rate in the general population or assess the full clinical spectrum of COVID-19 in the entire patient population.

However, viral nucleic acid testing may have a high rate of false-negative results.[15,16] False negatives may be caused by several factors such as the magnitude of the viral load at the early stage of infection, varying quality of test kits, the type and quality of biospecimen taken, and assay errors. A fraction of patients may have lower respiratory tract infections, which might affect their viral testing if a nasopharyngeal swab sample is used.[17] Of note, nucleic acid testing focuses on only parts of the viral genome, and that means that the virus *per se* is not the biospecimen. Thus, in most cases, the nucleic acid test cannot rule out a disease caused by another coronavirus except in the case of those panels designed with specific procedures.[14] Because of the limitation of viral testing, individuals' history of exposure, clinical symptoms, computed tomography, or X-rays should also be considered to diagnose suspect COVID-19.[18]

7.2.2 Population-based Infection Rate

The estimate of an infection rate can be obtained from several outbreaks in which personal protective measures are unlikely to have been taken by individuals. The first estimated infection rate was on the *Diamond Princess* cruise ship after an index case with COVID-19 was identified.[19] Passengers and crew members remained in a closed environment and later were quarantined together, so all individuals on the cruise ship may be assumed to have been exposed to the confirmed infection source. Altogether, 19% (712 out of 3711) of passengers and crew members tested positive for the SARS-CoV-2 viral RNA, of whom 46% (331 out of 712) were asymptomatic, according to the statistics of the United States Centers for Disease Control and Prevention. The infection rate was close to the estimates from the crew members of the USS *Theodore Roosevelt* (17%) and obstetric patients in New York (15%), but was lower than estimates of 24–69% in homeless shelters in Boston and Los Angeles, inmates in prison in four states of the United States, as well as the elderly in a nursing home facility in Washington State (Table 7.1).

The prevalence of infection appears to be lower in children exposed to a confirmed infection. In a population-based assessment of 1391 children with a median age of 6.7 years,[20] only 12.3% of them were found to be infected with SARS-CoV-2 after being exposed to persons with confirmed or suspected COVID-19. Even though the exposure is likely more intensive, the children appear to have a lower infection rate than adults (17–69%).

In addition, the infection rate of SARS-CoV-2 seems much lower in the general population without known history of exposure to a confirmed infection. Community-based screenings for SARS-CoV-2 infection have estimated an

Table 7.1 Infection rate after close exposure in different outbreaks.[19,11,108,37]

Authors	Population	Place	Exposed	Cases		Asymptomatic rate (%)
				n	%	
Mizumoto *et al.*	Adults	*Diamond Princess* cruise ship, Japan	3711	712	19.2	46.5
US Navy	Crew member	USS *Theodore Roosevelt*	4954	856	17.3	60.0
Sutton *et al.*	Obstetric patients	New York, NY, USA	214	33	15.4	87.9
Chou *et al.*	Homeless	Los Angeles, CA, USA	178	43	24.2	63.8
Baggett *et al.*	Homeless	Boston, MA, USA	408	147	36.0	87.8
Arons *et al.*	Elderly	Nursing home, Washington State, USA	89	57	64.0	58.7[a]
So *et al.*	Inmates	USA: Alaska, North Carolina, Virginia, Ohio	4963	3277	69.8	96.0
Oran and Topol	Crew member	Aircraft, Charles De Gaulle airport, France	1760	1046	54.9	50.0
Ing *et al.*	Cruise ship	Argentina	217	128	59.0	81.3
Lu *et al.*	Children	Wuhan, China	1391	171	12.3	22.8

[a]Rate of presymptomatic patients; the majority developed symptoms later.

infection rate of 0.8% in Ireland, and 1.7% and 1.8% in the residents of Indiana and San Francisco, CA, respectively.[11] The infection rate in community residents was estimated at 1.2–2.6% in Italy.[21]

7.2.3 Antibody Tests

Antibody testing detects immunoglobins (Ig; antibodies) specific to the SARS-CoV-2 antigen in the body. In response to the invasion of viral antigens, the host first initiates its innate immune response (*i.e.* innate immune cells) and activates adaptive immune response (*i.e.* T-cell activation and B-cell response) to fight the infection through destroying the infected cells and by antibodies (mainly by B-cells) that neutralize the pathogen. Some of these antibodies may provide long-term or even lifelong protective immunity to a re-infection by the same pathogen. Serum antibodies (IgM, IgA, and IgG) tested against SARS-CoV[22] and SARS-CoV-2[23–26] for evidence of early infection have been inconsistent. Approximately 100 manufacturers have developed variable serological tests.[27] On 18 April 2020, the FDA issued an umbrella EUA for multiple SARS-CoV-2 antibody tests, including lateral flow or enzyme-linked immunosorbent assay tests. However, on 21 July 2020, the FDA issued a revocation of the umbrella EUA and now plans to issue individual EUAs.[28] Concerns have been raised about the clinical utility of the serological tests.[27,29] Still, antibody tests can detect recently and even more temporally distal SARS-CoV-2 infections by searching specifically for IgG and IgM[25] as well as IgA.[30]

The population-based seroprevalence of antibody IgG and IgM is valuable for public health strategies and helps describe the distribution of the SARS-CoV-2 infection for a study of risk factors. Therefore, a well-designed serological survey would be necessary to specify the epidemiology of COVID-19 following the pandemic. A survey has been conducted in multiple selected samples in China of the seroprevalence of IgG and IgM antibodies against the SARS-CoV-2 antigens.[31] In confirmed patients with COVID-19 ($n = 242$), the seropositive rate reached 95% for IgG on day 16 and for IgM on day 20 following symptom onset. The seropositive IgG in patients with COVID-19 is consistent with that obtained in patients in the United States.[32] Therefore, IgG can be used as a biomarker to confirm that a recent viral infection has occurred. In the same study, immunoassays on antibodies were assessed in a few community-based samples.[31] The highest seropositive IgG, 3.2–3.9%, was found among healthcare workers, outpatients, and hotel staff workers. However, the IgM positives were relatively lower than the IgG in most samples, except for hemodialysis patients (1.9%) and community residents (0.3%), for whom IgG- and IgM-positive rates were similar. Seropositive IgG has been reported at 1.79% in a US community-based sample.[32]

Population-based seroprevalence surveys have recently provided some robust estimates of the infection rate in the COVID-19 pandemic. A nationwide seroprevalence survey of nearly 36 000 households ($n = 51\,958$) conducted in Spain, one of Europe's most severely affected countries,[33] provides the first estimate of seroprevalence from a large-scale randomly selected sample. The seroprevalence of IgG antibody was ~5% in the country as a whole, but manifested considerable regional variability. In the central region around Madrid, the epicenter of COVID-19 in Spain, the seroprevalence was as high as 10–13%, whereas Barcelona had ~7%, and other areas had 1–3%. Another study, conducted in Switzerland over five consecutive weeks with randomly selected samples, reported a seroprevalence of IgG at 5–11%.[34]

These low rates of seropositive IgG or IgM may indicate a reduced level of exposure to the sources of infection in the populations, probably due to the preventive measures implemented there. They may also indicate that a high proportion of people are still susceptible and that the community is at risk of a future outbreak unless or until a certain proportion of susceptible people have been vaccinated to produce effective "herd immunity". However, it seems unlikely that the natural "herd immunity" will be achievable, because of the use of testing tools yielding quick results and the use of personal protective equipment, as well as information sharing that enables adequate contact tracing, leading to isolation and self-quarantine as needed.

7.3 Clinical Spectrum of COVID-19

The heterogeneity of the clinical spectrum of COVID-19 must be understood in order to fully characterize its clinical epidemiology. The estimates of epidemiological parameters can also be used for mathematical modeling that describes the mode and dynamics of transmission for an infectious disease. As a new disease, defining the epidemiology of COVID-19 will inevitably require more comprehensive research efforts that use population-based representative samples. The numbers of exposures and infections in the population are fundamental for estimating related epidemiological parameters, such as infectivity, transmission, and transmissibility.

7.3.1 Classification by Clinical Type

The clinical types of COVID-19 can be classified as mild, moderate, severe, and critical. As a new disease, there are no long-established guidelines for COVID-19. There has been a trial version of a diagnosis and treatment protocol for coronavirus pneumonia,[18] which has been revised multiple times as knowledge and evidence have accrued. Criteria for diagnosing COVID-19 in children have also been proposed.[35] In addition to positive viral tests and clinical symptoms, definitions of these clinical types are as follows:

a. Mild: clinical symptoms are mild, and there is no sign of pneumonia on imaging.
b. Moderate: an individual presents with a fever and clinical symptoms; radiological findings indicate pneumonia.
c. Severe: individual adults meet any of the following criteria: respiratory distress (≥30 breaths per minute); oxygen saturation <93% at rest;

arterial partial pressure of oxygen/fraction of inspired oxygen ≤300 mmHg, which needs to be adjusted if individuals live at high altitude >1000 m. Infants and children have different diagnostic criteria for severe cases.[35]

d. Critical: cases meet any of the following criteria: respiratory failure that requires medical ventilation; shock; and presentation with other organ failures that require treatment in an intensive care unit (ICU).

The clinical development of COVID-19 can be divided into several stages with multiple outcomes. Exposure to the pathogen is required for an individual to be infected with SARS-CoV-2. Strictly speaking, factors associated with the likelihood of exposure to the pathogen may determine the risk of COVID-19 in general populations. Usually, medical and healthcare professionals or people working in service sectors are at a high risk of infection with SARS-CoV-2, because of their increased likelihood of exposure to confirmed or potential infections. As mentioned previously, population-based surveys have shown that healthcare workers, outpatients, and hotel staff workers have the highest seropositivity of IgG.[31,33]

7.3.2 Asymptomatic Cases

An individual's immune system may modify the course of clinical development after infection by the pathogen. When susceptible people (*i.e.* those without specific immunity) are infected with SARS-CoV-2, the host's body may initiate its innate immune system and then activate its adaptive immune response to fight the viral infection. For some people, this immune response may result in the eradication of the viral pathogen and the development of protective immunity without any symptoms. Individuals who are infected, but remain asymptomatic are unlikely to be identified as cases unless a related test is performed, such as viral nucleic acid testing or a subsequent antibody seropositive result. At present, a case of asymptomatic infection is defined as an individual who tests positive for SARS-CoV-2 but lacks any noted clinical symptoms or abnormalities detected by chest radiological imaging.[35] In the early stage of infection with SARS-CoV-2, some people who test positive for the SARS-CoV-2 viral genome are probably presymptomatic, but may develop symptoms later. Therefore, a follow-up should be conducted for those presymptomatic people to determine whether they remain so.

Asymptomatic infections may account for a considerable proportion of the infections or cases with COVID-19. In the outbreak on the *Diamond Princess* cruise ship, 46.7% of cases were asymptomatic (Table 7.1). A review of 16 out of 17 reports showed that asymptomatic patients might account for between 40–90% of cases.[11] These estimates have been mostly based on outbreaks in special populations, in which some participants may have been predominantly younger to middle-aged people such as crew members or the military. In addition, some infected people may develop mild symptoms but then soon recover with routine or even no medical treatment. According to the statistics of 44 672 confirmed cases in China, mild or moderate cases may account for 81% of total confirmed cases,[36] but it is unclear whether this estimation included entire asymptomatic patients, some of whom might not have had a chance to undergo viral testing.

It should be noted that there are distinct differences between asymptomatic and presymptomatic cases. As defined above, an asymptomatic case is an individual who has a positive nucleic acid test, but lacks any symptoms. Presymptomatic cases are individuals who are incubating the disease and have no symptoms at the time of viral testing, but who develop symptoms later. In a point-prevalence survey of residents in a nursing home facility,[37] 27 of 46 individuals who tested positive showed no symptoms, but 24 of the 27 developed symptoms a few days later, thereby demonstrating only a 6.5% asymptomatic rate among those elderly residents tested positive. A community-based screening in China identified 178 individuals who tested positive, and an additional 60 who lacked any symptoms within 14 days of quarantine.[10] However, continued observation of this population revealed that 23 of the 60 individuals developed symptoms, whereas 37 remained asymptomatic. Some individuals might simply require a more extended incubation period to have an onset of symptoms. In this population-based screening of individuals with suspected exposure and subsequent quarantine, the asymptomatic rate was estimated at 20% (37 out of 178). This estimate is supported by a recent nationwide survey in Spain, indicating that 22–36% of seropositive cases were asymptomatic.[33]

7.3.3 Severe Cases and Comorbidity

As noted earlier, symptomatic cases can be classified as mild, moderate, severe, or critical. Some patients may first present for care with severe symptoms, leading to immediate hospitalization, and then be diagnosed with COVID-19 through viral testing. This pattern was seen in the early epidemic in many countries or regions where the viral testing capacity was limited, and testing was only performed for individuals with severe symptoms who were suspected of infection. In this setting, most of the identified patients were in a severe or critical condition. In other settings, infected individuals identified by active viral testing in suspected people who had close contact with individuals with confirmed COVID-19 may have a high proportion of mild or moderate cases. It was estimated that 13.8% were severe, and 4.7% were critical cases, while mild or moderate cases accounted for 81% of overall patients.[36] These estimates might not have fully considered all asymptomatic patients, who were unknown in the community since population-based widespread testing was not performed.

Individuals who develop severe and critical cases of COVID-19 predominantly have underlying chronic health conditions. In a clinical study of 1099 patients from a national sample in China (median age 47 years, interquartile range (IQR) 35–58 years), in which severe cases were defined at hospital admission by the American Thoracic Society guideline for community-acquired pneumonia,[38] severe cases seemed to be more likely to have comorbidities (39%) than non-severe patients (21%). Specifically, severe cases had about a two- or three-fold higher likelihood of presenting with diabetes (16.2% *versus* 5.7%), hypertension (23.7% *versus* 13.7%), and coronary heart disease (5.8% *versus* 1.8%), and a five-fold higher likelihood of having chronic obstructive pulmonary disease (COPD). Conversely, they were 75% less likely to have a hepatitis B viral infection relative to non-severe patients. Of note, the overall prevalence of hypertension and diabetes in patients with COVID-19 seemed not to be highly elevated, as compared to the prevalence (47%) in the middle-aged population.[39] Since these comorbidities of cardiovascular disease and hypertension are prevalent and associated with older age,[40] the elderly who are infected with SARS-CoV-2 are likely to be severely or critically ill and may have a high fatality rate.[41] However, it is important to note that hospital-based patients constitute a highly selected population. Therefore, findings from an observational study of this nature are likely to be confounded or subject to selection bias.[42]

7.3.4 Clinical and Immunological Response

The severity of individuals infected with SARS-CoV-2 may vary. Some people may lack any clinical response (*i.e.* symptoms) but can have viral shedding and immunological response such as development of antibodies. The COVID-19 subtype may have implications for the transmission of SARS-CoV-2. Compared with mild cases, severe cases of COVID-19 tend to have a high viral load and a longer viral shedding period.[43] A recent study[10] showed that patients with positive viral tests and clinical symptoms had a median duration of viral shedding of 19 days (IQR 15–26 days), but the period for asymptomatic patients was 14 days (IQR 9–22 days). There were no significant differences in the RT-PCR cycle threshold value in the two target genes tested (ORF1b and N). IgG antibody testing positivity was also similar at 83.8% and 81.1% for symptomatic and asymptomatic patients, respectively. However, the mean levels of IgG seemed higher in symptomatic patients both in their acute and convalescent phase, which was defined as 8 weeks after patients were discharged from the hospital. More surprisingly, in the follow-up testing period that was based on an 80% response rate, >90% of patients demonstrated a decline in the level of IgG during their convalescent relative to their acute phase. The IgG was expected to produce long-term immunity to the viral pathogen. This finding provides additional evidence that a herd immunity strategy to combat COVID-19 may be dubious, and also that herd immunity has been shown to affect prognostic outcomes.[44] These evidences may provide information on the utility and viability of developing a vaccine that produces long-term or lifelong immunity.

In addition, the study performed a systematic examination of cytokine biomarkers in two different types of patients.[10] 18 serum cytokines were significantly elevated in symptomatic relative to asymptomatic patients. The top five cytokines include tumor necrosis factor-related apoptosis-inducing ligand (TRAIL), macrophage colony-stimulating factor, growth-regulated oncogene-α (GRO-α), granulocyte colony-stimulating factor, and interleukin (IL)6, of which IL6 has been proposed as an inflammatory marker for severe cases and a poor prognosis of COVID-19. However, compared with non-infected individuals, asymptomatic patients have elevated plasma levels of several other cytokines, including stem cell factor, IL-13, IL-12, and leukemia inhibitory factor.

Previous clinical studies of hospital-based patients have indicated that critically ill patients tend to have increased white blood cells (*e.g.* neutrophils), lymphopenia, and changes in many other clinical parameters in laboratory tests such as elevated levels of C-reactive protein.[38,45] However, the clinical utility of these parameters is uncertain. A regression analysis showed that they were not significantly associated with severity or death after some covariates were adjusted, but older age remained significantly associated with critical illness or death,[45] which was likely caused by underlying chronic conditions or compromised immune systems.

7.4 Regional Variation in Clinical Presentation

7.4.1 Clinical Manifestation

The clinical presentation of hospitalized patients appears to be heterogeneous. A study examined the difference in clinical symptoms and chest CTs between patients from nine hospitals[46] and found that fever was the most frequent symptom (84.8%). However, there was a substantial disparity between patients in Wuhan (92%, n = 338) relative to those outside the city (78%, n = 286). Dyspnea was also more common among Wuhan patients than outside (21% *versus* 3.8%). These regional differences in symptoms were likely due to the overwhelming number of patients who sought treatment at medical facilities in Wuhan, which may have delayed hospital admission in the epicenter. Additional heterogeneity in the clinical presentation seems to relate to various neurological manifestations; acute cerebrovascular disorder, impaired consciousness, cranial nerve manifestation, and autoimmune diseases were often reported in severe cases with COVID-19.[47]

A review of 67 studies showed that SARS-CoV-2 and other human coronaviruses have been associated with multiple neurological disorders, including polyneuropathy, stroke, and encephalitis.[48] Patients with COVID-19 tend to have headaches, myalgia, fatigue and dizziness, and chemosensory dysfunctions, suggesting that the nervous system is involved. The coronaviruses SARS-CoV-2, SARS-CoV and HCoV-NL63 share the angiotensin-converting enzyme 2 (ACE2) receptor, which facilitates the entry of the virus into the host and is expressed in the brain renin–angiotensin system in transgenic mice.[49] Mice in which the *Ace2* gene has been deleted have shown diverse phenotypes, including hypertension,[50] suggesting that ACE2 may regulate blood pressure or disease involving the autonomous nervous system.

7.4.2 Chemosensory Dysfunction

Chemosensory dysfunction is prevalent in patients with mild to moderate COVID-19. In a few related studies,[51–57] two larger-sized studies have reported a consistently high rate of chemosensory dysfunction in ~80% of patients with mild or moderate COVID-19 in European countries (Table 7.2). However, one study showed that the combined disturbance of olfactory and gustatory function was ~79%,[57] and isolated olfactory and gustatory dysfunction were 8.9% and 12.1%, respectively.[57] A study of the neurological manifestations in 214 patients with COVID-9 in China found that olfactory and gustatory dysfunction were 7.1% and 6.3%, respectively.[55] Understanding this considerable difference in chemosensory disturbances between populations may require a formal investigation. Recently, a national representative seroprevalence survey in Spain found that among 7273 individuals with anosmia, the seropositive rate was 15–19%, much higher than the 5% seropositive rate in the country as a while, suggesting that chemosensory dysfunction is associated with the infection with SARS-CoV-2.

The pathogenesis of chemosensory dysfunction is not clear, but might be associated with the receptors to which the viral pathogen SARS-CoV-2 binds. As mentioned previously, ACE2 is one of the significant receptors for the viral binding. ACE2 inhibitors and angiotensin II blockers may have a side-effect of altered chemosensitivity.[58,59] This side-effect typically disappears after drug discontinuation. Recent evidence shows that loss of smell might be due to the key genes (*ACE2* and *TMPRSS2*) expressed in non-neuronal support cells, stem cells, and perivascular cells.[60] SARS-CoV-2 also uses sialic acids linked to host cell surface gangliosides,[61] highly important molecules in immunology and which have been considered possible therapeutics for neurodegenerative disorders.[62] Sialic acid is a fundamental component of the salivary mucin that protects the glycoprotein conveying gustatory molecules

Table 7.2 Chemosensory dysfunction in patients with COVID-19.[54–60]

Author	Regions	Study type	Subjects (n)	Anosmia	Dysgeusia
Lechien	Europe	Multicenter	417	85.6%	88.0%
Yan	US	Single center	59	68%	71%
Luers	Germany	Single center	73	74%	69%
Vaira	Italy	Multicenter	256	79.3%	79.3%[a]
Passali	Italy	Hospital	NA	25–29%	25–29%
Giacomelli	Italy	Single center	59	>33.9%	>33.9%
Mao	Asia	Single center	214	7.1%	6.3%

[a]For olfactory and gustatory dysfunction, the isolated olfactory disorder was 8.6%, and isolated gustatory disorder was 12.1%.

inside the taste pores. When the host is infected with SARS-CoV-2, sialic acid reduction may increase the gustatory threshold.[59] Moreover, an *in silico* study indicates that the Spike protein of SARS-CoV-2 could use a dual strategy by binding ACE2 and interact with the sialic acid receptors of the host cells in the upper airways,[63] which might be a possible explanation for the observation of a high rate of disturbance in both olfactory and gustatory function in patients with COVID-19.[57]

In addition, inflammatory responses could contribute to the development of olfactory dysfunction in patients with COVID-19. Individuals infected with SARS-CoV-2 may increase the production of proinflammatory cytokines in the nasal epithelium, and then undergo nasal olfactory epithelium destruction, which could lead to olfactory dysfunction. A recent study of three cases with anosmia COVID-19 and three healthy controls has shown a trend that the IL-1B and tumor necrosis factor-α are elevated in the olfactory epithelium by infection with SARS-CoV-2.[64] However, further investigation with a larger sample is needed.

7.4.3 Common Symptoms

Clinical manifestation of symptoms can be used to make an early diagnosis of COVID-19, especially when there is limited viral testing availability. Early diagnosis can help initiate timely medical treatment of infected people and prompt preventive health measures to protect susceptible people. However, clinical manifestation varies greatly, depending on how or what medical information is collected at presentation. Case records may include a systematic review of a patient's multiple systems, but the collection of detailed information on clinical symptoms often varies with studies. The following subsections describe related studies conducted in three significant epicenters of the COVID-19 pandemic, namely China and other Asian countries, the United States, and Europe; the reports selected related to clinical manifestations in their respective earliest cohorts of patients and followed one large-sample study in each region.

7.4.3.1 Asia

A study reported clinical presentations in the first 41 patients in Wuhan, China,[65] of whom 73% were men and whose median age was 49 years. The most common symptoms at the onset of disease were fever (98%), cough (76%), myalgia or fatigue (44%), and sputum production (28%). Approximately 55% of patients developed dyspnea within a median of 8 days (IQR 5–13 days), and 63% had lymphopenia. All patients had pneumonia with abnormal imaging on chest CTs. 29% of these patients developed acute respiratory distress syndrome (ARDS); 32% were treated in ICUs; and 15% died. This case-fatality rate was relatively high in the first cohort, considering that the majority of patients were middle-aged.

Additional clinical studies seem to report consistent clinical symptoms in patients with COVID-19. One study was based on 214 patients in Wuhan, China,[55] of whom there were 88 severe and 126 non-severe cases; women accounted for 59% of these patients, and individuals aged >50 years constituted 58% of them. Another study included 1099 patients in a national sample,[38] of whom the majority were male and had a mean age of 47 years. Both reports showed that fever and cough were the main symptoms (Figure 7.1). Additionally, patients from Wuhan showed more symptoms related to the nervous system (36%) and anorexia (32%), while in the national sample, the more prevalent symptoms were fatigue (38%) and sputum production (34%). In addition, a report described 62 patients from seven hospitals in Zhejiang Province.[66] This cohort of patients seemed younger, with a median age of 41 years, and only one among them was admitted to the ICU. The most common symptoms in this group were cough (81%), fever (77%), expectoration (56%), myalgia or fatigue (52%), and headache (32%). The median duration from exposure to onset of symptoms was 4 days. All these patients seemed to be relatively mild cases.

In southeast Asia, a report described the first 11 patients hospitalized in Thailand.[67] Nine had a cough, malaise, and sore throat in the first 4 days after admission, and a few patients had a fever (>38 °C) in the first 2 days. Another study reported the first 21 confirmed cases with COVID-19 in India.[68] Their mean age was 40 years; 62% were men; and 62% had traveled within the past 30 days, of whom two-thirds had traveled to Italy. The most common symptoms reported were fever (43%) and cough (43%), followed by sore throat, headache, and shortness of breath.

In northeast Asia, South Korea reported the first 28 confirmed COVID-19 cases with hospitalization,[69] with a mean age of 40 years (range 20–73 years), of which 15 were men. The most common symptoms were cough (29%),

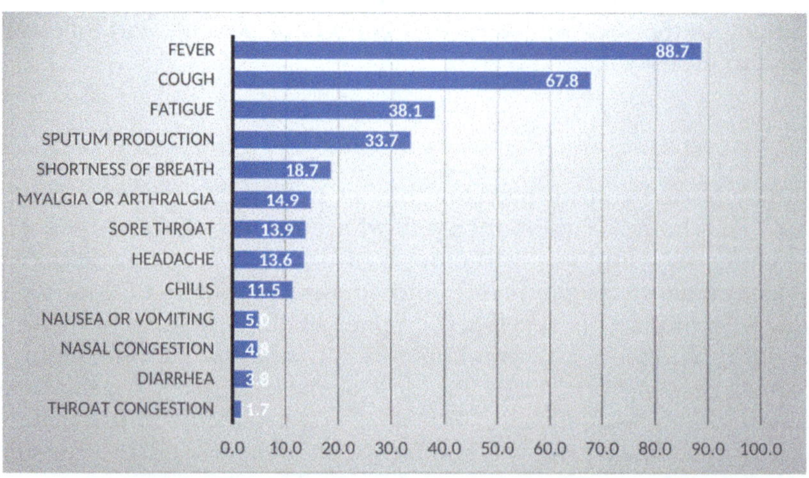

Figure 7.1 Clinical presentations of 1099 patients with COVID-19 in China.[55,38]

sore throat (29%), fever (25%), and headache (25%). The symptomatic rate seemed to be lower in Korea, probably due to their more active viral testing strategies, which have been considered a model for other countries actively seeking to identify individuals infected with SARS-CoV-2. Patients are therefore identified early, with milder and fewer symptoms. A study reported the clinical presentations of 70 patients with COVID-19 in the outbreak on the *Diamond Princess* cruise ship,[70] including those of Japanese (46%), other Asian (34%), and European ancestry (20%), with a median age of 67 years. Of these, 60% (43 out of 70) had pneumonia. The common symptoms were fever (64%), cough (54%), fatigue (24%), shortness of breath (17%), and diarrhea (14%).

7.4.3.2 USA

Patients with COVID-19 may show differences in clinical manifestations by age and severity. In the first 12 patients reported in the United States,[71] all had mild to moderately severe illness, their median age was 53 years, and 67% were men. The most common symptoms were cough (67%), fever (58%), fatigue (42%), and headache (25%). Another study reported 16 hospitalized elderly patients,[72] with a median age of 67 years, of whom 75% were men. The common symptoms were a history of fever (94%), dry cough (88%), dyspnea (81%), and fatigue (50%), which seemed more prevalent than that in the first cohort, probably because of their older age.

A clinical study retrospectively analyzed a hospitalized case series of 1000 patients with COVID-19 in New York City.[73] These patients were classified into three groups: those first seen at the emergency department (n = 150), those in hospital (but not in the ICU; n = 614), and those in the ICU (n = 236), with a median age of 63 years (range 50–75 years), and of whom 60% were men. This cohort of patients was ethnically diverse, including Hispanics or Latinx (25%), others (22%), not specified (19%), Black (18%), White (14%), and Asian (2%).

The data show that cough and fever were the main symptoms (>70%) across all three groups of patients, but that dyspnea was prevalent (73%) in patients who were in the ICU (Figure 7.2). Altogether, 46% of patients in the ICU died and 14% of others died in hospital.

7.4.3.3 Europe

Two early reports in Europe showed similar clinical symptoms in older adults and the elderly based on each country's first cohort of patients. At the beginning of the outbreak in Europe, a study reported the clinical presentations and outcomes of the first cohort of 46 patients with COVID-19 from a single hospital in northern Italy.[74] Most of the sample were men (65%), and elderly patients (age >65 years) accounted for 54% of cases. The common symptoms were fever (87%), cough (33%), and dyspnea (22%).

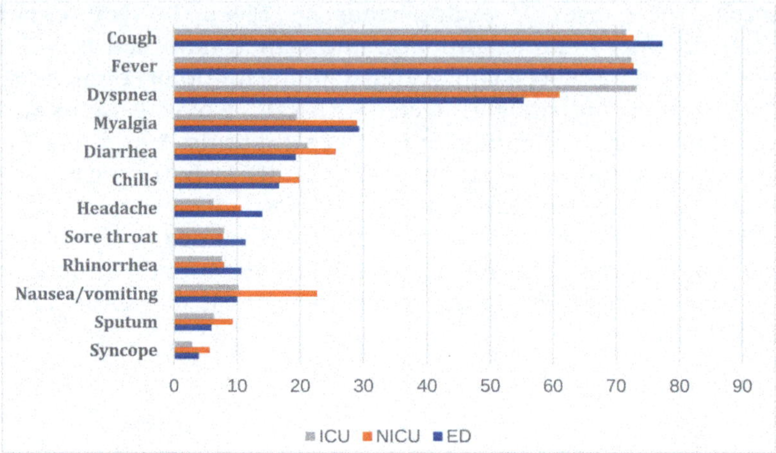

Figure 7.2 Clinical presentation of 1000 older patients with COVID-19 in New York City, USA. NICU; hospitalized, but not in the ICU, ED: emergency department.

A study described the first cohort of 96 patients diagnosed with COVID-19 in the United Kingdom;[75] the median age was 75 years (IQR 59–82 years); and 63% were men. Common symptoms were cough (74%), fever (72%), shortness of breath (43%), and myalgia (14%). At the time of the report, 21% of patients had died, 43% were discharged, and 37% remained in the hospital for treatment.

Later, a large clinical study reported the clinical characteristics of 1420 patients with COVID-19.[40] The study samples included patients in three different age groups: younger (15–39 years, 56%, $n = 793$), middle-aged (40–59 years, 38%, $n = 551$), and older (≥60 years, $n = 76$). All were mild to moderate cases. Data were collected from 18 hospital centers in five countries in Europe. The clinical symptoms specified were different across age groups. Interestingly, the major symptoms among younger patients were loss of smell, nasal obstruction, headache, and rhinorrhea, cough, and myalgia (Figure 7.3), whereas headache, loss of smell, myalgia, cough, and nasal obstruction were the major symptoms in the middle-aged subsample. Older patients mainly had a fever, loss of appetite, and cough, consistent with those patients of similar age in New York and the other first cohorts in Italy and the UK.

In summary, while the clinical symptoms of patients with COVID-19 vary with age, fever and cough were two consistent symptoms for older patients. Neurological symptoms were common, particularly among younger patients and those with mild to moderate severity. People with younger age or whose cases were detected early may have had relatively mild and fewer symptoms, predicting a better prognostic outcome. Older patients in Europe and the United States had a 20% death rate.[40,72]

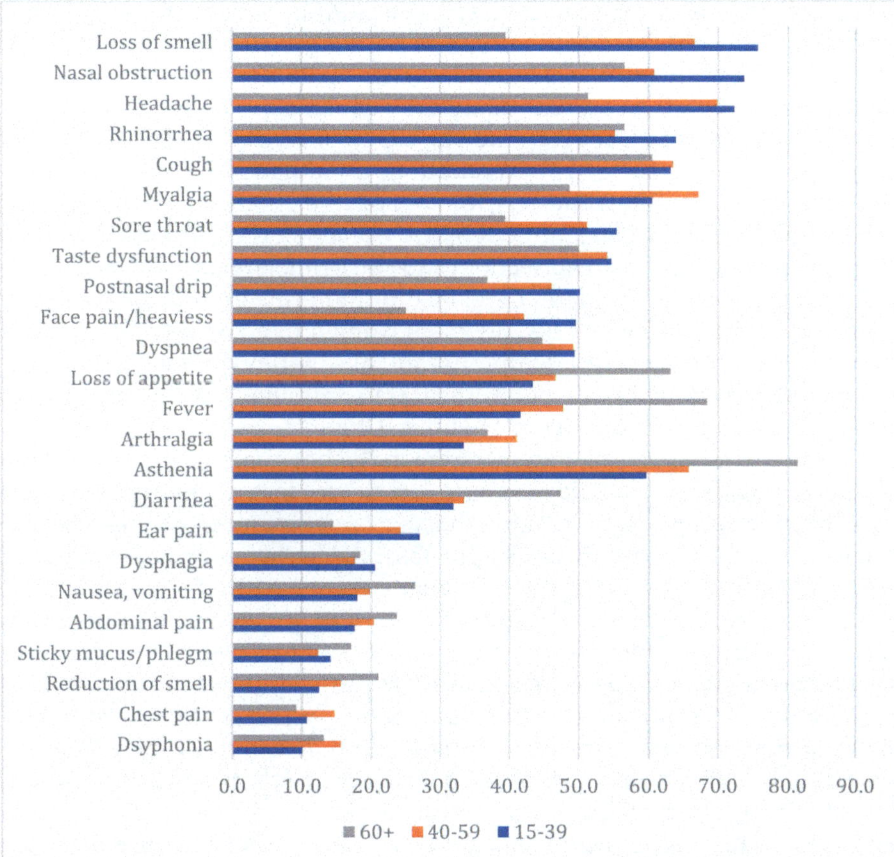

Figure 7.3 Clinical presentation of 1420 patients with COVID-19 by age in Europe. Data source: 18 hospital centers in five countries, including France (Paris, Marseille), Italy (Milan, Verona, Naples, Genova, Florence, and Forli), Spain (Sevilla, Santiago de Compostela, San Sebastian), Belgium (Mons, Brussels, Charleroi, Saint Ghislain) and Switzerland (Geneva).

7.5 Population Distribution

7.5.1 Pregnancy and Pregnant Women

Pneumonia caused by infectious agents is a prevalent and significant cause of death in pregnant women. During the 1918–1919 influenza pandemic, the case-fatality rate was estimated at 27% in pregnant women and was even higher among those in the third trimester.[76] Viral infections during the perinatal period are shown to harm pregnancy outcome and fetal development.[77] A 2004 study reported on 12 pregnant women in Hong Kong where were aged 27–44 years,[78] comprising five health workers and seven others in the community, who were infected with the SARS-CoV virus. These patients' diagnoses were confirmed with RT-PCR or antibody testing and chest X-ray or CT evidence for pneumonia. All patients with SARS pneumonia had a fever and myalgia, and other presentations such as malaise (11 out of 12), chills and rigor (11 out of 12), cough (nine out of 12), headache (six out of 12), and lymphopenia (eight out of 12). In addition, three patients died (one in the first and two in the third trimester); four out of seven in the first trimester had a spontaneous miscarriage; four out of five women in the third trimester had a delivery through a caesarean section; and two women recovered and continued their pregnancies, but later developed intrauterine growth restrictions.

During late gestational weeks, pregnant women infected with SARS-CoV-2 seem to develop fewer severe symptoms and have little evidence of intrauterine infections and maternal–fetal vertical transmission. A study reported six pregnant patients with COVID-19 in Hubei, China,[79] aged 25–31 years and at 38–41 gestational weeks; all patients were in the hospital for delivery and did not have a fever, but some did develop a low-degree fever 24 hours after delivery. Five patients had chest imaging that featured ground-glass opacities. No other symptoms or complications were observed in the pregnant women and newborns. Another report of three pregnant women with COVID-19,[80] at late gestational weeks and in the hospital for delivery, showed no abnormalities in either mothers or their newborns, but one patient developed a low-grade fever before delivery, and two did so after delivery. All three newborns tested negative for the SARS-CoV-2 infection. A retrospective review on nine hospitalized pregnant women with COVID-19,[81] all delivered in their third trimester through a caesarean section, showed that mothers and newborns did not exhibit any abnormalities. Various biospecimens from mothers and newborns tested for nucleic acids of SARS-CoV-2 showed no evidence for intrauterine vertical transmission. Other reports did not show any complications in pregnant women with COVID-19 and newborns, and all newborns tested negative.[82,83]

However, some pregnant women may develop symptoms and have preterm delivery after being infected with SARS-CoV-2. A study described 77 hospitalized pregnant women from 12 maternity hospitals in Italy,[84] of whom 14 (18%) had severe disease, and all had higher gestational body mass indices and increased heart and respiratory rates. Two-thirds of the 77 patients were in the third trimester and 84% had symptoms; nine women (13%) had a preterm delivery, and all premature newborns were admitted to the neonatal ICU.

7.5.2 Infants and Pediatric Patients

While there is no evidence for maternal–fetal vertical transmission, infants are susceptible to SARS-CoV-2. Infants are likely to be infected through close contact with infected family members. Nine infants were diagnosed with COVID-19 under 12 months,[85] five of whom had at least two family members infected with SARS-CoV-2; four had a fever, two had upper respiratory symptoms, and one was asymptomatic. All infected infants were confirmed positive with at least two tests. A 15-day-old neonate, reportedly the youngest patient with COVID-19, presented with fever, lethargy, and respiratory distress, but without a cough.[86] All infected infants seemed to have recovered well.

Children account for a small proportion of patients with COVID-19, probably due to their reduced likelihood of both exposure and testing, because they tend to be asymptomatic or have milder symptoms. Based on a nationwide sample of 1011 patients with COVID-19, a study showed that children aged <14 years accounted for ~1% of overall patients.[87] Later, a national sample of 72 314 COVID-19 patients estimated that <1% were aged <10 years.[88] In the general transmission dynamics of infectious diseases, an infection may start in the community and be transmitted to infants or preschool children through close contact with family members.[89] While an infected school-age child could spread the infection to other children or cause an outbreak in a school, this seems to have been less likely during the pandemic to date, and few school outbreaks have been reported. The proportion of cases with COVID-19 in children seems lower in China and Italy than those in Australia and Korea,[90] probably due to the respective testing policies.

Population-based data suggest that children are less likely to develop COVID-19 after exposure to the viral pathogen. A study has shown that *ACE2* gene expression is significantly lower in children aged <10 years than in those aged 10–17 years, 18–25 years, and ≥25 years.[91]

In a population-based assessment of 1390 children with a median age of 6.7 years in China,[20] 12% (171 out of 1391) were found to be infected with SARS-CoV-2 after exposure to people with confirmed or suspected COVID-19 (Table 7.1). This infection rate in children was lower than the 19% on the *Diamond Princess* cruise ship[19] and much lower than the 64% in the nursing home facility in Washington State.[92] Of note, children are generally likely to have more intensive exposure to infected family members than occurs among passengers on a cruise ship and a community of adult residents.

Of the 171 confirmed cases in children in the study described,[20] the most common symptoms noted were cough (48%), pharyngeal erythema (46%), and fever (42%) during the illness; the common radiological findings were bilateral ground-glass opacities (33%). Of three patients who required ICU care, all had co-existing conditions, including hydronephrosis, leukemia, or intussusception. Additionally, six of 171 patients (4%) had lymphopenia, and among 39 (23%) asymptomatic cases, 27 children showed no abnormalities in chest radiological imaging.

7.5.3 Asymptomatic Cases and Clinical Severity in Children

Children and adolescents with COVID-19 have less severe inflammatory biomarkers and clinical manifestations than do adults. The asymptomatic rate is estimated at 30–50%. In a report on 25 children tested positive for SARS-CoV-2, with a median age of 11 years (range 0.6–17 years), all of which were secondary to a family cluster outbreak, eight were asymptomatic, four were very mild, and 13 were moderate. However, all children manifested abnormal chest CT imaging. The most common symptoms were cough (13 out of 25) and fever (six out of 25), and the majority of children had normal lab tests, such as white blood cell counts at hospital admission (21 out of 25) and $CD8^+$ T-lymphocyte counts (23 out of 25). In a report of 36 children with COVID-19,[93] 17 were asymptomatic and the others were symptomatic, mainly manifesting fever, cough, or diarrhea. It has also been reported that many children infected with other coronaviruses such as SARS-CoV and MERS-CoV were asymptomatic.[94]

Children with COVID-19 seem less likely to be severe or critical cases. Based on all reported cases of 2143 children (731 confirmed and 1412 suspected) with COVID-19 in China,[35] a study found that the proportion of combined severe and critical cases constituted 2.9% (21 out of 731) of all confirmed cases and 7.4% (104 out of 1412) of all suspected cases. The difference in the rate of severe cases in children between the two groups of cases seemed to be associated with the proportion of the sample from Hubei, in which patients in children from Hubei accounted for 31% and 53.3% of confirmed and suspected cases, respectively.

As described earlier, a trial version of diagnostic criteria for COVID-19 in adults has been developed,[18] but children might require a different set of diagnostic criteria. In one study,[35] a suspected case of COVID-19 in a child was confirmed when they manifested any two of the following three criteria: (1) a fever or respiratory or digestive symptom or fatigue; (2) lab tests indicating that white blood cell counts were normal or that there was a decreased level of lymphocyte count or an increased level of C-reactive protein; and (3) abnormal X-ray imaging. A confirmed case was determined when viral testing indicated a positive for SARS-CoV-2, with either RT-PCR or genome sequencing in a nasopharyngeal swab or blood sample.

Even when children are critically ill with COVID-19, the fatality rate seems to be relatively low. A study examined the clinical presentations and treatment outcomes in critically ill children who tested positive for SARS-CoV-2 in a single tertiary children's hospital in the United States,[95] including 67 patients (21 outpatients and 46 inpatients) aged from 1 month to 21 years. Of 13 patients who were critically ill and who were admitted to the pediatric ICU, 10 developed ARDS, four were treated with hydroxychloroquine, six with remdesivir, six with methylprednisolone, and 11 with antibiotics. On day 14, one patient had died, and three were still hospitalized, and all others were discharged from the hospital.

7.5.4 Older Adults

Older adults seem to have a high risk of developing COVID-19 after being exposed to a confirmed infection. After an index case confirmed at a nursing facility in King County of Washington State in the United States,[92] 57 of 89 (64%) residents there tested positive 23 days later (Table 7.1).

Older patients are more likely to become severely or critically ill after infection with SARS-CoV-2. A study examined 203 adult patients and found that the elderly (age >65 years) with COVID-19 had a higher fatality rate at 35% (19 out of 55) than the 5% (seven out of 148) seen in younger adults,[96] probably because the older people had a higher prevalence of comorbidities including hypertension (38% *versus* 15%), diabetes (22% *versus* 3%), cardiovascular diseases (20% *versus* 3%), and COPD (13% *versus* 1%). Critical cases comprised 44% of the sample, which were far more common in the elderly than the 6.8% in the younger adults. The elderly cases with COVID-19 tended to have abnormal lab tests, including total white blood cell counts >10 000 (18% *versus* 3%), lymphocyte count <1000 (82% *versus* 49%), and level of IL6 >7 pg mL^{-1} (40% *versus* 15%). In France, as of 15 March 2020, people aged >75 years accounted for 20% of all confirmed cases, but 79% of the deaths.[97]

7.5.5 Sex Differences

Sex difference have been observed in prognostic outcomes in patients with COVID-19. A report noted a difference in death rate between males (11 out of 86, 13%) and females (6 out of 82, 7%),[98] based on 168 consecutive severe patients, and there were also significant differences between males and females in the symptom of headache (7% *versus* 20%), neutrophil/lymphocyte ratio (4% *versus* 3%), and mean level of C-reactive protein (60% *versus* 29%). According to the mortality surveillance system, males had a significantly excessive mortality rate in Italy across all age groups, from younger patients to the elderly,[99] especially in northern Italy.

The immune response to SARS-CoV-2 differs by sex in patients with COVID-19. In patients with mild or moderate COVID-19,[100] males tend to have elevated levels of plasma innate immune cytokines, chemokines, and monocytes, whereas female patients tend to mount a more robust T-cell activation. Males and females differ in their immune response to foreign antigens and their autoimmune response, mediated by differences in their innate and adaptive immune responses.[101] It has been argued that sex differences in susceptibility to COVID-19 are probably caused by differences in innate immunity, steroid hormones, and genes related to immune response on X-chromosomes, which can explain why women infected with SARS-CoV-2 have lower levels of viral load and inflammation, together with higher levels of activation in immune cells and antibody production.[102]

In addition, a sex bias has been noted for disease susceptibility to the predecessor coronaviruses that caused SARS and MERS.[103] In a mouse model,[104] male mice had a greater susceptibility than females to infection with

SARS-CoV, particularly in subjects at an advanced age. This enhanced susceptibility was accompanied in male mice with increased virus titer, vascular leakage, and alveolar edema, concurrent with more accumulated inflammatory monocyte macrophages and neutrophil in the mice's lungs. In addition, female mice treated with estrogen receptor antagonists were characterized by increased mortality following infection with SARS-CoV, suggesting that estrogen signaling may have a protective effect on mice.

Of note, recent reports show that women are more likely to be infected than men[105] because of upregulated ACE2 caused by estrogen and some genes on the X chromosome that are incompletely silenced. It is important to note that sex differences were not observed in the seroprevalence that was based on a large national representative sample in Spain.[33] Therefore, observed sex differences in the risk of death and infection need to be rigorously investigated to control the effects of potential confounding factors or selection bias, especially given that both gender and sex may unequally affect underlying chronic health conditions as well as the chance of exposure.

7.5.6 Occupation

Occupation affects the risk of developing COVID-19, probably due to the chance of exposure. Healthcare and medical workers are at a high risk of exposure to infected people and therefore are at increased risk of infection with SARS-CoV-2. 2 weeks after the first case of COVID-19 in the Netherlands was confirmed, 127 cases were reported. Nine of these cases were healthcare workers from two Dutch teaching hospitals, and eight had no travel history.[106] This evidence may have prompted a cross-sectional survey conducted in these two hospitals.[106] Among 1353 of 9705 healthcare workers who reported fever or respiratory symptoms, 86 (6.4%) were confirmed with COVID-19, but only three had histories of exposure to an inpatient who was a confirmed COVID-19 case.

China reported 1688 healthcare workers with a confirmed diagnosis of COVID-19.[36] Most cases (60%) were in Wuhan, probably due to the early overwhelming outbreak. Severe and critical cases accounted for 39% of patients by early January 2020. The rates of severe and critical cases then seemed to reduce over time (Table 7.3). A report on 30 medical and health professionals with COVID-19 in one hospital[107] showed that the common symptoms were cough (83%), fever (77%), fatigue (70%), headache (53%), and dyspnea (47%).

It is likely that other occupations in the public service sectors, such as hotel workers, may also be more likely to be exposed to a source of infection and may have a higher likelihood of developing COVID-19. The rate of infection of healthcare workers in the United States was 7.3%, compared with 0.4% among non-healthcare workers.[108] This is consistent with findings from a recent national seroprevalence survey in Spain where healthcare workers had the highest infection rate at 10%,[33] compared with 5% overall.

Table 7.3 Confirmed and severe cases of COVID-19 in medical workers by three different locations.[36]

	Wuhan			Hubei[a]			China[b]		
	Confirmed cases	Severe and critical cases	%	Confirmed cases	Severe and critical cases	%	Confirmed cases	Severe and critical cases	%
1–10 January 2020	18	7	38.9	1	1	100.0	1	1	100.0
11–20 January 2020	233	52	22.3	48	8	16.7	29	1	3.4
21–31 January 2020	656	110	16.8	250	29	11.6	130	10	7.7
After 1 February 2020	173	22	12.7	95	3	3.2	54	3	5.6
Total	1080	191	17.7	394	41	10.4	214	15	7.0

[a]After excluding Wuhan.
[b]After excluding Hubei.

7.6 Case-fatality Rate

An understanding of case-fatality rates can elucidate the severity and challenges that a disease presents and emphasize the urgency of developing treatment strategies. The case-fatality rate of COVID-19 at a specific time point is the proportion of the cumulative number of deaths over the number of cases identified or reported. Case-fatality rates vary significantly around the world. 65 countries reported >6000 cases with COVID-19 (https://coronavirus.jhu.edu/, accessed on 7 June 2020). The countries with the highest fatality rates (Table 7.4) were Belgium (16.2%), France (15.3%), Italy (14.4%), the United Kingdom (14.1%), the Netherlands (12.6%), Mexico (11.7%), Spain (11.2%), and Sweden (10.4%). Notably, all the countries except Mexico are in Europe, which has an aging population. The United States reported a fatality rate of 5.7%; that of other counties were China (5.5%), Japan (5.4%), Brazil (5.3%), Iran (4.8%), and Germany (4.8%). A variety of factors in each country may determine these comparatively low fatality rates.

However, one must be cautious when comparing the fatality rate across countries, because several factors determine the fatality calculation. First, viral testing policy, new case identification, demographic characteristics (*e.g.* the proportion of the elderly and the prevalence of chronic health conditions), and public transportation systems are important determinants of fatality. Second, the cause-of-death classification may constitute a factor in estimating a country's case-fatality rate, insofar as some people who died during the pandemic had significant underlying health conditions. In addition, variations in the fatality rate may depend on a country's strategies for prevention, its healthcare system, its capacity for and the availability of medical care, and the overall capability of a region to respond to the epidemic.

We examined the updated data from another source (https://ncov2019.live/data, accessed 31 August 2020), except that the United States where active testing strategies have been implemented lately, the case-fatality rate has dropped from 5.7 to 3%, ranking 42nd among all countries where more than 100 deceased cases with COVID-19 were reported. The case-fatality pattern remained similar, especially for the top 10 countries with high case-fatality rates. Of note, while Hungary reported <6000 cases, which were not included in Table 7.4, the fatality rate was reported at 10.3%.

As older adults have a high risk of developing COVID-19, the age of a country's population will have a significant impact on its case-fatality rate. In the early stage of the pandemic, the fatality rate was reported at 2% in China, according to the 44 568 confirmed cases as of 11 February 2020. Italy reported a higher fatality rate of 7% as of 17 March 2020.[41] When examining age-specific fatality, we can see that the age-specific case-fatality rate for people aged <70 years in China and Italy was similar, but that there was a significant difference in the elderly aged >70 years (Figure 7.4). By contrast, the fatality rate on the *Diamond Princess* cruise ship was 14% for individuals aged 80–89 years.[109]

Table 7.4 Number of confirmed cases, deaths, and fatality by country.

Rank	Country	Cases	Death	Fatality
1	Belgium	59 226	9595	16.20
2	France	191 102	29 158	15.26
3	Italy	234 998	33 899	14.43
4	United Kingdom	287 621	40 625	14.12
5	The Netherlands	47 780	6032	12.62
6	Mexico	117 103	13 699	11.70
7	Spain	241 550	27 136	11.23
8	Sweden	44 730	4659	10.42
9	Ecuador	43 120	3621	8.40
10	Canada	97 176	7877	8.11
11	Algeria	10 154	707	6.96
12	Ireland	25 201	1679	6.66
13	Romania	20 479	1333	6.51
14	Switzerland	30 965	1921	6.20
15	Indonesia	31 186	1851	5.94
16	Sudan	6081	359	5.90
17	USA	1 938 931	110 481	5.70
18	China	84 187	4638	5.51
19	Japan	17 039	917	5.38
20	Brazil	672 846	35 930	5.34
21	Denmark	12 148	589	4.85
22	Iran	171 789	8281	4.82
23	Germany	185 750	8685	4.68
24	Finland	6981	323	4.63
25	The Philippines	21 895	1003	4.58
26	Poland	26 561	1157	4.36
27	Portugal	34 693	1479	4.26
28	Honduras	6155	250	4.06
29	Austria	16 902	672	3.98
30	Egypt	34 079	1237	3.63
31	Moldova	9700	341	3.52
32	Bolivia	13 358	454	3.40
33	Czechia	9628	327	3.40
34	Guatemala	6792	230	3.39
35	Colombia	38 149	1265	3.32
36	Argentina	22 794	664	2.91
37	Ukraine	27 599	796	2.88
38	Nigeria	12 486	354	2.84
39	India	257 506	7207	2.80
40	Iraq	12 366	346	2.80
41	Norway	8547	238	2.78
42	Peru	196 515	5465	2.78
43	Turkey	170 132	4692	2.76
44	Dominican Republic	19 600	538	2.74
45	Cameroon	7908	212	2.68
46	Morocco	8224	208	2.53
47	Panama	16 004	386	2.41
48	South Korea	11 776	273	2.32
49	Serbia	11 823	249	2.11
50	South Africa	48 285	998	2.07
51	Pakistan	98 943	2002	2.02

(*continued*)

Table 7.4 (*continued*)

Rank	Country	Cases	Death	Fatality
52	Afghanistan	20 342	357	1.75
53	Israel	17 863	298	1.67
54	Armenia	13 130	200	1.52
55	Malaysia	8322	117	1.41
56	Australia	7260	102	1.40
57	Bangladesh	65 769	888	1.35
58	Russia	467 073	5851	1.25
59	Chile	134 150	1637	1.22
60	Azerbaijan	7553	88	1.17
61	Kuwait	31 848	264	0.83
62	United Arab Emirates	38 808	276	0.71
63	Saudi Arabia	101 914	712	0.70
64	Belarus	48 630	269	0.55
65	Ghana	9638	44	0.46
66	Oman	16 882	75	0.44
67	Kazakhstan	12 694	53	0.42
68	Bahrain	14 763	26	0.18
69	Qatar	68 790	54	0.08
70	Singapore	37 910	25	0.07

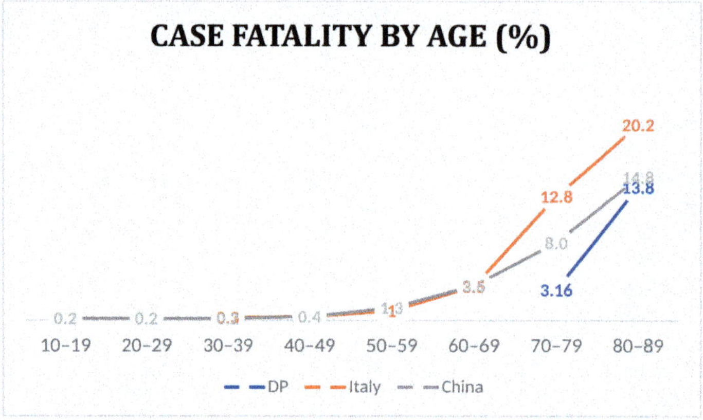

Figure 7.4 Case-fatality rate of people on the *Diamond Princess* (DP) cruise ship, in Italy and in China. *Note*: The total number of deaths were 301 on the *Diamond Princess* cruise ship, the total number of deaths were 1625 from Italy as of 17 March 2020, and 1023 deaths from China as of 11 February 2020.[41,109]

Comorbidities appear to increase the fatality rate in adults significantly. In China, the case-fatality rate was higher in individuals with cardiovascular disease (13%), diabetes (9%), hypertension (8%), chronic respiratory disease (8%), and cancer (8%), relative to those without chronic conditions (1%).[110] This implies that chronic health conditions or advanced

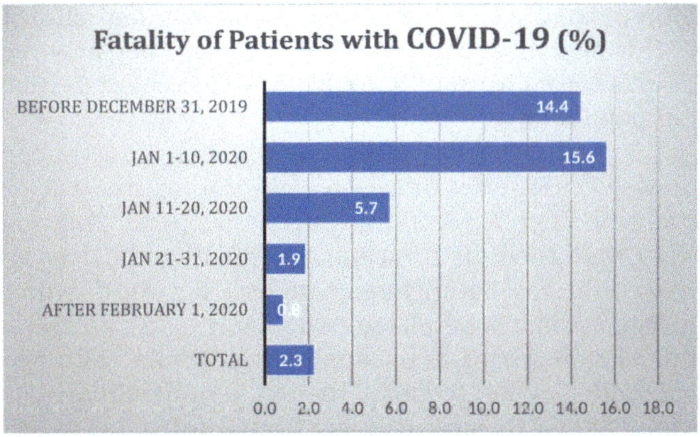

Figure 7.5 Case-fatality rate of patients with COVID-19 in the early stage of the pandemic in China. Data source: Total 44 568 cases and 1008 death from *China CDC Weekly*.[36]

age may increase the fatality risk by approximately four- to nine-fold. On this basis, 45% of adults in the United States are estimated to be at an increased risk of COVID-19, which varies from 20% among those aged 18–29 years to 80% in those aged >80 years, due to the presence of chronic health conditions.[111]

Fatality is affected by underlying health conditions in patients with COVID-19, but it also may reflect the capacity and experience of the local healthcare system to respond to the epidemic within a country. According to the 44 672 confirmed cases by the China Center for Disease Control and Prevention,[36] the fatality rate was 2.9% (n = 33 367) among patients from Hubei and 0.4% (n = 11 305) in patients outside the area. Individuals exposed in Wuhan experienced a fatality rate of 2.8% relative to 1.2% among those living elsewhere. The fatality rate has declined continuously over the period of the epidemic (Figure 7.5). The fatality trend was consistent with that calculated based on all reported patients (n = 82 719) and deaths (n = 4632) in China.[112] Overall, the fatality rate was higher in Wuhan than in other areas of Hubei and elsewhere, partly due to the limited capacity for prompt treatment of critical patients in hospitals at the time when medical facilities were overwhelmed.

7.7 Executive Summary and Conclusions

- Understanding the epidemiology and clinical epidemiology of COVID-19 is critical for effective public health strategies and conducting clinical studies and trials to test vaccines and new therapeutics, which are relied on to control or reduce the health and life cost of individuals infected with the severe acute respiratory syndrome virus (SARS-CoV-2).

- Clearly, the related parameters of epidemiology and clinical epidemiology of this new disease are not defined, and they may require some formal studies that are population-based or use a representative sample of patient populations.
- This chapter aims to review current research, mostly conducted in hospital-based case series or selected clinical reports. The descriptions focused on the rate of infection, clinical spectrum, and manifestation of COVID-19 and their distribution by populations such as age, sex, and occupation, and case-fatality rates across the world, which can help obtain rough estimates of related parameters.
- According to viral testing of nasopharyngeal swabs from people in outbreaks of SARS-CoV-2 infections, the rate of infection in several closed populations after exposure to a confirmed infection source was estimated at 15–30% in general, 24–36% in homeless people, and 70% in inmates; the infection rate was higher in older adults (64%) and lower in children (12%). However, a population-based large-scale serological survey in Italy showed that the infection rate was as high as 10–13% in Italy, one of the epicenters in Europe.
- Individuals infected with SARS-CoV-2 may be asymptomatic (40–60%), while those who develop symptoms can be diagnosed as mild or moderate (80%), severe (14%), or critically ill (5%). In selected reports, severe cases with COVID-19 presented with underlying chronic health conditions, including hypertension, cardiovascular diseases, and coronary heart disease. Both asymptomatic and symptomatic patients could be serologically positive for IgG (>80%) in the acute phase of infection, but the mean level of immunoglobin appears to decline in the infection's convalescent phase.
- Selected reports showed that common symptoms were fever, dry cough, dyspnea, and fatigue, but symptoms differed by region and other factors. Neurological manifestations were common in younger patients with mild to moderate symptoms.
- Infectivity and severity of COVID-19 appear different by demographics such as age, sex, and occupation. Pregnant women and infants are not immune to this new viral pathogen, but there was no convincing evidence of maternal–fetal transmission. Children with COVID-19 are less likely to have severe illness, while older people are more likely to be severe cases, probably due to the underlying chronic conditions and/or compromised immune systems. Seroprevalence surveys support the observation that people in healthcare and service sectors may have an increased likelihood of infection.
- Beyond the existing evidence obtained in the ongoing pandemic, more rigorous studies are needed to dissect the etiology of COVID-19, which may involve both the genomic structure of the pathogen, host-expressing receptors, immunity, and chronic health conditions. These etiological factors may provide important evidence for developing or conducting precision prevention and therapeutic applications.

References

1. P. Zhou, X.-L. Yang, X.-G. Wang, B. Hu, L. Zhang and W. Zhang, *et al.*, A pneumonia outbreak associated with a new coronavirus of probable bat origin, *Nature*, 2020, **579**(7798), 270–273.
2. J.-M. Kim, Y.-S. Chung, H. J. Jo, N.-J. Lee, M. S. Kim and S. H. Woo, *et al.*, Identification of Coronavirus Isolated from a Patient in Korea with COVID-19, *Osong Public Health Res. Perspect.*, 2020, **11**(1), 3–7.
3. K. G. Andersen, A. Rambaut, W. I. Lipkin, E. C. Holmes and R. F. Garry, The proximal origin of SARS-CoV2, *Nat. Med.*, 2020, **26**, 450–452.
4. C. I. Paules, H. D. Marston and A. S. Fauci, Coronavirus infections—more than just the common cold, *JAMA*, 2020, **323**(8), 707–708.
5. M. Lipsitch, D. L. Swerdlow and L. Finelli, Defining the Epidemiology of Covid-19 - Studies Needed, *N. Engl. J. Med.*, 2020, **382**(13), 1194–1196.
6. N. W. Furukawa, J. T. Brooks and J. Sobel, Evidence Supporting Transmission of Severe Acute Respiratory Syndrome Coronavirus 2 While Presymptomatic or Asymptomatic, *Emerging Infect. Dis.*, 2020, **26**(7), e201595.
7. C. Egloff, C. Vauloup-Fellous, O. Picone, L. Mandelbrot and P. Roques, Evidence and possible mechanisms of rare maternal-fetal transmission of SARS-CoV-2, *J. Clin. Virol.*, 2020, **128**, 104447.
8. M. Gandhi, D. S. Yokoe and D. V. Havlir, Asymptomatic Transmission, the Achilles' Heel of Current Strategies to Control Covid-19, *N. Engl. J. Med.*, 2020, **382**(22), 2158–2160.
9. Y. Bai, L. Yao, T. Wei, F. Tian, D.-Y. Jin and L. Chen, *et al.*, Presumed Asymptomatic Carrier Transmission of COVID-19, *JAMA*, 2020, **323**(14), 1406–1407.
10. Q. X. Long, X. J. Tang, Q. L. Shi, Q. Li, H. J. Deng and J. Yuan, *et al.*, Clinical and immunological assessment of asymptomatic SARS-CoV-2 infections, *Nat. Med.*, 2020, **26**, 1200–1204.
11. D. P. Oran and E. J. Topol, Prevalence of Asymptomatic SARS-CoV-2 Infection: A Narrative Review, *Ann. Intern. Med.*, 2020, **173**, 362–367.
12. The National Health Commission, *Guidance to the Control and Treatment for COVID-19*, 2020, http://wwwnhcgovcn/xcs/zhengcwj/202002/a5d6f7b8c48c451c87dba14889b30147/files/3514cb996ae24e2faf65953b4ecd0df4pdf.
13. Centers for Disease Control and Prevention, *CDC 2019-Novel Coronavirus (2019-nCov) Real-Time RT-PCR DIagnostic Panel*, 2020, https://wwwfdagov/media/134922/download.
14. B. Cao, Y. Wang, D. Wen, W. Liu, J. Wang and G. Fan, *et al.*, A Trial of Lopinavir–Ritonavir in Adults Hospitalized with Severe Covid-19, *N. Engl. J. Med.*, 2020, **382**, 1787–1799.
15. J. V. Waller, P. Kaur, A. Tucker, K. K. Lin, M. J. Diaz and T. S. Henry, *et al.*, Diagnostic Tools for Coronavirus Disease (COVID-19): Comparing CT and RT-PCR Viral Nucleic Acid Testing, *AJR, Am. J. Roentgenol.*, 2020, 1–5.
16. Y. Pan, L. Long, D. Zhang, T. Yuan, S. Cui and P. Yang, *et al.*, Potential False-Negative Nucleic Acid Testing Results for Severe Acute Respiratory Syndrome Coronavirus 2 from Thermal Inactivation of Samples with Low Viral Loads, *Clin. Chem.*, 2020, **66**(6), 794–801.

17. P. Gautret, J. C. Lagier, P. Parola, V. T. Hoang, L. Meddeb and M. Mailhe, *et al.*, Hydroxychloroquine and azithromycin as a treatment of COVID-19: results of an open-label non-randomized clinical trial, *Int. J. Antimicrob. Agents*, 2020, 105949.
18. The National Health Commission, Diagnosis and Treatment Protocol for Novel Coronavirus Pneumonia (Trial Version 7), *Chin. Med. J. (Engl).*, 2020, **133**(9), 1087–1095.
19. K. Mizumoto, K. Kagaya, A. Zarebski and G. Chowell, Estimating the asymptomatic proportion of coronavirus disease 2019 (COVID-19) cases on board the Diamond Princess cruise ship, Yokohama, Japan, 2020, *Eurosurveillance*, 2020, **25**(10), 2000180.
20. X. Lu, L. Zhang, H. Du, J. Zhang, Y. Y. Li and J. Qu, *et al.*, SARS-CoV-2 Infection in Children, *N. Engl. J. Med.*, 2020, **382**(17), 1663–1665.
21. E. Lavezzo, E. Franchin, C. Ciavarella, G. Cuomo-Dannenburg, L. Barzon and C. Del Vecchio, *et al.*, Suppression of COVID-19 outbreak in the municipality of Vo, Italy, *Nature.*, 2020, **584**, 425–429.
22. P. R. Hsueh, L. M. Huang, P. J. Chen, C. L. Kao and P. C. Yang, Chronological evolution of IgM, IgA, IgG and neutralisation antibodies after infection with SARS-associated coronavirus, *Clin. Microbiol. Infect.*, 2004, **10**(12), 1062–1066.
23. J. Zhao, Q. Yuan, H. Wang, W. Liu, X. Liao and Y. Su, *et al.*, Antibody responses to SARS-CoV-2 in patients of novel coronavirus disease 2019, *Clin. Infect. Dis.*, 2020, **71**(16), 2027–2034.
24. M. Yan, Y. Zheng, Y. Sun, L. Wang, L. Luan and J. Liu, *et al.*, Analysis of the diagnostic value of serum specific antibody testing for coronavirus disease 2019, *J. Med. Virol.*, 2021, **93**, 441–447.
25. Z. Li, Y. Yi, X. Luo, N. Xiong, Y. Liu and S. Li, *et al.*, Development and clinical application of a rapid IgM-IgG combined antibody test for SARS-CoV-2 infection diagnosis, *J. Med. Virol.*, 2020, **92**(9), 1518–1524.
26. N. Kohmer, S. Westhaus, C. Ruhl, S. Ciesek and H. F. Rabenau, Brief clinical evaluation of six high-throughput SARS-CoV-2 IgG antibody assays, *J. Clin. Virol.*, 2020, **129**, 104480.
27. E. S. Theel, P. Slev, S. Wheeler, M. R. Couturier, S. J. Wong and K. Kadkhoda, The Role of Antibody Testing for SARS-CoV-2: Is There One? *J. Clin. Microbiol.*, 2020, **58**(8), e00797–20.
28. The Food and Drug Administration, *SARS-CoV-2 Antibody Tests-Revocation of EUA*, 2020, https://wwwfdagov/media/140351/download.
29. M. Andersson, N. Low, N. French, T. Greenhalgh, K. Jeffery and A. Brent, *et al.*, Rapid roll out of SARS-CoV-2 antibody testing-a concern, *BMJ [Br. Med. J.]*, 2020, **369**, m2420.
30. H.-q. Yu, B.-q. Sun, Z.-f. Fang, J.-c. Zhao, X.-y. Liu and Y.-m. Li, *et al.*, Distinct features of SARS-CoV-2-specific IgA response in COVID-19 patients, *Eur. Respir. J.*, 2020, 2001526.
31. X. Xu, J. Sun, S. Nie, H. Li, Y. Kong and M. Liang, *et al.*, Seroprevalence of immunoglobulin M and G antibodies against SARS-CoV-2 in China, *Nat. Med.*, 2020, **26**, 1193–1195.

32. A. Bryan, G. Pepper, M. H. Wener, S. L. Fink, C. Morishima and A. Chaudhary, *et al.*, Performance Characteristics of the Abbott Architect SARS-CoV-2 IgG Assay and Seroprevalence in Boise, Idaho, *J. Clin. Microbiol.*, 2020, **58**(8), e00941–20.
33. M. Pollan, B. Perez-Gomez, R. Pastor-Barriuso, J. Oteo, M. A. Hernan and M. Perez-Olmeda, *et al.*, Prevalence of SARS-CoV-2 in Spain (ENE-COVID): a nationwide, population-based seroepidemiological study, *Lancet*, 2020, **96**(10250), 535–544.
34. S. Stringhini, A. Wisniak, G. Piumatti, A. S. Azman, S. A. Lauer and H. Baysson, *et al.*, Seroprevalence of anti-SARS-CoV-2 IgG antibodies in Geneva, Switzerland (SEROCoV-POP): a population-based study, *Lancet*, 2020.
35. Y. Dong, X. Mo, Y. Hu, X. Qi, F. Jiang and Z. Jiang, *et al.*, Epidemiological characteristics of 2143 pediatric patients with 2019 coronavirus disease in China, *Pediatrics*, 2020, **58**(4), 712–714.
36. NCPERE Team, The Epidemiological Characteristics of an Outbreak of 2019 Novel Coronavirus Diseases (COVID-19) — China, 2020, *China CDC Wkly.*, 2020, **2**(8), 113–122.
37. M. Arons, *et al.*, Presymptomatic SARS-CoV-2 Infections and Transmission in a Skilled Nursing Facility, *N. Engl. J. Med.*, 2020, **382**(22), 2081–2090.
38. Guan, *et al.*, Clinical characteristics of coronavirus disease 2019 in China, *N. Engl. J. Med.*, 2020, **382**, 1708–1720.
39. C. Xia and J. Li, Socioeconomic status impact hyertensive risk and treatment among older adults in China, *Global Clin. Transl. Res.*, 2019, **1**(2), 62–68.
40. J. R. Lechien, C. M. Chiesa-Estomba, S. Place, Y. Van Laethem, P. Cabaraux and Q. Mat, *et al.*, Clinical and Epidemiological Characteristics of 1,420 European Patients with mild-to-moderate Coronavirus Disease 2019, *J. Intern. Med.*, 2020, **288**(3), 335–344.
41. G. Onder, G. Rezza and S. Brusaferro, Case-Fatality Rate and Characteristics of Patients Dying in Relation to COVID-19 in Italy, *JAMA*, 2020, **323**(18), 1775–1776.
42. F. Zhang and C. Hughes, A Caution in Association of ABO Blood Group with COVID-19, *Global Clin. Transl. Res.*, 2020, **2**(1), 51–53.
43. Y. Liu, L. M. Yan, L. Wan, T. X. Xiang, A. Le and J. M. Liu, *et al.*, Viral dynamics in mild and severe cases of COVID-19, *Lancet Infect. Dis.*, 2020, **20**(6), 656–657.
44. F. Jung, V. Krieger, F. T. Hufert and J. H. Kupper, Herd immunity or suppression strategy to combat COVID-19, *Clin. Hemorheol. Microcirc.*, 2020, **75**(1), 13–17.
45. F. Zhou, T. Yu, R. Du, G. Fan, Y. Liu and Z. Liu, *et al.*, Clinical course and risk factors for mortality of adult inpatients with COVID-19 in Wuhan, China: a retrospective cohort study, *Lancet*, 2020, **395**(10229), 1054–1062.
46. J. H. Park, W. Jang, S. W. Kim, J. Lee, Y. S. Lim and C. G. Cho, *et al.*, The Clinical Manifestations and Chest Computed Tomography Findings of Coronavirus Disease 2019 (COVID-19) Patients in China: A Proportion Meta-Analysis, *Clin. Exp. Otorhinolaryngol.*, 2020, **13**(2), 95–105.

47. K. Vonck, I. Garrez, V. De Herdt, D. Hemelsoet, G. Laureys and R. Raedt, *et al.*, Neurological manifestations and neuro-invasive mechanisms of the severe acute respiratory syndrome coronavirus type 2, *Eur. J. Neurol.*, 2020, **27**(8), 1578–1587.
48. V. Montalvan, J. Lee, T. Bueso, J. De Toledo and K. Rivas, Neurological manifestations of COVID-19 and other coronavirus infections: A systematic review, *Clin. Neurol. Neurosurg.*, 2020, **194**, 105921.
49. M. F. Doobay, L. S. Talman, T. D. Obr, X. Tian, R. L. Davisson and E. Lazartigues, Differential expression of neuronal ACE2 in transgenic mice with overexpression of the brain renin-angiotensin system, *Am. J. Physiol.: Regul., Integr. Comp. Physiol.*, 2007, **292**(1), R373–R381.
50. N. Alenina and M. Bader, ACE2 in Brain Physiology and Pathophysiology: Evidence from Transgenic Animal Models, *Neurochem. Res.*, 2019, **44**(6), 1323–1329.
51. J. C. Luers, A. C. Rokohl, N. Loreck, P. A. Wawer Matos, M. Augustin and F. Dewald, *et al.*, Olfactory and Gustatory Dysfunction in Coronavirus Disease 19 (COVID-19), *Clin. Infect. Dis.*, 2020, **71**(16), 2262–2264.
52. C. H. Yan, F. Faraji, D. P. Prajapati, C. E. Boone and A. S. DeConde, Association of chemosensory dysfunction and COVID-19 in patients presenting with influenza-like symptoms, *Int. Forum Allergy Rhinol.*, 2020, **10**(7), 806–813.
53. J. R. Lechien, C. M. Chiesa-Estomba, D. R. De Siati, M. Horoi, S. D. Le Bon and A. Rodriguez, *et al.*, Olfactory and gustatory dysfunctions as a clinical presentation of mild-to-moderate forms of the coronavirus disease (COVID-19): a multicenter European study, *Eur. Arch. Oto-Rhino-Laryngol.*, 2020, **277**(8), 2251–2261.
54. A. Giacomelli, L. Pezzati, F. Conti, D. Bernacchia, M. Siano and L. Oreni, Self-reported olfactory and taste disorders in SARS-CoV-2 patients: a cross-sectional study [published online March 26, 2020], *Clin. Infect. Dis.*, 2020, **71**(15), 889–890.
55. L. Mao, H. Jin, M. Wang, Y. Hu, S. Chen and Q. He, *et al.*, Neurologic Manifestations of Hospitalized Patients With Coronavirus Disease 2019 in Wuhan, China, *JAMA Neurol.*, 2020, **77**(6), 683–690.
56. G. C. Passali and A. R. Bentivoglio, Comment to the article "Olfactory and gustatory dysfunctions as a clinical presentation of mild-to-moderate forms of the coronavirus disease (COVID-19): a multicenter European study", *Eur. Arch. Oto-Rhino-Laryngol.*, 2020, **277**(8), 2391–2392.
57. L. A. Vaira, C. Hopkins, G. Salzano, M. Petrocelli, A. Melis and M. Cucurullo, *et al.*, Olfactory and Gustatory Function Impairment in COVID-19 Patients: Italian Objective Multicenter-Study, *Head Neck*, 2020, **42**, 1560–1569.
58. S. Tsuruoka, M. Wakaumi, N. Araki, T. Ioka, K. Sugimoto and A. Fujimura, Comparative study of taste disturbance by losartan and perindopril in healthy volunteers, *J. Clin. Pharmacol.*, 2005, **45**(11), 1319–1323.
59. L. A. Vaira, G. Salzano, A. G. Fois, P. Piombino and G. De Riu, Potential pathogenesis of ageusia and anosmia in COVID-19 patients, *Int. Forum Allergy Rhinol.*, 2020, **10**(9), 1103–1104.

60. D. H. Brann, T. Tsukahara, C. Weinreb, M. Lipovsek, K. Van den Berge and B. Gong, *et al.*, Non-neuronal expression of SARS-CoV-2 entry genes in the olfactory system suggests mechanisms underlying COVID-19-associated anosmia, *Sci. Adv.*, 2020, **6**(31), eabc5801.
61. J. Fantini, C. Di Scala, H. Chahinian and N. Yahi, Structural and molecular modelling studies reveal a new mechanism of action of chloroquine and hydroxychloroquine against SARS-CoV-2 infection, *Int. J. Antimicrob. Agents*, 2020, **55**(5), 105960.
62. I. Mocchetti, Exogenous gangliosides, neuronal plasticity and repair, and the neurotrophins, *Cell. Mol. Life Sci.*, 2005, **62**(19–20), 2283–2294.
63. E. Milanetti, M. Miotto, L. D. Rienzo, M. Monti, G. Gosti and G. Ruocco, *et al.*, In-Silico evidence for two receptors based strategy of SARS-CoV-2, *Front. Mol. Biosci.*, 2021, **8**, 690655.
64. A. Torabi, E. Mohammadbagheri, N. Akbari Dilmaghani, A. H. Bayat, M. Fathi and K. Vakili, *et al.*, Proinflammatory Cytokines in the Olfactory Mucosa Result in COVID-19 Induced Anosmia, *ACS Chem. Neurosci.*, 2020, **11**(13), 1909–1913.
65. C. Huang, Y. Wang, X. Li, L. Ren, J. Zhao and Y. Hu, *et al.*, Clinical features of patients infected with 2019 novel coronavirus in Wuhan, China, *Lancet*, 2020, **395**(10223), 497–506.
66. X. W. Xu, X. X. Wu, X. G. Jiang, K. J. Xu, L. J. Ying and C. L. Ma, *et al.*, Clinical findings in a group of patients infected with the 2019 novel coronavirus (SARS-Cov-2) outside of Wuhan, China: retrospective case series, *BMJ [Br. Med. J.]*, 2020, **368**, m606.
67. W. A. Pongpirul, J. A. Mott, J. V. Woodring, T. M. Uyeki, J. R. MacArthur and A. Vachiraphan, *et al.*, Clinical Characteristics of Patients Hospitalized with Coronavirus Disease, Thailand, *Emerging Infect. Dis.*, 2020, **26**(7), 1580–1585.
68. N. Gupta, S. Agrawal, P. Ish, S. Mishra, R. Gaind and G. Usha, *et al.*, Clinical and epidemiologic profile of the initial COVID-19 patients at a tertiary care centre in India, *Monaldi Arch. Chest Dis.*, 2020, **90**(1), DOI: 10.4081/monaldi.2020.1294.
69. Covid-19 National Emergency Response Center, Epidemiology and Case Management Team and Korea Centers for Disease Control and Prevention, Early Epidemiological and Clinical Characteristics of 28 Cases of Coronavirus Disease in South Korea, *Osong Public Health Res. Perspect.*, 2020, **11**(1), 8–14.
70. H. Kato, H. Shimizu, Y. Shibue, T. Hosoda, K. Iwabuchi and K. Nagamine, *et al.*, Clinical course of 2019 novel coronavirus disease (COVID-19) in individuals present during the outbreak on the Diamond Princess cruise ship, *J. Infect. Chemother.*, 2020, **26**(8), 865–869.
71. Covid-Investigation Team, Clinical and virologic characteristics of the first 12 patients with coronavirus disease 2019 (COVID-19) in the United States, *Nat. Med.*, 2020, **26**, 861–868.
72. S. Aggarwal, N. Garcia-Telles, G. Aggarwal, C. Lavie, G. Lippi and B. M. Henry, Clinical features, laboratory characteristics, and outcomes of patients hospitalized with coronavirus disease 2019 (COVID-19): Early report from the United States, *Diagnosis*, 2020, **7**(2), 91–96.

73. M. G. Argenziano, S. L. Bruce, C. L. Slater, J. R. Tiao, M. R. Baldwin and R. G. Barr, *et al.*, Characterization and clinical course of 1000 patients with coronavirus disease 2019 in New York: retrospective case series, *BMJ [Br. Med. J.]*, 2020, **369**, m1996.
74. M. Colaneri, P. Sacchi, V. Zuccaro, S. Biscarini, M. Sachs and S. Roda, *et al.*, Clinical characteristics of coronavirus disease (COVID-19) early findings from a teaching hospital in Pavia, North Italy, 21 to 28 February 2020, *Eurosurveillance*, 2020, **25**(16), 2000460.
75. J. Tomlins, F. Hamilton, S. Gunning, C. Sheehy, E. Moran and A. MacGowan, Clinical features of 95 sequential hospitalised patients with novel coronavirus 2019 disease (COVID-19), the first UK cohort, *J. Infect.*, 2020, **81**(2), e59–61.
76. D. A. Schwartz and A. L. Graham, Potential Maternal and Infant Outcomes from (Wuhan) Coronavirus 2019-nCoV Infecting Pregnant Women: Lessons from SARS, MERS, and Other Human Coronavirus Infections, *Viruses*, 2020, **12**(2), 194.
77. C. Eppes, M. Rac, J. Dunn, J. Versalovic, K. O. Murray and M. A. Suter, *et al.*, Testing for Zika virus infection in pregnancy: key concepts to deal with an emerging epidemic, *Am. J. Obstet. Gynecol.*, 2017, **216**(3), 209–225.
78. S. F. Wong, K. M. Chow, T. N. Leung, W. F. Ng, T. K. Ng and C. C. Shek, *et al.*, Pregnancy and perinatal outcomes of women with severe acute respiratory syndrome, *Am. J. Obstet. Gynecol.*, 2004, **191**(1), 292–297.
79. S. Chen, E. Liao, D. Cao, Y. Gao, G. Sun and Y. Shao, Clinical analysis of pregnant women with 2019 novel coronavirus pneumonia, *J. Med. Virol.*, 2020, **92**(9), 1556–1561.
80. S. Chen, B. Huang, D. J. Luo, X. Li, F. Yang and Y. Zhao, *et al.*, [Pregnancy with new coronavirus infection: clinical characteristics and placental pathological analysis of three cases], *Zhonghua Binglixue Zazhi*, 2020, **49**(5), 418–423.
81. H. Chen, J. Guo, C. Wang, F. Luo, X. Yu and W. Zhang, *et al.*, Clinical characteristics and intrauterine vertical transmission potential of COVID-19 infection in nine pregnant women: a retrospective review of medical records, *Lancet*, 2020, **395**(10226), 809–815.
82. A. Pereira, S. Cruz-Melguizo, M. Adrien, L. Fuentes, E. Marin and T. Perez-Medina, Clinical course of Coronavirus Disease-2019 (COVID-19) in pregnancy, *Acta Obstet. Gynecol. Scand.*, 2020, **99**(7), 839–847.
83. Y. Liu, H. Chen, K. Tang and Y. Guo, Clinical manifestations and outcome of SARS-CoV-2 infection during pregnancy, *J. Infect.*, 2020, **82**(6), e9–e10.
84. V. M. Savasi, F. Parisi, L. Patane, E. Ferrazzi, L. Frigerio and A. Pellegrino, *et al.*, Clinical Findings and Disease Severity in Hospitalized Pregnant Women With Coronavirus Disease 2019 (COVID-19), *Obstet. Gynecol.*, 2020, **136**(2), 252–258.
85. M. Wei, J. Yuan, Y. Liu, T. Fu, X. Yu and Z. J. Zhang, Novel Coronavirus Infection in Hospitalized Infants Under 1 Year of Age in China, *JAMA*, 2020, **323**(13), 1313–1314.
86. M. Kamali Aghdam, N. Jafari and K. Eftekhari, Novel coronavirus in a 15-day-old neonate with clinical signs of sepsis, a case report, *Infect. Dis.*, 2020, **52**(6), 427–429.

87. W.-j. Guan, Z.-y. Ni, Y. Hu, W.-h. Liang, C.-q. Ou and J.-x. He, *et al.*, Clinical Characteristics of Coronavirus Disease 2019 in China, *N. Engl. J. Med.*, 2020, **382**, 1708–1720.
88. Z. Wu and J. M. McGoogan, Characteristics of and Important Lessons From the Coronavirus Disease 2019 (COVID-19) Outbreak in China: Summary of a Report of 72314 Cases From the Chinese Center for Disease Control and Prevention, *JAMA*, 2020, **323**(13), 1239–1242.
89. Q. Cao, Y. C. Chen, C. L. Chen and C. H. Chiu, SARS-CoV-2 infection in children: Transmission dynamics and clinical characteristics, *J. Formosan Med. Assoc.*, 2020, **119**(3), 670–673.
90. S. H. Choi, H. W. Kim, J. M. Kang, D. H. Kim and E. Y. Cho, Epidemiology and clinical features of coronavirus disease 2019 in children, *Clin. Exp. Pediatr.*, 2020, **63**(4), 125–132.
91. S. Bunyavanich, A. Do and A. Vicencio, Nasal Gene Expression of Angiotensin-Converting Enzyme 2 in Children and Adults, *JAMA*, 2020, **323**(23), 2427–2429.
92. M. M. Arons, K. M. Hatfield, S. C. Reddy, A. Kimball, A. James and J. R. Jacobs, *et al.*, Presymptomatic SARS-CoV-2 Infections and Transmission in a Skilled Nursing Facility, *N. Engl. J. Med.*, 2020, **382**(22), 2081–2090.
93. H. Qiu, J. Wu, L. Hong, Y. Luo, Q. Song and D. Chen, Clinical and epidemiological features of 36 children with coronavirus disease 2019 (COVID-19) in Zhejiang, China: an observational cohort study, *Lancet Infect. Dis.*, 2020, **20**(6), 689–696.
94. P. Zimmermann and N. Curtis, Coronavirus Infections in Children Including COVID-19: An Overview of the Epidemiology, Clinical Features, Diagnosis, Treatment and Prevention Options in Children, *Pediatr. Infect. Dis. J.*, 2020, **39**(5), 355–368.
95. J. Y. Chao, K. R. Derespina, B. C. Herold, D. L. Goldman, M. Aldrich and J. Weingarten, *et al.*, Clinical Characteristics and Outcomes of Hospitalized and Critically Ill Children and Adolescents with Coronavirus Disease 2019 (COVID-19) at a Tertiary Care Medical Center in New York City, *J. Pediatr.*, 2020, **223**, 14–19.e2.
96. T. Chen, Z. Dai, P. Mo, X. Li, Z. Ma and S. Song, *et al.*, Clinical characteristics and outcomes of older patients with coronavirus disease 2019 (COVID-19) in Wuhan, China (2019): a single-centered, retrospective study, *J. Gerontol., Ser. A*, 2020, **75**(9), 1788–1795.
97. J. F. Etard, P. Vanhems, L. Atlani-Duault and R. Ecochard, Potential lethal outbreak of coronavirus disease (COVID-19) among the elderly in retirement homes and long-term facilities, France, March 2020, *Eurosurveillance*, 2020, **25**(15), 2000448.
98. Y. Meng, P. Wu, W. Lu, K. Liu, K. Ma and L. Huang, *et al.*, Sex-specific clinical characteristics and prognosis of coronavirus disease-19 infection in Wuhan, China: A retrospective study of 168 severe patients, *PLoS Pathog.*, 2020, **16**(4), e1008520.

99. P. Michelozzi, F. de'Donato, M. Scortichini, M. De Sario, F. Noccioli and P. Rossi, *et al.*, Mortality impacts of the coronavirus disease (COVID-19) outbreak by sex and age: rapid mortality surveillance system, Italy, 1 February to 18 April 2020, *Euro Surveill*, 2020, **25**(19), 2000620.
100. T. Takahashi, P. Wong, M. Ellingson, C. Lucas, J. Klein and B. Israelow, *et al.*, Sex differences in immune responses to SARS-CoV-2 that underlie disease outcomes, *Nature*, 2020, **588**, 315–320.
101. S. L. Klein and K. L. Flanagan, Sex differences in immune responses, *Nat. Rev. Immunol.*, 2016, **16**(10), 626–638.
102. P. Conti and A. Younes, Coronavirus COV-19/SARS-CoV-2 affects women less than men: clinical response to viral infection, *J. Biol. Regul. Homeostatic Agents*, 2020, **34**(2), 339–343.
103. A. Jansen, M. Chiew, F. Konings, C. K. Lee and L. Ailan, Sex matters - a preliminary analysis of Middle East respiratory syndrome in the Republic of Korea, 2015, *West. Pac. Surveill. Response J.*, 2015, **6**(3), 68–71.
104. R. Channappanavar, C. Fett, M. Mack, P. P. Ten Eyck, D. K. Meyerholz and S. Perlman, Sex-Based Differences in Susceptibility to Severe Acute Respiratory Syndrome Coronavirus Infection, *J. Immunol.*, 2017, **198**(10), 4046–4053.
105. M. C. Gagliardi, P. Tieri, E. Ortona and A. Ruggieri, ACE2 expression and sex disparity in COVID-19, *Cell Death Discovery*, 2020, **6**, 37.
106. M. F. Q. Kluytmans-van den Bergh, A. G. M. Buiting, S. D. Pas, R. G. Bentvelsen, W. van den Bijllaardt and A. J. G. van Oudheusden, *et al.*, Prevalence and Clinical Presentation of Health Care Workers With Symptoms of Coronavirus Disease 2019 in 2 Dutch Hospitals During an Early Phase of the Pandemic, *JAMA Network Open*, 2020, **3**(5), e209673.
107. M. Liu, P. He, H. G. Liu, X. J. Wang, F. J. Li and S. Chen, *et al.*, [Clinical characteristics of 30 medical workers infected with new coronavirus pneumonia], *Zhonghua Jiehe He Huxi Zazhi*, 2020, **43**(3), 209–214.
108. E. S. Barrett, D. B. Horton, J. Roy, M. L. Gennaro, A. Brooks and J. Tischfield, *et al.*, Prevalence of SARS-CoV-2 infection in previously undiagnosed health care workers at the onset of the U.S. COVID-19 epidemic, *BMC Infect Dis*, 2020, **20**, 853.
109. T. W. Russell, J. Hellewell, C. I. Jarvis, K. van Zandvoort, S. Abbott and R. Ratnayake, *et al.*, Estimating the infection and case fatality ratio for coronavirus disease (COVID-19) using age-adjusted data from the outbreak on the Diamond Princess cruise ship, February 2020, *Euro Surveill*, 2020, **25**(12), 2000256.
110. The Word Health Organization, *Report of the WHO–China Joint Mission on Coronavirus Disease 2019 (COVID-19)*, 28 February 2020 (cited 23 June 2020), https://wwwwhoint/docs/default-source/coronaviruse/who-china-joint-mission-on-covid-19-final-reportpdf.
111. M. L. Adams, D. L. Katz and J. Grandpre, Population-Based Estimates ofChronicConditionsAffectingRiskforComplicationsfromCoronavirus Disease, United States, *Emerging Infect. Dis.*, 2020, **26**(8), 1831–1833.
112. X. Deng, J. Yang, W. Wang, X. Wang, J. Zhou and Z. Chen, *et al.*, Case fatality risk of the first pandemic wave of novel coronavirus disease 2019 (COVID-19) in China, *Clin. Infect. Dis.*, 2020, **73**(1), e79–e85.

CHAPTER 8

SARS-CoV-2 Genomics and Host Cellular Susceptibility Factors of COVID-19

FENGYU ZHANG*[a,b] AND MICHAEL D. WATERS[c]

[a]Global Clinical and Translational Research Institute, Bethesda, MD, USA; [b]Beijing Huilongguan Hospital and Peking University Huilongguan Clinical Institute, Beijing, China; [c]Michael Waters Consulting USA, Hillsborough, NC, USA
*E-mail: zhangfy@gcatresearch.org

8.1 Introduction

Coronavirus infectious disease 2019 (COVID-19) is a new respiratory disease with a rapid transmission and high infectivity. Some individuals who become infected may develop acute respiratory disease syndrome (ARDS) and a cytokine storm,[1,2] the sequelae of which can be life-threatening. The pathogen has been defined and classified through whole-genome sequencing as the second severe acute respiratory syndrome coronavirus (SARS-CoV-2) and is 80% identical to the earlier severe acute respiratory syndrome virus (SARS-CoV). SARS-CoV-2 has a high sequence similarity (96%) with a bat coronavirus, but its natural host remains uncertain.[3,4] SARS-CoV was more lethal and has not been reported to cause additional infections in nearly two decades, while SARS-CoV-2 has a rapid human-to-human transmission mode that

The Coronavirus Pandemic and the Future Volume 1: Virology, Epidemiology, Translational Toxicology and Therapeutics
Edited by Michael D. Waters, Alok Dhawan, Tim Marrs, Diana Anderson, Stafford Warren and Claude L. Hughes

Published by the Royal Society of Chemistry, www.rsc.org

enabled it within a short period to cause the ongoing pandemic and global public health emergency. SARS-CoV-2 and COVID-19 will be of great challenge to human medicine and public health in years to come, requiring a significant amount of time and effort to research and develop appropriate prevention and therapeutic measures.

SARS-CoV-2 is the sole etiological factor for COVID-19, but the risk of developing the disease and clinical manifestations are affected by pathogen–host interaction. Exogenous factors or activities associated with a higher likelihood of exposure to the pathogen and endogenous host factors that modulate the viral entry into the host cells may sequentially contribute to the disease risk and finally determine the status of infection and its severity COVID-19. Understanding the pathogen genomic and genetic factors, viral replication cycle, and human host factors may help to develop vaccines and drugs to prevent and treat individuals with COVID-19. This article aims to describe the genomics and genetics of SARS-CoV-2, its replication, and host cellular factors necessary for the viral entry into the host cell to cause the infection and further disease development. It will not consider host factors that may influence the sequelae of the infection and clinical outcomes.

8.2 Coronavirus and Human Diseases

8.2.1 Classification of Coronaviruses

Coronaviruses are enveloped viruses with a single-strand and positive-sense RNA genome that can infect mammals and birds. They are classified into three groups based on genetic and serological evidence. Group I includes human coronavirus (HCoV) 229E and NL63, transmissible gastroenteritis virus of swine (TGEV), and porcine epidemic diarrhea virus (PEDV). Group II includes murine hepatitis virus (MHV), bovine coronavirus, HCoV OC43, SARS-CoV, the Middle East respiratory syndrome (MERS-CoV), porcine hemagglutinating encephalitis virus (PHEV), rat sialodacryoadenitis virus,[5] canine corona-virus, and equine coronavirus.[6] Group III includes avian infectious bronchitis virus (IBV). These coronaviruses have been shown to cause respiratory, enteric, hepatic, and neurological diseases in humans and animals. Even domestic animals are arguably susceptible to SARS-CoV-2.[7,8]

Coronaviruses are currently grouped into four genera with gene sources in bats and avian species (Table 8.1). According to the International Committee on Taxonomy of Viruses, the coronaviruses belong to the family Coronaviridae under the order Nidovirales. They are *Alphacoronavirus*, *Betacoronavirus*, *Gammacoronavirus*, and *Deltacoronavirus* genera.[9] Animal surveillance indicates that bat coronaviruses mainly have gene sources of *Alphacoronavirus* and *Betacoronavirus*, accounting for most coronaviruses.[10] In contrast, avian coronaviruses are gene sources of *Gammacoronavirus* and *Deltacoronavirus*.[11] The coronaviruses can infect different animal species and replicate in the plasma of host cells after infection.

Table 8.1 Classification of coronaviruses. Reproduced from ref. 186 with permission from Global Clinical and Translational Research, Copyright © 2020.

Genus	Abbreviation	Full name	Human CoV
Alphacoronavirus	FIPV	Feline infectious peritonitis virus	
	PRCV	Porcine respiratory coronavirus	
	TGEV	Transmissible gastroenteritis virus of swine	
	Rh-BatCoV HKU2	*Rhinolophus* bat coronavirus HKU2	
	SDAS-CoV	Swine acute diarrhea syndrome coronavirus	
	HCoV-NL63	Human coronavirus NL63	Human
	HCoV-229E	Human coronavirus 229E	Human
	Mi-Bat-CoV 1A	*Miniopterus* bat coronavirus 1A	
	Mi-BatCov HKU8	*Miniopterus* bat coronavirus HKU8	
	Sc-BatCoV 512	*Scotophilus* bat coronavirus 512	
	PEDV	Porcine epidemic diarrhea virus	
Betacoronavirus	HCoV-OC43	Human coronavirus OC43	Human
	ECoV	Equine coronavirus	
	PHEV	Porcine hemagglutinating encephalomyelitis virus	
	CRCOV	Canine respiratory coronavirus	
	BCoV	Bovine coronavirus	
	MHV	Murine herpesvirus	
		Mouse hepatitis virus	
	HCoV-HKU1	Human coronavirus HKU1	Human
	SARS-CoV	Severe acute respiratory syndrome coronavirus	Human
	SARS-CoV-2	Severe acute respiratory syndrome coronavirus 2	Human
	MERS-CoV	Middle East respiratory syndrome coronavirus	Human
	btCoV-HKU4	Bat coronavirus HKU4	
	btCoV-HKU5	Bat coronavirus HKU5	
	btCoV-HKU9	Bat coronavirus HKU9	
Gammacoronavirus	BWCOV SW1	Beluga whale coronavirus SW1	
	TCoV	Turkey coronavirus	
	IBV	Avian infectious bronchitis virus	
Deltacoronavirus	Bu-CoV-HKU11	Bubaline coronavirus	
	MunCoV HKU13	Munia coronavirus	
	ThCoV HKU12	Thrush coronavirus	

8.2.2 Human Coronaviruses and Diseases

Human coronavirus is a coronavirus that can cause disease in humans and has a person-to-person transmission mode. So far, seven HCoVs have been identified (Table 8.1). Of the seven HCoVs, OC43, 229E, NL63, and HKU1 often cause mild diseases, including the common cold. They are called nonsevere acute respiratory syndrome-like coronavirus. The other three coronaviruses are novel, zoonotic, highly pathogenic, and can cause severe infectious diseases in humans.

The first two HCoVs (OC43 and 229E) were isolated and caused common cold and minor upper respiratory infections in the 1960s.[12,13] Seroepidemiological studies detected outbreaks of disease with a 229E-like virus and other coronaviruses such as OC38 and OC43.[14,15] OC43 and related strains may account for a considerable proportion of common cold or respiratory tract infection.[14] The outbreak of a severe viral respiratory disease in 2002 led to the identification of SARS-CoV.[16] The infection with SARS-CoV can cause humans to develop flu-like symptoms and pneumonia after an incubation period of 4–6 days. A total of 8098 people were diagnosed with SARS-CoV infection; 774 people died, yielding a case fatality rate of 8.7%.

Since the SARS outbreak, several new HCoVs have been identified. In 2004, HCoV-NL63 was isolated from a 7-month-old child with bronchitis,[17] and it was identified later as a causative agent for a lower respiratory tract disease.[18] A year later, HCoV-HKU1 was isolated in nasopharyngeal biospecimens and identified as a causative agent for an indexed patient with pneumonia in Hong Kong.[19] In 2012, MERS-CoV was identified as a novel coronavirus causing SARS-like respiratory disease in Saudi Arabia,[20] leading to the first outbreak of MERS with more than 2080 laboratory-confirmed cases and 722 related deaths, a fatality rate of 34.7%. In late 2019, the SARS-CoV-2 virus was isolated from seven patients with pneumonia admitted to an intensive care unit in Wuhan, China.[3]

8.2.3 The Natural Origin of HCoVs

All seven HCoVs have natural animal origins of either bat or rodents.[10] Of the three HCoVs that can cause the acute severe respiratory syndrome, SARS-CoV was transmitted to humans by the civet cat, while MERS-CoV was transmitted by the dromedary camel, and SARS-CoV-2 was likely transmitted by pangolins. HCoVs are related in structure and replication to neuroinvasive animal coronavirus, including PHEV, feline corona-virus (FCoV), and mouse hepatitis virus. All can invade the central nervous system and cause neuropathology.[21] HCoV 229E can infect human lung fibroblasts, and the cellular receptor, human amino-peptidase N (APN, also known CD13),[22] a zinc-dependent metalloprotease, which is also a receptor for FCoV serotype II, TGEV, PEDV, and the canine coronavirus that causes disease in animals.[23]

8.3 Genomic Structure and Genetics

8.3.1 Viral Genomic Structure

Coronaviruses are named for their large spike projections from the envelope, giving the virus a crown-like shape (Figure 8.1A). It contains the nucleocapsid, consisting of the genome RNA packaged inside the helical capsid formed by the phosphorylated nucleocapsid (N) protein, enclosed with nonstructural proteins inside the envelope of a phospholipid lipid bilayer derived from the membrane of host cells.[24] The envelope also consists of structural proteins, including spike (S) glycoprotein, hemagglutinin-esterase (HE), membrane (M) protein, and the enveloped (E) protein, plus the glycan groups and ORF7a accessory protein.[24]

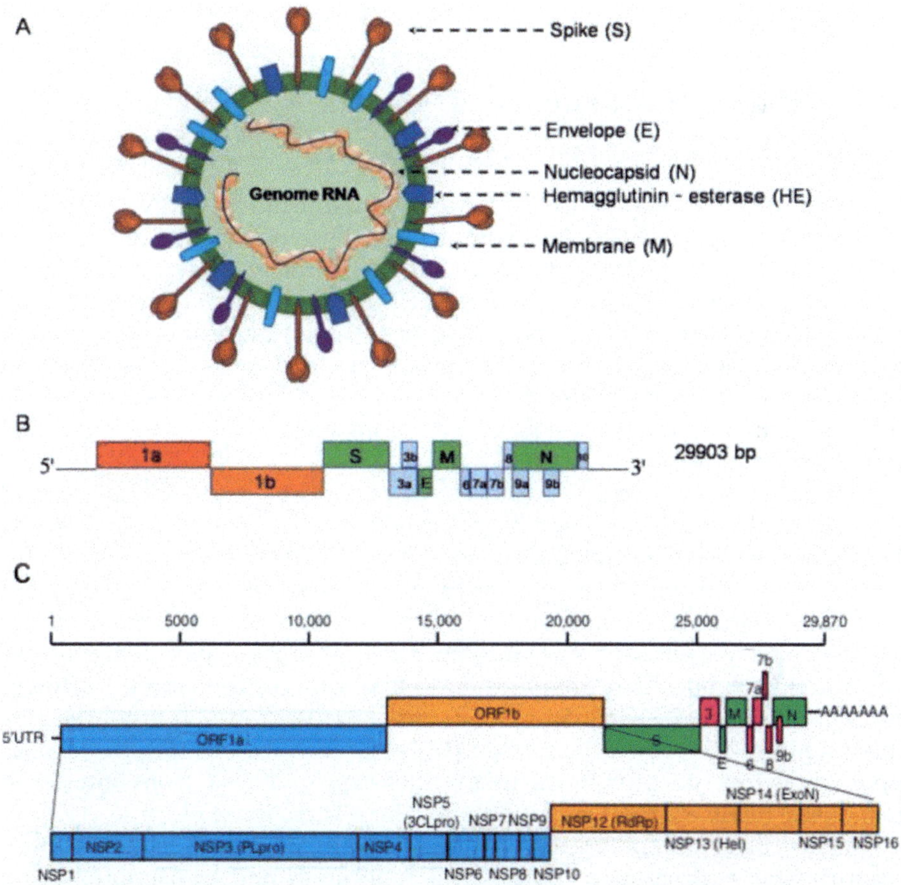

Figure 8.1 Schematic presentation of SARS-CoV-2 (A) virion structure, (B) RNA genome structure. Reproduced from ref. 45, https://doi.org/10.3390/v12040372, under the terms of the CC BY 4.0 license, https://creativecommons.org/licenses/by/4.0/. (C) Encoded proteins. Reproduced from ref. 36, https://doi.org/10.1038/s41467-020-17665-9, under the terms of the CC BY 4.0 license, http://creativecommons.org/licenses/by/4.0/.

The genome of SARS-CoV-2 is ~29.9 kb, comprising 14 open reading frames (ORFs), and it encodes 27 proteins (Figure 8.1B and C). It consists of a 5′ untranslated region (5′UTR), a replicase (ORF1ab), S, E, M, N, a 3′UTR, and other ORFs that have yet to be characterized,[25] but which probably encode accessory proteins required for viral replication inside the host cells. The viral particles include 16 nonstructural proteins (nsp1–nsp16) which are required for viral replication and modulate pathogenesis, four structural proteins (S, E, M, N) which are important for viral subtyping and response to vaccines, and other accessory proteins, including 3a, 3b, 6, 7a, 7b, 8, 9b, 9c, and 10.[26,27] While SARS-CoV-2 has a high similarity in genomic structure with two other betacoronaviruses that cause SARS and MERS in humans, the latter two diseases seem to be more severe in terms of case fatality, MERS, even more so than SARS. However, they lack the rapid human-to-human transmission characteristic of COVID-19.

8.3.2 The Replicase Locus and Proteins

8.3.2.1 Nonstructural Proteins

The replicase locus of SARS-CoV-2 contains two substantial ORFs, ORF1a and ORF1b, which span ~20 kb and two-thirds of the genome. ORF1a encodes a polyprotein (pp) 1a, while ORF1a and ORF1b together encode a C-terminally extended frameshift protein, pp1ab. The ORF1a/b locus encodes 16 nonstructural proteins, in which pp1a and pp1ab are first produced and then further processed *via* auto proteolysis by the viral-encoded polymerase in the virion into nsp1–nsp16 (Table 8.2). The replicase locus encoded nonstructural proteins (nsp1–nsp16) have all functions required for replication and transcription in the virus life cycle.[28]

Two coronavirus proteases (nsp5 and nsp3) play an essential role in converting the pp1a and pp1ab into nsp1–nsp16.[29] Protein nsp5 is the 3-chymotrypsin-like protease enzyme (3CL^{pro}), the main protease (M^{pro}), which cleaves the nsp4–nsp11 of pp1a and nsp4–nsp16 part of pp1ab, and digests the polyproteins at least in 11 conserved sites, first through the autolytic cleavage of itself from pp1a and pp1ab. Nsp3 is the papain-like protease enzyme (PL^{pro}), and its primary function is to process the viral polyprotein pp1a to release nsp1–nsp3. SARS-CoV-2 protease PL^{pro} preferentially cleaves the ubiquitin-like protein interferon (IFN)-stimulated gene 15 (ISG15), and, upon infection, it contributes to the cleavage of ISG15 from interferon responsive factor 3 (IRF3) and attenuates the IFN response.[30]

Nonstructural proteins are also involved in the innate host response, anchoring the viral replication complex, viral replication and transcription, and mRNA processing.[28] Viral replication and transcription are mediated by the replication–transcriptase complex (RTC), formed by several nonstructural proteins (nsp7–nsp16), N protein, and multiple host proteins in the infected cell. The RNA-dependent RNA polymerase (RdRP, nsp12), helicase (nsp13), nsp7, and nsp8 are central enzymes for RNA processing.[28,29] Nonstructural proteins

Table 8.2 Proteins encoded by the SARS-CoV-2 RNA genome.[a,b]

Genes	Protein	Amino acids	Functions
ORF1ab	nsp1	~100	A potentially major virulence factor in the cytoplasm of infected cells that may interfere with the cellular translational machinery and shut down the host mRNA translation through ribosome-binding.[31] SARS-CoV nsp1 promotes viral gene expression and immune-evasion in part by interfering with the virus- and IFN-mediated signaling.[54] It might be a potential drug target.
	nsp2	~628	SARS-CoV nsp2 interacts with host factors prohibitin 1 (PHB1) and PHB2, two evolutionarily conserved proteins with cellular functions of cell cycle progression, migration, cellular differentiation, and apoptosis, and mitochondrial biogenesis.[57] It is likely involved in altering the host cell environment. Nsp2 and S may have a stimulating effect on IFN induction.[36] Nsp2 is involved in the ER Ca^{2+}signaling and mitochondrial biogenesis.[58]
	nsp3	~1945	A papain-like protease (PL^{pro}), its primary function is to process the viral polyprotein to release nsp1-nsp3. PL^{pro} protease enzyme also has the function of de-ubiquitinating and de-ISGylating (interferon-stimulated gene 15, ISG15) from host cell proteins and aids in viral evasion of the host innate immune response;[30,33] interacting with nsp4 and nsp6.
	nsp4	~500	Is involved in the formation of double-membrane cytoplasmic vesicles.[28] Interaction with nsp3 is required for viral replication. Loss of nsp3–nsp4 interaction eliminates viral replication. Nsp4 protein interactors are involved in ER homeostasis.[58]
	nsp5	~306	The main protease (M^{pro}, also $3CL^{pro}$) cleaves at least 11 conserved sites in polyproteins pp1a and pp1ab to release nsp4–nsp16. It is necessary for viral replication and its survival in the host cell.[59]
	nsp6	~290	Induces double-membrane vesicles in infected cells with nsp3 and nsp4 in SARS-CoV;[60] nsp6 also limits autophagosome expansion, induced directly by nsp6 or indirectly by starvation or chemical inhibition of mTOR signaling.[61]
	nsp7	~83	Is formed with nsp8 as a co-factor for the RaRP (nsp12) and may have processivity or RNA primase function.[29,62]
	nsp8	~198	Is formed with nsp7 as a co-factor for the RdRP (nsp12) and may have processivity or RNA primase function.[62] Mutation of specific residues in nsp8 is lethal to SARS-CoV by impacting RNA synthesis.[29]

(*continued*)

Table 8.2 (*continued*)

Genes	Protein	Amino acids	Functions
	nsp9	~113	Is believed to mediate viral replication, overall virulence, and viral genomic RNA reproduction.[35]
	nsp10	~139	High sequence similarity with SARS-CoV nsp10 binds to nsp14 and nsp16 to stimulate their respective 3′-5′ exoribonuclease and 2′-*O*-methyltransferase activities building the viral mRNA capping machinery.[29,63]
	nsp11	~13–23	A pp1a cleavage product at the boundary of nsp10 and nsp11, and with function unknown.[29]
	nsp12	~932	RdRP forms a complex with nsp7 and nsp8 and performs replication and transcription of the viral genome.[29,64] It has more than 95% identity to the SARS-CoV polymerase, which is inhibited by the nucleoside analog remdesivir. The active site of SARS-CoV-2 RdRP is formed by conserved catalytic motifs A–G, of which A–E are located within the palm subdomain, and F and G are within the finger subdomain.[65,66]
	nsp13	~601	A helicase forming the replicase–transcriptase complex with nsp12 for viral genome replication is involved in viral mRNA capping and associated with nucleoprotein in membranous complexes.[67]
	nsp14	~527	It has both 3′-5′ exoribonuclease and N7-guanine methyltransferase (viral mRNA capping) activities and interacts with nsp10.[29,68]
	nsp15	~346	An endoribonuclease necessary for viral replication and its survival in the host cell. Loss of nsp15 affects both viral replication and pathogenesis.[69]
	nsp16	~298	It interacts with and is activated by nsp10. Nsp16 and nsp10 form a complex to catalyze the methylation of viral RNA cap.[70] Nsp16 may also work against host cell antiviral sensors.[29]
ORF2	S	~1273	The spike full-length protein precursor is cleaved into glycosylated subunits, S1, binding to the host's receptor ACE2, and S2 to mediate viral and host membrane fusion.[71]
ORF3a	ORF3a	~275	A transmembrane protein that forms a viroporin 3a in SARS-CoV can modify the host's cell membrane to facilitate viral release from infected cells. Full-length protein E and 3a are required for maximal replication and virulence, and a virus lacking both proteins is not viable.[47] ORF3a interacts with accessory protein 7a, M, S, and E. ORF3a also activates both NF-κB and the NLRP3 inflammasome,[72] contributing to the generation of the cytokine storm. It induces apoptosis in the host cell.[73]
ORF3b	ORF3b	~22	Along with N and ORF6, ORF3b appears to block induction of IFN-I, and the variant is present in other related viral genomes in bats and pangolins.[74]

ORF4	E	~75	A small membrane protein involved in viral assembly, budding, envelope formation, and pathogenesis. It is viroporin and forms a homopentameric ion channel; it interacts with M, N, 3a, and 7.[75]
ORF5	M	~222	A multimembrane-spanning protein on the viral envelope and the most abundant structural component is conserved and highly similar to its SARS-CoV counterpart. It interacts with N for virion RNA packaging and with accessory proteins 3a and 7a. It may aid in proliferation, replication, and immune-evasion.[76]
ORF6	ORF6	~61	It might be associated with the virulence of SARS-CoV as an antagonist of type I IFNs, and it is involved in the viral escape from the host's innate immune system.[36]
ORF7a	ORF7a	~121	A type I membrane protein interacts with bone marrow stromal antigen 2 (BST-2) in SARS-CoV. BST-2 binds virions to the host's plasma membrane. ORF7a binding inhibits BST-2 glycosylation and interferes with this restriction activity. It interacts with S, M, E, and ORF3a in SARS-CoV.[77]
RF7b	ORF7b	~43	It appears to be a viral attenuation factor for SARS-CoV and may influence the human infectivity of SARS-CoV-2.[78]
ORF8	ORF8	~121	It might be a luminal ER membrane-associated protein that may trigger ATF6 activation and affect the unfolded protein response (UPR). It may be involved in the human infectivity of SARS-CoV-2.[79]
ORF9a	N	~419	Binds viral genomic RNA and forms a helical ribonucleocapsid, is involved in genome protection, viral RNA replication, virion assembly, and immune-evasion (including IFN-I suppression), and interacts with M and nsp3 proteins.[80]
ORF9b	ORF9b	~97	Overlaps with N gene in some strains of SARS-CoV-2 isolated in the United States.[46] In SARS-CoV, the encoded protein localizes to mitochondria. It affects mitochondrial morphology and function and suppresses the innate immunity of host cell IFN response by targeting mitochondria and the MAVS/TRAF3/TRAF6 signaling.[81]
ORF9c	ORF9c	~70	Overlaps with N gene in some strains of SARS-CoV-2 isolated in the United States.[46] The encoded protein is associated with membrane-associated proteins in multiple organelles, including ER, Golgi, mitochondria, and surface membrane, which suppress antiviral response in the cell through downregulating IFN signaling and antigen processing.[82]
ORF10	ORF10	~38	ORF10 is unique to SARS-CoV-2 among HCoV and encodes a peptide of only 38 amino acids,[46] which interacts with factors in the CUL2 RING E3 ligase complex and thus may modulate ubiquitination.[68]

[a]Source: https://www.genetex.com/MarketingMaterial/Index/SARS-CoV-2_Genome_and_Proteome.
[b]Nsp: nonstructural proteins encoded by ORF1a and ORF1b gene; S: spike glycoprotein encoded by ORF2; E: envelope protein encoded by ORF4; M: membrane protein encoded by ORF5; nucleocapsid protein encoded by ORF9a; the RNA-dependent RNA polymerase (RdRP)

nsp3, nsp4, and nsp6 are involved in building the double-membrane vesicles required for RNA replication in the cytoplasm. The nonstructural proteins nsp1 and nsp2 are likely to modulate the host cell response through interacting with host cellular factors to promote viral gene expression and immune-evasion in part by interfering with the IFN-mediated signaling.[31]

Computational analysis has indicated that some segments within nsp2 and nsp3 have no homology with other coronaviruses such as SARS and bat SARS-like coronavirus.[32] These mutations in nsp2 and nsp3 might provide indications for examining the difference between SARS-CoV-2 and other coronaviruses in infectivity and disease clinical features. The primary function of nsp3 is to process polyproteins to release nsp1–nsp3,[33] indicating its viral replication role. In mice, nsp9 was found to be essential for the replication of SARS-CoV[34] as well as other coronaviruses[35]

Other nonstructural proteins are involved in the antiviral response of the host immune system. In screening 23 proteins of SARS-CoV-2 including 15 nonstructural proteins (nsp1–nsp10, nsp12–nsp16), four structural proteins, and four accessory proteins (ORF3, ORF6, ORF7, ORF8), Lei *et al.* found that nsp1, nsp3, nsp12–nsp14, ORF3, ORF6, and M proteins might inhibit Sendai virus-induced IFN-β promoter activation, but nsp2 and S proteins showed the opposite effect.[36]

8.3.2.2 Structural Proteins

The downstream ORFs of the RNA genome encode four structural proteins required for assembling a new RNA genome during replication (Table 8.3). The S protein is the largest structural protein encoded by the RNA genome. The receptor-binding domain of the S protein strongly binds to the human angiotensin-converting enzyme (ACE)2 receptor.[37] Therefore, S protein may be a target for inhibition of viral entry and development of antibody-based therapeutics to prevent the disease. The S protein has two functional units, S1 (bulb part) for receptor binding and S2 (stalk part) for membrane fusion; their functions may need other human receptors or proteases such as transmembrane protease, serine 2 (TMPRSS2) and furin to activate and facilitate the viral entry into host cells. They are discussed in Section 8.5 (Host Cellular Factors).

The envelope (E) protein is a small membrane protein that is anchored on the viral envelope and plays a role in the virion assembly and has additional effects on the infected cells of the host. A SARS-CoV lacking E protein adapted to infect mice *in vivo* seemed to be attenuated and could serve as a promising vaccine candidate.[38] The E protein alters coronavirus virulence by regulating the cell stress response and apoptosis during host cell infection.[39] In mice, the SARS-CoV E protein PDZ-binding motif was involved with the immunopathology, probably through interacting with syntenin and P38 MARK activation.[40]

The membrane (M) protein is a multimembrane-spanning protein and the most abundant in the viral envelope. It provides structural support for the virion and plays a role in the assembly and construction of virion

Table 8.3 Host cellular factors involved in viral attachment and entry of coronaviruses.

Host factor	Description	Virus name abbreviation	Role for viral protein
APN[22]	Human or porcine aminopeptidase N	229E, [TGEV]	Cell surface receptor
CTSL[95]	Cathepsin L	SARS, SARS-CoV-2	Mediates cleavage of the S1
CTSB[71]	Cathepsin B	SARS-CoV-2	Cleaves S protein in the cell
CD147[96]	Basigin (BSG)	SARS-CoV-2	A receptor on the host cell
ACE2[97]	Angiotensin-converting enzyme 2	SARS-CoV, CoV-2, NL63	Receptor
DDP4[98]	Dipeptidyl peptidase-4, CD26	MERS, 229E	Receptor
9-*O*-ac-sia[99]	9-*O*-acetylated sialic acid	OC43, HKU1	Receptor
Furin[100]	PCSK3	SARS-CoV-2, MERS, [IBV]	Cleaves and activates S protein
GRP78[101]	Binding immunoglobin protein	SARS-CoV-2, MERS	Receptor binding
LY6E[102]	Lymphocyte antigen 6E	All HCoVs	Restrict viral entry
Sialic acid[103]	In the brain and neuronal transmission	SARS-CoV-2, MERS	Receptor
TMPRSS11D[104]	Transmembrane serine protease 11D	SARS-CoV-2, SARS, 229E	Cleaves and activates S protein
TMPRSS2[71]	Transmembrane serine protease 2D	SARS-CoV-2	Cleaves and activates S protein
VCP[93]	Valosin-contacting protein	229E, [IBV]	Cleaves and activates S protein

particles. M proteins can have interactions among themselves and with other structural proteins.

N protein complexes with the genome RNA to form the viral nucleocapsid within the viral membrane and interact with the M protein in the assembly. The N protein may play a role in replicating RNA. Antibodies to N and S protein have been considered targets to diagnose infection with coronavirus.[41] The N protein, along with other accessory proteins (*e.g.* ORF6 and ORF8), has been shown to inhibit the type 1 IFN signaling pathway.[42]

8.3.2.3 Accessory Proteins

The viral genome of SARS-CoV-2 encodes accessory proteins including ORF3a, ORF3b, ORF6, ORF7a, ORF7b, ORF8, ORF9b, ORF9c, and ORF10.[27,43–46] They may play a critical role in determining virulence and modulating host immune response to induce pathogenicity caused by the viral pathogen. In the case of SARS-CoV, ORF3a partnering with E protein may play a critical role in maximal replication and virulence, and a coronavirus lacking both proteins is not viable.[47] ORF3a activates both nuclear factor (NF)-κB and the NLRP3 (nucleotide-binding oligoerization domain (NOD), leucine-rich repeat (LRR) and pyrin domain containing 3) inflammasome,[48] promoting the secretion of anti-inflammatory cytokines IL1B and IL18. Mutations in functional domains of ORF3a are linked to virulence, infectivity, and virus release.[49] By contrast, ORF3b may not code for a functional protein in all coronaviruses, as immunohistochemical analysis of tissue biopsies of patients with SARS failed to detect its presence *in vivo*.[46]

In addition, ORF6, ORF8, and N proteins can inhibit the type 1 IFN signaling pathway,[42] a key component of the antiviral response of the host innate immune system. ORF6 may play a critical role in inhibiting both IFN-β production and downstream signaling.[36] As mentioned earlier, ORF7a is also on the envelope, and large deletions have been detected in ORF7a, resulting in in-frame gene fusions.[50] SARS-CoV ORF7a interacts with the ribosomal transport proteins and has been shown to inhibit cellular translation and activate p38 MARK.[51]

8.3.3 Genomic Mutations

The coronavirus consists of an RNA genome that is naturally subject to a rapid genetic mutation. Mutations may change the dynamics of transmissibility and pathogenicity of SARS-CoV-2. Therefore, monitoring possible mutations is important in developing strategies for prevention and medications. With a bioinformatics investigation of the structural genomics and existing database of the related-SARS viruses, Srinivasan *et al.* provided structural genomics and interactome roadmaps of the SARS-CoV-2 virus.[43] In the nonstructural, structural, and accessory proteins encoded by SARS-CoV-2 RNA genome, a large novel insert was found between two putative functional domains in nsp3, and four inserts found in S protein were unique to SARS-CoV-2. Of note, the nsp3 is a PL^{pro} that processes the viral polyprotein to release nsp1–nsp3, and it may facilitate the viral evasion of host immune response.[52]

Another study focused on detecting mutations within the strains of SARS-CoV-2 to understand the mode of transmission and clinical manifestations. With nearly 5000 complete sequences of the SARS-CoV-2 genome, mostly from the United States and Europe, Zhao *et al.* performed a phylogenetic and structural analysis, aiming to identify the human adaption related mutation.[53] While noting a lower genetic distance of these genomic sequences, they found 11 residuals with high-frequency substitutions, located at nsp2, nsp5 (3CL-protease), nsp6, nsp12 (RdRP), nsp13 (helicase), ORF3a, ORF8, and structural proteins N and S.[53] These are essential proteins for viral replication. Furthermore, deletion of amino acids in SARS-COV-2 nsp1 protein could affect the regulation of viral replication and the host's gene expression. Nsp1 as leader protein is the most critical determinant of viral replication and the host's antiviral innate immune response, mainly through reducing the host's IFN-α expression. In SARS-CoV, residuals in nsp1 were found to affect the nsp1 inhibition of host gene expressions or antiviral signaling.[54] Benedetti *et al.* found that deletion of three amino acids in SARS-CoV-2 nsp1 protein could affect the protein's C-terminal structure, which is important for regulating viral replication and the negative effect of host gene expression.[55] A mutation was detected in the SARS-CoV-2 nsp6,[56] and mutations in ORF3a are found in six functional domains of SARS-CoV-2, probably linked to virulence and infectivity, ion channel formation, and virus release.[49]

8.4 Viral Replication

8.4.1 Attachment and Cell Entry

The viral attachment and host cell entry is the first step of viral replication in which the S glycoprotein binds the receptors (*e.g.*, ACE2) on the surface of host cells.[71] The subunit S1 binding specific receptor causes a drastic conformational change in the subunit S2 and then leads to a fusion between the virus envelope and the host cell's cellular membrane. Many proteases, including the host protease, TMPRSS2, can accomplish a cleavage of S2 protein to facilitate the viral fusion with the host cell membrane.[71,83] Also, SARS-CoV-2 S glycoprotein harbors a cleavage site at the boundary between S1 and S2 units,[84,85] activated by furin, a serine endoprotease, for facilitating the fusion of cell membrane. TMPRSS2 and furin are essential for proteolytic activation of the SARS-CoV-2 S protein in human airway cells.[86]

The furin cleavage site in the S protein is not found in SARS-CoV,[85] which, together with the receptor-binding domain (RBD) of the S protein, might contribute to the high infectivity of SARS-CoV-2. The furin cleavage site is polybasic, allowing an effective cleavage by furin and other proteases, and polybasic furin cleavage sites are often found in hemagglutinin proteins in avian and influenza viruses. In addition, the inserted proline-arginine-arginine-alamine (PRRA) at this site is predicted to add O-linked glycans to three variants, which flank the polybasic cleavage site.[88] This addition of O-linked glycans may enable S protein to bind additional receptors, as described later. In addition, the S protein RBD of SARS-CoV-2 that has an optimal binding affinity to ACE2 is structurally different from that of SARS-CoV.[88] The SARS-CoV-2 binding affinity to ACE2 is more than 10-fold higher than SARS-CoV.[89,90] Therefore, the feature of the furin cleavage site and different RBD sequences likely constitute an effective viral attachment and entry, leading a rapid transmission.

The virus can also enter the host cell through direct endocytosis and then, through a cathepsin-mediated process, release the viral RNA genome into plasma.[91] Other host proteins (*e.g.* CD147, GRP78) and host proteases (*e.g.* cathepsin L/B, trypsin) might be involved with the viral entry of SARS-CoV-2 into the host cell,[91] which may establish an infection and affect the infectivity and clinical outcomes.

8.4.2 Translation and RNA Replication

The RNA genome's translation and replication start after the virus disassembles within the host cells and releases its nucleocapsid and viral RNA into the host cell cytosol (Figure 8.2). The nucleocapsid undergoes an uncoating process to release the genomic RNA. The host ribosomes then translate the viral RNA to produce nonstructural proteins. This translation process starts with the replicase locus where the ORF1a and 1b are first translated to produce replicase polyproteins pp1a and pp1ab, which are further processed through proteolysis by viral-encoded proteases into nonstructural proteins (nsp1–nsp16). Some nonstructural proteins involved in forming the RTC include the RNA-dependent RNA polymerase (RdRP, also known as nsp12), which uses nsp7 and nsp8 as auxiliary factors and nsp13, the helicase. RdRP, as a critical component, catalyzes the synthesis of viral RNA and plays a central role in the viral replication of SARS-CoV-2.[66] The nsp13 replicase is a critical component for viral replication and has the highest sequence conservation across the coronavirus family.[92]

RdRP transcribes the full-length positive strand of viral genomic RNA to synthesize negative-strand genomic and subgenomic RNA template for positive-strand new genomic RNA and subgenomic RNA transcripts. The most abundant transcripts are N, S, 7a, 3a, 8, M, E, and 7b.[27] The subgenomic transcripts are translated to produce structural (S, E, M, N) and accessory proteins and genomic RNA to form the viral particle.

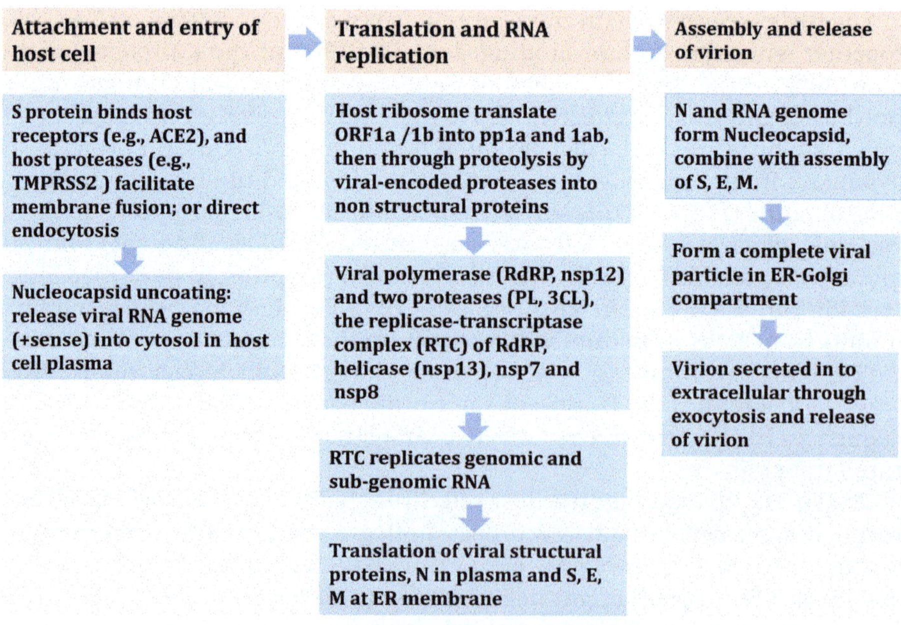

Figure 8.2 SARS-CoV-2 infection, viral genome replication and translation.

8.4.3 Assembly and Release of Virion

The S, E, and M proteins enter the endoplasmic reticulum (ER), and the N protein combines with the new positive-strand genomic RNA to form a nucleocapsid complex. They merge into a complete viral particle in the ER–Golgi apparatus compartment and are secreted into extracellular through exocytosis. The new virions are then released outside the host cell.

8.5 Host Cellular Factors

Coronaviruses can infect a variety of species, some of which may impact human life. Besides the ongoing SARS-CoV-2 pandemic which has an unprecedented impact on human history, coronaviruses can infect major domestic animals such as pigs, cattle, and chickens, significantly impacting economic activities and daily human life. More importantly, because they process an RNA genome, coronaviruses are subject to high recombination and, therefore, more likely to have a host switching. For example, the three HCoVs that can cause severe diseases to humans (SARS, MERS, and COVID-19) have migrated from a natural host such as a bat to a domestic host before infecting humans and causing diseases. The current pandemic of COVID-19 indicates the ability of coronaviruses to undergo host switching and cause human-to-human transmission.

Host cellular factors involved in the infection cycle of HCoVs are essential for understanding the viral infectivity and transmission and developing effective preventive and therapeutic strategies. Several host factors are involved with coronavirus, mainly HCoV, infection (Table 8.3), that effect viral attachment and entry into host cells. In addition, a genome-wide screen of small interfering RNA (siRNA) reveals that vasolin-containing protein (VCP) plays a role in the early stage of infection with IBV, an avian infectious bronchitis virus in human cells.[93]

With the technological advances in large-scale molecular biology and computational and bioinformatics tools, additional host factors have been rapidly indicated to affect the entry of viral SARS-CoV-2 into human host cells (Table 8.3). Besides the main receptor, ACE2 and TMPRSS2 are known to be involved with infections of other HCoVs, additional receptors and proteases, or even different virus routes by which SARS-CoV-2 entering host cells are discovered.[94]

Host factors associated with susceptibility to COVID-19 are divided into three sequential stages. They include (1) host cell exposure to the pathogen, (2) host cellular factors (host receptors and proteases) that modulate pathogen entering into the host cell, and (3) host response that is determined by individual-specific immunity and immune system response. Unlike the other six HCoVs that can cause human infections, SARS-COV-2 can be affected by more host receptors or cellular factors that can establish an attachment and/or facilitate viral entry into the host cell. Identifying new receptors and host cellular factors might help prevent SARS-CoV-2 infections after exposure by altering their expression.[105]

8.5.1 ACE2

ACE2 is a zinc-containing metalloenzyme located on the surface of endothelial and other cells[106] and moderately to highly expressed in the human upper respiratory tract, small intestine, colon and rectum, gallbladder, kidney, pancreas, testis, and fallopian tube.[107] It serves as a receptor for SARS-CoV-2 and SARS-CoV[108,109] and NL63.[95] The SARS-CoV-2 S protein RBD and ACE2 have a compact conformation, which indicates an increased binding affinity.[97,110] The SARS-CoV-2 virus has been isolated in human airway epithelial cells in patients with COVID-19. The *ACE2* gene is highly expressed in the epithelial cells of the oral mucosa,[111] other human tissues such as the gastrointestinal tract, liver and gallbladder, kidney, urine, and male bladder tissues, and heart muscle. ACE2 has genetic heterogeneity,[112] which may lead to a variation in infectivity among different populations.

Animal studies show that a high dose of an ACE inhibitor (ACEi) or angiotensin-receptor blockers (ARB) – two common antihypertensive drugs – may increase the level of ACE2.[113] It is speculated that continuing use of ACEi or ARBs during the pandemic, may elevate ACE2, which might increase the risk of SARS-CoV-2 infection and viral load and lung injury in humans.[114] However, so far, several observational studies have not provided any supporting evidence. A large prospective study of more than 8 million adult participants showed that using ACEi might reduce the risk of COVID-19, but did not significantly increase the risk for admissions to an intensive care unit.[115] Another case-population study in Spain showed that the use of ACEi did not increase severe COVID-19, but reduced the likelihood of required hospital admission in patients with diabetes.[116] However, black Africans who used ACEi were observed with an increased risk of COVID-19. In fact, other studies in animals showed that SARS-CoV-induced acute lung injury could be attenuated by blocking the renin–angiotensin pathway.[117]

The role of ACE2 in developing COVID-19 has not been determined and needs to be further studied. ACE2 is a receptor bound by the S protein of three HCoVs that cause disease in humans: SARS-CoV-2, SARS-CoV, and HCoV-NL63. However, the infectivity, disease severity, and transmission mode of SARS-CoV-2 are markedly different, indicating that the ACE2 receptor is necessary for establishing an infection, but not sufficient to account for all the features of SARS-CoV-2 infection. In addition, a protein expression study of various human cells and tissues indicates that ACE2 expression is limited in lungs, with no or low expression in a subset of cells.[118] The infection might be established in the upper respiratory tract where ACE2 protein has a medium level of expression in the nasopharynx and bronchus, according to the human protein atlas.[107] There might be additional features of SARS-CoV-2 or pathogen–host interactions that influence its differences from SARS-CoV and HCoV-NL63 infection. Current clinical observations show that severe COVID-19 is associated with underlying health conditions and advanced age.

ACE/ACE2 is important for maintaining physiological hemostasis and has implications for chronic health conditions.[119,120] The functions of ACE and ACE2 counter each other in the classical renin-angiotensin system (RAS). ACE, moderately expressed in lung, kidney, gastrointestinal tract, and reproductive tissues,[107] is a central component of the RAS that controls blood pressure by regulating the body's volume of fluids. It converts the inactive pro-hormone angiotensin I (Ang I, angiotensin 1–10), a decapeptide, into the active vasoconstrictor angiotensin II (Ang II, angiotensin 1–8), by removing two C-terminal residues. Ang II is a highly active octapeptide that, through binding angiotensin type 1 receptor (AT1R), causes vasoconstriction and other proinflammatory activities, which may cause cell injuries (Figure 8.3). Therefore, ACEi and ARBs aim to interfere with the Ang II/AT1R pathway to reduce blood pressure and protect cells from damage.

Conversely, the endogenous ACE2 counters the effects induced by ACE in the classical RAS.[121] ACE2 was discovered as a first homolog gene of ACE and exhibits more than 60% similarity with the ACE sequence. ACE2 converts Ang II to a vasodilator Ang 1–7, a heptapeptide, which, through binding the Mas 1 oncogene product receptor (Mas1R), stimulates nitric oxide synthase to produce a counter-regulatory response to Ang II-mediated effects. Therefore, Ang 1–7/Mas1R causes vasodilation and anti-inflammatory activities that may exhibit a cell protective effect (Figure 8.3). In addition, ACE2 has been shown to metabolize Ang I to Ang 1–9, a biologically active peptide that may act through angiotensin type 2 receptor (AT2R). Ang 1–9/AT2R may serve as an antagonist of the classical RAS signaling to produce cell protection.[122]

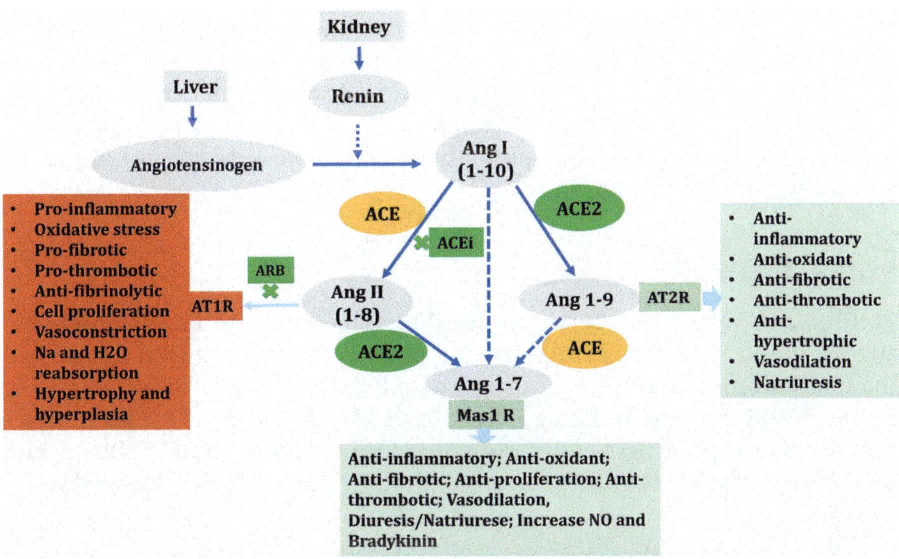

Figure 8.3 The role of ACE/ACE2 in the renin-angiotensin system (RAS).

An imbalance of ACE and ACE2 indicates a disturbance in the RAS system. The Ang II/AT1R pathway can activate immune cells and trigger the proliferation of splenic lymphocytes,[123] inducing inflammation, which has been indicated in the pathological effect of RAS dysfunction in cardiovascular disease. ACE/ACE2 imbalance has been considered to play a key role in the inflammatory processes in cardiovascular, cardiac hypertrophy, pulmonary hypertension, lung injury, sepsis, and acute pancreatitis.[120] Since SARS-CoV-2 utilizes ACE2 receptors to enter the host cells, it may deplete ACE2 receptors and cause an increase in ACE/ACE2 ratio and RAS dysfunction, leading to a worsening outcome in individuals infected with SARS-CoV-2.[124] In a report of 23 188 patients with COVID-19, 69% of patients were found to have hypertension, suggesting that RAS system imbalance may increase susceptibility to SARS-CoV-2.

ACE2 depletion might be associated with the severity of COVID-19. In a mouse model of acute lung injury, ACE2 deletion causes more severe disease, suggesting a protective role for ACE2 in lung tissues.[117] When SARS-CoV-2 S protein binds to ACE2, the depletion may reduce availability for its normal protective functions. Therefore, in turn, it may result in acute respiratory failure and other organ-associated damage,[117] commonly observed in severe COVID-19.[125] Depletion of ACE2 may thus play a causative role in the lung injury caused by SARS-CoV and SARS-CoV-2, as plasma Ang II is elevated in patients with critical COVID-19 compared to mild cases.[126] A high ratio of ACE to ACE2 may cause harm to the lungs and other organs due to the dysfunction of RAS. Furthermore, ACE and ACE2 might have a role in the kallikrein–kinin system. This humoral pathway modulates the blood coagulation system and endothelial cell growth and angiogenesis; all are related to the common pathogenesis of severe COVID-19.[127]

Of note, MERS-CoV also causes a similar lung disease without targeting ACE2 as a receptor, so other viral molecules and host factors must be involved in causing the severe lung diseases associated with SARS-CoV-2 infections. The effects of SARS-CoV-2 on the respiratory system and other major functioning organ systems in the human body are borne out by multi-organ histopathological manifestations.[128–131]

8.5.2 TMPRSS2

TMPRSS2 is an enzyme belonging to the serine protease family and which cleaves the S2 subunit of SARS-CoV-2 S protein to facilitate viral fusion with the host cell membrane.[71,84,85] As well as the SARS-CoV-2 virus, other viruses such as SARS-CoV and influenza virus H1N1 also depend on TMPRSS2 activation for entering the host cell.[132] An animal model shows that TMPRSS2 contributes to virus spread and immunopathology in the airways after infection with SARS-CoV-2.[133]

The involvement of TMPRSS2 might support some observations on the sex bias in COVID-19.[134] The *TMPRSS2* gene was initially identified in prostate cancer and is upregulated in prostate cancer cell lines in response to androgens.[135] It also participates in TMPRSS2-ERG gene fusion, the most common gene fusion events in solid tumors, presenting 50% of prostate cancer in European ancestry. Since TMPRSS2 is critical for SARS-CoV-2 viral infection and is upregulated by androgens, a sex bias in the severity of COVID-19 should be investigated with this gene. TMPRSS2 has been proposed as a potential biomarker for the study of COVID-19.[136]

TMPRSS2 inhibitors have been available[135] and suggested for use in possibly preventing SARS-CoV-2 infection. Primary nasal epithelial cells from human COVID-19 patients treated with azithromycin exhibited TMPRSS11D and TMPRSS2 as top gene downregulation compared with controls, although these were based on few samples.[137]

8.5.3 Furin

Furin is a protease enzyme encoded by a gene located in oncogene FES's upstream region (FURIN). It is a subtilisin pro-protein peptidase that cleaves and processes latent precursor proteins into their biologically active products. The spike glycoprotein of SARS-CoV-2 is found to harbor a polybasic cleavage site at S1/S2 subunits required for its efficient cleavage by furin, which is essential for S protein-mediated cell–cell fusion and viral entry into human lung cells.[138] However, the cleavage site is lacking in other SARS-like coronaviruses.[100] Increased cell–cell fusion by this cleavage site at optimized conditions might contribute to the rapid transmission mode of SARS-CoV-2 that is more competent than other SARS-like viruses.

Current evidence indicates that the SARS-CoV-2 enters the host cell by its S protein binding to ACE2 receptor through its cleavage at site S1/S2, first by furin and then at site S2' by TMPRSS2; both are essential for the proteolytic activation of the virus.[86] In a study that examined the expression of three genes, the co-expression of *FURIN* with *TMPRSS2* was more significant than with *ACE2* in the primary lung cells.[139] The antiviral activity of various TMPRSS2 inhibitors combined with a furin inhibitor is more efficient than any single inhibitor alone in human airway epithelial cells.[86] In addition, furin might be involved in the pathogenesis of SARS-CoV-2, as it is expressed by T-cells, which are essential for maintaining the peripheral immune tolerance and regulating cell-mediated immunity.[140] Pharmacological inhibition of furin in humans might slow viral entry into host cells and reduce the infection by SARS-CoV-2.

8.5.4 DPP4

Dipeptidyl peptidase 4 (DPP4) is an aminopeptidase protein expressed on the surface of several cell types and which is involved with signal transduction, immune regulation, and apoptosis. It is a serine protease known to cleave substrates such as growth hormones, chemokines, neuropeptides, and vasoactive peptides and inactivate them after cleavage.[141,142] Aminopeptidase N (APN) is a receptor for HCoV-229E and TGEV.[143,144] DPP4 is an abundant and ubiquitous protein that plays functional roles in metabolism, the immune and endocrine responses, bone marrow mobilization, cancer growth, and cell adhesion.[145] More specifically, its activity regulates glucose homeostasis and inflammation. DPP4 has been identified as a functional cellular receptor for MERS-CoV.[98] Recently, a molecular docking analysis has suggested that many residuals on DPP4 interact with the S1 domain of SARS-CoV-2.[146] The coronavirus enters and hijacks the host cells primarily through the S1 domain of the S protein interacting with host proteins such as DPP4, which cleaves the amino-terminal dipeptide to activate T-cells, a key immunoregulatory factor in viral infection.[146,147]

DPP4 might explain the severe COVID-19 response that is downstream of virion binding to the ACE2 receptor. Plasma DPP4 is higher in individuals with obesity, metabolic syndrome, and type 2 diabetes, and older adults, conditions that are associated with severity and unfavorable outcomes in COVID-19.[145,148] In human tissues, DPP4 is strongly expressed on the surface of epithelial cells in the liver, kidney, gut, lungs, and pancreas, and is present on capillary endothelial cells.[145] These are consistent with the multi-organ pathophysiology of severe COVID-19 and MERS; additionally, ACE2 is minimally or not expressed in the lung tissues and is not a receptor for MERS-CoV. Because DPP4 inhibitors have been approved for treating patients with type 2 diabetes, they might help prevent SARS-CoV-2 infection and disease progression.[149]

8.5.5 CD147

CD147 has been demonstrated to be a novel receptor for the SARS-CoV-2 S protein.[150] It is an extracellular matrix metalloproteinase inducer (EMMPRIN) encoded by the gene *basigin* (*BSG*) and is a transmembrane glycoprotein of the immunoglobin superfamily. CD147 also acts as a receptor for *Plasmodium falciparum*, a pathogen that causes malaria by invading red blood cells.[151] Drugs interfering with S/CD147 binding may inhibit pathogen invasion and protect other cells, including progenitor/stem cells. Humanized anti-CD147 antibodies can block host cell invasion by the human malaria parasite entirely, and can significantly block host cell invasion by SARS-CoV-2 *in vitro*[152] and human patients with COVID-19.[153] Patients treated with anti-CD147 antibodies have significantly reduced time to viral testing negative (median 3 *versus* 13 days) and improved clinical outcomes, compared with other hospitalized patients.[153] In addition, SARS-CoV N protein may bind to HAb18G/CD147 to invade the host cell.[154]

CD147 acts as the principal upstream stimulator of matrix metalloproteinases (MMPs), closely related to MMPs. Their expression is often increased in inflammation and on the surface of tumor cells. MMP1 and MMP7 are potential peripheral blood biomarkers for idiopathic pulmonary fibrosis,[155] which is observed in severe COVID-19.[156] Anti-CD147 antibodies can inhibit transcription growth factor-β induced proliferation and the differentiation of fibroblasts into myofibroblasts, which are originated from resident fibroblasts, pericytes, and resident stem cells.[152] Thus, anti-CD147 antibodies could also be used to prevent pulmonary fibrosis in the early stage of COVID-19.

In a gene expression study of SARS-CoV-2 related molecules, *ACE2* and *TMPRSS2* were co-expressed in the epithelial cells of lung and skin, while *CD147* and *CD26* (DPP4) were expressed in both epithelial and immune cells,[157] which might imply alternative routes of infection for severe COVID-19.

8.5.6 GRP78

The glucose-regulated protein 78 (GRP78), a binding immunoglobin protein encoded by *HSPA5*, is probably a receptor for SARS-CoV-2. GRP78 is a master chaperone protein widely expressed in human tissues and synthesized markedly under cell stress conditions that lead to an accumulation of unfolding polypeptides in the ER. It has been shown to play a role in infections with several viruses, including MERS-CoV, Zika, Ebola, and Dengue, and cancer aggressiveness.[158] Molecular docking analysis predicts that four spike protein regions can bind to the substrate-binding domains of GRP78 in recognition of the host cell receptor.[101] mRNA expression of *GRP78* in the blood of patients with SARS-CoV-2 pneumonia was found multi-fold higher than other pneumonia patients.[159] Further molecular docking analysis found that withaferin A, artemisinin, curcumin, and andrographolide had optimal interaction (*e.g.* low binding energies) with the GRP78 receptors.[160] These four natural compounds have been demonstrated with the efficacy of immunomodulation, anti-inflammation, and anti-anagenesis.

8.5.7 Cathepsins

Cathepsins are a group of endosomal and lysosomal proteases that exhibit endo- and exopeptidase activities. SARS-CoV-2 may enter host cells through endocytosis depending on low-pH environments, in which phosphatidylinositol-3 phosphate-5-kinase (PIKfyve) and cathepsin L (CTSL) are critical for endocytosis.[161] Molecules related to endocytosis might affect the virus entering *via* this route. A genome-wide loss-of-function screening identified top host factors that may reduce SARS-CoV-2 entry into human alveolar epithelial cells. These top factors are located at endosomal entry (*e.g.* ACE2, RAB7A), spike cleavage and membrane fusion (CTSL), and endosome recycling (*e.g.* VSP35), and they can be grouped into different pathways, including the vacuolar APTase proton pump, retromer, and commander complex.[162]

8.5.8 Ly6E

Ly6E is a protein belonging to the lymphocyte antigen 6 (Ly6) family and is encoded in humans by the *LY6E* gene. Its primary function is immune regulation, including T-cell activation, proliferation, and development. Ly6E[163] and other IFN-induced cellular proteins such as IFITMs (interferon-induced transmembrane proteins), GILT (gamma-interferon-inducible lysosomal thiol reductase), ADAP2 (ArfGAP domain-containing protein 2), 25HC (25-hydroxycholesterol) are found to enhance or inhibit the entry of various viruses, including coronaviruses, through viral and host interaction.[164–168] Recently, Ly6E was found to inhibit the entry of all HCoVs, including SARS-CoV-2, but through a distinct mechanism from IFITMs.[102] LY6E also might mediate the transport of adeno-associated virus across the human blood–brain barrier.[169]

Ly6E also plays a role in oncogenesis. Four lymphocyte antigen 6 family members (Ly6D, Ly6E, Ly6H, and Ly6K) were observed to have increased mRNA expression in tumor tissues of bladder, brain and central nervous system, breast, esophageal, gastric, pancreatic, colorectal, kidney, lung, melanoma, and prostate cancer compared with normal tissues;[170] over-expression of Ly6E was associated with poor prognostic outcomes of lung, gastric, and breast cancer.

8.5.9 L-SIGN and DC-SIGN

A preprint manuscript posted on 5 November 2020 reported a lung cDNA screening. It indicated that the N-terminal domain of SARS-CoV-2 S protein mediates viral infection through membrane fusion by associating with L-SIGN, a liver/lymph-node specific intracellular adhesion molecule (ICAM)-3-grabbing integrin; and DC-SIGN, a dendritic cell-specific ICAM-3-grabbing nonintegrin.[94] L-SIGN is a liver-specific receptor for hepatitis virus C (HCV), and both L-SIGN and DC-SIGN play a role in HCV infection and immunity.[171]

8.5.10 Sialic Acid

Sialic acid is a monosaccharide widely expressed on the surface of vertebrate cells. N-acetylneuraminic acid (Neu5Ac) is the predominant sialic acid found in human cells and many animal cells. It has been implicated in host cell entry of various viruses, including adenovirus serotype 37 and enterovirus serotype 70[172] and MERS-CoV.[173] Two HCoVs (OC43 and HKU1) in the same group as SARS-CoV-2 use the 9-*O*-acetylated-sialic acid as a receptor for viral entry.[99]

Recent *in silico* studies indicate that the S protein of SARS-CoV-2 also binds to sialic acid linked to host cell surface gangliosides as a cellular receptor for viral entry,[103,174] and hydroxychloroquine was found to block SARS-CoV-2 binding to gangliosides.[174] It is speculated that binding to the sialic receptor could contribute to the "cytokine storm" that frequently occurs in severe COVID-19 and MERS.[175] In addition, sialic acid is a fundamental component of the salivary mucin that protects the glycoproteins. SARS-CoV-2 infection through binding this receptor may reduce sialic acid availability in salivary mucin, causing an increase in gustatory threshold and then gustatory problems. Anosmia and dysgeusia are demonstrated to be prevalent in patients with mild COVID-19, both at a similar rate,[134] suggesting that SARS-CoV-2 might use a dual receptor-based strategy for viral entry.[103] Furthermore, the sialic acid receptor might be involved in gastrointestinal SARS-CoV-2 infection, which may add a fecal–oral route of infection.[176]

8.5.11 Plasmin and Other Proteases

Other proteases are likely to promote viral invasion to host cells and are associated with cytokine storm, ARDS, and multiorgan injuries.[177–179] Plasmin has been discussed as a protease of SARS-CoV-2 S protein to facilitate the binding to ACE2 receptor and viral entry and membrane fusion.[178] Elevated plasmin may associate with excessive D-dimer in ARDS and chronic conditions such as hypertension and diabetes, often in severe COVID-19.[178] Patients with COVID-19-induced ARDS showed a significantly higher level of fibrin/fibrinogen degradation product and D-dimer relative to non-COVID-19-induced ARDS,[180] and D-dimer level was strongly associated with COVID-19 fatality.[181] Therefore, high plasmin levels are likely to play an important role in determining disease severity. As such, related inhibitors should have more therapeutic implications for severe patients.[182,183] Of note, plasmin is a co-activator of NF-κB signaling and plasminogens mediate efferocytosis for the inflammation resolution,[184] there might be a complex molecular network involved with plasmin in response to the viral pathogen and its sequelae.

In summary, while the genomic structure in SARS-CoV-2 is highly similar to SARS-CoV, the difference in S protein variants might determine that more host cellular factors are involved in viral entry (Table 8.4); in addition,

Table 8.4 Host cellular factors for SARS-CoV-2 with other HCoVs.[a]

	Betacoronavirus					Alphacoronavirus	
	SARS-CoV-2	SARS	MERS	OC43	HKU1	299E	NL63
ACE2	++	++					++
CD147	++						
CSTB	++						
CSTL	++	++					
DC-SIGN	++						
DPP4	++		++			++	
Furin	**		**				
GRP78	++		++				
L-SIGN	++						
Ly6E	++	++	++	++	++	++	++
Plasmin	**						
Sialic acid	++		++				
TMPRSS11D	**	**					
TMPRSS2	**						

[a]++: Receptor; **: proteases.

a distinct route of endocytosis is indicated. These S protein differences may affect the kinetics of infection and clinical manifestations, and therefore the potential therapeutic approaches. Nonstructural proteins play some essential roles in viral replication, and their interaction with host intracellular molecules may affect the pathogenesis of COVID-19 and lead to variation in clinical manifestations. Therefore, it is critical to conduct timely formal clinical and molecular studies in human patients to understand how viral and host cellular factors influence viral infection and disease progression. Genome-wide screening for host factors in patients with COVID-19 would be a promising approach to define molecular-susceptible individuals and develop preventive therapeutics for early treatment for the population at risk exposure.

In addition, an *in vitro* screening study of human cell lines may help identify additional cellular factors at a different stage of the viral replication cycle that could indicate therapeutic targets. Given that COVID-19 generally has an incubation period of 4–14 days, it would be valuable to provide preventive therapeutics to those exposed to infection sources who may have a high likelihood of developing the disease. Hopefully, with newly approved or existing drugs that target the critical molecules of receptors and proteases, the quarantine burden could be reduced before effective vaccines are widely available for vaccination.

In summary, while the genomic structure in SARS-CoV-2 is highly similar to SARS-CoV, the difference in S protein variants might determine that more host cellular factors are involved in viral entry (Table 8.4); in addition, a distinct roue of endocytosis is indicated. These S protein differences may affect the kinetics of infection and clinical manifcstations, and therefore the potential therapeutic approaches. Nonstructural proteins play some essential roles in viral replication, and their interaction with host intracellular molecules may affect the pathogenesis of COVID-19 and lead to variation in clinical manifestations. Therefore, it is critical to conduct timely formal clinical and molecular studies in human patients to understand how viral and host cellular factors influence the viral infection and disease progression. Genome-wide screening for host factors in patients with COVID-19 would be a promising approach to define molecular-susceptible individuals and develop preventive therapeutics for early treatment for the population at risk of exposure.

8.6 Executive Summary

- SARS-CoV-2 is an enveloped coronavirus with a single strand and positive-sense RNA genome; it can infect mammals and birds.
- The SARS-CoV-2 RNA genome is ~29.9 kb, comprising 14 open reading frames (ORFs) that include a 5′ untranslated region, a replicase (ORF1ab), S, E, M, N, a 3′ untranslated region, and other open reading frames. The RNA genome encodes 27 proteins, including structural

proteins (S, E, M, N), 16 nonstructural proteins (nsp1–16), and accessory proteins. nsps play an essential role in viral replication and immune-invasion in the host cells.

- The establishment of SARS-CoV-2 infection is divided into three steps: spike protein binding the host cellular factors for viral entry to the host cell; viral RNA translation and transcription; and virion assembly and release out the infected cells. Host cellular factors are essential in determining infectivity, while viral nonstructural proteins affect the viral replication and reproduction of the virions. The S protein is subject to genetic mutations, which may affect virulence.
- Host cellular factors include the receptor protein (ACE2) on the host cell surfaces and the two main proteases, TMPRSS and furin. Proteases cleave spike protein sites to facilitate viral binding and fusion with the cell membrane. The furin cleavage site is unique to SARS-CoV-2. These factors are considered to build one of the main routes of viral entry to the host cell, but hardly explain the rapid transmission and high infectivity of SARS-CoV-2.
- Mounting evidence supports that additional receptors and proteases are involved in the S protein binding, activation, and fusion with the host cell membrane. Host cellular factors include CD147, GRP78, cathepsins B/L, DPP4, L-SIGN, DC-SIGN, sialic acid, Ly6E, and plasmin(ogen). There is even an additional direct route through endocytosis.
- The features of S protein and additional multiple host cellular factors together determine the characteristic of SARS-CoV-2, with high infectivity, rapid transmission, and a spectrum of clinical severity, which, together, are distinct from infection with any of the other human coronaviruses.

This chapter was initially published as journal article[185] and is granted to be included as a chapter of this book.

References

1. X. Zhang, Y. Tan, Y. Ling, G. Lu, F. Liu and Z. Yi, *et al.*, Viral and host factors related to the clinical outcome of COVID-19, *Nature*, 2020, **583**(7816), 437–440.
2. W. J. Guan, Z. Y. Ni, Y. Hu, W. H. Liang, C. Q. Ou and J. X. He, *et al.*, Clinical Characteristics of Coronavirus Disease 2019 in China, *N. Engl. J. Med.*, 2020, **382**(18), 1708–1720.
3. P. Zhou, X. Yang, X. Wang, B. Hu, L. Zhang and W. Zhang, *et al.*, A pneumonia outbreak associated with a new coronavirus of probable bat origin, *Nature*, 2020, **579**(7798), 270–273.
4. J. Kim, Y. Chung, H. Jo, N. Lee, M. Kim and S. Woo, *et al.*, Identification of Coronavirus Isolated from a Patient in Korea with COVID-19, *Osong Public Health Res. Perspect.*, 2020, **11**(1), 3–7.

5. A. Kojima, F. Fujinami, K. Doi, A. Yasoshima and A. Okaniwa, Isolation and properties of sialodacryoadenitis virus of rats, *Jikken Dobutsu*, 1980, **29**(4), 409–418.
6. L. Vijgen, E. Keyaerts, P. Lemey, P. Maes, K. Van Reeth and H. Nauwynck, *et al.*, Evolutionary history of the closely related group 2 coronaviruses: porcine hemagglutinating encephalomyelitis virus, bovine coronavirus, and human coronavirus OC43, *J. Virol.*, 2006, **80**(14), 7270–7274.
7. A. E. Stout, N. M. Andre, J. A. Jaimes, J. K. Millet and G. R. Whittaker, Coronaviruses in cats and other companion animals: Where does SARS-CoV-2/COVID-19 fit? *Vet. Microbiol.*, 2020, **247**, 108777.
8. Q. Zhang, H. Zhang, K. Huang, Y. Yang, X. Hui and J. Gao, *et al.*, A serological survey of SARS-CoV-2 in cat in Wuhan, *Emerging Microbes Infect.*, 2020, **9**(1), 2013–2019.
9. T. S. Fung and D. X. Liu, Human Coronavirus: Host-Pathogen Interaction, *Annu. Rev. Microbiol.*, 2019, **73**, 529–557.
10. Y. Fan, K. Zhao, Z.-L. Shi and P. Zhou, Bat coronaviruses in China, *Viruses*, 2019, **11**(3), 210.
11. P. C. Woo, S. K. Lau, C. S. Lam, C. C. Lau, A. K. Tsang and J. H. Lau, *et al.*, Discovery of seven novel Mammalian and avian coronaviruses in the genus deltacoronavirus supports bat coronaviruses as the gene source of alphacoronavirus and betacoronavirus and avian coronaviruses as the gene source of gammacoronavirus and deltacoronavirus, *J. Virol.*, 2012, **86**(7), 3995–4008.
12. D. Hamre, D. A. Kindig and J. Mann, Growth and intracellular development of a new respiratory virus, *J. Virol.*, 1967, **1**(4), 810–816.
13. D. A. Tyrrell and M. L. Bynoe, Cultivation of a Novel Type of Common-Cold Virus in Organ Cultures, *Br. Med. J.*, 1965, **1**(5448), 1467–1470.
14. K. McIntosh, A. Z. Kapikian, H. C. Turner, J. W. Hartley, R. H. Parrott and R. M. Chanock, Seroepidemiologic studies of coronavirus infection in adults and children, *Am. J. Epidemiol.*, 1970, **91**(6), 585–592.
15. J. J. Cavallaro and A. S. Monto, Community-wide outbreak of infection with a 229E-like coronavirus in Tecumseh, Michigan, *J. Infect. Dis.*, 1970, **122**(4), 272–279.
16. W. Li, Z. Shi, M. Yu, W. Ren, C. Smith and J. H. Epstein, *et al.*, Bats are natural reservoirs of SARS-like coronaviruses, *Science*, 2005, **310**(5748), 676–679.
17. L. van der Hoek, K. Pyrc, M. F. Jebbink, W. Vermeulen-Oost, R. J. Berkhout and K. C. Wolthers, *et al.*, Identification of a new human coronavirus, *Nat. Med.*, 2004, **10**(4), 368–373.
18. K. E. Arden, M. D. Nissen, T. P. Sloots and I. M. Mackay, New human coronavirus, HCoV-NL63, associated with severe lower respiratory tract disease in Australia, *J. Med. Virol.*, 2005, **75**(3), 455–462.
19. P. C. Woo, S. K. Lau, C. M. Chu, K. H. Chan, H. W. Tsoi and Y. Huang, *et al.*, Characterization and complete genome sequence of a novel

coronavirus, coronavirus HKU1, from patients with pneumonia, *J. Virol.*, 2005, **79**(2), 884–895.
20. R. J. de Groot, S. C. Baker, R. S. Baric, C. S. Brown, C. Drosten and L. Enjuanes, *et al.*, Middle East respiratory syndrome coronavirus (MERS-CoV): announcement of the Coronavirus Study Group, *J. Virol.*, 2013, **87**(14), 7790–7792.
21. M. Desforges, A. Le Coupanec, J. K. Stodola, M. Meessen-Pinard and P. J. Talbot, Human coronaviruses: viral and cellular factors involved in neuroinvasiveness and neuropathogenesis, *Virus Res.*, 2014, **194**, 145–158.
22. C. L. Yeager, R. A. Ashmun, R. K. Williams, C. B. Cardellichio, L. H. Shapiro and A. T. Look, *et al.*, Human aminopeptidase N is a receptor for human coronavirus 229E, *Nature*, 1992, **357**(6377), 420–422.
23. T. Cui, S. Theuns, J. Xie, W. Van den Broeck and H. J. Nauwynck, Role of Porcine Aminopeptidase N and Sialic Acids in Porcine Coronavirus Infections in Primary Porcine Enterocytes, *Viruses*, 2020, **12**(4), 402.
24. J. M. Parks and J. C. Smith, How to discover antiviral drugs quickly, *N. Engl. J. Med.*, 2020, **382**(23), 2261–2264.
25. S. J. R. da Silva, C. T. Alves da Silva, R. P. G. Mendes and L. Pena, Role of nonstructural proteins in the pathogenesis of SARS-CoV-2, *J. Med. Virol.*, 2020, **92**(9), 1427–1429.
26. M. Uddin, F. Mustafa, T. A. Rizvi, T. Loney, H. A. Suwaidi and A. H. H. Al-Marzouqi, *et al.*, SARS-CoV-2/COVID-19: Viral Genomics, Epidemiology, Vaccines, and Therapeutic Interventions, *Viruses*, 2020, **12**(5), 526.
27. D. Kim, J. Y. Lee, J. S. Yang, J. W. Kim, V. N. Kim and H. Chang, The Architecture of SARS-CoV-2 Transcriptome, *Cell*, 2020, **181**(4), 914–921.e10.
28. F. Robson, K. S. Khan, T. K. Le, C. Paris, S. Demirbag and P. Barfuss, *et al.*, Coronavirus RNA Proofreading: Molecular Basis and Therapeutic Targeting, *Mol. Cell*, 2020, **79**(5), 710–727.
29. E. J. Snijder, E. Decroly and J. Ziebuhr, The Nonstructural Proteins Directing Coronavirus RNA Synthesis and Processing, *Adv. Virus Res*, 2016, **96**, 59–126.
30. D. Shin, R. Mukherjee, D. Grewe, D. Bojkova, K. Baek and A. Bhattacharya, *et al.*, Papain-like protease regulates SARS-CoV-2 viral spread and innate immunity, *Nature*, 2020, **587**(7835), 657–662.
31. M. Thoms, R. Buschauer, M. Ameismeier, L. Koepke, T. Denk and M. Hirschenberger, *et al.*, Structural basis for translational shutdown and immune evasion by the Nsp1 protein of SARS-CoV-2, *Science*, 2020, **369**(6508), 1249–1255.
32. S. Angeletti, D. Benvenuto, M. Bianchi, M. Giovanetti, S. Pascarella and M. Ciccozzi, COVID-2019: The role of the nsp2 and nsp3 in its pathogenesis, *J. Med. Virol.*, 2020, **92**(6), 584–588.
33. Y. M. Baez-Santos, S. E. St John and A. D. Mesecar, The SARS-coronavirus papain-like protease: structure, function and inhibition by designed antiviral compounds, *Antiviral Res.*, 2015, **115**, 21–38.

34. M. Frieman, B. Yount, S. Agnihothram, C. Page, E. Donaldson and A. Roberts, *et al.*, Molecular determinants of severe acute respiratory syndrome coronavirus pathogenesis and virulence in young and aged mouse models of human disease, *J. Virol.*, 2012, **86**(2), 884–897.
35. D. R. Littler, B. S. Gully, R. N. Colson and J. Rossjohn, Crystal Structure of the SARS-CoV-2 Non-structural Protein 9, Nsp9, *iScience*, 2020, **23**(7), 101258.
36. X. Lei, X. Dong, R. Ma, W. Wang, X. Xiao and Z. Tian, *et al.*, Activation and evasion of type I interferon responses by SARS-CoV-2, *Nat. Commun.*, 2020, **11**(1), 3810.
37. W. Tai, L. He, X. Zhang, J. Pu, D. Voronin and S. Jiang, *et al.*, Characterization of the receptor-binding domain (RBD) of 2019 novel coronavirus: implication for development of RBD protein as a viral attachment inhibitor and vaccine, *Cell. Mol. Immunol.*, 2020, **17**(6), 613–620.
38. J. A. Regla-Nava, J. L. Nieto-Torres, J. M. Jimenez-Guardeño, R. Fernandez-Delgado, C. Fett and C. Castaño-Rodríguez, *et al.*, Severe acute respiratory syndrome coronaviruses with mutations in the E protein are attenuated and promising vaccine candidates, *J. Virol.*, 2015, **89**(7), 3870–3887.
39. M. L. DeDiego, J. L. Nieto-Torres, J. M. Jiménez-Guardeño, J. A. Regla-Nava, E. Alvarez and J. C. Oliveros, *et al.*, Severe acute respiratory syndrome coronavirus envelope protein regulates cell stress response and apoptosis, *PLoS Pathog.*, 2011, **7**(10), e1002315.
40. J. M. Jimenez-Guardeño, J. L. Nieto-Torres, M. L. DeDiego, J. A. Regla-Nava, R. Fernandez-Delgado and C. Castaño-Rodriguez, *et al.*, The PDZ-binding motif of severe acute respiratory syndrome coronavirus envelope protein is a determinant of viral pathogenesis, *PLoS Pathog.*, 2014, **10**(8), e1004320.
41. P. D. Burbelo, F. X. Riedo, C. Morishima, S. Rawlings, D. Smith and S. Das, *et al.*, Sensitivity in Detection of Antibodies to Nucleocapsid and Spike Proteins of Severe Acute Respiratory Syndrome Coronavirus 2 in Patients With Coronavirus Disease, *J. Infect. Dis.*, 2020, **22**(2), 206–213.
42. J. Y. Li, C. H. Liao, Q. Wang, Y. J. Tan, R. Luo and Y. Qiu, *et al.*, The ORF6, ORF8 and nucleocapsid proteins of SARS-CoV-2 inhibit type I interferon signaling pathway, *Virus Res.*, 2020, **286**, 198074.
43. S. Srinivasan, H. Cui, Z. Gao, M. Liu, S. Lu and W. Mkandawire, *et al.*, Structural Genomics of SARS-CoV-2 Indicates Evolutionary Conserved Functional Regions of Viral Proteins, *Viruses*, 2020, **12**(4), 360.
44. J. Diaz, SARS-CoV-2 Molecular Network Structure, *Front. Physiol.*, 2020, **11**, 870.
45. Y. Jin, H. Yang, W. Ji, W. Wu, S. Chen and W. Zhang, *et al.*, Virology, Epidemiology, Pathogenesis, and Control of COVID-19, *Viruses*, 2020, **12**(4), 372.
46. C. J. Michel, C. Mayer, O. Poch and J. D. Thompson, Characterization of accessory genes in coronavirus genomes, *Virol. J.*, 2020, **17**(1), 131.

47. C. Castano-Rodriguez, J. M. Honrubia, J. Gutierrez-Alvarez, M. L. DeDiego, J. L. Nieto-Torres and J. M. Jimenez-Guardeno, *et al.*, Role of Severe Acute Respiratory Syndrome Coronavirus Viroporins E, 3a, and 8a in Replication and Pathogenesis, *mBio*, 2018, **9**(3), e02325.
48. K. L. Siu, K. S. Yuen, C. Castaño-Rodriguez, Z. W. Ye, M. L. Yeung and S. Y. Fung, *et al.*, Severe acute respiratory syndrome coronavirus ORF3a protein activates the NLRP3 inflammasome by promoting TRAF3-dependent ubiquitination of ASC, *FASEB J.*, 2019, **33**(8), 8865–8877.
49. E. Issa, G. Merhi, B. Panossian, T. Salloum and S. Tokajian, SARS-CoV-2 and ORF3a: Nonsynonymous Mutations, Functional Domains, and Viral Pathogenesis, *mSystems*, 2020, **5**(3), e00266-20.
50. L. A. Holland, E. A. Kaelin, R. Maqsood, B. Estifanos, L. I. Wu and A. Varsani, *et al.*, An 81-Nucleotide Deletion in SARS-CoV-2 ORF7a Identified from Sentinel Surveillance in Arizona (January to March 2020), *J. Virol.*, 2020, **94**(14), e00711-20.
51. S. A. Kopecky-Bromberg, L. Martinez-Sobrido and P. Palese, 7a protein of severe acute respiratory syndrome coronavirus inhibits cellular protein synthesis and activates p38 mitogen-activated protein kinase, *J. Virol.*, 2006, **80**(2), 785–793.
52. Y. M. Báez-Santos, S. E. St John and A. D. Mesecar, The SARS-coronavirus papain-like protease: structure, function and inhibition by designed antiviral compounds, *Antiviral Res.*, 2015, **115**, 21–38.
53. J. Zhao, J. Sun, W. He, X. Ji, Q. Gao and X. Zhai, *et al.*, Snapshot of the evolution and mutation patterns of SARS-CoV-2, bioRxiv, 2020, DOI: 10.1101/2020.07.04.187435.
54. A. R. Jauregui, D. Savalia, V. K. Lowry, C. M. Farrell and M. G. Wathelet, Identification of residues of SARS-CoV nsp1 that differentially affect inhibition of gene expression and antiviral signaling, *PLoS One*, 2013, **8**(4), e62416.
55. F. Benedetti, G. A. Snyder, M. Giovanetti, S. Angeletti, R. C. Gallo and M. Ciccozzi, *et al.*, Emerging of a SARS-CoV-2 viral strain with a deletion in nsp1, *J. Transl. Med.*, 2020, **18**(1), 329.
56. D. Benvenuto, S. Angeletti, M. Giovanetti, M. Bianchi, S. Pascarella and R. Cauda, *et al.*, Evolutionary analysis of SARS-CoV-2: how mutation of Non-Structural Protein 6 (NSP6) could affect viral autophagy, *J. Infect.*, 2020, **81**(1), e24–e27.
57. C. T. Cornillez-Ty, L. Liao, J. R. Yates 3rd, P. Kuhn and M. J. Buchmeier, Severe acute respiratory syndrome coronavirus nonstructural protein 2 interacts with a host protein complex involved in mitochondrial biogenesis and intracellular signaling, *J. Virol.*, 2009, **83**(19), 10314–10318.
58. J. P. Davies, K. M. Almasy, E. F. McDonald and L. Plate. Comparative multiplexed interactomics of SARS-CoV-2 and homologous coronavirus non-structural proteins identifies unique and shared host-cell dependencies, *ACS Infect Dis*, 2020, **6**(12), 3174–3189.

59. C. C. Stobart, N. R. Sexton, H. Munjal, X. Lu, K. L. Molland and S. Tomar, *et al.*, Chimeric exchange of coronavirus nsp5 proteases (3CLpro) identifies common and divergent regulatory determinants of protease activity, *J. Virol.*, 2013, **87**(23), 12611–12618.
60. M. M. Angelini, M. Akhlaghpour, B. W. Neuman and M. J. Buchmeier, Severe acute respiratory syndrome coronavirus nonstructural proteins 3, 4, and 6 induce double-membrane vesicles, *mBio*, 2013, **4**(4), e00524-13.
61. E. M. Cottam, M. C. Whelband and T. Wileman, Coronavirus NSP6 restricts autophagosome expansion, *Autophagy*, 2014, **10**(8), 1426–1441.
62. R. N. Kirchdoerfer and A. B. Ward, Structure of the SARS-CoV nsp12 polymerase bound to nsp7 and nsp8 co-factors, *Nat. Commun.*, 2019, **10**(1), 2342.
63. A. Rogstam, M. Nyblom, S. Christensen, C. Sele, V. O. Talibov and T. Lindvall, *et al.*, Crystal Structure of Non-Structural Protein 10 from Severe Acute Respiratory Syndrome Coronavirus-2, *Int. J. Mol. Sci.*, 2020, **21**(19), 7375.
64. H. S. Hillen, G. Kokic, L. Farnung, C. Dienemann, D. Tegunov and P. Cramer, Structure of replicating SARS-CoV-2 polymerase, *Nature*, 2020, **584**(7819), 154–156.
65. Y. Jiang, W. Yin and H. E. Xu, RNA-dependent RNA polymerase: Structure, mechanism, and drug discovery for COVID-19, *Biochem. Biophys. Res. Commun.*, 2021, **538**, 47–53.
66. Y. Gao, L. Yan, Y. Huang, F. Liu, Y. Zhao and L. Cao, *et al.*, Structure of the RNA-dependent RNA polymerase from COVID-19 virus, *Science*, 2020, **368**(6492), 779–782.
67. K. J. Jang, S. Jeong, D. Y. Kang, N. Sp, Y. M. Yang and D. E. Kim, A high ATP concentration enhances the cooperative translocation of the SARS coronavirus helicase nsP13 in the unwinding of duplex RNA, *Sci. Rep.*, 2020, **10**(1), 4481.
68. D. E. Gordon, G. M. Jang, M. Bouhaddou, J. Xu, K. Obernier and K. M. White, *et al.*, A SARS-CoV-2 protein interaction map reveals targets for drug repurposing, *Nature*, 2020, **583**(7816), 459–468.
69. X. Deng, M. Hackbart, R. C. Mettelman, A. O'Brien, A. M. Mielech and G. Yi, *et al.*, Coronavirus nonstructural protein 15 mediates evasion of dsRNA sensors and limits apoptosis in macrophages, *Proc. Natl. Acad. Sci. U. S. A.*, 2017, **114**(21), E4251–E4260.
70. S. Lin, H. Chen, F. Ye, Z. Chen, F. Yang and Y. Zheng, *et al.*, Crystal structure of SARS-CoV-2 nsp10/nsp16 2'-O-methylase and its implication on antiviral drug design, *Signal Transduction Targeted Ther.*, 2020, **5**(1), 131.
71. M. Hoffmann, H. Kleine-Weber, S. Schroeder, N. Kruger, T. Herrler and S. Erichsen, *et al.*, SARS-CoV-2 Cell Entry Depends on ACE2 and TMPRSS2 and Is Blocked by a Clinically Proven Protease Inhibitor, *Cell*, 2020, **181**(2), 271–280 e8.

72. K. L. Siu, K. S. Yuen, C. Castano-Rodriguez, Z. W. Ye, M. L. Yeung and S. Y. Fung, *et al.*, Severe acute respiratory syndrome coronavirus ORF3a protein activates the NLRP3 inflammasome by promoting TRAF3-dependent ubiquitination of ASC, *FASEB J.*, 2019, **33**(8), 8865–8877.
73. Y. Ren, T. Shu, D. Wu, J. Mu, C. Wang and M. Huang, *et al.*, The ORF3a protein of SARS-CoV-2 induces apoptosis in cells, *Cell. Mol. Immunol.*, 2020, **17**(8), 881–883.
74. Y. Konno, I. Kimura, K. Uriu, M. Fukushi, T. Irie and Y. Koyanagi, *et al.*, SARS-CoV-2 ORF3b Is a Potent Interferon Antagonist Whose Activity Is Increased by a Naturally Occurring Elongation Variant, *Cell Rep.*, 2020, **32**(12), 108185.
75. D. Schoeman and B. C. Fielding, Coronavirus envelope protein: current knowledge, *Virol. J.*, 2019, **16**(1), 69.
76. S. Thomas, The Structure of the Membrane Protein of SARS-CoV-2 Resembles the Sugar Transporter SemiSWEET, *Pathog. Immun.*, 2020, **5**(1), 342–363.
77. J. K. Taylor, C. M. Coleman, S. Postel, J. M. Sisk, J. G. Bernbaum and T. Venkataraman, *et al.*, Severe Acute Respiratory Syndrome Coronavirus ORF7a Inhibits Bone Marrow Stromal Antigen 2 Virion Tethering through a Novel Mechanism of Glycosylation Interference, *J. Virol.*, 2015, **89**(23), 11820–11833.
78. S. Pfefferle, V. Krähling, V. Ditt, K. Grywna, E. Mühlberger and C. Drosten, Reverse genetic characterization of the natural genomic deletion in SARS-Coronavirus strain Frankfurt-1 open reading frame 7b reveals an attenuating function of the 7b protein in-vitro and in-vivo, *Virol. J.*, 2009, **6**, 131.
79. S. C. Sung, C. Y. Chao, K. S. Jeng, J. Y. Yang and M. M. Lai, The 8ab protein of SARS-CoV is a luminal ER membrane-associated protein and induces the activation of ATF6, *Virology*, 2009, **387**(2), 402–413.
80. J. Mu, J. Xu, L. Zhang, T. Shu, D. Wu and M. Huang, *et al.*, SARS-CoV-2-encoded nucleocapsid protein acts as a viral suppressor of RNA interference in cells, *Sci. China: Life Sci.*, 2020, **63**(9), 1–4.
81. C. S. Shi, H. Y. Qi, C. Boularan, N. N. Huang, M. Abu-Asab and J. H. Shelhamer, *et al.*, SARS-coronavirus open reading frame-9b suppresses innate immunity by targeting mitochondria and the MAVS/TRAF3/TRAF6 signalosome, *J. Immunol.*, 2014, **193**(6), 3080–3089.
82. A. Dominguez Andres, Y. Feng, A. R. Campos, J. Yin, C. C. Yang and B. James, *et al.*, SARS-CoV-2 ORF9c Is a Membrane-Associated Protein that Suppresses Antiviral Responses in Cells, bioRxiv, 2020, DOI: 10.1101/2020.08.18.256776.
83. M. Thunders and B. Delahunt, Gene of the month: TMPRSS2 (transmembrane serine protease 2), *J. Clin. Pathol.*, 2020, **73**(12), 773–776.
84. A. Walls, Y. Park, M. Tortorici, A. Wall, A. McGuire and D. Veesler, Structure, function, and antigenicity of the SARS-CoV-2 spike glycoprotein, *Cell*, 2020, **181**(2), 281–292.

85. T. S. Pillay, Gene of the month: the 2019-nCoV/SARS-CoV-2 novel coronavirus spike protein, *J. Clin. Pathol.*, 2020, **73**(7), 366–369.
86. D. Bestle, M. R. Heindl, H. Limburg, T. Van Lam van, O. Pilgram and H. Moulton, *et al.*, TMPRSS2 and furin are both essential for proteolytic activation of SARS-CoV-2 in human airway cells, *Life Sci. Alliance*, 2020, **3**(9), e202000786.
87. N. Nao, J. Yamagishi, H. Miyamoto, M. Igarashi, R. Manzoor and A. Ohnuma, *et al.*, Genetic Predisposition To Acquire a Polybasic Cleavage Site for Highly Pathogenic Avian Influenza Virus, Hemagglutinin, *mBio*, 2017, **8**(1), e02298-16.
88. K. G. Andersen, A. Rambaut, W. I. Lipkin, E. C. Holmes and R. F. Garry, The proximal origin of SARS-CoV-2, *Nat. Med.*, 2020, **26**(4), 450–452.
89. D. Wrapp, N. Wang, K. S. Corbett, J. A. Goldsmith, C. L. Hsieh and O. Abiona, *et al.*, Cryo-EM structure of the 2019-nCoV spike in the prefusion conformation, *Science*, 2020, **367**(6483), 1260–1263.
90. J. Lu and P. D. Sun, High affinity binding of SARS-CoV-2 spike protein enhances ACE2 carboxypeptidase activity, *J. Biolog. Chem.*, 2020, **295**, 18579–18588.
91. S. Joshi, M. Joshi and M. Degani, Tackling SARS-CoV-2: proposed targets and repurposed drugs, *Future Med. Chem.*, 2020, **12**(17), 1579–1601.
92. M. A. White, W. Lin and X. Cheng, Discovery of COVID-19 Inhibitors Targeting the SARS-CoV-2 Nsp13 Helicase, *J. Phys. Chem. Lett.*, 2020, **11**(21), 9144–9151.
93. H. H. Wong, P. Kumar, F. P. Tay, D. Moreau, D. X. Liu and F. Bard, Genome-Wide Screen Reveals Valosin-Containing Protein Requirement for Coronavirus Exit from Endosomes, *J. Virol.*, 2015, **89**(21), 11116–11128.
94. W. T. Soh, Y. Liu, E. E. Nakayama, C. Ono, S. Torii and H. Nakagami, *et al.*, The N-terminal domain of spike glycoprotein mediates SARS-CoV-2 infection by associating with L-SIGN and DC-SIGN, bioRxiv, 2020, DOI: 10.1101/2020.11.05.369264.
95. I. C. Huang, B. J. Bosch, F. Li, W. Li, K. H. Lee and S. Ghiran, *et al.*, SARS coronavirus, but not human coronavirus NL63, utilizes cathepsin L to infect ACE2-expressing cells, *J. Biol. Chem.*, 2006, **281**(6), 3198–3203.
96. K. Wang, W. Chen, Y.-S. Zhou, J.-Q. Lian, Z. Zhang and P. Du, *et al.*, SARS-CoV-2 invades host cells *via* a novel route: CD147-spike protein, bioRxiv. 2020, DOI: 10.1101/2020.03.14.988345.
97. J. Shang, G. Ye, K. Shi, Y. Wan, C. Luo and H. Aihara, *et al.*, Structural basis of receptor recognition by SARS-CoV-2, *Nature*, 2020, **581**(7807), 221–224.
98. V. S. Raj, H. Mou, S. L. Smits, D. H. Dekkers, M. A. Müller and R. Dijkman, *et al.*, Dipeptidyl peptidase 4 is a functional receptor for the emerging human coronavirus-EMC, *Nature*, 2013, **495**(7440), 251–254.
99. R. J. G. Hulswit, Y. Lang, M. J. G. Bakkers, W. Li, Z. Li and A. Schouten, *et al.*, Human coronaviruses OC43 and HKU1 bind to 9-O-acetylated sialic acids *via* a conserved receptor-binding site in spike protein domain A, *Proc. Natl. Acad. Sci. U. S. A.*, 2019, **116**(7), 2681–2690.

100. B. Coutard, C. Valle, X. de Lamballerie, B. Canard, N. G. Seidah and E. Decroly, The spike glycoprotein of the new coronavirus 2019-nCoV contains a furin-like cleavage site absent in CoV of the same clade, *Antiviral Res.*, 2020, **176**, 104742.
101. I. M. Ibrahim, D. H. Abdelmalek, M. E. Elshahat and A. A. Elfiky, COVID-19 spike-host cell receptor GRP78 binding site prediction, *J. Infect.*, 2020, **80**(5), 554–562.
102. X. Zhao, S. Zheng, D. Chen, M. Zheng, X. Li and G. Li, *et al.*, LY6E Restricts the Entry of Human Coronaviruses, Including the Currently Pandemic SARS-CoV-2, *J. Virol.*, 2020, e00562-20.
103. E. Milanetti, M. Miotto, L. Di Rienzo, M. Monti, G. Gosti and G. Ruocco, *In-Silico* evidence for two receptors based strategy of SARS-CoV-2, 2020, 24, https://arxiv.org/abs/2003.11107.
104. A. E. Renteria, L. Endam Mfuna, D. Adam, A. Filali-Mouhim, A. Maniakas and S. Rousseau, *et al.*, Azithromycin Downregulates Gene Expression of IL-1beta and Pathways Involving TMPRSS2 and TMPRSS11D Required by SARS-CoV-2, *Am. J. Respir. Cell Mol. Biol.*, 2020, **63**(5), 707–709.
105. M. Paulpandi, K. Kavithaa, A. Asaikkutty, V. Balachandar, N. Ayyadurai and N. Arul, Could host cell receptor alteration prevent SARS-CoV-2 viral entry? – Hype or hope, *Eur. Rev. Med. Pharmacol. Sci.*, 2020, **24**(8), 4554–4557.
106. A. J. Turner, ACE2 Cell Biology, Regulation, and Physiological Functions, *The Protective Arm of the Renin Angiotensin System (RAS)*, ed. T. Unger, U. M. Steckelings and R. A. S. dos Santos, Academic Press, Boston, 2015, ch. 25, pp. 185–189.
107. M. Uhlén, L. Fagerberg, B. M. Hallström, C. Lindskog, P. Oksvold and A. Mardinoglu, *et al.*, Proteomics. Tissue-based map of the human proteome, *Science*, 2015, **347**(6220), 1260419.
108. W. Li, M. J. Moore, N. Vasilieva, J. Sui, S. K. Wong and M. A. Berne, *et al.*, Angiotensin-converting enzyme 2 is a functional receptor for the SARS coronavirus, *Nature*, 2003, **426**(6965), 450–454.
109. F. Li, W. Li, M. Farzan and S. C. Harrison, Structure of SARS coronavirus spike receptor-binding domain complexed with receptor, *Science*, 2005, **309**(5742), 1864–1868.
110. Y. Wan, J. Shang, R. Graham, R. S. Baric and F. Li, Receptor Recognition by the Novel Coronavirus from Wuhan: an Analysis Based on Decade-Long Structural Studies of SARS Coronavirus, *J. Virol.*, 2020, **94**(7), e00127–20.
111. H. Xu, L. Zhong, J. Deng, J. Peng, H. Dan and X. Zeng, *et al.*, High expression of ACE2 receptor of 2019-nCoV on the epithelial cells of oral mucosa, *Int. J. Oral. Sci.*, 2020, **12**(1), 8.
112. F. Ali, M. Elserafy, M. H. Alkordi and M. Amin, ACE2 coding variants in different populations and their potential impact on SARS-CoV-2 binding affinity, *Biochem. Biophys. Rep.*, 2020, 100798.
113. C. Ferrario, J. Jessup, M. Chappell, D. Averill, K. Brosnihan and E. Tallant, *et al.*, Effect of Angiotensin-Converting Enzyme Inhibition and Angiotensin II Receptor Blockers on Cardiac Angiotensin-Converting Enzyme 2, *Circulation*, 2005, **111**(20), 2605–2610.

114. J. H. Diaz, Hypothesis: angiotensin-converting enzyme inhibitors and angiotensin receptor blockers may increase the risk of severe COVID-19, *J. Travel Med.*, **2020**, **27**(3), taaa041.
115. J. Hippisley-Cox, D. Young, C. Coupland, K. M. Channon, P. S. Tan and D. A. Harrison, *et al.*, Risk of severe COVID-19 disease with ACE inhibitors and angiotensin receptor blockers: cohort study including 8.3 million people, *Heart*, 2020, **106**, 1503–1511.
116. F. J. de Abajo, S. Rodriguez-Martin, V. Lerma, G. Mejia-Abril, M. Aguilar and A. Garcia-Luque, *et al.*, Use of renin-angiotensin-aldosterone system inhibitors and risk of COVID-19 requiring admission to hospital: a case-population study, *Lancet*, 2020, **395**(10238), 1705–1714.
117. K. Kuba, Y. Imai, S. Rao, H. Gao, F. Guo and B. Guan, *et al.*, A crucial role of angiotensin converting enzyme 2 (ACE2) in SARS coronavirus-induced lung injury, *Nat. Med.*, 2005, **11**(8), 875–879.
118. F. Hikmet, L. Méar, Å. Edvinsson, P. Micke, M. Uhlén and C. Lindskog, The protein expression profile of ACE2 in human tissues, *Mol. Syst. Biol.*, 2020, **16**(7), e9610.
119. M. C. Chappel and C. M. Ferrario, ACE and ACE2: their role to balance the expression of angiotensin II and angiotensin-(1–7), *Kidney Int.*, 2006, **70**(1), 8–10.
120. R. R. Gaddam, S. Chambers and M. Bhatia, ACE and ACE2 in inflammation: a tale of two enzymes, *Inflammation Allergy: Drug Targets*, 2014, **13**(4), 224–234.
121. R. A. S. Santos, G. Y. Oudit, T. Verano-Braga, G. Canta, U. M. Steckelings and M. Bader, The renin-angiotensin system: going beyond the classical paradigms, *Am. J. Physiol.: Heart Circ. Physiol.*, 2019, **316**(5), H958–H970.
122. M. P. Ocaranza, I. Godoy, J. E. Jalil, M. Varas, P. Collantes and M. Pinto, *et al.*, Enalapril attenuates downregulation of Angiotensin-converting enzyme 2 in the late phase of ventricular dysfunction in myocardial infarcted rat, *Hypertension*, 2006, **48**(4), 572–578.
123. C. Nataraj, M. I. Oliverio, R. B. Mannon, P. J. Mannon, L. P. Audoly and C. S. Amuchastegui, *et al.*, Angiotensin II regulates cellular immune responses through a calcineurin-dependent pathway, *J. Clin. Invest.*, 1999, **104**(12), 1693–1701.
124. P. Pagliaro and C. Penna, ACE/ACE2 Ratio: A Key Also in 2019 Coronavirus Disease (Covid-19)? *Front. Med.*, 2020, **7**, 335.
125. X. Qiu, Y. Tian, X. Jiang, J. Xu, Q. Zhang and J. Huang, Clinical Characteristics of Coronavirus Disease 19, *Global Clini .Transl. Res.*, 2020, **2**(3), 85–99.
126. Z. Wu, R. Hu, C. Zhang, W. Ren, A. Yu and X. Zhou, Elevation of plasma angiotensin II level is a potential pathogenesis for the critically ill COVID-19 patients, *Crit. Care*, 2020, **24**(1), 290.
127. M. K. Chung, S. Karnik, J. Saef, C. Bergmann, J. Barnard and M. M. Lederman, *et al.*, SARS-CoV-2 and ACE2: The biology and clinical data settling the ARB and ACEI controversy, *EBioMedicine*, 2020, **58**, 102907.

128. J. K. Achua, K. Y. Chu, E. Ibrahim, K. Khodamoradi, K. S. Delma and O. A. Iakymenko, *et al.*, Histopathology and Ultrastructural Findings of Fatal COVID-19 Infections on Testis, *World J. Mens Health*, 2021, **39**(1), 65–74.
129. B. T. Bradley, H. Maioli, R. Johnston, I. Chaudhry, S. L. Fink and H. Xu, *et al.*, Histopathology and ultrastructural findings of fatal COVID-19 infections in Washington State: a case series, *Lancet*, 2020, **396**(10247), 320–332.
130. A. Sonzogni, G. Previtali, M. Seghezzi, M. Grazia Alessio, A. Gianatti and L. Licini, *et al.*, Liver histopathology in severe COVID 19 respiratory failure is suggestive of vascular alterations, *Liver Int.*, 2020, **40**(9), 2110–2116.
131. M. Ackermann, S. E. Verleden, M. Kuehnel, A. Haverich, T. Welte and F. Laenger, *et al.*, Pulmonary Vascular Endothelialitis, Thrombosis, and Angiogenesis in Covid-19, *N. Engl. J. Med.*, 2020, **383**(2), 120–128.
132. C. Chaipan, D. Kobasa, S. Bertram, I. Glowacka, I. Steffen and T. S. Tsegaye, *et al.*, Proteolytic activation of the 1918 influenza virus hemagglutinin, *J. Virol.*, 2009, **83**(7), 3200–3211.
133. N. Iwata-Yoshikawa, T. Okamura, Y. Shimizu, H. Hasegawa, M. Takeda and N. Nagata, TMPRSS2 Contributes to Virus Spread and Immunopathology in the Airways of Murine Models after Coronavirus Infection, *J. Virol.*, 2019, **93**(6), e01815-18.
134. F. Zhang and C. Hughes, Clinical epidemiology of coronavirus disease 2019: defined on current research, *Global Clin. Transl. Res.*, 2020, **2**(3), 54–72.
135. K. H. Stopsack, L. A. Mucci, E. S. Antonarakis, P. S. Nelson and P. W. Kantoff, TMPRSS2 and COVID-19: Serendipity or Opportunity for Intervention? *Cancer Discovery*, 2020, **10**(6), 779–782.
136. J. D. Strope, D. C. Pharm and W. D. Figg, TMPRSS2: Potential Biomarker for COVID-19 Outcomes, *J. Clin. Pharmacol.*, 2020, **60**(7), 801–807.
137. A. E. Renteria, L. Mfuna Endam, D. Adam, A. Filali-Mouhim, A. Maniakas and S. Rousseau, *et al.*, Azithromycin Downregulates Gene Expression of IL-1beta and Pathways Involving TMPRSS2 and TMPRSS11D Required by SARS-CoV-2, *Am. J. Respir. Cell Mol. Biol.*, 2020, **63**(5), 707–709.
138. M. Hoffmann, H. Kleine-Weber and S. Pohlmann, A Multibasic Cleavage Site in the Spike Protein of SARS-CoV-2 Is Essential for Infection of Human Lung Cells, *Mol. Cell*, 2020, **78**(4), 779–784 e5.
139. S. Lukassen, R. L. Chua, T. Trefzer, N. C. Kahn, M. A. Schneider and T. Muley, *et al.*, SARS-CoV-2 receptor ACE2 and TMPRSS2 are primarily expressed in bronchial transient secretory cells, *EMBO J.*, 2020, **39**(10), e105114.
140. K. Drak Alsibai, Expression of angiotensin-converting enzyme 2 and proteases in COVID-19 patients: A potential role of cellular FURIN in the pathogenesis of SARS-CoV-2, *Med. Hypotheses*, 2020, **143**, 109893.
141. R. Mentlein, Dipeptidyl-peptidase IV (CD26)-role in the inactivation of regulatory peptides, *Regul. Pept.*, 1999, **85**(1), 9–24.

142. X. Chen, Biochemical properties of recombinant prolyl dipeptidases DPP-IV and DPP8, *Adv. Exp. Med. Biol.*, 2006, **575**, 27–32.
143. A. F. Kolb, J. Maile, A. Heister and S. G. Siddell, Characterization of functional domains in the human coronavirus HCV 229E receptor, *J. Gen. Virol.*, 1996, **77**(Pt 10), 2515–2521.
144. A. F. Kolb, A. Hegyi and S. G. Siddell, Identification of residues critical for the human coronavirus 229E receptor function of human aminopeptidase N, *J. Gen. Virol.*, 1997, **78**(Pt 11), 2795–2802.
145. M. F. Bassendine, S. H. Bridge, G. W. McCaughan and M. D. Gorrell, Covid-19 and co-morbidities: a role for Dipeptidyl Peptidase 4 (DPP4) in disease severity? *J. Diabetes*, 2020, **12**(9), 649–658.
146. N. Vankadari and J. A. Wilce, Emerging WuHan (COVID-19) coronavirus: glycan shield and structure prediction of spike glycoprotein and its interaction with human CD26, *Emerging Microbes Infect.*, 2020, **9**(1), 601–604.
147. C. Morimoto and S. F. Schlossman, The structure and function of CD26 in the T-cell immune response, *Immunol. Rev.*, 1998, **161**(1), 55–70.
148. I. Barchetta, M. G. Cavallo and M. G. Baroni, COVID-19 and diabetes: Is this association driven by the DPP4 receptor? Potential clinical and therapeutic implications, *Diabetes Res. Clin. Pract.*, 2020, **163**, 108165.
149. R. Strollo and P. Pozzilli, DPP4 inhibition: Preventing SARS-CoV-2 infection and/or progression of COVID-19? *Diabetes/Metab. Res. Rev.*, 2020, e3330.
150. K. Wang, W. Chen, Y.-S. Zhou, J.-Q. Lian, Z. Zhang and P. Du, *et al.*, CD147-spike protein is a novel route for SARS-CoV-2 infection to host cells, *Sig. Transduct. Target Ther.*, 2020, **5**, 283.
151. C. Crosnier, L. Y. Bustamante, S. J. Bartholdson, A. K. Bei, M. Theron and M. Uchikawa, *et al.*, Basigin is a receptor essential for erythrocyte invasion by Plasmodium falciparum, *Nature*, 2011, **480**(7378), 534–537.
152. H. Ulrich and M. M. Pillat, CD147 as a Target for COVID-19 Treatment: Suggested Effects of Azithromycin and Stem Cell Engagement, *Stem Cell Rev. Rep.*, 2020, **16**(3), 434–440.
153. H. Bian, Z.-H. Zheng, D. Wei, Z. Zhang, W.-Z. Kang and C.-Q. Hao, *et al.*, Meplazumab treats COVID-19 pneumonia: an open-labelled, concurrent controlled add-on clinical trial, medRxiv, 2020, DOI: 10.1101/2020.03.21.20040691.
154. Z. Chen, L. Mi, J. Xu, J. Yu, X. Wang and J. Jiang, *et al.*, Function of HAb18G/CD147 in invasion of host cells by severe acute respiratory syndrome coronavirus, *J. Infect. Dis.*, 2005, **191**(5), 755–760.
155. I. O. Rosas, T. J. Richards, K. Konishi, Y. Zhang, K. Gibson and A. E. Lokshin, *et al.*, MMP1 and MMP7 as potential peripheral blood biomarkers in idiopathic pulmonary fibrosis, *PLoS Med.*, 2008, **5**(4), e93.
156. P. M. George, A. U. Wells and R. G. Jenkins, Pulmonary fibrosis and COVID-19: the potential role for antifibrotic therapy, *Lancet Respir. Med.*, 2020, **8**(8), 807–815.

157. U. Radzikowska, M. Ding, G. Tan, D. Zhakparov, Y. Peng and P. Wawrzyniak, *et al.*, Distribution of ACE2, CD147, CD26, and other SARS-CoV-2 associated molecules in tissues and immune cells in health and in asthma, COPD, obesity, hypertension, and COVID-19 risk factors, *Allergy*, 2020, **75**(11), 2829–2845.
158. I. M. Ibrahim, D. H. Abdelmalek and A. A. Elfiky, GRP78: A cell's response to stress, *Life Sci.*, 2019, **226**, 156–163.
159. A. Palmeira, E. Sousa, A. Koseler, R. Sabirli, T. Goren and I. Turkcuer, *et al.*, Preliminary Virtual Screening Studies to Identify GRP78 Inhibitors Which May Interfere with SARS-CoV-2 Infection, *Pharmaceuticals*, 2020, **13**(6), 132.
160. H. V. Sudeep, K. Gouthamchandra and K. Shyamprasad, Molecular docking analysis of Withaferin A from Withania somnifera with the Glucose regulated protein 78 (GRP78) receptor and the SARS-CoV-2 main protease, *Bioinformation*, 2020, **16**(5), 411–417.
161. X. Ou, Y. Liu, X. Lei, P. Li, D. Mi and L. Ren, *et al.*, Characterization of spike glycoprotein of SARS-CoV-2 on virus entry and its immune cross-reactivity with SARS-CoV, *Nat. Commun.*, 2020, **11**(1), 1620.
162. Z. Daniloski, T. X. Jordan, H.-H. Wessels, D. A. Hoagland, S. Kasela and M. Legut, *et al.*, Identification of Required Host Factors for SARS-CoV-2 Infection in Human Cells, *Cell*, 2021, **184**(1), 92–105.e16.
163. J. Yu and S. L. Liu, Emerging Role of LY6E in Virus-Host Interactions, *Viruses*, 2019, **11**(11), 1020.
164. X. Zhao, J. Li, C. A. Winkler, P. An and J. T. Guo, IFITM Genes, Variants, and Their Roles in the Control and Pathogenesis of Viral Infections, *Front. Microbiol.*, 2018, **9**, 3228.
165. D. Chen, Z. Hou, D. Jiang, M. Zheng, G. Li and Y. Zhang, *et al.*, GILT restricts the cellular entry mediated by the envelope glycoproteins of SARS-CoV, Ebola virus and Lassa fever virus, *Emerging Microbes Infect.*, 2019, **8**(1), 1511–1523.
166. S. Y. Liu, R. Aliyari, K. Chikere, G. Li, M. D. Marsden and J. K. Smith, *et al.*, Interferon-inducible cholesterol-25-hydroxylase broadly inhibits viral entry by production of 25-hydroxycholesterol, *Immunity*, 2013, **38**(1), 92–105.
167. Q. Shu, N. J. Lennemann, S. N. Sarkar, Y. Sadovsky and C. B. Coyne, ADAP2 Is an Interferon Stimulated Gene That Restricts RNA Virus Entry, *PLoS Pathog.*, 2015, **11**(9), e1005150.
168. J. Yu and S. L. Liu, Emerging Role of LY6E in Virus-Host Interactions, *Viruses*, 2019, **11**(11), 1020.
169. A. M. Ille, E. Kishel, R. Bodea, A. Ille, H. Lamont and S. Amico-Ruvio, Protein LY6E as a candidate for mediating transport of adeno-associated virus across the human blood-brain barrier, *J. NeuroVirol.*, 2020, **26**, 769–778.
170. L. Luo, P. McGarvey, S. Madhavan, R. Kumar, Y. Gusev and G. Upadhyay, Distinct lymphocyte antigens 6 (Ly6) family members Ly6D, Ly6E, Ly6K and Ly6H drive tumorigenesis and clinical outcome, *Oncotarget*, 2016, **7**(10), 11165–11193.

171. J. P. Gardner, R. J. Durso, R. R. Arrigale, G. P. Donovan, P. J. Maddon and T. Dragic, *et al.*, L-SIGN (CD 209L) is a liver-specific capture receptor for hepatitis C virus, *Proc. Natl. Acad. Sci. U. S. A.*, 2003, **100**(8), 4498–4503.
172. S. Olofsson, U. Kumlin, K. Dimock and N. Arnberg, Avian influenza and sialic acid receptors: more than meets the eye? *Lancet Infect. Dis.*, 2005, **5**(3), 184–188.
173. W. Li, R. J. G. Hulswit, I. Widjaja, V. S. Raj, R. McBride and W. Peng, *et al.*, Identification of sialic acid-binding function for the Middle East respiratory syndrome coronavirus spike glycoprotein, *Proc. Natl. Acad. Sci. U. S. A.*, 2017, **114**(40), E8508–E8517.
174. J. Fantini, C. Di Scala, H. Chahinian and N. Yahi, Structural and molecular modelling studies reveal a new mechanism of action of chloroquine and hydroxychloroquine against SARS-CoV-2 infection, *Int. J. Antimicrob. Agents*, 2020, **55**(5), 105960.
175. P. Wielgat, K. Rogowski, K. Godlewska and H. Car, Coronaviruses: Is Sialic Acid a Gate to the Eye of Cytokine Storm? From the Entry to the Effects, *Cells*, 2020, **9**(9), 1963.
176. A. B. Engin, E. D. Engin and A. Engin, Dual function of sialic acid in gastrointestinal SARS-CoV-2 infection, *Environ. Toxicol. Pharmacol.*, 2020, **79**, 103436.
177. S. Seth, J. Batra and S. Srinivasan, COVID-19: Targeting Proteases in Viral Invasion and Host Immune Response, *Front. Mol. Biosci.*, 2020, **7**, 215.
178. H. L. Ji, R. Zhao, S. Matalon and M. A. Matthay, Elevated Plasmin(ogen) as a Common Risk Factor for COVID-19 Susceptibility, *Physiol. Rev.*, 2020, **100**(3), 1065–1075.
179. U. Kaur, S. S. Chakrabarti, B. Ojha, B. K. Pathak, A. Singh and L. Saso, *et al.*, Targeting host cell proteases to prevent SARS-CoV-2 invasion, *Curr. Drug Targets*, 2020, DOI: 10.2174/1389450121666200924113243.
180. Y. Umemura, K. Yamakawa, T. Kiguchi, T. Nishida, M. Kawada and S. Fujimi, Hematological Phenotype of COVID-19-Induced Coagulopathy: Far from Typical Sepsis-Induced Coagulopathy, *J. Clin. Med.*, 2020, **9**(9), 2875.
181. Y. Huang, X. Lyu, D. Li, L. Wang, Y. Wang and W. Zou, *et al.*, A cohort study of 676 patients indicates D-dimer is a critical risk factor for the mortality of COVID-19, *PLoS One*, 2020, **15**(11), e0242045.
182. B. Solun and Y. Shoenfeld, Inhibition of metalloproteinases in therapy for severe lung injury due to COVID-19, *Med. Drug Discovery*, 2020, **7**, 100052.
183. A. R. Thierry, Anti-protease Treatments Targeting Plasmin(ogen) and Neutrophil Elastase May Be Beneficial in Fighting COVID-19, *Physiol. Rev.*, 2020, **100**(4), 1597–1598.
184. B. Heissig, Y. Salama, S. Takahashi, T. Osada and K. Hattori, The multifaceted role of plasminogen in inflammation, *Cell. Signalling*, 2020, **75**, 109761.

185. F. Zhang and M. Walters, Pathogen genomics and host cellular susceptibility factors of COVID-19, *Global Clin. Transl. Res.*, 2020, **2**(4), 107–126.
186. F. Zhang and M. Waters, Correction: Pathogen Genomics and Host Cellular Susceptibility Factors, *Global Clin. Transl. Res.*, 2020, **3**(2), 35–36.

CHAPTER 9

Infection and Pathogenesis of SARS-CoV-2: an Immunological Perspective

SNEHA MOHANTY[a,b] AND PRABHANSHU TRIPATHI*[a]

[a]Bio-molecular Toxicology Lab, Food Drug and Chemical Toxicology Division, CSIR-Indian Institute of Toxicology Research, Lucknow-226001, India; [b]Academy of Scientific and Innovative Research (AcSIR), Ghaziabad-201002, India
*E-mail: p.tripathi1@iitr.res.in

9.1 Introduction

Coronaviruses constitute a large collection of viruses which are capable of infecting humans as well as animals. These viruses can infect the respiratory, gastrointestinal, renal, and nervous systems.[1,2] Prior to the current pandemic caused by severe acute respiratory syndrome coronavirus 2 (SARS-CoV-2), the 21st century has witnessed two other coronaviruses zoonotic outbreaks, caused by SARS-CoV in 2002 and Middle East respiratory syndrome (MERS-CoV) in 2012.[3] The SARS-CoV-2 pandemic originated when several people were exposed in the Huanan seafood market situated in Wuhan, China, in December 2019. Those people manifested a mysterious pneumonia with unknown etiology, and later the culprit was identified as a novel

The Coronavirus Pandemic and the Future Volume 1: Virology, Epidemiology, Translational Toxicology and Therapeutics
Edited by Michael D. Waters, Alok Dhawan, Tim Marrs, Diana Anderson, Stafford Warren and Claude L. Hughes

Published by the Royal Society of Chemistry, www.rsc.org

betacoronavirus, which was officially designated as SARS-CoV-2.[4,5] On 11 February 2020, the World Health Organization (WHO) named this disease "coronavirus disease 2019" (COVID-19), and subsequently, on 11 March 2020, announced COVID-19 as pandemic, due to high infectivity, and global morbidities and mortalities.[6]

Even though there are many uncertainties regarding SARS-CoV-2 transmittance in humans, it is speculated that the initial transmission to humans occurred *via* infected bats and pangolins, and which then expanded around the globe.[7] SARS-CoV-2 is transmitted from one person to another primarily either through aerosol microdroplets disseminated by sneezing or coughing or by direct interaction with an infected individual,[8,9] and the incubation period ranges from 2 to 14 days.[10] Symptoms range from mild to severe respiratory syndrome influenza-like illness with loss of taste and smell, lower respiratory tract symptoms, and severe gastrointestinal symptoms. In most cases, COVID-19 is mild; however, it can be fatal following the development of acute respiratory distress syndrome (ARDS), severe pneumonia, high fever, headache, and even multi-organ failure.[11] At the time of writing of this chapter, COVID-19 had infected 53 million (53 164 803) people and taken 1 million (1 300 576) lives all over the world, which has drastically affected public health and economies worldwide.

Despite the fact that SARS-CoV-2 has spread all around the world, severity and mortality vary in all regions. While some people develop severe COVID-19, others develop mild symptoms or even remain asymptomatic. A few of the possible reasons for the observed differential response include the nature of the infecting virus (wild or mutant), viral titer, and differential expression of angiotensin-converting enzyme 2 (ACE2).[12,13] An additional primary reason for such variation might be the immune system.

SARS-CoV-2 primarily uses the ACE2 receptor for entry into human cells.[14,15] The infected immune cell uses numerous pattern recognition receptors (PRRs) to detect the presence of SARS-CoV-2 while it replicates its RNA.[16] This recognition of foreign substances activates several downstream transcription factors, mainly interferon (IFN) and nuclear factor (NF-κB).[17] These transcription factors promote the activation of two antiviral processes. On one hand, cellular antiviral defences are engaged, mediated by induction of IFN-I and IFN-III transcription and consequent IFN-stimulated gene (ISG) upregulation.[18] On the other hand, specific subsets of leukocytes are recruited and coordinated, mainly by the secreted chemokines.[19,20] Thus, there are various pathways by which the body activates itself to respond to the viral infection. These immunological responses vary from one individual to another, which probably plays an important role in disease severity.

This chapter discusses how different immunological responses determine the severity of COVID-19 with consideration of factors such as age, food habits, and blood group. Lastly, we focus on selected ongoing clinical trials in various parts of world and their outcomes.

9.2 SARS-CoV-2 Virology

The viral taxonomic classification system assigns SARS-CoV-2 to the Nidovirales order, Coronaviridae family, *Betacoronavirus* (β-CoV) genus and *Sarbecovirus* (beta-CoV lineage B) subgenus, which is a new strain. Phylogenetic analysis demonstrated ~80% similarity with SARS-CoV and very close (>95%) similarity with coronaviruses found in Chinese bats, suggesting bats as a natural reservoir and possible originating source.[21,22]

9.2.1 Structure

The shape of SARS-CoV-2 has been found to be nearly spherical or somewhat polymorphic, with its diameter ranging from 80 nm to 160 nm.[23] The positive-sense RNA virion is enclosed in a lipid envelope. The RNA is associated with a phosphorylated capsid protein which forms the nucleocapsid (N) inside phospholipid bilayers which forms the envelope. The envelope seats the club shaped spike glycoprotein (S) and possibly the hemagglutinin-esterase (HE) protein. The S protein on the envelope appears as crown-like spikes, resembling the classic coronaviruses. Within the spikes in the envelope are located the membrane (M) protein and the envelope (E) protein.

9.2.2 Genome

The SARS-CoV-2 genome (GenBank no. MN908947) is 29.9 kb positive-sense single-stranded RNA (+ssRNA) encoding 9860 amino acids. The genome includes the 5′-untranslated region (5′-UTR), 3′-UTR, replicate open reading frame (ORF) and structural proteins. The genome contains no fewer than 10 ORFs, such as ORF1a/b, N, S, M, and E genes, and a few additional genes, for instance ORF3b and OFR8.[24] The ORF1a/b, *i.e.* the first ORF, occupies almost two-thirds of the RNA and encodes two large polyproteins (pps), namely pp1a and pp1ab. Furthermore, different proteases process these polyproteins to 16 non-structural proteins (NSPs) which are NSP1–NSP16 which form the replicase–transcriptase complex (RTC). The remaining one-third of the genome that contains other ORFs encodes the four principal structural (N, E, M, S) proteins, and various accessory proteins like 3a, 3b, p6, 7a, 7b, 8b, 9b.[5,25]

9.2.3 Proteome

The four structural proteins, N, E, M, and S participate in the formation of virion particles. Additionally, four other conserved NSPs, RNA-dependent RNA polymerase (RdRp, NSP12), 3CL-protease (NSP5), papain-like protease (NSP3), and helicase (NSP13)[26] are essential for RNA replication.[27,28]

The glycoprotein spikes (S) are 23 nm long and 7 nm wide.[23] The S proteins are viral fusion proteins, which facilitate the receptor binding that plays an immense role in selecting the host that is susceptible to infection.[29] The S proteins are trimers, with each one containing a *N*-glycosylated ectodomain (*N*-GE), a cleavable N-terminal signal peptide, a transmembrane region and one cytoplasmic tail. Upon protease action, the *N*-GE is cleaved into a variable S1 domain, which is involved in receptor binding, and a conserved S2 domain, which is associated with membrane fusion.[30]

The E protein, the smallest structural protein, ejects the virus from infected cells by acting as a viroporin that alters the cell membrane of the host. The M protein gives a distinct shape to the envelope by interacting with the N. The N protein, being a multifunctional protein, forms a complex with ssRNA. During the virion particle packaging, the N protein facilitates the interaction between M and N proteins thus enhancing the viral genome transcription. Furthermore, it weakens the host antiviral response [*e.g.* RNA interference (RNAi), interferon] by modulating the signal transduction pathways.[31] In addition to these structural proteins, the NSPs also play a vital role. The significant constituent of the replication/transcription machinery is NSP12, known as RdRp.[32]

9.2.4 Mutation

Frequent mutations have been evidenced in SARS-CoV-2, which shows that the virus is continuing to adapt to responses in humans. Extensive analysis of the viral genome in humans has identified frequent, independent mutations in 198 sites.[33] One of the most frequent mutations has been observed in the S protein, *i.e.* the receptor binding domain (RBD).[34] Mutation of the aspartate (D) located at position 614 in the RBD to glycine (G) (G614) has resulted in its stabilization, which increases the accessibility to the protease of the host cell.[35] Notably, it has been found that the mortality rate varies with frequency of the virus displaying mutation at G614 in different countries.[36] Therefore, mutation can have an intense impact on SARS-CoV-2 pathogenicity and disease severity. Significantly, a study has found that SARS-CoV-2 is evolving and different mutation patterns have been characterized in European, North American and Asian strains.[32] From these studies, it can be deduced that disease severity, which depends on mutation type, might be region-specific.

9.2.5 Transmission

As mentioned earlier, the primary mode of transmission of SARS-CoV-2 is through droplets and aerosols from oral and nasal passages that are generated while sneezing, coughing, and talking.[37] Other means of transmission include direct contact with COVID-19 patients and indirect contact with various surfaces, such as household items that have been associated with COVID-19 patients. Additionally, a study by Zhang *et al.* found that viral nucleotides were present in blood and anal swabs. They observed that the anal swab had a higher concentration of viral RNA than blood during the later stages of infection, suggesting the possibility of numerous ways of transmission of SARS-CoV-2.[38] Interestingly, Chen *et al.* found that COVID-19 severity was positively correlated with the levels of viral RNA in anal and blood swabs.[39] According to a study by Al-Rohaimi and Otaibi, the novel SARS-CoV-2 titer determines the rate of infection and severity of disease.[12]

There are various modes of transmission from one human to another, such as familial gatherings,[40] exposure to virus carriers without any symptoms,[41] and the three-phase breakout.[42] Additionally, the role of individuals who became "super-spreaders" can't be ignored in this pandemic.[43] Prominently, asymptomatic SARS-CoV-2 transmission has been evidenced, where infected persons without any symptoms could be sources of infection, during the incubation period.[44]

These studies demonstrate that SARS-CoV-2 can be spread through various routes such as the respiratory tract, alimentary tract, nasolacrimal system, oral–fecal, and body fluid, with the oral–fecal route implying high infection and viral shedding.

9.3 Pathogenesis

After SARS-CoV-2 comes in contact with an individual, it attaches to the upper respiratory airway, followed by entry into the host cells.

9.3.1 Attachment to the Host Cell

The interaction between the virus and a human cell starts when the S protein of the virus binds with the ACE2 receptor. A newer study has reported that the virus also penetrates the host cell by binding to cluster of differentiation 147 (CD147), also termed basigin (Bsg) or EMMPRIN.[45] Belonging to the immunoglobulin super family, CD147 is a transmembrane glycoprotein that is highly expressed in virus-infected cells, inflamed tissues, and tumor tissues. The spike protein binds with the CD147, thereby facilitating the viral invasion.[45] In addition to these, dendritic cells (DCs), intercellular adhesion molecule 3 (ICAM)-3, grabbing nonintegrin (DC-SIGN, *i.e.* CD209), and CD209L also act as SARS-CoV-2 receptors.[45–47]

9.3.2 Penetration of SARS-CoV-2 Inside the Host Cell

Following the attachment to host cells, the viral membrane fuses with the cell membrane of the host *via* an endosomal pathway and then enters the cell. For fusion, the S protein needs to be activated. Several host proteases including transmembrane protease serine protease (TMPRSS)-2, TMPRSS-4, trypsin, furin, cathepsins, or human airway trypsin-like protease facilitate priming cleavage between S1 and S2 and activate cleavage on the S2′ site.[29] S protein activation by TMPRSS2 protease is mandatory for attachment to the ACE2 receptor in the infection.[48]

The N-terminal domain of S1 structurally folds to present galectin (galactose-binding lectins) which binds with the sugar on the host cell membrane. On the other side, the C-terminal comain (CTD) of S1 assists in attachment to the ACE2. The CTD contains two subdomains: the primary one is composed of a five-stranded antiparallel β-sheet and the other is the receptor binding domain (RBD), which selects which receptor to bind.[30] The extended insertion of the RBD (between the β4 and β7 strands) contains some residues that are significant for receptor binding events. It has been observed that dimers of hACE2 and amino acid transporter B(0)AT1 are formed.[49] To this hACE2-B0AT1 dimer, two molecules of CTD are individually bound. The RBD augments the interaction by moving in a hinge-like conformation momentarily covering or uncovering the receptor binding elements.[49]

After cleavage, the S1 subunit separates from S2, the S2 subunit extends itself leading to a change in conformation from compressed to nail-shaped configuration. The S2 consists of fusion peptide (FP), an internal fusion peptide (IFP), a cleavage site (S2′), and two heptad-repeat domains before the transmembrane domain. Out of these, the FP and IFP are essential for entry of virus into the host cell.[50] Thus, the proteases cleave the S protein at priming and activating sites to release FP and IFP.[29] After this cleavage, the conformational change induces membrane fusion, in which the formerly generated peptide integrates itself into the host cell membrane.[30]

Significantly, S proteins have the ability to modulate the protease cleavage sites, as a result of which S protein can be cleaved by different proteases. This mechanism is one of numerous ways which SARS-CoV-2 uses to diversify the host cell membrane susceptible to its attack.[30]

Furin, a pro-protein convertase, is one of the extensively studied cell proteases in SARS-CoV infection. Only SARS-CoV-2, of all the betacoronaviruses, has the furin-like cleavage site (PRRAR↓SV).[51] Significantly, the lungs express furin to a large degree, reinforcing the fact that the human respiratory airway exhibits high susceptibility.[52,53] In addition, numerous studies on gene expression have reported the occurrence of ACE2 mainly in alveolar epithelial type II (AT2) cells of nasal epithelial cells (precisely, in ciliated cells and goblet cells) as well as in heart, lung, kidney, and adipose tissue, which accounts for the diverse pathogenicity of SARS-CoV-2.[54] Other than these, some immune cells like monocytes and macrophages can also express ACE2 (less than AT2 cells), making them vulnerable to SARS-CoV-2 attack.[55]

9.3.3 Assembly of the RTC

The viral lipid bilayer fuses with the host cell membrane through the endosomal pathway. Following this, the virus releases its ssRNA inside the cytoplasm of the host cell. The ssRNA utilizes the host cell ribosome to translate two polyproteins, pp1a and pp1ab, and structural proteins. Next, processing of polyproteins occurs by numerous protease cleavages to form a replicase–transcriptase complex (RTC). It is believed that the RTC comprises up to 16 viral subunits and several cellular proteins.[56]

9.3.4 Genome Replication, Transcription, and Translation

The cytoplasmic membranes are the sites of genome replication and transcription. The RTC mediates co-ordinated processes of both continuous and discontinuous transcription to produce sub-genomic mRNA.[54,57] RdRp functions to duplicate RNA. In addition to RdRp, the coronavirus replicase has been speculated to employ various RNA processing enzymes that either are extremely rare or absent in other RNA viruses. These RNA-processing enzymes include 3′-to-5′ exoribonuclease, putative sequence-specific endoribonuclease, ADP ribose 10-phosphatase and 20-*o*-ribose methyltransferase.[58,59] The resulting mRNA is then translated to related viral proteins.

9.3.5 Virion Assembly and Release

The replicated viral particles are further processed into the endoplasmic reticulum–Golgi intermediate compartment (ERGIC). In the ERGIC, the assembly of virions occurs. Next, the membranes coming from ERGIC are reorganized into double membrane vesicles by the NSPs. The replication and transcription of the virus genome occur in these vesicles.[54] Subsequently, the viral RNA and N protein aggregate to form the nucleocapsid. Finally, the virion particles are extricated from the host cell through exocytosis or extrusion.[60]

9.3.6 Incubation Period

The incubation period for SARS-CoV-2 infection is approximately 1–14 days, with the average incubation period being 5.2 days. It has been reported that 98% of symptomatic individuals develop symptoms within 11.5 days.[25]

9.4 Disease Progression

After SARS-CoV-2 enters into the host cell, it releases specific inflammatory mediators which stimulate macrophages.[48,61,62] The macrophages are activated and secrete chemokines (CCL2 and XCL10) and cytokines (interleukin (IL)-1, IL-6, tumor necrosis factor (TNF)-α) into the bloodstream. These secretions cause vasodilation and increase capillary permeability. As a result, the plasma leaks inside the interstitial spaces of the alveolar cells and accumulates around the alveoli,[43,63–67] compressing the alveolar space. This results in lowering of surfactant levels in AT2. Finally, the chain of events results in collapse of alveoli and gas exchange impairment.

Persistent inflammation impairs the activities of anticoagulant proteins such as antithrombin and tissue factor pathway inhibitor (TFPI) which completely disrupts the homeostatic equilibrium, leading to endothelial cell dysfunction, increased platelet activity, and thrombus formation.[67] Inflammatory cytokines including TNF-α, IL-1, and IL-6 activate coagulation by checking the fibrinolytic activity and stimulating prothrombic factors.[67–69] These processes lead to small-vessel vasculitis and microthrombosis.[70]

Simultaneously, inflammatory cytokine secretion increases what is well known as "cytokine storm".[71] Through CD^+ type 1 T-helper (Th1) cells, the inflammatory mediators increase the production and recruitment of macrophages and neutrophils using IL-17, IL-21, and IL-22.[66] As a result, large volumes of mucus and fluid accumulate in the alveoli leading to breathing difficulty, hypoxemia, and cough in the latter stages of disease.[72–74]

Body temperature is controlled by the hypothalamus. The released IL-1, IL-6, and TNF-α circulate in the blood affecting the hypothalamus.[43,63–67] The cytokines induce the secretion of Prostaglandin and Prostaglandin E2 (PGE2), leading to an increase in body temperature.

Autonomic sympathetic nerves can induce tachycardia in response to hypoxia. Ultimately, septic shock and multi-organ failure occurs due to these abnormal inflammatory responses.[75,76]

9.5 Immunology

9.5.1 Innate Immune Response

Innate immunity elicits the primary immune response upon infection. In the case of SARS-CoV-2 infection, the innate immune system elicits the first antiviral response; however, the characteristics of the innate immune system response to SARS-CoV-2 infection *per se* are not well understood. Nevertheless, because SARS-CoV-2 shares sequence homology with other coronaviruses and innate immune signalling is conserved, we can suggest the likely immune response towards SARS-CoV-2.

The response begins with the recognition of various pathogen-associated molecular patterns by the host cell's PRRs, such as the Toll-like receptors (TLRs) and retinoic acid-inducible gene I (RIG-I)-like receptors (RLRs) (a collective term used for RIG-I and melanoma differentiation-associated protein 5 (MDA5)).[77] These PRRs identify aberrant RNA structures often formed during virus replication.[16] This recognition activates a number of downstream transcription factors, such as NF-κB and interferon regulator factors (IRFs),[17] triggering the release of various cytokines. Among these cytokines, type I interferons (IFNα/β) are regarded as the most significant for defending against viral infections.[66] The secreted IFN binds to the IFN receptors and activates the Janus kinase-signal transducer and activator of transcription (JAK/STAT) pathway. This results in induction of transfer of the IFN-responsive transcriptional factors to the nucleus. These translocated transcriptional factors activate genes that have promoters consisting of IFN-stimulated response elements, leading to the expression of a group of IFN-stimulated genes (ISGs), which generates an antiviral state.[78] In addition, other pro-inflammatory cytokines, including IL-1, IL-6, IL-18, and TNF-α are secreted, which stimulate antiviral programs in virus-infected cells, leading to activation of an adaptive immune response.

The early presence of IFN-I and its proper localization can restrict the coronavirus infection efficiently.[79,80] Prior investigations (*in vitro*) have reported that IFN-I/III pre-treatment is effective against SARS-CoV-2, conceivably more so than for SARS-CoV-1.[81–84] Nevertheless, the specific ISGs that facilitate these antiviral programs are being investigated. A study by Pfaender *et al.* reported that the SARS-CoV-2 spike (S) protein-mediated membrane fusion is obstructed by lymphocyte antigen 6 complex locus E (LY6E), a product of ISG.[85]

9.5.1.1 *Evasion of Innate Sensing by SARS-CoV-2*

The coronaviruses, including SARS-CoV-2, attenuate the innate immune response, which promotes enhanced immunopathogenesis. However, due to the unavailability of detailed studies, we will discuss different ways of evasion, taking other coronaviruses into consideration.

9.5.1.1.1 Evasion of PRR Sensing. Coronavirus-mediated antagonism starts with the evasion of PRR recognition. The ssRNA of coronaviruses can be recognized by TLR7 or TLR8 and possibly RIG-I and protein kinase RNA (PKR). The coronaviruses replicate their ssRNA to form double-stranded (ds) RNA intermediates. These dsRNA intermediates are identified by cytosolic PKR, MDA5, RIG-I, and endosomal TLR3. Coronaviruses avoid PRR sensing by either evading recognition or contravening the PRR action (Figure 9.1).[86–91]

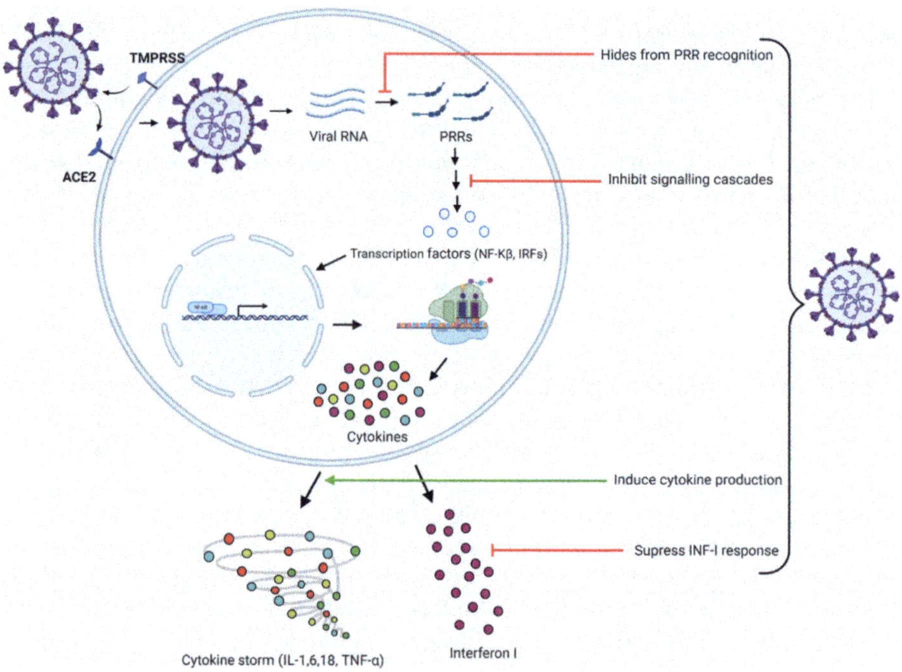

Figure 9.1 SARS-CoV-2 evading the innate immunity in the target host cell, showing how the innate immune system responds upon SARS-CoV-2 attack. After entry, the viral RNA structures are recognized by various pattern recognition receptors (PRRs). These activate signalling cascades that trigger various transcription factors resulting in release of cytokines. The right-hand side of the figure displays how SARS-CoV-2 protects itself from innate immunity. It evades PRR recognition and inhibits signalling cascades. It also induces cytokine production resulting in cytokine storm and inhibits the IFN-1 response which is significant in viral infection.

SARS-CoV-1 evades PRRs by shielding the dsRNA with membrane-bound compartments, which are formed at the time of replication.[91] In addition, the coronavirus NSPs (NSP10, 13, 14, 16) modify their RNA by methylation and guanosine-capping at the 5′ end,[86,87,90] thereby mimicking the mRNA of the host, which stimulates translation, obstructs degradation, and thereby escapes RLR sensing. Furthermore, coronaviruses evade MDA5 recognition by the use of NSP15, endoribonuclease, which cleaves 5′ polyuridines produced during viral replication.[88,89] A study has revealed that SARS-CoV-2 also inhibits the PRR recognition of viral RNA through the use of NSP15.[89] Other than these, coronaviruses have evolved to employ many other schemes to prevent PRR activation. However, to what extent SARS-CoV-2 uses these strategies to impede PRR activation is yet to be known.

Inhibition of Signalling Cascades. As described, TLR and RLR activation induces signalling cascades, which leads to the release of pro-inflammatory cytokines and IFN. Various proteomic researchers have confirmed the interactions between SARS-CoV-2 proteins and PRR signalling cascades. Indirect interactions between SARS-CoV-2 ORF9b and the signalling adaptor, mitochondrial antiviral-signalling protein (MAVS) *via* association with translocase of outer membrane (Tom70) has been demonstrated.[92] This finding is in accordance with previous studies demonstrating that SARS-CoV-1 ORF9b suppresses MAVS signalling.[93] In addition, SARS-CoV-2 NSP13 interacts with signalling intermediate TBK1 (TANK (TRAF-associated NF-κB activator) binding kinase 1) and NSP15 associates with RNF41 (ring finger protein 41, also known as E3 ubiquitin-protein ligase Nrdp1), which is an activator of IRF3 and TBK1, preventing the formation of the antiviral state.[92] Similarly, SARS-CoV-1 M protein and MERS-CoV ORF4b also inhibit the TBK1 signalling complex.[94,95]

Other proteins, such as SARS-CoV-1 papain-like protease (PLP), ORF3b, ORF6, and N protein, block IRF3 phosphorylation, as a result of which the nuclear translocation of IRF3 is blocked, leading to inhibition of induction of interferons.[96,97] However, the SARS-CoV-2 ORF6 sequence is only 69% similar to SARS-CoV-1, suggesting the non-functionality of ORF6 in SARS-CoV-2. In accordance with this, unlike in SARS-CoV-1 infection, SARS-CoV-2 infection fails to restrict STAT1 phosphorylation which contributes to the antiviral state in the host cell.[83]

Inhibition of IFN-I. Based on previous studies, various mechanisms have been adopted by coronaviruses to inhibit IFN-1 production and supress the subsequent signalling cascades which are closely related with disease severity.[66,98,99] Similar behaviors have been speculated for SARS-CoV-2 in suppressing the IFN-1 response. A study by Li *et al.* found that SARS-CoV-2 N, ORF6, and ORF8 proteins remarkably repressed the expression of ISG54, ISG56, and IFN-β mRNA in Sendai virus-induced HEK 293 T-cells (human embryonic kidney 293 cells).[100] In addition to this, SARS-CoV-2 ORF3b has been found to be a strong IFN antagonist, the activity of which is additionally enhanced by a naturally occurring elongation variant.[101]

These studies confirmed the prior reports on lack of strong IFN-I/III signatures from primary bronchial cells, infected cell lines, and a ferret model.[81] In the same study, it was found that at low multiplicity of infection (MOI), the virus did not act as a powerful IFN-I and -III inducer, which is the opposite to the phenotype seen when the MOI levels are high. These kinetics may contribute in exacerbating COVID-19 during the course of infection.[102] In another study, it was manifested that severe COVID-19 patients demonstrated exceptionally weakened IFN-I expression in comparison with moderate or mild cases.[103]

Therefore, the IFN-I levels might be an important signature in predicting the COVID-19 severity. However, in-depth investigations are essential to interpret the mechanism of IFN-I suppression and its effect on COVID-19 severity.

9.5.1.2 *Imbalanced Antiviral Defense and Pro-inflammatory Cytokine Production*

The coronaviruses are particularly adept at inhibiting the IFN-I pathway to escape innate immune sensing. This suggests that dysregulation of IFN-I is significant in SARS-CoV-2 pathogenicity. Many reports have demonstrated the importance of the timing in determining the role of IFN-I. If present early in the disease, it is protective, but it becomes pathological in later stages.[79,80] This effect might be due to IFN-induced ACE2 upregulation in epithelial cells of the airway, which might lead to an increase in the infected cell count.[104] In addition, while coronaviruses inhibit IFN signalling, they may activate other inflammatory pathways, thus enhancing pathogenicity.

A study by Blanco-Melo *et al.* has shown reduction in the levels of IFN-I and IFN-III and robust production of cytokines in response to SARS-CoV-2 infection. They found that monocyte-associated chemokines such as CCL2 and CCL8 that contribute to pathological inflammation were induced significantly.[81] In addition, Li *et al.* observed the interaction of NSP9 and NSP10 with NKRF, a NF-κB repressor, and also suggested that NSP9 and NSP10 might induce IL-6 and IL-8 production, which mediates strong chemotaxis of neutrophils and an extraordinary inflammatory response manifested in patients with COVID-19.[105] Altogether, these pro-inflammatory conditions could lead to the "cytokine storm" typically manifested in COVID-19 patients.

These studies suggest that if the host responds to the infection early, the antiviral defence might protect them; however, if it gets delayed, the IFN could contribute to the disease, making it more severe. Therefore, the interaction between antiviral and inflammatory activities is an important aspect of the immunopathogenesis of COVID-19.

9.5.1.3 *Macrophages*

Macrophages are monocyte-derived immune cells which are either found in a stationary form located in tissues or as mobile white blood cells at infection sites. On the basis of activation state and function, they are classified into the M1 type (classically activated macrophage) and M2 type (alternatively activated macrophage).[106,107]

Wang *et al.* investigated abnormalities in the lungs of two severe COVID-19 patients. They found that there was significant increase of macrophages in alveolar spaces with dispersed lymphocytes and neutrophils. They also examined secretion levels of several chemokines and inflammatory cytokines such as IL-6, IL-10, and TNF-α by alveolar macrophages (AMs). The macrophages manifested strong expression of IL-10 and moderate expression of IL-6 and TNF-α. Significantly, they found ACE2 receptor expression in AMs, which explains the former observation of cytokine secretion by macrophages.[108]

Numerous studies on peripheral blood mononuclear cell (PBMC) samples of symptomatic COVID-19 patients have evidenced remarkable influx of activated $CD4^+$ T-cells and $CD14^+HLA\text{-}DR^{lo}$ inflammatory monocytes (IMs) that produce granulocyte–macrophage colony-stimulating factor (GM-CSF).[109–111] Zhou *et al.* found that in intensive care unit (ICU) patients, the levels of inflammatory monocytes ($CD14^+CD16^+$) that were $GM\text{-}CSF^+IL\text{-}6^+$ increased. These monocytes had the ability to migrate into the lungs and could differentiate to monocyte-derived dendritic cells (DCs) or macrophages.[111] In contrast, Zhang *et al.* observed no changes in monocyte levels in PBMC samples of COVID-19 patients. However, they found substantial variation in morphology and marker expression of the inflammatory cells due to TNF-α, IL-6, and IL-10 secretion and increased CD11b, CD14, CD16, CD68, CD80, CD163, and CD206 expression on the surfaces of monocytes to generate the initial strong inflammatory response during COVID-19 progression.[110]

Similar results have been reported in peripheral blood of COVID-19 patients representing expansion of $CD14^+IL\text{-}1b^+$ monocytes,[112,113] IL-1b-associated inflammasome signatures,[114] and IFN–mitogen-activated protein kinase (MAPK)-driven adaptive immune responses.[115] Significantly, these immune responses correspond with the progression of the disease.

Liao *et al.* conducted a single-cell transcriptomic (scRNA-seq) study on the pulmonary tissues of severe COVID-19 patients. They manifested an increase of IMs and $Ficolin\text{-}1^+$ monocyte-derived macrophages at the cost of tissue-resident reparative AMs. In addition, the authors observed IFN signalling and recruitment of monocytes, which probably promote fast deterioration in alveolar patency and stimulate ARDS.[116]

In a study by Reyfman *et al.*, alveolar macrophages were classified into four groups on the basis of presence of FAB4, SPP1, and FCN1 markers. They arranged $FCN1^+$ macrophages derived from monocytes under groups 1 and 2. These $FCN1^+$ macrophages secrete high amounts of inflammatory chemokine and stimulate the ISGs, contributing to hyperinflammation.[117] Another study by Liao *et al.* demonstrated the predominance of the $FCN1^+$ macrophages in badly damaged lungs from patients manifesting ARDS, which explains the hyperinflammation in the lungs.[118] Group 3 comprised $SPP1^+$ macrophages that neutralize the inflammation by $FCN1^+$ cells and probably characterize a pro-fibrotic feature.[117] The group 4 $FABP4^+$ AMs were predominantly observed in mildly ill COVID-19 patients and were absent in severe patients.[118] From previous studies, it is evident that FABP4$^+$ AMs have a crucial role in lipid surfactant metabolization[119] promoted by GM-CSF cytokine,[120] which leads to the presumption that lack of AMs in severe COVID-19 patients might give rise to lung failure.

9.5.1.3.1 Macrophage Responses to SARS-CoV-2 Infection. Macrophages and monocytes accumulate and trigger cytokine secretion, leading to cases of uncontrolled cytokine storms. This results in the conversion of M1 AMs to the M2 type that additionally contribute to inflammatory injuries and respiratory tract fibrosis, especially in severe cases.[108,121] In another study on SARS-CoV-2 patients, it was found that a group of macrophages express genes correlated to fibrosis initiation and tissue repair similar to that manifested in liver cirrhosis.[122] Hence, the activated macrophages present in the lung may be involved in fibrotic complications and acute inflammation that are hallmark features in patients who progress to ARDS.[123] A detailed study on the mechanism of SARS-CoV-2 interference with monocyte/macrophage function and how it utilizes them for survival and the infection of other cells is essential for understanding SARS-CoV-2 pathogenicity.

9.5.1.4 Alveolar type II ATII Cells in the Innate Immune Response

Apart from immune cells, alveolar epithelial cells also contribute to the innate response generated by SARS-CoV-2 in the lung. Numerous studies have illustrated that SARS-CoV-2 particularly attacks ATII cells, which express high levels of ACE2 receptor on their surface and form 5–15% of the epithelial cells of the lung.[48,124–127] ATII have the ability to produce and release surfactant protein A (SPA). The role of SPA in surfactant metabolism and lung innate immunity is significant.[127] Interestingly, SARS-CoV-1 has the ability to spread in ATII cells until the cells undergo apoptosis or die, and afterwards the viral particles are disseminated to adjoining areas[128] which leads to spread of the infection. The ATII cells respond to SARS-CoV-2 infection, by expressing TLRs that lead to activation and secretion of inflammatory cytokines. ATII cells also activate or recruit other immune cells, including neutrophils and macrophages. In addition, ATII cells and AMs harmonize inflammatory responses that might cause lung damage connected with the disease.[124–126]

9.5.1.5 Neutrophils

Neutrophils are among the earliest immune cells to travel to infection sites. They are characterized by the presence of granules which contain enzymes and antimicrobial peptides used in killing micro-organisms. They form neutrophil extracellular traps (NETs) by releasing granule proteins and chromatin, resulting in the formation of extracellular fibres which trap the micro-organism, prevent spread and ensure their degradation.[129]

The importance of neutrophils was observed in a study by Liu *et al.*, who established that the neutrophil-to-lymphocyte ratio (NLR) is the most important prognostic feature in estimating disease severity. They found that patients bearing NLR ≥3.13 and aged ≥50 years manifested severe illness.[130] As the NLR ratio rises, the infiltration of neutrophils into the lungs increases, in contrast to the lower levels of lymphocytes in the periphery that characterize the immune dysregulation in severe cases.[131] A detailed study by Liu *et al.* suggested that the neutrophil to $CD8^{+}$T-cell ratio (N8R) should be measured as a more accurate prognostic indicator for determining the severity of disease in COVID-19 patients.[132]

NETs are considered a significant functional characteristic of neutrophils in SARS-CoV-2 infection. In patients with ARDS, high NET blood level was manifested, which is associated with higher severity and mortality.[133] Zuo *et al.* performed a study on 50 hospitalized patients and observed dramatic increases in the laboratory markers of NETosis which included myeloperoxidase (MPO)-DNA, cell-free DNA, and citrullinated histone H3 (CiteH3).[134] From these studies, it can be concluded that neutrophil signatures can be considered in estimations of disease severity.

9.5.1.6 *Dendritic Cells*

Dendritic cells are found in practically all tissues. Upon detection of tissue homeostatic imbalances, they process and present antigen to T-cells, establishing a bridge between innate and adaptive immunity.[135] Thus, they play a central role in both immune responses. A subset of DCs namely plasmacytoid DCs (pDCs) have the ability to generate large amounts of IFN-I.[136] Wang *et al.* performed a single-cell transcriptional profiling of 14 major peripheral immune cells. They observed that the monocyte-derived DC levels decreased in severe COVID-19 patients and their concentrations were later restored in convalescent patients.[137]

Zhou *et al.* conducted a thorough study on the effect of SARS-CoV-2 infection on DCs in acute patients and convalescent patients. They analysed the DC maturation upon cytokine treatment. They found that there was no significant increase of DC maturation in one out of four convalescent patients and three out of three acute patients. In addition, they observed that IFN- α was not induced in three out of four convalescent patients and three out of three acute patients and, in the meantime, there was no significant increase of IFN-β in three out of three acute patients, with a slight increase in three out of four convalescent patients. Next, they examined the functionality of DCs in stimulation of CD4 and CD8 T-cell proliferation. They found that convalescent patient-derived DCs were capable of inducing CD4 and CD8 proliferation and, in contrast, no induction was observed in acute patient-derived DCs. They concluded that acute patient-derived DCs were functionally impaired for IFN production and T-cell activation, thereby reducing the induction of the adaptive immune response. In addition, they found that there were significant changes in conventional DC (cDC) : pDC ratios and also established that high cDC : pDC ratios of ~50-fold might serve as a potential hallmark of severe disease.[138] Thus, DCs might be used as a new biomarker in estimations of disease severity. More studies on DCs are needed to understand the innate and adaptive immune response to SARS-CoV-2 infection.

9.5.1.7 Natural Killer Cells

Natural killer (NK) cells serve as the principal innate lymphocyte cells that facilitate antiviral responses. They exhibit activating and inhibitory receptors which regulate their cytotoxicity. NK cells kill cells infected with virus by upregulating virus-derived proteins and stress-inducible ligands. NK cell-activating receptors such as NKp46 recognize these molecules that result in cell lysis.[139–142]

Many studies on peripheral blood samples of COVID-19 patients have reduced NK cell numbers, which is associated with disease severity.[143–146] Interestingly, a recent scRNAseq analysis of lungs from COVID-19 patients and healthy donors by Liao *et al.* has revealed equal transcriptomic signature for NK cells in lungs from COVID-19 patients and healthy donors.[116]

It has been observed that most of the lung NK cells are migratory[147,148] and NK cell infiltration is mediated by CXCR3 during influenza infection.[149] *In vitro* studies have revealed increased CXCR3 ligands (CXCL9–11) in lung tissue infected with SARS-CoV-2,[150] and expansion of monocytes that produce CXCR3-ligand in the lungs of COVID-19 patients.[116] These studies suggest that the CXCR3 pathway might mediate recruitment of NK cells from the peripheral blood towards the lungs in COVID-19 patients.

9.5.1.7.1 Modification of NK Cells Upon SARS-CoV-2 Infection. Studies on peripheral blood samples of COVID-19 patients have shown that in NK cells there was a decrease in intracellular expression of CD107a, Ksp37, granulysin, and granzyme B, implying an weakened cytotoxicity, and diminished release of TNF-α, IFN-γ, and chemokines.[146,151]

Numerous other changes are also observed in NK cells upon SARS-CoV-2 infection. There was an increase of immune checkpoint NKG2A on $CD8^{+}$T and NK cells.[146] It is known that NKG2A binds with the non-classical human leukocyte antigen E HLA-E molecule and suppresses cytotoxicity,[152,153] which is closely associated with spread of HIV-1 infection.[154]

In addition, there is upregulation of the genes that code for the inhibitory receptors TIM3 and LAG3 in NK cells.[103,151] Thus, upon infection, the immune checkpoints on NK are increased, permitting the virus an opportunity to escape from NK cell cytotoxicity.

Higher plasma concentrations of IL-6 have been manifested in COVID-19 patients,[155] which can significantly be associated with lower NK cell numbers.[144,156] Previous *in vitro* studies have shown that upon stimulation of soluble IL-6 receptor and IL-6, the cytolytic functions (production of granzyme

B and perforin) of healthy donor NK cells are impaired, and this can be restored by treatment with immunosuppressive drug (see Sections 9.5.2.2.1 and 9.7.3.2) tocilizumab (IL-6R blockade).[157] Additionally, upregulation of TNF-α is observed in the plasma of COVID-19 patients.[155]

Thus, it is important to note that NK cell function is impaired in COVID-19 patients. However, the relation between the NK cell impairment and disease severity is yet to be investigated.

9.5.2 Adaptive Immune Response

After the innate response, the specific adaptive immune response comes into action. Upon virus attack, the specific cell-mediated immunity has the highest responsibility for defense, mainly the T-cells. Similar events take place in the specific case of SARS-CoV-2 attack. Lymphopenia is a well-known feature in the majority of COVID-19 patients, which is due to extensive lymphocyte (specifically T-cells) death.

According to Tan *et al.*, the time-lymphocyte model (TLM) can be considered to avoid probable over-/under-treatment incidence in hospitalized COVID-19 patients. In the TLM model, cases are considered as "moderate" if the blood lymphocyte percentage raises above 20% after 2 weeks of disease initiation, and "severe" if lymphocyte levels become lower than 20%. After the third week of disease in severe symptomatic patients, if the lymphocyte count increases more than 5%, then the situation might improve, otherwise the disease will lead to mortality in patients.[158]

9.5.2.1 B-cells

Antibody (Ab)-mediated immunity is very important for fighting infection. Not only is it responsible for viral clearance, but it also protects against reinfection by developing a memory response. In case of SARS-CoV-2 infection, the virus elicits a strong B-cell response, which is reflected by the global robust production of immunoglobulin (Ig) including IgA, IgM, IgG, and neutralizing IgG antibodies (nAbs) following infection.

COVID-19 is often detected 1–2 days after symptoms arise, peaks at approximately 4–6 days, and, in mild–moderate cases, usually resolves by day 18 after the onset of symptoms.[159] Mostly, seroconversion occurs at ~18 days and the IgG titer is found to be maximal at ~20 days after disease onset.[160]

Xiao *et al.* conducted research on the Ab profile in 34 COVID-19 patients for 7 weeks. They found that IgM and IgG were present in all the patients 2 weeks post symptom onset. At week 4, the IgM levels declined, in contrast to IgG, whose levels started increasing. The patients were IgM-negative within 5–7 weeks and correspondingly the IgG titre continued to rise. Thus, in most cases, the acute phase of infection lasted longer than 1 month.[161] Another investigation demonstrated the presence of IgM and IgA after 5 days and IgG post-14 days of symptoms onset.[162]

Numerous studies have detected that Abs usually bind either to the S protein or N protein of SARS-CoV-2.[163–165] The RBD (specifically R441 residue) of the S protein has high immunogenicity, and Abs that bind to the RBD can be effectively neutralizing, blocking the SARS-CoV-2–ACE2 interaction and preventing virus entry.[138,164,166] Most tested patients evidenced the presence of anti-RBD nAbs.[164–166] A study by Moderbacher *et al.* examined the Ab titers (IgA, IgM, IgG) against RBD on S protein in 43 COVID-19 patients. They found that IgG and IgA were present in most of the patients and IgM was less frequent. Additionally, they analysed Ab titers against complete S and N proteins, where they found similar results as in RBD. Furthermore, they performed neutralization assays to measure the functional Ab titers and found that the nAbs titers were correlated with RBD IgG and IgA, supporting the fact that RBD has maximum immunogenicity in comparison with S and N proteins.[168]

Numerous studies have reported robust evidence of the presence of SARS-CoV-2 RBD-specific IgG and nAb responses in severe COVID-19 patients.[135,165,166] Tan *et al.* observed in COVID-19 patients that a large amount of IgG and anti-nucleocapsid (NP) IgM after onset of symptoms was associated with severe disease.[167,168] Additionally, higher amounts of IgM, anti-S, and anti-NP IgG were correlated with poorer disease outcomes and older age.[169]

Another study found that there was no cross-reaction of RBD specific monoclonal antibodies between SARS-CoV-2 and MERS-CoV or SARS-CoV-1. Nor was there any neutralisation of SARS-CoV-2-infected plasma between SARS-CoV-2, MERS-CoV, or SARS-CoV-1,[164] suggesting virus specificity. In all, these studies reveal that Ab-mediated neutralization is established by binding of Ab to the epitopes within the RBD and is virus-specific.

Other than Ab levels, some researchers have proposed an Ab-dependent enhancement (ADE) theory to correlate previous exposure to coronaviruses with disease severity or death. They suggest that the previously elicited Abs might not be effective against SARS-CoV-2. It has been hypothesized that due to antigenic epitope heterogenicity, the pre-formed Abs do not neutralize the SARS-CoV-2 completely. In contrast, perhaps they form a complex with SARS-CoV-2 or virus-activated complement particles, which either bind to the Fc region or complement receptors on susceptible cells, leading to entry of virus inside the host cells.[169–171] Numerous *in vitro* and *in vivo* (mouse models) studies have demonstrated immune response dysregulation by ADE, which in severe patients could result in lymphopenia, inflammation-based injuries, and cytokine surge[170] in lungs or other organs.[79,172]

One study suggested that the presence of various epitopes on the SARS-CoV-2 S protein that were comparable, but not identical to SARS-CoV-1, might contribute towards ADE in case of SARS-CoV-2 infection.[173] Previous studies on the acute phase of SARS-CoV-1 infection have demonstrated the regulation of macrophage by anti-S-IgG where the secretion of transcription growth factor-β is reduced and production of pro-inflammatory cytokines is promoted, leading to acute lung injury.[174] Another study has observed few non-neutralizing Abs that bind to non-RBD regions of S protein which may lead to an ADE of SARS-CoV-1 infection, leading to harmful immune responses.[175] To date there is no evidence that Abs involved in ADE have led to harmful consequences. Therefore, further investigations on the role of ADE in SARS-CoV-2 are needed to see if there is a correlation with disease severity.

9.5.2.2 T-cells

The roles of T-cells are significant in viral infections. One type, the T-helper ($CD4^+$) cell, produces cytokines that stimulate B-cells to produce Abs, and a second T-cell type, known as the cytotoxic ($CD8^+$) T-cell, destroys virus-infected cells. However, dysregulation of T-cell responses results in immunopathology. Here, we discuss the T-cell response in COVID-19 and the mechanism by which the SARS-CoV-2 evades the T-cell immune response.

9.5.2.2.1 Lymphopenia. Lymphopenia has become a well-known hallmark of COVID-19. Many studies on the incidence of lymphopenia in COVID-19 have revealed a drastic reduction of peripheral $CD8^+$ and $CD4^+$ T-cell levels in moderate and severe cases.[146,156,176,177] Numerous studies have suggested the association of low $CD8^+$ T-cell count with severe COVID-19 pneumonia and high mortality.[178,179] In contrast, low $CD4^+$ T-cell count can be correlated with intensive care unit (ICU) admission.[180] At the same time, slightly increased or normal T-cell counts is a prognostic indicator of COVID-19 disease with mild symptoms.[181,182]

Various mechanisms are responsible for T-cell reduction during infection. Zhou *et al.* observed that reduced peripheral T-cell count in acute patients was probably associated with a decrease in $CD4^+$ T-cell proliferation and hyperactivation of $CD8^+$ T-cells along with migration of T-cells into the lungs during acute infection.[138] Another important factor is the level of inflammatory cytokines. It has been found that pro-inflammatory cytokine levels were lowered and T-cell frequencies were restored in convalescent patients compared with acute patients.[132,146,181,183,184] Furthermore, it has been reported that lymphopenia is probably associated with increased levels of serum TNF-α, IL-10, and IL-6.[74,184] Cytokines including TNF-α and IFN-I may constrain the T-cell recirculation in blood by attachment to endothelium and stimulation of T-cell retention in lymphoid organs.[185,186] Supporting this fact, tocilizumab, an IL-6 receptor antagonist has been shown to raise the circulating lymphocyte count[109] and is approved to treat cytokine release syndrome that is severe or life-threatening (https://www.cancer.gov/about-cancer/treatment/drugs/tocilizumab).

In addition, the recruitment of T-cells to infection sites might also decrease their levels in the peripheral blood compartment. Liao *et al.* performed a scRNA-seq analysis of bronchoalveolar lavage fluid of patients and evidenced a rise of $CD8^+$ T-cell infiltration.[116] Similarly, post-mortem examination of a COVID-19 patient who manifested severe ARDS witnessed predominant lymphocyte infiltration in the lungs.[187] However, another examination of post-mortem biopsies of four individuals only found the presence of neutrophilic infiltrate.[188] Therefore, more studies are essential to understand the cause and effect of lymphopenia in COVID-19 patients.

9.5.2.2.2 T-cell Status in SARS-CoV-2 Infection. Zhou *et al.* conducted a study on T-cell response in SARS-CoV-2 infection and witnessed N and S protein (RBD)-specific T-cell responses in acute patients.[138] In another study conducted on 12 patients, viral S-, M-, and N-specific robust T-cell responses were detected during recovery from mild disease, and only T-cell responses specific to N protein in one-third of patients after recovery.[189] A study on PBMC samples (14 days post-ICU admission) from patients having moderate to severe ARDS demonstrated the presence of virus-specific $CD4^+$ and $CD8^+$ T-cells in all patients with frequencies of approximately 1.4% and 1.3%, respectively. These cells were identified either as $CD8^+T_{EM}$ (effector memory) or $CD4^+T_{CM}$ (central memory) and T_{EMRA} (effector memory CD45RA) cells as per the expression status of CD45RA and CCR7.[190] In another study, S protein-specific T-cell analysis demonstrated strong IFN-γ, IL-2, and TNF-α induction and reduced levels of IL-1, IL-5, IL-9, IL-10, and IL-22. Furthermore, $CD4^+$ T-cell responses specific to S protein in 18 patients having mild or severe COVID-19 were also studied. The antiviral $CD4^+$ T-cells characterized the co-expression of CD154 and CD137. These antiviral cells were found in 83% of cases, which indicated higher Ki67, CD38, and HLA-DR expression implying recent activation. Significantly, low levels of S-reactive $CD4^+$ T-cells were detected in 34% of SARS-CoV-2 seronegative healthy donors. However, phenotypic markers of activation were absent in these $CD4^+$ T-cells. In addition, these cells were found to be C-terminal S protein epitope specific, which are also present in endemic coronaviruses. This suggests chances of cross-reactivity of CD4 memory T-cells in specific communities including children and young individuals who normally encounter high incidences of coronavirus infection which might convert to an augmented primary SARS-CoV-2 specific response.[191] Recently, a study on T-cell response specific for N, M, and S protein and SARS-CoV-2 orfeome (MP_R) found the presence of IFN-γ-producing cells and circulating T-follicular helper (T_{FH}) cells ($CD4^+$ T-cells that are essential for IgG and nAb responses) in acute patients. In case of $CD8^+$ T-cells, they reported the appearance of IFN-γ^+CD8^+ T-cells in a majority of cases.[168]

9.5.2.2.3 The T-cell Journey: from Protective to Destructive. A thorough examination by Zhou *et al.* of T-cell functional changes in acute patients found that the capacity of $CD4^+$ and $CD8^+$ T-cells for production of IFN-γ and IL-2 was reduced. Additionally, they observed that both central memory and effector memory $CD4^+$ T-cells had significant reductions in polyfunctionality for secreting TNF-α and IFN-γ in sox acute patients in comparison with convalescent patients and healthy donors. Likewise, effector memory and EMRA $CD8^+$ T-cells also displayed decreased polyfunctionality for releasing both TNF-α and IFN-γ release in six acute patients in comparison with convalescent patients and healthy donors. Additionally, effector memory and EMRA CD8 T-cells of acute patients showed a significant decrease in cytotoxicity for perforin and granzyme B production. This study concluded that in COVID-19 acute patients, the functions of both $CD4^+$ and $CD8^+$ T-cells become impaired.[138]

Zhou *et al.* performed a detailed study on T-cell status in COVID-19 patients. They observed that both $CD4^+$ T- and $CD8^+$ T-cells were CD38-, CD44-, and CD69-positive, which indicated that both were activated. Also, $CD4^+$ T-cells exhibited the OX40 marker, which reveals high secretion of cytokines, and $CD8^+$ T-cells exhibited 4-1BB ligand which indicates cytotoxic activity. Interestingly, they found an increase in $Tim^+PD\text{-}1^+$ co-expression in both T-cell types, which implies immunological exhaustion in ICU-admitted severe patients. In addition, these individuals had elevated levels of $IL\text{-}6^+$ $GM\text{-}CSF^+$ $CD4^+$ T-cells and $IFN\text{-}\gamma^+$ $GM\text{-}CSF^+$ Th1 cells, which generate a hyperinflammatory condition.[192] Another study reported a rise in the $CCR6^+$ Th17 population, which is associated with cytotoxic (T_c) T-cells as well as tissue injury.[187]

In addition, two studies have reported a reduction in levels of regulatory (Treg) cells in severe COVID-19 patients.[176,193] Since prior studies on Treg cells in mouse models have evidenced their involvement in resolving ARDS inflammation,[194] a loss of regulatory T-cells might mediate the development of immunopathology in lungs. Similarly, severe COVID-19 patients show a decrease in γδT-cells, a class of T-cells known to exhibit antiviral activities in influenza pneumonia,[195,196] suggesting lowered antiviral defense against SARS-CoV-2.[112,197] Therefore, the exact function of the $CD4^+$ T-cell as to whether it is a protective or a destructive cell remains a mystery.

9.5.2.2.4 T-cell Response and Disease Severity. In one non-severe patient, T-cell evaluation after 7 days of infection revealed a remarkable rise in the frequency as well as the function of $CD8^+$ T-cells specifically, less so in the case of $CD4^+$ T-cells. It was found that with a quick increase in immune cells, there was a rapid decline in SARS-CoV-2. Interestingly, SARS-CoV-2 was absent in nasopharyngeal, sputum, or fecal samples from the seventh day. Simultaneously, cytokine levels dropped with the rise in T-cell count. In addition, the pro-chemokines and inflammatory cytokines did not cause a cytokine storm during the symptomatic period of disease.[182]

Chen and co-workers compared the T-cell response in moderate and severe COVID-19 patients and found that there was a drastic decline in the count of $CD4^+$ and $CD8^+$ T-cell in severe cases along with decrease in secretion of Treg cells and IFN-γ specifically by $CD4^+$ T-cells. Correspondingly, severe cases manifested high levels of TNF-α, IL-10, and IL-2R, linking cytokine storm with disease severity.[176] Another study claimed the cause of reduction of $CD4^+$ T-cells in COVID-19 patients was due to a high rise in IL-6 secretion, which can cause T-cell death, leading to immunodeficiency.[198]

Zhou *et al.* conducted a study on T-cell responses in acute patients. They found N- and S-RBD specific T-cell responses in the acute patients with mild disease 2 weeks after symptom onset. However, no such response was witnessed in severe COVID-19 patients. They suggested that delays in T-cell responses might contribute to COVID-19 severity in acute SARS-CoV-2 infected patients.[138]

In a study involving a large population of hospitalized patients, ICU patients manifested overexpression of TNF-α, IL-6, IL-10 as compared with non-ICU cases, which usually have decreased T-cell numbers during the disease.[184] Similarly, in a study on a cohort of 522 COVID-19 patients, the ICU patients had significantly lower levels of T-cells than non-ICU patients and this decrease had more detrimental impact on older patients. Remarkably, the frequency of PD-1 expressing $CD8^+$ T-cells and Tim-3 expressing $CD4^+$ T-cells was much greater in ICU patients than in non-ICU patients. That increase was correlated with the presence of IL-10, which not only prevents the T-cell proliferation, but also stimulates a T-cell exhaustion process that is a state of T-cell dysfunction.[184]

A study by Moderbacher *et al.* on the relationship between cellular immunity and disease severity found that the SARS-CoV-2 specific $CD4^+$ and $CD8^+$ T-cells were associated with low-severity COVID-19. A fatal case demonstrated the absence of $CD4^+$ and $CD8^+$ T-cells 15 days after symptom onset. However, it is noteworthy that in the same case the nAb titers were high. In another case where the disease resolved without hospitalisation, there was absence of nAbs, but $CD4^+$ and $CD8^+$ T-cells were present, implying that cell-mediated immunity can limit the severity of COVID-19. The authors suggested that the coordination of the three branches of adaptive immunity instead of a partial response might limit the disease severity to a greater extent. Also, they proposed that CXCL10, a cytokine in plasma, can be used to detect T-cell impairment, which might help to estimate disease severity.[168]

Thus, it can be concluded that for severe COVID-19 patients, T-cells appear to become more activated, followed by their exhaustion, which is evident by the presence of inhibitory markers including PD1 and TIM3. This T-cell exhaustion leads to impairment in functionality and cytotoxicity. Additionally, these studies suggest that pro-inflammatory cytokine status, T-cell count, and delayed T-cell response might be associated with disease severity.

9.5.3 Cytokine Storm

Since the first studies on COVID-19, there has been clear evidence of the unmanageable inflammatory response caused by pro-inflammatory cytokines. This is responsible for severe damage in multiple tissues and organ failure that ultimately leads to death in severe COVID-19 patients.[187,199] ARDS has been reported to be the primary cause of death in COVID-19.[187]

From numerous studies, COVID-19 patients have witnessed increased levels of many peripheral blood immune mediators such as IL-2, IL-6, IL-7, IL-8, IL-9, IL-10, IL-1 receptor antagonist (IL-1RA), IL-Monokine induced by interferon-γ (MIG), IL-1β, IFN-γ, TNF-α, IP-10, CCL8 (MCP-2), MIP1-α, MIP1-β, MIP2-α (CXCL2), CXCL16, basic fibroblast growth factor (bFGF), MCP-1, granulocyte colony stimulating factor (G-CSF), GM-CSF, vascular endothelial growth factor (VEGF) and platelet-derived growth factor (PDGF). Among them, IL-2, IL-2R, IL-6, IL-7, IL-10, Interferon gamma-induced protein 10 (IP-10), MCP-1, MIP1-α, TNF-α, and G-CSF concentrations have been positively associated with intensity of disease signifying the contribution of hypercytokinemia towards inflammation and immunopathology.[81,155,176,193]

It has been found that during the first week of disease, severe COVID-19 patients manifest maximum cytokine levels that include INF γ, IL-2, IL-6, and IL-10.[132] Yang *et al.* found 48 cytokines in 53 plasma samples from patients having moderate/severe disease. They observed that IL-1RA, IP-10, and MCP-3 exhibited the highest prevalence in severe cases. They reported that these three cytokines were closely associated with COVID-19 severity.[200] Xiong *et al.* analysed the transcriptomes of bronchoalveolar lavage fluid and PBMC samples from 19 COVID-19 patients and focused on the connection between rise in secretion of cytokines such as CCL2 (MCP-1), CCL3 (MIP-1A), CCL4 (MIP1B), and CXCL10 (IP-10) and COVID-19 pathogenesis.[201] These studies demonstrate that the concentration of cytokines and chemokines are directly correlated with disease severity.

Chen *et al.* observed that a 10-fold rise in IL-6 in severe COVID-19 patients can contribute to cytokine storm which finally results in multi-organ failure.[202] Wang *et al.* conducted a study on 11 SARS-CoV-2 infected patients manifesting severe pneumonia and concluded that 72.7% of them suffered from cytokine release syndrome-like (CRSL) toxicity. Additionally, high IL-6 levels were observed in PBMCs, which were primarily responsible for the CRSL toxicity.[144] These studies suggest that IL-6 contributes significantly in causing the "cytokine super-cyclone".

9.5.3.1 IL-6 in Lymphopenia and Coagulation

IL-6 decreases the count of major histocompatibility complex class II molecules on the surface of monocytes, leading to monocyte hyperactivation, lymphopenia, and uncontrolled IL-6 production[109] (Figure 9.2).

It is known that inflammatory-induced coagulation activation can be triggered by pro-inflammatory cytokines. Various studies have revealed the anti-coagulant properties of TNF-α and IL-1. It has been shown that IL-6 activates coagulation pathways by interrupting in the production of thrombin.[203–205] Thrombin is responsible for clot formation and inflammation magnification through proteinase-activated receptor-1 (PAR-1). The anti-coagulants including antithrombin III, the protein C system, and tissue factor pathway inhibitor regulate thrombin production; however, in hyperinflammatory states, the impairment of all these functions occurs. There is evidence of pro-coagulant and anti-coagulant molecular imbalance which results in diffused intravascular coagulation and microthrombosis, and in due course leads to multi-organ dysfunction in severe COVID-19 cases.[206,207] Thus, the cytokine storm, specifically IL-6, leads the way to immune system dysfunction, contributing to and/or causing severe COVID-19.

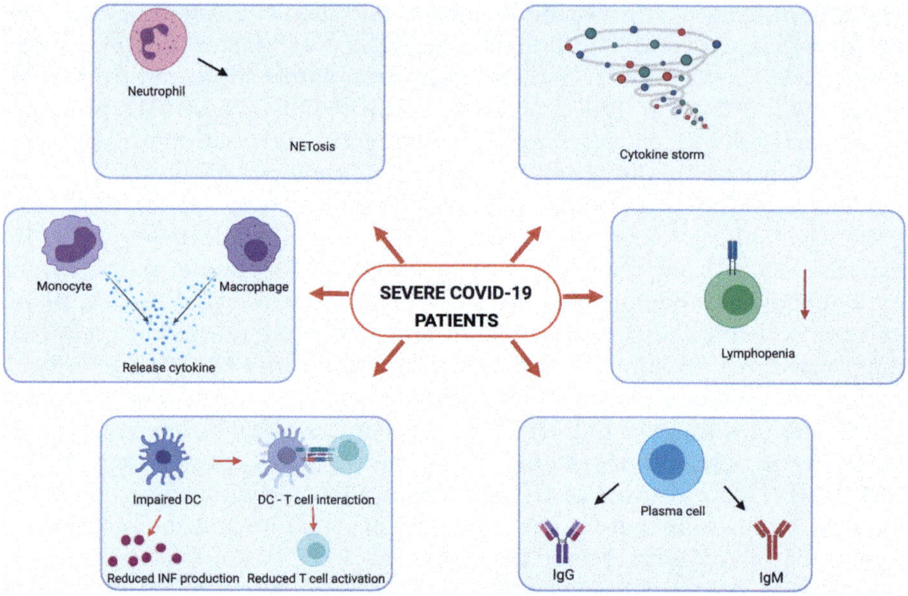

Figure 9.2 The characteristic features of the immune system as observed in severe COVID-19 patients: the formation of neutrophil extracellular traps that have been manifested in severe COVID-19 patients;[134] release of cytokines leading to cytokine storm is very well known in severe COVID-19 patients; impaired dendritic cells fail to induce the optimum adaptive immune response;[138] robust production of IgG and IgM were found in severe COVID-19 patients; [171,172] and lymphopenia is an important biomarker in severe COVID-19 patients.

9.6 Factors Affecting COVID-19 Severity

9.6.1 Human Genetic Variation

Recently, many studies have reported the impact of genetic variation in influencing the susceptibility of humans to SARS-CoV-2 infection and severity of COVID-19. It is well known that individuals with different racial and ethnic backgrounds display large allelic variation.[208–210] Prior studies on coronaviruses have highlighted the role of cellular receptors in incidence of disease. Thus, individuals with allelic variants in the genes coding for these cellular receptors such as ACE2 and TMPRSS2 may have differential expression and function of the genes, resulting in variation of SARS-CoV-2 infection susceptibility along with disease severity. These factors help explain why some individuals manifest mild COVID-19 and others very severe disease.

9.6.1.1 ACE2

As ACE2 is the primary cellular receptor for SARS-CoV-2 entry into the host cell, many researchers have investigated the correlation between ACE2 heterogeneity and COVID-19 incidence. Renieri *et al.* identified and characterized

33 ACE2 variants from approximately 7000 Italian individuals. They found a variant (N720D) that was adjacent to the TMPRSS2 cleavage site, and three other mutants (L351V, P389H, W69C) that were thought to modify conformation of ACE2 and alter the interaction with S (RBD) protein.[211] Stawiski and co-workers described a series of variants in ACE2 that apparently increase (I21V, S19P, E23K, F72V, H378R, K26R, N64K, Q102P, T27A, T92I) or decrease (D38V, D355N, D509Y, E35K, E37K, G352V, G326E, H34R K31R, K68E, M62V, N51S, N33I, Q388L, Y50F, Y83H) SARS-CoV-2 susceptibility based on the interaction with the RBD of S protein.[212] Another study on East Asian and European cohorts recognized two ACE2 variants (K26R and I468V) having the ability to decrease the ACE2 binding affinity for the RBD of S protein.[213] The differential interaction between ACE2 and S protein determines the viral attachment and internalization of SARS-CoV-2, which explains the variations in COVID-19 severity between individuals.

Additionally, the ACE2 polymorphism has been linked with COVID-19 severity in males. Some studies have reported that men express higher amounts of ACE2 than women,[214,215] in contrast to other studies reporting equivalent ACE2 expression in both sexes.[216,217] It is not clear whether the ACE2 expression density or pattern in different organs determine the disease severity or whether in fact it is a difference based on sex.[218–220] However, as the X chromosome codes for ACE2, males harbor only a single variant of ACE2 on all cells, in contrast to females, where X-inactivation leads to a mosaic pattern of ACE2 expression.[218] Thus, if the males carry ACE2 which has a strong affinity with the virus, they should be more susceptible to infection.

Pinto *et al.* examined 700 lung transcriptomic datasets of patients having comorbidities including chronic obstructive pulmonary disease (COPD), hypertension, and diabetes. These investigators observed higher ACE2 expression in these comorbid individuals in comparison with healthy individuals[221] which may explain their greater risk of severe COVID-19. In contrast to this, numerous studies have shown the association of higher ACE2 expression in prevention of lung injury and the decline in ACE2 expression with increase in age in animal models.[222–224] Therefore, detailed investigations are needed to understand the relationship between ACE2 expression regulation and underlying comorbidities and COVID-19 severity.

9.6.1.2 TMPRSS2

TMPRSS2 is the second protein that is associated with SARS-CoV-2 entry. However, fewer studies have been conducted on TMPRSS2 in relation to COVID-19 incidence. Russo *et al.* determined an intergenic single nucleotide polymorphism (SNP) that came to be concurrently related to increased TMPRSS2 expression. Thus, individuals harboring this SNP might have more susceptibility towards SARS-CoV-2 infection.[225] Another study on an Italian cohort identified various SNPs in TMPRSS2 that are associated with increased expression of TMPRSS2 receptors on the cell surface 112. Significantly, one of these variants was linked to an androgen-responsive enhancer, suggesting a probable mechanism whereby males are more susceptible to the virus.

However, studies conducted up until the writing of this chapter are not conclusive in establishing a relationship between ACE2 and TMPRSS2 receptors with COVID-19 severity.

9.6.2 Blood Group

Zhao *et al.* analysed ABO blood group distribution in 2173 COVID-19 patients. They observed that type A blood group persons were most prone to SARS-CoV-2 infection. On the contrary, O blood group was found to have the least chances of acquiring the infection.[226] Identical blood group distribution patterns were observed in another study.[227]

Additionally, Dai reported that COVID-19 patients with blood group A and cardiovascular diseases (particularly hypertension), are more likely to develop severe disease.[228] A study on 1980 COVID-19 patients from Italy and Spain found two genetic regions to be associated with respiratory failure. One of them was present at 9q34.2 inside the locus of ABO blood grouping, which demonstrated that blood group A individuals have higher risk (meta analysis result OR (odds ratio) 1.45, $P = 1.48 \times 10^{-4}$) of acquiring the infection, in contrast to blood group O individuals having lesser probabilities (OR 0.65, $P = 1.06 \times 10^{-5}$) of infection.[229]

There might be several reasons for such discrepancies. A study found that individuals with blood group A have been associated with higher risk of chronic diseases such as gastric cancer, coronary heart disease, and venous deep thromboembolism.[230] Moreover, genome-wide association studies have shown that genetic variations at ABO loci are found to be associated with increase in plasma lipid levels and inflammatory markers such as intercellular adhesion molecule 1,[231–233] which might increase the chances of developing cardiovascular diseases.[234,235] With the increased risk of cardiovascular diseases, individuals are at a higher risk of developing severe COVID-19 symptoms. Looking into these aspects, Zhang and Hughes have suggested further investigations regarding the correlation between ABO blood grouping and COVID-19.[236]

9.6.3 Gender

In comparison with females, males may be more susceptible to viral infection on the basis of differences in hormones, the immune system, and sex chromosomes. Consistent with this, numerous studies have shown that male COVID-19 patients develop more severe disease and experience higher mortality rates in comparison to females.[43,155,206] Gender-based data from 26 countries have revealed higher case fatality rates in males than females.[237] The higher death rate in males can be partly understood due to the existence of underlying diseases (such as diabetes, cardiovascular disease, chronic lung disease), habits (drinking alcohol and smoking), and occupational exposure.[237]

Sex chromosomes and hormones such as estrogens, progesterone, and androgens modulate the innate immune responses in respiratory viral infections.[238] In viral infections, disease incidence and severity in males and females depend on different copy numbers of X-linked genes, steroid hormone production by the gonads, and the existence of disease susceptibility genes.[239,240] The X-linked gene in females has a mosaic pattern of distribution. Because the X chromosome is known to harbor many immune-related genes, females tend to generate stronger innate and adaptive immune responses than males,[241] resulting in faster recovery. Zeng *et al.* conducted a study on 331 COVID-19 patients and observed that production of IgG was stronger in females than males in the early phase of the disease. Additionally, the female patients generated comparatively higher levels of IgG than male patients during the severe phase of the disease.[242] However, these strong immune responses may also make women more susceptible to inflammation and autoimmune diseases. Thus, in-depth studies are needed to ascertain a higher immunological response and lower COVID-19 severity in women.

Another biological difference that can be correlated with sex differences is ACE2 receptors. Interestingly, a study has reported that the reproductive organs exhibit different densities of ACE2. The testes in males have higher ACE2 levels in comparison with the ovaries in females.[243] While ACE2 forms the gateway for SARS-CoV-2 entry, it is also known to have anti-inflammatory activities that are assumed to have a role in lung injury prevention. These two opposite functions might be attributed to variation of ACE2 protein location, that is either transmembrane or within the plasma membrane.[244] As stated previously, the X chromosome hosts the ACE2 gene, suggesting that higher ACE2 expression in females may lead to protection against more severe disease compared to males.[245] Therefore, further investigation is essential to establish the correlation between sex and disease severity.

9.6.4 Age

Studies to date have reported a low incidence of COVID-19 in children.[8,42,155] Among COVID-19 patients aged <18 years, children aged <1 year seem to have the highest risk of developing severe disease.[246] Liu *et al.* analysed 366 hospitalized children (aged ≤16 years) from Wuhan. Among them, six children were detected positive for SARS-CoV-2 infection with a median age of 3 years (range 1–7 years). These children manifested moderate-to-severe respiratory illness, and one of them was admitted to a pediatric ICU.[247] Another study reported that clinical manifestations of COVID-19 in children may be less severe than in adult patients.[248] However, a recent study revealed that children are as likely as adults to transmit COVID-19.[249] Therefore, detection of COVID-19 in children should not be overlooked.

Elderly persons are more prone to SARS-CoV-2 attack, and have a higher risk of developing severe COVID-19 as well as a higher mortality rate.[43]

Consistent with this, an age-dependent study reported that the probability of developing COVID-19 symptoms ranged from 20% in those aged <10 years to $>70\%$ in older adults.[250] Liu *et al.* compared the clinical features of hospitalized elderly and young/middle aged COVID-19 patients. They observed that out of 56 patients, 12 elderly patients had chronic diseases, mainly diabetes and cardiovascular and cerebrovascular diseases, which was comparatively a much higher prevalence than in young and middle-aged patients. Additionally, they observed a higher mortality rate in elderly patients than in young and middle-aged patients. They also found that the elderly are likely to develop more severe disease.[251]

Another study on 783 COVID-19 Italian patients by Borghesi *et al.* showed that males aged ≥ 50 years and females aged ≥ 80 years have the highest risk of developing severe pneumonia in COVID-19.[252] This clearly displays the close association between age and COVID-19 symptoms in males as well as females.

9.6.5 Comorbidities

It has been found that severe COVID-19 patients who are admitted to the ICU frequently have underlying comorbidities. Among severe COVID-19 patients, approximately 75% have chronic comorbidities.[138] The most common underlying diseases has been found to be hypertension (30%), diabetes (19%), and cardiovascular diseases (8%).[206] Other underlying diseases include COPD, pharyngeal pain, abdominal pain, dizziness, anorexia, dyspnea, malignancy, cerebrovascular diseases, chronic kidney diseases, and hepatitis B infections. Complications including ARDS,[92] shock, arrhythmia and multiorgan failure,[8,253] are more likely to arise in these patients with comorbidities.

9.6.5.1 Hypertension

Hypertension has shown to increase the mortality rate of COVID-19 patients 2.5-fold.[254] Serum samples of hypertensive patients have shown high levels of IL-2, IL8, IL-1α, MCP-1, TNF-α, IFN-γ, epidermal growth factor, and VEGF.[255] Moreover, hypertensive patients exhibit comparatively lower levels of the anti-inflammatory cytokines IL-10 and IL-4, which suggests an irregular Th1, Th2, and Th17 immune response caused by hypertension.[255,256] Interestingly, COVID-19 patients also manifest increased IL-1, IL-2, IL-6, IL-8, TNF-α, and MCP-1 levels, suggesting a correlation between hypertension, inflammatory cytokines, and COVID-19 severity.

Furthermore, the involvement of the renin–angiotensin system in the pathogenesis of hypertension is well known and ACE-inhibitors and type 1 angiotensin II receptor blockers (ARBs) inhibitors are primarily used for treating hypertensive COVID-19 patients.[257] This leads to increased ACE2 expression. Thus, hypertensive patients are at a higher risk of SARS-CoV-2 infection. Additionally, computational network analysis determined IL-6, IL-10, CCL2, CXCL8, MMP-9, VEGF-A, and TNF to be the major genes in hypertensive patients predisposing to a having higher risk of SARS-CoV-2 infection.[258]

Thus, hypertensive patients, being more prone to SARS-CoV-2 attack, when infected by SARS-CoV-2 display comparatively very high inflammation leading to an intensified cytokine storm that worsens the COVID-19 symptoms.

9.6.5.2 Diabetes

Type 2 diabetes is characterized by increased pro-inflammatory cytokines and hyperimmune activation.[259] These pro-inflammatory cytokines alter insulin resistance and sensitivity, glucose metabolism, and enhance the synthesis of liver acute phase proteins and release of inflammatory mediators.[260] The predominant cytokines that are involved in immune pathological processes in individuals with type 2 diabetes are IL-1β, IL-6, TNF-α, and MCP-1.[261] In type 1 diabetes, high levels of IL-1α, IL-1β, IL-6, IL-10, IL-17A, IL-23, and TNF-α are found.[262–264] In type 1 diabetes, an inflammatory cascade leads to activation and infiltration of macrophages, T-lymphocytes, and other immune cells, along with secretion of inflammatory mediators in the pancreas.[265]

The concentration of inflammatory cytokines increases in COVID-19 diabetic patients. Hyperglycemia, elevated coagulation rate, and increased secretion of pro-inflammatory cytokines[266] all enhance the COVID-19 severity in diabetic patients. Type 1 or type 2 diabetic patients are administered ARBs and ACE inhibitors for treatment of COVID-19, which might facilitate a rise in ACE2 expression.[257]

A study found that COVID-19 patients with diabetes exhibit a two-fold higher risk of becoming infected with SARS-CoV-2 and of dying, in comparison with non-diabetics.[263] One study reported that two pregnant women with comorbid gestational diabetes mellitus succumbed to death due to SARS-CoV-2 infection.[267]

Thus, it can be deduced that the increased pro-inflammatory cytokines levels in diabetic COVID-19 patients further augment the cytokine storm, making the disease more severe.

9.6.5.3 COPD

In due course, chronic exposure to irritants, environmental pollutants, or smoking leads to COPD. The incidence of COPD in COVID-19 patients is reported to be only 3.2%.[268] However, COPD as an underlying disease has a tremendous impact on mortality rate and severity, increasing them by 63% and 60%, respectively[269] and worsening the COVID-19 effects by 5.9 times,[270] although stratification of these data is needed on the basis of cause and stages of COPD. A study reported that the frequency of smoking in COVID-19 patients is 9% and that current smokers in comparison with non-smokers are more prone to COVID-19 associated complications.[269]

COPD is mainly characterized by airflow obstruction, lung parenchymal inflammation, chronic bronchitis, and pulmonary emphysema.[271] As the epithelial cells become damaged, there is an infiltration of $CD8^+$ T-cells, neutrophils, and monocytes to lung parenchyma and peripheral airways, in addition to secretion of pro-inflammatory cytokines which enter into the systemic circulation, resulting in systemic inflammation.[272] Plasma and sputum samples of COVID-19 patients with COPD have shown increased levels of some prominent cytokines including IL-1β, IL-6, IL-8, IL-18, VEGF, and TNF-α.[273–276] In addition to these, a study reported upregulation of ACE2 receptors in the epithelium of the lower airway of COPD patients, conceivably explaining the increased susceptibility of these patients.[277] Altogether, the augmented cytokine production that leads to inflammation in COPD patients, when infected with SARS-CoV-2, is multiplied and results in increased disease severity in COVID-19 patients compared with COPD alone.

In conclusion, these comorbidities share a hyperinflammation profile similar to SARS-CoV-2 infection and greatly enhance the cytokine storm associated with COVID-19.

9.6.6 Diet and Microbiota

Nutrition plays a crucial role in shaping the health of an individual. Diets comprising high amounts of sugars, refined carbohydrates, and saturated fatty acids result in medical disorders such as obesity, hypertension, and diabetes, which increase the likelihood of severe COVID-19. The Western diet contains a high quantity of sugars, refined carbohydrates, and saturated fats, and lower amounts of fiber, antioxidants, and unsaturated fats.[278] Specifically, saturated fatty acids (SFA) activate the innate immune system *via* TLR4 expressed in macrophages, DCs, and neutrophils, which triggers inflammatory cascades releasing inflammatory cytokines.[279] For example, a study in mice fed a diet high in SFA demonstrated increased macrophage infiltration into the lung and mainly the alveoli,[280] which is also evident in COVID-19 pathology.[187] Furthermore, high levels of SFA inhibit the function of B- and T-cell lymphocytes in the adaptive immune system, by increasing oxidative stress.[281]

It has been well established that diet profoundly influences the composition of gut microbiota. In turn, the gut microbiota immensely affects the host's health *via* the host's immune system. A study has reported that various bacterial fragments and metabolites digest the dietary fiber and generate short-chain fatty acids in blood, which inhibit allergic pulmonary inflammation.[282] It has been shown that the RNA of SARS-CoV-2 is present in stools of COVID-19 patients, which suggests a relationship between pulmonary and intestinal infection.

Interestingly, imbalance in the gut microbiota, *i.e.* dysbiosis, has been manifested in immunocompromised patients and older people with comorbidities such as diabetes and cardiovascular diseases who develop severe COVID-19. Therefore, gut microbiota dysbiosis might also influence COVID-19 severity in patients.

There are several pre-biotics such as fructo-oligosaccharides, wheat bran, and galactosaccharides, which increase the concentration of butyrate resulting in reduction of inflammation.[283] A study has shown that healthy diets have an anti-inflammatory effect, even in obese individuals.[284] Moreover, a study claimed that dietary therapy and herbal medicines might be effective in the treatment of COVID-19 patients.[285] Therefore, consuming a healthy diet leading to a balanced gut microbiota may be important in reducing the probability of the development of severe COVID-19.

9.6.7 BCG Vaccine

Bacillus Calmette-Guérin (BCG) vaccine is a live attenuated strain procured from a bacterium named *Mycobacterium bovis*. It is administered in newborns for protection against tuberculosis (TB) and diseases caused by other non-tuberculous mycobacteria like Buruli ulcer[286] *Mycobacterium ulcerans*. Previous reports have revealed that BCG vaccine also confers protection against several DNA and RNA viruses and diseases such as influenza virus, herpes virus,[287] leprosy,[288,289] encephalomyocarditis[290] and bladder cancer.[291] A study also reported that the sub-Saharan African children aged <5 years who were administered BCG vaccine were less susceptible to malaria.[292] Moreover, BCG has a defensive role in respiratory tract infections not only in children,[293,294] but also in elderly people by increasing IL-10 and IFN-γ levels.[295]

BCG is known to establish non-specific innate immunity against pathogens *via* an innate immune mechanism known as "trained immunity". In this state, the innate immune system recalls its past encountered pathogens, that is, "immunological recall" through epigenetic reprogramming.[296] After vaccination, several innate immune cells, *e.g.* NK cells, monocytes, and macrophages undergo histone epigenetic modification including acetylation and methylation to elevate IL-6, IL-1β, and TNF-α expression. Upon exposure to pathogens, these trained immune cells are activated to augment production of pro-inflammatory cytokines, thereby amplifying host protection.[297,298] In addition, BCG further induces the adaptive immune system, promoting greater host immunity.[299]

As per these observations, BCG vaccine has the potential to protect against SARS-CoV-2.[300] Miller *et al.* compared the morbidity and mortality rate of COVID-19 in different countries having/not having BCG vaccination policies. They observed that countries with universal BCG vaccination policies in place were less severely affected by COVID-19 than those countries (such as USA, Italy) lacking universal BCG policies. It was suggested that the BCG vaccination was responsible for reducing morbidity and mortality rates.[301] Additionally, another study reported that the incidence of COVID-19 was low in countries having high BCG vaccination coverage, indicating some degree of protection in tuberculosis-endemic areas.[302] Ozdemir and co-workers demonstrated that a BCG-vaccinated population manifested comparatively lesser morbidity, mild disease, and a decreased mortality rate compared to a non-vaccinated population.[303] It has been estimated that BCG vaccination may confer a secondary innate immune response after 1 month that could result in lowering of virus load and improvement of antiviral responses to SARS-CoV-2 infection, leading to milder disease and fast recovery.[304]

Even though numerous studies validate the prediction of BCG vaccination stimulating the non-specific innate immune response against SARS-CoV-2,[301,305] there have been differences in statistical analysis, data extraction techniques, and confounding variables regarding BCG vaccine-mediated protection and the degree of protection currently remains an open question.[306,307] At present, according to the clinicaltrials.gov website, approximately 22 clinical trials are ongoing to estimate BCG vaccine induced immunity against SARS-CoV-2 infection.

9.7 Therapeutics

Several of the novel COVID-19 therapeutics specifically focus on immunomodulation, where various features of innate and adaptive immune response are targeted. Leading the battle against COVID-19, some of the significant mechanism used to obstruct SARS-CoV-2 infection include drugs preventing entry of SARS-CoV-2, arresting immunosuppressive molecules, and drugs inhibiting chemokine, cytokine, and JAK-STAT cascades.[308] The experience with some of these drugs in the treatment of COVID-19 is described in this section.[68]

9.7.1 Chloroquine and Hydroxychloroquine

Chloroquine (CQ) and hydroxychloroquine (HCQ) are commonly used as antimalarial drugs and as immunomodulatory therapeutics in rheumatological diseases. They have also been used as antivirals, including for influenza, Chikungunya, and SARS-CoV-1,[309,310] and this led to their use for SARS-CoV-2.

9.7.1.1 Assessment of HCQ Effectiveness

As of December 2020, numerous controlled clinical trials have been conducted and many more are ongoing. From the completed studies, the efficacy of CQ and HCQ in treating SARS-CoV-2 infection remains controversial.

A clinical trial approved by the French National Agency for Drug safety (ANSM) (EU Clinical Trial Register 2020-000890-25) was conducted on 6 March 2020. The trial included 20 patients in a single-arm protocol in which 600 mg of HCQ alone or in combination with azithromycin was administered daily and nasal swabs were tested daily for viral load. Investigators observed that with the combination of HCQ and azithromycin there was a reduction in viral load on day 6, suggesting efficacy for virus elimination.[311] A follow-up trial was conducted on 80 patients who were administered HCQ along with azithromycin. In 93% of cases, the polymerase chain reaction (PCR) results were negative on the eighth day, and by 10 days of treatment, 81.3% of patients were discharged. It is noteworthy that there were no control groups in either of the trials.[312]

A double-blind randomized controlled trial (RCT) (Chinese Clinical Trial Register ChiCTR2000029559) on 62 patients analysed HCQ monotherapy for mild COVID-19.[313] Patients were given 400 mg of HCQ daily for 5 days. By the sixth day, the HCQ-treated patients had clinical improvement on average 1 day earlier than the controls. Moreover, there was no progression to severe disease in HCQ-treated patients compared with controls, of whom four patients developed severe disease. A smaller RCT (Clinicaltrials.gov NCT04261517) was performed on 30 mild COVID-19 patients who were treated with 400 mg HCQ for 7 days. On the seventh day, there were no significant differences in the number of patients exhibiting negative PCR results, time to fever resolution, median duration of hospitalization, or disease progression on chest computed tomography (CT).[314] The largest RCT (Chinese Clinical Trial Register ChiCTR2000029868), an open-label trial of HCQ, recruited 150 mild COVID-19 patients. They also observed no notable differences between groups based on negative SARS-CoV-2 RT-PCR results or the rate of symptom resolution on day 28; in contrast, they found more adverse effects, large, but non-serious, in the HCQ-treated patients.[315] A meta-analysis by Shamshirian *et al.* considered the studies mentioned herein and found there was no clinical benefits of HCQ in COVID-19 patients receiving standard of care.[316]

Few studies have been performed to have evaluate the effectiveness of HCQ in severe cases. In one study, 11 patients were administered 600 mg HCQ with azithromycin for 10 days. After 5 days, the clinical status of several patients had deteriorated and one patient had died. On the tenth day, out of 10 patients, eight displayed positive PCR results.[317] Another ongoing double-blind RCT (Clinicaltrials.gov NCT04323527) on 81 severe COVID-19 patients divided into two groups were given low-dose (450 mg per day for 5 days) or high-dose HCQ (600 mg twice a day for 10 days).[318] The high-dose protocol was stopped because of poor safety outcomes. Investigators noticed that on the fourth day there were no significant differences in RT-PCR negative reports or any difference in requirement for mechanical ventilation on day 6. Overall, these clinical trials have not displayed any clear benefit of HCQ treatment together with azithromycin for COVID-19 patients.

9.7.1.2 *HCQ and Cardiac Arrhythmia Risk*

Recently, numerous studies found that HCQ/CQ treatment of COVID-19 patients observed an average increase of QTc interval by ~28–35 ms,[319–321] thus raising the risk of occurrence of malignant arrhythmias (*torsades de pointes*). It is noteworthy that electrocardiographic data from previous studies have shown that COVID-19 patients even without the use of HCQ/CQ have a prolonged QTc intervals[319–322] with a baseline mean 425–455 ms range in comparison to healthy volunteers having a 400 ms QTc interval.[323,324] Significantly, the study found that up to 20% of COVID-19 patients observed QTc prolongation in the range of ~500 ms, which is considered as a limit for the discontinuation of treatment.[325] In view of these findings, HCQ/CQ use might induce life-threatening arrhythmias, specifically in cases of COVID-19 patients with pre-existing cardiovascular diseases.

Importantly, the WHO announced the termination of the HCQ arm of the Solidarity trial to find an effective treatment for COVID-19 on 17 June 2020, as it was found that HCQ usage did not decrease the mortality rate of hospitalized COVID-19 patients.

9.7.2 Corticosteroids

Corticosteroids, due to their anti-inflammatory effect, are administered as an adjuvant therapy for cytokine storm and ARDS. However, this immunosuppression becomes a bane when it interferes with the progression of immunity against SARS-CoV-2. This was evident in a meta-analysis study on 5270 patients with SARS-CoV-1, SARS-CoV-2, and MERS-CoV, where the mortality increased upon corticosteroid treatment.[326] Additionally, there are reports claiming that corticosteroid treatment in SARS and MERS patients delayed viral clearance and increased complications.[327] Due to these findings, the WHO has advised against the application of systemic corticosteroids for treating COVID-19 patients. However, a meta-analysis of 2636 COVID-19 patients, including a group of patients manifesting ARDS demonstrated no difference in mortality upon corticosteroid treatment.[328] Another group found no difference in the viral clearance duration between the patients who were administered oral methylprednisolone (MPS) (mild cases) and intravenous MPS (severe cases) and the patients who were not given either.[329]

Retrospective studies in China observed that patients who did not receive corticosteroids were less likely to be admitted to an ICU.[8] Patients manifesting ARDS who were administered MPS showed a decrease in mortality risk.[65] In contrast, another retrospective study showed that patients who were given corticosteroids had more comorbidities and more possibilities of being admitted to the ICU.[330] However, a small study conducted on 31 patients did not find any correlation between corticosteroids treatment and symptom duration, viral clearance duration, or hospital stay length.[331] Another study of adjuvant corticosteroid treatment on 244 severe COVID-19 patients observed no effect of corticosteroids on 28-day mortality; a subgroup analysis of patients manifesting ARDS showed no significant relationship between corticosteroid treatment and clinical symptoms.[332] Alarmingly, investigators found that mortality risk increased significantly in association with increase in corticosteroids dosage.

A retrospective study of 46 patients, 26 of whom were administered early, low-dose MPS showed significant improvement in oxygen saturation and chest CT, time on supplemental oxygen therapy, and time to fever resolution.[333] Based on the anecdotal evidence, few reports have supported the usage of early[334] and low-dose, short-term administration.[335]

Dexamethasone, a glucocorticoid, has been used in the latest RECOVERY trial by a team of researchers at Oxford University (UK). They administered 6 mg of dexamethasone to 2104 COVID-19 patients once daily for

10 days and compared the results with 4321 COVID-19 patients who were not administered dexamethasone. The average age of the patients was 66.1 years and 36% of patients were female. They found that use of dexamethasone decreased 28-day mortality by 20% in patients requiring supplementary oxygen and by 35% in patients requiring mechanical ventilation. No side-effects were manifested, as the dose of dexamethasone was restricted to a low to medium dose. The positive effects of dexamethasone were clearly evident when the patients were given dexamethasone for more than 7 days after initial onset of symptoms. However, no effect of dexamethasone was seen in patients with mild symptoms, which clearly indicates that corticosteroids should be administered to severe COVID-19 patients.[336] In another study, it was found that corticosteroids not only lowered the mortality rate in severe COVID-19 patients, but also improved oxygen status, prevented the worsening of ventilation parameters, and shortened duration of hospital stay.[337]

However, dexamethasone use also has some side-effects such as headache, mood changes, and change in appetite. Furthermore, long-term usage, *i.e.* more than a week, could result in arrhythmias. Therefore, the administration of high-dose dexamethasone to people with comorbidities such as diabetes and cardiovascular diseases should be undertaken carefully.[338] The use of dexamethasone in COVID-19 patients with comorbidities might deteriorate their condition.

In view of these findings, WHO has appreciated the preliminary results regarding the usage of dexamethasone in treating COVID-19 patients.[338] As of the writing of this chapter, 13 clinical trials are ongoing for the usage of dexamethasone in the treatment of COVID-19, according to the clinicaltrials.gov website.

9.7.3 Cytokine Inhibition

9.7.3.1 IFN

As discussed earlier, the first host response after the entry of SARS-CoV-2 is release of type I and III IFNs. IFNs produce antiviral effects by modulating the innate immune response.[339] Following the binding of IFN to its receptor present on the cell surface, they regulate gene transcription *via* the JAK-STAT signalling pathway. Because the type I IFN receptors are expressed extensively, type I IFN signalling is associated with antiviral effects as well as activation of immune cells that accentuate pathogenesis. Type III IFN signalling is limited to specific immune and epithelial cells. They only exhibit antiviral effects and do not exacerbate pathogenic inflammation.[340,341]

Recently, a study illustrated that SARS-CoV-2 can obstruct type I IFN activity by inhibiting the STAT-1 intracellular signalling pathway.[342] Similarly, another study showed that the type I IFN signalling pathway is impaired upon SARS-CoV-2 infection; *in vitro* studies revealed that type I IFN might be more effective against SARS-CoV-2 than SARS-CoV-1.[81] In view of these findings, a few clinical trials on type I/III IFN are being conducted. One of them is the evaluation of the efficacy of recombinant IFN-α1β in treating COVID-19 patients of Wuhan (NCT04293887). Other trials include clinicaltrials.gov NCT04343768, NCT04343976, and NCT04331899.

9.7.3.2 Interleukin

As mentioned earlier, COVID-19 severity is related to the levels of inflammatory cytokines, mainly IL-6.[74,184,343–346] Therefore, many clinical trials are been conducted on monoclonal antibodies (Abs) that target the IL-6 signalling pathways.[345] Currently, numerous tests are being conducted on anti-IL-6 Abs such as tocilizumab (TCZ) (Actemra), sarilumab (Kevzara) and siltuximab (Sylvant) for their efficiency in COVID-19 treatment.

A study on a cohort of 20 COVID-19 patients administered TCZ (400 mg), together with methylprednisolone, lopinavir, and oxygen therapy (Chinese Clinical Trial Register ChiCTR2000029765) reported improvement in symptoms with no consequent pulmonary infections.[346] Another study on 30 severe, rapidly deteriorating COVID-19-related pneumonia demonstrated the correlation between the usage of TCZ and a decrease in ICU admission and mechanical ventilation.[347]

Additionally, a study was conducted on TCZ administration in 100 treated patients (88 male, 12 female, median age 62 years), of whom 43 patients were treated in the ICU and 57 patients were treated in a general ward due to unavailability of ICU beds. After 10 days, respiratory conditions were either stabilized or improved in 77 patients, of whom 15 patients were discharged and 61 patients displayed a remarkable clearance of diffused bilateral opacities in chest X-ray images. However, in 23 patients, the respiratory conditions deteriorated and 20 of these patients died. The investigators concluded that there was a quick and sustained response upon usage of TCZ with substantial improvement in clinical symptoms in many of these patients.[348]

Moreno-Pérez *et al.*[349] described the consequence of TCZ administration in 77 patients with comorbidities and manifesting severe pneumonia. With 83 days of TCZ usage follow up, they witnessed a decrease in the mortality rate of severe COVID-19 patients to about 12.9% in comparison with 21% (2653 patients)[350] and 28.2% (191 patients).[206] Moreover, early TCZ response was associated with shorter median hospital stay and lower comorbidities, resulting in an increased survival rate.

A study by Rodríguez-Baño *et al.*[351] estimated the correlation between TCZ and corticosteroids with the possibility of intubation or death in 778 COVID-19 patients with hyperinflammatory state from 60 Spanish hospitals. Four groups of patients were described on the basis of treatment including TCZ,[88] intermediate–high dose of corticosteroids,[117] pulse dose of corticosteroids,[78] combination therapy of TCZ and corticosteroids,[151] and no treatment.[345] Their clinical and laboratory data showed that TCZ was associated with lower hazard ratios for intubation or death in COVID-19 patients with hyperinflammation. Furthermore, another interesting study was conducted on a 50-year-old man who was COVID-19 positive with severe pneumonia and moderate ARDS. Upon admission, he was immediately given HCQ, azithromycin, and antibiotics and his respiratory symptoms worsened. After that, he was administered steroids along with intravenous TCZ and evidenced lower levels of IL-6 with gradual improvement in respiratory symptoms and lung function. The authors suggested that TCZ may block downstream signal transduction by binding with membrane-bound IL-6 receptor (mIL-6R) and soluble IL-6R (sIL-6R), resulting in reduction of cytokine storm and clinical improvement in the condition of COVID-19 patients.[352]

Sinha *et al.* conducted an observational study on administration of IL-6 receptor inhibitor (IL6ri) (sarilumab or TCZ) in 225 COVID-19 patients. The patients were divided into two groups based on fraction of inspired oxygen ($F\text{iO}_2$) requirement. One group was termed "stage IIB", who required ≤45% $F\text{iO}_2$, and the second group was termed "stage III", who required >45% $F\text{iO}_2$. They concluded that early administration of IL6ri before requirement of >45% $F\text{iO}_2$ improved the clinical outcomes of COVID-19 patients.[353]

These studies clearly show the efficacy of TCZ in treatment of COVID-19. Currently, 53 clinical trials are ongoing to assess the efficacy of TCZ in COVID-19 patients, as per the clinicaltrials.gov website.

9.7.3.3 *JAK-STAT Inhibitors*

JAK-STAT inhibitors antagonize many pro-inflammatory cytokines that cause the cytokine storms in COVID-19. A noteworthy JAK-STAT inhibitor is baricitinib. It is not only known to inhibit the JAK1/2 (Janus Kinase 1/2), but also inhibits AAK1 (AP2-associated protein kinase 1) which is known to promote endocytosis of SARS-CoV-2. Thus, this drug inhibits the viral entry into the target cell.[354] Numerous clinical trials are ongoing on the use of several JAK inhibitors such as ruxolitinib, baricitinib, and tofacitinib for the treatment of COVID-19. Currently, 13 clinical trials are ongoing for use of baricitinib in the treatment of COVID-19.

9.7.3.4 *Nod-like Receptor Family Pyrin Domain-containing-3 (NLRP3) Inflammasome Inhibitors*

The NLRP3 inflammasome plays a major role in regulating the innate immune response. Upon invasion of pathogens including RNA viruses, NLRP3 activates caspase-1, which in turn releases pro-inflammatory cytokines IL-18 and IL-1β, which mediate the cytokine storm. One of the NLRP3 inflammasome inhibitors is anakinra, which is being used in 26 clinical trials registered for COVID-19 as per the clinicaltrials.gov website. Colchicine is another NLRP3 inflammasome inhibitor that obstructs the release of inflammatory chemokines.[355] Currently, 24 clinical trials are ongoing on the efficacy of colchicine for COVID-19 treatment as per clinicaltrials.gov website.

9.7.4 Convalescent Plasma Therapy

Generally, plasma antibodies destroy a pathogen either by direct binding and neutralization or by indirectly triggering immune responses including phagocytosis, complement-dependent cytotoxicity, or antibody-dependent cellular cytotoxicity.[356,357] The plasma of COVID-19 patients consists of active specific neutralizing Abs (nAbs) that bind with SARS-CoV-2 and non-nAbs. This binding might enhance prophylaxis as well as improve the condition of infected patients. Convalescent plasma or hyperimmune immunoglobulin involves elevated Ab titers that decrease the viral load, resulting in reduced mortality in addition to shortened hospital stays, thus protecting individuals from viral infection or preventing the development of disease.[358,359] Plasma therapy is generally used for treatment during emergencies such as pandemics, when the resources or time for preparation of pooled Ig is limited. Convalescent plasma therapy eliminates free virus together with infected cells, reduces symptoms such as fever, sputum, dry cough, pulmonary damage, and muscle pain, and reduces viremia and decreases the chances of new infections (*via* nAbs). Additionally, it increases the blood oxygen saturation and lymphocyte ratio without any adverse effects.[360–363] Therefore, in the current pandemic, convalescent plasma therapy is being actively employed as a quick immunological therapy during the early stages of COVID-19,[364] to enhance immunity in infected patients, preventing COVID-19 progression, and increasing the success rate of therapeutics.[360] In addition to this, convalescent plasma therapy is used as a prophylactic in individuals who are likely to get infected with SARS-CoV-2 including individuals with comorbidities and healthcare workers.

The United States Food and Drug Administration has given a green light for convalescent plasma therapy use against SARS-CoV-2 infection. In addition, they have recommended plasma donation by COVID-19 patients after 14 days of recovery to other individuals helping them to recover from the disease. It has been suggested that each plasma donor can supply an adequate amount of plasma needed for four patients.

Numerous studies have been carried out to investigate the efficacy of convalescent plasma therapy in COVID-19 patients. Some of the more significant studies are presented in Table 9.1. From these studies, it is clear that convalescent plasma therapy has the potential to improve disease conditions. However, more RCTs are essential to further validate the use of convalescent plasma therapy. Currently, 137 clinical trials are ongoing worldwide to estimate the effectiveness of convalescent plasma therapy in COVID-19 patients.

9.7.5 Cell Therapy

At the present time, the modulation of host hyperimmune responses and SARS-CoV-2 pathogenicity is undergoing a paradigm shift from drug treatment to cellular host-directed therapies. Few clinical studies are investigating cell therapies in treating COVID-19, including application of NK

Table 9.1 Clinical studies on the effect of convalescent plasma (CP) therapy on COVID-19 patients.[a]

Title of study	Patient characteristics	Outcomes	Ref.
Effectiveness of convalescent plasma therapy in severe COVID-19 patients	10 severe patients (aged 34–78 years)	Clinical symptoms disappeared after 3 days Improvement in chest CT Lymphocyte counts increased SO_2 raised in all patients nAbs increased in five patients	365
Treatment with convalescent plasma for critically ill patients with severe acute respiratory syndrome coronavirus 2 infection	4 critical patients (aged 31–73 years)	Viral load decreased Chest CT improved Clinical improvement with no adverse reactions associated with CP Results must be confirmed by randomized studies	366
Use of convalescent plasma therapy in two COVID-19 patients with acute respiratory distress syndrome in Korea	2 severe patients (aged 67 and 71 years)	Viral load declined Chest CT improved Fever and oxygen demand decreased CRP and IL-6 levels reduced to normal Improvement in both patients after the CP therapy along with systemic corticosteroid	367
Treatment of 5 critically ill patients with COVID-19 with convalescent plasma	5 severe patients (aged 30–70 years)	Negative viral loads in 12 days Normalisation of body temperature in 4 out of 5 patients within 3 days SOFA score reduced Increase in nAb titers	361
Convalescent plasma to treat COVID-19: Chinese strategy and experiences	1 moderate patient, 2 critical patients	Clinical improvement in 2 patients	368
Effect of convalescent plasma therapy on time to clinical improvement in patients with severe and life-threatening COVID-19	103 patients (median age-70 years) divided into groups (52 CP group and 51 control group)	Clinical improvement in CP-treated group was 51.9%, compared with 43.1% in control group Mortality rate difference in both the groups was not significant	369
Clinical efficacy of convalescent plasma for treatment of COVID-19 infections: Results of a multicenter clinical study	189 patients divided into 2 groups (115 CP therapy and 75 control group)	98% of CP-therapy group were discharged in comparison to 79% in control group Shorter duration of hospitalization in CP therapy group (9.54 days) in comparison with control group (12.88 days) 7% intubation requirement in CP therapy group; 20% in control group	370

[a] SO_2: Oxygen saturation; CRP: C-reactive protein; SOFA: Sequential Organ Failure Assessment (higher score representing more illness).

cells to strengthen the immune response of the host (clinicaltrials.gov NCT04280224), or the use of mesenchymal stem cells (MSCs), that stimulate immune suppression and tissue regeneration.[371]

MSCs are non-hematopoietic cells that originate from bone marrow, dental pulp, lung, adipose tissue, umbilical cord tissue, and placenta. They exhibit immunoregulatory and anti-inflammatory functions, facilitate damaged tissue regeneration, and prevent tissue fibrosis. Currently, MSCs are being implemented as a potential therapy to reduce the mortality rate in SARS-CoV-2 infected patients.

The first MSC study was performed on a seriously ill 65-year-old female manifesting severe pneumonia, hypertension, moderate anemia, respiratory failure, and multiple organ failure, who was given three infusions of umbilical cord MSCs (UCMSCs, 5×10^7 cells per infusion), within an interval of 3 days. 2 days after the third infusion, the patient showed a negative result for SARS-CoV-2. Moreover, her pneumonia was resolved post-UCMSC administration and no adverse effects were manifested.[372] Another study reported that intravenous injection of clinical-grade human MSCs in COVID-19 patients resulted in improvement in clinical outcomes. These findings demonstrate the potential of MSCs in reducing the aberrant inflammatory responses against SARS-CoV-2 resulting in clinical improvement for COVID-19 patients, especially in severe cases.[373] However, more studies are needed to validate the efficacy of MSCs. Currently, 57 clinical trials are ongoing as per the clinicaltrials.gov website.

9.8 Conclusion

The significant morbidity of COVID-19 as well as its associated mortality clearly signifies the infectiousness of the novel SARS-CoV-2 virus. Its worldwide spread and growing death rates demand an urgency to fully understand the mechanism of SARS-CoV-2 pathogenicity. Epidemiological observations have documented the wide spectrum of disease severity in different individuals, consistent with differences in host immune responses. The various factors determining the variable immune responses remain a mystery, which can be solved by further detailed investigations. Until that promise is fulfilled, the best way to avoid the immunological maelstrom that SARS-CoV-2 infection can cause is to avoid infectivity and transmission by following adequate personal and public health preventive measures.

9.9 Executive Summary

- SARS-CoV-2 protects itself from the elicited immune response by various means, such as evading pattern recognition receptor sensing, evading the innate immune response, and inhibiting interferon production.
- Interferon-1 is a key player in COVID-19 severity. If present early it is protective, but if delayed, a "cytokine super-cyclone" may result in severe COVID-19.

- Several immune responses can be used to determine the severity of COVID-19 such as cytokine storm, formation of neutrophil extracellular traps, impaired dendritic cells, lymphopenia, and robust antibody production.
- There are other factors that may influence COVID-19 severity, such as genetic variation, blood group, gender, age, comorbidities, and gut microbiota.
- Many clinical trials have been carried out and many more are ongoing, yet optimal therapeutics for COVID-19 remain to be established.

References

1. J. Cui, F. Li and Z.-L. Shi, Origin and evolution of pathogenic coronaviruses, *Nat. Rev. Microbiol.*, 2019, **17**(3), 181–192.
2. S. Hassan, F. Sheikh, S. Jamal, J. Ezeh and A. Akhtar, Coronavirus (COVID-19): A Review of Clinical Features, Diagnosis, and Treatment, *Cureus*, 2020, **12**(3), e7355.
3. J. Guarner, *Three Emerging Coronaviruses in Two Decades: The Story of SARS, MERS, and Now COVID-19*, Oxford University Press US, 2020.
4. S. A. Hassan, F. N. Sheikh, S. Jamal, J. K. Ezeh and A. Akhtar, Coronavirus (COVID-19): a review of clinical features, diagnosis, and treatment, *Cureus*, 2020, **12**(3), e7355.
5. R. Lu, X. Zhao, J. Li, P. Niu, B. Yang and H. Wu, *et al.*, Genomic characterisation and epidemiology of 2019 novel coronavirus: implications for virus origins and receptor binding, *Lancet*, 2020, **395**(10224), 565–574.
6. Y. Chen, Q. Liu and D. Guo, Emerging coronaviruses: genome structure, replication, and pathogenesis, *J. Med. Virol.*, 2020, **92**(4), 418–423.
7. C. Del Rio and P. N. Malani, COVID-19—new insights on a rapidly changing epidemic, *JAMA*, 2020, **323**(14), 1339–1340.
8. D. Wang, B. Hu, C. Hu, F. Zhu, X. Liu and J. Zhang, *et al.*, Clinical characteristics of 138 hospitalized patients with 2019 novel coronavirus–infected pneumonia in Wuhan, China, *JAMA*, 2020, **323**(11), 1061–1069.
9. Y. Wang, Y. Wang, Y. Chen and Q. Qin, Unique epidemiological and clinical features of the emerging 2019 novel coronavirus pneumonia (COVID-19) implicate special control measures, *J. Med. Virol.*, 2020, **92**(6), 568–576.
10. S. Khan, R. Siddique, M. A. Shereen, A. Ali, J. Liu and Q. Bai, *et al.*, Emergence of a novel coronavirus, severe acute respiratory syndrome coronavirus 2: biology and therapeutic options, *J. Clin. Microbiol.*, 2020, **58**(5), e01297–20.
11. T. Singhal, A review of coronavirus disease-2019 (COVID-19), *Indian J. Pediatr.*, 2020, 1–6.
12. A. H. Al-Rohaimi and F. Al Otaibi, Novel SARS-CoV-2 outbreak and COVID19 disease; a systemic review on the global pandemic, *Genes Dis.*, 2020, **7**(4), 491–501.

13. A. Chafekar and B. C. Fielding, MERS-CoV: understanding the latest human coronavirus threat, *Viruses*, 2018, **10**(2), 93.
14. L. Bao, W. Deng and B. Huang, *et al.*, The pathogenicity of SARS-CoV-2 in hACE2 transgenic mice, *Nature*, 2020, **583**, 830–833.
15. Coronaviridae Study Group of the International Committee on Taxonomy of Viruses, The species *Severe acute respiratory syndrome-related coronavirus*: classifying 2019-nCoV and naming it SARS-CoV-2, *Nat. Microbiol.*, 2020, **5**, 536–544.
16. C. A. Janeway Jr. and R. Medzhitov, Innate Immune Recognition, *Annu. Rev. Immunol.*, 2002, **20**(1), 197–216.
17. S. Hur, Double-Stranded RNA Sensors and Modulators in Innate Immunity, *Annu. Rev. Immunol.*, 2019, **37**(1), 349–375.
18. H. M. Lazear, J. W. Schoggins and M. S. Diamond, Shared and distinct functions of type I and type III interferons, *Immunity*, 2019, **50**(4), 907–923.
19. A. E. I. Proudfoot, Chemokine receptors: multifaceted therapeutic targets, *Nat. Rev. Immunol.*, 2002, **2**(2), 106–115.
20. C. L. Sokol and A. D. Luster, The chemokine system in innate immunity, *Cold Spring Harbor Perspect. Biol.*, 2015, **7**(5), a016303.
21. M. Cevik, C. Bamford and A. Ho, COVID-19 pandemic–A focused review for clinicians, *Clin. Microbiol. Infect.*, 2020, **26**(7), 842–847.
22. S. Khan, H. Tombuloglu, S. E. Hassanein, S. Rehman, A. Bozkurt and E. Cevik, *et al.*, Coronavirus diseases 2019: Current biological situation and potential therapeutic perspective, *Eur. J. Pharmacol.*, 2020, 173447.
23. C. Liu, *et al.*, The architecture of inactivated SARS-CoV-2 with postfusion spikes revealed by cryo-EM and cryo-ET, *Structure*, 2020, **28**(11), 1218–1224.
24. Q. Han, Q. Lin, S. Jin and L. You, Coronavirus 2019-nCoV: A brief perspective from the front line, *J. Infect.*, 2020, **80**(4), 373–377.
25. J. Vallamkondu, A. John, W. Y. Wani, S. Ramadevi, K. K. Jella and P. H. Reddy, *et al.*, SARS-CoV-2 pathophysiology and assessment of coronaviruses in CNS diseases with a focus on therapeutic targets, *Biochim. Biophys. Acta, Mol. Basis Dis.*, 2020, 165889.
26. T. K. Warren, R. Jordan, M. K. Lo, A. S. Ray, R. L. Mackman and V. Soloveva, *et al.*, Therapeutic efficacy of the small molecule GS-5734 against Ebola virus in rhesus monkeys, *Nature*, 2016, **531**(7594), 381–385.
27. G. Li, E. De Clercq, *Therapeutic Options for the 2019 Novel Coronavirus (2019-nCoV)*, Nature Publishing Group, 2020.
28. S. Angeletti, D. Benvenuto, M. Bianchi, M. Giovanetti, S. Pascarella and M. Ciccozzi, COVID-2019: the role of the nsp2 and nsp3 in its pathogenesis, *J. Med. Virol.*, 2020, **92**(6), 584–588.
29. B. Coutard, C. Valle, X. de Lamballerie, B. Canard, N. G. Seidah and E. Decroly, The spike glycoprotein of the new coronavirus 2019-nCoV contains a furin-like cleavage site absent in CoV of the same clade, *Antiviral Res.*, 2020, **176**, 104742.

30. P. Sun, X. Lu, C. Xu, W. Sun and B. Pan, Understanding of COVID-19 based on current evidence, *J. Med. Virol.*, 2020, **92**(6), 548–551.
31. S. Sungsuwan, A. Jongkaewwattana and P. Jaru-Ampornpan, Nucleocapsid proteins from other swine enteric coronaviruses differentially modulate PEDV replication, *Virology*, 2020, **540**, 45–56.
32. M. Pachetti, B. Marini, F. Benedetti, F. Giudici, E. Mauro and P. Storici, *et al.*, Emerging SARS-CoV-2 mutation hot spots include a novel RNA-dependent-RNA polymerase variant, *J. Transl. Med.*, 2020, **18**, 1–9.
33. L. van Dorp, M. Acman, D. Richard, L. P. Shaw, C. E. Ford and L. Ormond, *et al.*, Emergence of genomic diversity and recurrent mutations in SARS-CoV-2, *Infect., Genet. Evol.*, 2020, 104351.
34. J. A. Sheikh, J. Singh, H. Singh, S. Jamal, M. Khubaib and S. Kohli, *et al.*, Emerging genetic diversity among clinical isolates of SARS-CoV-2: Lessons for today, *Infect., Genet. Evol.*, 2020, 104330.
35. E. Trucchi, P. Gratton, F. Mafessoni, S. Motta, F. Cicconardi, F. Mancia, G. Bertorelle, I. D'Annessa and D. Di Marino, Population Dynamics and Structural Effects at Short and Long Range Support the Hypothesis of the Selective Advantage of the G614 SARS-CoV-2 Spike Variant, *Mol. Biol. Evol.*, 2021, **38**(5), 1966–1979.
36. M. Becerra-Flores and T. Cardozo, SARS-CoV-2 viral spike G614 mutation exhibits higher case fatality rate, *Int. J. Clin. Pract.*, 2020, **74**(8), e13525.
37. K.-S. Yuen, Z.-W. Ye, S.-Y. Fung, C.-P. Chan and D.-Y. Jin, SARS-CoV-2 and COVID-19: The most important research questions, *Cell Biosci.*, 2020, **10**(1), 1–5.
38. W. Zhang, R.-H. Du, B. Li, X.-S. Zheng, X.-L. Yang and B. Hu, *et al.*, Molecular and serological investigation of 2019-nCoV infected patients: implication of multiple shedding routes, *Emerging Microbes Infect.*, 2020, **9**(1), 386–389.
39. W. Chen, Y. Lan, X. Yuan, X. Deng, Y. Li and X. Cai, *et al.*, Detectable 2019-nCoV viral RNA in blood is a strong indicator for the further clinical severity, *Emerging Microbes Infect.*, 2020, **9**(1), 469–473.
40. J. F.-W. Chan, S. Yuan, K.-H. Kok, K. K.-W. To, H. Chu and J. Yang, *et al.*, A familial cluster of pneumonia associated with the 2019 novel coronavirus indicating person-to-person transmission: a study of a family cluster, *Lancet*, 2020, **395**(10223), 514–523.
41. C. Rothe, M. Schunk, P. Sothmann, G. Bretzel, G. Froeschl and C. Wallrauch, *et al.*, Transmission of 2019-nCoV infection from an asymptomatic contact in Germany, *N. Engl. J. Med.*, 2020, **382**(10), 970–971.
42. Q. Li, X. Guan, P. Wu, X. Wang, L. Zhou, Y. Tong, R. Ren, K. S. M. Leung, E. H. Y. Lau, J. Y. Wong, X. Xing, N. Xiang, Y. Wu, C. Li, Q. Chen, D. Li, T. Liu, J. Zhao, M. Liu, W. Tu, C. Chen, L. Jin, R. Yang, Q. Wang, S. Zhou, R. Wang, H. Liu, Y. Luo, Y. Liu, G. Shao, H. Li, Z. Tao, Y. Yang, Z. Deng, B. Liu, Z. Ma, Y. Zhang, G. Shi, T. T. Y. Lam, J. T. Wu, G. F. Gao, B. J. Cowling, B. Yang, G. M. Leung and Z. Feng, Early Transmission Dynamics in Wuhan, China, of Novel Coronavirus-Infected Pneumonia, *N. Engl. J. Med.*, 2020, **382**(13), 1199–1207.

43. W.-j. Guan, Z.-y. Ni, Y. Hu, W.-h. Liang, C.-q. Ou and J.-x. He, *et al.*, Clinical characteristics of coronavirus disease 2019 in China, *N. Engl. J. Med.*, 2020, **382**(18), 1708–1720.
44. M. Peng, Outbreak of COVID-19: An emerging global pandemic threat, *Biomed. Pharmacother.*, 2020, 110499.
45. K. Wang, W. Chen and Z. Zhang, *et al.*, CD147-spike protein is a novel route for SARS-CoV-2 infection to host cells, *Signal Transduction Targeted Ther.*, 2020, **5**, 283.
46. A. Marzi, T. Gramberg, G. Simmons, P. Möller, A. J. Rennekamp and M. Krumbiegel, *et al.*, DC-SIGN and DC-SIGNR Interact with the Glycoprotein of Marburg Virus and the S Protein of Severe Acute Respiratory Syndrome Coronavirus, *J. Virol.*, 2004, **78**(21), 12090–12095.
47. G. Cai, X. Cui, X. Zhu, J. Zhou, A hint on the COVID-19 risk: population disparities in gene expression of three receptors of SARS-CoV, 2020.
48. M. Hoffmann, H. Kleine-Weber, S. Schroeder, N. Krüger, T. Herrler, S. Erichsen, T. S. Schiergens, G. Herrler, N. H. Wu, A. Nitsche, M. A. Müller, C. Drosten and S. Pöhlmann, SARS-CoV-2 Cell Entry Depends on ACE2 and TMPRSS2 and Is Blocked by a Clinically Proven Protease Inhibitor, *Cell*, 2020, **181**(2), 271–280.e8.
49. J. Yan, J. Guo, C. Fan, J. Juan, X. Yu, J. Li, L. Feng, C. Li, H. Chen, Y. Qiao, D. Lei, C. Wang, G. Xiong, F. Xiao, W. He, Q. Pang, X. Hu, S. Wang, D. Chen, Y. Zhang, L. C. Poon and H. Yang, Coronavirus disease 2019 in pregnant women: a report based on 116 cases, *Am. J. Obstet. Gynecol.*, 2020, **223**(1), 111.e1–111.e14.
50. G. Lu, Q. Wang and G. F. Gao, Bat-to-human: spike features determining 'host jump'of coronaviruses SARS-CoV, MERS-CoV, and beyond, *Trends Microbiol.*, 2015, **23**(8), 468–478.
51. M. Xie and Q. Chen, Insight into 2019 novel coronavirus - An updated interim review and lessons from SARS-CoV and MERS-CoV, *Int. J. Infect. Dis.*, 2020, **94**, 119–124.
52. G. Izaguirre, The proteolytic regulation of virus cell entry by furin and other proprotein convertases, *Viruses*, 2019, **11**(9), 837.
53. M. Moulard and E. Decroly, Maturation of HIV envelope glycoprotein precursors by cellular endoproteases, *Biochim. Biophys. Acta, Rev. Biomembr.*, 2000, **1469**(3), 121–132.
54. S. Jamwal, A. Gautam, J. Elsworth, M. Kumar, R. Chawla and P. Kumar, An updated insight into the molecular pathogenesis, secondary complications and potential therapeutics of COVID-19 pandemic, *Life Sci.*, 2020, 118105.
55. E. Prompetchara, C. Ketloy and T. Palaga, Immune responses in COVID-19 and potential vaccines: Lessons learned from SARS and MERS epidemic, *Asian Pac. J. Allergy Immunol.*, 2020, **38**(1), 1–9.
56. L. Mousavizadeh and S. Ghasemi, Genotype and phenotype of COVID-19: Their roles in pathogenesis, *J. Microbiol., Immunol. Infect.*, 2021, **54**(2), 159–163.

57. I. Sola, F. Almazan, S. Zuniga and L. Enjuanes, Continuous and discontinuous RNA synthesis in coronaviruses, *Annu. Rev. Virol.*, 2015, **2**, 265–288.
58. J. Ziebuhr, *The Coronavirus Replicase. Coronavirus Replication and Reverse Genetics*, Springer, 2005, pp. 57–94.
59. F. Almazán, M. L. DeDiego, C. Galán, D. Escors, E. Álvarez and J. Ortego, *et al.*, Construction of a severe acute respiratory syndrome coronavirus infectious cDNA clone and a replicon to study coronavirus RNA synthesis, *J. Virol.*, 2006, **80**(21), 10900–10906.
60. J. A. Siordia Jr, Epidemiology and clinical features of COVID-19: A review of current literature, *J. Clin. Virol.*, 2020, 104357.
61. C. J. Sigrist, A. Bridge and P. Le Mercier, A potential role for integrins in host cell entry by SARS-CoV-2, *Antiviral Res.*, 2020, **177**, 104759.
62. A. C. Walls, Y. J. Park, M. A. Tortorici, A. Wall, A. T. McGuire and D. Veesler, Structure, Function, and Antigenicity of the SARS-CoV-2 Spike Glycoprotein, *Cell*, 2020, **181**(2), 281–292.e6.
63. X. Yang, Y. Yu, J. Xu, H. Shu, J. Xia, H. Liu, Y. Wu, L. Zhang, Z. Yu, M. Fang, T. Yu, Y. Wang, S. Pan, X. Zou, S. Yuan and Y. Shang, Clinical course and outcomes of critically ill patients with SARS-CoV-2 pneumonia in Wuhan, China: a single-centered, retrospective, observational study, *Lancet Respir. Med.*, 2020, **8**(5), 475–481.
64. A. M. Zaki, S. van Boheemen, T. M. Bestebroer, A. D. M. E. Osterhaus and R. A. M. Fouchier, Isolation of a Novel Coronavirus from a Man with Pneumonia in Saudi Arabia, *N. Engl. J. Med.*, 2012, **367**(19), 1814–1820.
65. C. Wu, X. Chen, Y. Cai, J. Xia, X. Zhou, S. Xu, H. Huang, L. Zhang, X. Zhou, C. Du, Y. Zhang, J. Song, S. Wang, Y. Chao, Z. Yang, J. Xu, X. Zhou, D. Chen, W. Xiong, L. Xu, F. Zhou, J. Jiang, C. Bai, J. Zheng and Y. Song, Risk Factors Associated With Acute Respiratory Distress Syndrome and Death in Patients With Coronavirus Disease 2019 Pneumonia in Wuhan, China, *JAMA Intern. Med.*, 2020, **180**(7), 934–943.
66. E. de Wit, N. van Doremalen, D. Falzarano and V. J. Munster, SARS and MERS: recent insights into emerging coronaviruses, *Nat. Rev. Microbiol.*, 2016, **14**(8), 523–534.
67. C. T. Esmon, The interactions between inflammation and coagulation, *Br. J. Haematol.*, 2005, **131**(4), 417–430.
68. R. Lordan, A. Tsoupras and I. Zabetakis, Platelet activation and prothrombotic mediators at the nexus of inflammation and atherosclerosis: Potential role of antiplatelet agents, *Blood Rev.*, 2020, 100694.
69. M. Levi and T. van der Poll, Inflammation and coagulation, *Crit. Care Med.*, 2010, **38**(2 Suppl), S26–S34.
70. J. C. Song, G. Wang, W. Zhang, Y. Zhang, W. Q. Li and Z. Zhou, Chinese expert consensus on diagnosis and treatment of coagulation dysfunction in COVID-19, *Mil. Med. Res.*, 2020, **7**(1), 19.

71. N. W. Palm and R. Medzhitov, Not so fast: adaptive suppression of innate immunity, *Nat. Med.*, 2007, **13**(10), 1142–1144.
72. Y. Gao, T. Li, M. Han, X. Li, D. Wu, Y. Xu, Y. Zhu, Y. Liu, X. Wang and L. Wang, Diagnostic utility of clinical laboratory data determinations for patients with the severe COVID-19, *J. Med. Virol.*, 2020, **92**(7), 791–796.
73. P. Mehta, D. F. McAuley, M. Brown, E. Sanchez, R. S. Tattersall and J. J. Manson, *et al.*, COVID-19: consider cytokine storm syndromes and immunosuppression, *Lancet*, 2020, **395**(10229), 1033.
74. S. Wan, Q. Yi, S. Fan, J. Lv, X. Zhang and L. Guo, *et al.*, Characteristics of lymphocyte subsets and cytokines in peripheral blood of 123 hospitalized patients with 2019 novel coronavirus pneumonia (NCP), MedRxiv, 2020.
75. A. S. Cowburn, D. Macias, C. Summers, E. R. Chilvers and R. S. Johnson, Cardiovascular adaptation to hypoxia and the role of peripheral resistance, *Elife*, 2017, **6**, e28755.
76. R. A. Balk, Systemic inflammatory response syndrome (SIRS) Where did it come from and is it still relevant today? *Virulence*, 2014, **5**(1), 20–26.
77. Y. X. Lim, Y. L. Ng, J. P. Tam and D. X. Liu, Human coronaviruses: a review of virus–host interactions, *Diseases*, 2016, **4**(3), 26.
78. M. Catanzaro, F. Fagiani, M. Racchi, E. Corsini, S. Govoni and C. Lanni, Immune response in COVID-19: addressing a pharmacological challenge by targeting pathways triggered by SARS-CoV-2, *Signal Transduction Targeted Ther.*, 2020, **5**(1), 1–10.
79. R. Channappanavar, A. R. Fehr, R. Vijay, M. Mack, J. Zhao and D. K. Meyerholz, *et al.*, Dysregulated type I interferon and inflammatory monocyte-macrophage responses cause lethal pneumonia in SARS-CoV-infected mice, *Cell Host Microbe*, 2016, **19**(2), 181–193.
80. R. Channappanavar, A. R. Fehr, J. Zheng, C. Wohlford-Lenane, J. E. Abrahante and M. Mack, *et al.*, IFN-I response timing relative to virus replication determines MERS coronavirus infection outcomes, *J. Clin. Invest.*, 2019, **129**(9), 3625–3639.
81. D. Blanco-Melo, B. E. Nilsson-Payant, W. C. Liu, S. Uhl, D. Hoagland, R. Møller, T. X. Jordan, K. Oishi, M. Panis, D. Sachs, T. T. Wang, R. E. Schwartz, J. K. Lim, R. A. Albrecht and B. R. tenOever, Imbalanced Host Response to SARS-CoV-2 Drives Development of COVID-19, *Cell*, 2020, **181**(5), 1036–1045.e9.
82. E. Mantlo, N. Bukreyeva, J. Maruyama, S. Paessler and C. Huang, Antiviral activities of type I interferons to SARS-CoV-2 infection, *Antiviral Res.*, 2020, 104811.
83. K. G. Lokugamage, A. Hage and M. de Vries, *et al.*, Type I Interferon Susceptibility Distinguishes SARS-CoV-2 from SARS-CoV, *J. Virol.*, 2020, **94**(23), e01410-20.
84. M. L. Stanifer, C. Kee, M. Cortese, C. M. Zumaran, S. Triana, M. Mukenhirn, H. G. Kraeusslich, T. Alexandrov, R. Bartenschlager and S. Boulant, Critical Role of Type III Interferon in Controlling SARS-CoV-2 Infection in Human Intestinal Epithelial Cells, *Cell Rep.*, 2020, **32**(1), 107863.

85. S. Pfaender, K. B. Mar and E. Michailidis, *et al.*, LY6E impairs coronavirus fusion and confers immune control of viral disease, *Nat. Microbiol.*, 2020, **5**, 1330–1339.
86. M. Bouvet, C. Debarnot, I. Imbert, B. Selisko, E. J. Snijder and B. Canard, *et al.*, In vitro reconstitution of SARS-coronavirus mRNA cap methylation, *PLoS Pathog.*, 2010, **6**(4), e1000863.
87. Y. Chen, H. Cai, Ja. Pan, N. Xiang, P. Tien and T. Ahola, *et al.*, Functional screen reveals SARS coronavirus nonstructural protein nsp14 as a novel cap N7 methyltransferase, *Proc. Natl. Acad. Sci. U. S. A.*, 2009, **106**(9), 3484–3489.
88. X. Deng, M. Hackbart, R. C. Mettelman, A. O'Brien, A. M. Mielech and G. Yi, *et al.*, Coronavirus nonstructural protein 15 mediates evasion of dsRNA sensors and limits apoptosis in macrophages, *Proc. Natl. Acad. Sci. U. S. A.*, 2017, **114**(21), E4251–E4260.
89. M. Hackbart, X. Deng and S. C. Baker, Coronavirus endoribonuclease targets viral polyuridine sequences to evade activating host sensors, *Proc. Natl. Acad. Sci. U. S. A.*, 2020, **117**(14), 8094–8103.
90. K. A. Ivanov, V. Thiel, J. C. Dobbe, Y. van der Meer, E. J. Snijder and J. Ziebuhr, Multiple Enzymatic Activities Associated with Severe Acute Respiratory Syndrome Coronavirus Helicase, *J. Virol.*, 2004, **78**(11), 5619–5632.
91. K. Knoops, M. Kikkert, S. H. E. Worm, J. C. Zevenhoven-Dobbe, Y. van der Meer and A. J. Koster, *et al.*, SARS-Coronavirus Replication Is Supported by a Reticulovesicular Network of Modified Endoplasmic Reticulum, *PLoS Biol.*, 2008, **6**(9), e226.
92. D. E. Gordon, G. M. Jang, M. Bouhaddou, J. Xu, K. Obernier and K. M. White, *et al.*, A SARS-CoV-2 protein interaction map reveals targets for drug repurposing, *Nature*, 2020, 1–13.
93. C.-S. Shi, H.-Y. Qi, C. Boularan, N.-N. Huang, M. Abu-Asab and J. H. Shelhamer, *et al.*, SARS-Coronavirus Open Reading Frame-9b Suppresses Innate Immunity by Targeting Mitochondria and the MAVS/TRAF3/TRAF6 Signalosome, *J. Immunol.*, 2014, **193**(6), 3080–3089.
94. K.-L. Siu, K.-H. Kok, M.-H. J. Ng, V. K. Poon, K.-Y. Yuen and B.-J. Zheng, *et al.*, Severe acute respiratory syndrome coronavirus M protein inhibits type I interferon production by impeding the formation of TRAF3· TANK· TBK1/IKKϵ complex, *J. Biol. Chem.*, 2009, **284**(24), 16202–16209.
95. Y. Yang, F. Ye, N. Zhu, W. Wang, Y. Deng and Z. Zhao, *et al.*, Middle East respiratory syndrome coronavirus ORF4b protein inhibits type I interferon production through both cytoplasmic and nuclear targets, *Sci. Rep.*, 2015, **5**, 17554.
96. S. G. Devaraj, N. Wang, Z. Chen, Z. Chen, M. Tseng and N. Barretto, *et al.*, Regulation of IRF-3-dependent innate immunity by the papain-like protease domain of the severe acute respiratory syndrome coronavirus, *J. Biol. Chem.*, 2007, **282**(44), 32208–32221.
97. S. A. Kopecky-Bromberg, L. Martínez-Sobrido, M. Frieman, R. A. Baric and P. Palese, Severe acute respiratory syndrome coronavirus open reading frame (ORF) 3b, ORF 6, and nucleocapsid proteins function as interferon antagonists, *J. Virol.*, 2007, **81**(2), 548–557.

98. E. Kindler, V. Thiel and F. Weber, Interaction of SARS and MERS coronaviruses with the antiviral interferon response, *Adv. Virus Res.*, 2016, **96**, 219–243.
99. R. Channappanavar and S. Perlman, Pathogenic human coronavirus infections: causes and consequences of cytokine storm and immunopathology, *Semin. Immunopathol.*, 2017, **39**(5), 529–539.
100. J.-Y. Li, C.-H. Liao, Q. Wang, Y.-J. Tan, R. Luo and Y. Qiu, *et al.*, The ORF6, ORF8 and nucleocapsid proteins of SARS-CoV-2 inhibit type I interferon signaling pathway, *Virus Res.*, 2020, 198074.
101. Y. Konno, *et al.*, SARS-CoV-2 ORF3b Is a Potent Interferon Antagonist Whose Activity Is Increased by a Naturally Occurring Elongation Variant, *Cell Rep.*, 2020, **32**(12), 108185.
102. R. Wölfel, V. M. Corman, W. Guggemos, M. Seilmaier, S. Zange and M. A. Müller, *et al.*, Virological assessment of hospitalized patients with COVID-2019, *Nature*, 2020, **581**(7809), 465–469.
103. J. Hadjadj, N. Yatim, L. Barnabei, A. Corneau, J. Boussier and N. Smith, *et al.*, Impaired type I interferon activity and inflammatory responses in severe COVID-19 patients, *Science*, 2020, **369**(6504), 718–724.
104. C. G. K. Ziegler, S. J. Allon, S. K. Nyquist, I. M. Mbano, V. N. Miao, C. N. Tzouanas, Y. Cao, A. S. Yousif, J. Bals, B. M. Hauser, J. Feldman, C. Muus, M. H. Wadsworth 2nd, S. W. Kazer, T. K. Hughes, B. Doran, G. J. Gatter, M. Vukovic, F. Taliaferro, B. E. Mead, Z. Guo, J. P. Wang, D. Gras, M. Plaisant, M. Ansari, I. Angelidis, H. Adler, J. M. S. Sucre, C. J. Taylor, B. Lin, A. Waghray, V. Mitsialis, D. F. Dwyer, K. M. Buchheit, J. A. Boyce, N. A. Barrett, T. M. Laidlaw, S. L. Carroll, L. Colonna, V. Tkachev, C. W. Peterson, A. Yu, H. B. Zheng, H. P. Gidcon, C. G. Winchell, P. L. Lin, C. D. Bingle, S. B. Snapper, J. A. Kropski, F. J. Theis, H. B. Schiller, L. E. Zaragosi, P. Barbry, A. Leslie, H. P. Kiem, J. L. Flynn, S. M. Fortune, B. Berger, R. W. Finberg, L. S. Kean, M. Garber, A. G. Schmidt, D. Lingwood, A. K. Shalek, J. Ordovas-Montanes, HCA Lung Biological Network. Electronic address: lung-network@humancellatlas.org and HCA Lung Biological Network, SARS-CoV-2 Receptor ACE2 Is an Interferon-Stimulated Gene in Human Airway Epithelial Cells and Is Detected in Specific Cell Subsets across Tissues, *Cell*, 2020, **181**(5), 1016–1035.e19.
105. J. Li, M. Guo, X. Tian, X. Wang, X. Yang, P. Wu, C. Liu, Z. Xiao, Y. Qu, Y. Yin, C. Wang, Y. Zhang, Z. Zhu, Z. Liu, C. Peng, T. Zhu and Q. Liang, Virus-Host Interactome and Proteomic Survey Reveal Potential Virulence Factors Influencing SARS-CoV-2 Pathogenesis, *Med (N Y)*, 2021, **2**(1), 99–112.e7.
106. D. M. Mosser and J. P. Edwards, Exploring the full spectrum of macrophage activation, *Nat. Rev. Immunol.*, 2008, **8**(12), 958–969.
107. G. Germano, R. Frapolli, C. Belgiovine, A. Anselmo, S. Pesce and M. Liguori, *et al.*, Role of macrophage targeting in the antitumor activity of trabectedin, *Cancer Cell*, 2013, **23**(2), 249–262.
108. C. Wang, J. Xie, L. Zhao, X. Fei, H. Zhang and Y. Tan, *et al.*, Alveolar macrophage dysfunction and cytokine storm in the pathogenesis of two severe COVID-19 patients, *EBioMedicine*, 2020, **57**, 102833.

109. E. J. Giamarellos-Bourboulis, M. G. Netea, N. Rovina, K. Akinosoglou, A. Antoniadou, N. Antonakos, G. Damoraki, T. Gkavogianni, M. E. Adami, P. Katsaounou, M. Ntaganou, M. Kyriakopoulou, G. Dimopoulos, I. Koutsodimitropoulos, D. Velissaris, P. Koufargyris, A. Karageorgos, K. Katrini, V. Lekakis, M. Lupse, A. Kotsaki, G. Renieris, D. Theodoulou, V. Panou, E. Koukaki, N. Koulouris, C. Gogos and A. Koutsoukou, Complex Immune Dysregulation in COVID-19 Patients with Severe Respiratory Failure, *Cell Host Microbe*, 2020, **27**(6), 992–1000.e3.
110. D. Zhang, R. Guo, L. Lei, H. Liu, Y. Wang, Y. Wang, H. Qian, T. Dai, T. Zhang, Y. Lai, J. Wang, Z. Liu, T. Chen, A. He, M. O'Dwyer and J. Hu, Frontline Science: COVID-19 infection induces readily detectable morphologic and inflammation-related phenotypic changes in peripheral blood monocytes, *J. Leukoc. Biol.*, 2021 Jan, **109**(1), 13–22.
111. Y. Zhou and Yonggang, *et al.*, Pathogenic T-cells and inflammatory monocytes incite inflammatory storms in severe COVID-19 patients, *Natl. Sci. Rev.*, 2020, 7(6), 998–1002.
112. C. Guo, *et al.*, Single-cell analysis of two severe COVID-19 patients reveals a monocyte-associated and tocilizumab-responding cytokine storm, *Nat. Commun.*, 2020, **11**(1), 3924.
113. W. Wen, W. Su, H. Tang, W. Le, X. Zhang and Y. Zheng, *et al.*, Immune cell profiling of COVID-19 patients in the recovery stage by single-cell sequencing, *Cell Discovery*, 2020, **6**(1), 1–18.
114. E. Z. Ong, Y. F. Z. Chan, W. Y. Leong, N. M. Y. Lee, S. Kalimuddin, S. M. Haja Mohideen, K. S. Chan, A. T. Tan, A. Bertoletti, E. E. Ooi and J. G. H. Low, A Dynamic Immune Response Shapes COVID-19 Progression, *Cell Host Microbe*, 2020, **27**(6), 879–882.e2.
115. L. Huang, Y. Shi and B. Gong, *et al.*, Blood single cell immune profiling reveals the interferon-MAPK pathway mediated adaptive immune response for COVID-19, *medRxiv*, 2020, DOI: 10.1101/2020.03.15.20033472.
116. M. Liao, Y. Liu, J. Yuan, Y. Wen, G. Xu and J. Zhao, *et al.*, Single-cell landscape of bronchoalveolar immune cells in patients with COVID-19, *Nat. Med.*, 2020, 1–3.
117. P. A. Reyfman, J. M. Walter, N. Joshi, K. R. Anekalla, A. C. McQuattie-Pimentel and S. Chiu, *et al.*, Single-cell transcriptomic analysis of human lung provides insights into the pathobiology of pulmonary fibrosis, *Am. J. Respir. Crit. Care Med.*, 2019, **199**(12), 1517–1536.
118. M. Liao, Y. Liu and J. Yuan, *et al.*, Single-cell landscape of bronchoalveolar immune cells in patients with COVID-19, *Nat. Med.*, 2020, **26**, 842–844.
119. E. Evren, E. Ringqvist and T. Willinger, Origin and ontogeny of lung macrophages: from mice to humans, *Immunology*, 2020, **160**(2), 126–138.
120. C. Schneider, S. P. Nobs, M. Kurrer, H. Rehrauer, C. Thiele and M. Kopf, Induction of the nuclear receptor PPAR-γ by the cytokine GM-CSF is critical for the differentiation of fetal monocytes into alveolar macrophages, *Nat. Immunol.*, 2014, **15**(11), 1026–1037.

121. T. Hussell and T. J. Bell, Alveolar macrophages: plasticity in a tissue-specific context, *Nat. Rev. Immunol.*, 2014, **14**(2), 81–93.
122. P. Ramachandran, R. Dobie, J. R. Wilson-Kanamori, E. F. Dora, B. E. P. Henderson and N. T. Luu, *et al.*, Resolving the fibrotic niche of human liver cirrhosis at single-cell level, *Nature*, 2019, **575**(7783), 512–518.
123. N. E. Cabrera-Benitez, J. G. Laffey, M. Parotto, P. M. Spieth, J. Villar and H. Zhang, *et al.*, Mechanical Ventilation–associated Lung Fibrosis in Acute Respiratory Distress SyndromeA Significant Contributor to Poor Outcome, *Anesthesiology*, 2014, **121**(1), 189–198.
124. D. V. Pechkovsky, T. Goldmann, C. Ludwig, A. Prasse, E. Vollmer and J. Müller-Quernheim, *et al.*, CCR2 and CXCR3 agonistic chemokines are differently expressed and regulated in human alveolar epithelial cells type II, *Respir. Res.*, 2005, **6**(1), 75.
125. A. J. Thorley, D. Grandolfo, E. Lim, P. Goldstraw, A. Young and T. D. Tetley, Innate immune responses to bacterial ligands in the peripheral human lung–role of alveolar epithelial TLR expression and signalling, *PLoS One*, 2011, **6**(7), e21827.
126. O. D. Chuquimia, D. H. Petursdottir, N. Periolo and C. Fernández, Alveolar epithelial cells are critical in protection of the respiratory tract by secretion of factors able to modulate the activity of pulmonary macrophages and directly control bacterial growth, *Infect. Immun.*, 2013, **81**(1), 381–389.
127. P. Silveyra, Z. C. Chroneos, S. L. DiAngelo, N. J. Thomas, G. T. Noutsios and N. Tsotakos, *et al.*, Knockdown of Drosha in human alveolar type II cells alters expression of SP-A in culture: A pilot study, *Exp. Lung Res.*, 2014, **40**(7), 354–366.
128. Z. Qian, E. A. Travanty, L. Oko, K. Edeen, A. Berglund and J. Wang, *et al.*, Innate immune response of human alveolar type ii cells infected with severe acute respiratory syndrome–coronavirus, *Am. J. Respir. Cell Mol. Biol.*, 2013, **48**(6), 742–748.
129. V. Brinkmann, U. Reichard, C. Goosmann, B. Fauler, Y. Uhlemann and D. S. Weiss, *et al.*, Neutrophil extracellular traps kill bacteria, *Science*, 2004, **303**(5663), 1532–1535.
130. J. Liu, Y. Liu, P. Xiang, L. Pu, H. Xiong, C. Li, M. Zhang, J. Tan, Y. Xu, R. Song, M. Song, L. Wang, W. Zhang, B. Han, L. Yang, X. Wang, G. Zhou, T. Zhang, B. Li, Y. Wang, Z. Chen and X. Wang, Neutrophil-to-lymphocyte ratio predicts critical illness patients with 2019 coronavirus disease in the early stage,*J. Transl. Med.*, 2020, **18**(1), 206.
131. M. Hajivalili, M. Hosseini and M. Haji-Fatahaliha, Gaining insights on immune responses to the novel coronavirus, COVID-19 and therapeutic challenges, *Life Sci.*, 2020, 118058.
132. J. Liu, S. Li, J. Liu, B. Liang, X. Wang and H. Wang, *et al.*, Longitudinal characteristics of lymphocyte responses and cytokine profiles in the peripheral blood of SARS-CoV-2 infected patients, *EBioMedicine*, 2020, 102763.

133. B. J. Barnes, J. M. Adrover, A. Baxter-Stoltzfus, A. Borczuk, J. Cools-Lartigue, J. M. Crawford, J. Daßler-Plenker, P. Guerci, C. Huynh, J. S. Knight, M. Loda, M. R. Looney, F. McAllister, R. Rayes, S. Renaud, S. Rousseau, S. Salvatore, R. E. Schwartz, J. D. Spicer, C. C. Yost, A. Weber, Y. Zuo and M. Egeblad, Targeting potential drivers of COVID-19: Neutrophil extracellular traps, *J. Exp. Med.*, 2020, **217**(6), e20200652.
134. Y. Zuo, S. Yalavarthi, H. Shi, K. Gockman, M. Zuo, J. A. Madison, C. Blair, A. Weber, B. J. Barnes, M. Egeblad, R. J. Woods, Y. Kanthi and J. S. Knight, Neutrophil extracellular traps in COVID-19, *JCI Insight*, 2020, **5**(11), e138999.
135. T. A. Patente, *et al.*, Human Dendritic Cells: Their Heterogeneity and Clinical Application Potential in Cancer Immunotherapy, *Front. Immunol.*, 2019, **9**, 3176.
136. P. Fitzgerald-Bocarsly, J. Dai and S. Singh, Plasmacytoid dendritic cells and type I IFN: 50 years of convergent history, *Cytokine Growth Factor Rev.*, 2008, **19**(1), 3–19.
137. J. Y. Zhang, X. M. Wang and X. Xing, *et al.*, Single-cell landscape of immunological responses in patients with COVID-19, *Nat. Immunol.*, 2020, **21**, 1107–1118.
138. R. Zhou, K. K. To and Y. C. Wong, *et al.*, Acute SARS-CoV-2 Infection Impairs Dendritic Cell and T Cell Responses, *Immunity*, 2020, **53**(4), 864–877.e5.
139. A. Cerwenka and L. L. Lanier, Natural killer cells, viruses and cancer, *Nat. Rev. Immunol.*, 2001, **1**(1), 41–49.
140. M. Draghi, A. Pashine, B. Sanjanwala, K. Gendzekhadze, C. Cantoni and D. Cosman, *et al.*, NKp46 and NKG2D Recognition of Infected Dendritic Cells Is Necessary for NK Cell Activation in the Human Response to Influenza Infection, *J. Immunol.*, 2007, **178**(5), 2688–2698.
141. A. Duev-Cohen, Y. Bar-On and A. Glasner, *et al.*, The human 2B4 and NTB-A receptors bind the influenza viral hemagglutinin and co-stimulate NK cell cytotoxicity, *Oncotarget*, 2016, **7**(11), 13093–13105.
142. A. Glasner, A. Zurunic, T. Meningher, T. L. Rovis, P. Tsukerman and Y. Bar-On, *et al.*, Elucidating the mechanisms of influenza virus recognition by Ncr1, *PLoS One*, 2012, **7**(5), e36837.
143. C. Song, J. Xu, J. He and Y. Lu, COVID-19 early warning score: a multi-parameter screening tool to identify highly suspected patients, *medRxiv*, 2020, DOI: 10.1101/2020.03.05.20031906.
144. W. Wang, X. Liu and S. Wu, *et al.*, Definition and Risks of Cytokine Release Syndrome in 11 Critically Ill COVID-19 Patients With Pneumonia: Analysis of Disease Characteristics, *J. Infect. Dis.*, 2020, **222**(9), 1444–1451.
145. L. Yu, Y. Tong, G. Shen, A. Fu, Y. Lai and X. Zhou, *et al.*, Immunodepletion with Hypoxemia: A Potential High Risk Subtype of Coronavirus Disease 2019, medRxiv, 2020, DOI: 10.1101/2020.03.03.20030650.

172. T. Yoshikawa, T. Hill, K. Li, C. J. Peters and C.-T. K. Tseng, Severe acute respiratory syndrome (SARS) coronavirus-induced lung epithelial cytokines exacerbate SARS pathogenesis by modulating intrinsic functions of monocyte-derived macrophages and dendritic cells, *J. Virol.*, 2009, **83**(7), 3039–3048.
173. C. A. Donnelly, M. C. Fisher, C. Fraser, A. C. Ghani, S. Riley and N. M. Ferguson, *et al.*, Epidemiological and genetic analysis of severe acute respiratory syndrome, *Lancet Infect. Dis.*, 2004, **4**(11), 672–683.
174. L. Liu, Q. Wei, Q. Lin, J. Fang, H. Wang, H. Kwok, H. Tang, K. Nishiura, J. Peng, Z. Tan, T. Wu, K. W. Cheung, K. H. Chan, X. Alvarez, C. Qin, A. Lackner, S. Perlman, K. Y. Yuen and Z. Chen, Anti-spike IgG causes severe acute lung injury by skewing macrophage responses during acute SARS-CoV infection, *JCI Insight*, 2019, **4**(4), e123158.
175. L. Du, Y. He, Y. Zhou, S. Liu, B.-J. Zheng and S. Jiang, The spike protein of SARS-CoV — a target for vaccine and therapeutic development, *Nat. Rev. Microbiol.*, 2009, **7**(3), 226–236.
176. G. Chen, D. Wu, W. Guo, Y. Cao, D. Huang, H. Wang, T. Wang, X. Zhang, H. Chen, H. Yu, X. Zhang, M. Zhang, S. Wu, J. Song, T. Chen, M. Han, S. Li, X. Luo, J. Zhao and Q. Ning, Clinical and immunological features of severe and moderate coronavirus disease 2019, *J. Clin. Invest.*, 2020, **130**(5), 2620–2629.
177. S. Nie, X. Zhao, K. Zhao, Z. Zhang, Z. Zhang and Z. Zhang, Metabolic disturbances and inflammatory dysfunction predict severity of coronavirus disease 2019 (COVID-19): a retrospective study, medRxiv, 2020, DOI: 10.1101/2020.03.24.20042283.
178. B. Xu, C. Y. Fan, A. L. Wang, Y. L. Zou, Y. H. Yu, C. He, W. G. Xia, J. X. Zhang and Q. Miao, Suppressed T cell-mediated immunity in patients with COVID-19: A clinical retrospective study in Wuhan, China, *J. Infect.*, 2020, **81**(1), e51–e60.
179. R. H. Du, L. R. Liang, C. Q. Yang, W. Wang, T. Z. Cao, M. Li, G. Y. Guo, J. Du, C. L. Zheng, Q. Zhu, M. Hu, X. Y. Li, P. Peng and H. Z. Shi, Predictors of mortality for patients with COVID-19 pneumonia caused by SARS-CoV-2: a prospective cohort study, *Eur. Respir. J.*, 2020, **55**(5), 2000524.
180. J. Chen, T. Qi and L. Liu, *et al.*, Clinical progression of patients with COVID-19 in Shanghai, China, *J. Infect.*, 2020, **80**(5), e1–e6.
181. B. Liu, J. Han, X. Cheng, L. Yu, L. Zhang and W. Wang, *et al.*, Persistent SARS-CoV-2 presence is companied with defects in adaptive immune system in non-severe COVID-19 patients, medRxiv, 2020, DOI: 10.1101/2020.03.26.20044768.
182. I. Thevarajan, T. H. O. Nguyen, M. Koutsakos, J. Druce, L. Caly and C. E. van de Sandt, *et al.*, Breadth of concomitant immune responses prior to patient recovery: a case report of non-severe COVID-19, *Nat. Med.*, 2020, **26**(4), 453–455.
183. X. Chen, J. Ling, P. Mo, Y. Zhang, Q. Jiang and Z. Ma, *et al.*, Restoration of leukomonocyte counts is associated with viral clearance in COVID-19 hospitalized patients, MedRxiv, 2020, DOI: 10.1101/2020.03.03.20030437.

184. B. Diao, C. Wang, Y. Tan, X. Chen, Y. Liu and L. Ning, *et al.*, Reduction and functional exhaustion of T cells in patients with coronavirus disease 2019 (COVID-19), *Front. Immunol.*, 2020, **11**, 827.
185. E. Kamphuis, T. Junt, Z. Waibler, R. Forster and U. Kalinke, Type I interferons directly regulate lymphocyte recirculation and cause transient blood lymphopenia, *Blood*, 2006, **108**(10), 3253–3261.
186. L. R. Shiow, D. B. Rosen, N. Brdičková, Y. Xu, J. An and L. L. Lanier, *et al.*, CD69 acts downstream of interferon-α/β to inhibit S1P1 and lymphocyte egress from lymphoid organs, *Nature*, 2006, **440**(7083), 540–544.
187. Z. Xu, L. Shi, Y. Wang, J. Zhang, L. Huang and C. Zhang, *et al.*, Pathological findings of COVID-19 associated with acute respiratory distress syndrome, *Lancet Respir. Med.*, 2020, **8**(4), 420–422.
188. S. Tian, Y. Xiong, H. Liu, L. Niu, J. Guo and M. Liao, *et al.*, Pathological study of the 2019 novel coronavirus disease (COVID-19) through postmortem core biopsies, *Mod. Pathol.*, 2020, 1–8.
189. L. Ni, F. Ye, M. L. Cheng, Y. Feng, Y. Q. Deng, H. Zhao, P. Wei, J. Ge, M. Gou, X. Li, L. Sun, T. Cao, P. Wang, C. Zhou, R. Zhang, P. Liang, H. Guo, X. Wang, C. F. Qin, F. Chen and C. Dong, Detection of SARS-CoV-2-Specific Humoral and Cellular Immunity in COVID-19 Convalescent Individuals, *Immunity*, 2020, **52**(6), 971–977.e3.
190. D. Weiskopf, K. S. Schmitz, M. P. Raadsen, A. Grifoni, N. M. Okba and H. Endeman, *et al.*, Phenotype of SARS-CoV-2-specific T-cells in COVID-19 patients with acute respiratory distress syndrome, medRxiv, 2020.
191. J. Braun, L. Loyal and M. Frentsch, *et al.*, SARS-CoV-2-reactive T cells in healthy donors and patients with COVID-19, *Nature*, 2020, **587**, 270–274.
192. Y. Zhou, B. Fu, X. Zheng, D. Wang, C. Zhao and Y. Qi, *et al.*, Pathogenic T-cells and inflammatory monocytes incite inflammatory storms in severe COVID-19 patients, *Natl. Sci. Rev.*, 2020, **7**(6), 998–1002.
193. C. Qin, L. Zhou, Z. Hu, S. Zhang, S. Yang, Y. Tao, C. Xie, K. Ma, K. Shang, W. Wang and D. S. Tian, Dysregulation of Immune Response in Patients With Coronavirus 2019 (COVID-19) in Wuhan, China, *Clin. Infect. Dis.*, 2020, **71**(15), 762–768.
194. J. M. Walter, K. A. Helmin, H. Abdala-Valencia, R. G. Wunderink and B. D. Singer, Multidimensional assessment of alveolar T cells in critically ill patients, *JCI Insight*, 2018, **3**(17), e123287.
195. P. Dong, X. Ju, Y. Yan, S. Zhang, M. Cai and H. Wang, *et al.*, γδ T cells provide protective function in highly pathogenic avian H5N1 influenza A virus infection, *Front. Immunol.*, 2018, **9**, 2812.
196. J. Zheng, Y. Liu, Y.-L. Lau and W. Tu, γδ-T cells: an unpolished sword in human anti-infection immunity, *Cell. Mol. Immunol.*, 2013, **10**(1), 50–57.
197. L. Lei, H. Qian, X. Yang, X. Zhou, X. Zhang and D. Zhang, *et al.*, The phenotypic changes of γδ T cells in COVID-19 patients, *J. Cell. Mol. Med.*, 2020, **24**, 11603–11606.

198. Z. Zhao, J. Xie, M. Yin, Y. Yang, H. He and T. Jin, *et al.*, Clinical and laboratory profiles of 75 hospitalized patients with novel coronavirus disease 2019 in Hefei, China, MedRxiv, 2020, DOI: 10.1101/2020.03.01.20029785.
199. X. Li, M. Geng, Y. Peng, L. Meng and S. Lu, Molecular immune pathogenesis and diagnosis of COVID-19, *J. Pharm. Anal.*, 2020, **10**(2), 102–108.
200. Y. Yang, *et al.*, Plasma IP-10 and MCP-3 levels are highly associated with disease severity and predict the progression of COVID-19, *J. Allergy Clin. Immunol.*, 2020, **146**(1), 119–127.
201. Y. Xiong, Y. Liu, L. Cao, D. Wang, M. Guo and A. Jiang, *et al.*, Transcriptomic characteristics of bronchoalveolar lavage fluid and peripheral blood mononuclear cells in COVID-19 patients, *Emerging Microbes Infect.*, 2020, **9**(1), 761–770.
202. X. Chen, B. Zhao and Y. Qu, *et al.*, Detectable Serum Severe Acute Respiratory Syndrome Coronavirus 2 Viral Load (RNAemia) Is Closely Correlated With Drastically Elevated Interleukin 6 Level in Critically Ill Patients With Coronavirus Disease 2019, *Clin. Infect. Dis.*, 2020, **71**(8), 1937–1942.
203. M. Levi and T. van der Poll, Two-way interactions between inflammation and coagulation, *Trends Cardiovasc. Med.*, 2005, **15**(7), 254–259.
204. T. Van der Poll, M. Levi, C. E. Hack, H. Ten Cate, S. Van Deventer and A. Eerenberg, *et al.*, Elimination of interleukin 6 attenuates coagulation activation in experimental endotoxemia in chimpanzees, *J. Exp. Med.*, 1994, **179**(4), 1253–1259.
205. M. Levi, T. Van der Poll, H. Ten Cate and S. Van Deventer, The cytokine-mediated imbalance between coagulant and anticoagulant mechanisms in sepsis and endotoxaemia, *Eur. J. Clin. Invest.*, 1997, **27**(1), 3–9.
206. F. Zhou, T. Yu, R. Du, G. Fan, Y. Liu, Z. Liu, J. Xiang, Y. Wang, B. Song, X. Gu, L. Guan, Y. Wei, H. Li, X. Wu, J. Xu, S. Tu, Y. Zhang, H. Chen and B. Cao, Clinical course and risk factors for mortality of adult inpatients with COVID-19 in Wuhan, China: a retrospective cohort study, *Lancet*, 2020, **395**(10229), 1054–1062.
207. N. Tang, D. Li, X. Wang and Z. Sun, Abnormal coagulation parameters are associated with poor prognosis in patients with novel coronavirus pneumonia, *J. Thromb. Haemostasis*, 2020, **18**(4), 844–847.
208. A. L. Van Dyke, M. L. Cote, A. S. Wenzlaff, S. Land and A. G. Schwartz, Cytokine SNPs: comparison of allele frequencies by race and implications for future studies, *Cytokine*, 2009, **46**(2), 236–244.
209. M. Maiers, L. Gragert and W. Klitz, High-resolution HLA alleles and haplotypes in the United States population, *Hum. Immunol.*, 2007, **68**(9), 779–788.
210. S. Buhler and A. Sanchez-Mazas, HLA DNA sequence variation among human populations: molecular signatures of demographic and selective events, *PLoS One*, 2011, **6**(2), e14643.
211. E. Benetti, R. Tita, O. Spiga, A. Ciolfi, G. Birolo, A. Bruselles, G. Doddato, A. Giliberti, C. Marconi, F. Musacchia, T. Pippucci, A. Torella, A. Trezza, F. Valentino, M. Baldassarri, A. Brusco, R. Asselta, M. Bruttini,

S. Furini, M. Seri, V. Nigro, G. Matullo, M. Tartaglia, F. Mari, GEN-COVID Multicenter Study, A. Renieri and A. M. Pinto, ACE2 gene variants may underlie interindividual variability and susceptibility to COVID-19 in the Italian population, *Eur. J. Hum. Genet.*, 2020, **28**(11), 1602–1614.

212. K. Suryamohan, D. Diwanji and E. W. Stawiski, *et al.*, Human ACE2 receptor polymorphisms and altered susceptibility to SARS-CoV-2, *Commun. Biol.*, 2021, **4**, 475.
213. Q. Li, Z. Cao and P. Rahman, Genetic Variability of Human Angiotensin-Converting Enzyme 2 (hACE2) Among Various Ethnic Populations, *Mol. Genet. Genomic. Med.*, 2020, **8**(8), e1344.
214. T. Tukiainen, A.-C. Villani, A. Yen, M. A. Rivas, J. L. Marshall and R. Satija, *et al.*, Landscape of X chromosome inactivation across human tissues, *Nature*, 2017, **550**(7675), 244–248.
215. Y. Zhao, Z. Zhao, Y. Wang, Y. Zhou, Y. Ma and W. Zuo, Single-Cell RNA Expression Profiling of ACE2, the Receptor of SARS-CoV-2 [published correction appears in *Am. J. Respir. Crit. Care Med.*, 2021 Mar 15, **203**(6), 782], *Am. J. Respir. Crit. Care Med.*, 2020, **202**(5), 756–759.
216. R. Asselta, E. M. Paraboschi, A. Mantovani and S. Duga, ACE2 and TMPRSS2 variants and expression as candidates to sex and country differences in COVID-19 severity in Italy, *Aging*, 2020, **12**(11), 10087–10098.
217. G. Cai, Bulk and single-cell transcriptomics identify tobacco-use disparity in lung gene expression of ACE2, the receptor of 2019-nCov, MedRxiv, 2020, DOI: 10.20944/preprints202002.0051.v2.
218. B. Bakhshandeh, S. G. Sorboni and A. R. Javanmard, *et al.*, Variants in ACE2; potential influences on virus infection and COVID-19 severity, *Infect., Genet. Evol.*, 2021, **90**, 104773.
219. S. P. Iyer, J. Ensor, K. Anand, P. Hwu, V. Subbiah and C. Flowers, *et al.*, Higher mortality in men from COVID19 infection-understanding the factors that drive the differences between the biological sexes, medRxiv, 2020, DOI: 10.1101/2020.04.19.20062174.
220. C. Gebhard, V. Regitz-Zagrosek, H. K. Neuhauser, R. Morgan and S. L. Klein, Impact of sex and gender on COVID-19 outcomes in Europe, *Biol. Sex Differ.*, 2020, **11**(1), 1–13.
221. B. G. G. Pinto, A. E. R. Oliveira, Y. Singh, L. Jimenez, A. N. A. Gonçalves, R. L. T. Ogava, R. Creighton, J. P. Schatzmann Peron and H. I. Nakaya, ACE2 Expression Is Increased in the Lungs of Patients With Comorbidities Associated With Severe COVID-19, *J. Infect. Dis.*, 2020, **222**(4), 556–563.
222. L. Cristiani, E. Mancino and L. Matera, *et al.*, Will children reveal their secret? The coronavirus dilemma, *Eur. Respir. J.*, 2020, **55**(4), 2000749.
223. X. Xudong, C. Junzhu, W. Xingxiang, Z. Furong and L. Yanrong, Age-and gender-related difference of ACE2 expression in rat lung, *Life Sci.*, 2006, **78**(19), 2166–2171.
224. K. Kuba, Y. Imai, S. Rao, C. Jiang and J. M. Penninger, Lessons from SARS: control of acute lung failure by the SARS receptor ACE2, *J. Mol. Med.*, 2006, **84**(10), 814–820.

225. R. Russo, I. Andolfo, V. A. Lasorsa, A. Iolascon and M. Capasso, Genetic Analysis of the Coronavirus SARS-CoV-2 Host Protease *TMPRSS2* in Different Populations, *Front. Genet.*, 2020, **11**, 872.
226. J. Zhao, Y. Yang, H. Huang, D. Li, D. Gu, X. Lu, Z. Zhang, L. Liu, T. Liu, Y. Liu, Y. He, B. Sun, M. Wei, G. Yang, X. Wang, L. Zhang, X. Zhou, M. Xing and P. G. Wang, Relationship Between the ABO Blood Group and the Coronavirus Disease 2019 (COVID-19) Susceptibility, *Clin. Infect. Dis.*, 2021, **73**(2), 328–331.
227. M. Zietz, J. Zucker and N. P. Tatonetti, Testing the association between blood type and COVID-19 infection, intubation, and death, *medRxiv*, 2020, DOI: 2020.04.08.20058073.
228. X. Dai, ABO blood group predisposes to COVID-19 severity and cardiovascular diseases, *Eur. J. Prev. Cardiol.*, 2020, **27**(13), 1436–1437.
229. R. Russo, I. Andolfo, V. A. Lasorsa, A. Iolascon and M. Capasso, Genetic Analysis of the Coronavirus SARS-CoV-2 Host Protease *TMPRSS2* in Different Populations, *Front. Genet.*, 2020, **11**, 872.
230. G. M. Liumbruno and M. Franchini, Beyond immunohaematology: the role of the ABO blood group in human diseases, *Blood Transfus.*, 2013, **11**(4), 491–499.
231. G. Paré, D. I. Chasman, M. Kellogg, R. Y. Zee, N. Rifai and S. Badola, *et al.*, Novel association of ABO histo-blood group antigen with soluble ICAM-1: results of a genome-wide association study of 6,578 women, *PLoS Genet.*, 2008, **4**(7), e1000118.
232. M. Barbalic, J. Dupuis, A. Dehghan, J. C. Bis, R. C. Hoogeveen and R. B. Schnabel, *et al.*, Large-scale genomic studies reveal central role of ABO in sP-selectin and sICAM-1 levels, *Hum. Mol. Genet.*, 2010, **19**(9), 1863–1872.
233. L. Qi, M. C. Cornelis, P. Kraft, M. Jensen, R. M. van Dam and Q. Sun, *et al.*, Genetic variants in ABO blood group region, plasma soluble E-selectin levels and risk of type 2 diabetes, *Hum. Mol. Genet.*, 2010, **19**(9), 1856–1862.
234. M. He, B. Wolpin, K. Rexrode, J. E. Manson, E. Rimm and F. B. Hu, *et al.*, ABO blood group and risk of coronary heart disease in two prospective cohort studies, *Arterioscler., Thromb., Vasc. Biol.*, 2012, **32**(9), 2314–2320.
235. O. Wu, N. Bayoumi, M. A. Vickers and P. Clark, ABO(H) blood groups and vascular disease: a systematic review and meta-analysis, *J. Thromb. Haemostasis*, 2008, **6**(1), 62–69.
236. F. Zhang and C. Hughes, A Caution in Association of ABO Blood Group with COVID-19, *Global Clin. Transl. Res.*, 2020, **2**, 51–53.
237. G. Sharma, A. S. Volgman and E. D. Michos, Sex differences in mortality from COVID-19 pandemic: are men vulnerable and women protected? *JACC: Case Rep.*, 2020, **2**(9), 1407–1410.
238. S. Kadel and S. Kovats, Sex hormones regulate innate immune cells and promote sex differences in respiratory virus infection, *Front. Immunol.*, 2018, **9**, 1653.

239. S. L. Klein and K. L. Flanagan, Sex differences in immune responses, *Nat. Rev. Immunol.*, 2016, **16**(10), 626–638.
240. D. P. Robinson, S. A. Huber, M. Moussawi, B. Roberts, C. Teuscher and R. Watkins, *et al.*, Sex chromosome complement contributes to sex differences in coxsackievirus B3 but not influenza A virus pathogenesis, *Biol. Sex Differ.*, 2011, **2**(1), 8.
241. H. Schurz, M. Salie, G. Tromp, E. G. Hoal, C. J. Kinnear and M. Möller, The X chromosome and sex-specific effects in infectious disease susceptibility, *Hum. Genomics*, 2019, **13**(1), 2.
242. F. Zeng, C. Dai, P. Cai, J. Wang, L. Xu, J. Li, G. Hu, Z. Wang, F. Zheng and L. Wang, A comparison study of SARS-CoV-2 IgG antibody between male and female COVID-19 patients: A possible reason underlying different outcome between sex, *J. Med. Virol.*, 2020, **92**(10), 2050–2054.
243. L. Chen, X. Li, M. Chen, Y. Feng and C. Xiong, The ACE2 expression in human heart indicates new potential mechanism of heart injury among patients infected with SARS-CoV-2, *Cardiovasc. Res.*, 2020, **116**(6), 1097–1100.
244. D. Batlle, J. Wysocki and K. Satchell, Soluble angiotensin-converting enzyme 2: a potential approach for coronavirus infection therapy? *Clin. Sci.*, 2020, **134**(5), 543–545.
245. K. Bhatia, M. A. Zimmerman and J. C. Sullivan, Sex Differences in Angiotensin-Converting Enzyme Modulation of Ang (1–7) Levels in Normotensive WKY Rats, *Am. J. Hypertens.*, 2013, **26**(5), 591–598.
246. Y. Dong, X. Mo, Y. Hu, X. Qi, F. Jiang, Z. Jiang and S. Tong, Epidemiology of COVID-19 Among Children in China, *Pediatrics*, 2020, **145**(6), e20200702.
247. W. Liu, Q. Zhang, J. Chen, R. Xiang, H. Song and S. Shu, *et al.*, Detection of Covid-19 in Children in Early January 2020 in Wuhan, China, *N. Engl. J. Med.*, 2020, **382**(14), 1370–1371.
248. Y. Dong, X. Mo, Y. Hu, X. Qi, F. Jiang and Z. Jiang, *et al.*, Epidemiology of COVID-19 Among Children in China, *Pediatrics*, 2020, **145**(6), e20200702.
249. Q. Bi, Y. Wu, S. Mei, C. Ye, X. Zou, Z. Zhang, X. Liu, L. Wei, S. A. Truelove, T. Zhang, W. Gao, C. Cheng, X. Tang, X. Wu, Y. Wu, B. Sun, S. Huang, Y. Sun, J. Zhang, T. Ma, J. Lessler and T. Feng, Epidemiology and transmission of COVID-19 in 391 cases and 1286 of their close contacts in Shenzhen, China: a retrospective cohort study, *Lancet Infect. Dis.*, 2020, **20**(8), 911–919.
250. N. G. Davies, P. Klepac, Y. Liu, K. Prem, M. Jit and R. M. Eggo, *et al.*, Age-dependent effects in the transmission and control of COVID-19 epidemics, *Nat. Med.*, 2020, **26**, 1205–1211.
251. K. Liu, Y. Chen, R. Lin and K. Han, Clinical features of COVID-19 in elderly patients: A comparison with young and middle-aged patients, *J. Infect.*, 2020, **80**(6), e14–e18.

252. A. Borghesi, A. Zigliani, R. Masciullo, S. Golemi, P. Maculotti and D. Farina, *et al.*, Radiographic severity index in COVID-19 pneumonia: relationship to age and sex in 783 Italian patients, *Radiol. Med.*, 2020, 1.
253. W. J. Guan, W. H. Liang and Y. Zhao, *et al.*, Comorbidity and its impact on 1590 patients with COVID-19 in China: a nationwide analysis, *Eur. Respir. J.*, 2020, **55**(5), 2000547.
254. G. Lippi, J. Wong and B. M. Henry, Hypertension and its severity or mortality in Coronavirus Disease 2019 (COVID-19): a pooled analysis, *Pol. Arch. Intern. Med.*, 2020, **130**(4), 304–309.
255. S. R. Mirhafez, M. Mohebati, M. Feiz Disfani, M. Saberi Karimian, M. Ebrahimi and A. Avan, *et al.*, An imbalance in serum concentrations of inflammatory and anti-inflammatory cytokines in hypertension, *J. Am. Soc. Hypertens.*, 2014, **8**(9), 614–623.
256. Q. Ji, G. Cheng and N. Ma, *et al.*, Circulating Th1, Th2, and Th17 Levels in Hypertensive Patients, *Dis. Markers*, 2017, **2017**, 7146290.
257. L. Fang, G. Karakiulakis and M. Roth, Are patients with hypertension and diabetes mellitus at increased risk for COVID-19 infection? *Lancet Respir. Med.*, 2020, **8**(4), e21.
258. S. Kumar, COVID-19: A drug repurposing and biomarker identification by using comprehensive gene-disease associations through protein-protein interaction network analysis, Preprints 2020, DOI: 10.20944/preprints202003.0440.v1.
259. J. F. Navarro and C. Mora, Role of inflammation in diabetic complications, *Nephrol., Dial., Transplant.*, 2005, **20**(12), 2601–2604.
260. J. Pickup and M. Crook, Is type II diabetes mellitus a disease of the innate immune system? *Diabetologia*, 1998, **41**(10), 1241–1248.
261. K. Alexandraki, C. Piperi, C. Kalofoutis, J. Singh, A. Alaveras and A. Kalofoutis, Inflammatory process in type 2 diabetes: The role of cytokines, *Ann. N. Y. Acad. Sci.*, 2006, **1084**(1), 89–117.
262. N. Fatima, S. M. Faisal, S. Zubair, M. Ajmal, S. S. Siddiqui and S. Moin, *et al.*, Role of pro-inflammatory cytokines and biochemical markers in the pathogenesis of type 1 diabetes: correlation with age and glycemic condition in diabetic human subjects, *PLoS One*, 2016, **11**(8), e0161548.
263. M. Purohit, A. R. Yadav, Comparison of feature extraction techniques to recognize traffic rule violations using low processing embedded system, *2018 5th International Conference on Signal Processing and Integrated Networks (SPIN)*, 22–23 Feb 2018.
264. W. Gouda, L. Mageed, S. M. Abd El Dayem, E. Ashour and M. Afify, Evaluation of pro-inflammatory and anti-inflammatory cytokines in type 1 diabetes mellitus, *Bull. Natl. Res. Cent.*, 2018, **42**(1), 14.
265. L. Szablewski, Role of immune system in type 1 diabetes mellitus pathogenesis, *Int. Immunopharmacol.*, 2014, **22**(1), 182–191.
266. A. Hussain, B. Bhowmik and N. C. do Vale Moreira, COVID-19 and diabetes: Knowledge in progress, *Diabetes Res. Clin. Pract.*, 2020, **162**, 108142.

267. S. Hantoushzadeh, A. A. Shamshirsaz, A. Aleyasin, M. D. Seferovic, S. K. Aski, S. E. Arian, P. Pooransari, F. Ghotbizadeh, S. Aalipour, Z. Soleimani, M. Naemi, B. Molaei, R. Ahangari, M. Salehi, A. D. Oskoei, P. Pirozan, R. F. Darkhaneh, M. G. Laki, A. K. Farani, S. Atrak, M. M. Miri, M. Kouchek, S. Shojaei, F. Hadavand, F. Keikha, M. S. Hosseini, S. Borna, S. Ariana, M. Shariat, A. Fatemi, B. Nouri, S. M. Nekooghadam and K. Aagaard, Maternal death due to COVID-19, *Am. J. Obstet. Gynecol.*, 2020, **223**(1), 109.e1–109.e16.
268. J. Yang, Y. Zheng, X. Gou, K. Pu, Z. Chen, Q. Guo, R. Ji, H. Wang, Y. Wang and Y. Zhou, Prevalence of comorbidities and its effects in patients infected with SARS-CoV-2: a systematic review and meta-analysis, *Int. J. Infect. Dis.*, 2020, **94**, 91–95.
269. J. S. Alqahtani, T. Oyelade, A. M. Aldhahir, S. M. Alghamdi, M. Almehmadi and A. S. Alqahtani, *et al.*, Prevalence, severity and mortality associated with COPD and smoking in patients with COVID-19: a rapid systematic review and meta-analysis, *PLoS One*, 2020, **15**(5), e0233147.
270. B. Wang, R. Li, Z. Lu and Y. Huang, Does comorbidity increase the risk of patients with COVID-19: evidence from meta-analysis, *Aging*, 2020, **12**(7), 6049–6057.
271. A. Di Stefano, G. Caramori, F. L. M. Ricciardolo, A. Capelli, I. Adcock and C. Donner, Cellular and molecular mechanisms in chronic obstructive pulmonary disease: an overview, *Clin. Exp. Allergy*, 2004, **34**(8), 1156–1167.
272. C. Moermans, V. Heinen, M. Nguyen, M. Henket, J. Sele and M. Manise, *et al.*, Local and systemic cellular inflammation and cytokine release in chronic obstructive pulmonary disease, *Cytokine*, 2011, **56**(2), 298–304.
273. P. J. Barnes, The cytokine network in chronic obstructive pulmonary disease, *Am. J. Respir. Cell Mol. Biol.*, 2009, **41**(6), 631–638.
274. L. G. Franciosi, C. P. Page, B. R. Celli, M. Cazzola, M. J. Walker and M. Danhof, *et al.*, Markers of disease severity in chronic obstructive pulmonary disease, *Pulm. Pharmacol. Ther.*, 2006, **19**(3), 189–199.
275. P. J. Barnes and B. R. Celli, Systemic manifestations and comorbidities of COPD, *Eur. Respir. J.*, 2009, **33**(5), 1165–1185.
276. S. Selvarajah, I. Todd, P. J. Tighe, M. John, C. E. Bolton, T. Harrison and L. C. Fairclough, Multiple Circulating Cytokines Are Coelevated in Chronic Obstructive Pulmonary Disease, *Mediators Inflammation*, 2016, **2016**, 3604842.
277. J. M. Leung, C. X. Yang, A. Tam, T. Shaipanich, T.-L. Hackett and G. K. Singhera, *et al.*, ACE-2 expression in the small airway epithelia of smokers and COPD patients: implications for COVID-19, *Eur. Respir. J.*, 2020, **55**(5), 2000688.
278. L. Cordain, S. B. Eaton, A. Sebastian, N. Mann, S. Lindeberg and B. A. Watkins, *et al.*, Origins and evolution of the Western diet: health implications for the 21st century, *Am. J. Clin. Nutr.*, 2005, **81**(2), 341–354.
279. M. M. Rogero and P. C. Calder, Obesity, inflammation, toll-like receptor 4 and fatty acids, *Nutrients*, 2018, **10**(4), 432.

280. H. Tashiro, K. Takahashi, H. Sadamatsu, G. Kato, K. Kurata and S. Kimura, *et al.*, Saturated fatty acid increases lung macrophages and augments house dust mite-induced airway inflammation in mice fed with high-fat diet, *Inflammation*, 2017, **40**(3), 1072–1086.
281. W. D. Green and M. A. Beck, Obesity impairs the adaptive immune response to influenza virus, *Ann. Am. Thorac. Soc.*, 2017, **14**(Supplement 5), S406–S409.
282. A. Trompette, E. S. Gollwitzer, K. Yadava, A. K. Sichelstiel, N. Sprenger and C. Ngom-Bru, *et al.*, Gut microbiota metabolism of dietary fiber influences allergic airway disease and hematopoiesis, *Nat. Med.*, 2014, **20**(2), 159–166.
283. S. Anand and S. S. Mande, Diet, microbiota and gut-lung connection, *Front. Microbiol.*, 2018, **9**, 2147.
284. R. M. Connaughton, A. M. McMorrow, F. C. McGillicuddy, F. E. Lithander and H. M. Roche, Impact of anti-inflammatory nutrients on obesity-associated metabolic-inflammation from childhood through to adulthood, *Proc. Nutr. Soc.*, 2016, **75**(2), 115–124.
285. S. Panyod, C. T. Ho and L. Y. Sheen, Dietary therapy and herbal medicine for COVID-19 prevention: A review and perspective, *J. Tradit. Complement. Med.*, 2020, **10**(4), 420–427.
286. P. Zimmermann, A. Finn and N. Curtis, Does BCG Vaccination Protect Against Nontuberculous Mycobacterial Infection? A Systematic Review and Meta-Analysis, *J. Infect. Dis.*, 2018, **218**(5), 679–687.
287. S. Moorlag, R. Arts, R. Van Crevel and M. Netea, Non-specific effects of BCG vaccine on viral infections, *Clin. Microbiol. Infect.*, 2019, **25**(12), 1473–1478.
288. S. Luca and T. Mihaescu, History of BCG Vaccine, *Maedica (Buchar)*, 2013, **8**(1), 53–58.
289. P. E. M. Fine, BCG vaccination against tuberculosis and leprosy, *Br. Med. Bull.*, 1988, **44**(3), 691–703.
290. F. Floc'h and G. H. Werner, Increased resistance to virus infections of mice inoculated with BCG (Bacillus calmette-guérin), *Ann. Immunol. (Paris)*, 1976, **127**(2), 173–186.
291. O. Alhunaidi and A. R. Zlotta, The use of intravesical BCG in urothelial carcinoma of the bladder, *Ecancermedicalscience*, 2019, **13**, 905.
292. M. L. Berendsen, S. W. van Gijzel, J. Smits, Q. de Mast, P. Aaby and C. S. Benn, *et al.*, BCG vaccination is associated with reduced malaria prevalence in children under the age of 5 years in sub-Saharan Africa, *BMJ Global Health*, 2019, **4**(6), e001862.
293. L. G. Stensballe, E. Nante, I. P. Jensen, P.-E. Kofoed, A. Poulsen and H. Jensen, *et al.*, Acute lower respiratory tract infections and respiratory syncytial virus in infants in Guinea-Bissau: a beneficial effect of BCG vaccination for girls: Community based case–control study, *Vaccine*, 2005, **23**(10), 1251–1257.

294. M.-G. Hollm-Delgado, E. A. Stuart and R. E. Black, Acute Lower Respiratory Infection Among Bacille Calmette-Guérin (BCG)–Vaccinated Children, *Pediatrics*, 2014, **133**(1), e73–e81.
295. D. E. Wardhana, A. Sultana, V. Mandang and E. Jim, The efficacy of Bacillus Calmette-Guerin vaccinations for the prevention of acute upper respiratory tract infection in the elderly, *Acta Med Indones*, 2011, **43**(3), 185–190.
296. M. G. Netea and R. van Crevel, BCG-induced protection: Effects on innate immune memory, *Semin. Immunol.*, 2014, **26**(6), 512–517.
297. L. A. J. O'Neill and M. G. Netea, BCG-induced trained immunity: can it offer protection against COVID-19? *Nat. Rev. Immunol.*, 2020, **20**(6), 335–337.
298. C. Covián, A. Fernández-Fierro, A. Retamal-Díaz, F. E. Díaz, A. E. Vasquez and M. K.-L. Lay, *et al.*, BCG-induced cross-protection and development of trained immunity. Implication for vaccine design, *Front. Immunol.*, 2019, **10**, 2806.
299. J. I. Moliva, J. Turner and J. B. Torrelles, Immune responses to bacillus Calmette–Guérin vaccination: why do they fail to protect against Mycobacterium tuberculosis? *Front. Immunol.*, 2017, **8**, 407.
300. G. Redelman-Sidi, Could BCG be used to protect against COVID-19? *Nat. Rev. Neurol.*, 2020, **17**(6), 316–317.
301. A. Miller, M. J. Reandelar, K. Fasciglione, V. Roumenova, Y. Li, G. H. Otazu, Correlation between universal BCG vaccination policy and reduced morbidity and mortality for COVID-19: an epidemiological study, MedRxiv, 2020, DOI: 10.1101/2020.03.24.20042937.
302. M. Madan, *et al.*, TB infection and BCG vaccination: are we protected from COVID-19? *Public Health*, 2020, **185**, 91–92.
303. C. Ozdemir, U. C. Kucuksezer and Z. U. Tamay, Is BCG vaccination affecting the spread and severity of COVID-19? *Allergy*, 2020, **75**(7), 1824–1827.
304. R. J. Arts, S. J. Moorlag, B. Novakovic, Y. Li, S.-Y. Wang and M. Oosting, *et al.*, BCG vaccination protects against experimental viral infection in humans through the induction of cytokines associated with trained immunity, *Cell Host Microbe*, 2018, **23**(1), 89–100.
305. D. Klinger, I. Blass, N. Rappoport and M. Linial, Significantly improved COVID-19 outcomes in countries with higher BCG vaccination coverage: A multivariable analysis, *Vaccines*, 2020, **8**(3), 378.
306. S. Singh, BCG vaccines may not reduce COVID-19 mortality rates, medRxiv, 2020, DOI: 10.1101/2020.04.11.20062232.
307. M. Fukui, K. Kawaguchi, H. Matsuura, Does TB vaccination reduce COVID-19 infection?: No evidence from a regression discontinuity analysis. No Evidence from a Regression Discontinuity Analysis (9 April 2020), 2020, DOI: 10.2139/ssrn.3572090.
308. N. E. Ingraham, S. Lotfi-Emran, B. K. Thielen, K. Techar, R. S. Morris, S. G. Holtan, R. A. Dudley and C. J. Tignanelli, Immunomodulation in COVID-19, *Lancet Respir. Med.*, 2020, **8**(6), 544–546.

309. E. Keyaerts, L. Vijgen, P. Maes, J. Neyts and M. V. Ranst, In vitro inhibition of severe acute respiratory syndrome coronavirus by chloroquine, *Biochem. Biophys. Res. Commun.*, 2004, **323**(1), 264–268.
310. M. J. Vincent, E. Bergeron, S. Benjannet, B. R. Erickson, P. E. Rollin and T. G. Ksiazek, *et al.*, Chloroquine is a potent inhibitor of SARS coronavirus infection and spread, *Virology J.*, 2005, **2**(1), 69.
311. P. Gautret, J.-C. Lagier, P. Parola, V. T. Hoang, L. Meddeb and M. Mailhe, *et al.*, Hydroxychloroquine and azithromycin as a treatment of COVID-19: results of an open-label non-randomized clinical trial, *Int. J. Antimicrob. Agents*, 2020, **56**(1), 105949.
312. P. Gautret, J.-C. Lagier, P. Parola, V. T. Hoang, L. Meddeb and J. Sevestre, *et al.*, Clinical and microbiological effect of a combination of hydroxychloroquine and azithromycin in 80 COVID-19 patients with at least a six-day follow up: A pilot observational study, *Travel Med. Infect. Dis.*, 2020, **34**, 101663.
313. Z. Chen, J. Hu, Z. Zhang, S. Jiang, S. Han and D. Yan, *et al.*, Efficacy of hydroxychloroquine in patients with COVID-19: results of a randomized clinical trial. medRxiv, 2020, DOI: 10.1101/2020.03.22.20040758.
314. J. Chen, D. Liu, L. Liu, P. Liu, Q. Xu, L. Xia, Y. Ling, D. Huang, S. Song, D. Zhang, Z. Qian, T. Li, Y. Shen and H. Lu, A pilot study of hydroxychloroquine in treatment of patients with moderate COVID-19, *Zhejiang Daxue Xuebao, Yixueban*, 2020, **49**(2), 215–219.
315. W. Tang, Z. Cao, M. Han, Z. Wang, J. Chen, W. Sun, Y. Wu, W. Xiao, S. Liu, E. Chen, W. Chen, X. Wang, J. Yang, J. Lin, Q. Zhao, Y. Yan, Z. Xie, D. Li, Y. Yang, L. Liu, J. Qu, G. Ning, G. Shi and Q. Xie, Hydroxychloroquine in patients with mainly mild to moderate coronavirus disease 2019: open label, randomised controlled trial, *BMJ*, 2020, **369**, m1849.
316. A. Shamshirian, A. Hessami, K. Heydari, R. A. Navaei, M. A. Ebrahimzadeh, G. W. Yip, R. Ghasemian, M. Sedaghat, H. Baradaran, S. M. Yazdi, E. Aboufazeli, H. Jafarpour, E. Dadgostar, B. Tirandazi, R. Sadeghnezhad, K. Karimifar, A. Eftekhari and D. Shamshirian, The Role of Hydroxychloroquine in COVID-19: A Systematic Review and Meta-Analysis, *Ann. Acad. Med. Singap*, 2020, **49**(10), 789–800.
317. J. M. Molina, C. Delaugerre, J. Le Goff, B. Mela-Lima, D. Ponscarme and L. Goldwirt, *et al.*, No evidence of rapid antiviral clearance or clinical benefit with the combination of hydroxychloroquine and azithromycin in patients with severe COVID-19 infection, *Med. Mal. Infect.*, 2020, **50**(4), 384.
318. M. G. S. Borba, F. F. A. Val, V. S. Sampaio, M. A. A. Alexandre, G. C. Melo and M. Brito, *et al.*, Effect of High vs Low Doses of Chloroquine Diphosphate as Adjunctive Therapy for Patients Hospitalized With Severe Acute Respiratory Syndrome Coronavirus 2 (SARS-CoV-2) Infection: A Randomized Clinical Trial, *JAMA Network Open*, 2020, **3**(4), e208857.

319. M. P. H. van den Broek, J. E. Möhlmann, B. G. S. Abeln, M. Liebregts, V. F. van Dijk and E. M. W. van de Garde, Chloroquine-induced QTc prolongation in COVID-19 patients, *Neth. Heart J.*, 2020, **28**(7–8), 406–409.
320. E. Chorin, L. Wadhwani, S. Magnani, M. Dai, E. Shulman and C. Nadeau-Routhier, *et al.*, QT interval prolongation and torsade de pointes in patients with COVID-19 treated with hydroxychloroquine/azithromycin, *Heart Rhythm*, 2020, **17**(9), 1425–1433.
321. F. S. Sinkeler, F. A. Berger, H. J. Muntinga and M. Jansen, The risk of QTc-interval prolongation in COVID-19 patients treated with chloroquine, *Neth. Heart J.*, 2020, **28**(7–8), 418–423.
322. M. G. S. Borba, F. F. A. Val, V. S. Sampaio, M. A. A. Alexandre, G. C. Melo and M. Brito, *et al.*, Effect of High vs Low Doses of Chloroquine Diphosphate as Adjunctive Therapy for Patients Hospitalized With Severe Acute Respiratory Syndrome Coronavirus 2 (SARS-CoV-2) Infection: A Randomized Clinical Trial, *JAMA Network Open*, 2020, **3**(4), e208857.
323. F. Mzayek, H. Deng, F. J. Mather, E. C. Wasilevich, H. Liu and C. M. Hadi, *et al.*, Randomized dose-ranging controlled trial of AQ-13, a candidate antimalarial, and chloroquine in healthy volunteers, *PLoS Clin. Trials*, 2007, **2**(1), e6.
324. A. S. Vink, B. Neumann, K. V. V. Lieve, M. F. Sinner, N. Hofman and S. El Kadi, *et al.*, Determination and Interpretation of the QT Interval, *Circulation*, 2018, **138**(21), 2345–2358.
325. Y. Szekely, Y. Lichter, B. A. Shrkihe, H. Bruck, H. S. Oster and S. Viskin, Chloroquine-induced torsades de pointes in a patient with coronavirus disease 2019, *Heart Rhythm*, 2020, **17**(9), 1452–1455.
326. Z. Yang, J. Liu, Y. Zhou, X. Zhao, Q. Zhao and J. Liu, The effect of corticosteroid treatment on patients with coronavirus infection: a systematic review and meta-analysis, *J. Infect.*, 2020, **81**(1), e13–e20.
327. J. M. Sanders, M. L. Monogue, T. Z. Jodlowski and J. B. Cutrell, Pharmacologic Treatments for Coronavirus Disease 2019 (COVID-19): A Review, *JAMA*, 2020, **323**(18), 1824–1836.
328. K. K. Gangopadhyay, J. J. Mukherjee, B. Sinha and S. Ghosal, The role of corticosteroids in the management of critically ill patients with coronavirus disease 2019 (COVID-19): A meta-analysis, medRxiv, 2020, DOI: 10.1101/2020.04.17.20069773.
329. X. Fang, Q. Mei, T. Yang, L. Li, Y. Wang and F. Tong, *et al.*, Low-dose corticosteroid therapy does not delay viral clearance in patients with COVID-19, *J. Infect.*, 2020, **81**(1), 147–178.
330. D. Wang, J. Wang, Q. Jiang, J. Yang, J. Li and C. Gao, *et al.*, No Clear Benefit to the Use of Corticosteroid as Treatment in Adult Patients with Coronavirus Disease 2019 : A Retrospective Cohort Study, medRxiv, 2020, DOI: 10.1101/2020.04.21.20066258.
331. L. Zha, S. Li, L. Pan, B. Tefsen, Y. Li and N. French, *et al.*, Corticosteroid treatment of patients with coronavirus disease 2019 (COVID-19), *Med. J. Aust.*, 2020, **212**(9), 416–420.

332. X. Lu, T. Chen, Y. Wang, J. Wang and F. Yan, Adjuvant corticosteroid therapy for critically ill patients with COVID-19, *Crit. Care*, 2020, **24**(1), 241.
333. Y. Wang, W. Jiang, Q. He, C. Wang, B. Wang and P. Zhou, *et al.*, Early, low-dose and short-term application of corticosteroid treatment in patients with severe COVID-19 pneumonia: single-center experience from Wuhan, China, medRxiv, 2020, DOI: 10.1101/2020.03.06.20032342.
334. K.-Y. Lee, J.-W. Rhim and J.-H. Kang, Early preemptive immunomodulators (corticosteroids) for severe pneumonia patients infected with SARS-CoV-2, *Clin. Exp. Pediatr.*, 2020, **63**(4), 117–118.
335. L. Shang, J. Zhao, Y. Hu, R. Du and B. Cao, On the use of corticosteroids for 2019-nCoV pneumonia, *Lancet*, 2020, **395**(10225), 683–684.
336. RECOVERY Collaborative Group, P. Horby, W. S. Lim, J. R. Emberson, M. Mafham, J. L. Bell, L. Linsell, N. Staplin, C. Brightling, A. Ustianowski, E. Elmahi, B. Prudon, C. Green, T. Felton, D. Chadwick, K. Rege, C. Fegan, L. C. Chappell, S. N. Faust, T. Jaki, K. Jeffery, A. Montgomery, K. Rowan, E. Juszczak, J. K. Baillie, R. Haynes and M. J. Landray, Dexamethasone in Hospitalized Patients with Covid-19, *N. Engl. J. Med.*, 2021, **384**(8), 693–704.
337. Y. Wang, W. Jiang, Q. He, C. Wang, B. Liu and P. Zhou, *et al.*, Early, low-dose and short-term application of corticosteroid treatment in patients with severe COVID-19 pneumonia: single-center experience from Wuhan, China, medRxiv, 2020, DOI: 10.1101/2020.03.06.20032342.
338. M. H. Ahmed and A. Hassan, Dexamethasone for the Treatment of Coronavirus Disease (COVID-19): a Review, *SN Compr. Clin. Med.*, 2020, 1–10.
339. W. M. Schneider, M. D. Chevillotte and C. M. Rice, Interferon-Stimulated Genes: A Complex Web of Host Defenses, *Annu. Rev. Immunol.*, 2014, **32**(1), 513–545.
340. E. Andreakos, M. Salagianni, I. E. Galani and O. Koltsida, Interferon-λs: Front-line guardians of immunity and homeostasis in the respiratory tract, *Front. Immunol.*, 2017, **8**, 1232.
341. L. Prokunina-Olsson, N. Alphonse, R. E. Dickenson, J. E. Durbin, J. S. Glenn, R. Hartmann, S. V. Kotenko, H. M. Lazear, T. R. O'Brien, C. Odendall, O. O. Onabajo, H. Piontkivska, D. M. Santer, N. C. Reich, A. Wack and I. Zanoni, COVID-19 and emerging viral infections: The case for interferon lambda, *J. Exp. Med.*, 2020, **217**(5), e20200653.
342. P. Mosaddeghi, *et al.*, Therapeutic Approaches for COVID-19 Based on the Interferon-mediated Immune Responses, *Curr. Signal Transduction Ther.*, 2021, **16**, 1.
343. Y. Chen, Z. Feng, B. Diao, R. Wang, G. Wang and C. Wang, *et al.*, The Novel Severe Acute Respiratory Syndrome Coronavirus 2 (SARS-CoV-2) Directly Decimates Human Spleens and Lymph Nodes, medRxiv, 2020, DOI: 10.1101/2020.03.27.20045427.

344. D. Wang, R. Li and J. Wang, *et al.*, Correlation analysis between disease severity and clinical and biochemical characteristics of 143 cases of COVID-19 in Wuhan, China: a descriptive study, *BMC Infect. Dis.*, 2020, **20**(1), 519.
345. J. B. Moore and C. H. June, Cytokine release syndrome in severe COVID-19, *Science*, 2020, **368**(6490), 473–474.
346. X. Xu, M. Han, T. Li, W. Sun, D. Wang and B. Fu, *et al.*, Effective treatment of severe COVID-19 patients with tocilizumab, *Proc. Natl. Acad. Sci. U. S. A.*, 2020, **117**(20), 10970–10975.
347. M. Roumier, R. Paule, G. Matthieu, A. Vallée and F. Ackermann, Interleukin-6 blockade for severe COVID-19, medrxiv, 2020, DOI: 10.1101/2020.04.20.20061861.
348. P. Toniati, S. Piva, M. Cattalini, E. Garrafa, F. Regola and F. Castelli, *et al.*, Tocilizumab for the treatment of severe COVID-19 pneumonia with hyperinflammatory syndrome and acute respiratory failure: A single center study of 100 patients in Brescia, Italy, *Autoimmun. Rev.*, 2020, 102568.
349. O. Moreno-Pérez, M. Andres, J.-M. Leon-Ramirez, J. Sánchez-Payá, J. C. Rodríguez and R. Sánchez, *et al.*, Experience with tocilizumab in severe COVID-19 pneumonia after 80 days of follow-up: A retrospective cohort study, *J. Autoimmun.*, 2020, 102523.
350. S. Richardson, J. S. Hirsch, M. Narasimhan, J. M. Crawford, T. McGinn, K. W. Davidson, the Northwell COVID-19 Research Consortium, D. P. Barnaby, L. B. Becker, J. D. Chelico, S. L. Cohen, J. Cookingham, K. Coppa, M. A. Diefenbach, A. J. Dominello, J. Duer-Hefele, L. Falzon, J. Gitlin, N. Hajizadeh, T. G. Harvin, D. A. Hirschwerk, E. J. Kim, Z. M. Kozel, L. M. Marrast, J. N. Mogavero, G. A. Osorio, M. Qiu and T. P. Zanos, Presenting Characteristics, Comorbidities, and Outcomes Among 5700 Patients Hospitalized With COVID-19 in the New York City Area, *JAMA*, 2020, **323**(20), 2052–2059.
351. J. Rodríguez-Baño, J. Pachón, J. Carratalà, P. Ryan, I. Jarrín, M. Yllescas, J. R. Arribas, J. Berenguer, SAM-COVID Study Group, Fundación SEIMC-GESIDA, Hospital Universitario La Paz, Hospital Universitario Gregorio Marañón, Hospital Infanta Leonor, Complejo Hospitalario Virgen de la Salud, Hospital Universitario Rafael Méndez, Hospital Universitario de Cruces, Hospital de Melilla, Hospital San Eloy de Barakaldo, Hospital Universitario Central de Asturias, Hospital Universitario Puerto Real, Hospital do Salnés, Hospital del Mar, Hospital Virgen de la Arrixaca, Hospital Clínico San Cecilio, Parc Sanitari Sant Joan de Déu, Hospital Josep Trueta, Hospital Dos De Maig - Consorci Sanitari Integral, Hospital Clínico Universitario de Valencia, Complejo Asistencial de Ávila, Hospital Universitario Marqués de Valdecilla, Hospital de Barcelona SCIAS, Hospital Álvaro Cunqueiro, Hospital Universitario Severo Ochoa, Hospital CIMA-Sanitas, Hospital La Inmaculada, Hospital de Guadalajara, Hospital Universitario Infanta Sofia, Hospital Comarcal de Blanes,

Hospital Universitario de Gran Canaria Dr Negrín, Hospital Son Espases, Complejo Hospitalario Universitario A Coruña, Hospital Costa del Sol, Hospital Clínico Universitario Lozano Blesa, Hospital Mutua de Terrassa, Hospital Universitario Virgen Macarena, Hospital Universitari de Bellvitge, Hospital Universitario y Politécnico La Fe, Hospital de Sabadell (Parc Tauli), Hospital Fundación Jiménez Díaz, Hospital Clínico Universitario de Valladolid, Hospital Son Llatzer, Hospital Universitario de Álava, Complejo Hospitalario Universitario Santa Lucía, Hospital General Universitario Reina Sofía, Complejo Hospitalario Universitario de Ferrol, Hospital Universitario los Arcos del Mar Menor, Hospital Universitario de Jerez, Hospital de Donostia, Hospital Juan Ramón Jiménez, Hospital Vega Baja, Hospital Puerta de Hierro, Hospital Universitario de Getafe, Hospital General de la Palma, Fundación Hospital de Calahorra, Hospital Alto Deba, Hospital Universitario de Jaén, Hospital de Palamós, Hospital Universitario de Valme, Hospital Universitario Virgen del Rocío, Hospital Universitario Ramón y Cajal, Hospital Universitario San Pedro and Hospital Regional de Málaga, Treatment with tocilizumab or corticosteroids for COVID-19 patients with hyperinflammatory state: a multicentre cohort study (SAM-COVID-19), *Clin. Microbiol. Infect.*, 2021, **27**(2), 244–252.
352. P. Sodani, L. Mucci, R. Girolimetti, S. Tedesco, F. Monaco and D. Campanozzi, *et al.*, Successful recovery from COVID-19 pneumonia after receiving baricitinib, tocilizumab, and remdesivir. A case report: Review of treatments and clinical role of computed tomography analysis, *Respir. Med. Case Rep.*, **2020**, 101115.
353. P. Sinha, A. Mostaghim, C. G. Bielick, A. McLaughlin, D. H. Hamer and L. M. Wetzler, *et al.*, Early administration of interleukin-6 inhibitors for patients with severe COVID-19 disease is associated with decreased intubation, reduced mortality, and increased discharge, *Int. J. Infect. Dis.*, 2020, **99**, 28–33.
354. P. Richardson, I. Griffin, C. Tucker, D. Smith, O. Oechsle and A. Phelan, *et al.*, Baricitinib as potential treatment for 2019-nCoV acute respiratory disease, *Lancet*, **2020**, **395**(10223), e30–e31.
355. S. G. Deftereos, *et al.*, The Greek study in the effects of colchicine in COvid-19 complications prevention (GRECCO-19 study): Rationale and study design, *Hell. J. Cardiol.*, 2020, **61**(1), 42–45.
356. E. A. Van Erp, W. Luytjes, G. Ferwerda and P. B. Van Kasteren, Fc-mediated antibody effector functions during respiratory syncytial virus infection and disease, *Front. Immunol.*, 2019, **10**, 548.
357. B. M. Gunn, W.-H. Yu, M. M. Karim, J. M. Brannan, A. S. Herbert and A. Z. Wec, *et al.*, A role for Fc function in therapeutic monoclonal antibody-mediated protection against Ebola virus, *Cell Host Microbe*, 2018, **24**(2), 221–233.
358. K.-M. Yeh, T.-S. Chiueh, L. K. Siu, J.-C. Lin, P. K. S. Chan and M.-Y. Peng, *et al.*, Experience of using convalescent plasma for severe acute respiratory syndrome among healthcare workers in a Taiwan hospital, *J. Antimicrob. Chemother.*, 2005, **56**(5), 919–922.

359. J. Mair-Jenkins, M. Saavedra-Campos, J. K. Baillie, P. Cleary, F.-M. Khaw and W. S. Lim, *et al.*, The Effectiveness of Convalescent Plasma and Hyperimmune Immunoglobulin for the Treatment of Severe Acute Respiratory Infections of Viral Etiology: A Systematic Review and Exploratory Meta-analysis, *J. Infect. Dis.*, 2014, **211**(1), 80–90.
360. P. K. Law, *Emergent Serum Therapy and Antibody Medicine to Counteract Sudden Attacks of COVID-19 and Other Pathogenic Epidemics*, Scientific Research Publishing, 2020.
361. C. Shen, Z. Wang, F. Zhao, Y. Yang, J. Li and J. Yuan, *et al.*, Treatment of 5 critically ill patients with COVID-19 with convalescent plasma, *JAMA*, 2020, **323**(16), 1582–1589.
362. K. Duan, *et al.*, Effectiveness of convalescent plasma therapy in severe COVID-19 patients, *Proc. Natl. Acad. Sci.*, 2020, **117**(17), 9490–9496.
363. L. Chen, J. Xiong, L. Bao and Y. Shi, Convalescent plasma as a potential therapy for COVID-19, *Lancet Infect. Dis.*, 2020, **20**(4), 398–400.
364. Y. Cheng, R. Wong, Y. Soo, W. Wong, C. Lee and M. Ng, *et al.*, Use of convalescent plasma therapy in SARS patients in Hong Kong, *Eur. J. Clin. Microbiol. Infect. Dis.*, 2005, **24**(1), 44–46.
365. K. Duan, B. Liu, C. Li, H. Zhang, T. Yu and J. Qu, *et al.*, Effectiveness of convalescent plasma therapy in severe COVID-19 patients, *Proc. Natl. Acad. Sci. U. S. A.*, 2020, **117**(17), 9490–9496.
366. B. Zhang, *et al.*, Treatment with convalescent plasma for critically ill patients with severe acute respiratory syndrome coronavirus 2 infection, *Chest*, 2020, **158**(1), e9–e13.
367. J. Y. Ahn, Y. Sohn, S. H. Lee, Y. Cho, J. H. Hyun, Y. J. Baek, S. J. Jeong, J. H. Kim, N. S. Ku, J. S. Yeom, J. Roh, M. Y. Ahn, B. S. Chin, Y. S. Kim, H. Lee, D. Yong, H. O. Kim, S. Kim and J. Y. Choi, Use of Convalescent Plasma Therapy in Two COVID-19 Patients with Acute Respiratory Distress Syndrome in Korea, *J. Korean Med. Sci.*, 2020, **35**(14), e149.
368. S. Pei, X. Yuan, Z. Z. Zhang, R. R. Yao, Y. Xie and M. M. Shen, *et al.*, Convalescent plasma to treat covid-19: Chinese strategy and experiences, medRxiv, 2020.
369. L. Li, W. Zhang, Y. Hu, X. Tong, S. Zheng, J. Yang, Y. Kong, L. Ren, Q. Wei, H. Mei, C. Hu, C. Tao, R. Yang, J. Wang, Y. Yu, Y. Guo, X. Wu, Z. Xu, L. Zeng, N. Xiong, L. Chen, J. Wang, N. Man, Y. Liu, H. Xu, E. Deng, X. Zhang, C. Li, C. Wang, S. Su, L. Zhang, J. Wang, Y. Wu and Z. Liu, Effect of Convalescent Plasma Therapy on Time to Clinical Improvement in Patients With Severe and Life-threatening COVID-19: A Randomized Clinical Trial, *JAMA*, 2020, **324**(5), 460–470.
370. H. Abolghasemi, P. Eshghi, A. M. Cheraghali, A. A. I. Fooladi, F. B. Moghaddam and S. Imanizadeh, *et al.*, Clinical efficacy of convalescent plasma for treatment of COVID-19 infections: Results of a multicenter clinical study, *Transfus. Apheresis Sci.*, 2020, 102875.
371. P. Saeedi, R. Halabian and A. A. Imani Fooladi, A revealing review of mesenchymal stem cells therapy, clinical perspectives and Modification strategies, *Stem Cell Invest.*, 2019, **6**, 34.

372. B. Liang, J. Chen, T. Li, H. Wu, W. Yang, Y. Li, J. Li, C. Yu, F. Nie, Z. Ma, M. Yang, M. Xiao, P. Nie, Y. Gao, C. Qian and M. Hu, Clinical remission of a critically ill COVID-19 patient treated by human umbilical cord mesenchymal stem cells: A case report, *Medicine*, 2020, **99**(31), e21429.
373. Z. Leng, R. Zhu, W. Hou, Y. Feng, Y. Yang and Q. Han, *et al.*, Transplantation of ACE2-mesenchymal stem cells improves the outcome of patients with COVID-19 pneumonia, *Aging Dis.*, 2020, **11**(2), 216.

CHAPTER 10

Neurological Manifestations in COVID-19

A. SAMHITA PANDA*[a], B. SAKSHAM JAIN[a], C. SANJIV SHARMA[a] AND D. SARBESH TIWARI[b]

[a]Department of Neurology, All India Institute of Medical Sciences, Jodhpur, Rajasthan, India; [b]Department of Diagnostic and Interventional Radiology, All India Institute of Medical Sciences, Jodhpur, Rajasthan, India
*E-mail: samhitapanda@yahoo.com

10.1 Introduction

Coronavirus disease 2019 (COVID-19), caused by the single-stranded, positively sensed RNA virus, the severe acute respiratory syndrome coronavirus 2 (SARS-CoV-2), has marked its presence worldwide since December 2019. While the majority of patients with COVID-19 usually present with mild to moderate respiratory symptoms, the lungs are the most severely affected organ and pulmonary manifestations of COVID-19 have been well described and are characterized by dyspnea and rapidly deteriorating respiratory failure and potentially life-threatening severe hypoxemia.[1,2] Extra-pulmonary manifestations such as diarrhea, myalgias, renal failure and low exercise tolerance, among others, have slowly come to fore during the pandemic.[3] The initially reported neurological symptoms were ageusia and anosmia and then acute hemorrhagic necrotising encephalopathy, meningoencephalitis and seizures.[4–6] Subsequently, a few case series reported an increasing incidence of strokes in patients with moderate to severe COVID-19.[5,7,8] As more

The Coronavirus Pandemic and the Future Volume 1: Virology, Epidemiology, Translational Toxicology and Therapeutics
Edited by Michael D. Waters, Alok Dhawan, Tim Marrs, Diana Anderson, Stafford Warren and Claude L. Hughes

Published by the Royal Society of Chemistry, www.rsc.org

and more literature has been garnered, a variety of disorders of the central nervous system (CNS) and/or peripheral nervous system (PNS) have been reported. Guillain-Barré syndrome, myelitis, white matter demyelination and acute disseminated encephalomyelitis (ADEM) have been reported.[5,9–11] The apparent neurotropism and neuroinvasiveness of SARS-CoV-2 may be responsible for the neurological manifestations. However, there is lack of clarity as to whether the neurological features are a direct or an indirect effect of the infection.[3,4]

10.2 Pathophysiology of Neurotropism of SARS-CoV-2

The SARS-CoV-2 virus has four structural proteins – spike protein (S), an envelope protein (E), a membrane protein (M), and the nucleocapsid protein (N). It is the S protein which is important for virus attachment, entry and fusion.[12] The ability of a virus to infect specific cells or tissues while not affecting others is referred to as viral tropism. SARS-CoV-2 recognizes the angiotensin-converting enzyme 2 (ACE2) as a host cell entry receptor and therefore the presence of ACE2 receptor in tissues determines viral cellular tropism in humans.[13] ACE2 is expressed in many organs, such as in airway epithelia, kidney cells, the small intestine, lung parenchyma and vascular endothelia throughout the body including widely throughout the CNS.[14,15] Paniz-Mondolfi and co-workers, in an autopsy-based study, demonstrated the presence of the SARS-CoV-2 virus in neural and capillary endothelial cells of the frontal lobe of the brain.[16] ACE2 is additionally found in the oral mucosa. The SARS-CoV-2 virus can cross the blood–brain barrier and enter the brain, thereby traveling to the brain by a hematogenous route. More recently, it has been reported that the brain also expresses ACE2 receptors on glial cells and neurons. This is most prominent in the brainstem, the paraventricular nucleus, nucleus tractus solitarius and the rostral ventrolateral medulla, which all play a role in cardiovascular regulation.[14,16]

Apart from ACE2, the transmembrane serine protease 2 ($TMPRSS2^+$) also aids in neuroinvasion. $TMPRSS2^+$ is expressed in the cerebral cortex, hippocampus, caudate and nasopharynx and dual-positive $ACE2^+$ $TMPRSS2^+$ cells in the oligodendrocytes.[13,17] Hence, it is possible that the virus may result in endotheliitis of brain vasculature and then spread to the $TMPRSS2^+$ as well as by a paracellular route to infect the oligodendrocytes and neurons. The SARS-CoV-2 virus also enters the brain through a retrograde transfer *via* the olfactory epithelium utilizing axonal transport in nerve fibers. Therefore, trans-synaptic spread through the olfactory nerves and bulb or the vagus nerve as a source of direct viral spread is also postulated.[4,17]

Therefore, viral neuroinvasion could possibly be achieved by several routes:

1. trans-synaptic transfer across infected neurons
2. entry *via* the olfactory nerve

3. infection of vascular endothelium
4. paracellular infection of oligodendrocytes and neurons
5. leukocyte migration across the blood–brain barrier.

10.2.1 Risk Factors Potentiating Neurotropism

Comorbidities such as diabetes and hypertension enhance ACE2 receptor expression in the brain and the neurotropism of SARS-CoV-2.[18] Competitive blockage of ACE2 by the SARS-CoV-2 virus additionally down-regulates ACE2 expression, leading to uncontrolled blood pressure. Olds and Kabbani raised the question of nicotine-associated neurological comorbidity in COVID-19 patients, depending on published evidence that the viral target receptor ACE2 is expressed in the brain and functionally interacts with nicotinic acetylcholine receptors (nAChR).[19] They considered neural cells and astrocytes (especially in the hypothalamus and brain stem) to be more prone to infection in smokers because nicotine stimulation of the nAChR was found to increase ACE2 expression within them.

10.2.2 Mechanisms for Neurological Complications

Multiple neuropathological mechanisms have been proposed to explain the neurological complications of SARS-CoV-2.

10.2.2.1 Pathogenesis of Thrombogenicity

The complexity of the pro-coagulant state related to COVID-19-associated strokes is yet to be clearly elucidated. The SARS-CoV-2 infection triggers a cascade of inflammation with release of mediators such as interleukin (IL)-6, high-sensitivity C-reactive protein (hs-CRP) and surrogate markers like ferritin.[20] An increased D-dimer level has also been noted in conjunction with cytokine storm in COVID-19.[21] Thus, the thrombogenicity may be considered as part of an exaggerated immune response to SARS-CoV-2 in severe patients. Additionally, vascular wall changes occur with COVID-19 infection, including endotheliitis and induction of apoptosis of endothelial cells, either by direct invasion or by pro-inflammatory markers.[21,22] This may further trigger macrophage and granulocyte recruitment, leading to cytokine release and thrombin formation. Severe respiratory dysfunction may release hypoxia-inducible transcription factors, further inducing a hypercoagulable state. The inflammatory storm may also accelerate the atherosclerotic process and lead to the disruption of plaques, making them unstable and ulcerated, thereby creating a nidus for thrombus formation. Therefore, SARS-CoV-2 can cause activation and upregulation of pro-thrombotic pathways, leading to widespread coagulopathy and microvascular and macrovascular complications in the brain secondary to endothelial cell damage, monocyte activation, up-regulation of tissue factors, and the release of microparticles with the presence of thrombocytopenia, elevated D-dimer and hs-CRP seen in severe COVID-19. The occurrence of stroke in the setting of COVID-19 is consistent with this mechanism. Based on our current understanding, COVID-19 patients should be stratified as high thrombotic risk if they have dyspnea, respiratory rate >24 breaths per minute, oxygen saturation <94%, elevated hs-CRP, rising D-dimer level and elevated fibrinogen.[23]

10.2.2.2 Cytokine Storm

SARS-CoV-2 can indirectly cause brain damage by triggering a cytokine storm, *i.e.* immunologic dysregulation.

Cytokine storm is associated with enhanced vascular hyperpermeability, coagulopathies and multisystem dysfunction. IL-6 is a key element of the cytokine storm alongside other pro-inflammatory cytokines such as IL-1β, IL-2, IL-8, IL-17, chemokine ligand 3, granulocyte-colony stimulating factor, granulocyte–macrophage colony-stimulating factor, the human interferon inducible protein and tumor necrosis factor-α.[24]

10.2.2.3 Hypoxic Brain Damage

Neurological manifestations may also be caused by hypoxia-related injuries, as alveolar and interstitial lung inflammation may lead to CNS hypoxia. This in turn may cause cerebral vasodilatation and cerebral edema.[25] Poor respiratory function and acute respiratory distress syndrome (ARDS) leading to hypoxemia may further lead to release of hypoxia-inducible transcription factors, further unbalancing the pro-coagulant and anticoagulant forces in the body.

10.2.2.4 Blood Pressure Alterations

Normally, ACE2 expression lowers the blood pressure. Blockage of ACE2 signaling by SARS-CoV-2 leads to wide variations in blood pressure, resulting in cerebrovascular accidents.

10.2.3 Pathogenesis of Neuronal Autoimmunity in COVID-19

At the present juncture, our understanding of the pathophysiology of CNS and PNS demyelination includes the hyperimmune activation and cytokine storm which results in breakdown of the blood–brain barrier without actual viral invasion.[4,17] This inflammation is mediated by interleukins (IL-6 and IL-8) released by monocytes and macrophages to stimulate other monocytes and both B- and T-lymphocytes, in addition to monocyte chemoattractant protein-1, a chemokine responsible for the transmigration of the monocytes across the blood–brain barrier.[13,24] Certain SARS-CoV-2 virus proteins are structurally similar to human proteins. This results in cross-reactivity and can explain post-infectious demyelinating syndromes seen with COVID-19. Hence, most of the demyelinating events of the nervous system are delayed, occurring in or after the second week, usually after severe COVID-19 infection.[10,11,26] Therefore, the COVID reverse transcriptase polymerase chain reaction tests (RT-PCR) have been negative in all these reports, both in serum and cerebrospinal fluid (CSF). However, there are rare reports of detection of coronavirus in the CSF in patients with demyelination.[27]

10.3 Neurological Manifestations of COVID-19

Neurological symptoms may be one of the important extra-pulmonary manifestations of COVID-19. Neurological manifestations have been observed in up to 36% of COVID-19 patients and may involve either the CNS or the PNS.[28] After the revelations of the first major study from Wuhan by Mao *et al.*, a number of other studies have demonstrated the spectrum of neurological manifestations of COVID infection.[4–6,16,25,28] The relative frequencies of the different neurological presentations are given in Table 10.1.

Table 10.1 Frequency of neurological symptoms in patients with COVID-19.

Main neurological symptoms	Frequency	Reference
Acute cerebrovascular disease	2.8%	Mao *et al.*[5]
Acute hemorrhagic necrotizing encephalopathy	Case report	Poyiadji *et al.*[9]
Ageusia	1.7%	Giacomelli *et al.*[57]
Ageusia and anosmia	8.5%	Giacomelli *et al.*[57]
Ageusia and hyposmia	3.4%	Giacomelli *et al.*[57]
Ataxia	0.5%	Mao *et al.*[5]
Dizziness	9.4%	Mao *et al.*[5]
Dysgeusia	8.5%	Giacomelli *et al.*[57]
Dysgeusia and anosmia	3.4%	Giacomelli *et al.*[57]
Dysgeusia and hyposmia	3.4%	Giacomelli *et al.*[57]
Headache	3.4%	Giacomelli *et al.*[57]
	13.1%	Mao *et al.*[5]
Hyposmia	5.1%	Mao *et al.*[5]
Impaired consciousness	7.5%	Mao *et al.*[5]
Ischemic stroke	2.8%	Mao *et al.*[5]
Meningitis/encephalitis	Case reports	Zuhorn *et al.*[44] Monti *et al.*[45] Fadakar *et al.*[46] Ayuso *et al.*[47]
Nerve pain	2.3%	Mao *et al.*[5]
Seizure	0.5%	Mao *et al.*[5]
Smell impairment	5.1%	Mao *et al.*[5]
Taste impairment	5.6%	Mao *et al.*[5]
Vision impairment	1.4%	Mao *et al.*[5]
Vomiting	4.0%	Zhou *et al.*[39]

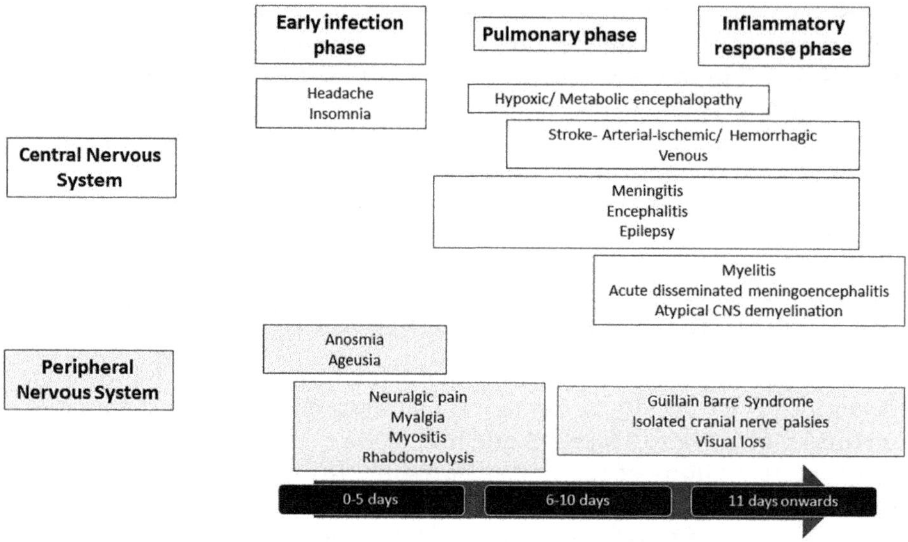

Figure 10.1 Timeline of neurological manifestations in COVID-19.

It is also noted that different types of neurological presentations are observed at different stages of the COVID-19 infection. Thus, while some features occur in the early infection phase and are relatively transient in nature with localized infection, more serious neurological symptoms may manifest in the pulmonary phase and the latter inflammatory response phase when the hypoxemia, thrombogenic state and cytokine storm kick in. These latter manifestations are associated with greater morbidity than isolated pulmonary involvement (Figure 10.1).

10.4 CNS Involvement in COVID-19

10.4.1 Vascular Manifestations

Stroke, whether arterial or venous, ischemic or hemorrhagic, have been observed more frequently with COVID-19 infection and contribute to significant morbidity. Ghannam *et al.* reported that 48.8% of neurological involvement in COVID-19 patients comprised cerebrovascular incidents, which consisted of 87.5% ischemic stroke, 5% cerebral venous thrombosis, 5% intraparenchymal hemorrhage and 2.5% subarachnoid hemorrhage.[29]

10.4.1.1 Ischemic Stroke

Out of all stroke types associated with COVID-19, ischemic stroke is probably the commonest. Acute and subacute cerebral infarction is reported in 2–6% of patients with COVID-19 infection and accounts for a quarter to a half of cases presenting with neurological symptoms.[7,30] In a study of 221 patients, Li *et al.* found that 11 had an acute ischemic stroke.[7] Beyrouti *et al.* noted elevated D-dimer levels ($\geq$1000 $\mu g\ L^{-1}$) in six severely affected patients with large cerebral infarcts, indicating a coagulopathy.[31] Similarly, ischemic strokes with large artery occlusions in multiple vascular territories have been noted in COVID-19.[32] The majority of ischemic stroke subtypes were large vessel occlusion (LVO), which consisted of 77% within a total pooled sample of 35 ischemic stroke cases.[32] This suggests that LVO in COVID-19 may occur either due to distal embolism from proximal vessels like aorta or subclavian arteries, or from a cardiogenic source. By contrast, Yaghi *et al.* reported that LVO consisted of only 6.2% out of a total of 32 ischemic stroke cases, and the majority of stroke subtypes were cryptogenic.[33]

The hypercoagulable state, vasculitis and cardiomyopathy appear to be the main mechanisms responsible for the occurrence of ischemic strokes in COVID-19.[20,31] Laboratory evidence of coagulopathy is noted in 20–55% of patients hospitalized with COVID-19, with raised D-dimer, mild thrombocytopenia and elevation of pro-thrombin time, activated partial thromboplastin time, fibrinogen and hs-CRP. In late-stage disease, fibrinogen levels may become depressed.[21,33] Antiphospholipid antibodies (anticardiolipin and anti-β-glycoprotein I antibodies) have also been documented in COVID-19 patients with multiple hemispheric infarcts.[34] The lupus anticoagulant was reported to be present in 45% of patients with COVID-19 *versus* only 10% with anticardiolipin antibody in a study on 50 patients in France.[35] The significance of these findings is unclear, as antiphospholipid antibodies were present in subjects with multiple other features of hypercoagulability, and the isolated finding of lupus anticoagulant was not associated with a high risk of thromboembolism. COVID-19 may stimulate the production of antiphospholipid antibody syndrome (APLA), although post-infectious APLA is transient and not usually associated with thrombosis. The pathogenic role of these antibodies in patients with COVID-19 is still uncertain.[34] As stated previously, endotheliitis by a direct local effect of SARS-CoV-2 on the vascular endothelial ACE2 receptors, as well as cytokine storm, additionally contribute to heightened thrombogenicity.[22] Direct invasion of virus in the myocardium may lead to injury and even death of cardiomyocytes (due to down-regulation of ACE2), with the simultaneous indirect effect of a systemic inflammatory response.[31,36] Furthermore, we have observed the presence of mobile thrombi in the aorta, subclavian, carotid or vertebral arteries, which are responsible for LVO. Similar observations have been observed by other authors with stroke as well as other embolic phenomena.[37,38]

Case Vignette 10.1

A 33-year-old male had fever and dry cough for 3 days, for which his RT-PCR tested positive for COVID-19 and he was advised to isolate at home. 10 days after onset, he developed weakness in the right side of the body with difficulty in speech. He was brought to hospital 2 days after the onset for altered sensorium. He did not have any known risk factors for stroke. CT head showed a large infarct in the left middle cerebral artery territory and CT angiography showed occlusion in left internal carotid artery with floating thrombus at the carotid bulb. The patient was treated with anticoagulants and had partial clinical improvement with change in modified Rankin scale score from 5 to 3 on follow-up.

10.4.1.2 Intracerebral Hemorrhage

Compared to ischemic stroke, intracerebral hemorrhage (ICH) is also observed with increased frequency in COVID-19. The incidence of ICH in COVID-19 patients ranges from 0.3% to 1.2%.[7] Most have been reported in active infection with severe pulmonary involvement in the presence of comorbidities. The reasons may be multifactorial. Pre-existing hypertension, ACE2 dysregulation, arteriopathy, coagulopathies and anticoagulant therapies are the probable cause of ICH. On one hand, direct effects of the SARS-CoV-2 virus on the ACE2 receptors and wide fluctuations in blood pressure may contribute. The indirect contribution of anticoagulation for treatment of the pro-coagulant state due to severe COVID-19 needs to be acknowledged. A total of 23 studies described ICH in 148 COVID-19 patients.[5,7] The majority of these patients with ICH were older males (65.8%), aged 31–78 years with comorbidities, the most common being systemic hypertension (54%). Hemorrhage involving multiple cranial compartments (MCH) was reported in 14 cases (9.5%), while single compartments were involved in the rest, with intraparenchymal hemorrhage (IPH) being the most common variety (62.6%), followed by subarachnoid hemorrhage (15.0%), subdural hemorrhage (11.6%) and intraventricular hemorrhage (1.4%). The most common location of ICH was the cerebral lobes (93.5%) followed by basal ganglia (5.4%) and cerebellum (1.1%). Half of these patients were on some form of anticoagulation. Importantly, the mortality rate in the COVID-19 patients with ICH was high, at approximately 48.6%.[5,39]

10.4.2 Seizures

Seizures are an uncommon manifestation in COVID-19. Mao and coworkers noted that among 214 patients admitted in intensive care units, seizures were recorded in only one patient.[5] Both direct and indirect mechanisms are implicated for seizures in COVID-19. Meningoencephalitis can be directly responsible for seizures. Alternatively, upregulation of components of the renin–angiotensin system (RAS) can occur in the hippocampus of patients with temporal lobe epilepsy.[40] The loss of ACE2 receptors leads to dysfunction of RAS and elevated production of angiotensin II, which results in a cascade of interactions that eventually leads to severe acute lung injury, vasoconstriction and oxidative processes that promote brain degeneration and possibly result in seizures. The overproduction of IL-6 further leads to neuroinflammation and seems to impact the inflammatory signaling on seizure thresholds.

No specific seizure type has been observed in patients with COVID-19. There may be either an exacerbation of a pre-existing seizure type or *de novo* seizures which may be of focal onset. A few authors had described new-onset seizures and either convulsive and nonconvulsive status epilepticus during the infection, which responded well to parenteral levetiracetam.[41]

Case Vignette 10.2

A 75-year-old female presented to the emergency room with complaints of recurrent episodes of jerking of the right upper limb with twitching of the right side of the face for 4 days. She very occasionally had seizures in the previous 2 years and was on irregular medications. She did not have any other febrile illness, headache, vomiting or focal weakness. On examination, the patient was in altered sensorium with right faciobrachial jerks occurring at regular intervals. No meningeal signs were noted. CT head was normal. Nasopharyngeal swab was positive for COVID PCR. The patient was treated with increasing doses of antiepileptic drugs, acyclovir and steroids and was finally controlled on three antiepileptic drugs.

10.4.3 Encephalitis

Encephalitis can occur in COVID-19 as a result of infection, parainfectious or post-infectious autoimmune processes. Very few cases of encephalitis due to SARS-CoV-2 infection have been reported to date.[42] The diagnosis of COVID-19-related encephalitis can be extremely challenging, as the definitive diagnosis of viral encephalitis largely depends on virus isolation from the CSF. This is difficult because SARS-CoV-2 dissemination is transient and its CSF titer may be extremely low. Domingues *et al.* claimed to isolate the first SARS-CoV-2 virus in cerebrospinal fluid (CSF).[42] Out of four cases with features consistent with encephalitis, only in two patients was the SARS-CoV-2 virus isolated in the CSF, and in one the virus was not identified in the throat swab. Neuroimaging is usually normal in patients with SARS-CoV-2 encephalitis. In one report of COVID-19-associated encephalitis, magnetic resonance imaging (MRI) revealed involvement of the limbic system. Electroencephalography has largely shown nonspecific findings.[43] Zuhorn *et al.* described parainfectious encephalitis with claustrum sign in a COVID-19-positive case.[44] *N*-Methyl-D-aspartate receptor encephalitis in a COVID-19 patient who presented with new-onset refractory status epilepticus has also been reported.[45] Rarely, cerebellitis can also occur in COVID-19.[46] Ayuso *et al.* reported a case of anti-GD1a antibody-positive Bickerstaff encephalitis in an elderly COVID-19 patient 8 days after constitutional symptoms.[47] The majority of patients with encephalitis have a good prognosis when treated early, and recover completely.

Case Vignette 10.3

A 12-year-old male presented with fever, headache and altered mental status for 15 days. Headache was new-onset, throbbing type, continuous, bifrontal and later becoming holocranial with associated projectile vomiting. On examination, the child was drowsy but arousable on repetitive verbal or painful stimuli. No cranial nerve or motor or sensory deficits were noted. On tandem walking, he had a tendency to fall towards the right side. No signs of meningeal irritation were observed. CT head revealed hydrocephalus with ill-defined area of hypodensity in right cerebellar hemisphere and downward displacement of cerebellar tonsil. COVID PCR was positive. In view of tonsillar herniation, bradycardia and increasing drowsiness, emergent external ventricular drainage was done. CSF was normal and negative for aerobic culture, fungal culture, india ink and neurotropic viruses. CSF was also negative for COVID RT-PCR. MRI brain revealed abnormal T2/FLAIR hyperintensity, diffusion restriction on diffusion-weighted imaging, post-contrast enhancement of bilateral cerebellar hemisphere (right > left) with mass effect (Figure 10.5). Steroid and acyclovir were given for 14 days with significant improvement.

10.4.4 Encephalopathy

Altered sensorium in severe COVID-19 ranges from confusion, delirium and stupor to coma. Delirium is often associated with systemic inflammation and prolonged hypoxia. Middle-aged and older adults with severe disease are more likely to be affected. Metabolic alterations may very rarely manifest with posterior reversible encephalopathy syndrome. Generally, in COVID-19, the presence of altered sensorium is associated with an increased risk of death.[42]

10.4.5 CNS Demyelination

Unlike other neurological manifestations noted through the COVID-19 pandemic, demyelinating illness of the CNS such as ADEM, acute necrotizing encephalopathy (ANE), myelitis or nonspecific white matter lesions have been reported sporadically.[9–11,26,48,49] In COVID-19, the demyelinating lesions have been observed to be multiple, small and scattered T2 hyperintense white-matter lesions with variable diffusion restriction and enhancement.[10,11,26] Occasional corpus callosal involvement has been noted.[26] Cases of lesional hemorrhage have been reported exceptionally.[7,9] We have encountered a patient presenting with a large expansive tumefactive demyelinating lesion without enhancement or hemorrhage in the COVID-19 setting.[50] Most of these reports revealed that the CNS demyelinating events were delayed, in or after the second week, usually after severe COVID-19 infection necessitating intubation.[10,11,26]

> **Case Vignette 10.4**
>
> A 35-year-old male presented with moderate to severe new-onset hemicranial headache without nausea or vomiting and simultaneous progressive left hemiparesis. The following day he had a left focal seizure and lapsed into altered sensorium. No antecedent fever, cough, sore throat, anosmia, behavioral change or visual symptoms were observed. No prior history of cranial neuropathy, speech, limb or bowel or bladder involvement was elicitable. Examination revealed the patient to be in deep altered sensorium with GCS-E2V2M4, left hemiparesis (Medical Research Council score 2/5) without pupillary asymmetry or papilledema. IL-6 was 3.33 (<4.4 pg mL^{-1}), ferritin 90.9 (22–322 ng mL^{-1}) and serum pro-calcitonin 0.14 (<0.02 ng mL^{-1}). COVID-19 PCR was positive. Noncontrast computed tomography of head revealed ill-defined hypodensities in the right fronto-insular region with involvement of right basal ganglia. MRI brain showed a large ill-defined tumefactive demyelinating lesion in the right fronto-parieto-insular region. After inadequate response to intravenous steroids, the patient underwent plasma exchange and subsequently improved completely.

10.4.6 Myelitis

Spinal cord involvement is a fairly uncommon manifestation of COVID-19. Chow *et al.* described acute myelitis in a 60-year-old patient.[51] The patient developed acute flaccid paraplegia with spinal sensory level at D10 and urinary incontinence. The patient was treated with intravenous corticosteroids and responded well. The cytokine storm and exaggerated inflammatory changes resulted in acute transverse myelitis. Another case of acute transverse myelitis in COVID-19 was reported by Chakraborty *et al.*[52] In another case, a young man with bilateral severe optic neuritis and myelitis, determined to be simultaneously SARS-CoV-2 and myelin oligodendrocyte glycoprotein antibody positive, was reported.[53]

10.4.7 Movement Disorders

There is a relative paucity of reports of patients presenting with *de novo* movement disorders with COVID-19. In a study of 601 patients, there were none.[54] Few scattered reports of hypokinetic rigid syndrome, tremor and myoclonus have been observed in the setting of encephalitis.[55,56] The presence of ACE2 receptors in the dopaminergic neurons as well as the microglia and astrocytes and the probable increased mRNA expression of IL-6 and other inflammatory markers may contribute. Additionally, the expression of the gene for ACE2 is co-regulated by the gene coding for dopa decarboxylase.

10.5 PNS Involvement in COVID-19

10.5.1 Loss of Smell and Taste Sensation

Anosmia (loss of sense of smell) and ageusia (loss of sense of taste) are the most frequent neurological manifestations of COVID-19 and appear early in the course of the illness with or without respiratory symptoms.[57] A study by Lechien *et al.* reported that out of 417 mild to moderate COVID-19 patients, 86% and 88% reported anosmia and ageusia, respectively.[58] This olfactory abnormality persisted after the resolution of other symptoms in 63.0% of patients. Though limited in number, studies from the Asia-Pacific region have demonstrated a lesser incidence of anosmia and ageusia,[25] but a direct comparison cannot be made across these studies due to absence of a uniform testing method and the development of a standardized questionnaire and testing method that would allow better inference among studies across different world regions. Sense of smell is more commonly involved than sense of taste, and both have been shown to occur either before, in unison with, or after the appearance of general symptoms. As far as the mechanism is concerned, there is still uncertainty regarding contribution of direct sensory neuronal damage *versus* inflammation due to the virus. Since it is an early manifestation, it is proposed that the utilization of ACE2 receptors in the olfactory epithelium by SARS-CoV-2 virus and its entry into the cells lead to neuronal damage.[4]

Case Vignette 10.5

A 48-year-old female presented with fever for a day followed by change of voice, sore throat and bifrontal headache. She tested positive for COVID RT-PCR. On day 4, she noticed a sudden inability to smell and following day she had complete loss of taste. Most of the previous symptoms subsided by day 8. The sense of taste to sour and bitter items was regained by day 10 with gradual recovery in other modalities. The sense of smell recovered last, over the following month.

10.5.2 Other Cranial Nerves

Isolated cranial neuropathies are quite rare in COVID-19. The few reports suggest the presence of an inflammatory reaction in the vessels supplying the nerve trunks. Wei *et al.* described a case of acute unilateral isolated oculomotor nerve palsy in a severely ill patient.[59] Another series of Bell's palsy was reported in COVID-19 patients from Brazil.[60] Blurring of unilateral vision or visual loss without field defects have been occasionally reported.[61] Most isolated cranial neuropathies respond well to steroids.

10.5.3 Guillain-Barré Syndrome and its Variants

Guillain-Barré syndrome is a frequently encountered neurological complication of COVID-19. Since the description of the first case of Guillain-Barré syndrome by Zhao *et al.* from China, nearly 20 more cases have since been described.[62,63] Almost all cases have shown the typical pattern of rapidly progressive, areflexic symmetrical limb weakness with sensory symptoms and two patients showed the Miller Fisher variant of the illness.[63] All these patients were positive for SARS-CoV-2 by nasal swab and chest computed tomography (CT) showed ground-glass opacities. The possible pathogenesis is a post-infectious autoimmune process triggering the complication.[28,63] Fortunately, most cases improved after being treated with intravenous immunoglobulin.

10.5.4 Neuromuscular Junction Disorders

Myasthenia gravis (MG) is an autoimmune disorder of neuromuscular transmission in which antibodies against the AChR and other similar proteins lead to the clinical spectrum of fluctuating muscle weakness involving ocular, bulbar and/or limb muscles (generalized MG). COVID-19 infection in MG can be challenging for multiple reasons: infections are known to trigger MG exacerbations/crises, MG patients may be at increased risk of such infections due to immunosuppressive medications and respiratory distress can be seen in both conditions, which can complicate identification and management. There are isolated reports of MG and COVID-19 infection and most have focused on increased severity of COVID-19 in MG patients.[64]

Case Vignette 10.6

A 40-year-male, previously diagnosed as AChR antibody positive MG since 2013, presented with increased weakness with worsening ptosis, ophthalmoplegia, bifacial weakness and dysphagia and dysarthria. He was on pyridostigmine 60 mg four times daily and had been stable on this medication. At admission, he tested positive for SARS-CoV-2, but was not found to have COVID-19-related respiratory issues. Hence, considering the present condition to be due to exacerbation of MG, he was treated with steroids and immunomodulation. Slight improvement in his symptoms was noted at 3 weeks. At follow-up after 1 month, the patient had significant improvement in Myasthenia Gravis Foundation of America grade from 4b to 2b.

10.5.5 Myalgia and Myositis

Myalgia is a relatively common presentation accompanying acute COVID-19 disease. Mao *et al.* reported that out of 214 COVID-19 patients, skeletal muscle injury occurred in 23 (10.7%).[5] However, as per later reports, it may be seen in approximately one-third of cases.[65,66] It is usually associated with febrile illness and upper respiratory symptoms and may involve isolated body parts such as the back, nape of neck, limbs or even buttocks. Creatine kinase, D-dimer, hs-CRP and lactate dehydrogenase levels were found to be elevated in patients with skeletal muscle injury. In addition to myalgia, myopathy with proximal limb weakness has been noted. It is not clear whether this was due to the direct effect of virus on muscle tissue or infection-mediated immune response causing elevated pro-inflammatory cytokines in serum resulting in skeletal muscle damage. Since no electrodiagnostic studies were done, it is difficult to distinguish whether this was due to critical illness myopathy and neuropathy in addition to skeletal muscle damage.[25,63] Rare forms of myositis proven by muscle biopsy in patients with COVID-19 have been reported.[67] A patient with COVID-19 infection and myopathy with biopsy evidence of virus-induced type I interferonopathy was described. Myxovirus resistance protein A is a type I interferon-inducible protein expressed in response to viral infection, including to SARS-CoV-2. Immunohistochemical analysis revealed abnormal expression of major histocompatibility complex class I antigen on the sarcolemma and sarcoplasm and abnormal presence of myxovirus resistance protein A on muscle fibers and capillaries without any virus in the muscle. Another important, though rare, phenomenon is rhabdomyolysis, which can be a heralding manifestation of COVID-19. A high index of suspicion for rhabdomyolysis should be kept in mind for any COVID-19 patient presenting with localized muscle pain or generalized weakness, because if not recognized and treated promptly, acute renal failure may set in.[25,28]

10.6 Neuroradiological Findings in COVID-19

Neuroimaging is usually required in 14–15% of the patients admitted with COVID-19 infection.[68] Among the imaged patients, ischemic and hemorrhagic stroke represent the most common neuroimaging findings.

10.6.1 Acute Ischemic Stroke

Stroke associated with COVID-19 infection tends to be severe, with higher National Institutes of Health Stroke Scale score, propensity for large vessel occlusion, multiterritorial involvement and a poor outcome (Figure 10.2).[33]

In addition, embolic stroke and small-vessel ischemic lesions are frequently encountered. COVID-19 infection is also associated with fulminant carotid and other cervical arterial thrombosis with mobile thrombus overlying mild atherosclerotic plaques leading to acute ischemic stroke (Figure 10.3).[69]

10.6.2 Acute Hemorrhagic Stroke

Hemorrhagic stroke has been reported with an incidence of 0.5–0.7% in patients with COVID-19 infection, less frequently than ischemic stroke. Lobar hemorrhage, subarachnoid hemorrhage, intraventricular hemorrhage and cerebral microhemorrhage are all described. A few cases of nonaneurysmal subarachnoid hemorrhage have also been reported.[70] The therapeutic or prophylactic anticoagulation regimen used in COVID-19 patients is also linked with an increased propensity for hemorrhagic stroke (Figure 10.4).[71]

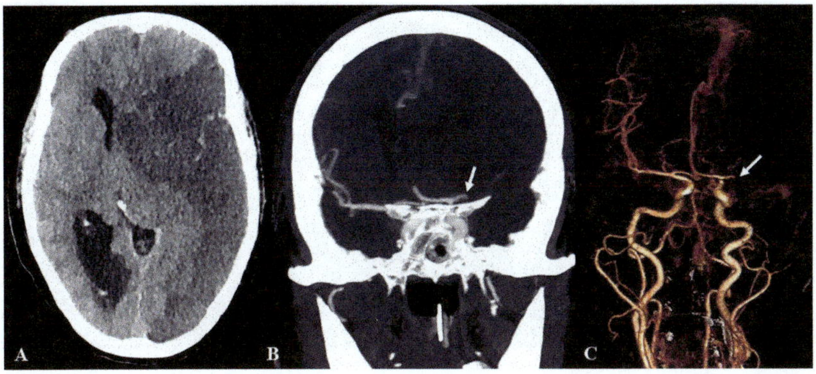

Figure 10.2 A 71-year-old COVID-19-positive patient presenting with right-sided hemiplegia of 48 hours duration. The axial noncontrast CT images (A) shows complete left middle cerebral artery (MCA) territory infarct with significant mass effect and midline shift. The maximum intensity projection image of CT angiogram (B) and volume rendered technique image (C) shows occlusion of the proximal M1 segment of left MCA (white arrow).

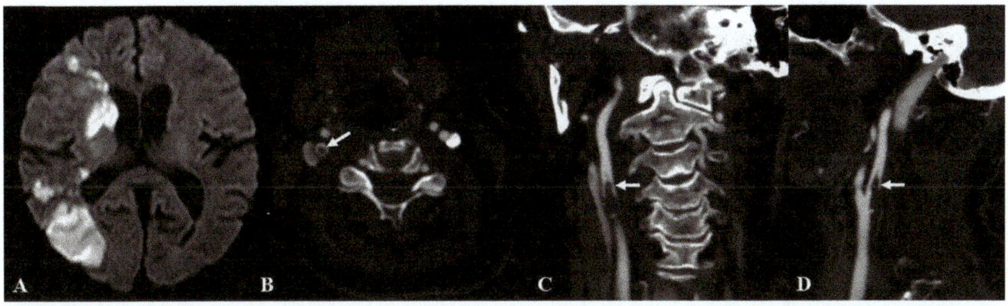

Figure 10.3 A 55-year-old COVID-19-positive patient with acute-onset left hemiparesis. The axial diffusion weighted image (A) shows patchy diffusion restricting acute infarct in the right middle cerebral artery territory. The axial (B), coronal (C) and sagittal (D) images of CT angiogram of neck vessels reveals a free-floating thrombus with plaque at the right carotid bulb extending into the proximal internal carotid artery (white arrow). These free-floating thrombi are frequently associated with COVID-19 infection.

Extensive white-matter microhemorrhages with predominant involvement of the juxtacortical and callosal white matter can be observed in critically ill patients with COVID-19. These microhemorrhages are thought to be related to chronic hypoxemia and disruption of the blood–brain barrier. Alternatively, these might be secondary to virus-induced thrombotic microangiopathy.[72]

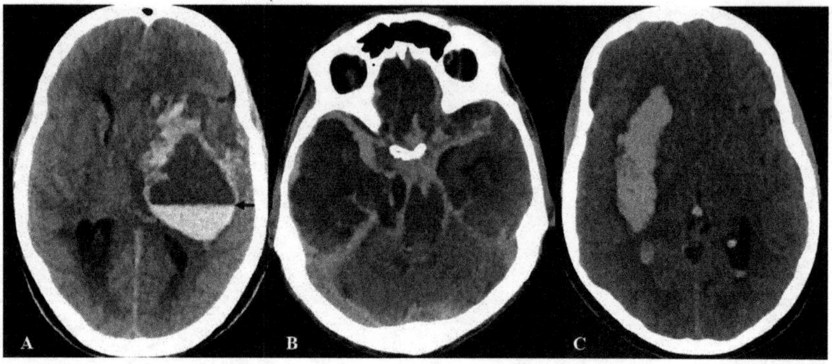

Figure 10.4 Hemorrhagic manifestations in COVID-19 infection in three different patients. Non-contrast CT head (A) shows anticoagulant-associated intraparenchymal hemorrhage at right ganglio-thalamic region with blood-fluid level (black arrow). Image (B) shows diffuse subarachnoid hemorrhage in the basal cisterns. No aneurysm was demonstrated in the CT angiogram (not shown). Image (C) shows large right ganglio-capsular hemorrhage with intraventricular extension and mass effect.

10.6.3 Acute Inflammatory/Infectious CNS Syndromes

Autoimmune encephalitis, ANE, myelitis and acute viral cerebellitis (Figure 10.5) are the range of acute inflammatory conditions associated with acute COVID-19 infections.[73,74]

ADEM-like presentation with scattered white matter lesions in a perivenular distribution and associated microhemorrhage has been reported in a few cases. In the case with hemorrhagic ANE, hemorrhagic lesions in the thalamus were noted.[9] Furthermore, cases with cortical–subcortical hemorrhagic lesions with restricted diffusion are described, causing what appears to be a posterior reversible encephalopathy syndrome.[75] Cytotoxic lesions of the corpus callosum are described in these patients, and are supposed to be related to the cytokine storm.[76]

10.6.4 Guillain-Barré Syndrome

Guillain-Barré syndrome is a significant parainfectious peripheral neurological manifestation of the COVID-19 infection. Neuroimaging studies are characterized by post-contrast enhancement of the cranial nerves and the cauda equina nerve roots. Leptomeningeal enhancement along the spinal cord and myelopathy are also reported features (Figure 10.6).[73]

10.6.5 Olfactory Bulb Involvement

Anosmia, a prominent early clinical manifestation of COVID-19 infection, correlates on MRI as T2 hyperintense edema of the olfactory bulbs and the gyrus rectus. In patients with prolonged anosmia, olfactory bulb atrophy is reported.[77]

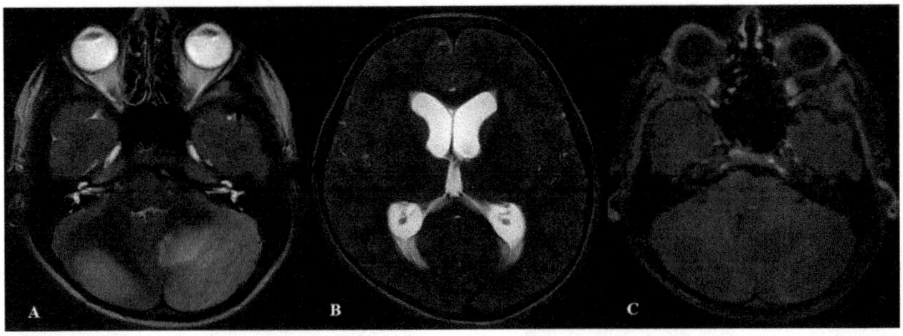

Figure 10.5 The axial T2 (A) image shows abnormal asymmetric hyperintensity involving bilateral cerebellar hemisphere (left > right) causing edematous change and effacement of the fourth ventricle. Note the dilatation of the supratentorial ventricular system (B). The involved cerebellar hemispheres show patchy folial enhancement on the post-contrast scan (C). The imaging features are suggestive of cerebellitis.

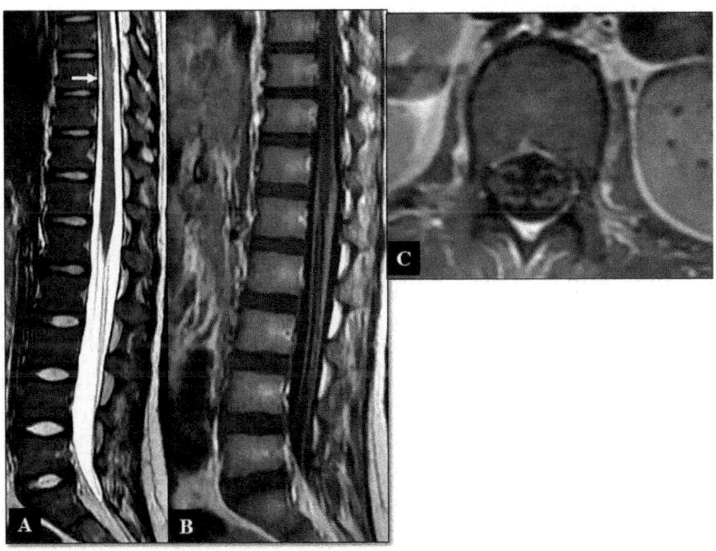

Figure 10.6 An 8-year-old female presenting with progressive paraparesis. Antibodies were positive for COVID-19 infection. The sagittal T2 image(A) shows abnormal hyperintense signal of the lower dorsal cord extending from D7 to D10 vertebra levels (white arrow) without any cord expansion. The post-contrast T1FS sagittal (B) and axial (C) images of the dorso-lumbar spine shows enhancement of the cauda equina nerve roots without enhancement of the lower dorsal cord. The imaging features are consistent with Guillain-Barré syndrome.

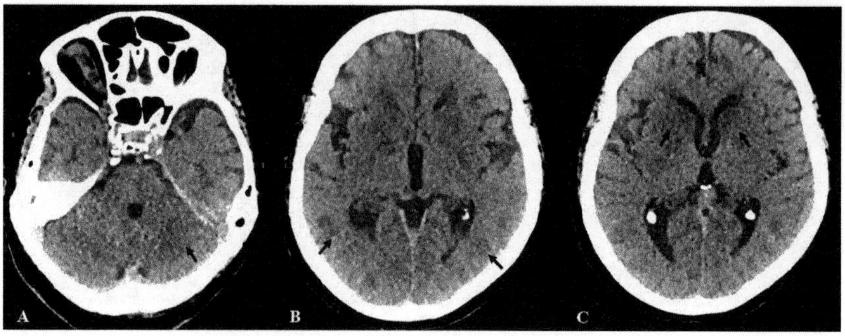

Figure 10.7 Critically ill COVID-19-positive patient presenting with diffuse hypoxic ischemic changes. The axial noncontrast CT scan images (A, B and C) show symmetrical hypodensities at bilateral cerebellar white matter, the subcortical white matter of the bilateral parieto-occipital cortex and bilateral basal ganglia (black arrows).

10.6.6 Global Hypoxia and Diffuse Post-hypoxic Leukoencephalopathy

Global hypoxic injury is a manifestation of severe COVID-19 infection and manifests on imaging as restricted diffusion of the basal ganglia, thalamus, hippocampi and the cortex (Figure 10.7).

Globus pallidus necrosis has also been reported, possibly secondary to severe hypoxia. Delayed post-hypoxic leukoencephalopathy manifests as symmetrical white matter T2 hyperintensity with restricted diffusion and relative sparing of the subcortical and infratentorial white matter.[78]

10.7 Treatment of Neurological Aspects of COVID-19

Along with standard management of COVID-19 as per the severity of involvement, neurological complications may need separate treatment depending on the type of involvement. Minor symptoms such as headache, dizziness and myalgia can be managed symptomatically. Meningitis and meningoencephalitis need appropriate evaluation to rule out other possibilities and empirical treatment with antibiotics and antiviral agents instituted until etiology is established. Antiepileptic medications should be given for seizures according to institutional protocol. Patients presenting as acute stroke also can be managed as per international/regional acute stroke guidelines. However, one should be cautious as coagulopathy and other systemic involvement like uremia and ARDS may be coexisting and can alter the clinical outcome. Since SARS-CoV-2 infection is linked to an increased incidence of thrombotic disease, ischemic strokes may be evaluated for the presence of thrombi and treated with anticoagulation. This has led to the modification of treatment strategies for patients with moderate and severe disease across centers worldwide. For complications arising out of aberrant immune responses such as Guillain-Barré syndrome, ANE, ADEM or other demyelination syndromes, intravenous steroids and immunoglobulins are the mainstay. For isolated cranial neuropathies, observation and follow-up alone may be all that is needed, as they tend to resolve spontaneously. Steroids or intravenous immunoglobulin can be given when appropriate on an individualized basis. Patients with myositis or myopathy respond well to steroids and recover well.

10.8 Outcomes

The general outcome of patients with stroke in COVID is poor, with mortality rates of up to 38% which specifically relate to the severity of COVID infection.[7] Risk factors for mortality include advanced age, presence of cardiovascular risk factors, ARDS, sepsis, high pro-inflammatory markers such as D-dimer, ferritin, hs-CRP, lymphopenia and especially multiorgan failure.[39] Mortality may be higher in hemorrhagic compared to ischemic stroke. However, LVO may contribute to large territorial infarcts and lead to rapid increase in brain edema. Strokes have also contributed to significant morbidity during the COVID-19 pandemic. However, the authors' experience with anticoagulation in the subset of ischemic stroke with mobile thrombi has been very encouraging with rapid improvement in physical deficits in a short period of time.[79]

The outcome of most of the other neurological manifestations are either transient, such as anosmia or the cranial neuropathies, or respond well to immunomodulation and specific management without significant sequelae. Therefore, timely recognition of these abnormalities is important to initiate appropriate treatment paradigms.

10.9 Future Perspectives

As humanity is trying to emerge knowledgeably through the COVID-19 pandemic, there seems more that eludes our eyes. More efforts and algorithms need to be evolved for risk stratification of stroke in COVID so as to prevent them altogether, mitigate their effect when they do occur and manage stroke patients appropriately so as to reduce their morbidity and mortality. This is a significant challenge and biomarkers for prognostication need to be sought. The effect of SARS-CoV-2 fetal or neonatal infection on neurodevelopmental outcomes need to be assessed. The possibilities of subtle autistic tendencies, learning disabilities and/or gene mutations on the incoming generations will unfold in the coming years.[80] Moreover, the neurological effects of new drugs and vaccines developed and approved for emergency use during the pandemic will be yet another story waiting to unfold.

List of Abbreviations

CNS	Central Nervous System
PNS	Peripheral Nervous System
ADEM	Acute disseminated encephalomyelitis
ACE2	Angiotensin-converting enzyme 2
TMPRSS2+	Transmembrane serine protease 2
nAChR	Nicotinic acetylcholine receptor
IL-6	Interleukin-6
hs-CRP	High-sensitivity C reactive protein
ARDS	Acute Respiratory Distress Syndrome
CSF	Cerebrospinal fluid
LVO	Large Vessel Occlusion
APLA	Antiphospholipid antibody syndrome
ICH	Intracerebral hemorrhage
RAS	Renin–angiotensin system
ANE	Acute hemorrhagic Necrotizing Encephalopathy
MG	Myasthenia Gravis

Acknowledgements

The authors would wish to acknowledge all the efforts of the doctors, nursing and healthcare staff involved in the management of patients with COVID-19 at the All India Institute of Medical Sciences, Jodhpur, and a special regard to all the patients who suffered through this infection.

References

1. World Health Organization, Coronavirus Disease 2019 (COVID-19) Situation Report, available at: https://www.who.int/emergencies/diseases/novel-coronavirus-2019/situation-reports, accessed 21 November 2020.
2. N. Zhu, D. Zhang, W. Wang, X. Li, B. Yang and J. Song, *et al.*, A novel coronavirus from patients with pneumonia in China, 2019, *N. Engl. J. Med.*, 2020, **382**, 727.
3. M. Al Samman, A. Caggiula, S. Ganguli, M. Misak and A. Pourmand, Non-respiratory presentations of COVID-19, a clinical review, *Am. J. Emerg. Med.*, 2020, **38**, 2444.
4. Z. Li, T. Liu, N. Yang, D. Han, X. Mi and Y. Li, *et al.*, Neurological manifestations of patients with COVID-19: potential routes of SARS-CoV-2 neuroinvasion from the periphery to the brain, *Front. Med.*, 2020, **14**, 533.
5. L. Mao, H. Jin, M. Wang, Y. Hu, S. Chen and Q. He, *et al.*, Neurologic manifestations of hospitalized patients with Coronavirus disease 2019 in Wuhan, China, *JAMA Neurol.*, 2020, **77**, 683.
6. F. J. Carod-Artal, Neurological complications of coronavirus and COVID-19, *Rev. Neurol.*, 2020, **70**, 311.

7. Y. Li, M. Li, M. Wang, Y. Zhou, J. Chang and Y. Xian, *et al.*, Acute cerebrovascular disease following COVID-19: a single center, retrospective, observational study, *Stroke Vasc. Neurol.*, 2020, **5**, 279.
8. A. A. Asadi-Pooya and L. Simani, Central nervous system manifestations of COVID-19: A systematic review, *J. Neurol. Sci.*, 2020, **413**, 116832.
9. N. Poyiadji, G. Shahin, D. Noujaim, M. Stone, S. Patel and B. Griffith, COVID-19-associated acute hemorrhagic necrotizing encephalopathy: CT and MRI features, *Radiology*, 2020, **296**, E119.
10. L. Zanin, G. Saraceno, P. P. Panciani, G. Renisi, L. Signorini and K. Migliorati, *et al.*, SARS-CoV-2 can induce brain and spine demyelinating lesions, *Acta Neurochir.*, 2020, **162**, 1491.
11. G. Brun, J.-F. Hak, S. Coze, E. Kaphan, J. Carvelli and N. Girard, *et al.*, COVID-19- White matter and globus pallidum lesions. Demyelination or small-vessel vasculitis? *Neurol. Neuroimmunol. Neuroinflamm.*, 2020, **7**, e777.
12. G. Nepal, J. H. Rehrig, G. S. Shrestha, Y. K. Shing, J. K. Yadav and R. Ojha, *et al.*, Neurological manifestations of COVID-19: a systematic review, *Crit. Care*, 2020, **24**, 421.
13. M. Hoffmann, H. Kleine-Weber, S. Schroeder, N. Kruger, T. Herrler and S. Erichsen, *et al.*, SARS-CoV-2 cell entry depends on ACE2 and TMPRSS2 and is blocked by a clinically proven protease inhibitor, *Cell*, 2020, **181**, 271.
14. R. Chen, K. Wang, J. Yu, Z. Chen, C. Wen and Z. Xu, The spatial and cell-type distribution of SARS-CoV-2 receptor ACE2 in human and mouse brain, *Front. Neurol.*, 2021, **11**, 573095.
15. M. Y. Li, L. Li, Y. Zhang and X. S. Wang, Expression of the SARS-CoV-2 cell receptor gene ACE2 in a wide variety of human tissues, *Infect. Dis. Poverty*, 2020, **9**, 45.
16. A. Paniz-Mondolfi, C. Bryce, Z. Grimes, R. E. Gordon, J. Reidy and J. Lednicky, *et al.*, Central nervous system involvement by severe acute respiratory syndrome coronavirus -2 (SARS-CoV-2), *J. Med. Virol.*, 2020, **92**, 699.
17. P. Guadarrama-Ortiz, J. A. Choreño-Parra, C. M. Sánchez-Martínez, F. J. Pacheco-Sánchez, A. I. Rodríguez-Nava and G. García-Quintero, Neurological Aspects of SARS-CoV-2 Infection: Mechanisms and Manifestations, *Front. Neurol.*, 2020, **11**, 1039.
18. V. G. Puelles, M. Lütgehetmann, M. T. Lindenmeyer, J. P. Sperhake, M. N. Wong and L. Allweiss, *et al.*, Multiorgan and renal tropism of SARS-CoV-2, *N. Engl. J. Med.*, 2020, **383**, 590.
19. J. L. Olds and N. Kabbani, Is nicotine exposure linked to cardiopulmonary vulnerability to COVID-19 in the general population? *FEBS J.*, 2020, **287**, 3651.
20. M. Panigada, N. Bottino, P. Tagliabue, G. Grasselli, C. Novembrino and V. Chantarangkul, *et al.*, Hypercoagulability of COVID-19 patients in intensive care unit: A report of thromboelastography findings and other parameters of hemostasis, *J. Thromb. Haemostasis*, 2020, **18**, 1738.

21. H. Han, L. Yang, R. Liu, F. Liu, K.-L. Wu and J. Li, *et al.*, Prominent changes in blood coagulation of patients with SARS-CoV-2 infection, *Clin. Chem. Lab. Med.*, 2020, **58**, 1116.
22. Z. Varga, A. J. Flammer, P. Steiger, M. Haberecker, R. Andermatt and A. S. Zinkernagel, *et al.*, Endothelial cell infection and endothelitis in COVID-19, *Lancet*, 2020, **395**, 1417.
23. J. S. Rico-Mesa, D. Rosas, A. Ahmadian-Tehrani, A. White, A. S. Anderson and R. Chilton, The role of anticoagulation in COVID-19-induced hypercoagulability, *Curr. Cardiol. Rep.*, 2020, **22**, 53.
24. C. Qin, L. Zhou, Z. Hu, S. Zhang, S. Yang and Y. Tao, *et al.*, Dysregulation of immune response in patients with COVID-19 in Wuhan, China, *Clin. Infect. Dis.*, 2020, **71**, 762.
25. A. Rahman, R. Niloofa, I. M. De Zoysa, A. D. Cooray, J. Kariyawasam and S. L. Seneviratne, Neurological manifestations in COVID-19: A narrative review, *SAGE Open Med.*, 2020, **8**, 2050312120957925.
26. A. Zoghi, M. Ramezani, M. Roozbeh, I. A. Darazan and M. A. Sahraian, A case of possible atypical demyelinating event of the central nervous system following COVID-19, *Mult. Scler. Relat. Disord.*, 2020, **44**, 102324.
27. E. Ann Yeh, A. Collins, M. E. Cohen, P. K. Duffner and H. Faden, Detection of coronavirus in the central nervous system of a child with acute disseminated encephalomyelitis, *Pediatrics*, 2004, **113**, e73.
28. R. K. Garg, Spectrum of Neurological Manifestations in Covid-19: A Review, *Neurol. India*, 2020, **68**, 560.
29. M. Ghannam, Q. Alshaer, M. Al-Chalabi, L. Zakarna, J. Robertson and G. Manousakis, Neurological involvement of coronavirus disease 2019: a systematic review, *J. Neurol.*, 2020, **267**, 3135.
30. M. F. Goldberg and M. F. Goldberg, Neuroradiologic manifestations of COVID-19: what the emergency radiologist needs to know, *Emerg. Radiol.*, 2020, **27**, 737.
31. R. Beyrouti, M. E. Adams, L. Benjamin, H. Cohen, S. F. Farmer and Y. Y. Goh, *et al.*, Characteristics of ischaemic stroke associated with COVID-19, *J. Neurol., Neurosurg. Psychiatry*, 2020, **91**, 889.
32. S. Escalard, B. Maïer, H. Redjem, F. Delvoye, S. Hebert and S. Smajda, *et al.*, Treatment of acute ischemic stroke due to large vessel occlusion with COVID-19: Experience from Paris, *Stroke*, 2020, **51**, 2540.
33. S. Yaghi, K. Ishida, J. Torres, B. M. Grory, E. Raz and K. Humbert, *et al.*, SARS2-CoV-2 and stroke in a New York healthcare system, *Stroke*, 2020, **51**, 2002.
34. Y. Zhang, M. Xiao, S. Zhang, P. Xia, W. Cao and W. Jiang, *et al.*, Coagulopathy and antiphospholipid antibodies in patients with COVID-19, *N. Engl. J. Med.*, 2020, **382**, e38.
35. I. Harzallah, A. Debliquis and B. Drenou, Lupus anticoagulant is frequent in patients with COVID-19, *J. Thromb. Haemostasis*, 2020, **18**, 2064.
36. M. Ackermann, S. E. Verleden, M. Kuehnel, A. Haverich, T. Welte and F. Laenger, *et al.*, Pulmonary vascular endothelialitis, thrombosis, and angiogenesis in COVID-19, *N. Engl. J. Med.*, 2020, **383**, 120.

37. M. de Carranza, D.-E. Salazar, J. Troya, R. Alcázar, C. Peña and E. Aragón, *et al.*, Aortic thrombus in patients with severe COVID19: review of three cases,*J. Thromb. Thrombolysis*, 2020, **51**(1), 237–242.
38. R. Silingardi, S. Gennai, M. Migliari, T. Covic and N. Leone, Acute limb ischemia in COVID-19 patients: Could aortic floating thrombus be the source of embolic complications?*J. Vasc. Surg.*, 2020, **72**, 1152.
39. F. Zhou, T. Yu, R. Du, G. Fan, Y. Liu and Z. Liu, *et al.*, Clinical course and risk factors for mortality of adult inpatients with COVID-19 in Wuhan, China: a retrospective cohort study, *Lancet*, 2020, **395**, 1054.
40. G. A. Arganaraz, A. C. Konno, S. R. Perosa, J. F. C. Santiago, M. A. Boim and D. B. Vidotti, *et al.*, The renin-angiotensin system is upregulated in the cortex and hippocampus of patients with temporal lobe epilepsy related to mesial temporal sclerosis, *Epilepsia*, 2008, **49**, 1348.
41. S. Somani, S. Pati, T. Gaston, A. Chitlangia and S. Agnihotri, De novo status epilepticus in patients with COVID-19, *Ann. Clin. Transl. Neurol.*, 2020, **7**, 1240.
42. R. B. Domingues, M. C. Mendes-Correa, F. B. V. de Moura Leite, E. C. Sabino, D. Z. Salarini and I. Claro, *et al.*, First case of SARS-COV-2 sequencing in cerebrospinal fluid of a patient with suspected demyelinating disease,*J. Neurol.*, 2020, **267**(11), 3154–3156.
43. P. Gélisse, A. O. Rossetti, P. Genton, A. Crespel and P. W. Kaplan, How to carry out and interpret EEG recordings in COVID-19 patients in ICU? *Clin. Neurophysiol.*, 2020, **131**, 2023.
44. F. Zuhorn, H. Omaimen, B. Ruprecht, C. Stellbrink, M. Rauch and A. Rogalewski, *et al.*, Parainfectious encephalitis in COVID-19: "The Claustrum Sign",*J. Neurol.*, 2021, **268**(6), 2031–2034.
45. G. Monti, G. Giovannini, A. Marudi, R. Bedin, A. Melegari and A. M. Simone, Anti-NMDA receptor encephalitis presenting as new onset refractory status epilepticus in COVID-19, *Seizure*, 2020, **81**, 18.
46. N. Fadakar, S. Ghaemmaghami, S. M. Masoompour, B. S. Yeganeh, A. Akbari and S. Hooshmandi, *et al.*, A first case of acute cerebellitis associated with coronavirus disease (COVID-19): a case report and literature review, *Cerebellum*, 2020, **19**, 911.
47. L. L. Ayuso, P. T. Rubio, B. R. F. do Rosário, M. L. G. Arroyo and F. Sierra-Hidalgo, Bickerstaff encephalitis after COVID-19, *J. Neurol.*, 2021, **268**(6), 2035–2037.
48. T. Parsons, S. Banks, C. Bae, J. Gelber, H. Alahmadi and M. Tichauer, COVID19associated acute disseminated encephalomyelitis (ADEM), *J. Neurol.*, 2020, **267**, 2799.
49. S. Abdia, A. Ghorbanib and F. Fatehic, The association of SARS-CoV-2 infection and acute disseminated encephalomyelitis without prominent clinical pulmonary symptoms,*J. Neurol. Sci.*, 2020, **416**, 117001.
50. S. Panda, P. Patel, S. Jain, S. Sharma, M. Vegda and A. Patel, *et al.*, Solitary large denovo tumefactive demyelination in COVID-19- is there a causal association?, *Neurology India*, in press.
51. C. C. N. Chow, J. Magnussen, J. Ip and Y. Su, Acute transverse myelitis in COVID-19 infection, *B. M. J. Case. Rep.*, 2020, **13**(8), e236720.

52. U. Chakraborty, A. Chandra, A. K. Ray and P. Biswas, COVID-19-associated acute transverse myelitis: a rare entity, *BMJ Case Rep.*, 2020, **13**, e238668.
53. S. Zhou, E. C. Jones-Lopez, D. J. Soneji, C. J. Azevedo and V. R. Patel, Myelin Oligodendrocyte glycoprotein antibody-associated optic neuritis and myelitis in COVID-19, *J. Neuroophthalmol.*, 2020, **40**, 398.
54. H. L. Geyer, D. M. Kaufman, R. K. Parihar and M. F. Mehler, Movement Disorders in COVID-19: Whither Art Thou? *Tremor Other Hyperkinet. Mov.*, 2020, **10**, 25.
55. A. Méndez-Guerrero, M. I. Laespada-García, A. Gómez-Grande, M. Ruiz-Ortiz, V. A. Blanco-Palmero and F. J. Azcarate-Diaz, *et al.*, Acute hypokinetic-rigid syndrome following SARS-CoV-2 infection, *Neurology*, 2020, **95**, e2109.
56. P. Rábano-Suárez, L. Bermejo-Guerrero, A. Méndez-Guerrero, J. Parra-Serrano, D. Toledo-Alfocea and D. Sánchez-Tejerina, *et al.*, Generalized myoclonus in COVID-19, *Neurology*, 2020, **95**, e767.
57. A. Giacomelli, L. Pezzati, F. Conti, D. Bernacchia, M. Siano and L. Oreni, *et al.*, Self-reported olfactory and taste disorders in SARS-CoV-2 patients: a cross-sectional study, *Clin. Infect. Dis.*, 2020, **71**, 889.
58. J. R. Lechien, C. M. ChiesaEstomba, D. R. De Siati, M. Horoi, S. D. Le Bon and A. Rodriguez, *et al.*, Olfactory and gustatory dysfunctions as a clinical presentation of mildtomoderate forms of the coronavirus disease (COVID19): A multicenter European study, *Eur. Arch. Otorhinolaryngol.*, 2020, **277**, 2251.
59. H. Wei, H. Yin, M. Huang and Z. Guo, The 2019 novel coronavirus pneumonia with onset of oculomotor nerve palsy: A case study, *J. Neurol.*, 2020, **267**, 1550.
60. M. A. Lima, M. T. T. Silva, C. N. Soares, R. Coutinho, H. S. Oliveira and L. Afonso, *et al.*, Peripheral facial nerve palsy associated with COVID-19, *J. Neurovirol.*, 2020, **26**, 941.
61. A. K. Tisdale and B. K. Chwalisz, Neuro-ophthalmic manifestations of coronavirus disease 19, *Curr. Opin. Ophthalmol.*, 2020, **31**(6), 489–494.
62. H. Zhao, D. Shen, H. Zhou, J. Liu and S. Chen, GuillainBarré syndrome associated with SARSCoV2 infection: Causality or coincidence? *Lancet Neurol.*, 2020, **19**, 383.
63. M. A. Ellul, L. Benjamin, B. Singh, S. Lant, B. D. Michael and A. Easton, *et al.*, Neurological associations of COVID-19, *Lancet Neurol.*, 2020, **19**, 767.
64. E. Aksoy and T. Oztutgan, COVID-19 Presentation in Association with Myasthenia Gravis: A Case Report and Review of the Literature, *Case Rep. Infect. Dis.*, 2020, **17**, 8845844.
65. A. C. Guidon and A. A. Amato, COVID-19 and neuromuscular disorders, *Neurology*, 2020, **94**, 959.
66. W. J. Guan, Z. Y. Ni, Y. Hu, W. H. Liang, C. Q. Ou and J. X. He, *et al.*, Clinical characteristics of coronavirus disease 2019 in China, *N. Engl. J. Med.*, 2020, **382**, 1708.

67. G. S. Manzano, J. K. Woods and A. A. Amato, COVID-19–associated myopathy caused by Type I Interferonopathy, *N. Engl. J. Med.*, 2020, **383**, 2389.
68. E. Lin, J. E. Lantos, S. B. Strauss, C. D. Phillips, T. R. Campion and B. B. Navi, *et al.*, Brain imaging of patients with COVID-19: findings at an academic institution during the height of the outbreak in New York City, *AJNR Am. J. Neuroradiol.*, 2020, **41**, 2001.
69. C. Esenwa, N. T. Cheng, E. Lipsitz, K. Hsu, R. Zampolin and A. Gersten, *et al.*, COVID-19-Associated Carotid Atherothrombosis and Stroke, *AJNR Am. J. Neuroradiol.*, 2020, **41**, 1993.
70. I. Cheruiyot, P. Sehmi, B. Ominde, P. Bundi, M. Mislani and B. Ngure, *et al.*, Intracranial hemorrhage in coronavirus disease 2019 (COVID-19) patients, *Neurol. Sci.*, 2020, 1–9.
71. S. Dogra, R. Jain, M. Cao, S. Bilaloglu, D. Zagzag and S. Hochman, *et al.*, Hemorrhagic stroke and anticoagulation in COVID-19, *J. Stroke Cerebrovasc. Dis.*, 2020, **29**, 104984.
72. A. Radmanesh, A. Derman, Y. W. Lui, E. Raz, J. P. Loh and M. Hagiwara, *et al.*, COVID-19–associated diffuse leukoencephalopathy and microhemorrhages, *Radiology*, 2020, **297**, E223.
73. S. Katal and A. Gholamrezanezhad, Neuroimaging findings in COVID-19: A narrative review, *Neurosci. Lett.*, 2021, **742**, 135529.
74. J. Finsterer, F. A. Scorza, C. A. Scorza and A. C. Fiorini, What the neuroradiologist should additionally consider in SARS-CoV-2 infection, *Emerg. Radiol.*, 2021, **28**(2), 437–438.
75. Y. Kaya, S. Kara, C. Akinci and A. S. Kocaman, Transient cortical blindness in COVID-19 pneumonia; a PRES-like syndrome: Case report, *J. Neurol. Sci.*, 2020, **413**, 116858.
76. P. Gaur, L. Dixon, B. Jones, H. Lyall and W. Jan, COVID-19-Associated Cytotoxic Lesions of the Corpus Callosum, *AJNR Am. J. Neuroradiol.*, 2020, **41**, 1905.
77. S. Klironomos, A. Tzortzakakis, A. Kits, C. Öhberg, E. Kollia and A. Ahoromazdae, *et al.*, Nervous system involvement in COVID-19: results from a retrospective consecutive neuroimaging cohort, *Radiology*, 2020, **297**, E324.
78. G. Moonis, C. G. Filippi, C. F. Kirsch, S. Mohan, E. G. Stein and J. A. Hirsch, *et al.*, The spectrum of neuroimaging findings on CT and MRI in adults with coronavirus disease (COVID-19), *Am. J. Roentgenol.*, 2021, **217**(4), 959–974.
79. S. Panda, S. Tiwari, J. Pannani, M. Vegda, A. Patel and S. Sharma, *et al.*, Large Vessel Occlusions By Free Floating Thrombi In Strokes During The COVID-19 Pandemic- A Single Center Observational Study, *Neurol. India*, in press.
80. S. Panda, Peep into the future of GEN C-19: the domino effect, *Postgrad. Med. J.*, 2021, **97**(1149), 415–416.

CHAPTER 11

COVID-19 and Coagulopathy

SHRUTI VASWANI[a], DEEPAK KUMAR[b], GOPAL KRISHANA BOHRA[b] AND ABHISHEK PUROHIT*[a]

[a]Department of Pathology and Laboratory Medicine, All India Institute of Medical Sciences, Jodhpur, Rajasthan, 342005, India; [b]Department of Medicine, All India Institute of Medical Sciences, Jodhpur, Rajasthan, 342005, India
*E-mail: purohitabhi80@gmail.com

11.1 Introduction

The 2019 novel coronavirus or, as it later came to be named, the severe acute respiratory syndrome coronavirus 2 (SARS-CoV-2), originated in Wuhan City of Hubei Province in China. In India, the first confirmed case of coronavirus disease 2019 (COVID-19) was reported on 30 January 2020, and since then the virus has spread across the country.[1] At the time of writing, more than 70 million cases of COVID-19 and more than 1.5 million deaths have been reported worldwide.[2,3]

The common presentation is with respiratory symptoms. However, symptoms related to other organ systems also have been reported.[4]

Several studies have shown many coagulation abnormalities in patients with COVID-19. Observations from clinical scenarios and autopsy-based findings have assisted in proposing various guidelines for patient management; however, many queries still remain unanswered and evidence-based clinical trials have not yet been completed in most areas.[5]

The Coronavirus Pandemic and the Future Volume 1: Virology, Epidemiology, Translational Toxicology and Therapeutics
Edited by Michael D. Waters, Alok Dhawan, Tim Marrs, Diana Anderson, Stafford Warren and Claude L. Hughes

Published by the Royal Society of Chemistry, www.rsc.org

Among many queries that need to be addressed are the mechanism(s) and pathophysiology underlying the development of a systemic coagulopathy characterized in most patients by symptoms of venous, arterial, and microvascular thrombosis.[5] Therefore, there is a need for early and effective predictors of clinical outcomes for risk stratification of COVID-19 patients.

COVID-19-associated coagulopathy (CAC)[6] raises questions about how to prevent or manage this state.

The International Society on Thrombosis and Haemostasis (ISTH), the American Society of Hematology (ASH), and the American College of Cardiology have given interim guidance on this topic.

In this chapter, we aim to provide a clinical overview on the pathogenesis, clinical features, investigations, and effective management of coagulation-related abnormalities in patients with COVID-19.

11.1.1 SARS-CoV-2

Human coronaviruses belong to the subfamily Coronavirineae in the Coronaviridae family of the order Nidovirales. The genome of coronavirus is comprised of single-stranded positive-sense RNA with a 5′ cap structure and a 3′-poly-A tail. The spikes seen on the viral surface are composed of homotrimers of S proteins and aid in binding to host receptors. Coronaviruses have as many as six open-reading frames, which encode for primary structural proteins of the virus including membrane, envelope, spike, and nucleocapsid. Membrane proteins are transmembrane proteins that give the virus its shape. The envelope protein is involved in virus assembly and pathogenesis. Spike proteins aid in attachment to host receptors. The nucleocapsid packages and encapsulates the genome into virions and antagonizes silencing.[7]

11.2 Pathogenesis

11.2.1 Virchow's Triad and COVID-19

The pathogenesis of coagulopathy in COVID-19 remains unclear. However, any coagulation abnormality in an individual can be explained by Virchow's triad[8] which is comprised of endothelial injury, stasis, and a hypercoagulable state.[8] Understanding the relationship of the components of Virchow's triad with COVID-19 might aid in appropriate management as well as coagulopathy prevention in these patients.

COVID-19 in the human body appears to directly attack endothelial cells by entering into the cell through the angiotensin converting enzyme-2 (ACE2) receptor, which is a type of metallopeptidase.[7] ACE2 mRNA is found in all major organs, including pharynx, stomach, small intestine, lungs, spleen, liver, brain, heart, and kidney, with binding to ACE2 receptors on endothelial cells, epithelial cells, and enterocytes. However, the protein expression of ACE2 mRNA is greatest in lungs, heart, veins, and arteries.[9]

Hamming *et al.* studied ACE2 receptor (ACE2R) expression in tissues obtained from COVID-19 patients undergoing diagnostic biopsies, surgically resected specimens, and clinical autopsies. High density of ACE2R expression was noted in endothelial cells from blood vessels in all tissues. Apart from endothelial cells, ACE2R expression was also noted in arterial smooth muscle cells of brain, which could point toward one aetiology of acute ischaemic stroke in these patients.[9]

A similar study by Chen *et al.* reported the distribution and density of ACE2R expression in the adult human heart. ACE2R expression was noted in eight major cell types in heart, including cardiac myocytes, fibroblasts, endothelial cells, pericytes, macrophages, T-cells, smooth muscle cells, and neuron-like cells. Among these cells, pericytes were identified to bear high levels of ACE2R expression. This discovery lead to the postulation that viral entry into these cells and injury may in turn lead to microvascular dysfunction *via* capillary endothelial cell injury.[10]

Ladikou *et al.* reported that SARS-CoV-2 entry damages the endothelium, which subsequently results in von Willebrand factor (vWF) release, mediating platelet aggregation and adhesion.[11]

It was also noted that infection with COVID-19 was associated with increased new blood vessel formation.[12]

Begbie *et al.* suggested that in addition to direct invasion by virus, endothelial injury was also caused by increased release of several cytokines and acute phase reactants in these patients.[13]

It has been observed that there is activation of both complement pathways (C5b-9 (membrane attack complex)), mannose-binding protein-associated serine protease 2, and C4d leading to further endothelial cell injury.[14]

Stasis can be explained by immobilisation in critically ill hospitalized patients (Figure 11.1).[8]

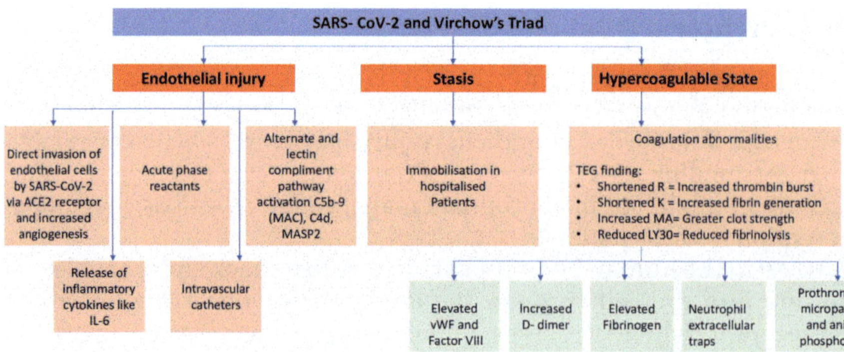

Figure 11.1 The pathogenesis of coagulopathy in COVID-19. Reproduced from ref. 8 with permission from Springer Nature, Copyright 2020.

Hypercoagulable state, the third component of Virchow's triad, is a result of increased levels of certain procoagulants such as vWF, anionic phospholipids, neutrophilic extracellular traps, prothrombotic microparticles, D-dimer, fibrinogen, and factor VIII.[15]

In a study by Panigada *et al.*, routine laboratory tests done in 24 severely ill COVID-19 patients reported several abnormalities, including normal or slightly prolonged prothrombin time (PT) and activated partial thromboplastin time (aPTT), normal or increased platelet count, and increased D-dimer and fibrinogen levels. Further testing also showed increased factor VIII activity and vWF antigens. Thromboelastography showed an increased early thrombin burst in 50% of patients, increased fibrin generation and greater clot strength in 83% of patients, and reduced fibrinolysis in 100% of patients.[16]

11.2.2 Disseminated Intravascular Coagulation and COVID-19

Disseminated intravascular coagulation (DIC) has been reported in some critically ill patients with COVID-19; however, there are many findings that differentiate CAC from DIC (Table 11.1).[8] In acute decompensated DIC, bleeding is a common presentation; however, CAC commonly presents as thrombosis. Markedly raised D-dimer levels a common finding in both CAC as well as acute decompensated DIC.[8,17]

Zhang *et al.* studied D-dimer levels on admission to predict in-hospital mortality in patients with COVID-19. They concluded that D-dimer >2.0 μg mL^{-1} on admission could effectively predict in-hospital mortality in patients with COVID-19, which indicated D-dimer could be an early and helpful marker to improve management of COVID-19 patients.[18]

Table 11.1 Distinguishing features between COVID-19-associated coagulopathy (CAC) and acute disseminated intravascular coagulation (DIC).[a] Reproduced from ref. 8 with permission from Springer Nature, Copyright © 2020, Springer Nature Switzerland AG.

	CAC	Acute DIC
Major clinical finding	Thrombosis	Bleeding
PT	Normal/prolonged	Prolonged
aPTT	Normal/prolonged	Prolonged
Platelet counts	Normal/increased/decreased	Decreased
Fibrinogen	Increased	Decreased
D-dimer	Increased	Increased
vWF and factor VIII activity	Increased	Increased
Antithrombin	Increased	Decreased
Anticardiolipin antibody	Positive	Normal
Protein C	Increased	Decreased

[a]aPTT: activated partial thromboplastin time; COVID-19: coronavirus disease 2019; PT: prothrombin time; vWF: von Willebrand factor.

Guan *et al.* analysed 1099 patients with COVID-19 and found that non-survivors had a significantly higher D-dimer (median 2.12 μg mL^{-1}) than that of survivors (median 0.61 μg mL^{-1}).[19]

Similarly, Tang *et al.* reported markedly high D-dimer levels in deaths with COVID-19.[20] Zhou *et al.* conducted a retrospective study on 191 patients with COVID-19 and found that D-dimer >1 μg mL^{-1} on admission was associated with in-hospital death (hazard ratio 18.42; 95% CI 2.64–128.55).[21] Huang and colleagues showed that D-dimer levels on admission were higher in intensive care unit (ICU) patients than in general ward patients (median 0.5 μg mL^{-1}).[22]

Thus, patients with markedly raised D-dimers (four-fold increase), even in the absence of symptoms, should be closely monitored.[18]

11.2.3 Viraemia and Disease Severity

Studies have shown that viraemia in COVID-19 correlates with severity of infection.[10] Zheng *et al.* detected viral RNA load in 96 COVID-19 patients in respiratory, serum, urine, and stool samples. Viral RNA was detected in all the sputum and saliva samples of patients, in 59% (n = 55) of stool samples and 41% (n = 39) of the serum samples. Only one urine sample was positive for viral RNA.[23] Hu *et al.* detected the time taken for negative conversion of viral RNA. The median time from the first day of a positive RNA test to consecutive negative tests was 14 days.[24]

11.2.4 Neutrophilic Extracellular Traps

Neutrophilic extracellular traps (NETs) are extracellular fibrillar networks that provide a high concentration of antimicrobial substances at sites of infection and prevent the spread of microbes by trapping them in the fibrils, a process known as NETosis. They are produced by neutrophils in response to infectious pathogens and inflammatory mediators.[25] These extracellular traps comprise circulating cell-free (cf)-DNA and histones. The process involves histone citrullination by peptidyl arginine deiminase-4, chromatin unwinding, breakdown of nuclear membranes, and cytolysis.[26,27]

Circulating cf-DNA in healthy individuals originates predominantly from apoptotic cells, whereas cell-free DNA/RNA released in the infection setting is the result of cellular and tissue necrosis, apoptosis, autophagy, or mitotic catastrophe.[28]

SARS-CoV-2 is an RNA virus and cell-free RNA of this virus has a clinical relevance in that it may serve as a cofactor for autoactivation of factor VII activating protease and initiate the extrinsic coagulation pathway.[29] Kannemeier *et al.* confirmed this in mice, in which extracellular RNA exposed following vascular injury provoked a thrombotic response.[30]

Zuo *et al.* reported high levels of circulating cf-DNA and DNA–myeloperoxidase complexes in patients with COVID-19. The levels correlated with the acuity of illness, inflammatory response, and need for mechanical ventilation.[31]

Gajsiewicz *et al.* reported that RNA structures such as hairpin-forming oligomers act as a procoagulant.[32] Phosphate groups and nucleic acids often tend to co-localize in high inflammatory environments and activate the coagulation pathway.[5]

In addition, NET-associated histones can induce platelet aggregation through toll-like receptors on platelets and other cells. Platelet signaling activates the major platelet adhesion receptor, integrin $\alpha IIb\beta 3$, which mediates platelet aggregation, as well granule release, phosphatidyl-serine exposure, FV/Va expression, and thrombin generation.[33–35]

11.2.5 Role of Cytokines in COVID-19

Studies have reported consistently raised interleukin (IL)-6 and −10 in almost all patients with COVID-19. IL-6 activates the IL-6 receptor, recruiting Janus kinases (JAK) transit cascade signal to activate signal transducer and activator of transcription 3 (STAT3).[36] IL-10 is an anti-inflammatory cytokine, which can be released by all immune cells in the body. Increased IL-10 levels block the activity of the nuclear factor-κB pathway to decrease the production of IL-6. However, raised levels of IL-10 also downregulate proliferation of immune cells (*e.g.* type 1 T-helper cells, natural killer cells, and CD8 T-cells), which delays the clearance of viruses.[37] Hence, IL-10 could be responsible for the normal levels of interferon (IFN)-γ, a cytokine for the clearance of viruses and the depletion of lymphocytes.[38] Additionally, SARS-CoV-2 infects airway epithelial cells and results in delayed release of chemokines such as CCL3, CCL5, CCL2, and CXCL10. Release of cytokines and chemokines leads to a poor prognosis.[39] The overproduction of early-response pro-inflammatory cytokines (tumour necrosis factor, IL-6, and IL-1β) leads to increased risk of vascular hyperpermeability, and in turn, imbalance in haemostasis.[40]

11.3 Clinical Autopsy Findings

Clinical autopsies were done on diseased patients with COVID-19 to aid in understanding the pathogenesis of COVID-19. Many studies reported macrovascular and microvascular thrombosis as a major histopathological finding in these patients.[41] Endotheliitis was considered as one of the major causes of endothelial injury leading to formation of these micro- and macrovascular thrombi.[5,8]

Monteil *et al.* engineered human blood vessels *in vitro* and showed that SARS-CoV-2 could invade these vessels *in vitro via* ACE2R on endothelial cells. They used human recombinant soluble ACE2 to block the invasion.[42]

One study reported involvement of endothelial cells by COVID-19. Diffuse lymphocytic endotheliitis and apoptotic bodies were noted in all the tissue specimens from deceased COVID-19 patients (Figure 11.2). In this study, core biopsies of lung, heart, small intestine, and kidneys were taken from deceased COVID-19 patients. Electron microscopy showed accumulation of viral inclusions in the endothelium from all the organs (Figure 11.2).[43]

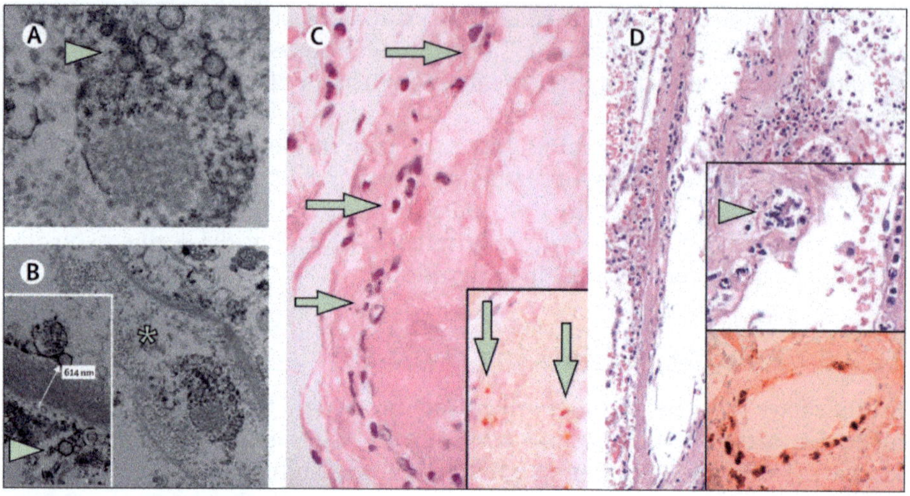

Figure 11.2 The mechanism(s) underlying a prolonged thrombophilic state following COVID-19 infection are unknown, but may be related to diffuse vascular endothelial cell infection, injury, and dysfunction. Viral particles and inclusion bodies are seen within endothelial cells of the glomerular capillary loops (A and B); in panel C, inflammatory cells are observed within capillaries serving the small intestine, while panel D demonstrates apoptotic endothelial cells and mononuclear cells. The available clinical and pathological evidence, coupled with strong and mature data that support a relationship between cell-free RNA, polyphosphate, and factor XI activation, raises a logical scientific premise that requires investigation. Reproduced from ref. 43 with permission from Elsevier, Copyright 2020.

Another study suggested that high IFN-1 expression in vascular endothelial cells caused reticular inclusions in these cells comprised of glycoproteins and phospholipids.[44]

As suggested by various pathologists globally, macro- and microvascular thrombosis with both red (erythrocytes, leukocytes, fibrin) and white thrombi (platelets and fibrin) was seen in almost all the organs. In addition, other findings such as endocardial thrombi, intravascular megakaryocytes, and viral particles in cells of adipose tissue and in platelets sequestered in the spleen was also noted.[5]

Fox *et al.* conducted clinical autopsies on four COVID-19-positive deceased patients. Grossly, lung parenchyma showed haemorrhage and oedema (in three out of four decedents) and microthrombi in sections of the peripheral parenchyma. Microscopic findings included interstitial lymphocytic infiltrates, accumulation of $CD4^+$ lymphocytes around thrombotic blood vessels, desquamated type-2 pneumocytes within the alveolar spaces, thickened alveolar capillaries with fibrin thrombi, entrapped neutrophils, and $CD61^+$ megakaryocytes. Cardiomegaly and right ventricular dilation suggested heart involvement. There was no evidence of coronary artery thrombosis in any of the four decedents.[41]

One of the studies also suggested the possibility of macrophage activation syndrome in COVID-19 patients; however, these patients did not meet all the criteria for diagnosis of haemophagocytic lymphohistiocytosis.[45]

The inhibition of $CD4^+$ T-cell IFN-γ production in COVID-19 pneumonia results in a hyper-inflammatory immunosuppressive state. The aetiology of macrophage activation syndrome has been proposed for this condition.[46]

11.4 Clinical Features

11.4.1 Features of Venous Thromboembolism

Venous thromboembolism (VTE) is characterized by deep vein thrombosis (DVT) and/or pulmonary embolism (PE) and is a common feature in patients with COVID-19. In spite of anticoagulant prophylaxis, it can be seen in most of the critically ill patients.[8] Among the studies done on COVID-19 and hypercoagulable state, two of the studies were on clinical autopsies. Older-aged male patients with obesity, smoking, and other chronic medical comorbidities were noted as being at risk of for thromboembolism.

In one study, clinical autopsies were conducted on 21 patients with COVID-19, of whom 45% showed alveolar microthrombi and 16.7% had glomerular microthrombi, which was similar to another study on 12 COVID-19 patients in which DVT was seen in 58% of patients with 33% mortality due to PE.[47,48] Markedly raised D-dimer levels (>20 μg mL^{-1}; normal value <0.5 μg mL^{-1}) were seen in 25% of these patients.

In another study of 184 critically ill COVID-19 patients, all patients received anticoagulant prophylaxis. In spite of the prophylaxis, VTE developed in 27% of patients[49] with 13% mortality rate.

In another series of 150 critically ill patients with COVID-19, PE was reported in 64 of the patients.[50] Extracorporeal circuit clotting was noted in 28 out of 29 patients receiving renal replacement therapy (RRT), and in two out of 12 patients undergoing extracorporeal membrane oxygenation. All of these patients received anticoagulation (therapeutic dose 30%; prophylactic dose 70%). This study also made comparison in COVID-19 patients with acute respiratory distress syndrome (ARDS) to a matched cohort of non-COVID-19 ARDS patients and found that the rate of PE was higher in COVID-19 patients (11.7% *versus* 2.1%; odds ratio (OR) 6.2, 95% CI 1.6–23.4; $p < 0.008$). Almost all patients (95%) had increased levels of fibrinogen and D-dimer with an increase in vWF activity, factor VIII, and vWF antigen.

In a similar study, PE was seen in 22 out of 107 ICU patients. By comparison, in two matched cohorts (one from the same time interval in the previous year and one from concurrent patients with influenza rather than COVID-19) 6.1% (absolute risk 14.4%, 95% CI 6.1–22.8%) and 7.5% (absolute risk 13.1%, 95% CI 1.9–24.3%) of patients, respectively, reported PE.[51] The study also showed that elevated plasma factor VIII, D-dimer, and vWF antigen were associated with high risk of PE.

In yet another study of 74 ICU patients, VTE was associated with higher death rate.[52] The incidence of VTE was higher in the critically ill ICU patients than the patients in general wards.

Another study conducted ultrasound screening of legs in 26 critically ill patients with COVID-19 who were all receiving either prophylactic (31%) or therapeutic (69%) doses of anticoagulation. Ultrasound reported VTE in 18 (69%) patients; the mortality rate was 12%.[53]

Two studies (314 ICU and 71 non-ICU patients, all of whom were on anticoagulation) found an incidence of VTE in the range 6.4–21%.[54,55]

11.4.2 Features of Arterial Thromboembolism

Although arterial thrombosis is less common than VTE in COVID-19, cases of acute limb ischaemia, myocardial infarction, and ischaemic stroke have been reported. In a case report, acute ischaemic stroke was noted in five COVID-19 patients aged <50 years during a 2-week period, as compared to an average of 0.73 patients every 2 weeks over the previous year.[56] The incidence rate of acute ischaemic stroke was noted to be between 2% and 3.7% in ICU patients[49,50] and 2% in non-ICU patients.[54]

In one of the studies, 20 patients with COVID-19 developed acute limb ischemia over a 3-month period. The incidence of acute limb ischemia was found to be significantly higher as compared to the previous year (16.3% in 2020 *versus* 1.8% in 2019; $p < 0.001$).[57] Out of the 20 patients, revascularization surgery was done in 17 patients, which was successful in 12 patients (71%). The mortality rate was up to 40% in this study, and was significantly higher in the older patients (81 ± 10 years *versus* 71 ± 5 years; $p = 0.008$). Postoperative heparin infusion resulted in significant improvement in survival in this small patient sample (57.1% *versus* 0%; $p = 0.42$).

Another study reported arterial thrombosis in two healthy patients (37-year-old and another 53-year-old) who were already on anticoagulant prophylaxis at the time of the event.[58] Both of the patients had markedly raised D-dimer levels. Two studies have reported mesenteric ischaemia and ST elevation myocardial infarction (STEMI) in COVID-19.[50,59]

11.5 Investigations

All patients with COVID-19 should undergo complete blood count, PT, aPTT, D-dimer, and fibrinogen levels. D-dimer levels have been reported to be directly proportional to the severity of COVID-19. These test reports were found to correlate with 28-day mortality based on a retrospective study in 447 patients [9]. D-dimer levels (OR 1.058, 95% CI 1.028–1.090) and PT (OR 1.107, 95% CI 1.008–1.215) correlated positively, while platelet count (OR 0.996, 95% CI 0.993–0.998) correlated negatively, with 28-day mortality.[60]

Some studies have reported various cut-off values of D-dimer levels to predict VTE in COVID-19 patients.[61] Cui *et al.* reported that a value of 1.5 μg mL^{-1} showed 85% sensitivity and 88.5% specificity in prediction of VTE with negative predictive value of 94.7%.[61] ASH guidelines suggested that a normal D-dimer could be sufficient to rule out PE in stable COVID-19 patients with moderate pretest probability, as calculated with Wells score.[43]

However, many COVID-19 patients present with high D-dimer levels because of underlying inflammatory process. Hence, ruling out VTE based on D-dimer alone is difficult in patients with high pretest probability.[62]

Due to the high risk of VTE, additional investigations such as computed tomography angiography (CTA) of the chest and ultrasonography and Doppler examination of the extremities should be considered, depending on clinical presentation. In patients with contraindications to CTA, a ventilation–perfusion (VQ) scan could be performed instead.[8]

In patients with contraindications to CTA as well as VQ scan, an echocardiogram with Doppler may provide valuable information including dilatatation/hypokinesia of the right ventricle with right heart strain and/or elevation of the right ventricular systolic pressure as a surrogate for pulmonary hypertension.[8]

Additional investigations should be carried out in patients with unexplained hypotension, worsening respiratory status, or other risk factors suspicious for thrombosis. Patients with prolonged PT/aPTT and/or severe thrombocytopenia require further work-up. Serum lactate dehydrogenase, ADAMTS13 activity, haptoglobin, and peripheral blood smear examination are recommended in case of suspected thrombotic microangiopathy. Antiphospholipid antibody testing is not recommended routinely, as it can be transiently present in viral infections.[8]

11.6 Management

The ISTH promulgated interim guidelines in March 2020 on recognition and management of coagulopathy in COVID-19 based on ISTH DIC score.[63] Similarly, ASH mentions management of CAC on its website.[64] A comparative analysis of these is shown in Table 11.2.

Due to lack of evidence-based studies, management of coagulopathy may be quite challenging. Immobilized ICU patients are known to be at higher risk of thrombosis (incidence 1.1% in non-ICU patients *versus* 79% in ICU patients).[64] Moores *et al.* recommended starting prophylaxis with low molecular weight heparin (LMWH) in all critically ill patients unless contraindicated.[65] Due to a higher number of patients presenting with VTE in spite of receiving anticoagulant prophylaxis, some recommend starting an intermediate dose (LMWH 30–40 mg subcutaneously twice daily) or full dose of anticoagulation in ICU patients with COVID-19.[49,66–68]

Most institutions have established guidelines to manage antithrombotic therapy. Balancing the risks of bleeding with the risk of thromboembolism is the mainstay of decision making regarding anticoagulant therapy.[8]

In one study, 786 out of 2773 (21%) patients with COVID-19 received anticoagulation and had an in-hospital mortality rate of 22.5% with a median

Table 11.2 Comparative analysis of academic society guidelines for COVID-19 thrombosis.[a]

Parameter	ASH[64]	ISTH[63]
Monitoring	Platelet count, PT, aPTT, D-dimer, and fibrinogen	PT, D-dimer, fibrinogen
Prophylaxis with LMWH	Recommended for all hospitalized COVID-19 patients	All patients (including non-ICU) who require hospitalization
Contraindications to prophylaxis	Active bleeding, platelet count $<25 \times 10^9$ L^{-1}, fibrinogen <0.5 g L^{-1}	Active bleeding, platelet count $<25 \times 10^9$ L^{-1}
Transfusion in bleeding patient	Transfuse platelets if count $<50 \times 10^9$ L^{-1}, give plasma if INR >1.8 and fibrinogen concentrate/cryoprecipitate if fibrinogen <1.5 g L^{-1}	Maintain platelet count $>50 \times 10^9$ L^{-1}, fibrinogen >2 g L^{-1} and INR <1.5
Transfusion in non-bleeding patient	No comment	Fibrinogen >2 g L and platelet count $>20 \times 10^9$ L^{-1}
Mechanical thromboprophylaxis	When pharmacological thromboprophylaxis contraindicated	When pharmacological thromboprophylaxis contraindicated Can be considered together in sick/immobile patients[b]
Therapeutic anticoagulation	VTE and atrial fibrillation	LMWH preferred, duration minimum 3 months
Extended anticoagulation		LMWH or DOAC for 14–30 days[b]
Screening for DVT		Not needed[b]

[a]aPTT: activated partial thromboplastin time; ASH: American Society of Hematology; DOAC: direct oral anticoagulants; DVT: deep vein thrombosis; ICU: intensive care unit; INR: international normalized ratio; ISTH: International Society on Thrombosis and Haemostasis; LMWH: low molecular weight heparin; PT: prothrombin time; VTE: venous thromboembolism.
[b]From new ISTH guidelines.[63]

survival of 21 days. In-hospital mortality rate was 22.8% in those who did not receive anticoagulation, with a median survival of 14 days.[63] In patients who did not receive anticoagulation, requirement of mechanical ventilation was significantly higher compared to those who received anticoagulation (29.8% *versus* 8.1%; $p < 0.001$). The in-hospital mortality rate was 62.7% with a median survival of 9 days in patients who were intubated but did not receive anticoagulation, as compared to a 29.1% in-hospital mortality rate with a median survival of 21 days in intubated patients who received anticoagulation.[63]

In another study, 99 out of 449 patients with severe COVID-19 received anticoagulation. A 28-day mortality reduction was noted in patients on heparin that had either a high D-dimer of more than six times the upper limit of normal (32.8% *versus* 52.4%; $p = 0.017$) or high sepsis-induced coagulopathy score (64.2% without heparin *versus* 40% with; $p = 0.029$).[60]

The recommended prophylactic dose of LMWH for patients with creatinine clearance (CrCl) >30 mL min^{-1} is 40 mg once daily, and for CrCl 15–30 mL min^{-1} is 30 mg once daily. Recommendations are to increase the dose by 30% for patients with a body mass index >40 kg m^{-2}. In patients receiving RRT or those with CrCl < 5 mL min^{-1}, unfractionated heparin must be given. Full-dose anticoagulant (*e.g.* LMWH 1 mg kg^{-1} every 12 h) is should be considered in patients with known VTE, similar to those without COVID-19. Full dose anticoagulant should also be considered for suspected VTE where confirmatory testing is not feasible[8] and also for patients with severe COVID-19, especially if they have high D-dimer, fibrinogen and/or other inflammatory markers. Most studies have shown that D-dimer levels in the range of 1.6–40 µg mL^{-1} correlate positively with risk for VTE.[54]

Therapeutic dose anticoagulation is also recommended in individuals with repeated clotting of extracorporeal circuits as well as central venous catheters and arterial lines.[8]

In individuals with any contraindication for heparin, alternative therapy may be considered. Tissue plasminogen activator may be used in cases of arterial thrombosis, such as acute STEMI when catheterization–laboratory intervention is decided against on clinical grounds and for acute limb ischaemia.[8]

Surgical consultation should be considered for patients with acute mesenteric ischaemia. In patients with severe acute renal failure, RRT may be required. Patients with known VTE should be given a minimum of 3 months of anticoagulation post-discharge, with appropriate follow-up.[8]

Some patients with recent surgery and/or trauma may need post-discharge prophylaxis even without known VTE. Rivaroxaban 10 mg daily for 31–39 days has been recommended for antithrombotic prophylaxis on an outpatient basis in these patients.[64] Equivalent doses of other similar oral anticoagulant drugs could likely be used effectively for this purpose as well.

11.7 Conclusion

The 2019 novel coronavirus has shown phenotypic variability in clinical presentation ranging from completely asymptomatic carriers to critically ill patients with high mortality. Coagulopathy in COVID-19 is a common presentation in critically ill patients. It manifests with elevated D-dimer and fibrinogen levels and is different from DIC in that it is usually manifested by thrombosis rather than bleeding, has a low occurrence of thrombocytopenia, a lack of severe thrombocytopenia, and is accompanied by only a mild prolongation of prothrombin time and activated partial thromboplastin time. Both venous and arterial thrombosis appear to be increased, especially in sick patients, and many of the thromboses may be subclinical in nature. Available evidence and guidelines recommend prophylactic anticoagulation with LMWH for all hospitalized patients. Academic societies have given guidelines on the management of patients with COVID-19. However, there is still an unclear understanding of the pathogenesis of this disease and therefore, high-quality evidence-based studies need to be encouraged.

References

1. T. Singhal, A Review of sCoronavirus Disease-2019 (COVID-19), *Indian J. Pediatr.*, 2020, **87**(4), 281–286.
2. World Health Organization, Situation Reports. Available at: https://www.who.int/emergencies/diseases/novel-coronavirus-2019/situation-reports/, accessed 10 December 2020.
3. Coronavirus Outbreak. Available at: https://www.worldometers.info/coronavirus/, accessed 27 October 2020.
4. N. Singhania, S. Bansal and G. Singhania, An atypical presentation of novel coronavirus disease 2019 (COVID-19), *Am. J. Med.*, 2020, **133**(7), e365–e366.
5. R. C. Becker, COVID-19 update: Covid-19-associated coagulopathy, *J. Thromb. Thrombolysis*, 2020, **50**(1), 54–67.
6. COVID-19 and coagulopathy: frequently asked questions, https://www.hemat.ology.org/covid-19/covid-19-and-coagulopathy, accessed December 2020.
7. R. C. Becker, Toward understanding the 2019 Coronavirus and its impact on the heart, *J. Thromb. Thrombolysis*, 2020, **50**(1), 33–42.
8. N. Singhania, S. Bansal, D. P. Nimmatoori, A. A. Ejaz, P. A. McCullough and G. Singhania, Current Overview on Hypercoagulability in COVID-19, *Am. J. Cardiovasc. Drugs*, 2020, **20**(5), 393–403.
9. I. Hamming, W. Timens, M. Bulthuis, A. Lely, G. Navis and H. van Goor, Tissue distribution of ACE2 protein, the functional receptor for SARS coronavirus. A first step in understanding SARS pathogenesis, *J. Pathol.*, 2004, **203**, 631–637.
10. W. Chen, Y. Lan, X. Yuan, X. Deng, Y. Li and X. Cai, *et al.*, Detectable 2019-nCoV viral RNA in blood is a strong indicator for the further clinical severity, *Emerging Microbes Infect.*, 2020, **9**, 469–473.
11. E. E. Ladikou, H. Sivaloganathan and K. M. Milne, *et al.*, Von Willebrand factor (vWF): marker of endothelial damage and thrombotic risk in COVID-19? *Clin. Med.*, 2020, e178–e182.
12. M. Ackermann, S. E. Verleden, M. Kuehnel, A. Haverich, T. Welte and F. Laenger, *et al.*, Pulmonary vascular endothelialitis, thrombosis, and angiogenesis in Covid-19, *N. Engl. J. Med.*, 2020, **383**, 120–128.
13. M. Begbie, C. Notley, S. Tinlin, L. Sawyer and D. Lillicrap, The factor VIII acute phase response requires the participation of NFκB and C/EBP, *Thromb. Haemostasis*, 2000, **84**, 216–222.
14. C. Magro, J. J. Mulvey, D. Berlin, G. Nuovo, S. Salvatore and J. Harp, *et al.*, Complement associated microvascular injury and thrombosis in the pathogenesis of severe COVID-19 infection: a report of five cases, *Transl. Res.*, 2020, **220**, 1–13.

15. M. Ranucci, A. Ballotta, U. Di Dedda, E. Bayshnikova, M. Dei Poli and M. Resta, *et al.*, The procoagulant pattern of patients with COVID-19 acute respiratory distress syndrome, *J. Thromb. Haemostasis*, 2020, **18**(7), 1747–1751.
16. M. Panigada, N. Bottino, P. Tagliabue, G. Grasselli, C. Novembrino and V. Chantarangkul, *et al.*, Hypercoagulability of COVID-19 patients in Intensive Care Unit. A report of thromboelastography findings and other parameters of hemostasis, *J. Thromb. Haemostasis*, 2020, **18**(7), 1738–1742.
17. M. Levi, C. H. Toh, J. Thachil and H. G. Watson, Guidelines for the diagnosis and management of disseminated intravascular coagulation, *Br. J. Haematol.*, 2009, **145**, 24–33.
18. L. Zhang, X. Yan, Q. Fan, H. Liu, X. Liu, Z. Liu and Z. Zhang, D-dimer levels on admission to predict in-hospital mortality in patients with Covid-19, *J. Thromb. Haemostasis*, 2020, **18**(6), 1324–1329.
19. W. J. Guan, Z. Y. Ni and Y. Hu, *et al.*, Clinical characteristics of coronavirus disease 2019 in China, *N. Engl. J. Med.*, 2020, **382**(18), 1708–1720.
20. N. Tang, D. Li, X. Wang and Z. Sun, Abnormal coagulation parameters are associated with poor prognosis in patients with novel coronavirus pneumonia, *J. Thromb. Haemostasis*, 2020, **18**, 844–847.
21. F. Zhou, T. Yu, R. Du, G. Fan, Y. Liu and Z. Liu, *et al.*, Clinical course and risk factors for mortality of adult inpatients with COVID-19 in Wuhan, China: a retrospective cohort study, *Lancet*, 2020, **395**(10229), 1054–1062.
22. C. Huang, Y. Wang and X. Li, *et al.*, Clinical features of patients infected with 2019 novel coronavirus in Wuhan, China, *Lancet*, 2020, **395**, 497–506.
23. S. Zheng, J. Fan, F. Yu, B. Feng, B. Lou and Q. Zou, *et al.*, Viral load dynamics and disease severity in patients infected with SARS-CoV-2 in Zhejiang province, China, January-March 2020: retrospective cohort study, *BMJ*, 2020, **369**, m1443.
24. X. Hu, Y. Xing, J. Jia, W. Ni, J. Liang, D. Zhao, X. Song, R. Gao and F. Jiang, Factors associated with negative conversion of viral RNA in patients hospitalized with COVID-19, *Sci. Total Environ.*, 2020, **728**, 138812.
25. A. J. McAdam, D. A. Milner and A. H. Sharpe, in *Inflammation and Repair*, ed. Kumar V., Abbas A. K. and Aster J. C., Elsevier Saunders, Philadelphia, 9th edn, 2014, p. 81.
26. A. Brill, T. A. Fuchs, A. S. Savchenko, G. M. Thomas, K. Martinod and S. F. De Meyer, *et al.*, Neutrophil extracellular traps promote deep vein thrombosis in mice, *J. Thromb. Haemostasis*, 2012, **10**, 136–144.
27. C. Thalin, M. Daleskog, S. P. Goransson, D. Schatzberg, J. Lasselin and A. C. Laska, *et al.*, Validation of an enzyme-linked immunosorbent assay for the quantification of citrullinated histone H3 as a marker for neutrophil extracellular traps in human plasma, *Immunol. Res.*, 2017, **65**, 706–712.
28. S. Jahr, H. Hentze, S. Englisch, D. Hardt, F. O. Fackelmayer, R. D. Hesch and R. Knippers, DNA fragments in the blood plasma of cancer patients: quantitations and evidence for their origin from apoptotic and necrotic cells, *Cancer Res.*, 2001 Feb 15, **61**(4), 1659–1665.

29. F. Nakazawa, C. Kannemeier and A. Shibamiya, *et al.*, Extracellular RNA is a natural cofactor for the (auto-)activation of Factor VII-activating protease (FSAP), *Biochem. J.*, 2005, **385**(Pt 3), 831–838.
30. C. Kannemeier, A. Shibamiya, F. Nakazawa, H. Trusheim, C. Ruppert and P. Markart, *et al.*, Extracellular RNA constitutes a natural procoagulant cofactor in blood coagulation, *Proc. Natl. Acad. Sci. U. S. A.*, 2007, **104**, 6388–6393.
31. Y. Zuo, S. Yalavarthi, H. Shi, K. Gockman, M. Zuo and J. A. Madison, *et al.*, Neutrophil extracellular traps in COVID-19, *JCI Insight*, 2020, **5**(11), e138999.
32. J. M. Gajsiewicz, S. A. Smith and J. H. Morrissey, Polyphosphate and RNA Differentially Modulate the Contact Pathway of Blood Clotting, *J. Biol. Chem.*, 2017, **292**(5), 1808–1814.
33. Y. Wang, L. Luo, O. Ö. Braun, J. Westman, R. Madhi and H. Herwald, *et al.*, Neutrophil extracellular trap-microparticle complexes enhance thrombin generation *via* the intrinsic pathway of coagulation in mice, *Sci. Rep.*, 2018, **8**(1), 4020.
34. Y. Wang, H. Gao, C. Shi, P. W. Erhardt, A. Pavlovsky and D. A. Soloviev, *et al.*, Leukocyte integrin Mac-1 regulates thrombosis *via* interaction with platelet GPIbalpha, *Nat. Commun.*, 2017, **8**, 15559.
35. Y. Wang, M. Li, S. Stadler, S. Correll, P. Li and D. Wang, *et al.*, Histone hypercitrullination mediates chromatin decondensation and neutrophil extracellular trap formation, *J. Cell Biol.*, 2009, **184**(2), 205–213.
36. T. Tanaka, M. Narazaki and T. Kishimoto, IL-6 in inflammation, immunity, and disease, *Cold Spring Harbor Perspect. Biol.*, 2014, **6**, a016295.
37. J. M. Rojas, M. Avia, V. Martín and N. Sevilla, IL-10: a multifunctional cytokine in viral infections, *J. Immunol. Res.*, 2017, **2017**, 6104054.
38. Y. Tang, J. Liu, D. Zhang, Z. Xu, J. Ji and C. Wen, Cytokine Storm in COVID-19: The Current Evidence and Treatment Strategies, *Front. Immunol.*, 2020, **11**, 1708.
39. Y.-T. Yen, F. Liao, C.-H. Hsiao, C.-L. Kao, Y.-C. Chen and B. A. Wu-Hsieh, Modeling the early events of severe acute respiratory syndrome coronavirus infection in vitro, *J. Virol.*, 2006, **80**, 2684–2693.
40. G. U. Meduri, G. Kohler, S. Headley, E. Tolley, F. Stentz and A. Postlethwaite, Inflammatory cytokines in the BAL of patients with ARDS. Persistent elevation over time predicts poor outcome, *Chest*, 1995, **108**(5), 1303–1314.
41. S. E. Fox, A. Akmatbekov, J. L. Harbert, G. Li, J. Quincy Brown and R. S. Vander Heide, Pulmonary and cardiac pathology in African American patients with COVID-19: an autopsy series from New Orleans, *Lancet Respir. Med.*, 2020, **8**(7), 681–686.
42. V. Monteil, H. Kwon, P. Prado, A. Hagelkrüys, R. A. Wimmer and M. Stahl, *et al.*, Inhibition of SARS-CoV-2 Infections in Engineered Human Tissues Using Clinical-Grade Soluble Human ACE2, *Cell*, 2020, **181**(4), 905–913.e7.

43. Z. Varga, A. J. Flammer, P. Steiger, M. Haberecker, R. Andermatt and A. S. Zinkernagel, *et al.*, Endothelial cell infection and endotheliitis in COVID-19, *Lancet*, 2020, **395**, 1417–1418.
44. M. Willicombe, J. Moss, L. Moran, P. Brookes, E. Santos-Nunez and A. G. McLean, *et al.*, Tubuloreticular inclusions in renal allografts associate with viral infections and donor-specific antibodies, *J. Am. Soc. Nephrol.*, 2016, **27**, 2188–2195.
45. P. Mehta, D. F. McAuley, M. Brown, E. Sanchez, R. S. Tattersall and J. J. Manson, COVID-19: consider cytokine storm syndromes and immunosuppression, *Lancet*, 2020, **395**, 1033–1034.
46. D. McGonagle, K. Sharif, A. O'Regan and C. Bridgewood, The role of cytokines including interleukin-6 in COVID-19 induced pneumonia and macrophage activation syndrome-like disease, *Autoimmun. Rev.*, 2020, **19**(6), 102537.
47. T. Menter, J. D. Haslbauer, R. Nienhold, S. Savic, H. Hopfer and N. Deigendesch, *et al.*, Post-mortem examination of COVID19 patients reveals diffuse alveolar damage with severe capillary congestion and variegated findings of lungs and other organs suggesting vascular dysfunction, *Histopathology*, 2020, **77**(2), 198–209.
48. D. Wichmann, J. P. Sperhake and M. Lütgehetmann, *et al.*, Autopsy Findings and Venous Thromboembolism in Patients With COVID-19: A Prospective Cohort Study, *Ann. Intern. Med.*, 2020, **173**(4), 268–277.
49. F. A. Klok, M. J. H. A. Kruip and N. J. M. van der Meer, *et al.*, Incidence of thrombotic complications in critically ill ICU patients with COVID-19, *Thromb. Res.*, 2020, **191**, 145–147.
50. J. Helms, C. Tacquard and F. Severac, *et al.*, High risk of thrombosis in patients with severe SARS-CoV-2 infection: a multicenter prospective cohort study, *Intensive Care Med.*, 2020, **46**(6), 1089–1098.
51. J. Poissy, J. Goutay, M. Caplan, E. Parmentier, T. Duburcq and F. Lassalle, *et al.*, Pulmonary Embolism in Patients With COVID-19: Awareness of an Increased Prevalence, *Circulation*, 2020, **142**(2), 184–186.
52. S. Middeldorp, M. Coppens, T. F. van Haaps, M. Foppen, A. P. Vlaar and M. C. A. Müller, *et al.*, Incidence of venous thromboembolism in hospitalized patients with COVID-19, *J. Thromb. Haemostasis*, 2020 Aug, **18**(8), 1995–2002.
53. J. F. Llitjos, M. Leclerc, C. Chochois, J. M. Monsallier, M. Ramakers and M. Auvray, *et al.*, High incidence of venous thromboembolic events in anticoagulated severe COVID-19 patients, *J. Thromb. Haemostasis*, 2020, **18**(7), 1743–1746.
54. J. F. Llitjos, M. Leclerc, C. Chochois, J. M. Monsallier, M. Ramakers, M. Auvray and K. Merouani, High incidence of venous thromboembolic events in anticoagulated severe COVID-19 patients, *J. Thromb. Haemostasis*, 2020 Jul, **18**(7), 1743–1746.
55. M. Artifoni, G. Danic and G. Gautier, *et al.*, Systematic assessment of venous thromboembolism in COVID-19 patients receiving thromboprophylaxis: incidence and role of D-dimer as predictive factors, *J. Thromb. Thrombolysis*, 2020, **50**(1), 211–216.

56. T. J. Oxley, J. Mocco, S. Majidi, C. P. Kellner, H. Shoirah and I. P. Singh, *et al.*, Large-vessel stroke as a presenting feature of Covid-19 in the young, *N. Engl. J. Med.*, 2020, **382**, e60.
57. R. Bellosta, L. Luzzani and G. Natalini, *et al.*, Acute limb ischemia in patients with COVID-19 pneumonia, *J. Vasc. Surg.*, 2020, **72**(6), 1864–1872.
58. P. Perini, B. Nabulsi, C. B. Massoni, M. Azzarone and A. Freyrie, Acute limb ischaemia in two young, non-atherosclerotic patients with COVID-19, *Lancet*, 2020, **395**, 1546.
59. G. G. Stefanini, M. Montorfano, D. Trabattoni, D. Andreini, G. Ferrante and M. Ancona, *et al.*, ST-Elevation Myocardial Infarction in Patients With COVID-19: Clinical and Angiographic Outcomes, *Circulation*, 2020, **141**(25), 2113–2116.
60. N. Tang, H. Bai, X. Chen, J. Gong, D. Li and Z. Sun, Anticoagulant treatment is associated with decreased mortality in severe coronavirus disease 2019 patients with coagulopathy, *J. Thromb. Haemostasis*, 2020, **18**(5), 1094–1099.
61. S. Cui, S. Chen, X. Li, S. Liu and F. Wang, Prevalence of venous thromboembolism in patients with severe novel coronavirus pneumonia, *J. Thromb. Haemostasis*, 2020, **18**(6), 1421–1424.
62. B. Yu, X. Li, J. Chen, M. Ouyang, H. Zhang and X. Zhao, *et al.*, Evaluation of variation in D-dimer levels among COVID-19 and bacterial pneumonia: a retrospective analysis, *J. Thromb. Thrombolysis*, 2020, 548–557.
63. J. Thachil, N. Tang, S. Gando, A. Falanga, M. Cattaneo and M. Levi, *et al.*, ISTH interim guidance on recognition and management of coagulopathy in COVID-19, *J. Thromb. Haemostasis*, **18**, 1023–1026.
64. https://www.hematology.org/covid-19/covid-19-and-coagulopathy, accessed December 2020.
65. L. K. Moores, T. Tritschler, S. Brosnahan, M. Carrier, J. F. Collen and K. Doerschug, *et al.*, Prevention, diagnosis, and treatment of VTE in patients with COVID-19, *Chest*, 2020, 1143–1163.
66. Clinical Reference Group SBAR, Therapies for COVID-19, https://www.bccdc.ca/Health-Professionals-Site/Documents/Guidelines_Unproven_Therapies_COVID-19.pdf, accessed 17 July 2020.
67. A. C. Spyropoulos, J. H. Levy, W. Ageno, J. M. Connors, B. J. Hunt and T. Iba, *et al.*, Scientific and Standardization Committee Communication: clinical guidance on the diagnosis, prevention and treatment of venous thromboembolism in hospitalized patients with COVID-19, *J. Thromb. Haemostasis*, 2020, 1859–1865.
68. British Thoracic Society Issues Guidance on VTE in Patients with COVID-19, https://brit-thoracic.org.uk/document-library/quality-improvement/covid-19/bts-guidance-on-venous-thromboembolic-disease-in-patients-with-covid-19/, accessed 17 July 2020.

CHAPTER 12

COVID-19 Endocrinopathies: Implications for Care

MAYA GOPALAKRISHNAN[a], BHARAT KUMAR[a], PRAKRATI YADAV[a] AND M. K. GARG*[b]

[a]Department of Medicine, All India Institute of Medical Sciences, Jodhpur, Rajasthan, India; [b]Department of Medicine & Endocrinology, All India Institute of Medical Sciences, Jodhpur, Rajasthan, 342005, India
*E-mail: mkgargs@gmail.com

12.1 Introduction

Coronavirus disease 2019 (COVID-19) caused by severe acute respiratory syndrome coronavirus-2 (SARS-CoV-2) is well-documented to be a multisystem disease in many cases. It is broadly accepted that this is due to the widespread presence of angiotensin-converting enzyme 2 (ACE-2) receptors in many tissues, and that the virus has a direct access point to infect various tissues.[1,2] As a part of the extrapulmonary manifestations of COVID-19, it is increasingly apparent that there is endocrine involvement.[3,4] Endocrine and immune systems function in a highly integrated manner; therefore, it is not surprising that endocrine dysfunction has been found in many COVID-19 patients, given that immune mediation is well-known to be a major component in COVID-19 pathogenesis.[1] In this chapter we aim to summarise the endocrine manifestations reported so far in COVID-19 and to explore mechanisms to decipher how SARS-CoV-2 may affect various endocrine organs.

The Coronavirus Pandemic and the Future Volume 1: Virology, Epidemiology, Translational Toxicology and Therapeutics
Edited by Michael D. Waters, Alok Dhawan, Tim Marrs, Diana Anderson, Stafford Warren and Claude L. Hughes

Published by the Royal Society of Chemistry, www.rsc.org

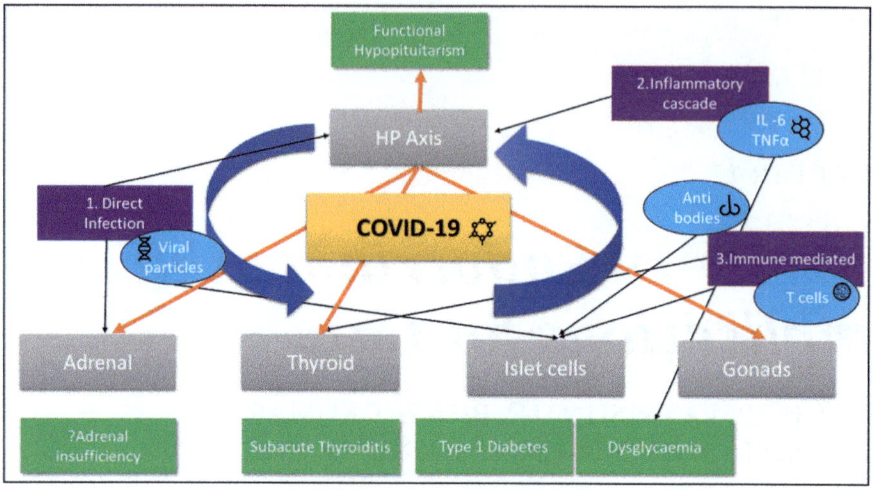

Figure 12.1 The multi-level interactions between SARS-CoV-2 and the endocrine system. Viral particles directly infect several endocrine tissues *via* ACE-2 receptors and cause cytopathology or apoptosis as in pancreas. Other mechanisms in COVID-19 are immune cell/antibody-mediated damage, as in subacute thyroiditis. A final mechanism is inflammatory cytokine-mediated dysregulation of the endocrine axis, which can cause functional hypopituitarism and relative adrenal insufficiency. Clinical features reported so far are marked in green.

There appear to be three main mechanisms of interaction between the endocrine system and SARS-CoV-2: direct viral infection of a particular endocrine gland; activation or suppression of the hypothalamo-pituitary axis (HPA) *via* inflammatory mediators; and immune-mediated glandular damage secondary to antibody formation or cell-mediated immunity. These mechanisms, their effectors and known clinical effects are summarized in Figure 12.1.

12.2 Hypothalamo-pituitary Axis

The HPA is central in stress responses during viral infections. The effect on HPA function can be directly virus-induced or secondary to inflammatory-related injury. Direct HPA injury is probably due to the ACE-2 receptor-mediated SARS-CoV-2 infection of neuronal cells causing cell edema and necrosis.[5] Direct viral damage may render the pituitary vulnerable to damage during acute illness or other conditions when perfusion is compromised. A case of gestational pituitary apoplexy with COVID-19 has been reported, supporting this pituitary vulnerability.[6]

The indirect effects on HPA are hypothesized to be mediated by cytokines such as interleukin (IL)-1, IL-6 and tumour necrosis factor (TNF)-α, which activate the HPA and subsequently raise serum cortisol levels.[7] In addition, inflammation may result in damage to the pituitary and hypothalamus, causing hypophysitis, as reported previously in SARS-CoV infection.[8]

Several inferences can also be attempted from previous experiences with reports of patients with SARS infection. SARS has been known to cause hypocortisolism and central hypothyroidism in long-term survivors.[8] Also, serum prolactin, luteinizing hormone (LH) and follicle stimulating hormone (FSH) have been shown to be elevated, while estrogen (E2), progesterone, and thyroid-stimulating hormone (TSH) are reduced in SARS-CoV infection.[9] This study had also demonstrated reduced expression of growth hormone, TSH, and adrenocorticotrophic hormone (ACTH), but increased prolactin, FSH and LH on pituitary autopsy samples of SARS-CoV-infected patients.[9] At least in theory, SARS-CoV-2 may have similar potential for HPA involvement.

While the role of growth hormone and insulin-like growth factor (IGF)-1 is currently unknown in COVID-19, IGF-1 has been known to modulate influenza A-mediated local lung injury in rats.[10] Inflammatory cytokine storm syndrome reported in COVID-19 is mediated by IL-6, which is known to supress IGF-1.[11] An intriguing cross-talk between IGF-1 and ACE-2 is the signal transducer and activator of transcription (STAT) pathway. Activation of ACE-2 receptors by cytokines through pan-JAK-STAT enhancers was demonstrated in a recent mRNA expression study.[12] The clinical relevance of this currently remains unknown.

In COVID-19, two studies have shown low normal TSH, tri-iodothyronine (T3), thyroxine (T4) suggesting a possible central hypothyroidism during acute infection.[4,13] This is probably secondary to cytokine mediation and is part of sick euthyroid syndrome, discussed below.

Several cases of severe hyponatraemia with possible syndrome of inappropriate antidiuretic hormone have been reported in the acute phase of COVID-19, suggesting posterior pituitary involvement, although several confounders were reported, including mechanical ventilation and coexistent diuretic use.[15,16] Long-term follow-up of patients recovering from COVID-19 is needed to estimate any long-term morbidity in relation to pituitary involvement.

12.3 Adrenals

The adrenals appear to be specifically vulnerable in COVID-19, due to various factors such as direct infection, inflammatory milieu, and a hypercoagulable state. The proclivity of SARS-CoV-2 to involve the adrenals directly is supported by the presence of ACE-2 receptors on the adrenal glands.[17] Stimulation of the HPA axis by cytokines results in increased perfusion of the adrenals, in turn increasing risk of adrenal haemorrhage.[18] Bilateral adrenal haemorrhage has been reported in the context of COVID-19.[19] Recently, an autopsy study of 28 severe COVID-19 patients also confirmed that 49% had features suggestive of adrenal haemorrhage.[20] Adrenal infarction secondary to coagulation abnormalities and prothrombotic state in COVID-19 has been reported.[21] A retrospective review of computed tomography findings also reported that 23% of patients were found to have imaging features of adrenal necrosis: a finding which the authors found to predict poor prognosis.[22]

In SARS-CoV, several viral peptides are structurally similar to ACTH. The host antibodies against these viral proteins might collaterally destroy ACTH, resulting in adrenal insufficiency.[23] There are reports of hypocortisolism in SARS survivors.[8] Similar molecular mimicry resulting in anti-ACTH antibody has been proposed, but not verified in SARS-CoV-2.

The basis for glucocorticoid use in COVID-19 stemmed from the ability of SARS-CoV-2 to cause a dysregulated stress response and consequent adrenal insufficiency. Apart from this, steroids may prove beneficial in the cytokine storm syndrome by limiting unregulated inflammation.[11,24] However, increased steroidogenesis results in immunomodulation towards the type 2 helper T-cell response, which may result in delay in viral clearing.[18] Additionally, steroids may have some local action on the pulmonary alveolar membrane, either due to their membrane-stabilising effects or other unknown mechanisms to prevent hyaline membrane formation in the lungs.[25]

Early results from the RECOVERY and other trials suggested significant benefits with steroid use in critically ill COVID-19 patients.[26] The trial demonstrated that dexamethasone reduced the risk of mortality by 17% in ventilated patients. Endocrine implications include risk of long-term adrenal suppression, infections and dysglycaemia due to exogenous steroids.[27] Furthermore, a large systematic review and metanalysis of trials (REACT) has confirmed the mortality benefit in using corticosteroids for treating COVID-19.[28]

Reports of adrenal insufficiency in SARS-CoV-2 infection are emerging, similar to the suppressed cortisol levels reported with SARS-CoV infection.[8] A recent report by Tan *et al.* has shown that severe COVID-19 patients mount a robust adrenal response and that baseline serum cortisol was a reliable predictor of mortality in COVID-19. The authors have suggested a cut-off of 27 $\mu g\ dL^{-1}$ (744 $nmol\ L^{-1}$) as a predictor of 15-day mortality.[3] Interestingly, the insufficient adrenal response is more in patients with mild and asymptomatic infection rather than severe COVID-19. These results suggest that the beneficial effects of steroids in critically ill COVID-19 patients is unlikely to be due to adrenal insufficiency and may possibly relate to local lung effects of steroids. It remains to be explored whether timing is critical to beneficial effects of glucocorticoid therapy and what other mechanisms may explain these beneficial effects of steroids.

12.4 Pancreas

The expression of ACE-2 receptors on exocrine pancreas and islets cells suggest that the pancreas is susceptible to SARS-CoV-2 infection. COVID-19 has been reported to cause exocrine pancreatic injury of varying severity.[29,30] Wang *et al.* reported pancreatic injury in 17% of 52 COVID-19 patients with elevated serum amylase or lipase levels, and two-thirds had abnormal blood glucose levels.[31]

Dysglycaemia is commonly reported in COVID-19.[32] A physiological model for SARS-CoV-2 using organoid derivatives from human pluripotent stem cells, demonstrated α- and β-cell death due to viral cytotoxic effects.[33] Direct islet cell injury by virus results in hyperglycaemia and acute diabetes. Glucocorticoids used for patients with severe COVID-19 disease may also result in hyperglycaemia.[34] Alteration in glucose metabolism due to stress response also results in hyperglycaemia.[35] Diabetes is a risk factor for severe COVID-19 disease and a predictor of morbidity and mortality. Type 1 diabetes patients experience two times the mortality compared to those without diabetes, and type 2 diabetes patients experience three times the mortality.[36–38] Inflammatory milieu and elevated levels of cytokines in diabetes may contribute to severe disease and acute respiratory distress syndrome (ARDS), thus establishing a bidirectional relationship between diabetes and COVID-19.[38]

SARS-CoV-2 infection has been reported to trigger type 1 diabetes mellitus with diabetic ketoacidosis and hyperosmolarity.[38] The possible mechanisms are either direct β-cell cytotoxicity or immune reaction.[39,40] Li *et al.* reported a 6.4% rate of ketosis without obvious cause among 658 patients.[41] Viral causes of ketosis-prone diabetes have been reported previously with other coronaviruses that bind to ACE-2 receptors.[42] A global database (CoviDiab Project) has been established to collect information on patients with COVID-19 and high blood-sugar.[43] Iacobellis *et al.* reported that admission hyperglycaemia was the best predictor of COVID-19 radiological findings.[32]

Clinical care in COVID-19 patients must consider these factors, and close monitoring of blood glucose levels is required in diabetics. Given the frequency of dysglycaemia in COVID-19, it may be prudent to employ blood glucose screening of all admitted COVID-19 patients to check for new-onset dysglycaemia. Consensus on management now favours insulin use with dipeptidyl peptidase 4 inhibitors while sodium–glucose cotransporter 2 inhibitors are not recommended in severe disease.

12.5 Thyroid Gland

There are multiple reports of thyroid involvement in COVID-19. ACE-2 receptor expression on thyroid tissue suggests the potential for direct SARS-CoV-2 infection. Wang *et al.* reported that serum levels of T3 and T4 in patients with SARS were significantly lower than those in the control group, and low T3 levels correlated positively with disease severity. Lower levels of TSH also present in these patients suggesting a central role.[44]

Leow *et al.* studied SARS survivors for hormonal derangements 3 months after recovery and noted that 6.7% of the patients had developed biochemical hypothyroidism. The majority of these patients (75%) had a central aetiology and only 25% developed primary hypothyroidism. Central hypocortisolism was also present in 66% of patients who had central hypothyroidism.[8]

Pituitary dysfunction in SARS patients was also confirmed by immunohistochemistry. Evaluation of the adenohypophysis from autopsies of five SARS patients in another study revealed that both the number and the immune-reactive intensity of TSH-positive cells were markedly decreased.[9] Beside pituitary dysfunction, direct involvement of thyroid in SARS was demonstrated with apoptosis and extensive injury to the follicular epithelium and parafollicular cells.[45]

Several studies have reported that COVID-19 can be associated with thyroid dysfunction due to a variety of mechanisms, including subacute thyroiditis, Hashimoto's thyroiditis, overt thyrotoxicosis or Graves' disease.[46–49]

Subacute thyroiditis (SAT)/De Quervain thyroiditis is an acute inflammatory self-limited thyroid disease caused by a viral or post-viral inflammatory process, usually presenting with fever, neck pain, and systemic symptoms. At present, there are approximately 16 reported cases of COVID-19-associated subacute thyroiditis reported in the literature.[46,47,50–60] The age range is 18–69 years with a 3 : 1 female : male ratio. These patients suffered from mild to moderate COVID-19 according to World Health Organization severity staging, and developed SAT during the acute illness or up to 42 days later.[61] The presenting symptoms were anterior neck pain (88%), fever (56%), features of thyrotoxicosis (62%), and most of them had clinically enlarged thyroid tenderness. Notably, one patient developed atrial fibrillation for which cardioversion was performed.[53] One patient did not have any symptoms of thyroiditis.[57] Mild to moderate biochemical thyrotoxicosis was present in all patients except one in whom thyroid function tests were not done.[47] TSH receptor antibodies and thyroid peroxidase antibodies were negative in all but one case, while thyroglobulin antibodies were positive in two patients. Inflammatory markers like erythrocyte sedimentation ratio and C-reactive protein were high in all cases. Thyroid ultrasonography showed typical features of thyroiditis and a thyroid scan revealed markedly diminished or absent uptake in patients. Approximately two-thirds of patients were given steroids; β-blockers and non-steroidal anti-inflammatory drugs (NSAIDs) were also used. Symptomatically all recovered, but in a quarter of patients, there was evidence of subclinical or overt hypothyroidism on follow-up. Longer follow-up is required to see whether they remain hypothyroid or recover completely. Muller *et al.* showed that the prevalence of thyroiditis is greater in severely ill COVID-19 patients than in non-severe COVID-19 patients.[62] High incidence of SAT in COVID-19 may be explained by the presence of mRNA encoding for the ACE-2 receptors, which is highly expressed in thyroid follicular cells, making them a potential target for SARS-CoV-2 entry.[63]

In a single-center retrospective study of 287 COVID-19 patients, thyrotoxicosis was present in 20% (10.8% overt), and hypothyroidism in 5.2% (overt in only two cases), suggesting that COVID-19 may be associated with high risk of thyrotoxicosis.[64] Two cases of Graves' disease also reported after COVID-19, one with a previous history of Graves' disease in remission for more than 30 years, and another with no previously known thyroid disease.[49]

A study in Wuhan, China, of 274 patients with COVID-19 revealed that TSH and free T3 were significantly less in patients who died than in survivors; here the alteration in the T4 levels was not statistically significant.[65] It is further supported by another prospective Chinese study of 100 patients wherein severely ill patients had lower levels of FT3 (free triiodotyronine), TSH (thyroid stimulating hormone) and FT3/FT4 (free thyroxine) than non-severely ill patients, and levels were lowest in those who died; levels of these hormones decreased with the clinical deterioration of the patient. In this study the authors recommended extra caution in patients with low FT3, as it was independently associated with all-cause mortality.[66]

Another single-center study of 191 patients by Lui *et al.* showed that 13% of predominantly mild to moderate COVID-19 patients had thyroid dysfunction, of which 10 patients had SAT and an equal number of patients had non-thyroidal illness syndrome (having isolated low FT3), and the authors suggested that low T3 may have prognostic significance.[67] The impact of hypothyroidism on COVID-19 was studied by van Gerwen *et al.* and they concluded that hypothyroidism is not associated with poor outcomes in COVID-19 patients.[68]

From the data available to date, we can say that thyroid dysfunction is fairly common in COVID-19, which may be due to direct effect of SARS-CoV-2 or antibody-mediated effects. SAT is the most common thyroid dysfunction, requiring only supportive treatment, and recovery occurs in most patients. In addition, we can conclude that low TSH and low T3 levels may be associated with disease severity and mortality. Longer follow-up studies are required to assess the long-term morbidity.

12.6 Gonads

Severity and case fatality of COVID-19 has been observed more in males as compared to females, similar to other coronaviruses.[69,70] ACE-2 receptors and cellular serine protease–TMPRSS-2 (transmembrane protease serine-2), which act as conduit to the virus are abundant in testes, with ACE-2 abundant on spermatogonia, Sertoli, and Leydig cells, while TMPRSS-2 is present on spermatogonia and spermatids.[71,72] This has led to the hypothesis that testes may act as a viral reservoir. Cases with epididymitis and testicular pain as symptom of COVID-19 have been reported.[73,74] In male patients with COVID-19, testes show significant seminiferous tubule injury, reduced Leydig cells and mild lymphocytic infiltration, although virus was not detected in 90% of cases.[71] Li *et al.* studied the impact of SARS-CoV-2 on male infertility and found impairment of spermatogenesis mainly due to elevated immune response in testes. There were increased seminal levels of IL-6, TNF-α and monocyte chemoattractant protein-1.[72] Semen analysis of men with COVID-19 has shown the presence of SARS-CoV-2, but its clinical implication or relationship to infertility is unknown.[75] Studies have shown low serum testosterone in COVID-19 patients compared to controls. Low testosterone levels may serve as a marker of COVID-19 disease and severity, at least in older males.[14] In these patients, serum LH was elevated, suggesting local testicular injury. Elevated levels of LH and lower ratio of testosterone to LH was demonstrated in men of reproductive age with COVID-19 compared to age-matched controls.[76] Based on these recent findings, the Italian Society of Andrology and Sexual Medicine suggests evaluation of gonadal function including semen analysis in patients who have recovered from COVID-19.[75]

The gene for ACE-2 receptors is on the X chromosome, which raises the possibility of different gender-specific activity due to X chromosome inactivation and imprinting; however, animal studies have shown that the differential ACE-2 activity in females is the result of female hormones and is independent of sex. In addition, this sexual dimorphism in ACE-2 activity is organ-specific and is not found in the lungs.[77] Estrogen and 17-β-estradiol can act on cellular subsets of the immune system by epigenetic mechanisms resulting in modulation of lymphocyte activity and number. This may explain why females were able to clear SARS-CoV-2 virus earlier than males. Montopoli *et al.* in their population-based study observed that men with prostate cancer on androgen deprivation therapies (ADT) are less likely to have COVID-19,[78] but Klein *et al.* found that ADT does not appear to be protective against COVID-19 in 1779 men with prostate cancer.[79] Several clinical trials for the use of estrogen and androgen deprivation therapies in reducing the severity of COVID-19 are ongoing.[59–62]

A case report of a patient with autoimmune polyendocrine syndrome type 1 with COVID-19 was reported wherein the patient experienced prolonged hospital stay and the authors recount issues in endocrine management.[80]

Table 12.1 summarises endocrine clinical features in COVID-19 and hypothesized mechanisms for these features.

Table 12.1 Clinical features of endocrine involvement in COVID-19 with possible mechanisms.

Gland involved	Suggested mechanism	Clinical implications
Hypothalamo-pituitary axis	Intact HPA	Robust adrenal response in critically ill
	Direct injury to hypothalamus and anterior pituitary cells	Basal serum cortisol may serve as severity and mortality marker
	Indirect injury due to cytokine storm and inflammatory mediators	Central hypothyroidism with low TSH, low T3 and low normal T4
	Anti-ACTH antibodies	Possible adrenal insufficiency ? Long-term adrenal insufficiency
Adrenal glands	Adrenal injury due to cytokine storm and inflammatory mediators	Possible adrenal insufficiency

Table 12.1 (*continued*)

Gland involved	Suggested mechanism	Clinical implications
Pancreas and islet cells	Direct cytopathic injury caused by virus in islet cells leading to altered functioning and hyperglycaemia	Development of new-onset diabetes in SARS-CoV-2 infection
	Stress response of the body	Poor diabetes control in previously diagnosed patients
	? Drug-induced hyperglycaemia	Elevated serum amylase and lipase
	Direct viral injury to pancreatic cells due to presence of ACE-2 receptors	Ketosis in the presence and absence of diabetes
	Possible autoimmune mechanism	Poor outcomes in type1 and type 2 diabetes patients
Thyroid gland	Direct injury by virus to thyroid gland	Atypical subacute thyroiditis
	Immune-mediated injury to thyroid gland	Hashimoto's thyroiditis Graves' disease
Gonads	Injury to testicular cells resulting in low testosterone and activation of hypothalamo–pituitary–gonadal axis	Elevated levels of LH and low serum testosterone levels Low serum testosterone may serve as severity marker in elderly males
	Presence of ACE-2 receptors on testes resulting in virus entry and presence in semen	Semen analysis of COVID-19 patients showed presence of SARS-CoV-2 Long-term implications unknown

12.7 Conclusion

COVID-19 is associated with endocrine dysfunction, which appears to be due to interactions at multiple levels. There are concerns for long-term metabolic and endocrine complications after SARS-CoV-2 infection, which need further exploration.[81]

References

1. S. Zaim, J. H. Chong, V. Sankaranarayanan and A. Harky, COVID-19 and Multiorgan Response, *Curr. Probl. Cardiol.*, 2020, **45**, 100618.
2. M. Noris, A. Benigni and G. Remuzzi, The case of Complement activation in COVID-19 multiorgan impact, *Kidney Int.*, 2020, **98**(2), 314–322.

3. T. Tan, B. Khoo, E. G. Mills, M. Phylactou, B. Patel and P. C. Eng, *et al.*, Association between high serum total cortisol concentrations and mortality from COVID-19, *Lancet Diabetes Endocrinol.*, 2020, **8**, 659–660.
4. W. Wang, X. Su, Y. Ding, W. Fan, J. Su and Z. Chen, *et al.*, Thyroid function abnormalities in COVID-19 patients, *Front. Endocrinol.*, 2021, **11**, 623792.
5. Y. Wu, X. Xu, Z. Chen, J. Duan, K. Hashimoto and L. Yang, *et al.*, Nervous system involvement after infection with COVID-19 and other coronaviruses, *Brain, Behav., Immun.*, 2020. Available from: http://www.sciencedirect.com/science/article/pii/S0889159120303573.
6. J. L. Chan, K. D. Gregory, S. S. Smithson, M. Naqvi and A. N. Mamelak, Pituitary apoplexy associated with acute COVID-19 infection and pregnancy, *Pituitary*, 2020, **23**, 716–720.
7. A. V. Turnbull and C. Rivier, Regulation of the HPA Axis by Cytokines, *Brain, Behav., Immun.*, 1995, **9**, 253–275.
8. M. K.-S. Leow, D. S.-K. Kwek, A. W.-K. Ng, K.-C. Ong, G. J.-L. Kaw and L. S.-U. Lee, Hypocortisolism in survivors of severe acute respiratory syndrome (SARS), *Clin. Endocrinol.*, 2005, **63**, 197–202.
9. L. Wei, S. Sun, J. Zhang, H. Zhu, Y. Xu and Q. Ma, *et al.*, Endocrine cells of the adenohypophysis in severe acute respiratory syndrome (SARS), *Biochem. Cell Biol.*, 2010, **88**, 723–730.
10. G. Li, L. Zhou, C. Zhang, Y. Shi, D. Dong and M. Bai, *et al.*, Insulin-Like Growth Factor 1 Regulates Acute Inflammatory Lung Injury Mediated by Influenza Virus Infection, *Front. Microbiol.*, 2019, cited 4 July 2020. 10 Available from: https://www.frontiersin.org/articles/10.3389/fmicb.2019.02541/full.
11. Q. Ye, B. Wang and J. Mao, The pathogenesis and treatment of the `Cytokine Storm' in COVID-19, *J. Infect.*, 2020, **80**, 607–613.
12. L. Hennighausen and H. K. Lee, Activation of the SARS-CoV-2 receptor Ace2 by cytokines through pan JAK-STAT enhancers during Pregnancy, *Cell Rep.*, 2020, **32**(13), 108199.
13. T. Chen, D. Wu, H. Chen, W. Yan, D. Yang and G. Chen, *et al.*, Clinical characteristics of 113 deceased patients with coronavirus disease 2019: retrospective study, *BMJ*, 2020, **368**, m1091.
14. D. Dworakowska and A. B. Grossman, Thyroid disease in the time of COVID-19, *Endocrine*, 2020, **68**, 471–474.
15. A. Naaraayan, S. Pant and S. Jesmajian, Severe Hyponatremic Encephalopathy in a Patient With COVID-19, *Mayo Clin. Proc.*, 2020, **95**, 2285–2286.
16. Z. Yousaf, S. D. Al-Shokri, H. Al-Soub and M. F. H. Mohamed, COVID-19-associated SIADH: a clue in the times of pandemic!, *Am. J. Physiol.*, 2020, **318**, E882–E885.
17. A. J. Turner and N. M. Hooper, 84 – Angiotensin-converting enzyme 2, in *Handbook of Proteolytic Enzymes*, ed. Barrett A. J., Rawlings N. D. and Woessner J. F., Academic Press, London, 2nd edn, 2004, pp. 349–51. Available from: http://www.sciencedirect.com/science/article/pii/B9780120796113500926.

18. W. F. Paolo Jr and J. D. Nosanchuk, Adrenal infections, *Int. J. Infect. Dis.*, 2006, **10**, 343–353.
19. J. Álvarez-Troncoso, M. Zapatero Larrauri, M. D. Montero Vega, R. Gil Vallano, E. Palmier Peláez and P. Martín Rojas-Marcos, *et al.*, Case Report: COVID-19 with Bilateral Adrenal Hemorrhage, *Am. J. Trop. Med. Hyg.*, 2020, **103**, 1156–1157.
20. M. Freire Santana, M. G. S. Borba, D. C. Baía-da-Silva, F. Val, M. A. A. Alexandre and J. D. Brito-Sousa, *et al.*, Case Report: Adrenal Pathology Findings in Severe COVID-19: An Autopsy Study, *Am. J. Trop. Med. Hyg.*, 2020, **103**, 1604–1607.
21. R. Kumar, T. Guruparan, S. Siddiqi, R. Sheth, M. Jacyna and M. Naghibi, *et al.*, A case of adrenal infarction in a patient with COVID 19 infection, *BJR Case Rep.*, 2020, cited 2 December 2020. 6 Available from: https://www.ncbi.nlm.nih.gov/pmc/articles/PMC7465735/.
22. P. Leyendecker, S. Ritter, M. Riou, A. Wackenthaler, F. Meziani and C. Roy, *et al.*, Acute adrenal infarction as an incidental CT finding and a potential prognosis factor in severe SARS-CoV-2 infection: a retrospective cohort analysis on 219 patients, *Eur. Radiol.*, 2020, 1–6.
23. R. Wheatland, Molecular mimicry of ACTH in SARS – implications for corticosteroid treatment and prophylaxis, *Med. Hypotheses*, 2004, **63**, 855–862.
24. L. Kolilekas, K. Loverdos, S. Giannakaki, L. Vlassi, A. Levounets and E. Zervas, *et al.*, Can steroids reverse the severe COVID-19 induced "cytokine storm", *J. Med. Virol.*, 2020, 2866–2869.
25. C. L. Hough, Should we ever give steroids to ARDS patients? *Clin. Chest Med.*, 2014, **35**, 781–795.
26. RECOVERY Collaborative Group, P. Horby, W. S. Lim, J. R. Emberson, M. Mafham and J. L. Bell, *et al.*, Dexamethasone in Hospitalized Patients with Covid-19, *N. Engl. J. Med.*, 2021, **384**, 693–704.
27. C. Scaroni, M. Armigliato and S. Cannavò, COVID-19 outbreak and steroids administration: are patients treated for Sars-Cov-2 at risk of adrenal insufficiency? *J. Endocrinol. Invest.*, 2020, **43**, 1035–1036.
28. WHO Rapid Evidence Appraisal for COVID-19 Therapies (REACT) Working Group, J. A. C. Sterne, S. Murthy, J. V. Diaz, A. S. Slutsky and J. Villar, *et al.*, Association Between Administration of Systemic Corticosteroids and Mortality Among Critically Ill Patients With COVID-19: A Meta-analysis, *JAMA*, 2020, **324**, 1330–1341.
29. J. McNabb-Baltar, D. X. Jin, A. S. Grover, W. D. Redd, J. C. Zhou and K. E. Hathorn, *et al.*, Lipase Elevation in Patients With COVID-19, *Am. J. Gastroenterol.*, 2020. Publish Ahead of Print. Available from: https://journals.lww.com/ajg/Fulltext/9000/Lipase_Elevation_in_Patients_With_COVID_19.99261.aspx.
30. A. Hadi, M. Werge, K. T. Kristiansen, U. G. Pedersen, J. G. Karstensen and S. Novovic, *et al.*, Coronavirus Disease-19 (COVID-19) associated with severe acute pancreatitis: Case report on three family members, *Pancreatology*, 2020, **20**, 665–667.

31. F. Wang, H. Wang, J. Fan, Y. Zhang, H. Wang and Q. Zhao, Pancreatic injury patterns in patients with COVID-19 pneumonia, *Gastroenterology*, 2020, **159**(1), 367–370.
32. G. Iacobellis, C. A. Penaherrera, L. E. Bermudez and E. Bernal Mizrachi, Admission hyperglycemia and radiological findings of SARS-CoV2 in patients with and without diabetes, *Diabetes Res. Clin. Pract.*, 2020, **164**, 108185.
33. L. Yang, Y. Han, B. E. Nilsson-Payant, V. Gupta, P. Wang and X. Duan, *et al.*, A Human Pluripotent Stem Cell-based Platform to Study SARS-CoV-2 Tropism and Model Virus Infection in Human Cells and Organoids, *Cell Stem Cell*, 2020, cited 26 June 2020. Available from: http://www.sciencedirect.com/science/article/pii/S1934590920302824.
34. S. Kwon, K. L. Hermayer and K. Hermayer, Glucocorticoid-Induced Hyperglycemia, *Am. J. Med. Sci.*, 2013, **345**, 274–277.
35. K. Chinsky, The Evolving Paradigm of Hyperglycemia and Critical Illness, *Chest*, 2004, **126**, 674–676.
36. O. Y. Bello-Chavolla, J. P. Bahena-López, N. E. Antonio-Villa, A. Vargas-Vázquez, A. González-Díaz and A. Márquez-Salinas, *et al.*, Predicting mortality due to SARS-CoV-2: A mechanistic score relating obesity and diabetes to COVID-19 outcomes in Mexico, *J. Clin. Endocrinol. Metab.*, 2020, **105**(8), dgaa346.
37. The Lancet Diabetes Endocrinology, COVID-19 and diabetes: a co-conspiracy? *Lancet Diabetes Endocrinol.*, 2020, **8**, 801.
38. F. Rubino, S. A. Amiel, P. Zimmet, G. Alberti, S. Bornstein and R. H. Eckel, *et al.*, New-Onset Diabetes in Covid-19, *N. Engl. J. Med.*, 2020, **383**, 789–790.
39. Y. J. Chee, S. J. H. Ng and E. Yeoh, Diabetic ketoacidosis precipitated by Covid-19 in a patient with newly diagnosed diabetes mellitus, *Diabetes Res. Clin. Pract.*, **2020**, **164**, 108166.
40. A. I. Heaney, G. D. Griffin and E. L. Simon, Newly diagnosed diabetes and diabetic ketoacidosis precipitated by COVID-19 infection, *Am. J. Emerg. Med.*, 2020, **38**(11), 2491.e3–2491.e4.
41. J. Li, X. Wang, J. Chen, X. Zuo, H. Zhang and A. Deng, COVID-19 infection may cause ketosis and ketoacidosis, *Diabetes, Obes. Metab.*, 2020, 1935–1941.
42. J.-K. Yang, S.-S. Lin, X.-J. Ji and L.-M. Guo, Binding of SARS coronavirus to its receptor damages islets and causes acute diabetes, *Acta Diabetol.*, 2010, **47**, 193–199.
43. CoviDiab Registry, cited 8 April 2021. Available from: https://covidiab.e-dendrite.com/.
44. W. Wang, Y. X. Ye and H. Yao, Evaluation and observation of serum thyroid hormone and parathyroid hormone in patients with severe acute respiratory syndrome, *J. Chin. Antituberc. Assoc.*, 2003, **25**, 232–234.
45. L. Wei, S. Sun, C.-H. Xu, J. Zhang, Y. Xu and H. Zhu, *et al.*, Pathology of the thyroid in severe acute respiratory syndrome, *Hum. Pathol.*, 2007, **38**, 95–102.

46. S. Ippolito, F. Dentali and M. L. Tanda, SARS-CoV-2: a potential trigger for subacute thyroiditis? Insights from a case report, *J. Endocrinol. Invest.*, 2020, 1171–1172.
47. A. Brancatella, D. Ricci, N. Viola, D. Sgrò, F. Santini and F. Latrofa, Subacute Thyroiditis After Sars-COV-2 Infection, *J. Clin. Endocrinol. Metab.*, 2020, **105**, dgaa276.
48. L. Y. Tee, S. Harjanto and B. H. Rosario, COVID-19 complicated by Hashimoto's thyroiditis, *Singapore Med. J.*, 2020, **62**(5), 265.
49. M. Mateu-Salat, E. Urgell and A. Chico, SARS-COV-2 as a trigger for autoimmune disease: report of two cases of Graves' disease after COVID-19, *J. Endocrinol. Invest.*, 2020, **43**, 1527–1528.
50. E. Asfuroglu Kalkan and I. Ates, A case of subacute thyroiditis associated with Covid-19 infection, *J. Endocrinol. Invest.*, 2020, **43**, 1173–1174.
51. R. M. Ruggeri, A. Campennì, M. Siracusa, G. Frazzetto and D. Gullo, Subacute thyroiditis in a patient infected with SARS-COV-2: an endocrine complication linked to the COVID-19 pandemic, *Hormones*, 2021, **20**(1), 219–221.
52. A. Khatri, E. Charlap and A. Kim, Subacute Thyroiditis from COVID-19 Infection: A Case Report and Review of Literature, *Eur. Thyroid J.*, 2020, **9**, 325–329.
53. A. Brancatella, D. Ricci, D. Cappellani, N. Viola, D. Sgrò and F. Santini, *et al.*, Is Subacute Thyroiditis an Underestimated Manifestation of SARS-CoV-2 Infection? Insights From a Case Series, *J. Clin. Endocrinol. Metab.*, 2020, **105**, dgaa537.
54. S. A. M. Mattar, S. J. Q. Koh, S. Rama Chandran and B. P. Z. Cherng, Subacute thyroiditis associated with COVID-19, *BMJ Case Rep.*, 2020, **13**, e237336.
55. E. Campos-Barrera, T. Alvarez-Cisneros and M. Davalos-Fuentes, Subacute Thyroiditis Associated with COVID-19, *Case Rep. Endocrinol.*, 2020, **2020**, 8891539.
56. M. D. J. San Juan, M. Q. V. Florencio and M. H. Joven, Subacute Thyroiditis in a Patient with Coronavirus Disease 2019, *AACE Clin. Case Rep.*, 2020, **6**, e361–e364.
57. R. Barahona San Millán, M. Tantinyà Daura, A. Hurtado Ganozá and M. Recasens Sala, Painless thyroiditis in SARS-CoV-2 infection, *Endocrinol. Diabetes Nutr.*, 2020, **68**(10), 757–758.
58. W. H. Chong, B. Shkolnik, B. Saha and S. Beegle, Subacute Thyroiditis in the Setting of Coronavirus Disease 2019, *Am. J. Med. Sci.*, 2020, **361**(3), 400–402.
59. R. Ruano, M. Zorzano-Martinez, A. Campos, F. Rius and M. Hernández, Subacute thyroiditis might be a complication triggered by SARS-CoV-2, *Endocrinol. Diabetes Nutr.*, 2020, **68**(10), 755–756.
60. M. Guven, Subacute Thyroiditis in the Course of Coronavirus Disease 2019: A Case Report, *J. Endocrinol. Metab.*, 2020, **10**, 110–112.
61. Clinical management of COVID-19, cited 12 December 2020. Available from: https://www.who.int/publications-detail-redirect/clinical-management-of-covid-19.

62. I. Muller, D. Cannavaro, D. Dazzi, D. Covelli, G. Mantovani and A. Muscatello, *et al.*, SARS-CoV-2-related atypical thyroiditis, *Lancet Diabetes Endocrinol.*, 2020, cited 6 August 2020. Available from: https://www.thelancet.com/journals/landia/article/PIIS2213-8587(20)30266-7/abstract.
63. M. Rotondi, F. Coperchini, G. Ricci, M. Denegri, L. Croce and S. T. Ngnitejeu, *et al.*, Detection of SARS-COV-2 receptor ACE-2 mRNA in thyroid cells: a clue for COVID-19-related subacute thyroiditis, *J. Endocrinol. Invest.*, 2020, **44**(5), 1085–1090.
64. A. Lania, M. T. Sandri, M. Cellini, M. Mirani, E. Lavezzi and G. Mazziotti, Thyrotoxicosis in patients with COVID-19: the THYRCOV study, *Eur. J. Endocrinol.*, 2020, **183**, 381–387.
65. G. Speer and P. Somogyi, Thyroid complications of SARS and coronavirus disease 2019 (COVID-19), *Endocr. J.*, 2021, **68**, 129–136.
66. W. Gao, W. Guo, Y. Guo, M. Shi, G. Dong and G. Wang, *et al.*, Thyroid hormone concentrations in severely or critically ill patients with COVID-19, *J. Endocrinol. Invest.*, 2020, **44**(5), 1031–1040.
67. D. T. W. Lui, C. H. Lee, W. S. Chow, A. C. H. Lee, A. R. Tam and C. H. Y. Fong, *et al.*, Thyroid Dysfunction in Relation to Immune Profile, Disease Status and Outcome in 191 Patients with COVID-19, *J. Clin. Endocrinol. Metab.*, 2020, **106**(2), e926–e935.
68. M. van Gerwen, M. Alsen, C. Little, J. Barlow, L. Naymagon and D. Tremblay, *et al.*, Outcomes of Patients With Hypothyroidism and COVID-19: A Retrospective Cohort Study, *Front. Endocrinol.*, 2020, **11**, 565.
69. J. Karlberg, D. S. Y. Chong and W. Y. Y. Lai, Do men have a higher case fatality rate of severe acute respiratory syndrome than women do? *Am. J. Epidemiol.*, 2004, **159**, 229–231.
70. C. Gebhard, V. Regitz-Zagrosek, H. K. Neuhauser, R. Morgan and S. L. Klein, Impact of sex and gender on COVID-19 outcomes in Europe, *Biol. Sex Differ.*, 2020, **11**, 29.
71. M. Yang, S. Chen, B. Huang, J.-M. Zhong, H. Su and Y.-J. Chen, *et al.*, Pathological Findings in the Testes of COVID-19 Patients: Clinical Implications, *Eur. Urol. Focus*, 2020, **6**, 1124–1129.
72. H. Li, X. Xiao, J. Zhang, M. I. Zafar, C. Wu and Y. Long, *et al.*, Impaired spermatogenesis in COVID-19 patients, *EClinicalMedicine*, 2020, cited 30 November 2020. 28 Available from: https://doi.org/10.1016/j.eclinm.2020.100604.
73. L. Gagliardi, C. Bertacca, C. Centenari, I. Merusi, E. Parolo and V. Ragazzo, *et al.*, Orchiepididymitis in a Boy With COVID-19, *Pediatr. Infect. Dis. J.*, 2020, **39**(8), e200–e202.
74. A. La Marca, S. Busani, V. Donno, G. Guaraldi, G. Ligabue and M. Girardis, Testicular pain as an unusual presentation of COVID-19: a brief review of SARS-CoV-2 and the testis, *Reprod. BioMed. Online*, 2020, **41**, 903–906.

75. G. Corona, E. Baldi, A. M. Isidori, D. Paoli, F. Pallotti and L. De Santis, *et al.*, SARS-CoV-2 infection, male fertility and sperm cryopreservation: a position statement of the Italian Society of Andrology and Sexual Medicine (SIAMS) (Società Italiana di Andrologia e Medicina della Sessualità), *J. Endocrinol. Invest.*, 2020, **43**, 1153–1157.
76. L. Ma, W. Xie, D. Li, L. Shi, G. Ye and Y. Mao, *et al.*, Evaluation of sex-related hormones and semen characteristics in reproductive-aged male COVID-19 patients,*J. Med. Virol.*, 2020, **93**(1), 456–462.
77. J. Liu, H. Ji, W. Zheng, X. Wu, J. J. Zhu and A. P. Arnold, *et al.*, Sex differences in renal angiotensin converting enzyme 2 (ACE2) activity are 17β-oestradiol-dependent and sex chromosome-independent, *Biol. Sex Differ.*, 2010, **1**(1), 6.
78. M. Montopoli, S. Zumerle, R. Vettor, M. Rugge, M. Zorzi and C. V. Catapano, *et al.*, Androgen-deprivation therapies for prostate cancer and risk of infection by SARS-CoV-2: a population-based study (N = 4532), *Ann. Oncol.*, 2020, **31**, 1040–1045.
79. A. Klein Eric, J. Li, A. Milinovich, J. D. Schold, S. Nima and M. W. Kattan, *et al.*, Androgen Deprivation Therapy in Men With Prostate Cancer Does Not Affect Risk of Infection With SARS-CoV-2,*J. Urol.*, 441–443.
80. G. Beccuti, L. Ghizzoni, V. Cambria, V. Codullo, P. Sacchi and E. Lovati, *et al.*, A COVID-19 pneumonia case report of autoimmune polyendocrine syndrome type 1 in Lombardy, Italy: letter to the editor, *J. Endocrinol. Invest.*, 2020, **43**(8), 1175–1177.
81. L. M. Mongioì, F. Barbagallo, R. A. Condorelli, R. Cannarella, A. Aversa and S. La Vignera, *et al.*, Possible long-term endocrine-metabolic complications in COVID-19: lesson from the SARS model, *Endocrine*, 2020, **68**(3), 467–470.

CHAPTER 13

Potential Risks of COVID-19 in Reproductive Health

KHALED HABAS[a], CLAUDE HUGHES[b,c] AND DIANA ANDERSON*[a]

[a]School of Chemistry & Biosciences, Faculty of Life Sciences, University of Bradford, Richmond Road, Bradford, West Yorkshire, UK; [b]Department of Obstetrics & Gynecology, Duke University Medical Center and Therapeutic Science & Strategy Unit, Durham, North Carolina, USA; [c]IQVIA, Durham, North Carolina, USA
*E-mail: d.anderson1@bradford.ac.uk

Highlights

- Pregnant women could be at higher risk of severe acute respiratory disease coronavirus 2 (SARS-CoV-2) infection compared to the general population.
- Pregnancy presents a unique immunological situation, characterized by increased vulnerability to some infectious diseases, including respiratory pathogens.
- Very limited evidence supports vertical transmission of COVID-19.
- Similar modes of entry for SARS-CoV-2 have been reported for cells in males and females.

The Coronavirus Pandemic and the Future Volume 1: Virology, Epidemiology, Translational Toxicology and Therapeutics
Edited by Michael D. Waters, Alok Dhawan, Tim Marrs, Diana Anderson, Stafford Warren and Claude L. Hughes

Published by the Royal Society of Chemistry, www.rsc.org

13.1 Introduction

Coronavirus disease 2019 (COVID-19) is an emerging acute infectious disease that was first reported in China in December 2019, and the World Health Organization declared the outbreak of COVID-19 a pandemic on 11 March 2020. COVID-19 results from infection with severe acute respiratory disease coronavirus 2 (SARS-CoV-2), which is a new virus to humans. However, information on diseases related to other highly pathogenic coronaviruses, *e.g.* severe acute respiratory syndrome (SARS) may provide insights into the effects of COVID-2019 during pregnancy. Moreover, previous data on SARS and Middle East respiratory syndrome (MERS) imply that clinical findings during pregnancy can range from no symptoms to severe disease and death. The main symptoms of COVID-19 are cough and fever, with more than 80% of hospitalized patients presenting with these symptoms.[1] The importance of understanding the underlying mechanisms of viral infection during pregnancy is obviously of great concern, since maternal infections could also adversely affect the foetus.[2] The risk and mode(s) of mother-to-child transmission of SARS-CoV-2, during pregnancy and labour and/or breastfeeding, is a key concern for the health and care of the infant. A very important question that remains unanswered is whether COVID-19 can be transmitted from a pregnant women to her foetus, a process named vertical transmission, and it remains to determine mechanisms of viral infection during pregnancy.[1,3–7]

13.2 COVID-19 and the Female Reproductive System

COVID-19 is caused by SARS-CoV-2, which shares 79% amino acid sequence homology with SARS; they may both enter into target host cells by the same cellular receptor, angiotensin-converting enzyme 2 (ACE2), and cellular protease, transmembrane protease serine (TMPRSS), due to similarity of structural analyses.[8,9] It has been reported that organs with a high expression of ACE2 and TMPRSS2 are more susceptible to infection, and the reproductive system has been reported as one of the potential target organs of COVID-19 infection.[10] Reis *et al.* demonstrated the presence of an angiotensin-(1–7)-Mas receptor–ACE2 axis and the ACE2 markers in all levels of follicles in the human ovary,[11] to be a part of the renin–angiotensin system, which plays a critical role in the female reproduction system, included steroidogenesis, folliculogenesis, oocyte maturation and ovulation.[11] ACE2 expression occurs in primordial, primary, secondary and antral follicles and corpora lutea of human ovaries.[11] ACE2 expression is also upregulated during the course of follicle development, and further increases after gonadotrophin stimulation.[12] However, a recent study reported that ACE2 had a very low expression in the different cell types, including syncytiotrophoblast (STB) cells, decidual stromal cells and epithelial glandular cells of the maternal–foetal interface.[13] This finding suggested that the potential effect of SARS-CoV-2 on early pregnancy could be due to the lack of susceptible cells at the interface between mother and child in early pregnancy and the rate of mother-to-infant transmission might be very low; SARS-COV-2 infection may alter the ACE2 expression at the interface with consequences on pregnancy loss.[13] ACE2 was found to reach higher levels in early gestation,[14] and is expressed in the primary and secondary decidual zone, and in luminal and glandular epithelial cells during early gestation.[15] During late gestation, ACE2 staining is visualized in the labyrinthine placenta.[16] The placenta is the main maternal–foetal interface during pregnancy, and virus receptor expression in the placenta may play a key role in permitting transmission of SARS-CoV-2.[1] Pringle *et al.* found that during early pregnancy, human placental renin–angiotensin–aldosterone system was upregulated and played a key role in the development of placental vasculature.[14] A study has shown that a high level of ACE2 mRNA was found in the placenta during early pregnancy, located in the STB and villous stroma, in addition to ACE2 controlling the release of Ang 1–7, which was valuable for maternal–foetal circulation, vasodilation and virus spread.[17]

13.3 Early Pregnancy

There is as yet no evidence of intrauterine infection caused by vertical transmission in women with COVID-19 in the third trimester, and the effect of SARS-CoV-2 on early pregnancy is still unidentified.[5] However, in general, pregnant women are particularly vulnerable to respiratory pathogens and are at heightened risk of severe pneumonia and severe disease related to other respiratory illnesses (*e.g.* 2009 H1N1 influenza),[18] because they are in an immunosuppressive state. Additionally, it is likely that physiological adaptive changes during pregnancy such as increased oxygen consumption and oedema of the respiratory tract mucosa reduce their tolerance to hypoxia.[1] It is noteworthy that the distinctive immunological adaptation of pregnancy that is essential for maintaining tolerance of the foetal semi-allograft[19] involves a state of transient suppressed immunity controlled *via* suppressing T-cell activity, which also predisposes pregnant women to viral infections.[20] The several physiological changes that occur in the immunological, respiratory and circulatory systems might worsen clinical outcomes when infected with a virus during pregnancy.[18] Nonetheless, there is not yet sufficient information to determine whether pregnancy increases vulnerability to COVID-19 compared with the general population.

Epidemiological evidence has shown that pregnant women are at a higher risk of severe illness and mortality from viral infections,[21,22] particularly during pandemics such as influenza, Ebola and Lassa fever.[23,24] Viral infections may predispose the pregnancy to preterm labour and preterm delivery by infection with other superimposed micro-organisms.[22,25] Consequently, understanding why pregnant women are at a higher risk is vital in order to design appropriate approaches for treatment as well as for prevention.[26] Furthermore, SARS and MERS viruses are known to be associated with severe complications during pregnancy, including the need for endotracheal intubation, admission to an intensive care unit (ICU), renal failure and death.[27,28]

SARS-CoV-2 infection could be related to poor pregnancy outcomes, such as spontaneous abortion, critical maternal illness and maternal death.[5] Wong and colleagues reported that four out of seven women infected with SARS had a spontaneous abortion in the first trimester.[28]

Potential foetal effects of COVID-19 and consequences for the neonate remain poorly understood. Most gestational infections reported from China have been during the third or late-second trimester, so whether first-trimester COVID-19 infection can induce birth defects or pregnancy loss is still unknown. Reported data indicated that some neonates born to mothers with COVID-19 during pregnancy were born preterm or of low birth weight, but whether these outcomes were causally associated with COVID-19 is unclear.[29] Recently, data from a case series of 18 women (19 infants) infected with COVID-19 in the third trimester of pregnancy recognized no laboratory evidence of vertical transmission.[1,30]

Transmission of COVID-19 from mother to infant might occur prenatally, perinatally or postnatally. Rasmussen and colleagues reported that most newborns tested after birth were negative for COVID-19.[18] However, symptomatic newborns of mothers infected with COVID-19 have been shown to have COVID-19 infection at a few days of life. Whether this was due to prenatal, perinatal or postnatal transmission is unidentified.[6] There has been no evidence so far of increased perinatal transmission during vaginal delivery.[6,31] It is important to note that at present and in relation to a vaginal birth, whenever vaginal secretions were tested for COVID-19, the results were negative.[32] It was reported that among six mothers with confirmed COVID-19, virus-specific antibodies were detected in neonatal blood sera samples. The immunoglobulin (Ig)G concentrations were elevated in five infants. IgG is passively transferred across the placenta from mother to foetus, beginning at the end of the second trimester, and reaches high levels at the time of birth.[33] IgM was detected in two infants; it is not usually transferred from mother to foetus because of its larger macromolecular structure. This suggests that IgM could have been produced by the infant if the virus had crossed the placenta.[30] In one instance, a newborn with elevated SARS-Cov-2 IgM and IgG at 2 hours after birth with negative nasopharyngeal swabs for reverse transcriptase polymerase chain reaction from 2 hours to 16 days after birth, no testing had been done on amniotic fluid, placenta, or cord blood.[34] These studies were not able to provide ultimate evidence for intrauterine transmission, because cross-reactivity and false-positive IgM test results can occur.[34]

It remains unclear whether SARS-CoV-2 can be shed into breastmilk and transmitted to a child through breastfeeding. Recent studies have found no evidence of SARS-CoV-2 in human breastmilk, but sample sizes were small.[35,36] It was reported that viral RNA has been detected in breastmilk samples from a single woman with COVID-19, and her infant tested positive, but whether the infant was infected through breastfeeding is unclear.[37] Further studies of milk samples from lactating women and potential virus transmission *via* breastfeeding are needed to develop recommendations on whether mothers with COVID-19 should breastfeed.

Physiological and mechanical changes in pregnancy increase susceptibility to infections in general, particularly when the cardiorespiratory system is affected, but there are no comparative data to determine whether pregnancy is a risk factor for severe SARS-CoV-2 pneumonia combinations with these risk factors. Maternal venous thromboembolism is a concern, since it is known that pregnancy induces a hypercoagulable state that predisposes pregnant women to blood clot formation. The latest evidence from patients with COVID-19 shows they face an increased risk of clotting.[38]

Another risk factor which has been revealed to increase maternal morbidity is smoking, which has been related to more severe disease in COVID-19 infection.[32] Data from China on preterm labour in 37 pregnant women with COVID-19 infection showed that the incidence of preterm birth was 19%.[39] However, it is unclear whether preterm labour in these women was iatrogenic or spontaneous.[1]

13.4 Clinical Presentation of COVID-19 Infection During Pregnancy

In pregnant women, the main presenting symptoms of COVID-19 are cough and fever, with more than 80% of hospitalized patients presenting with these symptoms.[1] Data from China showed that among 147 pregnant women, 8% had severe illness and 1% had serious infections, which were lower than the rates in the nonpregnant population, which were 14% with severe disease and 6% with critical illness.[40] A case series from China comprised mainly of women with third-trimester infection have revealed that clinical findings in pregnant women were similar to those seen in the general population.[18] Data from a Swedish study showed that pregnant and postpartum women infected with COVID-19 were five times more likely to be admitted to an ICU compared to nonpregnant women of a similar age.[41] Data from the United States showed that 43 pregnant women with COVID-19 presented for care at two hospitals in New York City. This case series did not include a nonpregnant control group; however, the proportion of pregnant women with severe disease was similar to that described in nonpregnant adults with COVID-19.[42]

Chen *et al.* (2020) reported the clinical features of nine pregnant women who tested positive for COVID-19 in the third trimester, which encompassed mainly fever and cough as well as other symptoms including malaise, myalgia, sore throat, diarrhoea and shortness of breath. Laboratory test results showed that most of the patients had lymphopenia and increased C-reactive protein; in addition, chest computed tomography scans showed multiple patchy ground-glass shadows in the lungs. Pregnancy complications that seemed to occur after the onset of COVID-19 infection were foetal distress in two out of nine patients and premature rupture of the membranes in two out of nine patients. None of the patients developed severe COVID-19 pneumonia or died.[1]

Ferrazzi *et al.* reported that among 42 affected pregnant women, 19 developed pneumonia, seven required oxygen and four required critical care. 18 out of these 19 women were delivered by Caesarean section, although in eight cases the indications were unrelated to COVID-19. Three babies tested positive for SARS-CoV-2. Two of these infants developed COVID-19 postnatally and had been breastfed without the mother using a mask. One infant delivered vaginally did not breastfeed, developed respiratory symptoms requiring ventilation for a day and tested positive for SARS-CoV-2. No mothers or babies died.[43]

In a United Kingdom national cohort of pregnant women admitted to hospital with COVID-19, factors associated with infection were identified and the outcomes were described, including transmission of infection for mothers and infants. In this prospective national population-based cohort study using the United Kingdom Obstetric Surveillance System, data showed that the incidence of admission to hospital with confirmed SARS-CoV-2 infection in pregnancy was 4.9 (95% CI 4.5–5.4) per 1000 maternities. 263 (56%) pregnant women admitted to hospital with SARS-CoV-2 infection in pregnancy were from black or other ethnic minority groups; 281 (69%) were overweight or obese; 175 (41%) were aged ≥35 years; and 145 (34%) had pre-existing comorbidities. 266 (62%) women gave birth or had a pregnancy loss; 196 (73%) gave birth at term. 41 (10%) women admitted to hospital needed respiratory support, and five (1%) women died. 12 (5%) out of 265 infants tested positive for SARS-CoV-2 RNA; six of them within the first 12 hours after birth. Most pregnant women admitted to hospital with COVID-19 were in the late second or third trimester, supporting guidance for continued social distancing measures in later pregnancy. Most had good outcomes; transmission of SARS-CoV-2 to infants was uncommon; and the high proportion of women from black or minority ethnic groups was notable.[44]

It was reported on 5 February 2020 by multiple media outlets that a newborn infant delivered to a mother who had tested positive for the virus during the epidemic in Wuhan had tested positive for SARS-CoV-2 at the Wuhan Children's Hospital in Hubei Province 30 hours following its birth.[4]

The Royal College of Obstetricians and Gynaecologists' guidance reported that COVID-19 in pregnancy caused serious morbidity, which occurred in two out of 12 women who required intensive care. Among those 12 women, nine were delivered by Caesarean section from 36 weeks onwards and two were preterm. There was no evidence of COVID-19 in the amniotic fluid, umbilical cord blood, neonatal throat swabs or breastmilk samples. Compared to SARS and MERS, COVID-19 appears to be less lethal, but there are a limited number of cases reported to date.[45]

It was reported from China that SARS-CoV-2 viral tests were carried out in a range of body fluids from mothers and babies with COVID-19 at RenMin Hospital of Wuhan University.[46] One neonate subsequently tested positive for serum conversion, but no others. One out of nine stool samples, none out of 13 vaginal samples and one out of three breastmilk samples were positive for COVID-19. Two preterm infants were diagnosed with pneumonia on chest X-ray. There was one biochemical pregnancy in the first trimester with a serum human chorionic gonadotrophin of 25.9 IU. 10 cases reverted to negative and no mothers or babies died.[46]

13.5 COVID-19 and the Male Reproductive System

Reproductive risks of COVID-19 are not restricted to women. There is modest evidence that men also suffer reproductive consequences from COVID-19, including potential destruction of testicular tissues.[47] Recent studies have recognized that higher rates of infection, morbidity and mortality in male patients are likely due to male genetic susceptibility.[48] Wang and Xu identified the main pathway for SARS-CoV-2 entry into cells to be by the viral spike (S) protein attaching to ACE2 and employing the cellular serine protease (TMPRSS2) for S-protein priming, known to be expressed in testicular tissue.[49] Shen *et al.* reported that the expression of ACE2 in testicular cells is age-related, with the highest expression documented in patients aged 30 years, which was higher than seen in men in their twenties; whereas patients aged 60 years revealed the lowest level of ACE2 expression.[50] This might suggest that young male patients are at higher risk of testicular impairment by COVID-19 than older patients.[51] Androgen receptor activity has also been considered a requirement for the transcription of the TMPRSS2 gene, as no other known TMPRSS2 gene promoter has been reported to exert the same action in humans.[52] The modulation of TMPRSS2 expression by testosterone has been suggested to contribute to the male predominance of COVID-19 infection.[53] Studies showed that ACE2 is a constitutive product of adult-type Leydig cells,[54] thus suggesting a role in testicular function and implying a possible involvement of the testes in COVID-19-infected patients, a factor which may affect testosterone secretion.[55]

A previous study reported that testes of six patients who died due to SARS in 2002 showed signs of orchitis and inflammatory infiltrates in the seminiferous tubules, and histopathological examination also showed IgG deposition in the seminiferous epithelium, interstitium, deteriorated germ cells and Sertoli cells.[56] These findings suggest that the testis could be a high-risk organ susceptible to attack by SARS-CoV-2. A recent report has shown the presence of SARS-CoV-2 in semen samples of six patients, including two subjects who were recovering from the clinical disease. This implies the possibility of prostate involvement, but this has not been investigated yet. Moreover, the RNA detected in semen could have been residual urinary presence.[57] SARS-CoV-2 infection may possibly affect sperm DNA and cause sperm DNA fragmentation through induction of oxidative stress following various direct oxidant-sensitive pathways, altering the reproductive endocrinological milieu.

It could also be through the course of COVID-19 treatment using certain broad-spectrum antiviral drugs.[58] Thus, disruption of functionality in male germ cells or supporting somatic cells might induce spermatogenic failure, resulting in male infertility following COVID-19 infection.

13.6 Conclusion

Taking account of the accumulating data, it appears that COVID-19 is less severe in pregnancy than the two earlier coronavirus infections, SARS and MERS. Further study is needed to evaluate the long-term effects of SARS-CoV-2 infection on female and male systems in relation to fertility, pregnancy outcomes and the growth and development of their offspring.

References

1. H. Chen, J. Guo, C. Wang, F. Luo, X. Yu and W. Zhang, *et al.*, Clinical characteristics and intrauterine vertical transmission potential of COVID-19 infection in nine pregnant women: a retrospective review of medical records, *Lancet*, 2020, **395**(10226), 809–815.
2. M.-T. Gervasi, R. Romero, G. Bracalente, T. Chaiworapongsa, O. Erez and Z. Dong, *et al.*, Viral invasion of the amniotic cavity (VIAC) in the midtrimester of pregnancy, *J. Matern.-Fetal Neonat. Med.*, 2012, **25**(10), 2002–2013.
3. Q. Li, X. Guan, P. Wu, X. Wang, L. Zhou and Y. Tong, *et al.*, Early transmission dynamics in Wuhan, China, of novel coronavirus–infected pneumonia, *N. Engl. J. Med.*, 2020, **382**(13), 1199–1207.
4. D. A. Schwartz and A. L. Graham, Potential maternal and infant outcomes from (Wuhan) coronavirus 2019-nCoV infecting pregnant women: lessons from SARS, MERS, and other human coronavirus infections, *Viruses*, 2020, **12**(2), 194.
5. X. Wang, Z. Zhou, J. Zhang, F. Zhu, Y. Tang and X. Shen, A case of 2019 Novel Coronavirus in a pregnant woman with preterm delivery, *Clin. Infect. Dis.*, 2020, **71**(15), 844–846.
6. L. Zeng, S. Xia, W. Yuan, K. Yan, F. Xiao and J. Shao, *et al.*, Neonatal early-onset infection with SARS-CoV-2 in 33 neonates born to mothers with COVID-19 in Wuhan, China, *JAMA Pediatr.*, 2020, **174**(7), 722–725.
7. N. Zhu, D. Zhang, W. Wang, X. Li, B. Yang and J. Song, *et al.*, A novel coronavirus from patients with pneumonia in China, 2019, *N. Engl. J. Med.*, 2020, **382**(8), 727–733.
8. R. Lu, X. Zhao, J. Li, P. Niu, B. Yang and H. Wu, *et al.*, Genomic characterisation and epidemiology of 2019 novel coronavirus: implications for virus origins and receptor binding, *Lancet*, 2020, **395**(10224), 565–574.
9. S. Lukassen, R. L. Chua, T. Trefzer, N. C. Kahn, M. A. Schneider and T. Muley, *et al.*, SARS-CoV-2 receptor ACE 2 and TMPRSS 2 are primarily expressed in bronchial transient secretory cells, *EMBO J.*, 2020, **39**(10), e105114.

10. X. Zou, K. Chen, J. Zou, P. Han, J. Hao and Z. Han, Single-cell RNA-seq data analysis on the receptor ACE2 expression reveals the potential risk of different human organs vulnerable to 2019-nCoV infection, *Front. Med.*, 2020, 1–8.
11. F. M. Reis, D. R. Bouissou, V. M. Pereira, A. F. Camargos, A. M. dos Reis and R. A. Santos, Angiotensin-(1-7), its receptor Mas, and the angiotensin-converting enzyme type 2 are expressed in the human ovary, *Fertil. Steril.*, 2011, **95**(1), 176–181.
12. V. M. Pereira, F. M. Reis, R. A. Santos, G. D. Cassali, S. H. Santos and K. Honorato-Sampaio, *et al.*, Gonadotropin stimulation increases the expression of angiotensin-(1–7) and Mas receptor in the rat ovary, *Reprod. Sci.*, 2009, **16**(12), 1165–1174.
13. Q.-L. Zheng, T. Duan and L.-P. Jin, Single-cell RNA expression profiling of ACE2 and AXL in the human maternal–fetal interface, *Reprod. Dev. Med.*, 2020, **4**(1), 7.
14. K. Pringle, M. Tadros, R. Callister and E. Lumbers, The expression and localization of the human placental prorenin/renin-angiotensin system throughout pregnancy: roles in trophoblast invasion and angiogenesis? *Placenta*, 2011, **32**(12), 956–962.
15. Y. Jing, L. Run-Qian, W. Hao-Ran, C. Hao-Ran, L. Ya-Bin and G. Yang, *et al.*, Potential influence of COVID-19/ACE2 on the female reproductive system, *Mol. Hum. Reprod.*, 2020, **26**(6), 367–373.
16. L. A. Neves, K. Stovall, J. Joyner, G. Valdés, P. E. Gallagher and C. M. Ferrario, *et al.*, ACE2 and ANG-(1–7) in the rat uterus during early and late gestation, *Am. J. Physiol.: Regul., Integr. Comp. Physiol.*, 2008, **294**(1), R151–R161.
17. S.-R. Isela, O.-V. Zendy, N.-M. Mónica, X.-R. Irene and A.-H. Omar, Changes in trophoblasts gene expression in response to perchlorate exposition, *Toxicol. In Vitro*, 2018, **50**, 328–335.
18. S. A. Rasmussen, J. C. Smulian, J. A. Lednicky, T. S. Wen and D. J. Jamieson, Coronavirus Disease 2019 (COVID-19) and pregnancy: what obstetricians need to know, *Am. J. Obstet. Gynecol.*, 2020, **222**(5), 415–426.
19. A. P. Weetman, Immunity, thyroid function and pregnancy: molecular mechanisms, *Nat. Rev. Endocrinol.*, 2010, **6**(6), 311–318.
20. M. Pazos, R. S. Sperling, T. M. Moran and T. A. Kraus, The influence of pregnancy on systemic immunity, *Immunol. Res.*, 2012, **54**(1–3), 254–261.
21. A. P. Kourtis, J. S. Read and D. J. Jamieson, Pregnancy and infection, *N. Engl. J. Med.*, 2014, **370**(23), 2211–2218.
22. J. Y. Kwon, R. Romero and G. Mor, New insights into the relationship between viral infection and pregnancy complications, *Am. J. Reprod. Immunol.*, 2014, **71**(5), 387–390.
23. D. J. Jamieson, T. M. Uyeki, W. M. Callaghan, D. Meaney-Delman and S. A. Rasmussen, What obstetrician–gynecologists should know about Ebola: a perspective from the Centers for Disease Control and Prevention, *Obstet. Gynecol.*, 2014, **124**(5), 1005–1010.

24. M. Price, S. Fisher-Hoch, R. Craven and J. McCormick, A prospective study of maternal and fetal outcome in acute Lassa fever infection during pregnancy, *Int. J. Gynecol. Obstet.*, 1989, **28**(3), 313.
25. R. Romero, J. Miranda, T. Chaiworapongsa, S. J. Korzeniewski, P. Chaemsaithong and F. Gotsch, *et al.*, Prevalence and clinical significance of sterile intra-amniotic inflammation in patients with preterm labor and intact membranes, *Am. J. Reprod. Immunol.*, 2014, **72**(5), 458–474.
26. M. Silasi, I. Cardenas, J. Y. Kwon, K. Racicot, P. Aldo and G. Mor, Viral infections during pregnancy, *Am. J. Reprod. Immunol.*, 2015, **73**(3), 199–213.
27. A. S. Omrani, J. A. Al-Tawfiq and Z. A. Memish, Middle East respiratory syndrome coronavirus (MERS-CoV): animal to human interaction, *Pathog. Global Health*, 2015, **109**(8), 354–362.
28. S. F. Wong, K. M. Chow, T. N. Leung, W. F. Ng, T. K. Ng and C. C. Shek, *et al.*, Pregnancy and perinatal outcomes of women with severe acute respiratory syndrome, *Am. J. Obstet. Gynecol.*, 2004, **191**(1), 292–297.
29. S. A. Rasmussen and D. J. Jamieson, Caring for women who are planning a pregnancy, pregnant, or postpartum during the COVID-19 pandemic, *JAMA*, 2020, **324**(2), 190–191.
30. H. Zeng, C. Xu, J. Fan, Y. Tang, Q. Deng and W. Zhang, *et al.*, Antibodies in infants born to mothers with COVID-19 pneumonia, *JAMA*, 2020, **323**(18), 1848–1849.
31. L. I. Zambrano, I. C. Fuentes-Barahona, D. A. Bejarano-Torres, C. Bustillo, G. Gonzales and G. Vallecillo-Chinchilla, *et al.*, A pregnant woman with COVID-19 in Central America, *Travel Med. Infect. Dis.*, 2020, **36**, 101639.
32. Royal College of Obstetricians and Gynecologists, *Coronavirus (COVID-19) Infection in Pregnancy. Information for Healthcare Professionals*, Royal College of Obstetricians and Gynaecologists, London, England, 2020.
33. P. F. Kohler and R. S. Farr, Elevation of cord over maternal IgG immunoglobulin: evidence for an active placental IgG transport, *Nature*, 1966, **210**(5040), 1070–1071.
34. D. W. Kimberlin and S. Stagno, Can SARS-CoV-2 infection be acquired in utero?: more definitive evidence is needed, *JAMA*, 2020, **323**(18), 1788–1789.
35. W.-j. Guan, Z.-y. Ni, Y. Hu, W.-h. Liang, C.-q. Ou and J.-x. He, *et al.*, Clinical characteristics of coronavirus disease 2019 in China, *N. Engl. J. Med.*, 2020, **382**(18), 1708–1720.
36. K. A. Lackey, R. M. Pace, J. E. Williams, L. Bode, S. M. Donovan and K. M. Järvinen, *et al.*, SARS-CoV-2 and human milk: What is the evidence? *Maternal Child Nutr.*, 2020, **16**(4), e13032.
37. R. Groß, C. Conzelmann, J. A. Müller, S. Stenger, K. Steinhart and F. Kirchhoff, *et al.*, Detection of SARS-CoV-2 in human breastmilk, *Lancet*, 2020, **395**(10239), 1757–1758.
38. B. Hunt, A. Retter, C. McClintock, *Practical Guidance for the Prevention of Thrombosis and Management of Coagulopathy and Disseminated Intravascular Coagulation of Patients Infected with COVID-19*, 2020.

39. L. Panahi, M. Amiri and S. Pouy, Risks of novel coronavirus disease (COVID-19) in pregnancy; a narrative review, *Arch. Acad. Emerg. Med.*, 2020, **8**(1), e34.
40. World Health Organization, *Report of the WHO-China Joint Mission on Coronavirus Disease 2019 (COVID-19)*, Geneva, Switzerland, WHO, 2020, p. 2020.
41. J. Collin, E. Byström, A. Carnahan and M. Ahrne, Pregnant and postpartum women with SARS-CoV-2 infection in intensive care in Sweden, *Acta Obstet. Gynecol. Scand.*, 2020, **99**(7), 819–822.
42. N. Breslin, C. Baptiste, R. Miller, K. Fuchs, D. Goffman and C. Gyamfi-Bannerman, *et al.*, COVID-19 in pregnancy: early lessons, *Am. J. Obstet. Gynecol. MFM*, 2020, 100111.
43. E. Ferrazzi, L. Frigerio, V. Savasi, P. Vergani, F. Prefumo and S. Barresi, *et al.*, Vaginal delivery in SARS-CoV-2-infected pregnant women in Northern Italy: a retrospective analysis, *BJOG*, 2020, **127**(9), 1116–1121.
44. M. Knight, K. Bunch, N. Vousden, E. Morris, N. Simpson and C. Gale, *et al.*, Characteristics and outcomes of pregnant women admitted to hospital with confirmed SARS-CoV-2 infection in UK: national population based cohort study, *BMJ*, 2020, **369**, m2107.
45. E. Mullins, D. Evans, R. Viner, P. O'Brien, E. Morris, Coronavirus in pregnancy and delivery: rapid review and expert consensus, *Ultrasound Obstet. Gynecol.*, 2020, **55**(5), 586.
46. Y. Wu, C. Liu, L. Dong, C. Zhang, Y. Chen and J. Liu, *et al.*, Coronavirus disease 2019 among pregnant Chinese women: case series data on the safety of vaginal birth and breastfeeding, *BJOG*, 2020, **127**(9), 1109–1115.
47. C. Fan, W. L. Lu, K. Li, Y. Ding and J. Wang, ACE2 expression in kidney and testis may cause kidney and testis damage after 2019-nCoV infection, *Front. Med.*, 2021, DOI: 10.3389/fmed.2020.563893.
48. M. L. Eisenberg, Coronavirus disease 2019 and men's reproductive health, *Fertil. Steril.*, 2020, **113**, 1154.
49. Z. Wang and X. Xu, scRNA-seq profiling of human testes reveals the presence of the ACE2 receptor, a target for SARS-CoV-2 infection in spermatogonia, Leydig and Sertoli cells, *Cells*, 2020, **9**(4), 920.
50. Q. Shen, X. Xiao, A. Aierken, W. Yue, X. Wu and M. Liao, *et al.*, The ACE2 expression in Sertoli cells and germ cells may cause male reproductive disorder after SARS-CoV-2 infection, *J. Cell. Mol. Med.*, 2020, **24**(16), 9472–9477.
51. A. Abobaker and A. A. Raba, Does COVID-19 affect male fertility? *World J. Urol.*, 2021, **39**(3), 975–976.
52. J. M. Lucas, C. Heinlein, T. Kim, S. A. Hernandez, M. S. Malik and L. D. True, *et al.*, The androgen-regulated protease TMPRSS2 activates a proteolytic cascade involving components of the tumor microenvironment and promotes prostate cancer metastasis, *Cancer Discov.*, 2014, **4**(11), 1310–1325.
53. K. H. Stopsack, L. A. Mucci, E. S. Antonarakis, P. S. Nelson and P. W. Kantoff, TMPRSS2 and COVID-19: serendipity or opportunity for intervention? *Cancer Discov.*, 2020, **10**(6), 779–782.

54. G. C. Douglas, M. K. O'Bryan, M. P. Hedger, D. K. Lee, M. A. Yarski and A. I. Smith, *et al.*, The novel angiotensin-converting enzyme (ACE) homolog, ACE2, is selectively expressed by adult Leydig cells of the testis, *Endocrinology*, 2004, **145**(10), 4703–4711.
55. P. Pozzilli and A. Lenzi, Commentary: testosterone, a key hormone in the context of COVID-19 pandemic, *Metab. Clin. Exp.*, 2020, **108**, 154252.
56. J. Xu, L. Qi, X. Chi, J. Yang, X. Wei and E. Gong, *et al.*, Orchitis: a complication of severe acute respiratory syndrome (SARS), *Biol. Reprod.*, 2006, **74**(2), 410–416.
57. C. Massarotti, A. Garolla, E. Maccarini, P. Scaruffi, S. Stigliani and P. Anserini, *et al.*, SARS-CoV-2 in the semen: where does it come from? *Andrology*, 2021, **9**(1), 39–41.
58. P. Sengupta and S. Dutta, Does SARS-CoV-2 infection cause sperm DNA fragmentation? Possible link with oxidative stress, *Eur. J. Contracept. Reprod. Health Care*, 2020, **25**(5), 405–406.

CHAPTER 14

Clinical Autopsy in COVID-19

DEEPIKA GUPTA[a], VIKARN VISHWAJEET*[a], AASMA NALWA[a], ABHISHEK PUROHIT[a] AND TANUJ KANCHAN[b]

[a]Department of Pathology and Lab Medicine, All India Institute of Medical Sciences, Jodhpur, Rajasthan, 342005, India; [b]Department of Forensic Medicine and Toxicology, All India Institute of Medical Sciences, Jodhpur, Rajasthan, 342005, India
*E-mail: vikarn.pmch@gmail.com

14.1 Introduction

The Coronaviridae family of viruses includes positive-sense, enveloped, single-stranded ribonucleic acid (RNA) viruses, which in humans typically cause mild respiratory diseases.[1] In December 2019, a member of this family, a novel *Betacoronavirus* (2019-nCoV) caused havoc in Wuhan, Hubei province of central China, with an outbreak of series of atypical respiratory infections, which has now spread rapidly across the rest of the world, infecting people in more than 218 countries. It was later named severe acute respiratory syndrome coronavirus 2 (SARS-CoV-2), as it was structurally homologous to the SARS-CoV, which was responsible for the pandemic of SARS in 2003.[2,3] Following the recommendations of the Emergency Committee, the Director-General of World Health Organization (WHO) declared the outbreak as a Public Health Emergency of International Concern on 30 January 2020 and was it declared a pandemic on 11 March 2020 by the WHO.[4] The first case in India was reported from Kerala on 30 January 2020; the patient was a student studying at Wuhan University. As of 23 November 2020, there are almost

The Coronavirus Pandemic and the Future Volume 1: Virology, Epidemiology, Translational Toxicology and Therapeutics
Edited by Michael D. Waters, Alok Dhawan, Tim Marrs, Diana Anderson, Stafford Warren and Claude L. Hughes

Published by the Royal Society of Chemistry, www.rsc.org

58.9 million cases worldwide; 9 140 312 in India, with approximately 1 393 678 reported deaths worldwide.[5]

14.1.1 Viral Transmission

It was considered that humans potentially acquired the pathogen from the seafood market *via* animal contact, although newer epidemiological data suggest that the virus might already have been circulating earlier.[6] The natural reservoir for Coronaviridae are certain bat species; pangolins might have served as intermediate hosts, and human infection probably occurred from them.[7] Human-to-human transmission is responsible for community spread and occurs by respiratory droplets, which occurs when a person is in close contact (within 1 m) with someone who is actively coughing or sneezing, and occurs through direct contact of mucous membranes of the host.[8–11] Transmission may also occur through fomites by direct or indirect contact through objects used on the infected person, such as thermometers, stethoscopes, blankets, towels, or bed sheets.[12] To date, airborne transmission (the presence of microbes within droplet nuclei, generally <5 μm in diameter) has not been reported for coronavirus disease 2019 (COVID-19), but may be possible in certain specific situations/procedures that generate aerosols, for example, bronchoscopy, open suctioning, endotracheal intubation, nebulization, manual ventilation, cardiopulmonary resuscitation, and tracheostomy.[9,12,13]

14.1.2 Pathogenesis

The coronaviruses are made up of four major structural proteins: the spike surface glycoprotein (S), membrane protein (M), small envelope protein (E), and nucleocapsid protein (N).[14] After the entry of the virus into the human body, SARS-CoV-2 infects the cells *via* the S protein by binding to the angiotensin-converting enzyme 2 (ACE2), which is highly expressed on conducting airway cells and multiple other human organs such as the heart (myocardial cells), kidney (proximal tubule cells), bladder, gastrointestinal tract, liver, kidney, spleen, lymph nodes, bone marrow, and brain.[15,16] The viral spike (S) glycoprotein extrudes from the surface of the virion, which is activated through cleavage by the enzyme furin and is further primed by host cell proteases *via* TMPRSS2 and enables attachment to ACE2, facilitating the fusion of viral and cellular membranes to enable entry into the host cell.[17] The virus replicates inside the cell and propagates along the conductive airway.[18] As the virus involves the upper airways, a greater immune response is mounted, leading to the release of interferons (IFN-β and IFN-λ) and C-X-C motif chemokine ligand 10 (CXCL-10) from the virus-infected cells.[19] This immune response is sufficient to contain the spread of infection in the majority of patients and they do not progress beyond this phase. The lower respiratory tract is involved in ~20% of all infected patients who progress to develop acute respiratory distress syndrome (ARDS).[20] This is due to the release of a large number of different inflammatory cytokines such as interleukins (IL-1, IL-2, IL-6, IL-7, IL-8, and IL-120), tumor necrosis factor (TNF)-α, IFN-λ, CXCL-10, monocyte chemoattractant protein-1, and macrophage inflammatory protein-1α, producing the so-called "cytokine storm", which acts as a chemoattractant for neutrophils, CD4 helper T-cells and CD8 cytotoxic T-cells, which then begin to get sequestered in the lung tissue, forming a double-edged sword, *i.e.* both helps to fight off the virus and causes lung injury leading to ARDS.

14.1.3 Clinical Features

The incubation period is 5–6 days, but can be up to 14 days.[20] During this time, the infected person (even if asymptomatic) might be contagious to healthy individuals.[21] There is no age preponderance; however, people in the 40–70 year age group are infected most commonly. Older age groups and the presence of comorbidities such as type 2 diabetes, hypertension, coronary artery disease, and obesity are risk factors for the development of severe disease.[22,23] As described by the Centers for Disease Control and Prevention (CDC), the symptoms range from mild to severe illness, which includes fever or chills, cough, shortness of breath or difficulty breathing, fatigue, muscle or body aches, headache, new loss of taste or smell, sore throat, congestion or runny nose, nausea or vomiting, and diarrhea.[24]

14.2 Importance of Autopsy

Autopsy in simple words means self-study of a dead body, which is carried out for medico-legal as well as clinical purposes. Clinical autopsy, also termed as the pathological autopsy, is done to diagnose the cause of death wherein an antemortem diagnosis could not be made. Clinical autopsy is also carried out to study the disease process despite knowing the cause of death, to enrich medical knowledge.[25] Given the significant risk of contagion, the initial indications discouraged autopsies in COVID-19 and were restrictive.[26,27] However, clinical postmortem examination followed by microscopic examination of the tissue sample is an invaluable tool in understanding the complexity of this newly emerging disease and clarifying viral invasion routes and distribution in different body parts, the framework of organ damage, its biological mechanism, and ultimately the prevention and control of this infectious disease.

14.3 Autopsy Protocols in Routine Practice

The Royal College of Pathologists (RCPath) has published a series of guidelines which are systematically developed statements to assist the decisions of practitioners, which enable pathologists to deal with non-forensic consent and coroners' postmortem examinations in a uniform pattern.[28] The College of American Pathologists (CAP) has also forwarded a Reporting Protocol for examining Gross Autopsy of Adult decedents. They intend to improve the completeness, clarity, and portability of autopsy reporting while paying attention to the wide range of practice settings in which the data in the report are generated and disseminated.[29]

14.4 Autopsy Protocols in COVID-19

As COVID-19 is an emerging infectious disease, in order to help healthcare workers, various researchers and healthcare organizations have published recommendations and guidelines on various aspects of management of the disease and how to proceed safely with suspected or confirmed COVID-19 postmortem examination. RCPath produced brief guidance in February 2020, to aid mortuary staff and pathologists with postmortem examination on a possible COVID-19 death.[26] To facilitate safe and effective autopsy performance in the COVID-19 setting, the CAP autopsy committee endorsed the recommendations in the CDC guidelines and amended their own guidelines regarding postmortem evaluation of suspected or confirmed COVID-19 decedents.[30]

14.4.1 Handling of the Dead Body in the Ward

Medical examiners should immediately notify their health department as dictated by the institutional policy in the event of the identification of a deceased person with known or suspected COVID-19.[31] Corpses need to be handled with paramount care to circumvent virus transmission. To date, there is no evidence regarding whether individuals have become infected from exposure to dead bodies infected with COVID-19; recommended safety guidelines should be strictly followed. The Advisory Committee on Dangerous Pathogens has categorized pathogens into four hazard groups (HG) depending upon their risk to humans; SARS-CoV-2 has been classified as a HG3 organism.[32]

Pre-autopsy risk assessment should be done on a case-by-case basis. This includes a complete review of the medical history, examination findings, and laboratory records. Written or verbal informed consent for an autopsy should be obtained from the next of kin, respecting their cultural and religious beliefs. In case the COVID-19 status of the deceased is unknown, testing should be done to ascertain the COVID-19 status before performing the autopsy. This has been referred to as a staged autopsy by the RCPath.[26,27]

14.4.2 Transportation to the Mortuary

Interim guidelines on the safe management of a dead body in the context of COVID-19 have been provided by WHO.[33]

- Personnel interacting with the body should follow standard precautions,[34,35] which include hand hygiene before and after interaction with the body and the environment, and the use of appropriate personal protection equipment (PPE) including gown and gloves, depending upon the level of interaction with the body.

- Any body fluids leaking from orifices should be contained. Facial protection, including the use of a face shield, goggles, and medical masks should be undertaken to avoid the risk of coming in direct contact with the body fluid.
- Movement and handling of the body should be minimal and there should be no haste in transferring the body.
- The body should be wrapped in a cloth while moving. In the case of body fluid leaks, body bags should be used.

14.4.3 Mortuary Handling

The healthcare worker preparing the body for the autopsy should wear proper PPE with a disposable impermeable gown, according to standard precautions. Embalming of the body is not advised, to prevent undue manipulation of the body. Individuals who are aged >60 years and are immunosuppressed should not be allowed in direct contact with the body.[33]

14.4.4 Prosector Precautions

A clinical autopsy is usually performed by a qualified pathologist who has at least a postgraduate diploma or degree. Autopsies should be performed using the minimum number of personnel necessary and their details should be noted for future follow-up.[30] Appropriately experienced and trained autopsy pathology trainees (as assessed by senior staff) may be involved in the autopsies with adequate supervision.[26] There is no specific infection risk to pregnant trainees, but they may not undertake autopsy work, which may be discussed with the deanery. The prosector should have complete knowledge of the potential hazards that may be encountered.

Universal precautions in autopsy dissection should be practised,[26] which include the following.

- Pathologists should wear PPE in the clean area. They should use double surgical gloves interposed with a layer of cut-proof Kevlar or neoprene synthetic mesh gloves, fluid-resistant gown with a waterproof apron over it, goggles or face shield, surgical cap, and rubber boots with metal-protected toe caps. National Institute for Occupational Safety and Health-approved disposable N-95 or powered, air-purifying respirators with high-efficiency particulate aerosol (HEPA) filters offer a high level of protection, particularly in the case of aerosol-generating procedures (AGPs).[35]
- Round-ended scissors, and PM40 blades with blunted points should be used.
- There should be a single operator within the body cavity at a time.
- Unfixed organs must be sliced carefully using a sponge to protect the hands from injury, and after fluid sampling, needles should not be resheathed and should be placed in a sharps bucket along with the syringe.

- Careful doffing of the PPE should be done and discarded in the appropriate laundry or waste receptacle. Immediately after doffing PPE, wash hands with soap and water for 20 seconds. An alcohol-based hand sanitizer that contains 60–95% alcohol may be used alternatively, if the hands are not visibly dirty.

14.4.5 Laboratory Biosafety Guidelines

It is important to make sure that health laboratories stick to recommended biosafety practices. The WHO and CDC have provided the following biosafety guidelines for general laboratory safety practices related to COVID-19.[36,37]

1. Site- and activity-specific risk assessment should be performed by responsible and trained personnel before all procedures in compliance with the relevant protocols (which include analyzing the number of personnel that can be adequately accommodated in the laboratory while sustaining complete social distancing, a one-way path for entry and separate for exit, assessment of cleaning and sanitizing of the equipment being used as well as the areas, review of emergency communication and operational plans).
2. Initial processing of all specimens should be done in a validated biological safety cabinet (BSC).
3. Non-propagative diagnostic laboratory work (such as nucleic acid amplification tests) should be conducted at a facility using procedures equivalent to biosafety level 2 (BSL-2).
4. Propagative work (for example virus culture or neutralization assays) should be conducted in a containment laboratory with inward directional airflow (BSL-3).
5. Point-of-care (POC) or near-POC assays can be performed on a bench without employing a BSC, keeping in view the local risk assessment.
6. Disinfectants that act against enveloped viruses should be used (such as hypochlorite, alcohol, hydrogen peroxide, quaternary ammonium compounds, and phenolic compounds).
7. Samples from suspected or confirmed patients should be transferred as UN3373, "Biological Substance Category B". Viral isolates or cultures should be transferred as Category A, UN2814, "infectious substance, affecting humans".
8. Each institution should have a COVID-19 health and safety plan for the protection of employees (educating the employees regarding the spread of the virus, reinforcing training, providing sick leave if needed and someone to contact to in need of emergency, access to PPE and disinfectants).
9. While handling and processing infectious specimens, standard microbiological practice and procedure should be followed.

14.4.6 Requirement of Dedicated Autopsy Facility, Including Engineering and Environmental Controls

The mortuary and the autopsy room for decedents known or suspected to be COVID-19 cases should be specially designed and conducted in airborne infection isolation rooms (AIIRs).[31] The requirements are as follows.

1. The room should be at negative pressure to surrounding areas, adequately ventilated, *i.e.* at least natural ventilation with ≥160 L s^{-1} per patient airflow with at least six air changes per hour (ACH) for existing structures and 12 ACH for renovated or new structures and controlled direction of airflow when using mechanical ventilation.[31,38] Modifications can be carried out according to the institute's bioengineering department and have air exhausted directly outside or use a HEPA filter.
2. If the use of an AIIR or HEPA unit is not possible, the procedure should be performed in the most protective environment possible.
3. An ample amount of space and a designated hand-washing basin must be provided.[36]
4. Doors should be labeled properly, kept closed except during entry and egress, and laboratory walls, floors, and furniture must be smooth for easy cleaning, impermeable to liquids, and chemical-resistant.
5. Human coronavirus can remain infectious on surfaces for up to 9 days and has been detected after up to 72 hours in experimental conditions.[35]
6. Environmental surfaces, where the body was prepared, should first be cleaned with soap and afterwards cleaned with a disinfectant with a minimum concentration of 0.1% (1000 ppm) sodium hypochlorite (bleach), or 70% ethanol for at least 1 minute.[35]
7. AGPs such as oscillating bone saw usage should be avoided. If used, attach a vacuum shroud to contain aerosols. Hand shears can be used as an alternative cutting tool.
8. First-aid facilities must be accessible.
9. Appropriate methods for waste decontamination must be available close to the laboratory.
10. Safety systems must cover fire and electrical emergencies.

14.5 Different Types of Autopsy

14.5.1 Complete Autopsy

A complete autopsy is defined as dissection and examination of a dead body which involves comprehensive external as well as internal examination of the whole body, comprising of removal of all thoraco-abdominal, pelvic, and neck organs along with the examination of the brain, followed by their dissection, to find the cause and sequence of events leading to death and substantiate the clinical diagnosis, and taking samples for additional specialized studies.

The autopsy should be performed with discernment of the deceased's clinical history, in a manner that thoroughly evaluates and documents if there were any comorbidities or co-infections that will allow adequate sample collection along with ancillary studies to understand the disease better.[30] Autopsy should begin within 1–2 hours, as samples for electron microscopy and molecular investigations on RNA and proteins give the best results within this time.[39] There must be an adequate amount of alcohol to sanitize instruments and surfaces, labeled containers for different organs filled with 10% buffered formalin, containers with glutaraldehyde for electron microscopy samples, and 70% alcohol solution for samples for extraction of DNA and RNA. All the equipment should be checked and placed before the procedure. The work of all personnel should be divided beforehand.

14.5.1.1 Autopsy Procedure

14.5.1.1.1 External Examination of the Body.

1. Place the body on the table.
2. Check the nutritional status, any fat loss, cutaneous lesions, any leakage from body orifices, oral dentition.
3. Mention any abnormality grossly detectable by bare eyes.

14.5.1.1.2 Internal Examination of the Body.

1. The “Y” shaped incision is made through the skin and subcutis for the examination of the neck and thoraco-abdominal cavity. It starts from the bilateral acromion processes, joins at the suprasternal notch, and is extended towards the xiphoid process of the sternum, and continued along the midline of the abdomen up to the pubis.
2. Open the thoracic cavity and look for any pleural effusion.
3. Open the pericardial sac and look for the presence of any fluid or blood or adhesions in the pericardial cavity. Assess the pericardial surfaces.
4. Open the abdominal cavity and look for free fluid.
5. Abdominal and thoracic organs can be removed *en bloc* by freeing the lateral and posterior attachment. The superior cut is done just above the level of epiglottis. The inferior cut is made at the level of the lower rectum.
6. Place the *en bloc* dissected organs in a tray.
7. Separate the abdominal and thoracic organs.
8. Individual organs should be weighed, and sliced at the desired thickness to allow permeation of formalin for fixation
9. The organs should be immersed in formalin-filled jars for a minimum of 2–3 days for adequate fixation.

Alternatively, the individual organs from thoracic and abdominal cavity may be eviscerated *in situ*.

14.5.1.1.3 Opening of the Skull.

1. The cut is given over the scalp starting from one mastoid process, passing through the vertex of the head, and extending to the other mastoid process, after which the scalp is reflected. Skull is opened with a saw.
2. After examination of the dura mater, the brain is eviscerated and weighed.
3. To prevent distortion, the entire brain is quickly suspended in an appropriate container by a string tied around the basilar artery without touching the sides and bottom and containing appropriate fixative (usually 20% formalin).
4. The cerebrum, brainstem, and cerebellum must be maintained in their anatomical continuity, or the brainstem and cerebellum usually are separated from the cerebrum by a sleek cut midway between the pons and the mammillary bodies.
5. Before placing the brain in formalin, take representative samples for electron microscopy and molecular testing.

14.5.2 Partial Autopsy

If an objection to complete autopsy is intended by the next of kin/family, a partial autopsy may be carried out, which is defined as an examination that refrains from any part of the complete autopsy definition and involves the examination of a specific part of the body, *e.g.* examination of chest cavity organs or brain examination.

Advantages:

1. Requires fewer resources and is cost-effective.
2. Fewer chances of transmission of an infectious agent to the autopsy personnel.

Disadvantages:

1. The cause of death may not be the organ that is examined.
2. Accuracy is potentially less than that of a complete autopsy, as partial autopsy gives partial results.

14.5.3 Minimally Invasive Autopsy

Minimally invasive autopsy (MIA) is a substantiated and safe approach to identify the cause of death that has been reviewed in low-resource settings.[40] The MIA consists of radiologically guided organ-directed needle biopsies from different organs such as the lung (most commonly involved in SARS-CoV-2 infection), heart, liver, spleen, kidney, or collection of various bodily fluids such as blood and/or cerebrospinal fluid samples.

Advantages:

1. There is no need to open up the entire body.
2. Cost-effective and can be performed in resource-poor countries.
3. Any trained professional can perform the technique.
4. Minimizes the risk of transmission of infectious disease as the procedure is non-invasive, therefore making it an ideal tool in the current COVID-19 pandemic.[27]

When the pathology is focal, the use of MIA as a blind tool is limited, and sometimes unable to identify clinically relevant findings. For example, MIA has insufficient diagnostic accuracy for pulmonary embolism, which is an important complication in COVID-19 patients.[41] Therefore, complete autopsies remain the gold standard.

14.6 Sample Acquisition

14.6.1 Histopathology Examination

The different organs are placed in labeled containers with 10% buffered formalin, in a volume that is approximately 10 times greater than the volume of tissue. Optimal fixation is 3 days (72 hours). Prolonged immersion in formalin (*e.g.* >2 weeks) can decrease the sensitivity of virus detection assays. CDC has provided guidelines regarding the safe preparation of the specimens for shipment.[31] After collection and properly securing the specimens in primary containers, they must be transferred from the autopsy room safely to laboratory staff who can process them for shipping.

1. Within the autopsy room, primary containers should be placed into a larger secondary container.
2. Subsequently, the secondary container should be placed into a resealable plastic bag.
3. The resealable plastic bag should then be placed into a biological specimen bag with absorbent material; and then can be transferred outside of the autopsy room.
4. Workers receiving the biological specimen bag outside the autopsy room should wear disposable nitrile gloves.

14.6.2 Molecular Testing (RT-PCR)

According to recommended guidelines given by the CDC,[31] if an autopsy is performed for a suspected case of SARS-CoV-2, collection of the following postmortem specimens is recommended:

1. Postmortem swab specimens for testing of SARS-CoV-2:
   a. nasopharyngeal (NP) swab for upper respiratory tract sampling
   b. lung swab from both lungs for lower respiratory tract sampling.

2. For other respiratory pathogens and postmortem testing, a separate specimen may be collected.

If an autopsy is conducted for a confirmed case of SARS-CoV-2, the following postmortem specimens should be collected.

Recommendations, if only NP swab is being collected:[31]

1. A negative pressure room is not required.
2. Wear non-sterile, nitrile, latex, or rubber gloves.
3. If there is a risk of injury to the hands, heavy-duty gloves should be preferred over nitrile gloves.
4. Wear a long-sleeved fluid-resistant or impermeable hospital isolation gown.
5. To protect the face, use a plastic face shield or a face mask.

14.6.2.1 Collection of Postmortem Swab Specimens for COVID-19 Testing During Autopsy[31]

An NP specimen is favored for testing of SARS-CoV-2 from upper respiratory tract swab collection. When NP swab collection is not possible, the following are admissible options:

1. An oropharyngeal (OP) specimen.
2. A nasal mid-turbinate swab.
3. An anterior nares (nasal swab) specimen.
4. Nasopharyngeal wash/aspirate or nasal aspirate specimen.

Types of swabs to be used for sample collection:

1. Synthetic fiber swabs with plastic shafts should be only used.
2. Use of calcium alginate swabs or swabs with wooden shafts is deterred, as they might contain virus-inactivating substances and inhibit polymerase chain reaction (PCR) testing.

Nasal swabs or nasal mid-turbinate swabs should be placed in a transport tube containing either viral transport medium, Amies transport medium, or sterile saline, immediately after collection.

If both NP and OP swabs are collected, they should be combined in a single tube to augment the sensitivity of the test, and limiting the wastage of resources.

14.6.2.1.1 Collection of Upper Respiratory Tract Specimen: NP Swab.

1. Insert flexible mini tip wire shaft swab through the nares parallel to the palate (not upwards) until resistance is encountered which indicates contact with the nasopharynx.
2. Gently rub and roll the swab.

3. The swab is kept in place for a few seconds to absorb the local secretions.
4. Remove it slowly while rotating the swab.

14.6.2.1.2 Collection of Lower Respiratory Tract Specimen: Lung Swabs.

1. One swab should be collected from each lung (right and left).
2. Options for collection of lung swabs depend on the individual institute practices:
   a. after the heart–lung block is taken out, insert one swab as far down into the tracheobronchial tree as possible on either side, or
   b. wipe the surface of each lung with a disinfectant containing iodine, clean the surface, let it dry and then with a sterile scalpel, make a nick in the lung parenchyma, and insert the swab for collecting the sample, from both sides.

14.6.2.2 Storage of Postmortem Swab Specimens[31]

After collection, specimens can be stored at 2–8 °C for up to 72 hours. If a delay in transferring the specimen or its testing is anticipated, store specimens at −70 °C or below.

14.6.3 For Electron Microscopy

The most common advantage of using electron microscopy is that it does not need organism-specific reagents for identifying the pathogenic agent. Sample collection should be completed within 3 hours of the presumed patient's death. The various organs from which electron microscopy samples for SARS-CoV-2 can be collected are bilateral lungs, heart (bilateral ventricles, the interventricular septum, and site of origin of the aortic tract and pulmonary artery), brain, liver (both lobes), spleen, kidneys, skeletal muscles (two to three different group of muscles), blood, and bone marrow (from the first–second rib).[39] The samples can be collected in 2.5% glutaraldehyde or half-strength Karnovsky fixative followed by postfixation in 1% osmium tetroxide, and then processed.

14.7 Pathological Examination

14.7.1 Gross Examination of Organs

14.7.1.1 Lung

The brunt of the disease is faced by the lungs. Grossly, the lungs are heavy, red-brown, and appear firm. There is a loss of normal parenchymal crepitancy. Most cases have patchy or diffuse areas of consolidation, correlating with the ground-glass opacities observed in the imaging studies. Frothy fluid exudating from the tissue can be observed on compression. A few studies have documented the presence of pulmonary thrombosis in large hilar vessels. Areas of intraparenchymal hemorrhage and infarction are also observed in occasional cases.[42] A lobar pattern of consolidation, multiple well-defined nodules, pleural effusion, and hilar lymphadenopathy are not characteristic features of SARS-CoV-2 and warrant exclusion of associated bacterial infections or other pathologies.[43] The tracheobronchial trees show congested and edematous mucosa, and may or may not contain mucous plugs.[44]

For diagnosing SARS-CoV-2 infection using electron microscopic and molecular techniques on the tissue, it is recommended to collect the specimen before keeping in the formalin container, as the viral antigens and nucleic acids may be distributed focally and the quantity of SARS-CoV-2 virus in the respiratory tissues can vary between different patients.[31] A minimum of three representative sections from different locations of lung parenchyma is recommended for collection. A minimum of two sections of the airway is taken to include the trachea and bronchi. The sample should be taken immediately after removal from the chest cavity to minimize possible viral contamination of uninvolved tissue.

14.7.1.2 Heart

Gross findings of heart in SARS-CoV-2 are described in only a few studies.[45–47] The two commonly described pathologies are cardiomegaly and right ventricular dilatation.[45,46] Associated findings, unrelated to SARS-CoV-2 viral infection, include atherosclerotic narrowing of the major cardiac vessels.[45]

14.7.1.3 Brain

Gross findings of the brain described in the literature varied from nonspecific changes such as edema and hyperemia to occasional cases with disseminated hemorrhagic lesions involving subcortical white matter. Territorial ischemic lesions and ischemic infarction due to thromboembolic events and scattered punctate intracerebral, midbrain, and subarachnoid hemorrhages are also reported.[48–50] No significant gross lesions were seen in a substantial number of cases.[51]

14.7.1.4 Other Organs

The liver shows variable congestion, ischemia, and steatosis. No additional specific gross lesions are described.[52] Pre-existing hepatic pathology such as cirrhosis, abscesses, and mass lesions may be encountered. The kidney may be swollen due to underlying acute tubular injury. Changes due to associated comorbid diseases such as renal parenchymal scarring in hypertensive renal disease may be noted.[53] In SARS-CoV-2, splenomegaly and infarction are described in the literature.[43] The gastrointestinal tract may show nonspecific congestion and edema of the mucosal surface.[51,54,55]

14.7.2 Light Microscopic Examination of Organs

14.7.2.1 Lungs

Microscopic findings in SARS-CoV-2 from autopsy studies started to emerge in the literature from February 2019.[56–58] The respiratory system is the most commonly and severely involved organ system, although the extent and distribution of involvement are variable.

The characteristic and most commonly encountered histological finding is diffuse alveolar damage (DAD), which is a microscopic counterpart of ARDS (Figure 14.1). The spectrum of DAD changes is observed from the early exudative stage, characterized by intra-alveolar edema, congested capillaries, type 1 pneumocyte necrosis, and formation of hyaline membranes, to the proliferative (organizing) phase represented by loose fibroblastic proliferation in alveolar spaces (Masson bodies), and type 2 pneumocyte hyperplasia. The fibrotic phase of DAD, seen as dense collagenous fibrosis and architectural remodeling, is not usually observed. Hyaline membrane, composed of cellular debris and plasma proteins, appears as a deep eosinophilic membrane along the walls of the alveoli and alveolar ducts on routine hematoxylin and eosin stain. Both the exudative and proliferative phases can be observed in the same patient, indicating that there is temporal and spatial heterogeneity of the pathologic lesions. Occasional studies have demonstrated proliferative changes even in the early stages of the disease onset.[51] These histological changes may be accompanied by non-specific features such as macrophage accumulation in alveoli, presence of multinucleated giant cells, squamous metaplasia, and interstitial infiltrates.

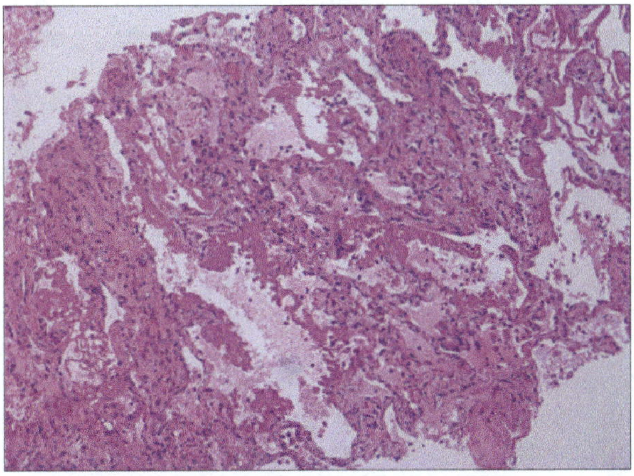

Figure 14.1 Microscopic examination from lung of a deceased COVID-19 patient shows deep eosinophilic hyaline membranes layering the alveolar lining and terminal airways (hematoxylin and eosin, 100×).

Xu *et al.* showed cytopathic effects in pneumocytes in the form of large nuclei, prominent nucleoli, and granular amphophilic cytoplasm.[56] The presence of fibrin balls within the alveolar spaces, sometimes labeled as acute fibrinous and organizing pattern (AFOP), has also been observed by some.[43] DAD and AFOP, though appearing to represent two different patterns of injury, are a continuum of lung injury.

One important finding highlighted in various studies is the presence of fresh fibrin thrombi in small pulmonary vessels such as alveolar capillaries and septal veins, reported in 35.6–80% of cases.[44,51,59] Microvascular injury has been proposed as the main pathogenic mechanism for the presence of these microthrombi. One study described characteristic vascular features, distinct from the ARDS, in the form of prominent endothelialitis, thrombosis, and neoangiogenesis. The diffuse presence of these vascular lesions is seen in SARS-CoV-2 infections and appears to be distinct from the H1N1 influenza infections.[60] The study also highlighted that microthrombi in smaller vessels were significantly more common in cases with SARS-CoV-2 infections when compared to those of H1N1 influenza infections.[60] A few studies pointed to the presence of coagulopathy as the possible causation of diffuse thrombosis and disseminated intravascular coagulation.[61,62]

Superimposed bronchopneumonia or aspiration pneumonia may also be seen in some cases. The main bronchi and bronchioles exhibited only mild non-specific changes in the form of mucosal congestion, excessive secretions, epithelial exfoliation, focal squamous metaplasia, and mild transmural monocytic inflammatory infiltrates.[43]

14.7.2.2 Heart

It has been hypothesized that SARS-CoV-2 may infect the pericytes and cause microvascular (thrombosis) and capillary endothelial cell dysfunction that could lead to degeneration and necrosis of individual cells without significant inflammation.[63] In a multicentre study comprising 21 patients, the authors emphasized two key pathological findings: multifocal lymphocytic myocarditis observed in 14% of cases, and diffuse interstitial macrophage accumulation without accompanying myocyte injury in 86% of cases.[64] However, it was not clear from the study that the myocarditis described was attributed to direct viral insult, as an ultrastructural study was not conducted in these cases. Furthermore, the diffuse macrophage infiltration in the interstitium may be attributed to systemic effects of increased proinflammatory cytokines such as IL-6 and TNF-α.

Other microscopic findings observed in the study included focal lymphocytic pericarditis, strain-induced acute myocyte injury in the right ventricle, and intramyocardial small vessel thrombi, each noted in four cases.[64] Apart from these findings, non-specific histological features such as myocyte hypertrophy and patchy interstitial fibrosis were seen in other studies.[65] Occasional cases of recent or remote myocardial ischemia, including infarction, have been noted, which appears to be less likely to be related to SARS-CoV-2 infection.[45–47]

14.7.2.3 Brain

Central nervous system (CNS) involvement in SARS-CoV-2 has been described in several studies; however, whether pathological findings in the brain result from viral infection or elevated inflammatory cytokines, sepsis, hypoxia, and multi-organ failure remain to be determined. A review summarizing the pathological findings in the brain postulated that the possible CNS pathology is related to ischemic/hypoxic encephalopathy, acute cerebrovascular events, and viral encephalitis. In addition, the authors considered, but strongly questioned, demyelinating events and the increased propensity of opportunistic infections as contributing events in the causation of CNS pathology, as strong evidence is still lacking in the literature.[48] Solomon *et al.* studied gross and microscopic findings in 18 brain specimens and noted the presence of acute hypoxic–ischemic injury in the frontal lobe, hippocampus, and cerebellum, arteriosclerosis with perivascular rarefaction, microglial nodules, and perivascular inflammation. No vasculitis or thrombi were noted in the study.[49] One case described the presence of perivascular hemorrhagic white matter lesions causing axonal injury, demyelination, and infiltration of macrophages. The finding raised a suspicion of microscopic thromboembolic events related to SARS-CoV-2 infection.[50] Several studies reported no significant microscopic alterations in a substantial proportion of cases.[51]

14.7.2.4 Kidney

Acute kidney injury occurs in a significant proportion (28–46%) of cases in hospitalized individuals.[53,66] The microscopic findings in renal parenchyma have been described by a few authors.[66–70] Acute tubular injury (ATI) was seen in all cases. ATI is characterized by loss of brush border epithelium, luminal ectasia, necrosis, and detachment of the tubular epithelial cells. ATI is attributed to a profound hemodynamic alteration in these patients and, thus, likely to be ischemic in origin. Findings ascribed to SARS-CoV-2-specific insults as well as changes associated with underlying comorbidities are reported in the glomerular compartment. A recent study described the presence of collapsing glomerulopathy, characterized by the collapse of glomerular tuft with overlying podocyte hyperplasia, in eight out of 13 indication kidney biopsies from COVID-19 patients.[67] The presence of collapsing glomerulopathy has also been described in other studies in patients of SARS-CoV-2. The authors postulated that the elevated cytokine levels, particularly interferons, might be the factor responsible for the development of collapsing glomerulopathy.[70] Apart from these lesions, other glomerular pathologies included the presence of segmental fibrin thrombi in glomerular tuft with associated endothelial injury and diffuse coagulative cortical necrosis in isolated cases. Age-related glomerulosclerosis or pre-existing glomerulopathies such as diabetic glomerulosclerosis, hypertensive nephrosclerosis, or other specific primary glomerular lesion are described. Congested glomerular capillary loops and margination of inflammatory cells along peritubular capillaries are other non-specific findings described.[66–70]

14.7.2.5 Liver

The morphological findings in the liver are usually related to indirect insults in form of ischemic changes, circulatory disturbances, or shock-induced changes. Hepatic injury caused by direct viral insult appears less likely. This was supported by the lack of ultrastructural demonstration of viral particles in the hepatocytes.[51] However, a study has shown the presence of SARS-CoV-2 within hepatic circulation in blood clots or endothelial cell cytoplasm, using *in situ* hybridization.[71] In our experience, frequently encountered histological findings include steatosis, centrilobular congestion with hepatocytic necrosis, Kupffer cell hyperplasia (with or without hemophagocytosis), and sinusoidal congestion with influx of neutrophils (Figure 14.2). Cholestasis is usually focal and intracytoplasmic. A mild degree of portal inflammation is seen. Lagana *et al.*, in their 40 postmortem cases, found steatosis in 75% of cases, lobular necroinflammation (50% of cases), portal inflammation (50% of cases), and cholestasis (38% of cases, mild and focal).[52] Although the authors report lobular necroinflammation in 50% of cases, it was of mild intensity in 80% of these cases, while none had submassive or massive hepatocytic necrosis. Portal inflammation was also described as minimally increased lymphomononuclear cells.[52] The vascular events were described in form of phlebosclerosis of the portal vein and occasional central vein, muscular hyperplasia of portal arterioles, and sinusoidal microthrombi in a smaller proportion of cases. In addition, the authors used the term "thrombotic bodies" to refer to aggregates of platelets in sinusoids and noticed these in two cases.[52]

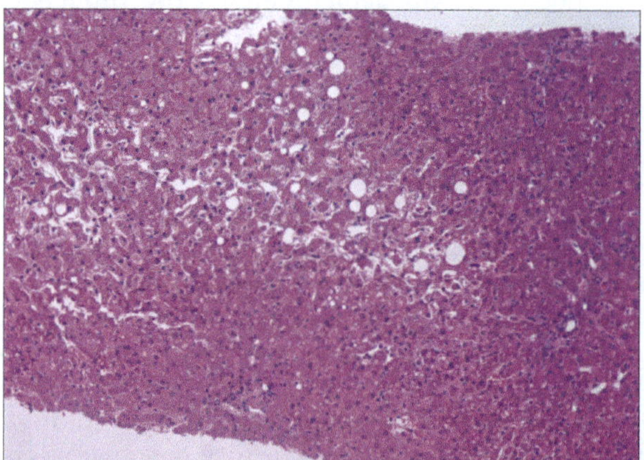

Figure 14.2 Histopathological findings from liver of a deceased COVID-19 patient shows sinusoidal congestion and macrovesicular steatosis in centrizonal areas. Note the absence of portal tract inflammation and lobular inflammation in this case (hematoxylin and eosin, 100×).

Another study also stressed the vascular alterations in form of increased thin-walled portal vascular channels, which was also seen in our experience (unpublished data).[71]

14.7.2.6 Pancreas

Occasional case reports of acute pancreatitis in SARS-CoV-2 have been described in the literature.[72] Studies describing light microscopic changes in the pancreas are scarce. A few autopsy reports documented no identifiable pathology in the exocrine compartment, while others revealed islet cell degeneration.[73,74]

14.7.2.7 Gastrointestinal Tract

A subset of patients often present with gastrointestinal manifestations such as diarrhea and vomiting. Autopsy studies describing the pathological changes in the luminal gastrointestinal tract are lacking. Prominent autolytic changes at the time of autopsy often prevent a detailed assessment of these organs. However, a few studies described non-specific features such as hemorrhages and non-specific inflammation in the mucosa.[51,55]

14.7.2.8 Spleen

The splenic tissue often shows depletion of white pulp which correlates with lymphopenia, a feature of viral infection. A recent study, summarizing the splenic pathology in six cases, noted the depletion of the white pulp with diminished mantle zone in all of them. Besides, red pulp expansion by lymphoplasmacytic cells infiltrate was observed.[43] T-lymphocyte destruction induced by viral invasion may be one of the factors responsible for lymphopenia. Red pulp hemorrhage has also been noted.[51]

14.7.2.9 Bone Marrow

Some studies have described the bone marrow findings in SARS-CoV-2 viral infection. Histiocytic prominence with hemophagocytosis was the most notable finding (Figure 14.3).[75,76] This feature has been described as a non-specific response to several viral infections, particularly in herpesviruses. A possible role of macrophage hyperactivation due to hyperproduction of interferon-γ, related to perforin suppression and reduced $CD8^+$ T-cell cytotoxicity, has been postulated in H5N1 influenza infections.[77] Additionally, increased number of the myeloid series of cells and lymphocytosis is described.[75]

14.7.2.10 Skin

Cutaneous manifestations are highly variable and non-specific. The frequency of skin lesions in SARS-CoV-2 ranges between 1.8% and 20.4%.[78] The different forms of clinical lesions include maculopapular eruptions, vesicular

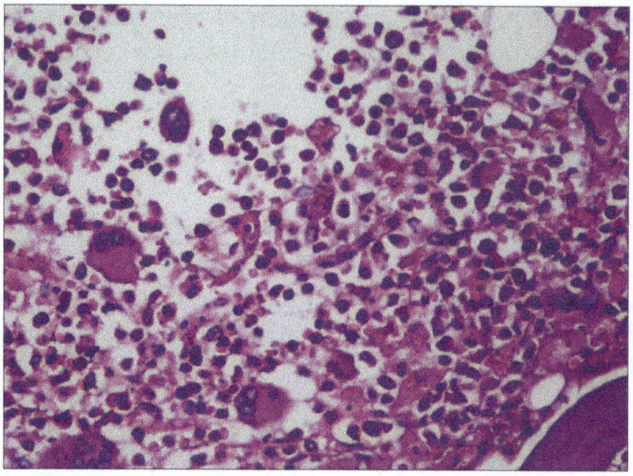

Figure 14.3 Bone marrow from a deceased COVID-19 patient shows prominent histiocytes with some of them showing hemophagocytosis (hematoxylin and eosin, 400×).

eruptions, urticarial lesions, and livedoid hemorrhagic lesions. The microscopic features also vary considerably, corresponding to the clinical appearance of the lesions. However, certain features are common to all the skin biopsies such as prominent dilated vascular channels with expanded endothelial layer, vessels stuffed with red blood cells, and perivascular inflammation comprising lymphocytes and eosinophils.[78–80]

14.7.2.11 Placenta

Universal testing of women for SARS-CoV-2 infection at the time of labor and delivery provided an opportunity to study placental pathology in COVID-19. Nearly 15.4% of women visiting hospitals for delivery tested positive for the virus; however, most of them (87.9%) were asymptomatic.[81] A few studies have described various histopathological parameters to be studied in the placenta of patients with SARS-CoV-2 infection. The studies reported a spectrum of pathological features including fetal vascular malperfusion (avascular villi, delayed villous maturation, mural fibrin in fetal vessels, and chorangiosis) as well as maternal vascular malperfusion (villous infarction, villous agglutination, decidual arteriopathy, subchorionic thrombi, and accelerated villous maturation) to a variable extent.[82–85] Smithgall *et al.* documented the presence of maternal vascular malperfusion in 38 (74.5%) out of 51 placentae, and the presence of subchorionic thrombi and villous agglutination in a significant number of SARS-CoV-2 placentae. The study documented the presence of fetal vascular malperfusion in 12 (23.5%) out of placentae. It was noted that there was an increased frequency of avascular villi, chorangiosis, and fetal thrombotic vasculopathy in the placentas of SARS-CoV-2-infected females compared to the healthy controls; however, this was not statistically significant.[82] These features provide indirect evidence of increased fetal stress in the virus-infected cases and warrants additional investigation. Apart from these, other features described are perivillous fibrin accumulation, increased syncytial knots, inflammatory changes (villitis, chorioamnionitis, funisitis), villous edema, and increased circulating nucleated red blood cells. No viral cytopathic changes were seen in any of the cases. Despite these myriad changes, no adverse pregnancy outcome or vertical transmission of the virus into the neonate was noted in the majority of the cases.[82–85]

14.7.3 Immunohistochemistry

Several authors used orthogonal approaches to demonstrate viral particles in pulmonary as well as in extra-pulmonary locations. These different approaches include immunohistochemistry (IHC), *in situ* hybridization, PCR, and electron microscopy. IHC offers a convenient tool because of easy accessibility and is less labor-intensive.

Antibodies directed against the spike protein and the nucleocapsid protein have been used for detecting SARS-CoV-2 by immunohistochemistry. Several clones with specificity against the nucleocapsid protein (4B21 and 6F10)and the spike protein (1A9) are available for immunohistochemical assays.[86] Expression of SARS-CoV-2 spike protein has been demonstrated in bronchial epithelial cells, pneumocytes (both type I and type II), and alveolar macrophages, within respiratory organs. Apart from these, positivity for this protein has also been noted in lymphocytes and submucosal glands of the trachea. Variable degrees of immunostaining of SARS-CoV-2 spike protein have been described in pneumocytes, with strong staining noted in areas with early features of DAD.[51] Yao *et al.* reported positive immunohistochemical staining in the type II pneumocytes and bronchial epithelium using an antibody against the nucleocapsid protein.[87]

Variable degrees of immunohistochemical expression of viral protein have also been documented in extra-pulmonary organs. One study performing IHC staining for spike protein showed focal granular cytoplasmic positivity in renal tubular epithelial cells. In this study, no expression was seen in the heart, liver, spleen, lymph node, or brain.[51]

14.7.4 *In Situ* Hybridization

In situ hybridization offers direct visualization of viral nucleic acids in specific cellular/tissue compartments. Probes are prepared to target the spike, nucleocapsid, and open reading frame 1ab (ORF1ab) genes. One study using a cell culture line (Vero cells) infected with a strain of SARS-CoV-2 showed strong membranous and cytoplasmic signals, regardless of target genes.[86] The study used RNAscope®-based assays on formalin-fixed paraffin-embedded (FFPE) cell pellets prepared from SARS-CoV-2-infected Vero monolayers.[86]

Liu *et al.*, using FFPE pellets of infected Vero 76 cells, developed a dual-staining approach using *in situ* hybridization as well as IHC to document the presence of SARS-CoV-2 RNA and protein.[88] Best Rocha *et al.*, in their eight lung samples with COVID-19, showed similar expression patterns in both IHC and *in situ* hybridization assays performed on FFPE tissue sections. In addition, the authors reported positive expression in cytotrophoblasts and syncytiotrophoblasts of the placenta, while the viral antigen expression was absent in kidney tissue.[89] Santoriello *et al.* performed SARS-CoV-2 *in situ* hybridization using two antisense probes against genes encoding spike protein in 10 autopsy cases and documented dot-like staining in renal tubular epithelial cells, although it was infrequently seen.[53]

14.7.5 Electron Microscopy

Electron microscopy is a valuable tool to detect emerging infections. Its role in the detection of viruses is well established in the literature. The first ultrastructural description of coronaviruses was described by Almeida and Tyrrell in 1965 from the culture of nasal epithelium and trachea. These viruses varied from 80 nm to 120 nm in size and were roundish with distinct surface projections 20 nm in length, which had a narrow stalk and head 10 nm across.[90] After the emergence of this recent pandemic, several studies have used transmission electron microscopy (TEM) to identify SARS-CoV-2 in different organs. The first electron microscopic demonstration of SARS-CoV-2 in lung tissue was done on bronchoalveolar lavage fluid (BAL) samples.[91] The authors infected the human airway epithelial cells with supernatant prepared from the BAL fluid and cultured. The cultured cells showing cytopathic effects were subjected to TEM, which revealed the typical virions. These viral particles varied from approximately 60 nm to 140 nm and were roughly spherical with distinct spikes, approximately 9 nm to 12 nm, giving an effect of the solar corona.[91]

SARS-CoV-2 is an enveloped RNA virus and, hence, has a membrane-bound nucleocapsid. When it infects cells, the spike protein gets attached to the ACE2 receptor on the cell surface and is sorted to the endocytic vesicles. Once inside, the spike protein is cleaved and activated and there is a fusion of the vesicle membrane and the virus envelope, resulting in the release of the nucleocapsid in the cytoplasm. Here the replicative cycle of the virus starts. This involves the generation of double-membrane vesicles, where the synthesis of viral proteins and genome occurs. The final assembly of these viral proteins and genomes into particles occurs and these viral particles, then, bud into the endoplasmic reticulum–Golgi intermediate compartment. As a result, the ultrastructural detection of viral particles should be done within a cellular structure resembling endoplasmic reticulum, Golgi apparatus, large vacuoles, and vesicles, also in the extracellular compartments.[92]

14.8 Molecular Testing in Autopsies

Novel viruses such as influenza A/H7N9 virus and Middle East respiratory syndrome coronavirus have caused disease outbreaks in the past, and the recent example is SARS-CoV-2. Molecular testing plays a crucial role in rapid identification and acts as a first step to mitigate the effects of future viral threats.

Molecular techniques identify the genomic and proteomic composition of the virus and the host response to the virus after infection. More than 1000 genomic sequences of SARS-CoV-2 have been identified and shared in the GenBank and Global Initiative on Sharing All Influenza Data (GISAID).[93,94]

Postmortem sampling can be done from the throat of the deceased patients. Studies have shown viral RNA in the pharynx of the deceased individuals up to 128 hours after death.[95] Swabs can be collected from the throat, lung parenchyma, bowel mucosa, brain, and other organs. The presence of SARS-CoV-2 RNA can be determined molecularly by quantitative real-time reverse transcriptase PCR (qRT-PCR). Selective amplification of the target nucleic acid from the samples can be done by use of forward and reverse primers for the ORF1a/b non-structural region, which is unique to SARS-CoV-2.[95,96] The results are interpreted as cycle threshold (Ct) values. Skok *et al.*,[95] in their study on postmortem viral dynamics, did not find any significant correlation of the individual Ct value with the degree of organ damage. However, their study demonstrated lower Ct values in the lungs, suggestive of higher viral RNA load.[95]

Autopsy tissues are fixed in formalin, processed and embedded in paraffin, and routinely sectioned and stained with haematoxylin and eosin stain. The formalin causes tissue fixation by cross-linking of tissue macromolecules. Paraffin-embedded tissue blocks can be preserved at room temperature for indefinite periods. However, this makes the extraction of DNA and RNA from the FFPE tissue challenging. RNA is extracted from the deparaffinized tissue by proteolytic digestion with Proteinase K followed by RNA extraction.[97] Various commercially available kits that are specifically dedicated to the extraction of RNA from the FFPE tissue are recommended for this purpose. Amplification by RT-PCR is done by target-specific probes which may include the SARS-CoV-2 specific RNA-dependent RNA polymerase (RdRp) gene in the ORF1a/b region, nucleocapsid genes N1 and N2, and envelope E genes.[96,98] The RdRp and E genes have high sensitivity.[96] The problem with PCR in the molecular analysis of samples is false-positive or false-negative results in autopsy samples of suspected cases.[99,100] Other issues are improper sample collection and RNA degradation due to improper storage. The US CDC has developed a diagnostic panel for SARS-CoV-2 RT-PCR. It uses a one-step qRT-PCR for the detection of viral loads.[101] It is essential to run both a positive and negative control to interpret the final results of qRT-PCR assay. nCoVPC is the commercial sequence provided by the CDC for positive control.[101]

Sekulic *et al.*[102] detected SARS-CoV-2 RNA from postmortem FFPE tissues from various organs with good results. They also detected a mutation consistent with a subset of the Western European Clade A2 a (C3037T, C14408T, A23403G) by next-generation sequencing.[102] Identification of the mutations in the viral genome is necessary to understand the pathobiology of the disease and subsequently vaccine development against the mutated hotspots in the genome.[103]

Since FFPE tissues are easy to store, the archived postmortem samples can be used for retrospective sequencing of the SARS-CoV-2 RNA for gaining a better understanding of the virus prevalence and epidemiology.

In conclusion, autopsy offers a unique opportunity to learn and expand the knowledge about the disease. Characterizing morphological changes in different organs at a microscopic level provides a prospect for clinicians to understand more about COVID-19 and seek effective management strategies.

References

1. V. M. Corman, J. Lienau and M. Witzenrath, Coronaviren als Ursache respiratorischer Infektionen [Coronaviruses as the cause of respiratory infections], *Internist*, 2019, **60**, 1136.
2. P. Zhou, X. L. Yang, X. G. Wang, B. Hu, L. Zhang, W. Zhang, H. R. Si, Y. Zhu, B. Li, C. L. Huang, H. D. Chen, J. Chen, Y. Luo, H. Guo, R. D. Jiang, M. Q. Liu, Y. Chen, X. R. Shen, X. Wang, X. S. Zheng, K. Zhao, Q. J. Chen, F. Deng, L. L. Liu, B. Yan, F. X. Zhan, Y. Y. Wang, G. F. Xiao and Z. L. Shi, A pneumonia outbreak associated with a new coronavirus of probable bat origin, *Nature*, 2020, **579**, 270.
3. T. G. Ksiazek, D. Erdman, C. S. Goldsmith, S. R. Zaki, T. Peret, S. Emery, S. Tong, C. Urbani, J. A. Comer, W. Lim, P. E. Rollin, S. F. Dowell, A. E. Ling, C. D. Humphrey, W. J. Shieh, J. Guarner, C. D. Paddock, P. Rota, B. Fields, J. DeRisi, J. Y. Yang, N. Cox, J. M. Hughes, J. W. LeDuc, W. J. Bellini, L. J. Anderson and SARS Working Group, A novel coronavirus associated with severe acute respiratory syndrome, *N. Engl. J. Med.* 2003, **348**, 1953.
4. *Rolling Updates on Coronavirus Disease (COVID-19)*, https://www.who.int/emergencies/diseases/novel-coronavirus-2019/events-as-they-happen, accessed 03 December 2020.
5. *COVID-19 Coronavirus Pandemic*, https://www.worldometers.info/coronavirus/, accessed 23 November 2020.
6. F. Calabrese, F. Pezzuto, F. Fortarezza, P. Hofman, I. Kern, A. Panizo, J. von der Thüsen, S. Timofeev, G. Gorkiewicz and F. Lunardi, Pulmonary pathology and COVID-19: lessons from autopsy. The experience of European Pulmonary Pathologists, *Virchows Arch.*, 2020, **477**, 359.
7. T. Zhang, Q. Wu and Z. Zhang, Probable pangolin origin of SARS-CoV-2 associated with the COVID-19 outbreak, *Curr. Biol.*, 2020, **30**, 1346.
8. Q. Li, X. Guan, P. Wu, X. Wang, L. Zhou, Y. Tong, R. Ren, K. S. M. Leung, E. H. Y. Lau, J. Y. Wong, X. Xing, N. Xiang, Y. Wu, C. Li, Q. Chen, D. Li, T. Liu, J. Zhao, M. Liu, W. Tu, C. Chen, L. Jin, R. Yang, Q. Wang, S. Zhou,

R. Wang, H. Liu, Y. Luo, Y. Liu, G. Shao, H. Li, Z. Tao, Y. Yang, Z. Deng, B. Liu, Z. Ma, Y. Zhang, G. Shi, T. T. Y. Lam, J. T. Wu, G. F. Gao, B. J. Cowling, B. Yang, G. M. Leung and Z. Feng, Early transmission dynamics in Wuhan, China, of novel coronavirus-infected pneumonia, *N. Engl. J. Med.*, 2020, **382**, 1199.

9. World Health Organization, https://www.who.int/newsroom/commentaries/detail/modes-oftransmission-of-virus-causing-COVID-19-implications-for-ipc-precautionrecommendations, accessed 23 November 2020.
10. J. F. Chan, S. Yuan, K. H. Kok, K. K. To, H. Chu, J. Yang, F. Xing, J. Liu, C. C. Yip, R. W. Poon, H. W. Tsoi, S. K. Lo, K. H. Chan, V. K. Poon, W. M. Chan, J. D. Ip, J. P. Cai, V. C. Cheng, H. Chen, C. K. Hui and K. Y. Yuen, A familial cluster of pneumonia associated with the 2019 novel coronavirus indicating person-to-person transmission: a study of a family cluster, *Lancet*, 2020, **395**, 514.
11. C. Huang, Y. Wang, X. Li, L. Ren, J. Zhao, Y. Hu, L. Zhang, G. Fan, J. Xu, X. Gu, Z. Cheng, T. Yu, J. Xia, Y. Wei, W. Wu, X. Xie, W. Yin, H. Li, M. Liu, Y. Xiao, H. Gao, L. Guo, J. Xie, G. Wang, R. Jiang, Z. Gao, Q. Jin, J. Wang and B. Cao, Clinical features of patients infected with 2019 novel coronavirus in Wuhan, China, *Lancet*, 2020, **395**, 497.
12. S. W. X. Ong, Y. K. Tan, P. Y. Chia, T. H. Lee, O. T. Ng, M. S. Y. Wong and K. Marimuthu, Air, Surface Environmental, and Personal Protective Equipment Contamination by Severe Acute Respiratory Syndrome Coronavirus 2 (SARS-CoV-2) From a Symptomatic Patient, *JAMA*, 2020, **323**, 1610.
13. R. M. Burke, C. M. Midgley, A. Dratch, M. Fenstersheib, T. Haupt, M. Holshue, I. Ghinai, M. C. Jarashow, J. Lo, T. D. McPherson, S. Rudman, S. Scott, A. J. Hall, A. M. Fry and M. A. Rolfes, Active Monitoring of Persons Exposed to Patients with Confirmed COVID-19 – United States, January–February 2020, *Morb. Mortal. Wkly. Rep.*, 2020, **69**, 245.
14. B. J. Bosch, R. van der Zee, C. A. de Haan and P. J. Rottier, The coronavirus spike protein is a class I virus fusion protein: structural and functional characterization of the fusion core complex, *J. Virol.*, 2003, **77**, 8801.
15. S. Tian, Y. Xiong, H. Liu, L. Niu, J. Guo, M. Liao and S. Y. Xiao, Pathological study of the 2019 novel coronavirus disease (COVID-19) through postmortem core biopsies, *Mod. Pathol.*, 2020, **33**, 1007.
16. X. Zou, K. Chen, J. Zou, P. Han, J. Hao and Z. Han, Single-cell RNA-seq data analysis on the receptor ACE2 expression reveals the potential risk of different human organs vulnerable to 2019-nCoV infection, *Front. Med.*, 2020, **14**, 185.
17. M. Hoffmann, H. Kleine-Weber, S. Schroeder, N. Krüger, T. Herrler, S. Erichsen, T. S. Schiergens, G. Herrler, N. H. Wu, A. Nitsche, M. A. Müller, C. Drosten and S. Pöhlmann, SARS-CoV-2 Cell Entry Depends on ACE2 and TMPRSS2 and Is Blocked by a Clinically Proven Protease Inhibitor, *Cell*, 2020, **181**, 271.

18. A. C. Sims, R. S. Baric, B. Yount, S. E. Burkett, P. L. Collins and R. J. Pickles, Severe acute respiratory syndrome coronavirus infection of human ciliated airway epithelia: role of ciliated cells in viral spread in the conducting airways of the lungs, *J. Virol.*, 2005, **79**, 15511–15524.
19. N. L. Tang, P. K. Chan, C. K. Wong, K. F. To, A. K. Wu, Y. M. Sung, D. S. Hui, J. J. Sung and C. W. Lam, Early enhanced expression of interferon-inducible protein-10 (CXCL-10) and other chemokines predicts adverse outcome in severe acute respiratory syndrome, *Clin. Chem.*, 2005, **51**, 2333.
20. A. Parasher, COVID-19: Current understanding of its pathophysiology, clinical presentation and treatment, *Postgrad. Med. J.*, 2021, **97**, 312–320.
21. World Health Organization, https://www.who.int/docs/defaultsource/coronaviruse/situationreports/20200402-sitrep-73-COVID-19.pdf?sfvrsn=5ae25bc7_6, accessed 23 November 2020.
22. F. Zhou, T. Yu, R. Du, G. Fan, Y. Liu, Z. Liu, J. Xiang, Y. Wang, B. Song, X. Gu, L. Guan, Y. Wei, H. Li, X. Wu, J. Xu, S. Tu, Y. Zhang, H. Chen and B. Cao, Clinical course and risk factors for mortality of adult inpatients with COVID-19 in Wuhan, China: a retrospective cohort study, *Lancet*, 2020, **395**, 1054.
23. C. Wu, X. Chen, Y. Cai, J. Xia, X. Zhou, S. Xu, H. Huang, L. Zhang, X. Zhou, C. Du, Y. Zhang, J. Song, S. Wang, Y. Chao, Z. Yang, J. Xu, X. Zhou, D. Chen, W. Xiong, L. Xu, F. Zhou, J. Jiang, C. Bai, J. Zheng and Y. Song, Risk Factors Associated With Acute Respiratory Distress Syndrome and Death in Patients With Coronavirus Disease 2019 Pneumonia in Wuhan, China, *JAMA Intern. Med.*, 2020, **180**, 934.
24. Centers for disease control and prevention, *Symptoms of Coronavirus*, http://www.cdc.gov/coronavirus/2019-ncov/symptoms-testing/symptoms.html, accessed 23 November 2020.
25. R. B. Kotabagi, S. C. Charat and D. Jayachandar, Clinical Autopsy vs Medicolegal Autopsy, *Med. J. Armed Forces India*, 2005, **61**, 258.
26. M. Osborn, S. Lucas, R. Stewart, B. Swift and E. Youd, The Royal College of Pathologists, Briefing on COVID-19, Autopsy Practice Relating to Possible Cases of COVID-19 (2019-nCov, Novel Coronavirus from China 2019/2020), https://www.rcpath.org/uploads/assets/d5e28baf-5789-4b0f-acecfe370eee6223/fe8fa85a-f004-4a0c-81ee4b2b9cd12cbf/Briefing-on-COVID-19-autopsy-Feb-2020.pdf, accessed 23 November 2020.
27. B. Hanley, S. B. Lucas, E. Youd, B. Swift and M. Osborn, Autopsy in suspected COVID-19 cases, *J. Clin. Pathol.*, 2020, **73**, 239.
28. The Royal College of Pathologists, *Autopsy Guideline Series*, https://www.rcpath.org/profession/guidelines/autopsy-guidelines-series.html, accessed 24 November 2020.
29. College of American Pathologists (CAP), *Reporting Protocol for the Examination of Gross Autopsy of Adult Decedents*, https://documents.cap.org/protocols/ap-autopsy-adult-20-02.pdf, accessed 24 November 2020.

30. College of American Pathologists (CAP), *Amended COVID-19 Autopsy Guideline Statement from the CAP Autopsy Committee*, https://documents.cap.org/documents/COVIDAutopsy-Statement-05may2020.pdf, accessed 24 November 2020.
31. Centers for Disease Control and Prevention, *Collection and Submission of Postmortem Specimens from Deceased Persons with Known or Suspected COVID-19*, https://www.cdc.gov/coronavirus/2019-ncov/hcp/guidance-postmortem-specimens.html, accessed 03 December 2020.
32. Health and Safety Executive Advisory Committee on Dangerous Pathogens, *The Approved List of Biological Agents*, https://www.hse.gov.uk/pubns/misc208.pdf, accessed 03 December 2020.
33. World Health Organization, *Infection Prevention and Control for the Safe Management of a Dead Body in the Context of COVID-19: Interim Guidance*, https://www.who.int/publications/i/item/infection-prevention-and-control-for-the-safe-management-of-a-dead-body-in-the-context-of-covid-19-interim-guidance, accessed 03 December 2020.
34. World Health Organization, *Standard Precautions in Healthcare*, https://www.who.int/publications-detail/standard-precautions-in-health-care. accessed 24 November, 2020.
35. World Health Organization, *Infection Prevention and Control during Health Care when COVID-19 is Suspected: Interim Guidance*, 19 March 2020, https://apps.who.int/iris/handle/10665/331495, accessed 24 November 2020.
36. World Health Organization, *Laboratory Biosafety Guidance Related to the Novel Coronavirus (2019-nCoV): Interim Guidance*, https://www.who.int/publications-detail/laboratory-biosafetyguidance-related-to-coronavirus-disease-2019-(covid-19), accessed 23 November 2020.
37. Centers for Disease Control and Prevention (CDC), *Guidance for General Laboratory Safety Practices during the COVID-19 Pandemic*, https://www.cdc.gov/coronavirus/2019-ncov/lab/lab-safety-practices.html, accessed 23 November 2020.
38. World Health Organization Publication/Guidelines, *Natural Ventilation for Infection Control in Health Care Settings*, https://www.who.int/water_sanitation_health/publications/natural_ventilation.pdf, accessed 23 November 2020.
39. L. Carpenito, M. D'Ercole, F. Porta, E. Di Blasi, P. Doi, G. R. Fagara, R. Rey and G. Bulfamante, The autopsy at the time of SARS-CoV-2: Protocol and lessons, *Ann. Diagn. Pathol.*, 2020, **48**, 151562.
40. P. Castillo, M. J. Martínez, E. Ussene, D. Jordao, L. Lovane, M. R. Ismail, C. Carrilho, C. Lorenzoni, F. Fernandes, R. Bene, A. Palhares, L. Ferreira, M. Lacerda, I. Mandomando, J. Vila, J. C. Hurtado, K. Munguambe, M. Maixenchs, A. Sanz, L. Quintó, E. Macete, P. Alonso, Q. Bassat, C. Menéndez and J. Ordi, Validity of a Minimally Invasive Autopsy for Cause of Death Determination in Adults in Mozambique: An Observational Study, *PLoS Med.*, 2016, **13**, e1002171.

41. V. D'Onofrio, E. Donders, M. E. Vanden Abeele, J. Dubois, R. Cartuyvels, R. Achten, M. Lammens, A. Dendooven, A. Driessen, L. Augsburg, J. Vanrusselt and J. Cox, The clinical value of minimal invasive autopsy in COVID-19 patients, *PLoS One*, 2020, **15**, e0242300.
42. S. F. Lax, K. Skok, P. Zechner, H. H. Kessler, N. Kaufmann, C. Koelblinger, K. Vander, U. Bargfrieder and M. Trauner, Pulmonary Arterial Thrombosis in COVID-19 With Fatal Outcome : Results From a Prospective, Single-Center, Clinicopathologic Case Series, *Ann. Intern. Med.*, 2020, **173**, 350.
43. L. M. Buja, D. A. Wolf, B. Zhao, B. Akkanti, M. McDonald, L. Lelenwa, N. Reilly, G. Ottaviani, M. T. Elghetany, D. O. Trujillo, G. M. Aisenberg, M. Madjid and B. Kar, The emerging spectrum of cardiopulmonary pathology of the coronavirus disease 2019 (COVID-19): Report of 3 autopsies from Houston, Texas, and review of autopsy findings from other United States cities, *Cardiovasc. Pathol.*, 2020, **48**, 107233.
44. L. M. Barton, E. J. Duval, E. Stroberg, S. Ghosh and S. Mukhopadhyay, COVID-19 Autopsies, Oklahoma, USA, *Am. J. Clin. Pathol.*, 2020, **153**, 725.
45. S. S. Elsoukkary, M. Mostyka, A. Dillard, D. R. Berman, L. X. Ma, A. Chadburn, R. K. Yantiss, J. Jessurun, S. V. Seshan, A. C. Borczuk and S. P. Salvatore, Autopsy Findings in 32 Patients with COVID-19: A Single-Institution Experience, *Pathobiology*, 2020, **17**, 1.
46. S. E. Fox, G. Li, A. Akmatbekov, J. L. Harbert, F. S. Lameira, J. Q. Brown and R. S. Vander Heide, Unexpected features of cardiac pathology in COVID-19 infection, *Circulation*, 2020, **142**, 1123.
47. S. E. Fox, A. Akmatbekov, J. L. Harbert, G. Li, J. Q. Brown and R. S. Vander Heide, Pulmonary and cardiac pathology in African Amcrican patients with COVID-19: an autopsy series from New Orleans, *Lancet Respir. Med.*, 2020, **8**, 681.
48. S. Al-Sarraj, C. Troakes, B. Hanley, M. Osborn, M. P. Richardson, M. Hotopf, E. Bullmore and I. P. Everall, Invited Review: The spectrum of neuropathology in COVID-19, *Neuropathol. Appl. Neurobiol.*, 2021, **47**, 3–16.
49. I. H. Solomon, E. Normandin, S. Bhattacharyya, S. S. Mukerji, K. Keller, A. S. Ali, G. Adams, J. L. Hornick, R. F. Padera Jr and P. Sabeti, Neuropathological Features of Covid-19, *N. Engl. J. Med.*, 2020, **383**, 989.
50. R. R. Reichard, K. B. Kashani, N. A. Boire, E. Constantopoulos, Y. Guo and C. F. Lucchinetti, Neuropathology of COVID-19: a spectrum of vascular and acute disseminated encephalomyelitis (ADEM)-like pathology, *Acta Neuropathol.*, 2020, **140**, 1.
51. B. T. Bradley, H. Maioli, R. Johnston, I. Chaudhry, S. L. Fink, H. Xu, B. Najafian, G. Deutsch, J. M. Lacy, T. Williams, N. Yarid and D. A. Marshall, Histopathology and ultrastructural findings of fatal COVID-19 infections in Washington State: a case series, *Lancet*, 2020, **396**, 320.
52. S. M. Lagana, S. Kudose, A. C. Iuga, M. J. Lee, L. Fazlollahi, H. E. Remotti, A. Del Portillo, S. De Michele, A. K. de Gonzalez, A. Saqi, P. Khairallah, A. M. Chong, H. Park, A. C. Uhlemann, J. H. Lefkowitch and E. C. Verna,

Hepatic pathology in patients dying of COVID-19: a series of 40 cases including clinical, histologic, and virologic data, *Mod. Pathol.*, 2020, **33**, 2147.
53. D. Santoriello, P. Khairallah, A. S. Bomback, K. Xu, S. Kudose, I. Batal, J. Barasch, J. Radhakrishnan, V. D'Agati and G. Markowitz, Postmortem Kidney Pathology Findings in Patients with COVID-19, *J. Am. Soc. Nephrol.*, 2020, **31**, 2158.
54. Q. Liu, R. S. Wang, G. Q. Qu, Y. Y. Wang, P. Liu, Y. Z. Zhu, G. Fei, L. Ren, Y. W. Zhou and L. Liu, Gross examination report of a COVID-19 death autopsy, *Fayixue Zazhi*, 2020, **36**, 21.
55. F. Xiao, M. Tang, X. Zheng, Y. Liu, X. Li and H. Shan, Evidence for Gastrointestinal Infection of SARS-CoV-2, *Gastroenterology*, 2020, **158**, 1831.
56. Z. Xu, L. Shi, Y. Wang, J. Zhang, L. Huang, C. Zhang, S. Liu, P. Zhao, H. Liu, L. Zhu, Y. Tai, C. Bai, T. Gao, J. Song, P. Xia, J. Dong, J. Zhao and F. S. Wang, Pathological findings of COVID-19 associated with acute respiratory distress syndrome, *Lancet Respir. Med.*, 2020, **8**, 420.
57. S. Tian, Y. Xiong, H. Liu, L. Niu, J. Guo, M. Liao and S. Y. Xiao, Pathological study of the 2019 novel coronavirus disease (COVID-19) through postmortem core biopsies, *Mod. Pathol.*, 2020, **33**, 1007.
58. C. Shao, H. Liu, L. Meng, L. Sun, Y. Wang, Z. Yue, H. Kong, H. Li, H. Weng, F. Lv and R. Jin, Evolution of severe acute respiratory syndrome coronavirus 2 RNA test results in a patient with fatal coronavirus disease 2019: a case report, *Hum. Pathol.*, 2020, **101**, 82.
59. A. N. Duarte-Neto, R. A. A. Monteiro, L. F. F. da Silva, D. M. A. C. Malheiros, E. P. de Oliveira, J. Theodoro-Filho, J. R. R. Pinho, M. S. Gomes-Gouvêa, A. P. M. Salles, I. R. S. de Oliveira, T. Mauad, P. H. N. Saldiva and M. Dolhnikoff, Pulmonary and systemic involvement in COVID-19 patients assessed with ultrasound-guided minimally invasive autopsy, *Histopathology*, 2020, **77**, 186.
60. M. Ackermann, S. E. Verleden, M. Kuehnel, A. Haverich, T. Welte, F. Laenger, A. Vanstapel, C. Werlein, H. Stark, A. Tzankov, W. W. Li, V. W. Li, S. J. Mentzer and D. Jonigk, Pulmonary Vascular Endothelialitis, Thrombosis, and Angiogenesis in Covid-19, *N. Engl. J. Med.*, 2020, **383**, 120.
61. D. Wichmann, J. P. Sperhake, M. Lütgehetmann, S. Steurer, C. Edler, A. Heinemann, F. Heinrich, H. Mushumba, I. Kniep, A. S. Schröder, C. Burdelski, G. de Heer, A. Nierhaus, D. Frings, S. Pfefferle, H. Becker, H. Bredereke-Wiedling, A. de Weerth, H. R. Paschen, S. Sheikhzadeh-Eggers, A. Stang, S. Schmiedel, C. Bokemeyer, M. M. Addo, M. Aepfelbacher, K. Püschel and S. Kluge, Autopsy Findings and Venous Thromboembolism in Patients With COVID-19: A Prospective Cohort Study, *Ann. Intern. Med.*, 2020, **173**, 268.
62. Z. Grimes, C. Bryce, E. M. Sordillo, R. E. Gordon, J. Reidy, A. E. Paniz Mondolfi and M. Fowkes, Fatal Pulmonary Thromboembolism in SARS-CoV-2-Infection, *Cardiovasc. Pathol.*, 2020, **48**, 107227.

63. L. Chen, X. Li, M. Chen, Y. Feng and C. Xiong, The ACE2 expression in human heart indicates new potential mechanism of heart injury among patients infected with SARS-CoV-2, *Cardiovasc. Res.*, 2020, **116**, 1097.
64. C. Basso, O. Leone, S. Rizzo, M. De Gaspari, A. C. van der Wal, M. C. Aubry, M. C. Bois, P. T. Lin, J. J. Maleszewski and J. R. Stone, Pathological features of COVID-19-associated myocardial injury: a multicentre cardiovascular pathology study, *Eur. Heart J.*, 2020, **41**, 3827.
65. G. Tavazzi, C. Pellegrini, M. Maurelli, M. Belliato, F. Sciutti, A. Bottazzi, P. A. Sepe, T. Resasco, R. Camporotondo, R. Bruno, F. Baldanti, S. Paolucci, S. Pelenghi, G. A. Iotti, F. Mojoli and E. Arbustini, Myocardial localization of coronavirus in COVID-19 cardiogenic shock, *Eur. J. Heart Fail.*, 2020, **22**, 911.
66. P. Sharma, N. N. Uppal, R. Wanchoo, H. H. Shah, Y. Yang, R. Parikh, Y. Khanin, V. Madireddy, C. P. Larsen, K. D. Jhaveri, V. Bijol and Northwell Nephrology COVID-19 Research Consortium, COVID-19-Associated Kidney Injury: A Case Series of Kidney Biopsy Findings, *J. Am. Soc. Nephrol.*, 2020, **31**, 1948.
67. S. H. Nasr, M. P. Alexander, L. D. Cornell, L. H. Herrera, M. E. Fidler, S. M. Said, P. Zhang, C. P. Larsen and S. Sethi, Kidney biopsy findings in patients with COVID-19, kidney injury, and proteinuria, *Am. J. Kidney Dis.*, 2021, **77**, 465–468.
68. S. Kudose, I. Batal, D. Santoriello, K. Xu, J. Barasch, Y. Peleg, P. Canetta, L. E. Ratner, M. Marasa, A. G. Gharavi, M. B. Stokes, G. S. Markowitz and V. D. D'Agati, Kidney Biopsy Findings in Patients with COVID-19, *J. Am. Soc. Nephrol.*, 2020, **31**, 1959.
69. H. Su, M. Yang, C. Wan, L. X. Yi, F. Tang, H. Y. Zhu, F. Yi, H. C. Yang, A. B. Fogo, X. Nie and C. Zhang, Renal histopathological analysis of 26 postmortem findings of patients with COVID-19 in China, *Kidney Int.*, 2020, **98**, 219.
70. C. P. Larsen, T. D. Bourne, J. D. Wilson, O. Saqqa and M. A. Sharshir, Collapsing Glomerulopathy in a Patient With COVID-19, *Kidney Int. Rep.*, 2020, **5**, 935.
71. A. Sonzogni, G. Previtali, M. Seghezzi, M. Grazia Alessio, A. Gianatti, L. Licini, D. Morotti, P. Zerbi, L. Carsana, R. Rossi, E. Lauri, A. Pellegrinelli and M. Nebuloni, Liver histopathology in severe COVID 19 respiratory failure is suggestive of vascular alterations, *Liver Int.*, 2020, **40**, 2110.
72. A. Hadi, M. Werge, K. T. Kristiansen, U. G. Pedersen, J. G. Karstensen, S. Novovic and L. L. Gluud, Coronavirus Disease-19 (COVID-19) associated with severe acute pancreatitis: Case report on three family members, *Pancreatology*, 2020, **20**, 665.
73. X. H. Yao, T. Y. Li, Z. C. He, Y. F. Ping, H. W. Liu, S. C. Yu, H. M. Mou, L. H. Wang, H. R. Zhang, W. J. Fu, T. Luo, F. Liu, Q. N. Guo, C. Chen, H. L. Xiao, H. T. Guo, S. Lin, D. F. Xiang, Y. Shi, G. Q. Pan, Q. R. Li, X. Huang, Y. Cui, X. Z. Liu, W. Tang, P. F. Pan, X. Q. Huang, Y. Q. Ding and X. W. Bian, [A pathological report of three COVID-19 cases by minimal invasive autopsies], *Zhonghua Binglixue Zazhi*, 2020, **49**, 411.

74. X. W. Bian and COVID-19 Pathology Team, Autopsy of COVID-19 victims in China, *Natl. Sci. Rev.*, 2020, **7**, 1414.
75. L. Prieto-Pérez, J. Fortes, C. Soto, Á. Vidal-González, M. Alonso-Riaño, M. Lafarga, M. J. Cortti, A. Lazaro-Garcia, R. Pérez-Tanoira, Á. Trascasa, A. Antonio, R. Córdoba, S. M. Rodríguez-Pinilla, O. Cedeño, G. Peces-Barba, I. Fernández-Ormaechea, M. J. Díez Medrano, M. López de Las Heras, A. Cabello, E. Petkova, B. Álvarez, I. Carrillo, A. M. Silva, M. Castellanos, S. Calpena, M. Valverde-Monge, D. Fresneda, R. Rubio-Martín, I. Cornejo, L. Astilleros Blanco de Cordova, S. de la Fuente, S. Recuero, M. Górgolas and M. A. Piris, Histiocytic hyperplasia with hemophagocytosis and acute alveolar damage in COVID-19 infection, *Mod. Pathol.*, 2020, **3**, 1.
76. T. Menter, J. D. Haslbauer, R. Nienhold, S. Savic, H. Hopfer, N. Deigendesch, S. Frank, D. Turek, N. Willi, H. Pargger, S. Bassetti, J. D. Leuppi, G. Cathomas, M. Tolnay, K. D. Mertz and A. Tzankov, Postmortem examination of COVID-19 patients reveals diffuse alveolar damage with severe capillary congestion and variegated findings in lungs and other organs suggesting vascular dysfunction, *Histopathology*, 2020, **77**, 198.
77. N. R. Maakaroun, A. Moanna, J. T. Jacob and H. Albrecht, Viral infections associated with haemophagocytic syndrome, *Rev. Med. Virol.*, 2010, **20**, 93.
78. G. Kaya, A. Kaya and J. H. Saurat, Clinical and Histopathological Features and Potential Pathological Mechanisms of Skin Lesions in COVID-19: Review of the Literature, *Dermatopathology*, 2020, **7**, 3.
79. C. Galván Casas, A. Català, G. Carretero Hernández, P. Rodríguez-Jiménez, D. Fernández-Nieto, A. Rodríguez-Villa Lario, I. Navarro Fernández, R. Ruiz-Villaverde, D. Falkenhain-López, M. Llamas Velasco, J. García-Gavín, O. Baniandrés, C. González-Cruz, V. Morillas-Lahuerta, X. Cubiró, I. Figueras Nart, G. Selda-Enriquez, J. Romaní, X. Fustà-Novell, A. Melian-Olivera, M. Roncero Riesco, P. Burgos-Blasco, J. Sola Ortigosa, M. Feito Rodriguez and I. García-Doval, Classification of the cutaneous manifestations of COVID-19: a rapid prospective nationwide consensus study in Spain with 375 cases, *Br. J. Dermatol.*, 2020, **183**, 71.
80. R. Gianotti, P. Zerbi and R. P. Dodiuk-Gad, Clinical and histopathological study of skin dermatoses in patients affected by COVID-19 infection in the Northern part of Italy, *J. Dermatol. Sci.*, 2020, **98**, 141.
81. D. Sutton, K. Fuchs, M. D'Alton and D. Goffman, Universal screening for SARS-CoV-2 in women admitted for delivery, *N. Engl. J. Med.*, 2020, **382**, 2163.
82. M. C. Smithgall, X. Liu-Jarin, D. Hamele-Bena, A. Cimic, M. Mourad, L. Debelenko and X. Chen, Third-trimester placentas of severe acute respiratory syndrome coronavirus 2 (SARS-CoV-2)-positive women: histomorphology, including viral immunohistochemistry and in-situ hybridization, *Histopathology*, 2020, **77**, 994.
83. R. N. Baergen and D. S. Heller, Placental Pathology in Covid-19 Positive Mothers: Preliminary Findings, *Pediatr. Dev. Pathol.*, 2020, **23**, 177.

84. E. D. Shanes, L. B. Mithal, S. Otero, H. A. Azad, E. S. Miller and J. A. Goldstein, Placental Pathology in COVID-19, *Am. J. Clin. Pathol.*, 2020, **154**, 23.
85. M. C. Sharps, D. J. L. Hayes, S. Lee, Z. Zou, C. A. Brady, Y. Almoghrabi, A. Kerby, K. K. Tamber, C. J. Jones, K. M. Adams Waldorf and A. E. P. Heazell, A structured review of placental morphology and histopathological lesions associated with SARS-CoV-2 infection, *Placenta*, 2020, **101**, 13.
86. M. Carossino, H. S. Ip, J. A. Richt, K. Shultz, K. Harper, A. T. Loynachan, F. Del Piero and U. B. R. Balasuriya, Detection of SARS-CoV-2 by RNAscope® *in situ* hybridization and immunohistochemistry techniques, *Arch. Virol.*, 2020, **165**, 2373.
87. X. H. Yao, Z. C. He, T. Y. Li, H. R. Zhang, Y. Wang, H. Mou, Q. Guo, S. C. Yu, Y. Ding, X. Liu, Y. F. Ping and X. W. Bian, Pathological evidence for residual SARS-CoV-2 in pulmonary tissues of a ready-for-discharge patient, *Cell Res.*, 2020, **30**, 541.
88. J. Liu, A. M. Babka, B. J. Kearney, S. R. Radoshitzky, J. H. Kuhn and X. Zeng, Molecular detection of SARS-CoV-2 in formalin-fixed, paraffin-embedded specimens, *JCI Insight*, 2020, **5**, e139042.
89. A. Best Rocha, E. Stroberg, L. M. Barton, E. J. Duval, S. Mukhopadhyay, N. Yarid, T. Caza, J. D. Wilson, D. J. Kenan, M. Kuperman, S. G. Sharma and C. P. Larsen, Detection of SARS-CoV-2 in formalin-fixed paraffin-embedded tissue sections using commercially available reagents, *Lab. Invest.*, 2020, **100**, 1485.
90. J. D. Almeida and D. A. Tyrrell, The morphology of three previously uncharacterized human respiratory viruses that grow in organ culture, *J. Gen. Virol.*, 1967, **1**, 175.
91. N. Zhu, D. Zhang, W. Wang, X. Li, B. Yang, J. Song, X. Zhao, B. Huang, W. Shi, R. Lu, P. Niu, F. Zhan, X. Ma, D. Wang, W. Xu, G. Wu, G. F. Gao, W. Tan and China Novel Coronavirus Investigating and Research Team, A Novel Coronavirus from Patients with Pneumonia in China, 2019, *N. Engl. J. Med.*, 2020, **382**, 727.
92. S. Akilesh, R. F. Nicosia, C. E. Alpers, M. Tretiakova, T. Y. Hsiang, M. Gale Jr and K. D. Smith, Characterizing viral infection by electron microscopy: Lessons from the COVID-19 pandemic, *Am. J. Pathol.*, 2021, **191**, 222–227.
93. Next SARS-CoV2 App, *GISAID*, https://www.gisaid.org/epiflu-applications/next-sars-cov2-app, Accessed 10 December 2020.
94. National Center for Biotechnology Information, *SARS-CoV-2 (Severe Acute Respiratory Syndrome Coronavirus 2) Sequences*, https://www.ncbi.nlm.nih.gov/genbank/sars-cov-2-seqs, accessed 10 December 2020.
95. K. Skok, E. Stelzl, M. Trauner, H. H. Kessler and S. F. Lax, Post-mortem viral dynamics and tropism in COVID-19 patients in correlation with organ damage, *Virchows Arch.*, 2021, **478**, 343–353.

96. B. Udugama, P. Kadhiresan, H. N. Kozlowski, A. Malekjahani, M. Osborne, V. Y. C. Li, H. Chen, S. Mubareka, J. B. Gubbay and W. C. W. Chan, Diagnosing COVID-19: the disease and tools for detection, *ACS Nano*, 2020, **14**, 3822.
97. S. Bonin and G. Stanta, in *Guidelines for Molecular Analysis in Archive Tissues*, ed. G. Stanta, Springer, Berlin, Heidelberg, 2011, vol. 1, pp. 57–61.
98. A. Ghosh and R. Colling, An overview of COVID-19 for diagnostic pathologists: clinicopathological correlation and diagnostic techniques, *Diagn. Histopathol.*, 2020, **26**, 529.
99. Y. Pan, L. Long, D. Zhang, T. Yuan, S. Cui, P. Yang, Q. Wang and S. Ren, Potential False-Negative Nucleic Acid Testing Results for Severe Acute Respiratory Syndrome Coronavirus 2 from Thermal Inactivation of Samples with Low Viral Loads, *Clin. Chem.*, 2020, **66**, 794.
100. A. Tahamtan and A. Ardebili, Real-time RT-PCR in COVID-19 detection: issues affecting the results, *Expert Rev. Mol. Diagn.*, 2020, **20**, 453.
101. Centers for Disease Control and Prevention, *CDC 2019-Novel Coronavirus (2019-nCoV) Real-Time RT-PCR Diagnostic Panel*, https://www.fda.gov/media/134922/download, accessed 10 December 2020.
102. M. Sekulic, H. Harper, B. G. Nezami, D. L. Shen, S. P. Sekulic, A. T. Koeth, C. V. Harding, H. Gilmore and N. Sadri, Molecular Detection of SARS-CoV-2 Infection in FFPE Samples and Histopathologic Findings in Fatal SARS-CoV-2 Cases, *Am. J. Clin. Pathol.*, 2020, **154**, 190.
103. P. Forster, L. Forster, C. Renfrew and M. Forster, Phylogenetic network analysis of SARS-CoV-2 genomes, *Proc. Natl. Acad. Sci. U. S. A.*, 2020, **117**, 9241.

CHAPTER 15

COVID-19 Pharmacotherapy – Present and Future

DEEPAK KUMAR[a], RAVI MANGALIA[a], MAHADEV MEENA[a], VARATHARAJAN SAKTHIVADIVEL[a], GOPAL KRISHNA BOHRA[a] AND STAFFORD WARREN*[a]

[a]Anne Arundel Medical Center, Annapolis, MD, USA
*E-mail: staffwarren@yahoo.com

15.1 Introduction

The battle of mankind against the viruses began in time immemorial. The alteration in human behavior that took place in the Neolithic period, nearly 12000 years ago, when humans began to stay in agricultural communities which were densely populated, is thought to be the inciting event. This allowed viruses to spread swiftly, following which they became endemic. Smallpox and measles are among the oldest of the viruses that infected humans. These supposedly transformed from viruses that were pathogens to other animals. They made their appearance thousands of years ago in Europe and North Africa. The Spanish conquests may have carried the viruses to humans in the New World. The resulting epidemics killed millions of the native population which had no natural resistance to these viruses. Influenza pandemics have been recorded since the late sixteenth century. They occurred with increasing frequency in successive periods. The most devastating of the influenza pandemics (the "Spanish Flu", in the years 1918–1919) is estimated to have caused 40–50 million deaths.[1]

In recent times, many viruses transmitted from animals have triggered outbreaks that have claimed thousands of lives. The Ebola outbreak in 2014,

The Coronavirus Pandemic and the Future Volume 1: Virology, Epidemiology, Translational Toxicology and Therapeutics
Edited by Michael D. Waters, Alok Dhawan, Tim Marrs, Diana Anderson, Stafford Warren and Claude L. Hughes

Published by the Royal Society of Chemistry, www.rsc.org

Marburg in 2005, the H1N1 influenza virus in 2009, and the viruses from the coronavirus family (severe acute respiratory syndrome coronavirus (SARS-CoV) and Middle East respiratory syndrome coronavirus (MERS-CoV)) have wreaked havoc in the form of numerous epidemics.

The search for an effective drug against severe acute respiratory syndrome coronavirus 2 (SARS-CoV-2) has proven to be an arduous task, as the process associated with drug development and subsequent approval is very complex. The drugs that are in active clinical use, as well as investigational and experimental agents including an array of antiviral agents and immunomodulators, are reviewed in this chapter – including those already approved for coronavirus disease 2019 (COVID-19), repurposed drugs already on the market for other indications, and new drug candidates undergoing trials as potential therapies. While vaccines are expected to dramatically reduce the spread of disease, three factors are likely to maintain the need for drug therapy indefinitely: 1) persons who refuse to be vaccinated or who still contract SARS-CoV-2 despite being vaccinated; 2) countries with limited or no access to a vaccine; and 3) viral mutations that limit the effectiveness of existing vaccines and that may produce more transmissible and/or more deadly variants. The urgent need for pharmacological treatment during the dire times of a pandemic may lead to inaccurate interpretation of outcomes for a particular therapy. This happens when the data are not carefully collected and interpreted, and can expose patients to unwanted and potentially serious side effects.

15.2 Antivirals

15.2.1 Remdesivir

Remdesivir is a broad-spectrum inhibitor of viral RNA-dependent RNA polymerase and was given emergency use authorization (EUA) by the United States Food and Drug Administration (FDA) in May 2020, for its use in confirmed or suspected cases including hospitalized adults and children with severe COVID-19. In October 2020, remdesivir was approved by the FDA for COVID-19 treatment.[2] Remdesivir is a pro-drug which is an adenosine nucleotide analog. It terminates the chain during viral RNA replication as it competes with the ATP for insertion into the developing RNA chains by RNA-dependent RNA polymerase.[3]

Gilead Sciences developed the drug to treat respiratory syncytial virus and hepatitis C in 2009. Remdesivir was determined to be ineffective for either disease; however, it showed some potential against other viruses such as Marburg and Ebola. Other studies suggested that remdesivir was effective against other members of the coronavirus family, such as MERS-CoV and SARS-CoV.[4] The earlier trials with remdesivir for patients with COVID-19 are listed in Table 15.1.

In the *New England Journal of Medicine* (*NEJM*) (5 November 2020),[5] the final report of the Adaptive COVID-19 Treatment Trial-1 (ACTT-1) trial was published with 1062 hospitalized patients with severe COVID-19 and evidence

Table 15.1 Initial trials of remdesivir against SARS-CoV-2.

Trial	Comments
ACTT-1 (Adaptive COVID-19 Treatment Trial-1) phase 3, randomized, double-blind, placebo controlled	The trial concluded that the 10 days of treatment with remdesivir led to faster recovery than similar patients who received placebo. A statistically significant difference was not seen in terms of mortality by day 15 or by day 29 compared to placebo[5]
A phase 3 randomized study to evaluate the safety and antiviral activity of remdesivir in participants with moderate COVID-19 compared to standard of care	The trial concluded that those with moderate COVID-19 who were treated with remdesivir for 5 days showed statistically significant improvement in clinical status compared with those receiving the standard of care. However, this finding was indecisive. The patients who were administered remdesivir for 10 days had no statistically significant improvement in clinical status when compared with the standard of care[6]
A phase 3 randomized study to evaluate the safety and antiviral activity of remdesivir in participants with severe COVID-19	The study showed no significant difference between 5 days and 10 days administration of remdesivir in severe COVID-19 patients not requiring mechanical ventilation[7]
A phase 3–4 multicentre, adaptive, randomized trial of the safety and efficacy of treatments of COVID-19 in hospitalized adults (SOLIDARITY trial)	The trial observed that the treatment with remdesivir, lopinavir/ritonavir, hydroxychloroquine and interferon-β-1A regimens had a little effect on hospitalized COVID-19 patients, as suggested by length of hospital stay, instigation of mechanical ventilation, and overall mortality[8]

of lower respiratory tract infection from 60 trial sites in North America, Europe, and Asia, randomized in a double-blind, placebo-controlled 1:1 trial of intravenous remdesivir, given in a dosage of 200 mg the first day then 100 mg per day for nine additional days *vs.* placebo. Those who received remdesivir had a median recovery time of 10 days as compared to 15 days in the placebo group ($p < 0.001$ by a log rank test) and after adjustment for disease severity. Kaplan–Meier estimates of mortality were 6.7% with remdesivir and 11.9% with placebo by day 15, and 11.4% with remdesivir *vs.* 15.2% with placebo by day 29 (95% confidence limits remdesivir 9–14.5 *vs.* placebo 12.3–18.6). Side effects were few and serious adverse side effects were reported more frequently in the placebo than in the remdesivir group. Then, surprisingly, 2 weeks later, the most recent World Health Organization (WHO) guidelines were released on 20 November 2020, and warned against using remdesivir for COVID-19 in hospitalized patients regardless of disease severity as a conditional recommendation "as there is no evidence that remdesivir improves survival and other outcomes in these patients". These guidelines were promulgated by an international guideline development group including 28 clinical care experts, one ethicist and four patient-partners.[9]

An interim report from the WHO Solidarity Trial Consortium was published in the *NEJM* on 11 February 2021. The trial involved 11 300 adults with COVID-19 in 405 hospitals in 30 countries randomized in an open-label format. Each patient received one of the four repurposed drugs: remdesivir (n = 2750), hydroxychloroquine (n = 954), lopinavir (without interferon (IFN), n = 1411), IFN (n = 2063, inclusive of 651 receiving IFN plus lopinavir), and 4088 received no trial drug and only hospital-specific standard care. The trial concluded that these four different repurposed drugs had little or no effect on hospitalized patients with COVID-19 as indicated by overall mortality, initiation of ventilation, and duration of hospital stay. Mortality at 28 days with remdesivir (which was administered for 10 days) compared to control was 10.9% *vs.* 11.2% (p = 0.50). Hospital length of stay was confounded by each drug delaying discharge by 1–3 days while it was being given. Whereas the report stated that there were no appreciable differences in discharge rates once the drug regimens were completed, it is possible that the 10 day duration of drug therapy may have obscured differences in the time to discharge. In the ACTT-1 trial, the proportion of patients not receiving high-flow oxygen or mechanical ventilation happened to be greater in the remdesivir group compared to placebo, which may have biased the final results in favor of the (lower severity) remdesivir group.[8] Spinner *et al.* found that a 5-day course of remdesivir improved clinical status by 11 days after the initiation of treatment but that a 10-day course did not.[6] Wang *et al.*, in a 237-person randomized double-blind placebo-controlled trial of patients with severe COVID-19 and pneumonia in Hubei, China, found no overall benefit in time to improvement except for a possible remdesivir benefit regarding time to recovery in the subgroup who received a diagnosis <10 days before the start of treatment.[10] As discussed by Harrington *et al.*, in a *NEJM* editorial (2 December 2020),[11] "Even without a reduction in in-hospital mortality, reducing the time to recovery and hospital discharge among patients who survive is important, both for patients and for stressed health care systems, and was the basis for the recent approval of remdesivir by the FDA."[11,12]

A double-blind randomized, placebo-controlled trial of 1033 patients hospitalized with COVID-19 (ACTT-2) at 67 trial sites in eight countries, compared the combination of remdesivir and baricitinib (an orally administered JAK 1 and 2 inhibitor) with a placebo control group in a 1:1 format. The results were reported on 4 March 2021 in the *NEJM*[13] and found that the combination reduced recovery time and accelerated improvement in clinical status, notably among those receiving high-flow oxygen or non-invasive ventilation (in this subset the difference in time to recovery was 10 *vs.* 18 days). The combination therapy was also associated with fewer adverse events. The 28 day mortality was 5.1% in the combination group and 7.8% in the control group (hazard ratio for death 0.65, 95% CI 0.39–1.09). The doses used were 200 mg remdesivir intravenously on day 1 then 100 mg daily intravenously for days 2–10 or until hospital discharge or death. Baricitinib was administered as a 2 mg tablet twice daily or administered through a nasogastric tube for intubated patients for 14 days or until discharge (the dose was reduced to 2 mg once daily for those patients with a glomerular filtration rate <60 mL min^{-1}). A matching placebo was administered on the same schedule as the active drug.

The current dosing status of remdesivir for adults and children (age ≥12 years and weighing ≥40 kg) for COVID-19 requiring hospitalization is 200 mg given by intravenous route on day 1, infused slowly over 30–120 minutes, which is followed by 100 mg once daily till day 5 (in the absence of clinical improvement with 5 days' course and in patients on mechanical ventilation or extracorporeal membrane oxygenation it can be administered up to 10 days). If a patient is otherwise ready for discharge prior to completion of the course, remdesivir can be discontinued.

More common adverse effects of remdesivir include reduction in estimated glomerular filtration rate (eGFR), impaired glucose tolerance, reduction in hemoglobin and lymphopenia. Those occurring with reduced frequency are an increase in prothrombin time, elevated liver transaminases, and raised bilirubin. Rare adverse effects include hypersensitivity reactions, generalized seizures and rash, which occur in less than 2% of those receiving remdesivir. The contraindication to its use includes an eGFR of <30 mL min^{-1} per 1.73 m^2. Increased alanine transaminases levels to more than 10 times of the upper normal limit should prompt discontinuation of remdesivir. It is not recommended to administer remdesivir with chloroquine or hydroxychloroquine, as they have demonstrated an antagonistic effect on its antiviral activity.[14]

15.2.2 Favipiravir

Favipiravir (6-fluoro-3-hydroxypyrazine-2-carboxamide) is another broad-spectrum pro-drug that interferes with viral replication by inhibiting viral RNA-dependent RNA polymerase[15] similar to remdesivir. It has the advantage that it is given orally and has already been approved in Japan to treat influenza. It has exhibited *in vitro* antiviral potential against SARS-CoV-2 and has been approved for emergency use in Italy. It has been used in Russia, Ukraine, Uzbekistan, Moldova, Kazakhstan, and was recently granted approval in Saudi Arabia and the United Arab Emirates.[16] The Indian authorities have allowed the drug for "restricted emergency use" in mild to moderate cases of COVD-19. However, there have been no large randomized controlled trials to evaluate the efficacy of this repurposed drug to treat COVID-19. In India, a 150-person randomized multicenter trial in hospitalized patients with mild to moderate COVID-19 administered 3600 mg daily in two divided doses on day 1, then 800 mg twice daily on days 2–14. They reported a 28.7% faster viral clearance in the favipiravir group than the standard of care group. 70% were judged "cured" by day 4 *vs.* 44% in the standard of care group.[16] A prospective randomized open-label trial of early *vs.* late favipiravir therapy in asymptomatic or mildly ill hospitalized patients to evaluate viral clearance by day 6 was carried out in 25 hospitals in Japan. 89 patients were enrolled and of these 69 were evaluable. There was a trend toward better viral

Table 15.2 A summary of two of the initial trials using favipiravir.

Trial	Comments
An adaptive, phase2/3, multicenter, randomized, open-label, comparative clinical study to assess efficacy and safety of favipiravir in hospitalized patients with COVID-19	Trial pilot data demonstrated promising results in moderate COVID-19 patients. It further suggested that in cases where the cytokine storm is suspected and signs of respiratory distress are obvious, favipiravir should be administered in combination with anti-inflammatory agents with an established role against COVID-19 [18]
Favipiravir *versus* Arbidol (umifenovir) for COVID-19: a randomized clinical trial	Trial involved moderate to critically sick adults with COVID-19 and compared favipiravir to umifenovir as supplement to standard of care. They observed that at day 7 the clinical recovery rate was significantly superior in patients who received favipiravir (71.4%) as opposed to umifenovir (55.8%) in moderate cases. Favipiravir was tolerated well and it also shortened the time to relief from cough and fever[19]

clearance in the early-treatment group, but with a wide confidence interval (66.7% *vs.* 56.1%, 95% CI 0.76–2.62%). Of 30 patients who had a fever on day 1, there was a trend toward more rapid defervescence in the favipiravir group, but again with a wide confidence interval (2.1 days *vs.* 3.2 days, 95% CI 0.81–4.35 days).[17] Two of the initial trials using favipiravir are summarized in Table 15.2.

In Japan, an observational study of favipiravir treatment in patients with mild to moderate COVID-19 (n = 2158), observed clinical improvement in up to 74% and 88% of the patients by the seventh and the fourteenth days of treatment, respectively. Those aged <60 years had better rates of clinical improvement.[17]

A randomized multicenter trial in hospitalized patients from Stanford Medical Center is ongoing in the United States (https://clinicaltrials.gov/ct2/show/NCT04358549).

In India, under the restricted emergency use label, the recommended dose of favipiravir for adults (aged >18 years) with mild to moderate COVID-19 is a loading dose of 3600 mg (1800 mg twice daily) on day 1 followed by 1600 mg (800 mg twice daily) from day 2 to 14.[20]

A limitation regarding the use of favipiravir is its side effect profile. It is reported to have teratogenic potential and embryotoxicity and cannot be administered to pregnant women. Other milder and usually non-limiting side effects include hyperuricemia, reduced neutrophil count, and hepatic transaminitis[21] and it is not indicated for patients with severe renal or hepatic disease.[20]

15.2.3 Molnupiravir (MK-4482)

Molnupiravir is another oral antiviral agent. It is a precursor of nucleoside derivative N4-hydroxycytidine and exerts its antiviral effects by introducing copying errors during viral RNA replication, with the expectation that mutations in the viral sequence will lead to its degredation. In Q4, 2020, the drug is entering phase 3 randomized, double-blind placebo-controlled trials for hospitalized (ClinicalTrials.gov identifier NCT04575584) and non-hospitalized COVID-19 patients (ClinicalTrials.gov identifier NCT04575597).

15.2.4 Umifenovir

This antiviral drug is approved in Russia and China for the treatment of influenza. The target of umifenovir is the spike glycoprotein on SARS-CoV-2.[22] A phase 3 trial combining umifenovir with favipiravir was started in May 2020, and revealed that the inclusion of umifenovir did not demonstrate any significant clinical advantage in COVID-19 patients[23] (see Table 15.2).

15.2.5 Lopinavir/Ritonavir

These are protease inhibitors used primarily for human immunodeficiency virus (HIV) infection. Lopinavir is a protease inhibitor and ritonavir is used in combination with it, as it prolongs the action of lopinavir by preventing its metabolism. Cao *et al.* reported no benefit in a 1 : 1 randomized trial of 199 patients with severe COVID-19 given 400 mg of lopinavir and 100 mg of ritonavir twice daily for 14 days.[24] The failure of this and other trials to demonstrate a clinical benefit in COVID-19 patients, as well as the unfavorable pharmacodynamics associated with this combination, have led to the recommendation against its use to treat COVID-19.[25]

In July 2020, thc SOLIDARITY trial discontinued use of the lopinavir/ritonavir combination. The interim results released in mid-October 2020 found lopinavir/ritonavir (plus or minus IFN) were of no significant benefit to hospitalized patients with COVID-19, as suggested by the length of hospital stay, initiation of ventilation, and overall mortality.[26]

15.3 Drugs with Antiviral Activity

15.3.1 Ivermectin

Ivermectin has been primarily used as an antiparasitic drug, but is also indicated for lice and scabies and has been shown to be effective against ticks. It has demonstrated effectiveness against onchocerciasis (river blindness), filariasis, ascariasis and strongyloidiasis. It has also demonstrated antiviral activity against a broad range of both RNA and DNA viruses, including influenza, HIV-1, West Nile virus, flaviviruses such as Zika and dengue, and alpha viruses such as Venezuelan equine encephalitis virus and Chikungunya, in addition to SARS-CoV-2.[27,28] It has a unique history in that it was discovered in the 1970s from avermectin, which was found in an unusual *Streptomyces* bacterium unique to the soil in southeast Japan, and isolated at the Kitasato Institute in Tokyo.[29] Originally introduced as a veterinary drug for animal health in 1981, it was first used in humans to treat river blindness in 1988 and is listed in the WHO's 2019 list of essential medicines.[30] Unlike the previously discussed medications, ivermectin has been used for prevention of infection among family members and close contacts as well as for treatment of different stages of COVID-19 in several small and medium-sized trials. In the >32 years that it has been used in humans, it has been well tolerated and caused few side effects. Guzzo *et al.* in 2002 found no serious side effects when doses up to 10 times the usual therapeutic doses were used in healthy volunteers (about 2.0 mg kg^{-1} *vs.* 0.2 mg kg^{-1} usual therapeutic dosage).[31] It has been used extensively and more widely than most drugs with more than 3.7 billion doses dispensed in Africa and Central and South America over the past >30 years to control tropical diseases.[31] Usually administered in only one or two doses orally for most diseases it treats and in generic formulation for many years, it is also inexpensive.

The mechanism of action against SARS-CoV-2 may well be multifactorial, through as many as three different actions that curtail viral replication and modulate the host inflammatory response.

Ivermectin has been reported to inhibit host importin-α (IMPα) as well as the IMPα/β-1 mediated nuclear transport, which is crucial in the process of cellular transportation that the virus requires after entering the cell membrane to enhance infection by entering the nucleus, replicating, and repressing the host's response.[27] A second mechanism may be by inhibition of lipopolysaccharide-induced production of inflammatory cytokines. Zhang *et al.*, in 2008, demonstrated reductions of tumor necrosis factor (TNF)-α, interleukin (IL)-6 and IL-1B and improved survival in a mouse model with ivermectin given after administration of a lethal dose of lipopolysaccharide.[32] The SARS-CoV-2 virus has a lipopolysaccharide coating. A third proposed mechanism is that complexes of ivermectin molecules may act as an ionophore that could lead to ionic transport across the viral membrane creating an ionic imbalance and consequent osmotic lysis.[28] While ivermectin exhibited potent *in vitro* antiviral activity with as much as a 5000-fold reduction in SARS-CoV-2 compared to placebo,[33] the very large equivalent dosage in humans would likely cause toxicity. However, a systematic review of experimental evidence for the antiviral effects of ivermectin in animals, with an *in silico* analysis of ivermectin's possible mode of action against SARS-CoV concluded that ivermectin was effective at the standard doses used to treat parasitic diseases *vs.* controls with a *p*-value of <0.00001.[34] This makes sense if the primary action of ivermectin against SARS-CoV-2 is to modify the host's response to the virus (host-directed effect) rather than by a direct virucidal or virustatic effect.

The studies thus far in humans have shown benefit with the same dosing that has an established safety profile and is used to treat human parasitic diseases. Ivermectin appears to accumulate at approximately 10× the concentration in lung tissue compared to serum,[33] which would be expected to increase its effect against respiratory viruses like SARS-CoV and SARS-CoV-2.

Clinical trials have been carried out in multiple countries, some assessing ivermectin's value as a preventive for exposed persons and others studying mild, moderate and/or severe COVID-19 disease. As of 26 February 2021, ivermectin is used countrywide in 15 countries and in certain regions of another 10 countries.[35] While the first trial using ivermectin for COVID-19 involved 280 patients with COVID-19 and was carried out in Broward County, FL, USA,[36] all of the subsequent completed trials have been in other countries and have been doctor-directed trials enrolling fewer than 500 patients. None have been sponsored by large pharmaceutical companies or government institutions like the United States National Institutes for Health (NIH), that would have the economic means to fund a trial of several thousand patients. The Broward County trial was observational, comprising 173 patients with COVID-19 treated with ivermectin at standard antiparasitic doses compared to 107 controls in four hospitals. The primary outcome was mortality and the study found lower mortality in the treated group (15% *vs.* 25.2%, $p = 0.03$). A secondary analysis was done using propensity matching in 196 patients, with similar results (13.3% mortality in treated *vs.* 24.5% in controls; odds ratio 0.47, 95% CI 0.22–0.99; $p < 0.05$) which was an 11.2% absolute risk reduction with a number needed to treat of 8.9 (95% CI 4.5–263).

Subsequent trials are described divided according to whether they addressed prevention or treatment of established disease.

15.3.1.1 Prevention Trials with Ivermectin

Three randomized controlled trials and four observational trials have been performed in various countries. Elgazzar, in Egypt, randomized 200 healthcare workers and household contacts to ivermectin 0.4 mg kg^{-1} (one dose), repeated at 1 week *vs.* controls, followed for 3 months and with both groups wearing personal protective equipment. He found that 2% of the ivermectin group contracted COVID-19 *vs.* 10% of the control group ($p = 0.05$).[132] A second randomized trial by Shouman, from a different university in Egypt, randomized 340 asymptomatic close family contacts to take 0.25 mg kg^{-1} ivermectin in one dose then repeated in 3 days *vs.* controls ($n = 228$ and $n = 112$, respectively). In 14 days, 7.4% of the ivermectin group had COVID-19 *vs.* 58.4% in the control arm.[37] The third randomized prevention trial was the IVERCOR study in Argentina (ClinicalTrials.gov NCT04529525) which randomized 234 healthcare workers in Argentina in a 1 : 1 fashion to a fixed 12 mg dosage of ivermectin (the equivalent of 0.2 mg kg^{-1} for a 60 kg person) once weekly plus six sprays of iota-carrageenan daily for 4 weeks. Iota-carrageenan is a sulfated polysaccharide widely used in the food industry for gelling, thickening, and stabilizing.[38] COVID-19 developed in 3.4% of the treated group *vs.* 21.4% in the controls ($p < 0.001$).[39]

Pooling the numbers from these three prevention trials with slightly different protocols (and overlooking what effect iota-carrageenan may or may not have had in the IVERCOR trial, as the results were in line with the two other trials) yields nearly a 3.5-fold reduction in disease: 5% *vs.* 29% for healthcare workers and close family contacts compared to controls during the duration of these three prevention studies (21 out of 240 *vs.* 94 out of 318, respectively).

Four observational prevention trials also used slightly different protocols, but in the same dosage range as the aforementioned randomized trials. Pooling the numbers in the observational trials for the development of COVID-19 in the treatment *vs.* control groups yielded a 30-fold reduction in disease: 1.8% *vs.* 55% (19 out of 1068 *vs.* 463 out of 846, respectively).[40–43] All seven of these prevention trials individually revealed statistically significant reductions in COVID-19 cases. Given the many different variables and limitations in combining randomized and observational trials, the difference in the development of COVID-19 between treated and control groups is nevertheless striking (2.1% of 1488 treated *vs.* 48% of 1164 control) among these exposed persons. However, until a larger randomized and controlled double blind study has been completed, there will not be consensus on whether ivermectin is an effective preventive drug in the doses currently prescribed for parasitic diseases.

15.3.1.2 Treatment Trials with Ivermectin

In a pre-print meta-analysis, Hill *et al.*[44] assessed 18 randomized clinical trials of patients with COVID-19 comprising 2282 patients. These studies were identified through systematic searches of PubMed, Embase, medRxiv, and trial registries (prevention studies, non-randomized, and case–control studies were excluded from the analysis). Ivermectin was associated with reduced inflammatory markers including C-reactive protein, ferritin, and D-dimer, and it was associated with faster viral clearance by polymerase chain reaction (PCR), which was found to be dose- and duration-dependent. In the six randomized clinical trials of moderate–severe disease, he reported there was a 75% reduction in mortality (relative risk 0.25, 95% CI 0.12–0.52; $p = 0.0002$). There were 14 out of 650 deaths (2.1%) on ivermectin *vs.* 57 out of 597 (9.5%) among controls. Reduced hospitalization duration and favorable clinical recovery were also noted in the ivermectin group.[44]

By 27 February 2021, 42 clinical trials (21 of which were randomized controlled trials) had been completed including approximately 15 000 patients including registered and unregistered studies (for prevention as well as treatment) according to Ōmura and co-workers,[35] and these studies were subjected to a meta-analysis after exclusion of certain biasing factors, finding that 83% showed improvement with early treatment, 51% improved with late-stage treatment, and 89% had prevention of illness by contact treatment with ivermectin, in doses comparable to what is recommended for parasitic diseases.[35] A more recent randomized double-blind dose-response and placebo-controlled study, not included in the above meta-analysis, was instructive regarding dose-response. It was reported by Babalola *et al.*[45] in 62 patients with mild–moderate COVID-19 randomized into three groups. One group received 6 mg orally every 84 hours for 2 weeks ($n = 20$); a second group received 12 mg by mouth every 84 hours for 2 weeks ($n = 20$); and the third group was the control arm. In addition, lopinavir/ritonavir and standard of care were given to all three groups. The time to SARS-CoV-2 negativity was substantially and dose-dependently reduced compared to the control arm (5.33 compared to 9.15 days with 95% CI 4.33–6.32 days and 5.68–12.62 days, respectively). In addition, the days to reverse transcriptase (RT)-PCR negativity were shortened by 4.5 days in the 12 mg group, compared to being shortened by only 3.15 days in the 6 mg group compared to the control arm. These differences were significant by ANOVA and suggested that a patient receiving the 12 mg dose would improve 2.38 times faster than the control group.[45]

For the treatment of mild COVID-19 in a younger cohort, one recent study reported in March 2021 found no significant benefit. This was a randomized placebo-controlled trial of ivermectin from Cali, Colombia,[46] in which 398 patients with mild COVID-19, average age 37 years (interquartile range 29–48 years), were randomized to ivermectin solution, rather than tablets, 0.3 mg kg^{-1} per day for 5 days. The number of patients with comorbid conditions was unusually low (21%), perhaps reflecting their young age (approximately two decades younger than most other COVID-19 trials) and the presence of only mild disease. The primary outcome was resolution of symptoms at 21 days, which was not significantly different between ivermectin and the control group (82% *vs.* 79%). Few patients in either group had deterioration of care of ≥2 points on an eight-point ordinal scale (2% ivermectin *vs.* 3.5% controls) or required escalation of care (2% ivermectin *vs.* 2% controls). Most of the patients in the study were at home and not under direct medical observation. There were no serious side effects attributed to the medication. This study suggests that in younger persons with few if any comorbidities, ivermectin may offer little benefit, while at the same time causing no serious side effects.[46]

Post-COVID-19, according to the NIH, approximately 30% of patients experience persistent symptoms such as alterations in taste and/or smell, fatigue, myalgia, and "brain fog". One small observational study of 33 patients 4–12 weeks post-COVID-19 with diagnosed "long COVID" were treated with 0.2–0.4 mg kg^{-1} ivermectin (the dose depending on whether the symptoms were mild or moderate) daily for 2 days. The authors reported that 87.9% of the patients had complete resolution of symptoms after the two doses and that another 7% had complete resolution after additional doses.[47] As this syndrome is not uncommon, can be disabling, and since there is no officially approved treatment for these sequelae of COVID-19, further studies are urgently needed. With over 130 million cases of COVID-19 reported worldwide as of the first week in April 2021, and assuming that as many as 30% may have "long COVID" symptoms, that means that in theory as many as 43 million or more persons may fall into this category with the uncertainty about symptom duration.

On 14 January 2021, the NIH upgraded their recommendation regarding ivermectin from "against" to a neutral stance, neither recommending for nor against ivermectin for the treatment of COVID-19. However, taking into account the latest preventive and therapeutic data, a strong case can be made that a large randomized double-blind controlled trial needs to be carried out, especially given the low cost and wide availability of this drug for other applications, which would simplify distribution if effectiveness against COVID-19 is confirmed. Initiating large randomized trials for inexpensive and generic medications is challenging and will likely require a non-profit organization such as the NIH or WHO to initiate such a trial.

15.3.2 Nitazoxanide

Nitazoxanide is an oral antiparasitic and an antiviral pro-drug. It is mainly used for treating diarrhea occurring due to *Cryptosporidium parvum* and *Giardia lamblia*. It has shown its inhibitory potential for an array of viruses *in vitro*, including MERS-CoV, SARS-CoV, and SARS-CoV-2.[48]

Two phase 3 trials for the prevention of COVID-19, are ongoing in groups that are at increased risk. The first includes elderly people residing in long-term care facilities to evaluate post-exposure prophylaxis.[49] The second trial involves healthcare workers and others and is evaluating nitazoxanide's role in pre- and post-exposure prophylaxis.[50]

A randomized, multicenter phase 3 study commenced in August 2020 for treating people (aged >12 years) having pyrexia and respiratory symptoms related to COVID-19. An analysis to evaluate the efficacy will be conducted in people having confirmed SARS-CoV-2 infection.[51]

15.3.3 Merimepodib

Merimepodib is an inhibitor of inosine-5′-monophosphate dehydrogenase, which leads to depletion of guanosine required by the viral polymerase during the process of viral replication.[52]

An oral antiviral phase 2 trial began in June 2020 with the purpose of assessing its efficacy and safety in patients with advanced COVID-19 when administered in combination with remdesivir, but was terminated when it failed to meet its primary endpoint.[53]

15.3.4 Niclosamide

Niclosamide is an FDA-approved antihelmintic agent used for treating tapeworm infestations. It has potential for use as an antiviral agent by two mechanisms, *i.e.* by blocking endocytosis and by acting as an S-phase kinase associated protein 2 inhibitor, reducing viral replication. It also has immunomodulatory action. Phase 2 and 3 trials in Brazil and Argentina are in process for mild–moderate COVID-19, to analyze the effectiveness and safety of its oral (ANA001) formulation.[54a,b] It has also been formulated in a nasal spray, but systematic studies with the spray have not yet been reported (Smyth, Hugh, Texas Pharmacy, Sept 29, 2020, accessed at pharmacy.utexas.edu).

15.3.5 Rintatolimod

Rintatolimod is a double-stranded RNA antiviral that binds to the Toll-like receptor 3 (TLR-3) and thereby protects and stimulates the innate immune pathway activated during viral infections. It is currently on the market under the brand name Ampligen and is used for chronic fatigue syndrome; in December 2020, it was granted orphan drug authorization by the FDA for pancreatic cancer.

A prospective trial in phase 1 and 2a, to treat cancer patients with mild or moderate COVID-19, is studying the adverse effects of the combination of IFNα-2b with rintatolimod. IFN-α clears the virus by stimulating and potentiating the immune response. The combination of rintatolimod and IFNα-2b may potentially stimulate the immune system to stop the multiplication and spreading of the virus.[55]

15.3.6 β-D-N4-Hydroxycytidine

β-D-N4-Hydroxycytidine is an interesting drug, in that it induces a high mutation rate in a viral genome, affecting the rate of particle release and particle infectivity. It has demonstrated efficacy against alpha viruses (and has been shown to inhibit murine hepatitis virus and MERS-CoV with minimal cytotoxicity).[56,57] It is an oral broad-spectrum antiviral and a ribonucleoside analog. It has led to improvement in lung function and reduction in virus titers when administered to infected mice with SARS-CoV or MERS-CoV.[58]

15.3.7 Bemcentinib

Bemcentinib is an oral selective AXL kinase inhibitor (AXL from the receptor tyrosine kinase family, over-expressed by many types of cancer cells). It has been reported in pre-clinical models to exhibit antiviral activity against numerous enveloped viruses, which include Ebola virus and Zika virus. Current research has shown it to be effective against SARS-CoV-2. A study in phase 2 to evaluate its use in those hospitalized for COVID-19 is ongoing as a component of the United Kingdom's Accelerating COVID-19 Research and Development (ACCORD) program.[59]

15.3.8 Plitidepsin

This drug is employed in Australia for treating multiple myeloma. It potentially targets eEF1A (α-subunit of eukaryotic elongation factor) and has been developed as an antitumor agent. The eEF1A is also important for multiplication and spread of the coronavirus.[60] A phase 1 trial of plitidepsin started in May 2020, with the objective of selecting appropriate dosing of plitidepsin for future phase 2/3 trials.[61]

15.3.9 Emetine Hydrochloride

Historically, emetine hydrochloride is a component of syrup of ipecac, which has been on the market for many years with a proven safety profile and is used for inducing emesis. Emetine has proven to be effective in several *in vitro* studies against many DNA and RNA viruses that include HIV, influenza, echovirus, cytomegalovirus and herpes simplex virus.[61] It has shown promise as an inhibitor of viral entry of genetically divergent coronaviruses and has demonstrated a viral blocking effect in Vero cells at low doses. Plans are in progress to begin trials for evaluating antiviral activity and safety of emetine in high-risk symptomatic adults having COVID-19 who are not in need of hospitalization.[62]

15.3.10 Trabedersen

Trabedersen is an antisense oligonucleotide that selectively targets and inhibits the expression of transforming growth factor (TGF)-β2. The proposed mechanisms by which it works in COVID-19 includes inhibition of viral replication, interfering with the cellular binding of virus, and by suppressing virus-induced pneumonia and progression of acute respiratory distress syndrome (ARDS).[63]

15.3.11 Stannous Protoporphyrin

Stannous protoporphyrin (active ingredient in RBT-9) has exhibited both antiviral and immunomodulatory action in preclinical studies. It is in a phase 2 trial for evaluation of its use in COVID-19 and has been developed particularly for benefiting patients who are at increased risk due to age or comorbidities.[64]

15.3.12 Antroquinonol

Antroquinonol is an antiviral and anti-inflammatory agent. It has been shown to inhibit viral RNA replication as well as viral protein synthesis, both *in vitro* and in animals. A phase 2 trial is evaluating its efficacy and safety in hospitalized patients with mild to moderate COVID-19 pneumonia.[65]

15.3.13 Apilimod Dimesylate

Apilimod dimesylate acts by inhibiting the lipid kinase enzyme PIKfyve. It intervenes in the process of intracellular entry and trafficking of virus. A randomized controlled trial is in phase 2, evaluating the efficacy of apilimod in contrast to placebo for treating SARS-CoV-2 infection in adults on an outpatient basis.[66]

15.3.14 VIR-2703

VIR-2703 is an investigational small interfering RNA (siRNA) that targets the SARS-CoV-2 genome. The siRNA silences messenger RNA that encodes for disease-causing proteins.[67] It is not yet in clinical trials.

15.3.15 Colchicine

Colchicine has been used for thousands of years, with the first recorded use in Egypt in 1550 BC for pain and swelling. It is administered orally and is primarily active in the inflammatory response of the innate rather than the adaptive immune system, but its influence is multifactorial in that it affects neutrophil and macrophage/monocyte cell migration, cytokine release, and intracellular trafficking.[68]

Colchicine inhibits neutrophil adhesion, extravasation, and recruitment by altering neutrophil L-selectin expression and E-selectin distribution as well as suppressing the release of the chemotactic agent leukotriene B4. It also alters neutrophil deformability and modulates leukocyte-mediated inflammatory activities including inhibiting leukocyte production of superoxides and release of various cytokines and pyrogens.[68]

A large randomized, double-blind (COLCORONA)[69] trial of 4488 non-hospitalized patients from the Montreal Heart Institute with COVID-19 diagnosed by PCR testing or clinical criteria were assigned to receive either colchicine 0.5 mg twice daily for 30 days or placebo. Of the 4150 patients with PCR-confirmed COVID-19, the composite primary endpoint of death or hospitalization occurred in 4.6% of those treated *vs.* 6.0% of controls (odds ratio 0.75, 95% CI 0.57–0.99; $p = 0.04$). The exact same odds ratio held for fewer hospitalizations for treated patients and there were fewer patients needing mechanical ventilation (odds ratio 0.50, 95% CI 0.23–1.07) and fewer deaths (odds ratio 0.56, 95% CI 0.19–1.66) in the colchicine-treated patients. Pneumonia occurred in 2.9% of treated patients and 4.1% of controls ($p = 0.02$). The only significant side effect in the treated group was diarrhea, which occurred in 13.7% of the colchicine group *vs.* 7.3% of controls ($p < 0.0001$), but severe side effects were reported to be less frequent in the treated than the control group.[69] The time to treatment was not listed, which could be a potential confounder, but, overall, the improved outcome with colchicine points toward immune dysregulation contributing significantly toward morbidity and mortality from COVID-19, covered in more detail in the next section.

15.4 Immunomodulators

Immunomodulation by various methods is being explored, mostly by revisiting existing drugs. So far, corticosteroids have been the immunomodulators that have made a significant and generally accepted impact on mortality from COVID-19. The target is to blunt the inflammation caused by cytokine release which can lead to organ damage and death.[70] Rizk *et al.* reviewed and summarized the roles of innate immunity and adaptive immunity, along with immunomodulation by various agents including the interleukins, convalescent plasma, and JAK inhibitors.[71]

15.4.1 Corticosteroids

Corticosteroids modulate the dysregulated immune system and dampen the immune response. In the past, there have been a limited number of studies on the use of steroids as an adjunctive therapy in viral pneumonia, but their status in the treatment of viral pneumonias remains controversial to date and corticosteroids are not employed routinely for treating viral pneumonias.[72] Their benefits in septic shock resulting from bacterial pneumonia may be due to the blunting of the immune reaction to released toxins. Alternatively, using corticosteroids may potentially cause harm through their immunosuppressant effects. Furthermore, there is no evidence of their success in providing benefits in other viral epidemics, such as those of influenza, SARS, and MERS.[73]

Various proposed mechanisms of immunopathology associated with the human coronaviruses prompted researchers to look into the role of corticosteroids. The virus invades alveolar and airway epithelial cells, stimulating the immune cells to produce massive amounts of cytokines and chemokines. The major ones include IL-1, IL-6, IL-8, IL-12, TNF-α, IFN-α, IFN-β, IFN-γ, and monocyte chemoattractant protein 1. A vicious cycle begins as these cytokines facilitate infiltration of additional inflammatory cells, activating a cascade to reproduce more cytokines. This process, sometimes referred to as "cytokine storm", inflicts injury to lungs and other organs.[74,75]

The UK RECOVERY trial evaluated the utility of steroids in COVID-19 in a controlled open-label randomized protocol to determine mortality at day 28 in hospitalized COVID-19 patients, who received a 10 day course of dexamethasone (6 mg per oral or intravenous once daily) along with the standard of care. One group was assigned to receive dexamethasone (n = 2104) plus standard of care, and the other was administered standard of care alone (n = 4321). The overall mortality was significantly lower (22.9%) in those who received dexamethasone compared to the standard of care group (25.7%). However, it was noted that the mortality benefit in the dexamethasone group was limited to those on invasive mechanical ventilation or who were receiving supplemental oxygen therapy but without mechanical ventilation. Those receiving no respiratory support (*i.e.* those on room air) at randomization had no mortality benefit and the results were consistent with possible harm in this subgroup. The results from this trial can be considered a landmark, as they provided clinicians with strong evidence to consider when the administration of corticosteroids might be advantageous.[76]

The publication of the RECOVERY results halted many trials that were set to assess the role of corticosteroids in COVID-19. The WHO undertook a meta-analysis, Rapid Evidence Appraisal for COVID-19 Therapies (REACT), from the data they assembled from seven trials with 1703 patients (the number of those enrolled who were administered corticosteroids was 678 and standard of care or placebo were 1025). The use of corticosteroids and the lowering of mortality had similar results for both hydrocortisone and dexamethasone. The mortality at day 28 was significantly lower in those who were administered glucocorticoids and the investigators concluded that it was a class effect.[77]

Recommendations regarding the dosing of corticosteroids were made by the WHO panel. One of these was the strong recommendation for the use of systemic corticosteroid therapy (oral or intravenous administration of 6 mg of dexamethasone once daily or intravenous 50 mg of hydrocortisone administered 8 hourly) for a period of 7–10 days to treat severe COVID-19. The panel recommended conditionally against the use of corticosteroids in non-severe COVID-19.[78]

15.4.2 Interleukin Inhibitors

Cytokine release in the form of interleukins, namely IL-1, IL-6, IL-12, and IL-18, TNF-α, and various other inflammatory mediators cause severe damage to lung tissue. The increased alveolar–capillary gas exchange due to increased pulmonary inflammatory response makes oxygenation difficult in severe COVID-19 pneumonia. To ameliorate the damage inflicted on lung tissue by this release of cytokines, interleukin inhibitors have been tried with varying success in severe COVID-19, as discussed later.

15.4.2.1 IL-6 Inhibitors

IL-6 is a pro-inflammatory cytokine, produced by various cells, including lymphocytes, monocytes, epithelial cells, and fibroblasts. Bronchial epithelial cells produce IL-6 after becoming infected by SARS-CoV-2 in a dose-dependent manner. This observation led researchers to hypothesize that IL-6 receptor blockade might interrupt this inflammatory cascade at a crucial stage which became the rationale for studying IL-6 inhibitors. The NIH guidelines in June 2020 observed that there was a paucity of data to suggest for or against their use in treating COVID-19.[79]

One of the first among the IL-6 inhibitors to enter the trials was sarilumab. One of the initial trials, phase 2/3 was started in March 2020 and was double-blind, multicenter, and had an adaptive design comprising two parts. The patients with severe pneumonia due to COVID-19 were incorporated in the first part which evaluated sarilumab's effect on pyrexia and the requirement for oxygen supplementation. A second part was designed to evaluate improvement in long-term outcomes, including supplemental oxygen requirement, need for mechanical ventilation, and prevention of mortality.[80]

In accordance with the analysis of data from the phase 2 trial, the design for phase 3 was altered for inclusion of sarilumab at a higher dose (400 mg) and investigators compared it with placebo in those who were critically ill (those requiring high-flow oxygenation or intensive care unit (ICU) admission or mechanical ventilation). There was no statistical significance observed in the critical patients on 400 mg sarilumab and on mechanical ventilation at baseline compared to placebo. Negative trends were observed in the group of severely ill patients without mechanical ventilation at baseline.[81] These results prompted the researchers to stop the US-based trial, which administered a higher dose (800 mg).[82]

Tocilizumab, a recombinant humanized monoclonal antibody, is an IL-6 inhibitor which has been widely evaluated so far with mixed results. It is used for rheumatoid arthritis, giant cell arteritis, cytokine release syndrome induced by chimeric antigen receptor T-cells (CAR-T), and systemic juvenile idiopathic arthritis. It binds competitively to the IL-6 soluble and membrane-bound IL-6 receptors and inhibits IL-6-mediated signaling through these receptors.[83]

The effectiveness of early administration of tocilizumab was assessed in hospitalized patients with COVID-19 with arterial oxygen tension ($P\text{aO}_2$)/inspiratory oxygen fraction ($F\text{IO}_2$) ratio between 200 and 300 mmHg. The trial did not show any benefit on disease progression when compared with the standard of care.[84]

Tocilizumab has been evaluated extensively in several phase 3 randomized controlled trials (EMPACTA, COVACTA and REMDACTA) that evaluated its efficacy and safety along with standard of care in adult hospitalized patients having severe pneumonia due to COVID-19 and compared it to placebo plus standard of care. The COVACTA, a phase 3 trial involving those with pneumonia due to COVID-19, observed that neither the primary endpoint of improvement in clinical status, nor the secondary endpoint of decreased mortality was achieved.[85,86] In the EMPACTA trial of tocilizumab in hospitalized patients with COVID-19 pneumonia who were not receiving mechanical ventilation, 389 patients were randomized in a 2 : 1 ratio. Tocilizumab reduced the likelihood of progression to the composite endpoint of mechanical ventilation or death by day 28 (12% *vs.* 19.3%; $p = 0.04$ by the log rank test). However, survival was not improved by tocilizumab (10.4% by day 28 in the tocilizumab group *vs.* 8.6% in the placebo group).[87]

While several observational studies have suggested a mortality benefit,[88–90] data from randomized clinical trials of tocilizumab to treat COVID-19 have overall not shown a mortality benefit.

15.4.2.2 IL-1 Inhibitors

IL-1 plays a crucial role in inflammatory responses. The drugs that have been evaluated as IL-1 inhibitors include anakinra and canakinumab. Anakinra is a human recombinant IL-1 receptor antagonist, whereas canakinumab is a monoclonal antibody targeting IL-1β. The FDA has approved anakinra to manage a variety of conditions including rheumatoid arthritis, neonatal-onset multisystem inflammatory disease and cryopyrin-associated periodic syndromes. The off-label use of anakinra is for macrophage activation syndrome and cytokine release syndrome mediated by severe CAR-T. The NIH guidelines, released in June 2020, noted that the data were insufficient to recommend for or against the use of IL-1 inhibitors such as anakinra for the treatment of COVID-19.[91]

Anakinra has been evaluated in several small studies. A retrospective Italian study was undertaken in COVID-19 patients who developed ARDS and were managed with non-invasive ventilation. It observed that anakinra administered at high doses was linked to reduced C-reactive protein levels, improved pulmonary function, decreased requirement for mechanical ventilation, and reduced mortality at 21 days. A meta-analysis of non-randomized cohort studies of anakinra for COVID-19 was reported on 5 February 2021 and included 184 patients. The overall mortality of treated patients was significantly lower than mortality in the control group (95% CI 0.14–0.48, $p < 0.0001$). No difference was found in adverse events or physical limitations at the time of hospital discharge.[92] While these results are encouraging, the numbers of patients treated thus far are small and confirmation awaits larger randomized controlled trials.

On 6 November 2020, Novartis released its interim results of canakinumab in the CAN-COVID trial, a multicenter, randomized, double-blind, placebo-controlled phase 3 trial in 454 patients hospitalized with COVID-19 pneumonia and cytokine release syndrome. The trial failed to show any survival benefit from canakinumab plus standard of care *vs.* standard of care alone and there was no reduction in the need for mechanical ventilation by day 29, despite no increase in adverse events compared to control patients (accessed at media.relations@novartis.com).

The recombinant interleukin inhibitor CYT107 (RevImmune) has been shown to increase T-cell production; several phase 2 trials have been completed in France, Belgium, and the UK to assess reconstitution in lymphopenic patients with COVID-19 and await further studies.

15.4.3 NAK and JAK Inhibitors

The kinase inhibitors have been proposed as a treatment for COVID-19 because they can prevent phosphorylation of key proteins involved in the signal transduction leading to immune activation and inflammation, *e.g.* by IL-6. Additionally, JAK inhibitors, particularly baricitinib, may have direct antiviral activity through interference with viral endocytosis, which would potentially prevent entry into and infection of susceptible cells.[93] The angiotensin-converting enzyme (ACE)2 receptors, expressed on type 2 alveolar epithelial cells, are the recognized portals of entrance in the case of SARS-CoV-2. The clathrin AP2-associated protein kinase-1 (AAK1) is a principal regulator of endocytosis. Therefore, to prevent the intracellular entry of the virus, AAK1 can be targeted so as to derange its function. Janus kinase (JAK) inhibitors have a high affinity for AAK1. Baricitinib, which is a JAK inhibitor and also a numb-associated kinase (NAK) inhibitor, can cause the inhibition of cytokine signaling that is implicated in the inflammatory response and diminish alveolar as well as systemic inflammation in cases of severe COVID-19. Baricitinib, pacritinib, fedratinib, tofacitinib, and ruxolitinib are JAK inhibitors that may diminish the effects of elevated cytokines and are being investigated in various trials.[94]

The National Institute of Allergies and Infectious Diseases' ACTT-2 study https://clinicaltrials.gov/ct2/show/NCT04280705 evaluated baricitinib (4 mg orally administered once daily for a period of 14 days) combined with remdesivir (100 mg administered intravenously once daily for a period of 10 days) compared to the administration of remdesivir alone. This was a double-blind randomized, placebo-controlled trial in 1033 hospitalized adult patients and the primary outcome was time to recovery, which was 1 day shorter in the combination baricitinib and remdesivir group (7 days *vs.* 8 days, $p = 0.03$). In addition, there was a 30% greater chance of recovery at day 15 and the mortality trended lower in the combination group (5.1% *vs.* 7.8%, odds ratio 0.65, 95% CI 0.39–1.09). Serious adverse events were less frequent in the combination group than the control group (16% *vs.* 21%, $p = 0.03$), as were new infections (5.9% *vs.* 11.2%, $p = 0.03$). The benefit was most pronounced in those patients receiving high-flow oxygen or noninvasive ventilation at the time of enrollment (10 days to recovery *vs.* 18 days for control patients). Walz *et al.* reviewed the results of 733 searched studies and did a meta-analysis of the effect of JAK inhibitors and type I IFNs in patients with COVID-19. They concluded that those who received JAK inhibitors had significantly reduced odds of mortality (odds ratio 0.12, $p < 0.001$), ICU admission (odds ratio 0.05, $p < 0.001$), and had significantly increased odds of hospital discharge when compared to the standard treatment group.[95] Review of data that came from the ACTT-2 trial and other studies earned an EUA from the FDA for baricitinib on 19 November 2020[96]

Ruxolitinib, an oral JAK inhibitor selective for JAK 1 and JAK 2 that is currently approved for myelofibrosis, polycythemia, and acute graft-*vs.*-host disease can, like baricitinib, modulate inflammatory responses *via* JAK1/JAK2 inhibition, and has exhibited inhibition of IL-6-induced STAT3 phosphorylation. It also has potential antiviral blocking effects, preventing SARS-CoV-2 entry and infection of lung cells and is currently in phase 2 trials.[96a]

Pacritinib, which is awaiting FDA approval for use in myelofibrosis, is a JAK2 inhibitor and colony stimulating factor-1 receptor (CSF-1R) inhibitor. A trial in phase 3 compared pacritinib with the standard of care and found that it blocked cytokine storm by IL-6 inhibition. The effects of macrophage activation syndrome can be ameliorated by pacritinib, as it is a CSF-1R inhibitor.[97] Further trials with JAK inhibitors, either alone or in combination with other drugs, hold promise.

15.4.4 Convalescent Plasma

Passive immunotherapy with convalescent plasma has been used for a variety of diseases since the nineteenth century, and a Nobel prize was awarded for the use of convalescent plasma for diphtheria in 1901.[98] Convalescent plasma is a source of neutralizing antibodies. There also may be certain other mechanisms by which convalescent plasma can be effective in COVID-19, including immune mechanisms such as phagocytosis, activation of the complement system, or antibody-dependent cell-mediated cytotoxicity. The collection of plasma is from eligible donors, post-recovery from COVID-19.[99]

Some, but not all observational studies have suggested a moderate benefit and these studies have confirmed its safety profile, but IgG titers, timing of the infusion, and severity of disease when treatment was begun varied, and there were other shortcomings and biases inherent in observational studies.[98]

The PLACID Trial[16] was a randomized controlled phase 2 trial that was conducted in 39 public and private hospitals across India and reported in October 2020. It evaluated the efficacy of convalescent plasma involving adults suffering from "moderate", RT-PCR-positive COVID-19. The trial started on 22 April 2020 and continued till 14 July 2020, enrolling 464 adults (aged ≥18 years) with moderate COVID-19 (determined by PaO_2/FIO_2, *i.e.* the ratio of partial pressure of oxygen in arterial blood divided by the fraction of oxygen in the inspired air ratio falling between 200 and 300 mmHg, or respiratory rate exceeding 24 breaths per minute and maintaining the oxygen saturation of ≤93% on room air), from which 235 were assigned to the intervention group that would receive standard of care in addition to the convalescent plasma and 229 were put in the group of controls that would receive only the standard-of-care treatment. Of note, as the authors state, "These criteria were similar to those for patients with severe illness in studies from other countries". The intervention group was administered 200 mL convalescent plasma, in two doses administered 24 hours apart. The difference that was observed in progression to more severe disease or mortality at day 28, among those with moderate COVID-19 treated with the plasma therapy in addition to the standard of care compared with those receiving only the standard of care, was not statistically significant. Development of severe pneumonia (decided by PaO_2/FIO_2 <100 mmHg) or all-cause mortality observed at 28 days after enrollment was not found to differ between the two groups; 19% and 18% of the participants in the intervention and the control groups, respectively.[100]

The question of antibody titer and disease severity was addressed in an observational study from the Mayo Clinic involving 4330 hospitalized adult patients in whom the 7 day mortality between patients who received low-titer plasma and high-titer plasma was not statistically different in the overall population, but for non-intubated patients, the 7 day mortality was lower in the high-titer compared to the low-titer group (14% *vs.* 11%, $p = 0.03$). The timing of treatment was also found to be important. Non-intubated patients aged <80 years, treated within 72 hours after diagnosis with high-titer plasma, had a lower 7 day mortality of 6% *vs.* 11.3% ($p = 0.0008$). Taken together, these findings indicate that early high-titer treatment might be effective, whereas early low-titer treatment might not, and that convalescent plasma might be ineffective for moderate and severe COVID-19. Early treatment was associated with a reduction in mortality.

Further clarification was provided by the studies of Libster *et al.*[101] and Simonovich *et al.*[102] in the 18 February 2021 edition of *NEJM*. Libster *et al.*[101] conducted a randomized, double-blind, placebo-controlled trial with convalescent plasma that had high IgG titers (>1 : 1000 in serum) in 160 persons within 72 hours of the onset of symptoms of COVID-19, with the primary endpoint being the development of severe disease (respiratory rate of ≥30 breaths per min or oxygen saturation of ≤93%). In the intention-to-treat group, severe respiratory disease developed in 16% *vs.* 31% in the controls (p = 0.03), with a relative risk reduction of 48%. A larger effect was noted using a modified intention-to-treat analysis after excluding six patients who developed severe disease before the infusion was begun (relative risk reduction of 60%). No adverse effects were noted.[101]

In contrast, Simonovich *et al.*[102] treated 228 hospitalized patients with severe COVID-19 pneumonia in a randomized controlled 2 : 1 ratio trial using high-titer convalescent plasma (median titer 1 : 3200). At 30 days, no significant difference was found in clinical outcomes between the convalescent plasma and control groups. Overall mortality was 10.96 in the convalescent plasma group and 11.43 in the control group. Neither adverse nor serious adverse events were different between the two groups,[102] agreeing with previous observational trials reporting an adequate safety profile for convalescent serum.[103]

The FDA granted an EUA on 23 August 2020 for investigational convalescent plasma for the treatment of COVID-19 in hospitalized patients. On 4 February 2021, they updated the EUA to "limit the authorization to the use of high titer COVID-19 convalescent plasma for the treatment of hospitalized patients early in the disease course". The Infectious Diseases Society of America and the Association for the Advancement of Blood and Biotherapies recommended in November 2020 that the use of convalescent plasma be limited to clinical trials and that critically ill patients and those in ICUs were not likely to benefit from convalescent plasma, and that when administered it should be used as early as possible, preferably within 3 days of onset.[98,104]

A further delineation regarding who should and who should not be given convalescent plasma was provided in mid-March 2021 by Joyner *et al.* and funded by the US Department of Health and Human Services from a retrospective analysis of 30-day mortality in 3082 patients hospitalized with COVID-19, characterized by whether they were given low- (n = 561), medium- (n = 2006), or high- (n = 515) titer convalescent plasma. These generally were high-risk patients in that approximately two-thirds of the patients in each of these three groups before treatment had severe or life-threatening COVID-19; approximately 60% of each group were in the ICU when treatment was begun; and approximately one-third of each group were on mechanical ventilation prior to convalescent plasma administration. The only subset that did not benefit was those patients on mechanical ventilation at the time that the infusion was begun. As reported in prior studies, higher-titer treatment was of greater benefit; mortality at 30 days was 22.3% in the high-titer group, 27.4% in the medium-titer group, and 29.6% in the low-titer group.[103]

A Cochrane review is being updated regularly as data related to the utility of plasma therapy in COVID-19 continue to emerge. However, based on current data, it would seem reasonable to administer high-titer convalescent plasma when available to hospitalized patients with COVID-19 who are not being mechanically ventilated, but have one or more risk factors for progression to more severe disease.

15.5 Monoclonal Antibodies

Monoclonal antibodies directed against SARS-CoV-2 can prove to be extremely potent therapeutic agents. Monoclonal antibodies directed against the spike proteins of SARS-CoV-2 are intended to prevent the virus from attaching to the cell and subsequently entering into it. On 21 November 2020, the FDA issued an EUA for the co-administration of casirivimab and imdevimab in laboratory-confirmed adult and pediatric patients (aged ≥12 years and weighing ≥40 kg) with mild to moderate COVID-19.[105] A randomized controlled trial recruited a total of 799 patients with mild to moderate COVID-19 who were not hospitalized. From this group of 799 patients, 266 were given a single intravenous infusion of 2400 mg of casirivimab and imdevimab (1200 mg each). A higher combined dose (8000 mg, comprised of 4000 mg of each) of the two drugs was administered to 267 patients and the remaining 266 received placebo. The primary endpoint was the reduction in viral load from baseline, which was considerably more in the two groups that received the monoclonal antibodies. It was found that the two dosing regimens were similar in efficacy in terms of the primary and secondary endpoints achieved.

Intravenous administration of this combination of monoclonal antibodies must be done together. The regimen is not to be used in the cases of hospitalization due to COVID-19 when receiving supplemental oxygen therapy, as no benefits were documented in such patients. Also, administration of the two monoclonal therapeutics to patients on mechanical ventilation and in those with need for high-flow oxygenation can rather prove harmful, as observed in the trial. The associated side effects include anaphylaxis, infusion-related reactions, fever, chills, flushing, itching, and hives.[106]

A more recent randomized, double-blind, placebo-controlled, single-infusion phase 2/3 trial of 577 patients at 49 US centers who tested positive for SARS-COV-2, the BLAZE-1 study, used bamlanivimab (also known as LY3819253 or LY-CoV55) alone and in combination with etesevimab (LY3832479 or LY-CoV016). These two compounds are potent spike-neutralizing monoclonal antibodies derived from two separate patients, one in North America and one in China, respectively. The reason for combining the two was that in pre-clinical experiments, etesevimab was shown to bind a different epitope from bamlanivimab and to neutralize resistant variants with mutations in the epitope bound by bamlanivimab. The study randomized outpatients aged ≥18 years who tested positive for SARS-CoV-2, and who had one or more mild or moderate symptoms, presenting within 3 days of their first positive test. A 1 hour infusion of one of three different doses of bamlanivimab, placebo, or bamlanivimab plus etesevimab was administered in the outpatient setting. The three doses of bamlanivimab were 700 mg, 2800 mg, and 7000 mg, and in the combined group, 2800 mg of bamlanivimab and 2800 mg of etesevimab were administered. The final analysis included results for the five treatment groups. The primary endpoint of the study was the change in SARS-CoV-2 log viral load from baseline to day 11, and there were nine pre-specified secondary outcomes, including five on symptoms and one on outcome. No statistically significant difference was found between the three doses of bamlanivimab and placebo in log viral load, but combination therapy resulted in a significant reduction in viral load of −0.57 ($p = 0.01$) by day 11. No deaths occurred during the study in any of the groups, but immediate hypersensitivity reactions were reported in nine patients (six bamlanivimab, two combination therapy, and one control). Among the secondary outcomes, 10 out of 84 endpoints reached statistical significance. Among these were COVID-19-related hospitalizations or emergency department visits, which were 5.8% for controls, and for bamlanivimab, 1% for the 700 mg, 1.9% for the 2800 mg, and 2% for the 7000 mg groups and 0.9% for the combination group. Further ongoing trials will focus on assessment of clinical benefit of anti-spike-neutralizing antibodies in patients with COVID-19 as a primary endpoint.[107]

In the meantime, the combination of these two monoclonal antibodies was given an EUA by the FDA in the US on 9 February 2021 for outpatients (or inpatients admitted for a non-COVID-19 diagnosis) with mild–moderate COVID-19 who are at high risk of progression as defined by EUA criteria (B11a) with the recommendation for dosages of 700 mg of bamlanivamab and 1400 mg of etesevimab to be given in combination by infusion as soon as possible after symptom onset. However, the dosage of bamlanivamib was lower than in the BLAZE-1 trial based on extrapolation of *in vitro* data and pharmacokinetic/pharmacodynamic modeling (NIH.COVID-19@service.govdelivery.com). A subsequent study of 314 patients published in March 2021 using bamlanivamib 7000 mg in conjunction with remdesivir as a single 1 hour infusion in hospitalized COVID-19 patients without end-organ failure did not demonstrate efficacy.[106a,b]

Monoclonal antibody combination therapy, such as bamlanivamab plus etesevimab or casirivimab plus imdevimab, directed against the spike protein is therefore promising for treatment of mild–moderate COVID-19 early in the course of the disease with proven efficacy in reducing viral load and reducing disease severity when administered early after onset of symptoms.

15.6 Ancillary Nutritional Therapies

15.6.1 Vitamin D

Vitamin D-deficient individuals are at an elevated risk of COVID-19 related mortality.[108] The immunomodulatory effects of vitamin D on various bacterial and viral infections have been noticed in various studies. Its role is yet to be determined in SARS-CoV-2 infection. The influence of vitamin D on expression of various genes involved in the immune system (innate immunity, adaptive immunity) and the downstream inflammatory cascade affects the susceptibility to bacterial and viral infections, as well as their severity.[109] Vitamin D deficiency in COVID-19 is associated with increased disease severity. This can be related to augmented pro-inflammatory cytokine milieu due to vitamin D deficiency. SARS-CoV-2 entry into the alveolar cells occurs *via* ACE2 receptors on the cells. The downregulation of the expression of ACE2 is the proposed mechanism by which vitamin D may prevent the virus from entering cells.[110] It is very possible that vitamin D supplementation may reduce the likelihood of becoming infected with SARS-CoV-2, or may cause an early viral clearance in vitamin D-deficient individuals. It has been observed that vitamin D levels >30 ng mL^{-1} are associated with a decline in viral infection, disease severity and associated deaths.[111] A public health study by Nurshad *et al.* examined the correlation of vitamin D levels with COVID-19 cases and deaths in 20 European countries as of 20 May 2020. A significant negative correlation (p = 0.033) was observed between mean vitamin D levels and COVID-19 cases per 1 million population. However, the correlation of vitamin D with COVID-19 deaths in those countries was not significant.

A randomized, placebo-controlled study conducted in India explored the impact of vitamin D supplementation at high doses on viral clearance in SARS-CoV-2 in RT-PCR-positive patients. It concluded that high-dose oral supplementation aiming at vitamin D levels >50 ng mL^{-1} was associated with faster viral RNA negativity in a majority of asymptomatic individuals with vitamin D deficiency along with a significant lowering of inflammatory markers. Achieving SARS-CoV-2 RNA negativity by supplementing cholecalciferol (vitamin D3) would therefore appear useful in lowering transmission rates in vitamin D-deficient persons.[112]

However, studies have not been uniformly positive for vitamin D supplementation in patients with COVID-19. The largest published multicenter, double-blind, randomized placebo-controlled trial conducted in two centers in Brazil in 240 patients (mean age 58 years) given a single dose of 200 000 IU of vitamin D3 or placebo who were hospitalized with severe COVID-19 conducted between June and October 2020, found no reduction in length of hospital stay, mechanical ventilation requirement, mortality, or any other relevant outcome in those who received vitamin D supplementation. This was despite a significant mean rise in serum 25-hydroxyvitamin D level of 24 ng mL^{-1} (from 21 ng mL^{-1} to 44 ng mL^{-1}; p = 0.001). No adverse effects were noted from this high dosage of vitamin D.[113] The accompanying editorial pointed out that the study excluded critically ill patients (patients admitted to the ICU or on mechanical ventilation) so the results can only be generalized to moderately ill, but not critically ill patients.[114] The issue becomes more complex when determining the diagnosis of vitamin D deficiency, as 25-hydroxyvitamin D levels that are low do not necessarily indicate vitamin D deficiency in those persons who have congenitally low levels of vitamin D binding globulin, as is often found in persons with African ancestry[115] or in persons who have liver disease. In normal non-pregnant persons, 85% of 25-hydroxyvitamin D is bound to vitamin D binding protein (DBP), made in the liver. Approximately 15% is bound to albumin and 0.03% is free.[116] Anything that interferes with liver function may reduce the serum level of vitamin D without necessarily reducing free serum vitamin D or making the person vitamin D deficient. Measuring free vitamin D may therefore be a more accurate method of determining vitamin D deficiency, but typically 25-hydroxyvitamin D is measured instead and is subject to these limitations.[116]

Further clarification was given from an Indian study by Jain and colleagues who studied 154 patients with COVID-19 over a 6-week period, 90 of whom were judged to be vitamin D deficient (very low mean 25-OH vitamin D level of 14.35). Serum levels of inflammatory markers were found to be higher in vitamin D-deficient COVID-19 patients (IL-6 19.34 *vs.* 12.18 pg mL^{-1}; serum ferritin 319.17 *vs.* 186 ng mL^{-1}; and TNF-α 13.26 *vs.* 11.87 pg mL^{-1}). The fatality rate was 21% in the vitamin D-deficient patients *vs.* 3.1% in those judged not to be vitamin D deficient.[117] A logical way to reconcile these divergent results is to postulate that if a person is vitamin D deficient when they contract COVID-19, mortality and morbidity are increased and supplementation with vitamin D appears to confer benefit, whereas if a person is not already vitamin D deficient, supplementation will not likely improve outcome. Vitamin D as measured by the 25-hydroxyvitamin D level may decrease in sick hospitalized patients due to decreased nutritional intake and/or concomitant medications from a decrease in DBP production in the liver, but not reflect vitamin D deficiency and in such patients, supplementing a person with a low 25-hydroxyvitamin D but normal free vitamin D level might be expected to confer little or no benefit.

15.6.2 Vitamin C

Vitamin C significantly impacts immunity, probably by playing a significant part in the functioning of phagocytes, conversion of T-lymphocytes and production of interferons.[118]

A healthy person requires only 0.1 g of vitamin C per day. A critically ill patient might require higher doses in the range of 1–4 g per day. Along with being an essential nutrient, vitamin C is cheap and safe. There has been limited systematic study of vitamin C supplementation in persons with COVID-19. A randomized controlled pilot study by Zhang *et al.* in China administered high-dose vitamin C intravenously to 56 critically ill COVID-19 patients and failed to find a significant mortality benefit at 28 days (p = 0.27), but during the 7 day treatment period, the vitamin C group had a steady rise in the PaO_2/FIO_2 ratio (229 *vs.* 151, $p = 0.01$) and IL-6 was lower than in the control group (19.42 *vs.* 158.00, $p = 0.04$ on day 7). Therefore, further investigation of vitamin C's potential impact on COVID-19 must be encouraged.[119a]

However, in the COVID A to Z trial from the Cleveland Clinic discussed, in more detail in section 15.6.3, outpatient treatment with 8000 mg of vitamin C daily in divided doses neither shortened the symptom duration for the 28 day study duration nor impacted hospitalization or death rate, although the baseline level of these vitamins in the study population was not measured and persons with nutritional deficiencies of any type may have been under-represented in the trial.[120] As of 11 February 2021, the NIH recommended that "There are insufficient data to recommend either for or against the use of vitamin C for the treatment of COVID-19".[121]

15.6.3 Zinc

Zinc is an essential trace mineral required to maintain natural barriers such as the respiratory and gastrointestinal lining epithelium, thereby preventing entry of pathogens. It is also necessary as a co-factor for various enzymes as well as for keeping the immune system intact. Zinc is known to play a role in immune function *via* antibody and white blood cell production[122] and zinc supplementation has been suggested to increase the ability of polymorphonuclear cells to fight infection. There is also evidence that zinc deficiency increases pro-inflammatory cytokines and decreases the production of antibodies. Zinc has been implicated in coronavirus biology.[123] ACE2 is a zinc metalloprotease that is important for cellular entry of coronavirus.[124] Prior studies involving SARS-CoV have shown that zinc can inhibit its ribonucleic acid polymerase.[123] However, the biological activity of zinc against viruses may require ionophores, such as pyrithione, to block viral replication.[119b] Deficiency of zinc predisposes to infection by SARS-CoV-2 and is associated with adverse outcomes in COVID-19. Those with zinc deficiency may benefit from its supplementation.[125] However, no causal relationship was observed between zinc and survival in COVID-19 patients requiring hospitalization.[126] Outpatient oral zinc supplementation was evaluated in the multicenter, single health system (Cleveland Clinic), randomized, factorial open-label COVID A to Z trial of 214 patients (aged ≥18 years) randomized in a 1 : 1 : 1 : 1 design to either zinc gluconate 50 mg daily at bedtime, 8000 mg of vitamin C in divided doses, both vitamin C and zinc in combination, or usual care without any of the study medicines. Symptom questionnaires were completed weekly for 1 month and the number of days to 50% symptom resolution was not found to be statistically significantly different between the four groups; the study was "stopped for futility". There was also no difference in hospitalizations or death rates between the four groups.[120] Using hydroxychloroquine as an ionophore to facilitate zinc entry into cells, a multicenter, non-randomized, retrospective cohort study of patients admitted to four New York City hospitals with SARS-CoV-2 infection was carried out and included 1006 who received zinc plus hydroxychloroquine, 2467 who received hydroxychloroquine without zinc, and 1097 who received zinc alone. In a Cox regression analysis that adjusted for confounders, treatment with zinc plus hydroxychloroquine was associated with a significantly reduced risk of in-hospital death (p = 0.023), but treatment with zinc alone did not affect mortality (p = 0.296), and hydroxychloroquine alone appeared to be harmful (p = 0.001).[127] However, a small randomized trial of 191 patients from three academic medical centers in Egypt with laboratory-confirmed severe COVID-19 randomized to receive 220 mg of zinc twice daily plus hydroxychloroquine for a 5 day course or hydroxychloroquine alone, and matched for age and gender revealed no significant difference between the two arms of the trial in terms of recovery at 28 days (p = 0.969), the need for mechanical ventilation (p = 0.537), or overall mortality (p = 0.986).[128] As of 11 February 2021, the NIH recommendation states that "There are insufficient data to recommend for or against the use of zinc for the treatment of COVID-19.[129]

15.7 Anticoagulants

COVID-19 predisposes to arterial as well as venous thrombosis. Thrombotic complications have occurred with increasing frequency in the form of venous thromboembolism (VTE), myocardial infarction, ischemic stroke, and peripheral arterial thrombosis. Autopsy findings have revealed thromboembolism in various organs including the lungs, kidneys, and the heart. Multisystem organ dysfunction syndrome, which occurs frequently in critically ill COVID-19 patients, can be partly attributed to thrombosis.

Elevated levels of D-dimer, fibrinogen, and factor VIII have been observed in patients with COVID-19. Decreased serum antithrombin along with lower protein C and protein S levels, in addition to detection of antiphospholipid antibodies, have also been noticed. These abnormalities are more pronounced in seriously ill patients. In COVD-19, the pathophysiology of thromboembolism differs from other viral diseases as it may be related to viral-mediated endothelial inflammation and more platelet activation. This occurs in conjunction with hypercoagulability associated with acquired antiphospholipid antibodies, increase in the levels of coagulation factors, and lower concentrations of endogenous anticoagulants. SARS-CoV-2 is histopathologically distinct from the other respiratory viruses, as it directly invades the endothelial cells with perivascular cuffing of T-cells, causes abnormal activation of macrophages, immune cell depletion, and thrombotic microangiopathy, along with accompanying angiogenesis.[130]

Low molecular weight heparin (LMWH) is to be administered to all adults with COVID-19 who require hospitalization. LMWH is preferred over unfractionated heparin. Fondaparinux is advised for patients who have a history of, or who develop, heparin-induced thrombocytopenia. The institutional guidelines should direct dosage adjustment based on body weight. The use of mechanical thromboprophylaxis is advocated in cases where pharmacological anticoagulation is contraindicated. There are no recommendations for the combined use of pharmacological and mechanical prophylaxis. Many institutions follow an intermediate-intensity dosing (LMWH administered twice daily in usual doses), or based on local experience some the institutes may go with therapeutic-intensity dosing for thromboprophylaxis, despite there being a lack of published quality evidence. In some COVID-19 patients, microvascular thrombosis has been recognized as a cause of type 1 (hypoxemic) respiratory failure. Findings revealed in autopsies include microvascular thrombosis, thrombosis involving large vessels, pulmonary thrombosis, and hemorrhages. In critically ill patients with COVID-19 who are suspected of having VTE, the status of anticoagulation dosing is yet to be determined. The dose-wise effects of heparin on various clinical consequences are being evaluated in numerous randomized controlled trials. The incidence of asymptomatic VTE, 30–42 days post-discharge for COVID-19 patients has ranged between 0 and 0.6%. Various trials have been investigating the need for thromboprophylaxis after discharge.[131]

The COVID-19 pandemic has raised unprecedented challenges for humans. The entire process of scientific discovery, trials, and the research associated with it is ever-expanding and poses a daunting task for researchers and drug developers. The fast-tracking of treatments has become imperative in these times of pandemic emergency. Initiatives by the FDA such as the Coronavirus Treatment Acceleration Program and the Accelerating COVID-19 Therapeutic Interventions and Vaccines program pursued by the NIH, along with the various efforts across the globe, demonstrate that we must be united in moving toward the goal of ending this pandemic. In this chapter, we have tried to review and summarize potential therapeutic interventions, encompassing the approved, proposed, repurposed, and experimental ones (Table 15.3). It leaves a lot to be desired, as the research is growing rapidly and new findings and recommendations are emerging nearly every week.

15.8 Executive Summary†

1. The antiviral intravenous drug remdesivir, approved by the FDA for treatment of COVID-19 in adults and children aged ≥12 years, likely shortens disease duration without a proven effect on mortality. Its beneficial effect may be limited to the first 10 days of symptomatic disease.
2. Remdesivir in combination with the JAK/NAK inhibitor and the repurposed drug baricitinib shortens recovery time and accelerates clinical improvement, especially in those patients on oxygen or non-invasive ventilation, and this drug combination may reduce mortality.

†The pharmacotherapy reviewed in this chapter has been updated as of Nov 1, 2021.

Table 15.3 Summary of various therapeutic classes.

Drug class	Examples
Antiviral drugs	Remdesivir, favipiravir, molnupiravir, umifenovir
Other drugs with antiviral properties	Nitazoxanide, ivermectin, niclosamide, rintatolimod, bemcentinib, antroquinonol, brequinar, AT-527, VIR-2703, plitidepsin, emetine hydrochloride, trabedersen, apilimod dimesylate
Immunomodulatory drugs	1. Corticosteroids: dexamethasone, methylprednisolone, hydrocortisone 2. IL-6 inhibitors: tocilizumab, siltuximab, sarilumab 3. T-cell co-stimulatory immunomodulator: abatacept 4. TNF inhibitor: infliximab 5. Chemokine blocker: cenicriviroc 6. IL-1 inhibitor: anakinra 7. IL-7 inhibitor: C-YT107 8. JAK inhibitors: baricitinib, fedratinib, ruxolitinib, pacritinib
Ancillary nutritional therapy	Vitamin D, vitamin C, zinc, selenium
Anticoagulants and antithrombotics	Low molecular weight heparin, unfractionated heparin, rivaroxaban, AB201
Investigational immunotherapies	Brilacidin, Allogenic natural killer cells, ADX-629, CD24Fc, bucillamine, peginterferon lambda, inhaled interferon β1a
Investigational antibody-directed therapies	Bamlanivimab, Casirivimab, Imdevimab, VIR-7831, VIR-7832, JS-016, COVI-SHIELD, STI-1499, AZD7442, foralumab intranasal
Additional investigational drugs	Neurokinin-1 receptor antagonists: tradipitant, aprepitant Colony-stimulating factors: sargramostim, gimsilumab, mavrilimumab, lenzilumab Mesenchymal stem cells: Remestemcel-L, PLX-PAD, BM-Allo.MSC Phosphodiesterase inhibitors: ibudilast, apremilast Others: ifenprodil, eculizumab, ravulizumab, aviptadil, ATYR1923, BIO-11006, dociparstat sodium, opaganib, intranasal vazegepant, selinexor
Therapies deemed ineffective	Hydroxychloroquine, chloroquine, doxycycline, lopinavir/ritonavir, azithromycin

3. Favipiravir, another antiviral drug, given orally, and similar in mechanism to remdesivir, has been approved for use in several countries for COVID-19, but has not yet undergone large randomized trials to confirm efficacy. It appears to shorten hospital course and hasten viral clearance and, like remdesivir, is more likely to be beneficial earlier in the course of disease. It is unlikely to benefit patients with severe COVID-19; it is also embryotoxic and cannot be given when pregnancy is a consideration.

4. The repurposed antihelminthic drug ivermectin in several smaller trials has demonstrated efficacy in randomized placebo-controlled trials both in preventing COVID-19 in contacts as well as for the treatment of mild, moderate, and severe disease, with evidence supporting faster viral clearance and clinical recovery as well as a reduction in mortality. It is unclear if it provides benefit in younger patients with mild disease who have few if any comorbidities. However, it has the advantage of a proven safety profile in the doses used to treat parasitic diseases, is readily available worldwide, and is inexpensive. It may have multiple mechanisms of action against COVID-19. One small study suggested benefit for post-COVID "long hauler" syndrome, which bears further investigation, as there is no known treatment for this condition at present.
5. Lopinavir, ritonavir, hydroxychloroquine, and umifenovir have failed efficacy testing to treat COVID-19.
6. Dexamethasone and hydrocortisone reduce mortality in patients with severe COVID-19, but are not indicated for milder forms of the disease.
7. The IL-6 inhibitor tocilizumab and the IL-1 inhibitor canakinumab have not demonstrated a survival benefit in the setting of severe COVID-19, but the IL-1 inhibitor anakinra has shown promise in observational trials and is now awaiting the outcome of randomized controlled trials.
8. NAK and JAK inhibitors prevent signal transduction that leads to immune activation and baricitinib has received an EUA from the FDA for use in adults and in children aged ≥2 years, in combination with thc antiviral drug remdesivir. In the only randomized controlled trial reported thus far, this combination improved clinical recovery with trending lower mortality in hospitalized COVID-19 patients (see #2 above).
9. Convalescent plasma was granted an EUA by the FDA for use in hospitalized COVID-19 patients early in the course of their disease, using high-titer plasma, given preferably within the first 72 hours of symptoms. It is probably ineffective against severe COVID-19, at least when the disease has progressed to the point of requiring mechanical ventilation.
10. Monoclonal antibody combinations have been shown to decrease viral load, but may be of value only in mild–moderate cases early in the course of disease of COVID-19, and they have not been of value for severe disease.
11. Drugs with novel and different potential mechanisms of action against COVID-19 are in clinical trials and include the antihelminth drugs nitazoxanide and niclosamide, the guanine-depletor merimepodib, the TLR-3 receptor binder rintatolimod, the viral genome mutagens molnupiravir and β-D-N4-hydroxycytidine, the AXL kinase inhibitor bemcentinib, the antitumor eEFIA inhibitor plitidepsin, the *in vitro* viral blocker of cell entry emetine hydrochloride, the inhibitor of TGF-β2

trabedersen, the dual-action antivirals and immunomodulators stannous protoporphyrin and antroquinonol, the lipid kinase enzyme PIKfyve inhibitor apilimod dimesylate, and the longstanding gout prevention and anti-inflammatory drug colchicine. Colchicine is the only one of these drugs that in a moderately large randomized controlled trial was found to reduce hospitalizations, development of pneumonia, and to reduce deaths in hospitalized patients with COVID-19.

12. Ancillary therapy with vitamins C and D and zinc may be beneficial for patients deficient in these nutrients at the time of infection, but there is no evidence that supplementation in the absence of deficiency alters the disease trajectory. However, all three play a role in immune competency under normal circumstances.
13. Anticoagulation is an accepted and beneficial treatment for hospitalized patients with COVID-19 due to the microangiopathy and vascular inflammation with resulting thrombosis resulting from SARS-COV-2 infection.
14. Drug trials are being released nearly every week and further exciting developments are likely ahead.

References

1. L. M. Van Blerkom, Role of viruses in human evolution, *Am. J. Phys. Anthropol.*, 2003, (Suppl 37), 14–46.
2. Coronavirus (COVID-19) Update: FDA Issues Emergency Use Authorization for Potential COVID-19 Treatment, FDA, 2020. Available from: http://www.fda.gov/news-events/press-announcements/coronavirus-covid-19-update-fda-issues-emergency-use-authorization-potential-covid-19-treatment. Last accessed on 10 December 2020.
3. R. T. Eastman, J. S. Roth, K. R. Brimacombe, A. Simeonov, M. Shen and S. Patnaik, *et al.*, Remdesivir: A Review of Its Discovery and Development Leading to Emergency Use Authorization for Treatment of COVID-19, *ACS Cent. Sci.*, 2020, **6**(5), 672–683.
4. Final report confirms remdesivir benefits for COVID-19 [Internet], National Institutes of Health (NIH), 2020, Available from: https://www.nih.gov/news-events/nih-research-matters/final-report-confirms-remdesivir-benefits-covid-19. Last accessed on 12 December 2020.
5. J. H. Beigel, K. M. Tomashek, L. E. Dodd, A. K. Mehta, B. S. Zingman and A. C. Kalil, *et al.*, Remdesivir for the Treatment of Covid-19 - Final Report, *N. Engl. J. Med.*, 2020, **383**(19), 1813–1826.
6. C. D. Spinner, R. L. Gottlieb, G. J. Criner, J. R. Arribas López, A. M. Cattelan and A. Soriano Viladomiu, *et al.*, Effect of Remdesivir *vs.* Standard Care on Clinical Status at 11 Days in Patients With Moderate COVID-19: A Randomized Clinical Trial, *JAMA*, 2020, **324**(11), 1048–1057.
7. J. D. Goldman, D. C. B. Lye, D. S. Hui, K. M. Marks, R. Bruno and R. Montejano, *et al.*, Remdesivir for 5 or 10 Days in Patients with Severe Covid-19, *N. Engl. J. Med.*, 2020, **383**(19), 1827–1837.

8. WHO Solidarity Trial Consortium, H. Pan, R. Peto, A.-M. Henao-Restrepo, M.-P. Preziosi and V. Sathiyamoorthy, *et al.*, Repurposed Antiviral Drugs for Covid-19 - Interim WHO Solidarity Trial Results, *N. Engl. J. Med.*, 2021, 497–511.
9. WHO recommends against the use of remdesivir in COVID-19 patients, Available from: https://www.who.int/news-room/feature-stories/detail/who-recommends-against-the-use-of-remdesivir-in-covid-19-patients. Last accessed on 16 December 2020.
10. Y. Wang, D. Zhang, G. Du, R. Du, J. Zhao, Y. Jin, S. Fu, L. Gao, Z. Cheng, Q. Lu, Y. Hu, G. Luo, K. Wang, Y. Lu, H. Li, S. Wang, S. Ruan, C. Yang, C. Mei, Y. Wang, D. Ding, F. Wu, X. Tang, X. Ye, Y. Ye, B. Liu, J. Yang, W. Yin, A. Wang, G. Fan, F. Zhou, Z. Liu, X. Gu, J. Xu, L. Shang, Y. Zhang, L. Cao, T. Guo, Y. Wan, H. Qin, Y. Jiang, T. Jaki, F. G. Hayden, P. W. Horby, B. Cao and C. Wang, Remdesivir in adults with severe COVID-19: a randomised, double-blind, placebo-controlled, multicentre trial, *Lancet*, 2020, **395**, 1569–1578.
11. D. P. Harrington, L. R. Baden and J. W. Hogan, A Large, Simple Trial Leading to Complex Questions, *N. Engl. J. Med.*, 2020, **384**, 576–577.
12. E. J. Rubin and D. L. Longo, SARS-CoV-2 Vaccination — An Ounce (Actually, Much Less) of Prevention, *N. Engl. J. Med.*, 2020, **383**, 2677–2678.
13. A. C. Kalil, *et al.*, Baricitinib plus Remdesivir for Hospitalized Adults with Covid-19, *N. Engl. J. Med.*, 2020, **384**(9), 795–807.
14. Veklury (remdesivir) dosing, indications, interactions, adverse effects, and more Available from: https://reference.medscape.com/drug/veklury-remdesivir-4000090. Last accessed on 16 December 2020.
15. A. Shannon, B. Selisko, N. Le, J. Huchting, F. Touret, G. Piorkowski, V. Fattorini, F. Ferron, E. Decroly, C. Meier, B. Coutard, O. Peersen and B. Canard, Favipiravir strikes the SARS-CoV-2 at its Achilles heel, the RNA polymerase, *bioRxiv*, 2020, DOI: 10.1101/2020.05.15.098731
16. A. Agarwal, A. Mukherjee, G. Kumar, P. Chatterjee, T. Bhatnagar and P. Malhotra, Convalescent plasma in the management of moderate covid-19 in adults in India: open label phase II multicentre randomised controlled trial (PLACID Trial), *BMJ*, 2020, **371**, m3939.
17. Y. Doi, M. Hibino, R. Hase, M. Yamamoto, Y. Kasamatsu, M. Hirose, Y. Mutoh, Y. Homma, M. Terada, T. Ogawa, F. Kashizaki, T. Yokoyama, H. Koba, H. Kasahara, K. Yokota, H. Kato, J. Yoshida, T. Kita, Y. Kato, T. Kamio, N. Kodama, Y. Uchida, N. Ikeda, M. Shinoda, A. Nakagawa, H. Nakatsumi, T. Horiguchi, M. Iwata, A. Matsuyama, S. Banno, T. Koseki, M. Teramachi, M. Miyata, S. Tajima, T. Maeki, E. Nakayama, S. Taniguchi, C. K. Lim, M. Saijo, T. Imai, H. Yoshida, D. Kabata, A. Shintani, Y. Yuzawa and M. Kondo, A Prospective, Randomized, Open-Label Trial of Early *versus* Late Favipiravir Therapy in Hospitalized Patients with COVID-19, *Antimicrob. Agents Chemother.*, 2020, **64**, e01897-20.
18. A. A. Ivashchenko, K. A. Dmitriev, N. V. Vostokova, V. N. Azarova, A. A. Blinow, A. N. Egorova, I. G. Gordeev, A. P. Ilin, R. N. Karapetian, D. V. Kravchenko, N. V. Lomakin, E. A. Merkulova, N. A. Papazova, E. P. Pavlikova, N. P. Savchuk, E. N. Simakina, T. A. Sitdekov, E. A. Smolyarchuk,

E. G. Tikhomolova, E. V. Yakubova and A. V. Ivachtchenko, AVIFAVIR for Treatment of Patients With Moderate Coronavirus Disease 2019 (COVID-19): Interim Results of a Phase II/III Multicenter Randomized Clinical Trial, *Clin. Infect. Dis.*, 2021, **73**, 531–534.

19. C. Chen, Y. Zhang, J. Huang, P. Yin, Z. Cheng, J. Wu, S. Chen, Y. Zhang, B. Chen, M. Lu, Y. Luo, L. Ju, J. Zhang and X. Wang, Favipiravir *versus* Arbidol for COVID-19: A Randomized Clinical Trial, *medRxiv*, 2020, DOI: 2020.03.17.20037432.
20. Glenmark, Formulations Global Presence, Glenmark, 2020, http://www.glenmarkpharma.com/formulations/formulations-global-presence (not secure).
21. V. Pilkington, T. Pepperrell and A. Hill, A review of the safety of favipiravir – a potential treatment in the COVID-19 pandemic?, *J. Virus Erad.*, 2020, **6**, 45–51.
22. N. Vankadari, Arbidol: A potential antiviral drug for the treatment of SARS-CoV-2 by blocking trimerization of the spike glycoprotein, *Int. J. Antimicrob. Agents*, 2020, **56**, 105998.
23. Combination of antiviral drugs shows 'no superior outcome' for Covid treatment: Firm, *Hindustan Times*, 2020. Available from: https://www.hindustantimes.com/india-news/combination-of-antiviral-drugs-shows-no-superior-outcome-for-covid-treatment-firm/story-CS4ySHTJj37EhPZoKKP7jJ.html. Last accessed on 14 December 2020.
24. B. Cao, Y. Wang, D. Wen, W. Liu, J. Wang, G. Fan, L. Ruan, B. Song, Y. Cai, M. Wei, X. Li, J. Xia, N. Chen, J. Xiang, T. Yu, T. Bai, X. Xie, L. Zhang, C. Li, Y. Yuan, H. Chen, H. Li, H. Huang, S. Tu, F. Gong, Y. Liu, Y. Wei, C. Dong, F. Zhou, X. Gu, J. Xu, Z. Liu, Y. Zhang, H. Li, L. Shang, K. Wang, K. Li, X. Zhou, X. Dong, Z. Qu, S. Lu, X. Hu, S. Ruan, S. Luo, J. Wu, L. Peng, F. Cheng, L. Pan, J. Zou, C. Jia, J. Wang, X. Liu, S. Wang, X. Wu, Q. Ge, J. He, H. Zhan, F. Qiu, L. Guo, C. Huang, T. Jaki, F. G. Hayden, P. W. Horby, D. Zhang and C. Wang, A Trial of Lopinavir–Ritonavir in Adults Hospitalized with Severe Covid-19, *N. Engl. J. Med.*, 2020, **382**, 1787–1799.
25. A. Bhimraj, R. L. Morgan, A. H. Shumaker, V. Lavergne, L. Baden, V. C. Cheng, *et al.*, COVID-19 Guideline, Part 1: Treatment and Management. Available from: https://www.idsociety.org/practice-guideline/covid-19-guideline-treatment-and-management. Last accessed on 18 December 2020.
26. "Solidarity" clinical trial for COVID-19 treatments. Available from: https://www.who.int/emergencies/diseases/novel-coronavirus-2019/global-research-on-novel-coronavirus-2019-ncov/solidarity-clinical-trial-for-covid-19-treatments. Last accessed on 18 December 2020.
27. D. A. Jans and K. M. Wagstaff, Ivermectin as a Broad-Spectrum Host-Directed Antiviral: The Real Deal? *Cells*, 2020, **9**, 2100.
28. E. Rizzo, Ivermectin, antiviral properties and COVID-19: a possible new mechanism of action, *Naunyn-Schmiedeberg's Arch. Pharmacol.*, 2020, **393**, 1153–1156.

29. A. Crump, Ivermectin: enigmatic multifaceted 'wonder'drug continues to surprise and exceed expectations, *J. Antibiot. Res.*, 2017, **70**, 495–505.
30. A. Zumla, J. F. Chan, E. I. Azhar, D. S. Hui and K. Y. Yuen, 'Coronaviruses - drug discovery and therapeutic options, *Nat. Rev. Drug Discovery*, 2016, **15**, 327–347.
31. C. A. Guzzo, C. I. Furtek, A. G. Porras, C. Chen, R. Tipping, C. M. Clineschmidt, D. G. Sciberras, J. Y. Hsieh and K. C. Lasseter, Safety, tolerability, and pharmacokinetics of escalating high doses of ivermectin in healthy adult subjects, *J. Clin. Pharmacol.*, 2002, **42**, 1122–1133.
32. X. Zhang, Y. Song, X. Ci, N. An, Y. Ju, H. Li, X. Wang, C. Han, J. Cui and X. Deng, Ivermectin inhibits LPS-induced production of inflammatory cytokines and improves LPS-induced survival in mice, *Inflammation Res.*, 2008, **57**, 524–529.
33. L. Caly, J. D. Druce, M. G. Catton, D. A. Jans and K. M. Wagstaff, The FDA-approved drug ivermectin inhibits the replication of SARS-CoV-2 *in vitro*, *Antiviral Res.*, 2020, **178**, 104787.
34. R. T. Kinobe and L. Owens, A systematic review of experimental evidence for antiviral effects of ivermectin and an in-silico analysis of ivermectin's possible mode of action against SARS-CoV-2, *Fundam. Clin. Pharmacol.*, 2021, 260–276.
35. M. Yagisawa, P. J. Foster, H. Hanaki and S. Ōmura, Global trends in clinical studies of ivermectin in COVID-19, *Jpn. J. Antibiot.*, 2021, **74–1**, 44–95.
36. J. C. Rajter, M. S. Sherman, N. Fatteh, F. Vogel, J. Sacks and J. J. Rajter, Use of Ivermectin Is Associated With Lower Mortality in Hospitalized Patients With Coronavirus Disease 2019: The Ivermectin in COVID Nineteen Study, *Chest*, 2021, **159**, 85–92.
37. W. Shouman, *et al.*, Use of Ivermectin as a potential chemoprophylaxis for COVID-19 in Egypt : A Randomised clinical trial, *J. Clin. Diagn. Res.*, 2021, **15**(2), OC27–OC32.
38. M. Krempel, K. Griffin, and H. Khouryieh, 13. Hydrocolloids as Emulsifiers and Stabilizers in Beverage Preservation, in *Preservatives and Preservation Approaches in Beverages*, ed. A. M. Grumezescu and A. M. Holban, Academic Press, 2019, pp. 427–465.
39. J. Vallejos, *et al.*, Ivermectin to prevent hospitalizations in patients with COVID-19 (IVERCOR-COVID19) a randomized, double-blind, placebo-controlled trial, *BMC Infect. Dis.*, 2021, **21**(1), 635.
40. M. Alam, *et al.*, A Case Series of 100 COVID-19 Positive Patients Treated with Combination of Ivermectin and Doxycycline, *J. Bangladesh Coll. Phys. Surg.*, 2020, **38**, 10–15.
41. H. Carvallo, R. Hirsch and M. Farinella, Safety and efficacy of the combined use of ivermectin, dexamethasone, enoxaparin and aspirin against covid 19, *medRxiv*, 2020, DOI: 10.1101/2020.09.10.20191619.
42. H. Carvallo, R. Hirsch and P. Alkis, *et al.*, Study of the efficacy and safety of topical ivermectin + iona-Carrageenan in the prophylaxis against COVID-19 in health personnel, *J. Biomed. Res. Clin. Invest.*, 2020, **2**, 1007.

43. P. Behera, B. K. Patro, A. K. Singh, P. D. Chandanshive, S. R. Ravi Kumar, S. K. Pradhan, S. S. K. Pentapati, G. Batmanabane, B. M. Padhy, S. Bal, S. R. Singh and R. R. Mohanty, Role of ivermectin in the prevention of COVID-19 infection among healthcare workers in India: A matched case-control study, *medRxiv*, 2020, DOI: 10.1101/2020.10.29.20222661.
44. A. Hill, A. Abdulamir, S. Ahmed, A. Asghar, O. E. Babalola, R. Basri, A. Z. K. Chachar, A. T. M. Chowdhury, A. Elgazzar, L. Ellis, J. Falconer, A. Garratt, B. M. Hany, H. A. Hashim, W. U. Haque, A. Hayat, S. He, R. Jamshidian, W. A. Khan, R. Kirti, A. Krolewiecki, C. Lanusse, J. Levi, R. Mahmud, S. A. Mangat, K. McCann, A. Mohan, M. S. Niaee, N. Okumuş, V. Pilkington, C. S. Podder, A. Qavi, H. Raad, A. Samaha, H. Mouawia, M. S. Rezai, S. Sasank, V. Spoorthi, T. Suri, J. Wang and H. Wentzel, Preliminary meta analysis of randomized trials of ivermectin to treat SARS CoV 2 infection, *Res. Sq.*, 2021, DOI: 10.21203/rs.3.rs-148845/v1.
45. O. E. Babalola, *et al.*, Ivermectin shows clinical benefits in mild to moderate COVID19: A randomised controlled double-blind, dose-response study in Lagos, *QJM*, 2021, 780–788.
46. E. Lopez-Medina, *et al.*, Effect of Ivermectin on Time to Resolution of Symptoms Among Adults With Mild COVID-19: A Randomized Clinical Trial, *JAMA*, 2021, 1426–1435.
47. G. Aguirre Chang, *et al.*, *COVID-19 Post-Aguda O Prolongada: Tratamiento Con Ivermectina De Pacientes con Sintomas Persistentes*, 2020.
48. J. Meneses Calderon, M. d. R. Figueroa Flores, L. Paniagua Coria, J. C. Briones Garduno, J. Meneses Figueroa, M. J. Vargas Contretas, L. de la Cruz Avila, S. Diaz Meza, R. Ramirez Chacon, S. Padmanabhan and H. Mendieta Zeron, Nitazoxanide against COVID-19 in three explorative scenarios, *J. Infect. Dev. Ctries.*, 2020, **14**, 982–986.
49. Romark Laboratories L.C., A Randomized, Double-Blind, Placebo Controlled Trial to Evaluate the Efficacy and Safety of Nitazoxanide (NTZ) for Pre-and Post Exposure Prophylaxis of COVID-19 and Other Viral Respiratory Illnesses (VRI) in Healthcare Workers and Others at Increased Risk of SARS-CoV-2 Infection. clinicaltrials.gov. October 2020. Report No: NCT04359680. Available from: https://clinicaltrials.gov/ct2/show/NCT04359680. Last accessed on 18 December 2020.
50. Romark Laboratories L.C., A Randomized, Double-Blind, Placebo Controlled Trial to Evaluate the Efficacy and Safety of Nitazoxanide (NTZ) for Post-Exposure Prophylaxis of COVID-19 and Other Viral Respiratory Illnesses in Elderly Residents of Long-Term Care Facilities (LTCF). clinicaltrials.gov, 2020, Report No.: NCT04343248. Available from: https://clinicaltrials.gov/ct2/show/NCT04343248. Last accessed on 13 December 2020.
51. Romark Initiates New Phase 3 Clinical Trial Of NT-300 For The Treatment Of COVID-19, BioSpace, Available from: https://www.biospace.com/article/romark-initiates-new-phase-3-clinical-trial-of-nt-300-for-the-treatment-of-covid-19/. Last accessed on 14 December 2020.

52. N. Bukreyeva, E. K. Mantlo, R. A. Sattler, C. Huang, S. Paessler and J. Zeldis, The IMPDH inhibitor merimepodib suppresses SARS-CoV-2 replication *in vitro*, *bioRxiv*, 2020, DOI: 2020.04.07.028589.
53. ViralClear Pharmaceuticals, Inc, A Phase 2, Randomized, Double-Blind, Placebo-Controlled Study of the Efficacy and Safety of Oral Merimepodib in Combination With Intravenous Remdesivir in Adult Patients With Advanced Coronavirus Disease 2019 (COVID-19), clinicaltrials.gov, December 2020. Report No: NCT04410354. Available from: https://clinicaltrials.gov/ct2/show/NCT04410354. Last accessed on 18 December 2020.
54. (a) ANA Therapeutics, A Phase 2/3 Randomized and Placebo-Controlled Study of ANA001 in Moderate COVID-19 Patients, clinicaltrials.gov, November 2020. Report No: NCT04603924. Available from: https://clinicaltrials.gov/ct2/show/NCT04603924. Last accessed on 18 December 2020; (b) S. Pindiprolu and S. H. Pindiprolu, Plausible mechanisms of Niclosamide as an antiviral agent against COVID-19, *Med. Hypotheses*, 2020, **140**, 109765.
55. Rintatolimod and IFN Alpha-2b for the Treatment of Mild or Moderate COVID-19 Infection in Cancer Patients – Full Text View, ClinicalTrials.gov. Available from: https://clinicaltrials.gov/ct2/show/NCT04379518. Last accessed on 18 December 2020.
56. N. Urakova, T. Strive and M. Frese, RNA-Dependent RNA Polymerases of Both Virulent and Benign Rabbit Caliciviruses Induce Striking Rearrangement of Golgi Membranes, *PLoS One*, 2017, **12**(1), e0169913.
57. M. L. Agostini, *et al.*, Small-Molecule Antiviral beta-D-N(4)-Hydroxycytidine Inhibits a Proofreading-Intact Coronavirus with a High Genetic Barrier to Resistance, *J. Virol.*, 2019, **93**(24) .
58. T. P. Sheahan, A. C. Sims, S. Zhou, R. L. Graham, A. J. Pruijssers and M. L. Agostini, *et al.*, An orally bioavailable broad-spectrum antiviral inhibits SARS-CoV-2 in human airway epithelial cell cultures and multiple coronaviruses in mice, *Sci. Transl. Med.*, 2020, **12**(541), eabb5883.
59. ASA B. Update On Bemcentinib's Involvement in Potential Treatment of COVID-19 Pandemic. Available from: https://www.prnewswire.com/news-releases/update-on-bemcentinibs-involvement-in-potential-treatment-of-covid-19-pandemic-301101363.html. Last accessed on 18 December 2020.
60. K. El Bairi, D. Trapani, A. Petrillo, C. Le Page, H. Zbakh and B. Daniele, *et al.*, Repurposing anticancer drugs for the management of COVID-19, *Eur. J. Cancer*, 2020, **141**, 40–61.
61. PharmaMar, Multicenter, Randomized, Parallel and Proof of Concept Study to Evaluate the Safety Profile of Three Doses of Plitidepsin in Patients With COVID-19 Requiring Hospitalization, clinicaltrials.gov, October 2020. Report No: NCT04382066. Available from: https://clinicaltrials.gov/ct2/show/NCT04382066. Last accessed on 12 December 2020.

62. Acer Therapeutics to Develop Emetine as Potential COVID-19 Treatment in Collaboration with National Center for Advancing Translational Sciences, One of the National Institutes of Health, Acer Therapeutics, 2020. Available from: https://www.acertx.com/2020/05/11/acer-therapeutics-to-develop-emetine-as-potential-covid-19-treatment-in-collaboration-with-national-center-for-advancing-translational-sciences-one-of-the-national-institutes-of-health/. Last accessed on 12 December 2020.
63. Mateon to Develop its OT-101, a Phase 3 Clinical Drug Candidate, Initially Against COVID-19, Mateon Therapeutics. Available from: http://investor.mateon.com/news-releases/news-release-details/mateon-develop-its-ot-101-phase-3-clinical-drug-candidate/. Last accessed on 12 December 2020.
64. Renibus Therapeutics, Inc, A Phase 2, Randomized, Placebo-Controlled Study to Evaluate the Effect of RBT-9 on Progression of COVID-19 in High-Risk Individuals (The PREVENT Study), clinicaltrials.gov, September 2020. Report No: NCT04364763. Available from: https://clinicaltrials.gov/ct2/show/NCT04364763. Last accessed on 13 December 2020.
65. Golden Biotechnology Corporation, A Phase 2 Randomized, Double-blind, Placebo-controlled, Proof of Concept Study to Evaluate the Safety and Efficacy of Antroquinonol in Hospitalized Patients With Mild to Moderate Pneumonia Due to COVID-19, clinicaltrials.gov, December 2020. Report No: NCT04523181. Available from: https://clinicaltrials.gov/ct2/show/NCT04523181. Last accessed on 13 December 2020.
66. A Study of LAM-002A for the Prevention of Progression of COVID-19, Full Text View – ClinicalTrials.gov. Available from: https://clinicaltrials.gov/ct2/show/NCT04446377. Last accessed on 13 December 2020.
67. Vir and Alnylam Identify RNAi Therapeutic Development Candidate, VIR-2703 (ALN-COV), Targeting SARS-CoV-2 for the Treatment of COVID-19, Vir. Available from: https://investors.vir.bio/news-releases/news-release-details/vir-and-alnylam-identify-rnai-therapeutic-development-candidate/. Last accessed on 10 December 2020.
68. A. Slobodnick, *et al.*, Update on colchicine, 2017, *Rheumatology*, 2018, **57**(suppl. 1), i4–i11.
69. J.-C. Tardif, *et al.*, Efficacy of Colchicine in Non-Hospitalized Patients with COVID-19. medRxiv, 2021, DOI: 2021.01.26.21250494.
70. D. C. Fajgenbaum and C. H. June, Cytokine Storm, *N. Engl. J. Med.*, 2020, **383**(23), 2255–2273.
71. J. G. Rizk, K. Kalantar-Zadeh, M. R. Mehra, C. J. Lavie, Y. Rizk and D. N. Forthal, Pharmaco-Immunomodulatory Therapy in COVID-19, *Drugs*, 2020, **80**(13), 1267–1292.
72. Clinical management of COVID-19. Available from: https://www.who.int/publications-detail-redirect/clinical-management-of-covid-19. Last accessed on 10 December 2020.

73. C. D. Russell, J. E. Millar and J. K. Baillie, Clinical evidence does not support corticosteroid treatment for 2019-nCoV lung injury, *Lancet*, 2020, **395**(10223), 473–475.
74. D. C. Fajgenbaum and C. H. June, Cytokine Storm, *N. Engl. J. Med.*, 2020, **383**(23), 2255–2273.
75. R. Channappanavar and S. Perlman, Pathogenic human coronavirus infections: causes and consequences of cytokine storm and immunopathology, *Semin. Immunopathol.*, 2017, **39**(5), 529–539.
76. RECOVERY Collaborative Group, P. Horby, W. S. Lim, J. R. Emberson, M. Mafham and J. L. Bell, *et al.*, Dexamethasone in Hospitalized Patients with Covid-19 - Preliminary Report, *N. Engl. J. Med.*, 2020, **17**.
77. The WHO Rapid Evidence Appraisal for COVID-19 Therapies (REACT) Working Group, J. A. C. Sterne, S. Murthy, J. V. Diaz, A. S. Slutsky and J. Villar, *et al.*, Association Between Administration of Systemic Corticosteroids and Mortality Among Critically Ill Patients With COVID-19: A Meta-analysis, *JAMA*, 2020, **324**(13), 1330.
78. Corticosteroids for COVID-19. Available from: https://www.who.int/publications-detail-redirect/WHO-2019-nCoV-Corticosteroids-2020.1. Last accessed on 14 December 2020.
79. Interleukin-6 Inhibitors, COVID-19 Treatment Guidelines. Available from: https://www.covid19treatmentguidelines.nih.gov/immune-based-therapy/immunomodulators/interleukin-6-inhibitors/. Last accessed on 08 December 2020.
80. Regeneron and Sanofi Begin Global Kevzara® (sarilumab) Clinical Trial Program in Patients with Severe COVID-19 Regeneron Pharmaceuticals Inc. Available from: https://investor.regeneron.com/news-releases/news-release-details/regeneron-and-sanofi-begin-global-kevzarar-sarilumab-clinical/. Last accessed on 08 December 2020.
81. Regeneron and Sanofi Provide Update on U.S. Phase 2/3 Adaptive-Designed Trial of Kevzara® (sarilumab) in Hospitalized COVID-19 Patients, Regeneron Pharmaceuticals Inc. Available from: https://investor.regeneron.com/news-releases/news-release-details/regeneron-and-sanofi-provide-update-us-phase-23-adaptive/. Last accessed on 10 December 2020.
82. Sanofi: Press Releases, Thursday, 2 July 2020. https://www.sanofi.com/en/media-room/press-releases/2020/2020-07-02-22-30-00. Available from: https://www.sanofi.com/media-room/press-releases/2020/2020-07-02 22-30-00-2057183. Last accessed on 10 December 2020.
83. S.-N. J. Song and K. Yoshizaki, Tocilizumab for treating rheumatoid arthritis: an evaluation of pharmacokinetics/pharmacodynamics and clinical efficacy, *Expert Opin. Drug Metab. Toxicol.*, 2015, **11**(2), 307–316.
84. C. Salvarani, G. Dolci, M. Massari, D. F. Merlo, S. Cavuto and L. Savoldi, *et al.*, Effect of Tocilizumab *vs.* Standard Care on Clinical Worsening in Patients Hospitalized With COVID-19 Pneumonia: A Randomized Clinical Trial, *JAMA Intern. Med.*, 2020, **20**, e206615.

85. J. H. Stone, *et al.*, Efficacy of Tocilizumab in Patients Hospitalized with Covid-19, *N. Engl. J. Med.*, 2020, **383**(24), 2333–2344.
86. Roche's phase III EMPACTA study showed Actemra/RoActemra reduced the likelihood of needing mechanical ventilation in hospitalised patients with COVID-19 associated pneumonia. Available from: https://www.roche.com/investors/updates/inv-update-2020-09-18.htm. Last accessed on 10 December 2020.
87. C. Salama, *et.al.*, *NEJM*, 7 January 2021, https://www.nejm.org/doi/full/10.1056/NEJMoa2030340.
88. N. Biran, *et al.*, Tocilizumab among patients with COVID-19 in the intensive care unit: a multicentre observational study, *Lancet Rheumatol.*, 2020, **2**(10), e603–e612.
89. E. C. Somers, *et al.*, Tocilizumab for Treatment of Mechanically Ventilated Patients With COVID-19, *Clin. Infect. Dis.*, 2020.
90. G. Guaraldi, *et al.*, Tocilizumab for severe COVID-19 pneumonia - Authors' reply, *Lancet Rheumatol.*, 2020, **2**(11), e660–e661.
91. Interleukin-1 Inhibitors, COVID-19 Treatment Guidelines. Available from: https://www.covid19treatmentguidelines.nih.gov/immune-based-therapy/immunomodulators/interleukin-1-inhibitors/. Last accessed on 10 December 2020.
92. L. Pasin, *et al.*, Anakinra for patients with COVID-19: a meta-analysis of non-randomized cohort studies, *Eur. J. Intern. Med.*, 2021.
93. https://www.treatmentguidelinesforcovid-19.nih.gov accessed 29th September 2021.
94. J. Stebbing, A. Phelan, I. Griffin, C. Tucker, O. Oechsle and D. Smith, *et al.*, COVID-19: combining antiviral and anti-inflammatory treatments, *Lancet Infect. Dis.*, 2020, **20**(4), 400–402.
95. L. Walz, *et al.*, Janus Kinase-Inhibitor and Type I Interferon Ability to Produce Favorable Clinical Outcomes in COVID-19 Patients: A Systematic Review and Meta-Analysis. medRxiv, 2020, DOI: 2020.08.10.20172189.
96. (a) NIH, *Fourth Iteration of COVID-19 Treatment Trial Underway*, National Institutes of Health (NIH), 2020. Available from: https://www.nih.gov/news-events/news-releases/fourth-iteration-covid-19-treatment-trial-underway. Last accessed on 04 December 2020; (b) NIH Guidelines Updated 11 February 2021, accessed www.covid19treatmentguidelines.nih.gov.
97. Corp C.B, CTI BioPharma Announces Enrollment of First Patient in COVID-19 PRE-VENT Phase 3 Clinical Trial. Available from: https://www.prnewswire.com/news-releases/cti-biopharma-announces-enrollment-of-first-patient-in-covid-19-pre-vent-phase-3-clinical-trial-301068348.html. Last accessed on 05 December 2020.
98. L. M. Katz, (A Little) Clarity on Convalescent Plasma for Covid-19, *N. Engl. J. Med.*, 2021, **384**(7), 666–668.
99. M. Rojas, Y. Rodríguez, D. M. Monsalve, Y. Acosta-Ampudia, B. Camacho and J. E. Gallo, *et al.*, Convalescent plasma in Covid-19: Possible mechanisms of action, *Autoimmun. Rev.*, 2020, **19**(7), 102554.

100. A. Agarwal, A. Mukherjee, G. Kumar, P. Chatterjee, T. Bhatnagar and P. Malhotra, Convalescent plasma in the management of moderate covid-19 in adults in India: open label phase II multicentre randomised controlled trial (PLACID Trial), *BMJ*, 2020, **371**, m3939.
101. R. Libster, *et al.*, Early High-Titer Plasma Therapy to Prevent Severe Covid-19 in Older Adults, *N. Engl. J. Med.*, 2021, **384**(7), 610–618.
102. V. A. Simonovich, *et al.*, A Randomized Trial of Convalescent Plasma in Covid-19 Severe Pneumonia, *N. Engl. J. Med.*, 2020, **384**(7), 619–629.
103. M. J. Joyner, *et al.*, Safety Update: COVID-19 Convalescent Plasma in 20,000 Hospitalized Patients, *Mayo Clin. Proc.*, 2020, **95**(9), 1888–1897.
104. P. Marks, FDA In Brief: FDA Updates Emergency Use Authorization for COVID-19 Convalescent Plasma to Reflect New Data.
105. Commissioner O of the Coronavirus (COVID-19) Update: FDA Authorizes Monoclonal Antibodies for Treatment of COVID-19, FDA, 2020. Available from: https://www.fda.gov/news-events/press-announcements/coronavirus-covid-19-update-fda-authorizes-monoclonal-antibodies-treatment-covid-19. Last accessed on 23 December 2020.
106. (a) Statement on Casirivimab Plus Imdevimab EUA, COVID-19 Treatment Guidelines. Available from: https://www.covid19treatmentguidelines.nih.gov/statement-on-casirivimab-plus-imdevimab-eua/. Last accessed on 23 December 2020; (b) ACTIV-3/TICO LY-CoV555 Study Group, A Neutralizing Monoclonal Antibody for Hospitalized Patients with Covid-19, *N. Engl. J. Med.*, 2021, **384**, 905–914.
107. R. L. Gottlieb, *et al.*, Effect of Bamlanivimab as Monotherapy or in Combination With Etesevimab on Viral Load in Patients With Mild to Moderate COVID-19: A Randomized Clinical Trial, *JAMA*, 2021, **325**(7), 632–644.
108. É. Lansiaux, P. P. Pébaÿ, J.-L. Picard and J. Forget, Covid-19 and vit-d: Disease mortality negatively correlates with sunlight exposure, *Spat. Spatio-temporal Epidemiol.*, 2020, **35**, 100362.
109. M. T. Zdrenghea, H. Makrinioti, C. Bagacean, A. Bush, S. L. Johnston and L. A. Stanciu, Vitamin D modulation of innate immune responses to respiratory viral infections, *Rev. Med. Virol.*, 2017, **27**(1).
110. J. F. Arboleda and S. Urcuqui-Inchima, Vitamin D Supplementation: A Potential Approach for Coronavirus/COVID-19 Therapeutics? *Front. Immunol.*, 2020, **11**, 1523.
111. Z. Maghbooli, M. A. Sahraian, M. Ebrahimi, M. Pazoki, S. Kafan and H. M. Tabriz, *et al.*, Vitamin D sufficiency, a serum 25-hydroxyvitamin D at least 30 ng/mL reduced risk for adverse clinical outcomes in patients with COVID-19 infection, *PLoS One*, 2020, **15**(9), e0239799.
112. A. Rastogi, A. Bhansali, N. Khare, V. Suri, N. Yaddanapudi and N. Sachdeva, *et al.*, Short term, high-dose vitamin D supplementation for COVID-19 disease: a randomised, placebo-controlled, study (SHADE study), *Postgrad. Med. J.*, 2020, **12**.

113. I. H. Murai, *et al.*, Effect of Vitamin D_3 Supplementation *vs.* Placebo on Hospital Length of Stay in Patients with Severe COVID-19: A Multicenter, Double-blind, Randomized Controlled Trial. medRxiv, 2020, DOI: 2020.11.16.20232397.
114. D. E. Leaf and A. A. Ginde, Vitamin D3 to Treat COVID-19: Different Disease, Same Answer, *JAMA*, 2021, **325**(11), 1047–1048.
115. C. E. Powe, *et al.*, Vitamin D–binding protein and vitamin D status of black Americans and white Americans, *N. Engl. J. Med.*, 2013, **369**(21), 1991–2000.
116. D. D. Bikle and J. Schwartz, Vitamin D binding protein, total and free vitamin D levels in different physiological and pathophysiological conditions, *Front. Endocrinol.*, 2019, **10**, 317.
117. N. Ali, Role of vitamin D in preventing of COVID-19 infection, progression and severity, *J. Infect. Public Health*, 2020, **13**(10), 1373–1380.
118. H. Hemilä, Vitamin C and SARS coronavirus, *J. Antimicrob. Chemother.*, 2003, **52**(6), 1049–1050.
119. (a) H. Hemilä and E. Chalker, Vitamin C as a Possible Therapy for COVID-19, *Infect. Chemother.*, 2020, **52**(2), 222–223; (b) B. M. Krenn, E. Gaudernak, B. Holzer, K. Lanke, F. J. Van Kuppeveld and J. Seipelt, Antiviral activity of the zinc ionophores pyrithione and hinokitiol against picornavirus infections, *J. Virol.*, 2009, **83**(1), 58–64.
120. S. Thomas, *et al.*, Effect of High-Dose Zinc and Ascorbic Acid Supplementation *vs.* Usual Care on Symptom Length and Reduction Among Ambulatory Patients With SARS-CoV-2 Infection: The COVID A to Z Randomized Clinical Trial, *JAMA Network Open*, 2021, **4**(2), e210369.
121. https://www.covid19treatmentguidelines.nih.gov accessed 29th September, 2021.
122. N. Z. Gammoh and L. Rink, Zinc in Infection and Inflammation, *Nutrients*, 2017, **9**(6), 624.
123. A. J. W. te Velthuis, *et al.*, Zn(2+) inhibits coronavirus and arterivirus RNA polymerase activity in vitro and zinc ionophores block the replication of these viruses in cell culture, *PLoS Pathog.*, 2010, **6**(11), e1001176.
124. W. Li, *et al.*, Angiotensin-converting enzyme 2 is a functional receptor for the SARS coronavirus, *Nature*, 2003, **426**(6965), 450–454.
125. I. Wessels, B. Rolles and L. Rink, The Potential Impact of Zinc Supplementation on COVID-19 Pathogenesis, *Front. Immunol.*, 2020, **11**, 1712.
126. J. S. Yao, J. A. Paguio, E. C. Dee, H. C. Tan, A. Moulick and C. Milazzo, *et al.*, The Minimal Effect of Zinc on the Survival of Hospitalized Patients With COVID-19: An Observational Study, *Chest*, 2020; S0012-3692(20)31961-9.
127. J. A. Frontera, *et al.*, Treatment with Zinc is Associated with Reduced In-Hospital Mortality Among COVID-19 Patients: A Multi-Center Cohort Study, *Res. Sq.*, 2020.
128. S. Abd-Elsalam, *et al.*, Do Zinc Supplements Enhance the Clinical Efficacy of Hydroxychloroquine?: a Randomized, Multicenter Trial, *Biol. Trace Elem. Res.*, 2020, 1–5.

129. https://www.treatmentguidelinesforcovid-19.nih.gov accessed 29th September 2021.
130. G. Piazza and D. A. Morrow, Diagnosis, Management, and Pathophysiology of Arterial and Venous Thrombosis in COVID-19, *JAMA*, 2020, 2548–2549.
131. COVID-19 and VTE-Anticoagulation. Hematology.org. Available from: https://www.hematology.org:443/covid-19/covid-19-and-vte-anticoagulation. Last accessed on 04 December 2020.
132. A. Elgazzar, B. Hany, S. A. Youssef, M. Hafez, H. Moussa and A. Eltaweel, Efficacy and safety of ivermectin for treatment and prophylaxis of COVID-19 pandemic, *Research Square* (preprint), 2021.

CHAPTER 16

Knowledge from Human Relevant Cell, Tissue and Mathematics-based Methods as Key Tools for Understanding COVID-19 Dynamics, Kinetics, Symptoms, Risk Factors and Non-conventional Treatments

SANDRA COECKE*[a], AMALIA MUNOZ[b], VITO D'ALESSANDRO[c], FRANCESCA DE BERNARDI[d], PIETRO ROMEO[e], FELIPE TORRES[f], GEORGINA HARRIS[g] AND SURAT PARVATAM[h]

[a]European Commission Joint Research Centre (JRC), Ispra, Italy; [b]European Commission Joint Research Centre (JRC), Geel, Belgium; [c]University of Salento, Institute of Nanotechnology CNR-NANOTEC, Lecce, Italy; [d]Department of Biotechnology and Life Sciences, University of Insubria, Circolo Hospital and Macchi Foundation, Varese, Italy; [e]Centre for Applied Research on Biophysical Stimulation of Musculoskeletal Tissues, Istituto Ortopedico Galeazzi, Milano, Italy; [f]Cell Regeneration Medicine Organization, Bogotá, Colombia; [g]Frontiers Media SA, Lausanne, Switzerland; [h]Centre for Predictive Human Model Systems, Atal Incubation Centre-Centre for Cellular and Molecular Biology (AIC-CCMB), Hyderabad, India
*E-mail: Sandra.COECKE@ec.europa.eu

The Coronavirus Pandemic and the Future Volume 1: Virology, Epidemiology, Translational Toxicology and Therapeutics
Edited by Michael D. Waters, Alok Dhawan, Tim Marrs, Diana Anderson, Stafford Warren and Claude L. Hughes

Published by the Royal Society of Chemistry, www.rsc.org

16.1 Cell, Tissue and Mathematical-based Methods Used for COVID-19 Research

Coronavirus disease 2019 (COVID-19), caused by the severe acute respiratory syndrome coronavirus-2 (SARS-CoV-2) belonging to the *Betacoronavirus* genus, emerged in the human population in late 2019 and has spread *via* human-to-human transmission to most countries in the world.

The variability of the symptoms, ranging from: (1) asymptomatic or pre-symptomatic, (2) mild illness, (3) moderate illness, (4) severe illness and (5) critical illness (Figure 16.1), requires a deep understanding of the pathogenesis of COVID-19 to be able to establish effective diagnostic tests and subsequent therapeutic interventions consistent with the observed features (disease symptoms) and the stage of the disease in each patient having with their own genetic make-up and related vulnerabilities and risk factors.

It is necessary to understand the dynamics and kinetics of SARS-CoV-2 in detail, which includes the ways of entry into the organism *via* the external body barriers (*e.g.* nose, lung, eye, intestine), its distribution and passage through internal body barriers (*e.g.* placenta barrier, blood–brain barrier) into the various human body organs and its subsequent viral dynamics (Figure 16.1).

Furthermore, the cellular immune response and specific downstream cellular responses can contribute to the multi-organ dysfunction. Mechanistic understanding on modulators of the immune response, cell homeostasis balance, and specific risk factors is critical to understand the systems biology processes underlying the multi-organ effects, and will and will improve diagnosis,[2,3] prevention and therapeutic strategies such as using immunomodulators and cytokine-directed therapies.

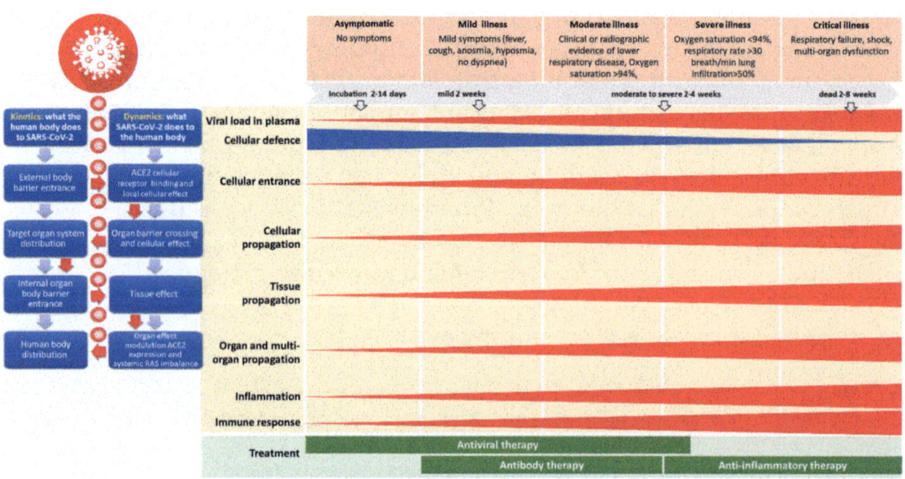

Figure 16.1 Relationships between cellular kinetics, dynamics and variability of symptoms of COVID-19 disease.[1]

The infection cycle of the SARS-CoV-2 virus begins with its cellular entry. Using the S glycoprotein, it attaches itself to a surface receptor of the host cell. The place of entry *via* the external body barriers (Figure 16.2) and the host cell receptor(s) and its (their) distribution determine which type of cells, tissues and organ systems get infected. The specificity of the S protein to a particular receptor influences viral tropism. SARS-CoV uses the human ACE2 receptor as points of entry. The ACE2 enzyme is a pivotal component of the renin–angiotensin system and exerts its physiological functions by modulating the levels of angiotensin II (Ang II) and Ang-(1–7).[5]

The often-recurrent paradigm for biomedical research and drug testing postulates that cell and tissue-based methods (*in vitro*, Latin for "within the glass") and mathematical-based methods (*in silico*, pseudo-Latin for "*in silicon*", relating to the use of silicon material for computer chips) inform animal studies that will subsequently inform human studies.

In vitro studies are conducted using components of a whole organism that have been isolated from their usual biological surroundings, such as tissue or organ fragment, organ explant, dissociated cells or primary isolated cells from different organs of the human body. "*In silico*" is an expression meaning "performed on computer or *via* computer simulation" in reference to biological experiments.

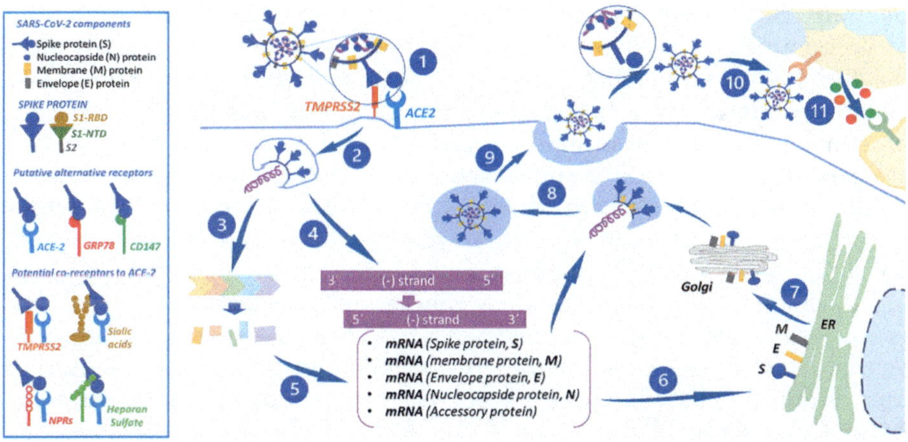

Figure 16.2 Access, multiplication and release of COVID-19 particles in human cells. The process is divided into the following steps: (1) attachment to the host cell; (2) viral–host membrane fusion and release of viral RNA; (3) translation and cleavage of polyproteins; (4) translation of full-length (−) strand by replication complex; (5) transcription and replication of viral RNA genome; (6) translation of structural and accessory proteins; (7) trafficking of newly synthesized proteins from the endoplasmic reticulum (ER) to the Golgi; (8) assembly of mature virion in building vesicle; (9) mature virion release *via* endocytosis; (10) presentation of SARS-CoV-2 antigens to antigen presenting cells (APCs); (11) release of cytokines (enhanced pro-inflammatory response).[4] Adapted from ref. 4 with permission from John Wiley & Sons, Copyright © 2021 Wiley-VCH GmbH.

Both "*in vitro*" testing methods and "*in silico*" methods have been used to characterize, for instance, specific adsorption, distribution, metabolism and excretion processes of drugs or general chemicals (kinetic processes, "what the body does with the drug or chemical") or adverse effects of drugs and chemicals on specific target organ systems of the human body (dynamic processes, "what the drug or chemical does with the human body") inside a living organism.[6]

Similarly, these types of kinetic and dynamic processes can be studied, keeping in mind that the SARS-CoV-2 virus is a foreign entity entering the human body (Figure 16.1).

When SARS-CoV-2 research work is performed *in vitro*, it happens outside of a living organism and its normal biological context. *In vitro* cell and tissue culture work provide a controlled environment for generating SARS-CoV-2 related experimental mechanistic data. Since the *in vitro* work occurs outside of a living organism, results must be considered carefully and need to be interpreted knowing the system biological differences in comparison with the intact organism, or the target organ, tissue or cells systems of that organism one tries to mimic.

In vitro and *in silico* method-based experimental studies can permit species-specific comparison and extrapolation, and provide simple, standardized, harmonized, higher throughput, and specific mechanistic investigations. Such studies can facilitate understanding the modes of action of compounds foreign to the human body (xenobiotics) and invaders (bacteria, viruses, *etc.*) that are difficult or impossible to investigate in the whole organism. Just as studies in whole animals have been carried out pre-clinically to inform human clinical trials, *in vitro* and *in silico* studies are used to replace, reduce and/or refine (3Rs[7]) experimental animal investigations, or even better, to directly inform human biological processes which cannot be studied in animals due to species differences. Indeed, SARS-CoV-2 has been shown to provoke highly heterogeneous responses in individuals, and as such the new generation of well-chosen *in vitro* and *in silico* methods can give the necessary data to diagnose the pathological features of COVID-19 disease (kinetics and dynamics) and also be used to understand the clinical symptoms, its risk factors and potential novel alternative treatments.

The primary disadvantage of *in vitro* and *in silico* experimental studies is the difficulty to extrapolate the results of the *in vitro* and *in silico* work back to the biology of the intact organism. With the new life sciences, scientists have moved to integrate the knowledge obtained with these new-approach methodologies,[8,9] thus contributing to accelerating knowledge on organismal and systems biology, which is of critical importance in COVID-19 research.

An animal method driven approach is not always the ideal methodology to represent human diversity, genetic make-up, polymorphism, risk factors, overlapping open reading frames (ORFs), ORF1a and ORF1b and vulnerabilities, nor can it provide a complete insight on this very challenging new viral disease which causes a wide variety of adverse effects in the human body, both pulmonary and extra-pulmonary, with subsequent multi-organ system effects.

SARS-CoV-2 animal studies have employed mammalian species such as bats, cats, ferrets, hamsters, mice (including genetically modified organism (GMO) *e.g.* human angiotensin-converting enzyme 2 (hACE2) mice due to ACE2 species difference), mink and non-human primates.

The animal studies looked at aspects such as:

- virus replication in the upper and lower respiratory tract and other organs
- clinical signs such as fever, nasal discharge, labored breathing
- pneumonia and details on bilateral lung involvement, ground-glass opacities, focal oedema, inflammation, or acute respiratory distress syndrome (ARDS)
- SARS-CoV-2 transmission
- immunology, such as aspects of seroconversion, neutralizing antibody titres, T-cell immunity or pro-inflammatory cytokines
- demographics, such as differential clinical effects between by females and males or younger and older patients.[10–12]

The contribution and the impact of using the available human-relevant novel methodological approaches like *in vitro* and *in silico* methods or the novel generation of clinical methods using directly patient-derived cell and tissue material – within this new global research area – is yet to be properly quantified. This quantification is important since biomedical research and drug discovery and development includes two steps of knowledge transfer, and it is critical to evaluate the effectiveness of both in order to properly benefit from the added value of COVID-19 research, compliant with the 3R principles (replacement, reduction and refinement) as described in the European Directive 2010/63/EU.

In general, biomedical research utilizing *in vitro* and *in silico* data is not the norm. In contrast to animal research community publications, the human medical community tends to refer more frequently to available mechanistic knowledge from *in vitro* and *in silico* methods as knowledge support for understanding disease and its dynamics and kinetics. Although, in the case of COVID-19, circumstances are diverse. Due to the pandemic and its impact on population and economy, it seems that the Scopus literature findings follow a similar pattern. One can argue that if the animal research community is not citing *in vitro* and *in silico* papers on COVID-19, this research might be of limited use. However, that is inconsistent with their substantial use by the human medical community, which cites these research findings more than research based on animal studies. Additionally, this lack of transferability of knowledge between the animal and the human medical research communities is further evidenced by the fact that, in general, most citations received by animal research papers are referring to other animal-based studies, rather than to human medical papers.

Cross-community research on SARS-CoV-2 is essential to understand its detailed pathophysiology and mechanisms of infection, the host–virus interaction, replication kinetics and targets for therapeutic screening. As such, *in vitro* cell- and tissue-based models that can faithfully reproduce the viral life cycle and reproduce the pathology of COVID-19 are required. This requires *Good In vitro Method Practices*[13,14] and more complex technologically innovative two- (2D) and three-dimensional (3D) cultures to comprehend disease features and other life-science questions including the biology of the SARS-CoV-2 virus.

As such, with advances of genetic engineering, tissue engineering and novel technologies, *in vitro* target organ system models have emerged and are used as more realistic *in vitro* models, allowing for instance the construction of complex cytoarchitecture, with better representation of cell heterogeneity, extracellular matrix (ECM) composition, and functionality of native tissues.[15] 3D *in vitro* models consist of scaffold-free (spheroids and organoids) or scaffold-based (3D scaffolding and 3D bioprinting) systems used to study infectivity, replication kinetics and host–viral interactions of SARS-CoV-2, showing increased physiological relevance as compared to 2D models.

Different types of *in vitro* cellular 2D- and 3D-derived test systems are reported in COVID-19 research papers (Figure 16.3), *e.g.*

- tissue or organ fragment
- organ explant
- dissociated cells
- primary cell culture
- continuous or finite cell lines
- immortalized cell lines

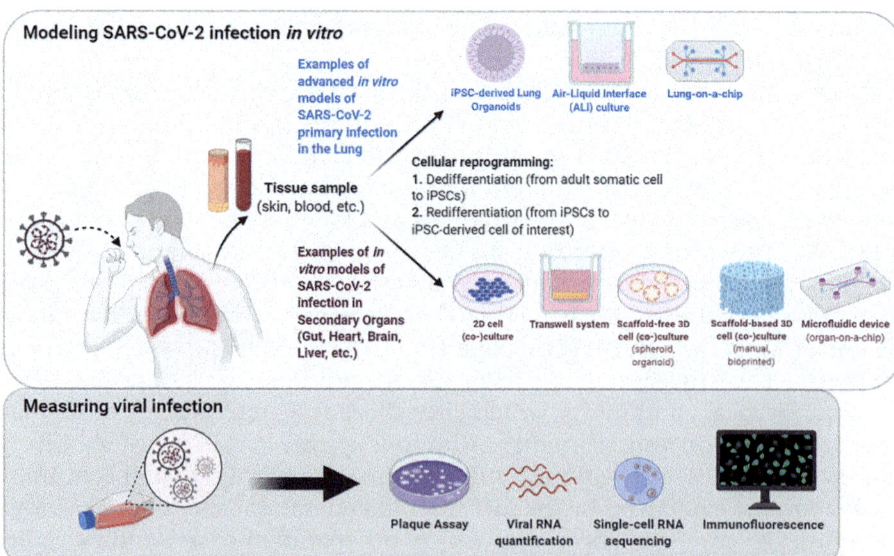

Figure 16.3 More complex 3D culture system like spheroids, organoids, organs-on-a-chip, 3D-scaffold or 3D-bioprinting cell and tissue-based systems.[18,21,22] Created in BioRender.com.

- human stem cells and human-derived induced pluripotent 2D and 3D stem (hiPSC) cell cultures (see Figure 16.3)
- complex 3D culture systems like spheroids, organoids, organs-on-a-chip, 3D-scaffold or 3D-bioprinting cell and tissue-based systems (see Figure 16.3)
- re-differentiated cells
- sub-cellular fractions like cytosol and microsomes or subcellular organelles
- proteins
- cellular-derived genetic material.

It was demonstrated that SARS-CoV-2 is able to infect a number of the above human 2D- or 3D-derived cell systems. Some examples are listed hereing[10,11,16–22] and further detailed in this chapter for various applications in COVID-19 related to the dynamics, kinetics, symptoms, risk factors and treatments.

Some generic examples of reported 2D *in vitro* models (Figure 16.3) for SARS-CoV-2 research are:

- primary cell cultures in 2D set-ups from various target organ systems
- human peripheral blood mononuclear cells studying the immune response of SARS-CoV-2
- U251 derived from a malignant glioblastoma tumor by explant technique
- BEAS-2B derived from human bronchial tracheal cells
- A549 derived from explanted cultures of human lung adenocarcinoma
- Calu-3 derived from human bronchial submucosal glands
- Huh-7 derived from human liver carcinoma
- Caco-2 derived from human colorectal adenocarcinoma
- HUVEC derived from human endothelium of veins from the umbilical cord
- HT-29 derived from human primary colon adenocarcinoma
- HEK293T derived human embryonic kidney cells that expresses a mutant version of the SV40 large T antigen
- HuH7 derived from human liver epithelial-like tumorigenic cells.

Some generic examples of reported 3D *in vitro* models (Figure 16.3) for SARS-CoV-2 research are:

- human intestine-on-chip-device
- human induced pluripotent stem cell (iPSC)-derived BrainSphere neurons
- primary human airway epithelial cells derived from primary nasal (hAECN) or bronchial (hAECB) origin that can be cultured at an air–liquid interface (ALI)
- human embryonic, adult stem cell or human iPSC-derived gut, heart, kidney, pancreatic, liver, *etc.* organoids
- human blood vessel organoids, *etc.*

The application of *in silico* mathematical modelling and simulation in COVID-19 research may be used to understand the dynamics, kinetics, transmission, symptoms, risk factors, prevention strategies and treatments of the COVID-19 disease.

In silico methods can also be used to actively design better vaccine prototypes, support decision making, decrease experimental costs and time and eventually improve success rates of therapeutic and vaccine trials. *In silico* clinical trials, for design and testing of therapeutic interventions, can accelerate and speed-up the therapy and vaccine discovery pipeline, predicting any therapeutic failure and very importantly minimizing undesired effects, a key focus for the toxicological research community. Also, *in silico* machine-learning models are demonstrating efficacy in predicting critical COVID-19 adverse effects defined as mechanical ventilation, multi-organ failure, admission to the intensive care unit (ICU) and/or death. Hence, artificial intelligence may be applied for accurate risk prediction of patients with COVID-19, to optimize patients triage and in-hospital allocation, better prioritization of medical resources and improved overall management of the COVID-19 pandemic,[23] which is an avenue to further explore with the data scientist community.

Mathematical approaches are also of critical importance for understanding the dynamics and kinetics of SARS-CoV-2 infection. Some generic examples of how *in silico* mathematical models are used for SARS-CoV-2 mechanistic research are:[24,25]

- SARS-CoV-2 host tropism and analysis of the main cellular factors
- *in silico* prediction of molecular targets by systems network pharmacology and bioinformatic gene expression analysis
- protein–protein interaction network construction and key SARS-CoV-2 pathways
- *in silico* molecular docking of key SARS-CoV-2 targets
- studying host selectivity of SARS-CoV-2 components, spike protein, putative alternative receptors and putative co-receptors of ACE2.

16.2 Human Body Barriers and Cell and Mathematics-based Methods as Tools for Understanding COVID-19 Infection Kinetics

The human body is enclosed by the external skin and mucosa barriers, which are the cellular barriers that separate the internal human body milieu and the external non-sterile environment (Figure 16.4). Additional cellular internal barriers, such as the blood–brain and placental barriers, define protected niches within the human body. In addition to their physiological roles, these host barriers provide both physical and immune defense against SARS-CoV-2 infection. The high contagion rate of SARS-CoV-2 occurs through direct transmission (coughing, sneezing and inhalation of viral particles) or through

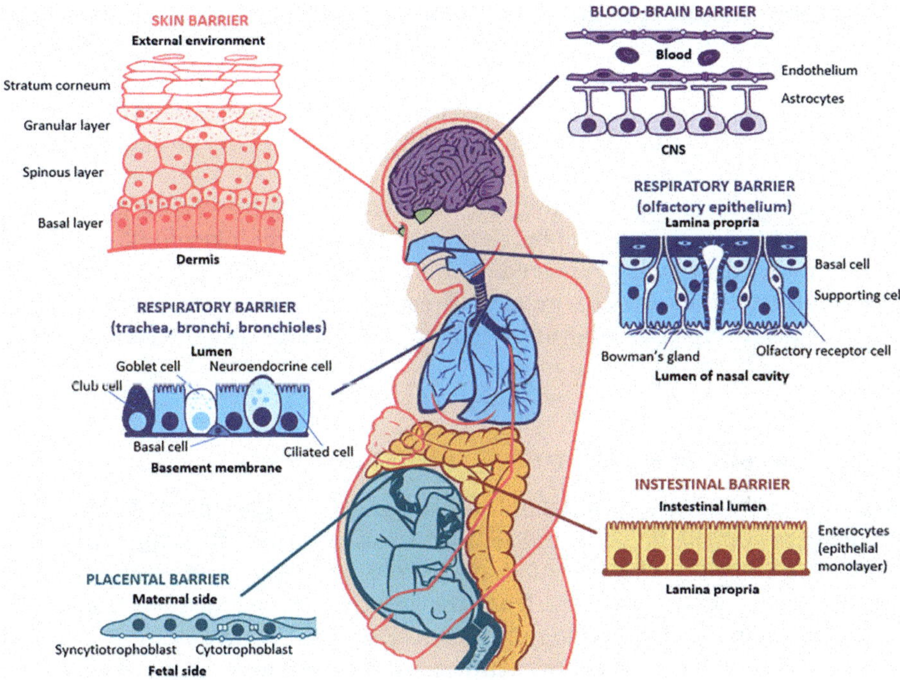

Figure 16.4 Human external and internal body barriers. Image credit: Alessia Nuvoloni, ARTSTATION, Civita Castellana, Italy.

contact (oral, nasal or ocular mucous membranes) that occur between humans, thus favoring its rapid spread around the world by efficiently crossing the respiratory, intestinal and blood retinal human body barriers. Like many pathogens, SARS-CoV-2 has a cellular mechanism to penetrate these barriers *via* cellular entry, resulting in a SARS-CoV-2 invasion of host barrier cells, disruption of barrier integrity and systemic dissemination and invasion of deeper organs and tissues. The skin and linings of the mouth, pharynx, esophagus, urethra and vagina are protective epithelia. The protective epithelia prevent exchanges between the external and internal environments and protect the regions subjected to chemical and mechanical stress. These epithelia are layered tissues, composed of numerous overlapping cell layers. Their resistance is pronounced thanks to the presence of keratin, a protein that is also found in hair and nails. Since protective epithelia, such as the skin, are exposed to chemical irritants and pathogens, their cells die and are replaced by new ones quite frequently. It has therefore been of interest to study whether a certain human barrier could be involved or not in SARS-CoV-2 infection, and to what extent several cellular types play a role in this process. One of the first evidences on the role of barriers in COVID-19 comes from a study using a single-cell RNA-sequencing approach across various barrier tissues and model organisms to identify the potential initial cellular targets of SARS-CoV-2 infection, showing that ACE2 and TMPRSS2, the main host protein involved in the initial viral entry, are present in different cell types at different levels.[26] Indeed, ACE2 has been found in type II pneumocytes in the lung, absorptive enterocytes within the gut, and goblet secretory cells of the nasal mucosa, while ACE2 and TMPRSS2 co-expression in respiratory tissues has been found only among a rare subset of epithelial cells.

A review of some representative examples of SARS-CoV-2 interactions with human body barriers, including the external respiratory (lung and nose), intestinal, eye and cutaneous barriers and the internal, placental, testis and blood–brain barriers, is presented in this section, including a discussion on how SARS-CoV-2 adheres to, invades, breaches, or compromises these barriers and its subsequent dynamic effects on the 11 target organ systems in the dynamics section.[27]

16.2.1 Respiratory Barrier

The most robust *in vitro* model of the respiratory epithelium is based on primary airway epithelial cells and is used to study molecular and functional aspects of SARS-CoV-2 virus infection *in vitro*. Primary nasal epithelial cells can non-invasively be isolated from the inferior surface of the middle turbinate of nostrils with cytology brushes. Progress in modelling the upper respiratory tract led to the establishment of well-differentiated cell cultures. In an ALI culture system, airway epithelial cells grow on a porous membrane, and their basal surface is in contact with culture medium in the basolateral chamber; after reaching confluency, the removal of apical medium and the exposure of the apical side of the cells to air mimic the conditions found in the airways. Airway epithelial cells grown at the ALI are now widely used to study airway disease mechanisms or for drug discovery, as they are robust surrogates of the structure and barrier functions of native respiratory epithelium.[28]

16.2.2 Intestinal Barrier

Several *in vitro* models that simulate the barrier function of the intestinal epithelial layer have been developed and applied to assess gut infection.

To evaluate absorption and permeability across the intestinal membrane, 2D monolayer culture systems have been developed. The main representative example is the Transwell system, which is composed of a well plate and a well insert. The bottom of the insert is a semipermeable membrane on which intestinal epithelial cells can be cultured to form cell layers. The compounds to be tested are added to the apical side of the cell layer, then the absorption and permeability can be assessed by analyzing the amount of compound in the basolateral side. In most cases, Caco-2 cells, an intestinal epithelial cell line derived from colon carcinoma, are cultured on the membrane to form cell layers. Interestingly, SARS-CoV-2 showed high replication capacity in Caco-2 cells.[29] However, the Transwell system is a very simplified approximation of the real *in vivo* situation, because it lacks the gut epithelium distinctive 3D structures, consisting of villi and crypts, that not only increase the surface area, but also affect the physiology of gut epithelial cells, such as enzymatic activity and the expression of specific membrane proteins in the cells. For this reason, the development of 3D *in vitro* structures assumes considerable importance in the study of human physiopathology.

It has been shown that SARS-CoV-2 can infect human intestinal organoids, or "mini-gut", resulting in the production of large amounts of infective virus particles in the intestine.[30] This evidence demonstrates that intestinal organoids can serve as a model to understand SARS-CoV-2 biology and infectivity in the gut, suggesting that SARS-CoV-2 can take advantage of this and infect intestinal enterocytes.

To overcome conventional static cell-culture limitations, physiological parameters, such as mechanical movement and flow of medium, have been integrated into culture systems using microtechnology known as organ-on-a-chip technology. For example, microfluidics-based flow environments have been applied to cell culture, using top and bottom microchannels separated by a porous membrane. It has been shown that, by culturing Caco-2 cells on this membrane, is possible to measure the absorption permeability and the functionality of the intestinal wall.[31] Also, a microfluidic cell culture device in which Caco-2 cells grew into a 3D structure on the porous membrane forming villus-like complexes as well as tight junctions with expression of occluding under fluidic conditions has been developed.[32] As previously reported, organs-on-chips are also versatile tools to study organ interactions, by combining 3D cultures of different organs and allowing them to share biological signaling. Given the causal link between SARS-CoV-2 infection, gut alterations and the disruption of the gut–lung axis,[33] the use of multi-organ-on-a-chip could represent a good option to study the effect of supportive treatment strategies against the viral infection in a more physiologically relevant way.

16.2.3 Anterior Segment Barrier

The eyes are an important entrance port and replication site of SARS-CoV-2.[34] If the virus first contacts the conjunctiva of the patient's eyes, or the hand touches the virus and rubs the eye, the virus will invade the patient's eye conjunctiva, infect and reproduce, causing eye swelling which can even lead to conjunctivitis. The replicated virus may pass through the tear fluid to patient's nasolacrimal duct and enter the respiratory infection pathway when coughing, or enter the digestive tract infection pathway when swallowing food.

16.2.4 Blood–Retinal Barrier

The cells that line the blood–retinal barrier, the retinal endothelium, and retinal pigment epithelium are highly permissive and susceptible to SARS-CoV-2-induced microvascular alterations and cell death. Retinal vein diameter seems directly correlated with disease severity.[35]

Another potential role of the eye in viral transmission of SARS-CoV-2 could be as a transport conduit to respiratory tract tissues. When ocular secretions and tears are drained through the nasolacrimal duct to the inferior meatus of the nose, viral particles may be transported from the ocular surface to the airway tissue, causing respiratory infection and diseases.

16.2.5 Skin Barrier

It has also been suggested that injured skin barrier can act as a reservoir for SARS-CoV-2. ACE2 receptors have been detected in the basal cell layer of the epidermis and hair follicles, eccrine glands, blood vessels and capillaries, sebaceous glands, and its surrounding smooth muscle cells. Patients with injured skin can be exposed to the virus *via* respiratory droplets, saliva or contaminated surfaces. This can provide an opportunity for the virus to survive for a while after binding to ACE2 receptors expressed in skin cells. This can further facilitate cellular viral entry and replication and protects SARS-CoV-2 from washing and antiseptics. However, further studies requiring isolation of the virus from active skin lesions would be necessary to prove this hypothesis.

In a study by Hirose *et al.*, the researchers developed an *ex vivo* human skin model using 1 day old abdominal skin autopsy samples.[36] The model was designed such that the skin samples did not deteriorate because of drying even after long-term incubation. Using this model, the authors found that SARS-CoV-2 could survive on human skin for more than 9 hours; however, more studies would be required to study whether the virus could cross the skin barrier.

16.2.6 Blood–Brain Barrier

To answer the question of whether human brain cells can be infected with SARS-CoV-2, some experimental *in vitro* models have been challenged. For instance, it has been found that a 3D hiPSC-derived organotypic brain model, known as BrainSpheres, can be infected by SARS-CoV-2.[17]

Considering these findings, there is huge interest in modelling the blood–brain barrier (BBB) *in vitro*; thus, several cellular models *in vitro* have been developed.[37] BBB permeability is a key determinant for central nervous system (CNS) exposure, and it must be considered when evaluating the dynamics and kinetics of SARS-CoV-2 brain penetration. According to a report of the European Centre for the Validation of Alternative Methods, there are a number of minimal requirements for any BBB model to be useful, such as the presence of restrictive paracellular permeability, the possession of a physiologically realistic cell architecture, the expression of the functional transporter mechanisms present *in vivo* and ease of culture.[38] A negative effect, due to a pro-inflammatory response of the endothelial cells, of SARS-CoV-2 spike subunit S1 on BBB integrity has been demonstrated, in experiments performed using primary human brain microvascular endothelial cells in both confluent cell monolayers and 3D microfluidic models of the BBB.[39]

16.2.7 Placental Barrier

It could be useful to resort to pre-clinical *in vitro* models of the placental barrier consisting of multi-nucleated layer of syncytiotrophoblast and vascular endothelial cells. In this context, an *in vitro* model of the human placental barrier has been developed by co-culturing syncytialized trophoblastic cells and vascular endothelial cells on contralateral sides of ECM-coated Transwell inserts under physiologically relevant oxygen levels.[40] In fact, physiological oxygen levels in the human placenta are approximately 3–8%, substantially lower than the 21% of oxygen typical of the atmospheric air. Moreover, a 1 day protocol for the production of a 3D *in vitro* placental barrier model using primary human trophoblasts and containing blood capillary networks which can be used to examine the mechanism of feto-maternal dialogue during pregnancy has been reported.[41] It is also possible to replicate the microarchitecture of the placental barrier by using micro-engineered devices that mimic the structural and functional complexity of this specialized tissue *in vitro*. For example, a multi-layered microfluidic system has been created that enables co-culture of human trophoblast cells and human fetal endothelial cells in a physiologically relevant spatial arrangement to replicate the characteristic architecture of the human placental barrier observed *in vivo*.[42]

16.2.8 Blood–Testis Barrier

In recent years, there has been increasing interest in using Sertoli cells as a model to study blood–testis barrier dynamics.[43] Sertoli cells cultured at high density on extracellular matrix-impregnated permeable supports have the ability to form a polarized monolayer that closely mimics the blood–testis barrier *in vivo*, both structurally and functionally. By maintaining the monolayers in appropriate culture chambers, the cells can develop a functional tight junction permeability barrier and are able to achieve the separation of apical from basal compartments.[44] Although 2D culture methods have provided us with much information on testicular biology, they often fail to mimic organ-specific physiology, leading to inappropriate and biologically irrelevant cell–cell interactions.[45] Organ culture methods were applied to address the lack of 3D cell–cell interactions of 2D culture. In organ culture, small testicular tissue fragments rather than single cells are placed in cultures.[46] Although organ culture models have been used successfully to study testicular biology and toxicity in rodents, the system does not allow the study of testicular morphogenesis.[47] The rising interest in hiPSC-derived organoid culture has led to the development of tissue-specific organ-like structures to investigate testicular physiology *in vitro*.[48] Organoids can be used as a physiologically more relevant model system to study cell–cell interactions, development and tissue morphogenesis.[49] These testicular organoids might be useful in answering scientific questions about the link between the SARS-CoV-2 infection and the altered regulation and function of the spermatogonial stem-cell niche as well as germ cell proliferation and differentiation in the presence of the viral particles.

16.3 Tissue-specific Distribution Patterns

A detailed characterization of receptor(s) and protease(s) involved in SARS-CoV-2 entry is of major importance to understand SARS-CoV-2 cellular entrance, distribution patterns and the pulmonary and extra-pulmonary COVID-19 disease features. Indeed, while in the early phase of the pandemic, SARS-CoV-2 was believed to behave like other respiratory viruses leading to ARDS, it is now very clear that it is a very unusual pathogen that provokes manifestations outside the respiratory apparatus, leading to fatal outcomes in vulnerable people. Although research on receptors and proteases has highly benefited from accumulated knowledge on other human coronaviruses, notably the closely related SARS-CoV, much remains to be learned in the context of SARS-CoV-2 infection. Numerous data published support a main role of ACE2 as for SARS-CoV. However, there is still some doubt on the hypothesis as ACE2 as the unique SARS-CoV-2 cellular entrance port by the observation that ACE2 is at most poorly expressed in the respiratory epithelium. Whether other receptors than ACE2 or facilitators as TMPRSS2, such as the extracellular matrix metalloproteinase inducer CD147 (BSG-Basigin), furin, CD209L/L-SIGN and a related protein, CD209/DSIGN, neuropilins (NRP1), heparan sulfate or sialic acids exposed at the surface of target cells could be play a role should be further confirmed.[50–53]

With COVID-19 emerging as a disease with vascular consequences and effects on many body target organ systems rather than just a respiratory syndrome, the role and qualitative and quantitative expression of the SARS-CoV-2 of entry receptor(s) is key to evaluate target organ system and tissue distribution and the related local pathogenic effects. Infection of endothelial cells is consistent with the role and expression of ACE2.[51]

However, more solid data, notably relying on post-mortem analyses or from patient-derived *in vitro* cell and tissue cultures, are urgently needed to validate this observation and determine whether it is anecdotic or more widespread among COVID-19 patients. Such data will help to elucidate the role of the modulation of the entry receptors in patients with comorbidities associated with severe symptoms and fatal outcome of COVID-19.

ACE2 in SARS-CoV-2 infection and its impact on the renin–angiotensin system (RAS), kallikrein–kinin system (KKS) and coagulation systems is another area that can be further mechanistically investigated with *in vitro* cell and tissue culture tools to elucidate the inflammatory state of severe COVID-19. The KKS is an intricate endogenous pathway involved in several physiological and pathological cascades in the brain. Due to the pathological effects of kinins in blood vessels and tissues, their formation and degradation are tightly controlled. Their components have been related to several central nervous system diseases such as stroke, Alzheimer's disease, Parkinson's disease, multiple sclerosis, epilepsy and others provoking some symptoms similar to those observed in COVID-19 disease. The RAS is an important blood pressure regulator and controls both sodium and water intake. AngII is a potent vasoconstrictor molecule and ACE enzyme is the major enzyme responsible for its release. AngII acts mainly on the AT1 receptor, with involvement in several systemic and neurological disorders. Brain RAS has been associated with physiological pathways, but is also associated with brain disorders.[54,55]

The tissue-wise distribution patterns of SARS-CoV-2 receptors in the human body can help in predicting the complications and early warning signs for the affected organs.[56]

ACE2 is expressed in the nasal and pulmonary respiratory airway epithelium at low levels compared to the intestine, kidney, heart and pancreas. However, in nasal and bronchial tissues, ACE2 is mainly expressed by ciliated, club and goblet cells. ACE2 is highly expressed in the sustentacular cells of the olfactory epithelium and the pericytes of the olfactory bulb. Low levels are also observed in the liver. It is also found in type-2 pneumocytes of alveoli and in endothelial cells of pulmonary capillaries.

Based on these findings studies looked at the tissue distribution of SARS-CoV-2 receptors and facilitators, including ACE2 and TMPRSS2, but also others such as the ECM MMP inducer CD147 (Basigin), furin CD209L/L-SIGN encoded by CLEC4M and a related protein, CD209/DSIGN, NRP1, heparan sulfate or sialic acids and BSG[57,58] as mentioned earlier.

Tools such as Gene ORGANizer (a phenotypic tool which directly links human genes to the organ systems they affect), the Human Protein Atlas and the PubMed Gene database and literature survey are used to identify potentially high-risk organs vulnerable to COVID-19 infection such as lungs, intestine, kidney, testis, heart, liver, placenta, hematopoietic tissue and nerve tissue.

16.4 Human Pulmonary and Extra-pulmonary Target Organ Systems Cell and Mathematics-based Methods as Tools for Understanding COVID-19 Infection Dynamics

To describe the infection dynamics of SARS-CoV-2, the understanding of the role of the key processes, pathways and subcellular components following SARS-CoV-2 coronavirus cellular entrance *via* the ACE2 enzyme receptor is critical.

The ACE2 enzyme itself is a pivotal component of the RAS exerting its physiological functions by modulating the levels of angiotensin II (Ang II) and Ang-(1–7) on human body organ systems. RAS plays an important role in inflammation and fibrosis. Figure 16.5 shows the RAS modulatory role of cellular functions and molecular signaling pathways and their key role in COVID-19 pulmonary and extra-pulmonary system multi-organ system effects.

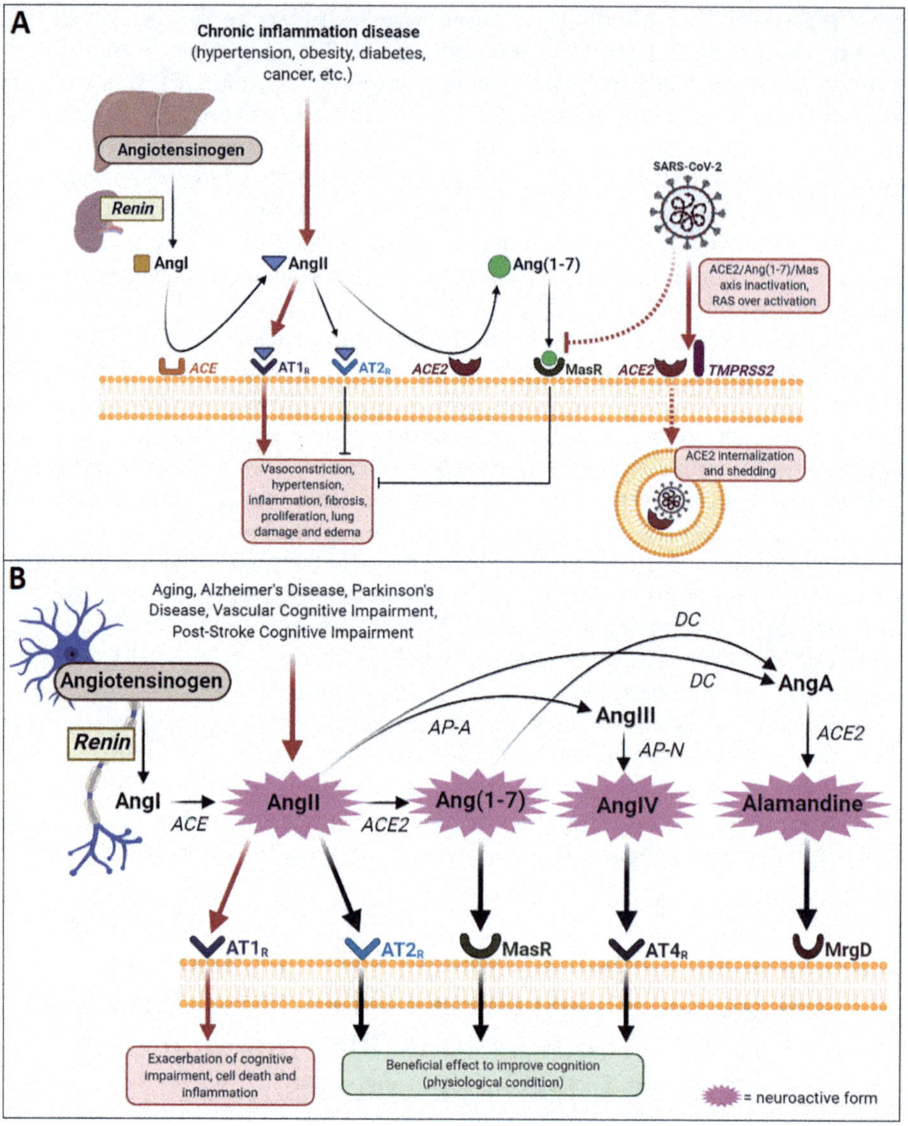

Figure 16.5 A) SARS-CoV-2 induces the ACE2 internalization and shedding, leading to the inactivation of the ACE2/Ang (1–7)/Mas axis and the human body cells expressing ACE2 as entrance port for SARS-CoV-2 lose their natural protection against cell and tissue injury, inflammation and fibrosis. RAS modulation is critical in balancing the COVID-19 disease towards either defense or progression.[61] B) Molecular RAS cascades connected to SARS-CoV-2 infection of the cerebral vasculature. Endotheliopathy resulting from direct Sars-CoV-2 infection of the cerebral vasculature can modulate ACE2/AT1R/MasR signaling pathways. Coronavirus S protein binding of ACE2 downregulates ACE2 protein expression that may attenuate the neuroprotective effects of Angiotensin-1,7 signaling. Created in BioRender.com.

The classical axis of the RAS is formed by ACE, Ang II and AT1 activating several cellular functions and molecular signaling pathways related to tissue injury, inflammation and fibrosis.

In sharp contrast, the RAS axis composed by ACE2, angiotensin-(1–7) and Mas receptor exerts opposite effects in relation to inflammatory response and tissue fibrosis. The Mas receptor has been identified as the protein transducing the vasodilator and antiproliferative actions of Ang-(1–7). The Mas protein is a G-protein coupled receptor originally linked to modulation of growth-regulating pathways involved in oncogenic effects and high expression in human eye, testis, hypothalamus and amygdala.[59]

For many years, modulators of RAS have been described to be essential for the control of essential hypertension. However, RAS modulators (including SARS-CoV-2 and chronic inflammation) have pleiotropic properties independent of their hypotensive effects, such as cognitive impairment of multiple etiologies. The spike glycoprotein S on the virion surface docking onto the ACE2 dimer (S1 and S2) (Figure 16.2) is an essential step in the process of SARS-CoV-2 infection in any cell of the human body organ system. It involves downregulation of ACE2 expression with systemic RAS imbalance and promotion of multi-organ damage.

Within the brain, different components of RAS have been extensively studied in the context of neuroprotection and cognition. Interestingly, a crosstalk between the RAS and other systems such as cholinergic, dopaminergic, and adrenergic systems have been demonstrated.

The expression and function of the different receptor subtypes within the RAS, such as:

- angiotensin II type I receptor (AT1R)
- angiotensin II type II receptor (AT2R)
- angiotensin IV receptor (AT4R), Mas receptor (MasR)
- and Mas-related-G protein-coupled receptor (MrgD) (Figure 16.5)

on different cell types within the brain can be modulated by RAS modulators. As such, the importance of RAS on cognition and different disease conditions caused by SARS-CoV-2 infection is an avenue central to understanding the overall COVID-19 disease etiology.[60]

16.4.1 Potential SARS-CoV-2 Target Organ Systems

To better understand how downregulation of ACE2 expression with systemic RAS imbalance and other related mechanisms lowering cellular defense can lead to promotion of pulmonary and extra-pulmonary multi-organ damage in the human body, the human body has been divided into 11 potential SARS-CoV-2 target organ systems (Table 16.1 and Figure 16.6).[5]

Table 16.1 SARS-CoV-2 target organ systems, functions, target organs, tissues, cell types of the human body and a listing of some of the molecular and cellular effects observed and the examples of the cell-based observation methods and some of the COVID-19 symptoms and laboratory findings reported[118–123] (cutaneous),[124] (neuro),[125] (systematic review),[126] (smell and taste systematic review),[127,128] (central nervous system systematic review).[129]

		COVID-19 relevant			
System (function)	Organs/ tissues	Cell types	Molecular and cellular effects	Examples of cell-based methods	Symptoms laboratory findings
1. Respiratory (oxygen/carbon dioxide exchange)	Nasal cavities	Epithelial cells (lining the nasal turbinate, ethmoid sinus, trachea and bronchus)	ACE2 is robustly expressed in the motile cilia of the respiratory tract[62]	Immunohistochemistry using a panel of ACE2 antibodies on human tissue microarrays	Dry cough (31–80%)
				Human nasal tissue specimen collection	Nasal obstruction (0–4%)
	Pharynx			Tissue microarrays	Sore throat (0–5%)
	Larynx			Chromogenic immunohistochemistry	
	Trachea			IMCD3 cell culture, transfection, and staining[62]	
	Respiratory barrier	Ciliated cell, globet cell, basal cell, basal lamina			
	Lung	Alveolar epithelial cell type II (AEC II)	ACE2 proteins are most significantly expressed on the surface of alveolar epithelial cells[63]	Serum levels of pro-inflammatory cytokines[66–68]	Dyspnea (30–66%)
				CT scan	ARDS (29–67%)
				Histopathology from deceased	Pneumothorax (2%)
				In vitro cell-derived human airway organoids and *ex vivo* bronchus cultures[69]	Haemoptysis (2%)

			Inflammatory cytokine storm[64,65]	Primary human airway epithelial cells derived from primary nasal (hAECN) or bronchial (hAECB) origin that can be cultured at an ALI BEAS-2B derived from human bronchial tracheal cells A549 derived from explanted cultures of human lung adenocarcinoma Calu-3 derived from human bronchial submucosal glands	D-dimer levels were markedly higher in severe cases
2. Cardiovascular (oxygen and nutrients transport; waste products elimination)	Heart Cardiovascular tissue	Cardiac myocytes, fibroblasts, endothelial cells, smooth-muscle cells: ACE2 has high expression	a. Direct viral injury: isolation of the virus from myocardial tissue has been reported in a few autopsy studies[70–72] b. Inflammatory infiltrates without myocardial evidence of SARS-CoV-2 [73,74] c. Direct viral infection of the endothelium and accompanying inflammation[75] d. Systemic inflammatory response syndrome (cytokine storm) is another putative mechanism of myocardial injury[76]	Human engineered 3D heart tissue model to demonstrate that SARS-CoV-2 selectively infects cardiomyocytes provoking cell death and loss of contractility[77]	Chest tightness (11%) In severe cases: cardiac injury (12–23%) (elevated creatine kinase, creatinine kinase MB)

(*continued*)

Table 16.1 (*continued*)

System (function)	Organs/ tissues	COVID-19 relevant: Cell types	Molecular and cellular effects	Examples of cell-based methods	Symptoms laboratory findings
3. Digestive (food processing; nutrient absorption)	Oesophagus Intestinal tract Liver Pancreas Salivary glands Stomach Bowel	Differentiated enterocytes,[78] intestinal glandular cells, gastric, duodenal, rectal epithelial cells, glandular enterocytes (presence of ACE2)	a. Virus-mediated direct tissue damage[78–81]	Histopathology from deceased	Anorexia (14–30%)
				Confocal and electron microscopy[78]	Diarrhea (9–13%)
				Histopathology from biopsy[80] Human primary gut epithelial cells used to establish small intestinal organoid in 3D culture[21,78]	Nausea (3–6%)
				Huh-7 derived from human liver carcinoma Caco-2 derived from human colorectal adenocarcinoma	Vomiting (3–5%)
				HT-29 derived from human primary colon adenocarcinoma HuH7 derived from human liver epithelial-like tumorigenic cells	Abdominal pain (2–6%)
				Human intestine-on-chip-device 3D liver and cholangiocytes for chemokine, IL-17, TNF and NF-κB signaling[22]	In severe cases: liver dysfunction (16–37%) with alanine aminotransferase and aspartate aminotransferase levels significantly higher
	Intestinal barrier	Enterocytes, basal lamina			

4. Urinary (waste elimination; blood volume and pH regulation)	Kidneys Urethra Urinary Bladder	Renal cells (presence of ACE2 receptors)	a. Direct infection of renal cells[70,82,83] b. Endothelial damage (glomerular capillary endothelial cells viral inclusion)[75] c. Cytokine storm[76,84] d. Immunocomplexes of viral antigen or virus-induced specific immunological effector mechanisms[85]	Kidney CT scan Histopathology (autopsy)[82] *In silico* analysis of publicly available data sets of single-cell RNA sequencing[70,82,83] Viral inclusion particles[75] Biopsy[85] HEK293T-derived human embryonic kidney cells that expresses a mutant version of the SV40 large T-antigen Human embryonic stem cell derived and iPSC-derived kidney organoids where single-cell RNA sequencing showed ACE2 expression in tubular-like cells and SCL3A1, SCL27A2, SCL5A12[20,86]	In severe cases: Acute kidney injury (7–29%) Electrolyte abnormalities (hyperkalemia, hyponatremia, hypernatremia) Proteinuria Hematuria Increased serum creatinine (10%) Metabolic acidosis

(*continued*)

Table 16.1 (*continued*)

		COVID-19 relevant			
System (function)	Organs/ tissues	Cell types	Molecular and cellular effects	Examples of cell-based methods	Symptoms laboratory findings
5. Nervous (organ systems activity coordination; stimuli response)	Brain Ears	Neurons, glial cells, endothelial cells, and arterial smooth muscles in the central nervous system	Hematogenous spread (2)- retrograde neuronal transport *via* the olfactory nerve	U251 derived from a malignant glioblastoma tumour by explant technique	Headache (6–13%) Dizziness (16%)
	Eyes Nerves		The cytokine cascade that follows potentiates the invasion of the virus into the brain[87]	Human induced pluripotent stem cell (iPSC)-derived BrainSphere neurons	Smell impairment (5–74%) Taste impairment (5–50%)
	Spinal cord			Pluripotent stem cells to produce a whole eye organoid model of retina, retinal pigment epithelium, ciliary margin, iris lens and cornea[88]	In severe cases: acute cerebrovascular disease (5–9%) ischemic stroke (1%) epilepsy (1%)
	Barrier Nasal olfactory epithelium	Supporting cell/ sustentacular cell, olfactory receptor neuron, basal cell			
	Blood–brain barrier	Endothelial cells, pericytes, astrocytes			

6. Lymphatic/hematopoietic (tissue fluid to blood return; defence against foreign organisms)	Bone marrow Thymus Lymph nodes Lymphatic vessels Mucous membranes Spleen Tonsils			Human peripheral blood mononuclear cells studying the immune response of SARS-CoV-2 Human blood vessel organoids closely simulating human vascular capillary growth with a lumen, PDGFR− pericyte coverage CD31+ endothelial lining, and formation of a basal membrane[20]	Fever (98–100%) Lymphopenia in severe cases (70%) Procalcitonin tended to be higher in severe cases
7. Muscular (locomotion; heat production)	Muscles Tendons	Muscle stem cells (satellite cells) Myonuclei (muscle fibers) Vascular cells (*i.e.* endothelial cells, smooth muscle cells, pericytes) Macrophage, adaptive immune cells (B-, T-, or natural killer cells)	a. Vascular cells (endothelial cells, smooth muscle cells, pericytes), muscle stem cells (satellite cells), macrophages, adaptive immune cells (B-, T-, or natural killer cells), and myonuclei (muscle fiber) express TMPRSS2 b. Neuronal demyelination[89]	Autopsy[89] 3D human skeletal muscle culture model[90]	Myalgia (11–40%)

(*continued*)

Table 16.1 (*continued*)

System (function)	Organs/ tissues	COVID-19 relevant: Cell types	Molecular and cellular effects	Examples of cell-based methods	Symptoms laboratory findings
8. Integumentary (physical barrier; temperature control)	Hair	Keratinocytes	Diffuse and dense lymphoid infiltrates, along with signs of endothelial inflammation	Human histopathology[91,92]	Cutaneous manifestations (0–20%) Erythematous rash (15%)
	Skin	Keratinocytes, melanocytes, Langerhan's cells, basal lamina	Immune hypersensitivity	Transcriptomic data[94]	Urticaria (3%) Chickenpox-like vesicles (1%)
	Subcutaneous tissue		Endothelial inflammation[91–93]	*In vitro* organoid model of dermal papilla of human hair follicle[95]	Maculopapular eruptions (47%) Vesicular eruptions (9%)
	Skin barrier (stratum corneum, granular layer, spinus layer, basal layer)		Human skin keratinocytes and basal cells express more ACE2 than lung epithelial cells[94]		Erythema with vesicles or pustules (pseudo-chilblain) (19%) Livedo or necrosis (6%)

9. Reproductive (germ cell production (eggs and sperm); environment for growth of fetus (female))	External genitalia Mammary glands Ovaries Placenta Prostate gland Testes Uterus	Female: oocytes Epithelial cells uterus and vagina Placental villi[5]	a. ACE2 is widely expressed in the ovary, uterus, vagina and placenta. b. modulating the levels of angiotensin II (Ang II) and Ang[5,52]	GeneCards database[96] Bgee[5,97] HUVEC derived from human endothelium of veins from the umbilical cord	Infertility?
		Placental barrier: syncytiotrophoblast cytotrophoblast			
		Male: spermatogonia, Leiden cells, Sertoli cells[98]	a. ACE2 is widely expressed in the testis b. Modulating the levels of angiotensin[98]	Microarray profiling assay Autopsy[99] Potential to use *in vitro* testicular organogenesis from human fetal gonads to produce fertilization-competent spermatids to study SARS-CoV-2 effects[100]	Infertility Orchitis Penetrate the barriers and induce testicular dysfunctions
		Blood testis barrier: Sertoli cell, Leydig cell, basal lamina			
10. Skeletal (body support; internal organ protection; mineral storage)	Bones Cartilage	Osteoclast Osteoblast	Increased bone resorption and bone loss: including (ACE2)-dependent bone resorption, inflammatory cytokines storm, immunosuppression and hypoxaemia[101]	RNA sequencing identifying skeletal muscle, synovium and cortical bone as potential sites of direct SARS-CoV-2 infection[89]	Arthralgia (0–1%)
	Ligaments	Chondrocytes	Chondrocytes express ACE2	*In vitro* model of mesenchymal condensation during chondrogenic development[69]	

Continued

Table 16.1 (*continued*)

System (function)	Organs/ tissues	COVID-19 relevant			
		Cell types	Molecular and cellular effects	Examples of cell-based methods	Symptoms laboratory findings
11. Endocrine (body function regulation (hormones))	Adrenal gland		Ref. 102–104	Potential for studying dynamics of calcium-regulated parathyroid hormone. Secretion *in vitro* to investigate the relationship between the Parathyroid cell cycle and the calcium-regulated parathyroid hormone secretion[105]	"An acute adrenal insufficiency may also be due to a thrombotic event at the adrenal level in COVID-19 patients. This could cause an acute adrenal insufficiency with impaired hormone production with consequent shock and worsening of the possibility of reacting to severe respiratory distress. A timely screening for pituitary–adrenal axis function and identification of this condition could allow adequate replacement therapy avoiding severe shock"[106]
	Parathyroid gland				
	Pituitary gland				
	Thymus				
	Thyroid gland		a. ACE2 and TMPRSS2 expression levels are high in the thyroid gland and more than in the lungs[107] b. Thyroid damage may be secondary to a hypothalamic–pituitary system virus insult leading to thyroid disconnection[107,108]	Anatomopathological studies	Thyrotoxicosis: fatigue, palpitations, inappetence, sweating, insomnia, anxiety, tremor, weight loss

		Hypothyroidism: fatigue, stypsis
	SARS genomic sequence positive lymphocytes and monocytes in the vessel of the thyroid gland from a SARS autopsy[109,110]	Subacute thyroiditis
Physiological concentrations of L-thyroxine (T4) and 3,3′,5-triiodo-L-thyronine (T3) stimulate the production and release of cytokines	Immunohistochemistry Evaluation of adenohypophysis from autopsies[108]	
Thyroid hormones are capable to potentiate the antiviral action of IFN-γ. It is also of interest that some pathways (*i.e.* the cytokine and hyper-activation of Th1 cell responses) of immune responses to virus infection are observed in thyroid disorders (*i.e.* classical autoimmune thyroid diseases, IFN-α-related thyroid disease, immune checkpoint inhibitor mediated thyroiditis, alemtuzumab-induced thyroid dysfunctions)[107]	*In vitro* models for assessing thyroid effects at TSAR "Test Methods Tracking"[111]	

Continued

Table 16.1 (*continued*)

System (function)	Organs/ tissues	COVID-19 relevant			
		Cell types	Molecular and cellular effects	Examples of cell-based methods	Symptoms laboratory findings
	Pancreas	Pancreatic cell	*Via* hematogenous spread: ACE 2 receptor expression is present in the pancreas	Quantitative expression map for ACE	Hyperglycemia
			ACE 2 expression is noted both in the exocrine pancreatic glands as well as in the islets[112]	RNA was extracted from 72 human tissues[114,115]	Ketoacidosis, including in patients with previously undiagnosed diabetes or no diabetes
			In pancreas, binding to ACE2 with subsequent damages of islets and reduced insulin release[113]	*In vitro* human pancreatic islets[116]	Euglycemic ketosis Severe illness in patients with pre-existing diabetes and/or obesity[117]

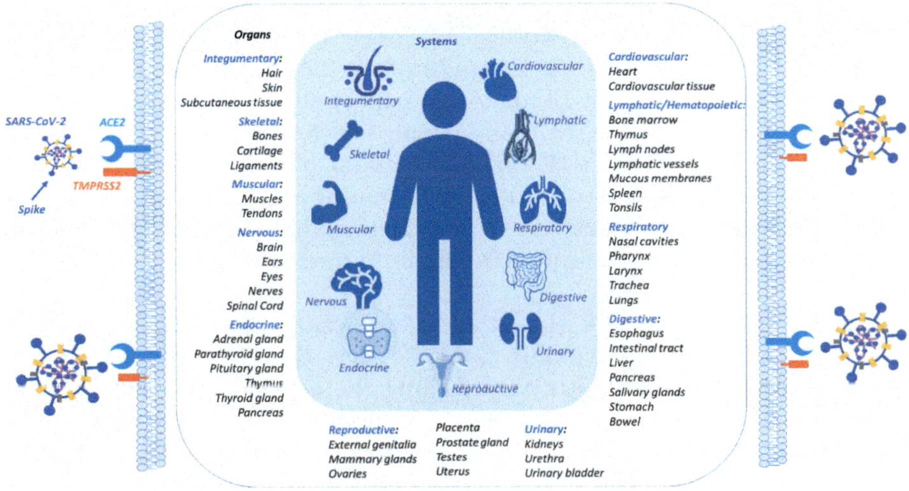

Figure 16.6 Human body target organ system containing several potential specific target organs where SARS-CoV-2 effects can be seen at the molecular, cellular, tissue or organ level.

Each target organ system contains several potential specific target organs where effects can be seen at the molecular, cellular, tissue or organ level. For the purpose of this chapter, an organ is defined as a unique anatomic structure consisting of groups of tissues that work in concert to perform specific functions. Furthermore, potential mechanisms of cellular and molecular interaction and signaling pathways, elicited by functional receptors, in major targeted tissues/organs (*e.g.* such as from the respiratory, digestive, lymphatic, cardiovascular, urinary, and nervous systems) are listed. Moreover, the potential involvement of these pathways and processes in evoking the onset and progression of COVID-19 symptoms in these organ systems following SARS-CoV-2 coronavirus cellular target organ entrance are discussed.

16.4.2 SARS-CoV-2 Target Organ Systems Relevant Cell Types, Molecular and Cellular Events, Methods and Symptoms

Table 16.1 includes the structures and functions of these 11 organ systems describing their function and organ composition as well as the COVID-19 relevant cell types, molecular and cellular events, methods used to identify the effects and the symptoms associated. Patients who develop severe/critical illness are reported to have more frequently acute and sub-chronic effects observed at target organ systems listed in Table 16.1.

16.4.2.1 Respiratory Organ Dynamics

16.4.2.1.1 Stage 1: Infection. From the dynamic point of view, the inhaled virus SARS-CoV-2 probably binds to epithelial cells in the nasal cavity and starts replicating. ACE2 is the main receptor for SARS-CoV-2 but alternative receptors and co-receptors (Figure 16.2) might play an additional role. *In vitro* data on with SARS-CoV-2 infection dynamics indicate that the ciliated cells are primary cells infected in the conducting airways. There is local propagation of the virus, but a limited innate immune response.

16.4.2.1.2 Stage 2: Upper Airway and Conducting Airway Response (Next Few Days). From the dynamic point of view the virus propagates and migrates down the respiratory tract along the conducting airways, and a more robust innate immune response is triggered. The level of signaling chemokine CXCL10 (or other innate response cytokine) may be predictive of the subsequent clinical course. SARS-CoV-2 virus infected epithelial cells are a major source of β and λ interferons. CXCL10 is an interferon-responsive gene that has an excellent signal-to-noise ratio in the alveolar type II cell response to both SARS-CoV and influenza and as such is a useful disease marker for COVID-19. Determining the host's innate immune response is used as a monitoring tool to improve predictions on the subsequent course of the disease and need for more aggressive monitoring.

16.4.2.1.3 Stage 3: Hypoxia and Progression to ARDS. From a dynamic point of view, SARS-CoV-2 reaches the gas exchange units of the lung and infects alveolar type II cells. Both SARS-CoV and influenza preferentially infect type II cells compared to type I cells. The infected alveolar units are mainly peripheral and subpleural. SARS-CoV propagates within type II cells and a large number of viral particles are released, leading the cells to undergo apoptosis and die. Subsequently, the released viral particles infect type II cells in adjacent units. Parts of the lungs loose most of their type II cells, and secondary pathway for epithelial regeneration are triggered. Normally, type II cells are the precursor cells for type I cells. The pathological result of COVID-19 is diffuse alveolar damage with fibrin rich hyaline membranes and a few multi-nucleated giant cells. The wound healing process leads to severe scarring and fibrosis. Recovery of fibrosis requires a vigorous innate and acquired immune response and epithelial regeneration.

Following the SARS-CoV-2 infection, macrophages are released as a response to inflammatory signals by type II cells. Cytokines are released by macrophages that in turn results in the release of more immune cells to the injury site. Cytokines also cause vasodilatation. Fluid accumulation in alveoli causes surfactant (a surface-active lipoprotein complex – phospholipoprotein – formed by type II alveolar cells) damage, and thus alveolar collapse which subsequently affects gas exchange and extensive release of cytokines occurs including interleukin (IL)-6, resulting in subsequent increase in the vascular permeability. This further leads to entry of a large number of blood cells and fluid into the lungs, and causes dyspnea and respiratory failure.[130]

An overactive immune defect (by localizing neutrophils and increasing cytokines) leads to an increase in free radicals, cell debris and proteases, and edema resulting from the increase in proteins in the interstitial space. Consequently, vasoconstriction occurs through platelet activation that further alters alveolar gas exchange with severe hypoxemia and tissue hypoxia, which will lead to multi organ failure.[131]

It has been described that, in severe COVID-19 infection, the dysregulated immune system responds by secreting cytokines in an uncontrolled manner leading to marked increases in cytokine release, or a "cytokine storm" syndrome (Moore and June 2020;[132] Zhang *et al.* 2020b[133]). The neutrophil-to-lymphocyte ratio is described as the leading indicator of cytokine storms (hypercytokinaemia), with increased neutrophil-to-lymphocyte ratio in the blood due to the elevated cytokine levels. SARS-CoV-2 can infect monocytes, macrophages, dendritic cells, and lymphocytes, which together play an indispensable role in the development of cytokine storm.[134]

Other authors suggest that the descriptor "cytokine storm" does not appropriately describe the milieu in COVID-19-induced organ dysfunction, since autopsy reports consistently describe widespread dissemination of SARS-CoV-2 throughout diverse target organ systems. Lymphopenia is common and prognostic. T-lymphocytes are directly susceptible to SARS-CoV-2 infection and are depleted in clinical COVID-19. A predominantly hypoimmune state with subsequent (direct) virus-mediated target organs system and tissue damage, and dysregulated inflammation is consistent with clinical and pathological abnormalities in COVID-19 and the high concentrations of circulating acute-phase reactants reported here.[135]

That SARS-CoV-2 cellular entrance and spread stimulates a proliferation of immune cells in the lung, that are part of the aberrant immune system activation, is being increasingly recognized as a feature of severe COVID-19.[136]

16.4.2.2 Cardiovascular Organ Dynamics

Cardiovascular dynamics are influenced by:

- direct viral injury: isolation of the virus from myocardial tissue has been reported in a few autopsy studies[70–72]
- inflammatory infiltrates without myocardial evidence of SARS-CoV-2[73,74]
- direct viral infection of the endothelium and accompanying inflammation[75]
- systemic inflammatory response syndrome (cytokine storm) is another putative mechanism of myocardial injury.[76]

A human engineered heart tissue model demonstrated that SARS-CoV-2 selectively infects cardiomyocytes. Viral infection is dependent on expression of ACE2 and endosomal cysteine proteases, suggesting an endosomal mechanism of cell entry. After infection with SARS-CoV-2, engineered tissues display typical features of myocarditis, including cardiomyocyte cell death, impaired cardiac contractility, and innate immune cell activation. Autopsy tissue obtained from individuals with COVID-19 myocarditis demonstrated these findings showing cardiomyocyte infection, cell death and macrophage-predominant immune cell infiltrate. 3D *in vitro* models permit mechanistic studies of human cardiomyocyte SARS-CoV-2 tropism and provides an experimental platform for interrogating and mitigating cardiac complications of COVID-19.[77]

16.4.2.3 Digestive Organ System Dynamics

Regarding the dynamics of the digestive system, SARS-CoV-2 virus-mediated direct digestive tissue damage has been reported.[78,79,81,137] Intestinal glandular cells, gastric, duodenal, rectal epithelial cells, glandular enterocytes all have ACE2 receptors.

SARS-CoV-2 directly invades the digestive tract through binding with ACE2 receptors in glandular cells of gastric, duodenal and rectal epithelial cells, as well as in enterocytes of small intestinal.[80]

ACE2 receptors at the enterocyte level (glandular cells of gastric, duodenal and distal enterocytes) may explain gastrointestinal involvement, resulting usually in malabsorption, unbalanced intestinal secretion and activated enteric nervous system. Relevant *in vitro* models are described in Section 16.2.2.[21,138]

When examining liver material from infected patients by electron microscopy, immunohistochemistry, terminal deoxynucleotidyl transferase nick end labeling staining (TUNEL assay) and pathological studies, conclusions were drawn that liver involvement is a crucial cause of hepatic impairment in COVID-19 patients. Indeed, SARS-CoV-2 causes conspicuous hepatic cytopathy, with massive apoptosis and presence of binuclear hepatocytes as predominant histological features.[139]

Post-mortem biopsy studies in patients with COVID-19 show moderate microvascular steatosis and mild lobular and portal activity, indicating that the injury could have been caused by either COVID-19 or drug-induced liver injury.[140]

SARS-CoV-2 infection and cholangiocyte damage have been studied *in vitro* with human liver ductal organoids as a new tool studying tropism and pathogenesis.[141]

Other 3D *in vitro* liver models consisting of scaffold-free (spheroids and organoids) or scaffold-based (3D scaffolding and 3D bioprinting) systems have been used to study infectivity, replication kinetics and host–viral interactions of SARS-CoV-2, showing increased physiological relevance as compared to 2D models.[22,88]

16.4.2.4 Urinary Organ System Dynamics

Possible dynamic mechanisms for kidney injury in COVID-19 include direct infection of the kidney as well as cytokine storm related to sepsis. Renal cells have ACE2 receptors, so a directly infection of renal cells is possible.[70,82,83] Endothelial damage is demonstrated by glomerular capillary endothelial cells viral inclusion.[75]

Cytokine storm has been reported in relation to the immunopathology of acute kidney injury. In fact, it has been speculated that this is an underlying mechanism of the clinical "viral sepsis" and multiple-organ dysfunction.[76,84]

Kidney organoids established from human embryonic stem cells or from induced pluripotent stem cells into 3D showing tubular structures, expressing different cell types (tubular-like cells and podicytes) and markers and the solute carrier SCL3A1 gene together with SCL27A2 and SCL5A12, have been used to study mechanistic effects such as the prediction about the effect of human recombinant soluble ACE2.[20,86]

Microfluidic devices are also an attractive option for COVID-19 disease modeling of kidney injury, offering the opportunity to utilize human cells, control experimental and environmental conditions, and combine with other on-chip devices. Kidney "spheroids", spontaneously organized cells in 3D that mimic tubule polarization and organization, have been in wide use since the 1980s, but are not amenable to shear stress conditions within the spheroid interior. To ensure shear flow conditions in the tubule "lumen" and maintain cells in a monolayer for simple visualization, the most common design for a kidney-mimicking microfluidic device is a sandwich of two chips of a non-porous material with a porous membrane between them. Instead of synthetic porous membranes, protein-derived hydrogels made of collagen or basement membrane extract allow for the study of extracellular matrix remodeling and 3D migration, both of which occur during fibrosis.[142]

For cell and tissue-based models, developed to model the urinary target organ system and the kidney, a challenge is to recapitulate blood clot or renin-angiotensin system, which is an important complex cascade of pathways that results in SARS-CoV-2 infection.

16.4.2.5 Nervous Organ System Dynamics

Dynamics range from directly infecting neurons (*e.g.* olfactory neuroepithelium) to underlying inflammatory immune responses and oxygen deprivation.

New-approach cell- and tissue-based methods are contributing to understanding the root cause of the neurological problems and giving insights into the understanding of symptoms and their temporary or permanent features such as cognitive impairments (*e.g.* ICU delirium).

SARS-CoV-2 may access the central nervous system *via* the neuropethelium, lamina ribrosa, and olfactory bulb or *via* retrograde axonal transport.

Sustentacular cells of the neuroepithelium display the highest expression of ACE2. This may account for the symptoms of altered or loss of sense of taste (hypogeusia: reduced ability to taste sweet, sour, bitter, or salty things; ageusia: loss of sense of taste) or smell (hyposmia: reduced ability to smell; anosmia: loss of sense of smell) frequently reported retrospectively in COVID-19 patients.

Other neurological manifestations support, at least, the neuroinvasion of COVID-19, perhaps by a hematogenous way reflecting the pro-inflammatory and pro-thrombotic cascade in the wake of cytokine storm as it affects brain vasculature and the BBB.[118]

The infection of the central nervous system and SARS-CoV-2 neuroinvasion, neuroinflammation, and BBB dysfunction may be implicated in the development of the observed disorders. Basic research is critical to understand the detailed mechanisms and pathway of infectivity underlying central nervous system pathogenesis. In order to understand why some patients develop such symptoms and others do not, mechanistic studies using advanced cell and tissue culture methods such as state-of-the-art 3D human induced pluripotent organotypic stem cell cultures, can give more insights into the infection dynamics and give more details on the onset of encephalitis and other neurological disorders.[17,143]

16.4.2.6 Lymphatic/Hematopoietic Organ System Symptoms and Dynamics

Regarding the dynamics, the inaccessibility of living bone marrow (BM) hampers the study of its pathophysiology under myelotoxic stress induced by SARS-CoV-2. Novel technologies like organ-on-a-chip systems offers the possibility to study such dynamic effects. Vascularized human BM-on-a-chip (BM chip) that support the differentiation and maturation of multiple blood cell lineages, facilitate CD34+ cell maintenance and recapitulate aspects of BM injury, are a novel elegant tool to use to study SARS-CoV-2 dynamics. The chips comprise a fluidic channel filled with a fibrin gel in which CD34+ cells and BM-derived stromal cells are co-cultured, a parallel channel lined by human vascular endothelium and perfused with culture medium, and a porous membrane separating the two channels.[21,144]

Induced pluripotent stem cells, differentiated to endothelial lineage and cultured in matrigel–collagen gel in 96-well plates to form human capillary organoids, were successfully infected with SARS-CoV-2. Matrigel is a gelatinous protein mixture derived from mouse tumor cells and is commonly used as a basement membrane matrix for stem cells because it maintains the stem cells in an undifferentiated state. The *in vitro* 3D model uses clinical-grade soluble human ACE2 and simulates human vascular capillary growth with a lumen, platelet-derived growth factor receptors (PDGFR+) pericyte coverage CD31+ endothelial lining, and formation of a basal membrane and models the early stages of infection.[20]

To avoid batch-to-batch variations of the matrigel-collagen gel with attendant lack of reproducibility, biologically relevant and chemically defined matrigel alternatives are being made available by several commercial suppliers, tailored for different cell types.[14]

16.4.2.7 Muscular Organ System Dynamics

For human skeletal muscle tissue, numerous cell types express TMPRSS2, including vascular cells such as endothelial cells, smooth muscle cells, pericytes, muscle stem cells (satellite cells), macrophages, adaptive immune cells

(B-, T- or natural killer cells), and myonuclei (muscle fibers). However, only smooth muscle cells and pericytes express ACE2 critical for the dynamics of muscular-skeletal manifestations.[89]

In COVID-19 patients, widespread muscle fiber atrophy was noted, with sporadic and focal muscle fiber necrosis and immune cell infiltration. Neuronal demyelination has also been reported, which may also contribute to muscle weakness and fatigue. In addition to potential direct viral infection, the cytokines and pro-inflammatory signaling molecules induced by the infection can lead to pathological changes in skeletal muscle tissue.[89] A subset of those pro-inflammatory cytokines is suggested to stimulate muscle atrophy and weakness during critical illness. Of these, three cytokines, tumor necrosis factor (TNF)-α, IL-1 and IL-6 are the most investigated in critically ill patients. *In vitro* studies have shown that incubation of cultured myotubes with clinically relevant doses of TNF-α caused progressive reduction in myotube diameter and total muscle protein content.[90]

16.4.2.8 Integumentary Organ System Dynamics

In vitro vascularized human skin equivalents as a novel *in vitro* model of skin fibrosis can be an avenue to explore to study SARS-CoV-2 chilblain-like lesions.[145] There is growing evidence that adipose tissue plays a key role in the aggravation of COVID-19 where adipocytes act as a reservoir for SARS-CoV-2 and increase viral load in obese or overweight individuals. Furthermore, release of substances that boost the inflammatory reaction by adipocytes into the bloodstream has been described. This knowledge is important to consider when obesity is at play as the lipid profile is already disrupted, which may lead to increased susceptibility to infection which if occurs will further alter the lipid profile inducing hyper-inflammation.[146,147]

16.4.2.9 Reproductive Organ System Dynamics

SARS-CoV-2 may infect the ovary, uterus, vagina and placenta through the ubiquitous expression of ACE2 important in reproductive dynamic processes.[5]

- ACE2 is most abundantly expressed in the ovary and also the expression level of *ACE2* in oocytes is relatively high. Therefore, the ovary and oocyte might be potential targets of SARS-CoV-2. Expression is also observed in uterus, vagina and placenta.[5]
- *ACE2* mRNA has been identified in the uterus of human after analyzing the data from the Human Protein Atlas[148,149] and GeneCards.[5]
- The Human Protein Atlas and GeneCards database show the presence of *ACE2* in female breasts.[5]
- Ang II, ACE2 and Ang-(1–7) regulate follicle development and ovulation, modulate luteal angiogenesis and degeneration, and also influence the regular changes in endometrial tissue and embryo development. Taking these functions into account, SARS-CoV-2 may disturb female reproductive functions through its interaction with ACE2.[5]

Male studies showed mapping of ACE2 in human testicular tissue using single-cell transcriptome resolution with ACE2 protein expression localized in spermatogonia, Leydig, and Sertoli cells. The testis has been found to have the highest level of ACE2 expression compared to other tissue.[150]

Increased blood viral load (as seen in SARS-CoV infection) can increase the vulnerability and risk of male gonads to viral infection and predominantly affect testicular functions. ACE2 can be utilized as a potent receptor by the virus to penetrate the testes. The rapid replication of the virus after entry may promote pyroptosis of the immune cells and apoptosis of the endothelial cells in the testes with subsequent release of inflammatory biomolecules including cytokines, chemokines and adhesion molecules. Also, inflammatory cytokines (TNF-α, IL-1β) may promote oxidative stress in the Sertoli cells and compromise blood–testis barrier integrity. A significant increase in cellular inflammation has been well correlated with cell death and multiple organ damage. Local or systemic infection can cause orchitis (an inflammatory lesion of the testicles) due to the hematogenous propagation of the pathogens. Orchitis has been reported to be a risk factor for male infertility, reported in 15% of infected or recovered patients undergoing fertility tests.[98]

Testicular autopsy of patients who died of SARS-CoV infection (a virus sharing similar pathogenicity with COVID-19 disease) revealed that there was inflammation of one or both testes resulting from the infection with substantial destruction and death of spermatogenic cells. Also, there is a significant reduction in sperm cells present in the germinal epithelium, macrophages infiltration into the tubular lumen and thickness of the basement membrane. Yang *et al.* have discussed pathological findings in the testes of COVID-19 patients and thcir clinical implications.[99]

A multitude of *in vitro* and *ex vivo* models to study reproduction have been made available and can be applied to the real needs of mechanistic understanding of SARS-CoV-2 on male and female reproduction.[151]

16.4.2.10 Skeletal Organ System Dynamics

COVID-19 dynamic processes may affect the skeletal system through multiple mechanisms[101] (Figure 16.7):

(1) ACE2 has been described as a critical player to induce bone resorption. ACE2 and MasR are expressed by osteoblasts and osteoclasts. The activation of ACE2/Ang-(1–7)/MasR axis can impact the bone resorptive milieu by inhibiting the expression of receptor activator of nuclear factor (NF)-κB ligand (RANKL)3. When COVID-19 targets ACE2, the ACE2 expression is downregulated, impacting the ACE2/Ang-(1–7)/MasR cascade, resulting in unbalanced bone homeostasis and acceleration of both osteoclastogenesis and osteoblastogenesis.

(2) High levels of inflammatory cytokines, such as IL-1β, IL-6, TNF-α, granulocyte colony-stimulating factor, Interferon gamma inducible

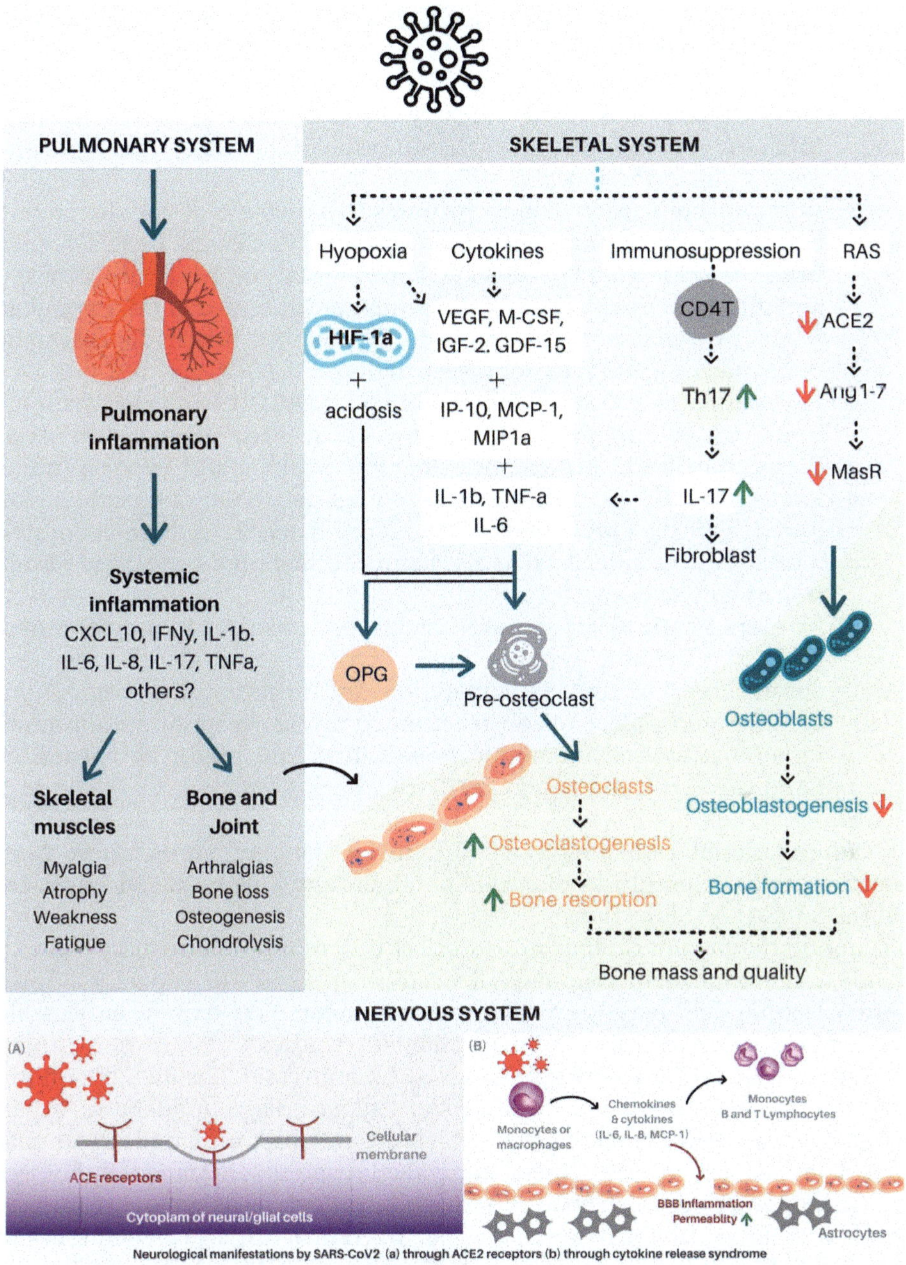

Figure 16.7 Potential immunopathogenesis in SARS-CoV-2 infection depicted for pulmonary, skeletal and nervous systems resulting from the cytokine cascade.

protein-10 (IP-10), macrophage chemotactic protein-1 (MCP-1), macrophage inflammatory protein-1α (MIP-1α) may play a role in different aspects of the osteoclastogenic pathway from osteoclast formation to enhance bone resorptive capacity *via* RANK/RANKL/OPG signaling. Inflammatory cytokines may recruit osteoclasts, trigger bone loss and resist bone formation.

(3) Immunosuppression also contributes to bone destruction. Reduction of lymphocytes in peripheral blood, mainly T-cells and B-cells, is seen in COVID-19 infected patients. In such a situation, T-cells are activated and differentiate into Th17 cells, which produce IL-17 and upregulate RANKL to facilitate osteoclast differentiation, which degrades the bone matrix, leading to bone destruction.

(4) Hypoxemia is a common symptom in COVID-19 infected patients needing high-level oxygen support with ICU care. Hypoxia stimulates the production of pro-osteoclastogenic cytokine and hypoxia inducible factor (HIF-1α) inhibits osteoprotegerin (OPG). The acidosis followed hypoxia upregulates RANKL and nuclear factor of activated T-cells cytoplasmic 1 (NFATc1), acting to facilitate osteoclast formation and bone destruction.

(5) Extra-traumatic osteonecrosis (sickle cell disease, hemoglobinopathies, autoimmune systemic lupus erythematosus, Gaucher disease, chronic renal failure, HIV infection and intravascular coagulation) are common aspects of obstruction of vascular supply, autoimmune inflammatory and metabolic impairment that might be caused by SARS-CoV-2 infection and COVID-19.[152]

Computational modelling and biochemical signaling studies have been used to predict musculoskeletal cellular targets and long-term consequences of the SARS-CoV-2 infection.[89]

Among the *in vitro* models of osteoclast differentiation, primary cell cultures and monocyte/macrophage-like cell linages with the capacity to form osteoclast-like cells have been developed. Engineered bone tissue is available and is being used for SARS-CoV-2 mechanistic studies.[21,153]

Several cells in the synovium express ACE2 and TMPRSS2, including fibroblasts, monocytes, B-cells and T-cells. For articular cartilage, proliferative, hypertrophic and effector chondrocytes (which are a subset of chondrocytes that appear to have a high level of metabolic activity) express ACE2, and only homeostatic chondrocytes (which appear to control circadian clock rhythm in cartilage) express TMPRSS2. In the meniscus, a small fraction of cartilage progenitors and regulatory fibrochondrocytes express ACE2, with no TMPRSS2 detected.[89]

16.4.2.11 Endocrine Organ System Dynamics

Pancreas-related dynamics are affected by the fact that pancreatic cells highly express ACE2, so a direct cytopathic effect of SARS-CoV-2 may appear[154] and an indirect systemic inflammatory and immune-mediated cellular responses results in organ damage.

ACE2 and TMPRSS2 expression levels are higher in thyroid glands than in lungs, therefore it is expected to affect thyroid related infection dynamics.[155] Abnormal immune responses and cytokine storm associated with COVID-19 may induce thyroid gland inflammation. Two mechanisms (*i.e.* indirect and direct) might account for the changes in the thyroid gland and hypothalamic–pituitary–thyroid axis (HPT axis).[89,107,156–158]

16.4.3 SARS-CoV-2 Cellular Mechanisms

SARS-CoV-2 enter cells *via* the ACE2 receptor, a critical actor itself in RAS-driven cellular modulation using the serine protease TMPRSS2 (transmembrane protease, serine 2). Following receptor binding, proteolytic cleavage of the viral S protein by TMPRRS2 exposes a fusion peptide signal that permits mixing of viral and human membranes and release of viral RNA into the cellular cytoplasm. Once the viral RNA has access to the host cell cytoplasm (Figure 16.2), translation of viral proteins and replication of viral RNA can occur, ultimately leading to the assembly of virions. Once inside the host cell, the viral replicase protein complexes are translated and assembled, the viral RNA is synthetized, replicated and subsequently encapsulated, resulting in the formation of the mature virus. Following assembly, virions are carried to the cell membrane into intracellular vesicles and released by exocytosis.

Several proteins encoded by SARS-CoV-2 viral RNA can also interact with various human cellular proteins to disrupt their function. Among the human proteins, processes, pathways and subcellular components predicted to be targeted by SARS-CoV-2 proteins are those involved with the following.

1. Intracellular vesicle trafficking: SARS-CoV-2 exploits most of the cell machinery aimed at producing and releasing extracellular vesicles, which in turn seem to be crucial components in the pathogenesis of virus infection.[159]
2. Ubiquitin ligases: the N-terminal of SARS-CoV-2 spike protein contains a PPxY L-domain motif that is known to hijack host cell WW-domain of Nedd4 E3 ubiquitin ligases and ultimately the endosomal sorting complex required for transport (ESCRT) complex to enhance virus budding and spread. Importantly, this motif is absent in SARS-CoV, and could explain why SARS-CoV-2 is more contagious than SARS-CoV, and can lead to severe and/or long-term systemic effects on pulmonary and extra-pulmonary organs.[160]
3. Nuclear transport: to successfully establish infection, SARS-CoV-2 overcomes the IFN-mediated antiviral response using viral accessory protein Orf6, capable of inhibiting STAT1 nuclear translocation to block IFN signaling in the cells of body organ systems.[161]
4. Cytoskeletal stability: the cytoskeleton is an intricate network involved in controlling cell shape, cargo transport, signal transduction and cell division. Dynamic interactions of SARS-CoV-2 between actin filaments, microtubules and intermediate filaments of the host cell plays an essential role for cytoskeleton in mediating the outcome of host–virus

interactions. Disruption of host cytoskeleton homeostasis and modification state is tightly connected to a variety of organ systems cellular pathological processes, such as defective cytokinesis, demyelinating, cilia loss and neuron necrosis.[161]

5. Mitochondrial respiration: the cellular innate immune system of a SARS-CoV-2 infected host target organ cell fighting the viral infection strongly relies on its membrane-bound cell organelles (mitochondrion, singular) that generate most of the chemical energy, adenosine triphosphate (ATP) needed to power the host cell's biochemical reactions. SARS-CoV-2 can evade host immune defense mechanisms by hijacking mitochondrial pathways, as recently suggested and the dysfunctional mitochondria could contribute to severe disease progression.[162]
6. Inflammatory signaling: the cytokines and signaling molecules, induced by SARS-CoV-2 infection, include C-X-C motif chemokine 10 (CXCL10), IFN-γ, IL-1β, IL-6, IL-8, IL-17 and TNF-α.[89]

16.4.4 Immunity Dynamics Following SARS-CoV-2 Infection

Immunity to viral infection is caused by a variety of specific and non-specific mechanisms. The activation of different immune functions and the duration and magnitude of the immune response depend on how SARS-CoV-2 interacts with host cells. SARS-CoV-2 can cause cytolytic, steady-state, latent and/or integrated infection depending on the cell type and the related cellular viral load. Organ system damage is also linked to how the virus spreads (by local, primary hematogenous, secondary hematogenous and/or nervous system spread). Therefore, viral antigens may be present in different parts of the body depending on the route of spread and phase of infection. Local infections at surfaces such as the mucosa can elicit humoral (IgA) and local cell-mediated immune responses, but not necessarily systemic immunity. The host has multiple immune defense functions that can eliminate virus and/or viral disease.

Humoral immunity occurs when the SARS-CoV-2 virus and/or virus-infected cells can stimulate B-lymphocytes to produce antibody (specific for viral antigens). Antibody neutralization is most effective when the virus is present in large fluid spaces (*e.g.* serum) or on moist surfaces (*e.g.* the gastrointestinal and respiratory tracts). IgG, IgM and IgA have all been shown to exert antiviral activity. Antibody can neutralize virus by (1) blocking virus–host cell interactions or (2) recognizing viral antigens on virus-infected cells, which can lead to antibody-dependent cytotoxic cells or complement-mediated lysis. IgG antibodies are responsible for most antiviral activity in serum, while IgA is the most important antibody when viruses infect mucosal surfaces.

The term cell-mediated immunity refers to (1) the recognition and/or killing of virus and virus-infected cells by leukocytes and (2) the production of different soluble factors (cytokines) by these cells when stimulated by virus or virus-infected cells. Cytotoxic T-lymphocytes, natural killer cells and antiviral macrophages can recognize and kill virus-infected cells. Helper T-cells can recognize virus-infected cells and produce a number of important cytokines. Cytokines produced by monocytes (monokines), T-cells and natural killer cells (lymphokines) play important roles in regulating immune functions and developing antiviral immune functions. A number of cytokines, including IL-6, IL-1β, TNF-α and IFN-γ, have been frequently reported to be elevated in COVID-19. A putative systemic outcome due to this effect is known as cytokine release syndrome, also called "cytokine storm" (Figure 16.7).

16.5 Risk Factors

Numerous studies, including reviews of existing PubMed literature and retrospective cohort studies show that certain genetic, cellular and organ-level phenotype and factors lead to an increased risk for both progression of the disease and increased mortality (Table 16.2). Identification of these factors

Table 16.2 Genetic, molecular, cellular, tissue and organism cell-based methodologies that have been used to identify and validate SARS-CoV-2 risk factors.

Genetic	Cellular	Tissue/organ	Organism
HLA haplotype[164]	ACE2 levels[167,168]	Pulmonary fibrosis[182]	Higher age
GA variant at 3p21.31 locus[163]	TMRPSS2 levels[169]		Chronic lung disease
ABO blood group[163]	Coagulopathy[170]	Myocardial infarction[183]	Heart disease
ACE2 polymorphism[165]	Pro-NET-otic state[172]	Kidney injury[184]	Diabetes
TMRPSS2 polymorphism[165]	Presence of D-dimers[170]	Arrhythmia[185]	Severe obesity (BMI >40 kg m^{-2})
TLR polymorphism[165]	Inflammation[173]	Cytokine-mediated brain damage	Kidney/liver diseases
	Apoptosis[180]		
CCR5 polymorphism[176]	IL-6, CRP levels[171]	Sepsis[186]	
IL6 polymorphism[177]	Low mitochondrial fitness[175]	Necrosis of adrenal gland	
1. effect of various cytokine SNPs on miRNAs (*in silico* work)[178]	SARS-CoV-2 viral loads[181]		
2. the heme oxygenase[179]			
Methodologies used			
Genome-wide association studies	Single-cell RNA sequencing	ECG	
Whole-exome sequencing	Biomarkers for cellular injury, cell death in serum	MRI	
ANNOVAR (ANNOtate VARiation)	D-dimer assay	CT scan	
	Cryo-Electron microscopy		
	Histopathology	Autopsy	
	Immunohistochemistry	Radiography	
	GO analysis (up-/down-regulation of genes)	Meta-analysis and correlation studies	

is important to predict the severity of disease and to make policy decisions regarding restrictions of movement and contact prohibition.

In this section, we review risk factors detected at various levels of biological organization, such as genetic, molecular, cellular, tissue and various *in silico* and cell-based methodologies that have been used to identify and validate those risk factors. The analysis of methodologies will also help in designing *in vitro* tests to identify the risks in various human populations.

16.5.1 Genetic Factors

Genome-wide association studies for severe COVID-19 have been conducted indicating potential molecular and genetic mechanisms involved in the disease. Associations between risk of severe COVID-19, a multigene locus at 3p21.31 and the ABO blood group locus at 9q34.2 has been reported. Patients with blood group A were reported to have an increased risk of severe COVID-19, and those with blood group O have been reported to have a decreased risk. The frequency of the rs11385942 insertion–deletion GA or G variant at locus 3p21.31 has been reported more in patients who received mechanical ventilation, indicating that this risk allele is associated with more severe forms of COVID-19. There are six different candidate genes at 3p21.31 locus, and LZFL1, and the proteins involved in protein trafficking to primary cilia, which have been highlighted as targets to consider.[163]

Regional distribution of frequent human leukocyte antigen (HLA) haplotypes has been studied.[164] However, to understand whether HLA haplotypes could be associated with disease susceptibility, global networks with patient information from varying geographic areas need to be built.

The entry of SARS-CoV-2 into the cells depends for the major part on the host factor ACE2 and the transmembrane serine protease TMPRSS2 for SARS-CoV-2 spike (S) protein priming. ACE2 is encoded on the X-chromosome and is involved in the catalysis of angiotensin II to angiotensin-(1–7) which is a vasodilator. It has been reported that unique DNA polymorphisms exists in ACE2 and TMPRSS2 in different populations which might lead to variations in genetic susceptibility towards SARS-CoV-2. For example, certain variants have been found to inhibit the interaction between ACE2 and spike protein.[165]

Toll-like receptors (TLR) are a family of immune sensor proteins and have a crucial role in the cytokine expression in the immune system. In addition, TLR genes show distinct population distribution pattern and are also subject to selection pressure and could thus play a key role in the differential susceptibility to SARS-CoV-2 infection.[165] It has been shown that patients with severe form of COVID-19 were associated with a rare putative loss-of-function variant of X-chromosomal TLR7.[166]

While the analysis of gene polymorphisms is mainly carried out by whole-exome sequencing, many of the annotation and association studies were carried out using genome wide association studies and ANNOVAR, a bioinformatics tool for functional annotation of genetic variants.

16.5.2 Cellular Factors

Apart from polymorphisms, a genome-wide association study showed that the expression levels of ACE2 itself is significantly associated with a higher risk for severe COVID-19 infection.[167,168]

Male sex, cardiovascular disease and diabetes, all of which have been reported to be risk factors for severe COVID-19, were also associated with higher levels of ACE2. Similarly, the levels of TMPRSS2 have also been proposed as a biomarker for SARS-CoV-2.[169]

A coagulopathy has been reported in up to 50% of patients with severe COVID-19 manifestations. An increase in D-dimer is the most significant change in coagulation parameters in severe COVID-19 patients, and progressively increasing values can be used as a prognostic parameter indicating a worse outcome.[170]

In a study conducted to develop a model which could predict the risk to a fatal outcome in COVID-19 patients, a prospective study was conducted and the study found that high IL-6 level, C-reactive protein (CRP) level, lactate dehydrogenase level, ferritin level, D-dimer level, neutrophil count and neutrophil-to-lymphocyte ratio were all predictive of mortality in COVID-19.[171]

Neutrophil extracellular traps (NETs) are extracellular webs of chromatin, microbicidal proteins, and oxidant enzymes released by neutrophils in response to infections. A study found correlation between high serum levels of NET and hospitalization of COVID patients.[172] However, further studies are required to estimate the predictive power and risk associated with circulating NETs.

Systemic inflammation is often associated with development of atherosclerosis, type 2 diabetes, and hypertension, which are well-known comorbidities that can increase the risk for adverse outcomes associated with COVID-19.[173]

Studies show that SARS-CoV-2 infection pathogenesis is related to oxidative stress in patients in response to infection.[174] As mitochondria are the hub of cellular oxidative homeostasis, the increased inflammatory/oxidative state during COVID-19 infection may lead to mitochondrial dysfunction further causing platelet damage and apoptosis. The interaction of dysfunctional platelets with coagulation cascades can also aggravate clotting and thrombus formation. With the critical role of mitochondria in inflammation and oxidative stress, low mitochondrial fitness has also been suggested as a risk factor for severe COVID-19.[175]

A study found that SARS-CoV-2 plasma viremia is commonly detected in hospitalized individuals and symptomatic non-hospitalized outpatients diagnosed with COVID-19. In addition, they found that SARS-CoV-2 viral loads, especially within plasma, are associated with systemic inflammation, disease progression, and increased risk of death.

The expression levels of cellular factors were determined using single cell RNA sequencing, and distribution patterns were visualized using cryo-electron microscopy and histopathology. In many cases, cellular events, such as death, injury, or presence of D-dimers were assessed by detecting biomarkers for specific events.

16.5.3 Tissue/Organ Level

Pulmonary fibrosis is associated with permanent changes in lung tissue architecture and function, and it has been shown to complicate SARS-CoV-2 infection. Myocardial injury, visualized using biomarkers, is another risk factor which can exaggerate the clinical course and worsen the outcomes of COVID-19 infection.

Using multivariate analysis, a study showed association of acute kidney injury with not just disease severity, but also lymphopenia and D-dimer levels.

While bacterial infections are often the main cause of sepsis, viral sepsis, again monitored using biomarkers of sepsis, is often ignored and was shown to be a risk factor associated with COVID-19 mortality.

16.5.4 Organism Level

Table 16.2 includes the risk factors that have been reported by the Centers for Disease Control and Prevention. The factors that are commonly mentioned include higher age, male gender, obesity, diabetes, and diseases associated with liver, heart, renal, and respiratory system. Higher age is one of the significant and consistent risk factors, and this could be due to declining immune function at older age, increased likelihood of comorbidities or the associated medications.[187]

While ethnicity has been reported as one of the risk factors, currently not enough evidence of the impact of ethnicity on the risk associated with acquiring the infection and progression to worse clinical outcomes has been demonstrated.[188]

Both organ and organism level risk information can be obtained using whole organ-level structure and function abnormalities can be captured using electrocardiogram (ECG), computed tomography (CT) scans, magnetic resonance imaging (MRI) or histopathology of autopsy samples in some cases.

16.6 Vulnerabilities

Epidemics and pandemics, such as COVID-19, often disproportionately affect vulnerable populations. Thus, there is a need to identify the populations that are either vulnerable to increased risk of infection or inferior health outcomes, as this information can help in informing where additional resources are most needed. In this section, we discuss the vulnerabilities towards severe prognosis of COVID-19 in various populations, such as malnourished, pregnant and immunocompromised populations (Table 16.3).

Table 16.3 Cell- and tissue-based strategies used to identify or visualise the cellular or molecular effects associated with COVID-19 vulnerability.

Vulnerable population[a]	Cellular/tissue-level effects	Methodologies used to visualize cellular/molecular effects
Malnourished[189]	Immunosuppression Lymphopenia	
Population with high BMI[190]	Chronic inflammation Increased cytotoxic T-cells Increased effector T-cells	For B- and T-cell numbers: Biomarkers for inflammation LysiSpot assay[191] IFN-γ ELISpot[191] Flow cytometry Cytotoxic T-lymphocyte (CTL) activity assay[192] (CTLp) frequency assessment[192] Tetramers[193] Immunoscope[193] OT T-cell system[192]
Old age[194]	Limited capacity of the thymus to generate "naive" T-cells Reduced ability to regulate production of macrophages and neutrophils Immunosuppression (reduced number of B- and T-cells)	For macrophage number and activity: Quantitative phase imaging and Raman spectroscopy[195]
AIDS patients	Immune deficiency (progressive loss of CD4+ T-cells)	For inflammation: Inflammatory biomarkers
Patients on immuno-suppressants (such as corticosteroids)	Dose-dependent effects on: Reduced number of circulating B- and T-cells Suppression of lymphocyte activation and in the production of antibodies by B-cells	Placental breach: Single-cell RNA sequencing[196]
Pregnancy[196,197]	Anti-inflammatory Immunosuppression Placental breach	
Diabetes[198,199]	Reduced forced vital capacity Reduced forced expiration volume Inflammation	
Hypertension	Increased troponin[200] Cardiac injury[201] Fibrosis[201]	Tissue injury/fibrosis: Cardiovascular magnetic resonance[202] CT scan

[a]Known risk factors for severe COVID-19.

We also discuss the cell-based strategies used to identify or visualize the cellular or molecular effects associated with COVID-19 vulnerability.

16.6.1 Malnutrition

There is increasing evidence that suggests that COVID-19 may be associated with negative outcome in hypoalbuminergic patients and patients with high body mass index (BMI). Interestingly, malnutrition is associated with immunosuppression, while overnutrition is associated with chronic inflammation, both of which are known risk factors of severe COVID-19.

16.6.2 Immunocompromised Conditions

Patients suffering from HIV infection, those on immunosuppressant drugs or pregnant women are in an immunodeficient state which could lead to inabilities in establishing immune response towards external infections. However, there is still insufficient evidence to suggest whether reduced immune function is linked to more severe outcomes in COVID-19. The cellular responses in these conditions include suppression of circulating B- and T-cells and specifically CD4+ T-cells in cases of HIV infection. Studies are looking at the proteomic and the transcriptomic profiles using RNA sequencing from lung samples of fatal COVID-19 cases and the authors suggest that patients' deaths may be due to uncontrolled host inflammatory processes.[203]

It has been reported that immunosuppressive therapies are not associated with increased risk for COVID-19.[204]

Using single-cell RNA-sequencing data, placental cells were found to express ACE2 and S protein priming protease TMPRSS2 during first and second trimester, indicating that placenta is permissive to SARS-CoV-2. Additionally, the placental cells also had the mRNA for proteins that physically interact with SARS-CoV-2 in host cells.[196]

16.6.3 Presence of Comorbidities

Diabetes has been associated with reduced lung capacity, including forced vital capacity and forced expiratory volume in 1 second, which could increase the risk of severe COVID-19.[198]

There is also increasing evidence that suggests that diabetes is an inflammatory disease, where type 1 diabetes is characterized by autoimmune-mediated destruction of pancreatic β-cells.[205] Thus, existing inflammation in diabetes (detected using biomarkers for inflammation) may increase the risk of severe COVID-19.

Hypertension results in a number of pathophysiological changes in the cardiovascular system, including left ventricular hypertrophy and fibrosis.[201] These symptoms are also known risk factors for severe COVID-19, making the hypertensive heart particularly susceptible to SARS-CoV-2.

16.7 Non-conventional or Supportive Treatments or Approaches Based on Underlying Cellular Mechanisms

In this section, we list various molecular/cellular signaling pathways that are affected during COVID-19 infection, and treatment strategies that are based on effecting these pathways. We also discuss various *in silico*, cellular strategies and *in vitro* model systems that are being used to understand and assess the therapeutic potential of various therapeutic candidates.

The interruption of any stage of the viral life cycle can become a significant therapeutic approach for treating coronavirus-related diseases (Table 16.4).

A recent SARS-CoV-2-human protein–protein interaction analysis showed that SARS-CoV-2 contains many druggable proteins, each of which has several ligand binding sites. The most interesting coronavirus proteins are the S glycoprotein, proteases main protease (Mpro) and papain-like protease (PLpro), RNA dependent RNA polymerase (RdRP) and helicase. Broad-spectrum antivirals targeting the viruses and new alternative methods to identify damaged target organ systems are also used for identifying adverse target organ system features of the COVID-19 disease (Table 16.5).[6,206]

Based on the molecular, cellular and tissue-level effects observed during COVID-19 infection, various drug targets are being investigated that affect one or more of these pathways.

Table 16.4 Molecular/cellular signalling pathways that are affected during COVID-19 infection, and treatment strategies that are based on affecting these pathways.

Molecular/signalling effects	Treatment strategies
Binding of spike protein (S-RBD) and ACE2 host cell receptor	Decoy ACE2 receptors for virus attachment Blocking SARS-CoV-2 replication
Spike protein (S) priming by TMPRSS2	Blocking certain Serine proteases
Conversion of Ang I to Ang II by ACE2	ACE2 inhibitors
Activation of RAS signalling	RAS inhibitors
Activation of JAK/STAT pathway by IL-6	Downregulation of STAT3
Activation of MAPK pathway	MAPK inhibitors
Activation of NF-κB	Inhibition of NF-κB
Cellular/tissue effects	Treatment strategies
Pro-inflammatory cytokines (IL-6)	IL-6 receptor antagonists[207]
Venous thromboembolism	Low molecular weight heparin
Pulmonary embolism	
Hypercoagulability	
Pulmonary fibrosis	Antifibrotics (nintedanib)
Organ injury	Systemic steroids
Hypoxemia	Bronchodilators, mechanical ventilation

Table 16.5 Mechanistic methods used to study effect of interfering with stages of the SARS-CoV-2 viral life cycle at the molecular, cellular and tissue level as treatment strategies.

In silico	Cellular assays (to check drug mechanisms and efficacy)	Cell lines	Stem cells	Organoids	Organ-on-chip
Molecular docking and simulation	Cytokine biomarkers	Human airway epithelial cells	Mesenchymal stem cells	Lung organoid model	Human airway chip
Network pharmacology	Furin cleavage assay	*In vitro* SARS-CoV-2 replication inhibition assay using VeroE6 and Caco-2 cells[208]		Colon organoid model	
Pharmacokinetic–pharmacodynamic models	ELISpot cell-mediated immunity				
Simulating viral infection dynamics					
Tensor decomposition-based unsupervised feature extraction					

16.7.1 *In Silico* Strategies

16.7.1.1 *Molecular Docking and Simulation*

Docking studies have been performed over the binding pocket of COVID-19 to find potential small molecules, including approved drugs and compounds undergoing clinical trials to combat SARS-CoV-2.[209]

Docking studies have also been performed to determine the binding energies of compounds/drugs in isolation or in combination with other drugs. For example, a study showed that caulerpin and its derivatives could be used in combination with lopinavir, simeprevir, hydroxychloroquine, chloroquine, and amprenavir to disrupt the stability of SARS-CoV-2 receptor proteins and so increase the antiviral activity of these drugs.[210]

16.7.1.2 *Network Pharmacology*

Network pharmacology and bioinformatics analysis, including the Kyoto Encyclopedia of Genes and Genomes (KEGG) molecular interaction/reaction network diagram pathway, Gene Ontology (GO) enrichment analysis and the Protein–Protein Interactions Network have been evaluated.[211] The GO analysis results of SARS-CoV-2 differentially expressed genes revealed that the most significant terms were type I interferon signaling pathway, cellular response to type I interferon, response to type I interferon, defense response to the virus and mRNA binding involved in post-transcriptional gene silencing. Influenza A, measles, NOD-like receptor signaling pathway, hepatitis C, herpes simplex infection and cytokine–cytokine receptor interaction were the most enriched pathways in KEGG (a database resource for understanding high-level functional and biological systems from large-scale molecular data generated by high-throughput experimental techniques) enrichment analysis. Identification of drug targets using network pharmacology approaches by identifying compounds that show overlapping differentially expressed genes with SARS-CoV-2 is an avenue that is routinely applied.

16.7.1.3 *Pharmacokinetic–Pharmacodynamic Models*

A study by Garcia-Cremades *et al.* developed a mechanistic pharmacokinetic model to predict the SARS-CoV-2 rate of viral decline and QTc prolongation associated with various doses of hydroxychloroquine. The simulation showed that while hydroxychloroquine doses >400 mg predicted the rapid decline of viral loads, hydroxychloroquine doses >600 mg were also predicted to prolong QTc intervals.[212] Thus, the model helped to evaluate the benefit/risk analysis of high doses for a drug or a compound.

16.7.1.4 *Simulating Viral Infection Dynamics*

A study by Wang *et al.* built a mathematical model to study various features of SARS-CoV-2 infection by examining the interaction between virus, cells and immune responses.[213] The simulation showed that in the first stage, the viral load increases rapidly and reaches the peak, followed by a plateau phase, possibly due to lymphocytes as a secondary target of infection.

Then, the viral load decreases due to the emergence of adaptive immune responses. The model also showed that anti-inflammatory treatments or antiviral drugs combined with interferon reduced the duration of the viral plateau phase and the time to recovery. Thus, such models that help in understanding the infection dynamics might also help in designing appropriate treatment strategies.

16.7.1.5 Tensor Decomposition-based Unsupervised Feature Extraction for Drug Discovery

A study by Taguchi and Turki proposed a tensor decomposition-based unsupervised feature extraction method to predict drug candidate compounds without knowledge of known compounds.[214] Tensor decomposition-based unsupervised feature extraction was applied to the gene expression profiles of multiple lung cancer cell lines infected with SARS-CoV-2. Approximately 163 genes were identified as differentially expressed genes in SARS-CoV-2 infection, and drugs were screened based on the coincidence of differentially expressed genes between drug treatments and SARS-CoV-2 infection.

16.7.2 Fit-for-purpose Cellular Assays

Various bioassays have been devised to detect biomarkers, antibodies and therapeutic responses, and neutralization of the virus.

16.7.2.1 Cytokine Biomarkers

The release of pro-inflammatory cytokines post-SARS-CoV-2 infection has been shown to be responsible for the respiratory complications associated with the disease. Thus, cytokine biomarkers can be monitored in patients enrolled in SARS-CoV-2 trials, to understand mechanisms of action and effects of treatments.

16.7.2.2 Furin Cleavage Assay

SARS-CoV-2 uses furin, a protease highly expressed in the lung, to cleave S protein into S1 and S2, which then bind to the ACE2 receptor. Intracellular furin also helps in packaging new viral particles. Certain therapies work by blocking the interaction of S protein and furin or by targeting furin. For this class of therapeutic, furin cleavage assay can be used as a proof of concept/mechanism of action.

16.7.2.3 ELISpot Cell-mediated Immunity

Many therapies are focused on affecting the T-cell cytokine as that is one of the primary drivers of progression to severe COVID-19. T-cell responses can be assessed using enzyme-linked immune absorbent spot (ELISpot) assays.

This *in vitro* assay can screen responses in a cost-effective manner to an entire pathogen proteome and estimate memory response in patients.

16.7.3 Cell Lines

16.7.3.1 Human Airway Epithelial Cells

Human airway epithelial cells highly express ACE2 and TMPRSS2, the receptor that the virus uses to prime the S protein (spike protein of SARS-CoV-2). These cells also show cytopathic effects 96 hours after infection. Drug validation studies were conducted on differentiated ALI cultures of proximal airway epithelium.[215] ALI medium containing drugs was added to the cultures 3 hours prior to infection, then, 100 µL of SARS-CoV-2 viral inoculum was added, and after 2 hours of viral adsorption, cells were washed and fresh medium containing drugs was added. Remdesivir strongly suppressed viral infection/replication, indicating the potential relevance of this platform.

16.7.3.2 Vero-E6, Caco-2 Cell Lines

Vero-E6 cells, the most widely used cell lines to replicate and isolate SARS-CoV-2, are isolated from kidney epithelial cells of the African green monkey. Recently, the antiviral activity of 1520 approved drugs from Prestwick Chemical Libraries were tested on Vero-E6 and Caco-2 cells.[208] Of them, 11 compounds, including macrolide antibiotics, proton-pump inhibitors, anti-arrhythmic agents or central nervous system drugs showed antiviral potency with $2 < EC_{50} \leq 20$ µM.

16.7.4 Mesenchymal Stem Cells

The role of mesenchymal stem cells (MSCs) in promoting regeneration and suppressing inflammation at sites of injury have been well recognized. Many clinical trials have explored the role of MSCs to reduce inflammation in patients with osteoarthritis, multiple sclerosis, type 1 diabetes mellitus and influenza virus-induced lung injury. Two reports from Chinese hospitals document the outcome for patients receiving MSCs for SARS-CoV-2 infection. Both studies showed significant clinical improvement in the test group compared to the control group.[216] MSCs have been shown to have broad immunoregulatory properties through the interaction of immune cells in both innate and adaptive immune systems.[217] This can lead to immunosuppression of many effector activities, and reduction of the cytokine storm, indicating the potential role of MSCs in COVID-19 treatment.

Extracellular vesicles include both exosomes and microvesicles. The diameter of exosomes is <200 nm, while the diameter of microvesicles can reach up to 1000 nm. Exosomes secreted from MSCs have been studied in various model systems and have been shown to reduce edema in human injured lungs in an *ex vivo* lung perfusion model.[218] Thus, exosomes are also emerging as a therapeutic option for treatment of organ injury during COVID-19.[219]

16.7.5 3D Tissue/Organoids

16.7.5.1 Lung Organoid Model

Han *et al.* developed a lung organoid model using human pluripotent stem cells (hPSCs).[220] Various cell types, including alveolar type II (AT2) cells, alveolar type I (AT1) cells, stromal cells, proliferating cells, a low number of pulmonary neuroendocrine cells and airway epithelial cells could be identified in the lung organoid. Pos- SARS-CoV-2 infection, there was a robust production of chemokines and cytokines with little type I/III interferon signaling, which is also present during human COVID-19 infection. A high-throughput screen was performed using hPSC-derived lung organoids which identified United States Food and Drug Administration (FDA)-approved drug candidates, including imatinib and mycophenolic acid, as inhibitors of SARS-CoV-2 entry.

16.7.5.2 Colon Organoid Model

Han *et al.* developed hPSC-derived colonic organoids.[221] These organoids consisted of multiple colonic cell types – especially enterocytes, expressed ACE2 and were permissive to SARS-CoV-2 infection. The authors performed a high-throughput screen of FDA-approved drugs, including imatinib, mycophenolic acid, and quinacrine dihydrochloride. The treatment of these drugs at physiologically relevant levels inhibited the SARS-CoV-2 infection of colonic organoids, indicating these model systems as a valuable resource for drug screening to identify candidate COVID-19 therapeutics.

16.7.6 Organ-on-a-chip

Si *et al.* developed a human airway chip, which is a microfluidic device where primary human lung airway basal stems cells are grown under an ALI, along with the primary human lung endothelium which is grown on the opposite side.[222] The lung epithelium is also exposed to continuous fluid flow. The chip supports differentiation of the lung airway basal stem cells into a mucociliary pseudostratified epithelium and airway-specific cell types, such as ciliated cells, mucus-producing goblet cells, club cells and basal cells. It also showed establishment of continuous ZO1-containing tight junctions and cilia; permeability barrier properties; and mucus production similar to a human airway *in vivo*. This system was used to test inhibitory activities of seven clinically approved drugs, including chloroquine, arbidol, toremifene, clomiphene, amodiaquine, verapamil, and amiodarone. While all seven of these drugs inhibited infection by viral pseudoparticles expressing SARS-CoV-2 spike protein in cell lines (Huh-7 cells, Vero cells), only two of them inhibited infection by viral infection in airway chips, suggesting that they can be used in conjunction with cell-based screening assays to expedite drug repurposing in SARS-CoV-2 infection.

16.7.7 Diamagnetism as an Alternative or Integrating Cellular Therapy for COVID-19

Naturally occurring electric fields are not only important for cell-surface interactions, but are also pivotal for the normal development of the organism and its physiological functions. Selective control of cell function by applying specifically configured, weak, time-varying magnetic fields has added a therapeutic dimension to biology and medicine, *e.g.* diamagnetic therapies. Using diamagnetic therapy, there is a powerful effect on cellular features including specific cell-membrane mechanisms reducing edema and inflammatory processes and the damaged tissue is quickly repaired with an immediate analgesic effect. Field parameters for therapeutic pulsed electromagnetic field (PEMFs) were designed to induce voltages similar to those normally produced during dynamic mechanical deformation of, for example in the human, body connective tissues. It has been applied successfully for nerve regeneration, wound healing, graft behavior, diabetes and myocardial and cerebral ischemia (heart attack and stroke), among other conditions.[223]

The biophysical stimulation induced by low-frequency high-intensity PEMF – diamagnetotherapy – has recently been demonstrated to be effective in the treatment of lung fibrosis associated to autoimmune diseases as well as to fibrosis in post-COVID-19 pneumonia.

In this dysregulation of the immune system, diamagnetotherapy ameliorated the functionality of respiratory muscles in treated patients, improved dyspnea and oxygen saturation, indicative of a better performance of the lung. In these cases, a double effect in lung parenchyma and in respiratory muscles has been evaluated, also considering that in post-intensive care syndrome, patients report muscle weakness, impaired mobility and balance and joint stiffness while the associated neuropathy and myopathy are a consequence of SARS-CoV-2 infection and worsen morbidity and mortality. Diamagnetoteraphy also offers the possibility to treat epilepsy, cognitive impairment, or motor imbalance, including the problems of the peripheral nervous system in COVID-19 patients.

The dynamic of diamagnetism moves liquids and solutes in the ECM and in the intracellular environment thanks to the effects of a high intensity magnetic field. This type of impulse, has be shown to reduce peripheral chronic lymphoedema and, potentially, offers the possibility to treat the lymphatic imbalance consequent to the prolonged COVID-19 related immobility.

High-intensity magnetic fields share the therapeutic and biological effects of conventional low intensity magnetic fields in terms of anti-inflammatory and regenerative effects.[224] Moreover, the high intensity of the magnetic field moves diamagnetic substances such water, ions, proteins and nanoparticles. This results in repulsive molecular movements that are able to increase the normal transmembrane flow of these substances, usually at the origin of the physiological fluctuations of the electrical potential in cells.

In the course of inflammation, variable intensities of the magnetic field inhibit the expression of IL-6, IL-1β and TNF-α at the gene and protein level in injured tissues.[225] This modulatory effect is enforced by the activation of the anti-inflammatory cytokine IL-10,[226] which normally supports the integrity of the ECM through the regulation of the metabolism of proteoglycans.[227] Diamagnetism-based treatments could be applied as alternative or integrating therapy to recover the structural effects of the immune dysregulation induced by COVID-19 at ECM level in various tissues.

The chronobiology of the inflammatory process relates to IL-1, IL-6 and TNF,[228] followed by physiological self-control and cellular self-defense of inflammation given by the response time of IL-10, and transforming growth factor (TGF)-β genetic expression.[229]

In COVID-19 patients, the rapid exaggerated and continuous increase in IL-6 can lead to a chronic activation of IL-17 with subsequent activation of specific multilevel antibodies, which may have autoimmune or degenerative sequelae. However, the physiological response to high levels of IL-6 given by the increase of IL-10 and TGF-β[230] should be taken into account if the over-expression of these last two molecules, and the return to normality of TGF-β activity is slow. The continuous activation of TGF-β can cause changes in the ECM,[231] making it less soluble and evoking an increase in secondary fibrosis in the inflamed areas. In the case of the post-COVID-19 patients, these tissues are identified by ACE receptors and through the stimulation given by chemotaxis and necrotaxis, these tissues change the inflammatory adaptation response of MMPs.[232]

The ECM takes within its sol–gel regulation the management of its remodeling and plasticity homeostasis, cell proliferation, migration and differentiation that are crucial for the cell survival.[233,234] By virtue of its structure and composition, the homeostatic functions of ECM include the storage of growth factors and their release, acting in this way as a filter. According to the local inflammatory state, this regulatory function is controlled by reactive oxygen species (ROS) and pH values which modify the heparan sulphate chains of proteoglycans or activate the MMPs, thus promoting alternating changes in the sol–gel structure of the ECM.[234,235] The sol phase within the ECM has a protease activity (hydrolysis of proteoglycans); the maximum pH in this phase is 7.35, with sympathicotonic activation. The gel phase has anti-protease activity restoring the matrix by means of the synthesis of matrix proteins. The pH peak in this phase is 7.45, so it is an alkaline phase with vagotonic activation.[236]

MMPs are 22 human proteolytic enzymes regulated by Th1 lymphocytes, and therefore carry out degradation effects of the ECM. MMPs are regulated by inhibitory tissue of metalloproteinases (TIMPs), which are four inhibitory proteins of MMPs controlled by Th2 lymphocytes, and therefore carry out protein deposition in the ECM; this inhibition is carried out 1 : 1 from TIMPs to MMPs.[237,238]

ROS overexpression and pH reduction create pathological inflammation like that seen in COVID-19 in the acute and semi-acute phases (sol phase or degradative and inflammatory phase). At the same time, after the acute phase, a low-grade inflammation is related to the expression and the secretion of IL-6, which indices as a regulatory control an increase in the secretion and expression of TGF-β1, creating ECM rigidity due to an increase in the deposits of the proteoglycans giving the fibrotic phenotype.[239]

The rigidity of the ECM modifies the intercellular junctions. In the case of COVID-19, the mucosa represents alterations in the intercellular junctions of lung and receptors of the respiratory and oropharyngeal mucosa. In the processes of traditional inflammatory control, IL-10 activates the STAT3 receptors that decrease the expression of Claudin-2 and increase the expression of Claudin-4. TGF-β activation of the ERK and SMAD2 receptors results in increased expression of Claudin-1, as observed in Crohn´s disease patients that have an alteration of the intestinal mucosa. These patients have been treated with low dose IL and sequential kinetic activation. Low dose IL is a treatment modality to provide interleukins that remain active thanks to a pharmaceutical preparation process known as "sequential kinetic activation" (SKA).[240–243]

An integrative treatment option with diamagnetic therapy is the implementation, penetration or implantation of therapeutic agents directly on the tissues under treatment. Due to the effect of diamagnetic repulsion movement, it is possible to introduce a product up to 8–10 cm deep (Figure 16.8).

Therapeutic agents with the greatest possibility of transport are interleukins, hormones, neuropeptides or growth factors at low dose, which have already shown adequate control of the processes of low-grade chronic

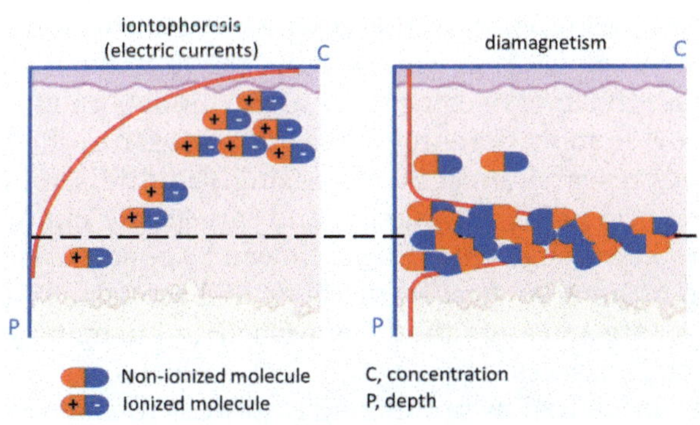

Figure 16.8 Distribution of pharmacological substances as a function of concentration-depth ratio (PC), induced respectively by electric currents (iontophoresis) or high intensity-PEMF. Note the more homogeneous and deeper distribution of a hydro-soluble (diamagnetic) therapeutic agent due to the high-intensity magnetic field. Image credit: Alessia Nuvoloni, ARTSTATION, Civita Castellana, Italy.

inflammation which causes COVID-19 inflammatory, autoimmune or degenerative sequelae observed in patients 2–3 weeks and up to 3–4 months after the acute phase.[244]

16.8 Concluding Remarks

COVID-19 research suggests that there is a need for changing the paradigm that postulates that cell-, tissue- and mathematical-based methods used for COVID-19 research inform animal studies that will subsequently inform human studies. *In vitro* models using 2D and 3D cell- and tissue-based systems and *in silico* methods can be used as powerful tools to accelerate SARS-CoV-2 studies and discoveries. These models are proven to efficiently mimic a variety of subcellular, cellular, tissue and organ physiological aspects of relevance for the understanding of SARS-CoV-2 related risk factors, vulnerabilities and potential therapies.

Due to the extraordinary crisis caused by COVID-19, medical researchers, cell biologists, life science experts, mathematical modelers and bioengineers across the world are actively collaborating to accelerate the rapid development of relevant cell, tissue and mathematical methods and approaches to gain detailed mechanistic knowledge of this new disease. This global collaborative knowledge sharing effort assists clinical approaches and treatments through the unravelling of the mechanistic understanding of SARS-CoV-2 to provide innovative solutions, such as the development of new diagnostic tools, and preventive and curative strategies.

The challenge is not only technological, but also cultural. The perspective that sees only clinical studies at the center of the cognitive process of a disease needs change, but it must include the integration of *in silico* and *in vitro* methods with clinical studies. It is therefore necessary to disseminate knowledge on all new methodologies. Clinical doctors should strive to reconstruct the complexity of the organism from the molecular, cellular, tissue, organ and whole-organism level and the researchers instead should strive to increase the level of complexity of the models and their integration in order to provide tools that are increasingly closer to the human organism. Biomedical research must be increasingly multidisciplinary and integrate advanced technological aspects and large-scale clinical research to allow for faster knowledge gathering followed by concrete actions.

Information technology, *in silico* methods and artificial intelligence could support a more targeted and rational approach and to manage the complexity of human body processes relatively quickly and at low cost. Furthermore, knowledge can be shared by optimizing resources thus reducing waste and overlap. The positive effect of the worldwide sharing of scientific knowledge has permitted the rapid understanding of COVID-19 and trying together to combat the spread of the disease and its consequences.

In vitro methods dilucidated different aspects of the cellular function independently from the context, the organism, tissue and organ within the cells operate in physiological conditions. This is an advantage, as it simplifies the model, but may result in a rough approximation of reality. For example, the variations linked to body kinetics including absorption, distribution, metabolism and excretion are a challenge to overcome, but progress is being made embracing new methodologies. For dealing with complex body dynamics and kinetics, advance technologies and increasingly complex dynamic systems such as 3D organoids and microphysiological systems have been developed.

Human biology is complex, and it is therefore important to remember that only integrated and complementary use of different experimental models and methods allows us to obtain useful information. The dissemination of this knowledge is fundamental and will allow an increasingly effective and multidisciplinary research to reduce the disease burden of society.

Acknowledgements

We acknowledge support on the figures from Alessia Nuvoloni, ARTSTATION, Civita Castellana, Italy. We also thank the following for their support on the overall review of the chapter: Michael D. Waters, Michael Waters Consulting USA, Hillsborough, NC, North Carolina, United States; Maddalena Querci, European Commission Joint Research Centre (JRC), Ispra, Italy; Gerard Bowe, European Commission Joint Research Centre (JRC), Ispra, Italy; and Laura Gribaldo, European Commission Joint Research Centre (JRC), Ispra, Italy.

References

1. R. T. Gandhi, J. B. Lynch and C. Del Rio, Mild or Moderate Covid-19, *N. Engl. J. Med.*, 2020, **383**(38), 1757–1766.
2. European Centre for Disease Prevention and Control, *Diagnostic Testing and Screening for SARS-CoV-2*, 2020. Available from: https://www.ecdc.europa.eu/en/covid-19/latest-evidence/diagnostic-testing.
3. Joint Research Centre, *The JRC Releases New Reference Materials for the Quality Control of SARS-CoV-2 Antibody Tests*, 2020.
4. S. B. Kadam, et al., SARS-CoV-2, the pandemic coronavirus: Molecular and structural insights, *J. Basic Microbiol.*, 2021, **61**(3), 180–202.
5. Y. Jing, *et al.*, Potential influence of COVID-19/ACE2 on the female reproductive system, *Mol. Hum. Reprod.*, 2020, **26**(6), 367–373.
6. F. Busquet, *et al.*, Harnessing the power of novel animal-free test methods for the development of COVID-19 drugs and vaccines, *Arch. Toxicol.*, 2020, **94**(6), 2263–2272.
7. Animals used for scientific purposes, Replacement, Reduction and Refinement – the "Three Rs". Available from: https://ec.europa.eu/environment/chemicals/lab_animals/3r/alternative_en.htm.
8. European Commission, *New Approach Methodologies in Regulatory Science Proceedings of a Scientific Workshop*, 2016.

9. A. Punt, *et al.*, New approach methodologies (NAMs) for human-relevant biokinetics predictions. Meeting the paradigm shift in toxicology towards an animal-free chemical risk assessment, *Altex*, 2020, **37**(4), 607–622.
10. D. S. Khoury, *et al.*, Measuring immunity to SARS-CoV-2 infection: comparing assays and animal models, *Nat. Rev. Immunol.*, 2020, **20**(12), 727–738.
11. K. Takayama, In vitro and Animal Models for SARS-CoV-2 research, *Trends Pharmacol. Sci.*, 2020, **41**(8), 513–517.
12. C. Munoz-Fontela, *et al.*, Animal models for COVID-19, *Nature*, 2020, **586**(7830), 509–515.
13. S. Coecke, M. Balls, G. Bowe, J. Davis, G. Gstraunthaler, T. Hartung, R. Hay, O. W. Merten, A. Price, L. Schechtman, G. Stacey and W. Stokes, Second ECVAM Task Force on Good Cell Culture Practice. Guidance on good cell culture practice. A report of the second ECVAM task force on good cell culture practice, ATLA, *Altern. Lab. Anim.*, 2005, **33**(3), 261–287.
14. OECD, *Guidance Document on Good In Vitro Method Practices (GIVIMP), OECD Series on Testing and Assessment*, OECD Publishing, vol. 286, 2018.
15. A. Aydin, *et al.*, Combating COVID-19 with tissue engineering: a review, *Emergent Mater.*, 2020, 1–21.
16. A. Pizzorno, *et al.*, Characterization and Treatment of SARS-CoV-2 in Nasal and Bronchial Human Airway Epithelia, *Cell Rep. Med.*, 2020, **1**(4), 100059.
17. C. K. Bullen, *et al.*, Infectability of human BrainSphere neurons suggests neurotropism of SARS-CoV-2, *Altex*, 2020, **37**(4), 665–671.
18. B. A. G. de Melo, et al., 3D culture models to study SARS-CoV-2 infectivity and antiviral candidates: From spheroids to bioprinting, *Biomed. J.*, 2021, **44**(1), 31–42.
19. S. Hao, et al., Long-Term Modeling of SARS-CoV-2 Infection of In vitro Cultured Polarized Human Airway Epithelium, *mBio*, 2020, **11**(6), e02852-20.
20. V. Monteil, *et al.*, Inhibition of SARS-CoV-2 Infections in Engineered Human Tissues Using Clinical-Grade Soluble Human ACE2, *Cell*, 2020, **181**(4), 905–913 e7.
21. J. Chakraborty, *et al.*, Bioengineered *in vitro* Tissue Models to Study SARS-CoV-2 Pathogenesis and Therapeutic Validation, *ACS Biomater. Sci. Eng.*, 2020, **6**(12), 6540–6555.
22. C. R. Simoneau and M. Ott, Modeling Multi-organ Infection by SARS-CoV-2 Using Stem Cell Technology, *Cell Stem Cell*, 2020, **27**(6), 859–868.
23. D. Assaf, *et al.*, Utilization of machine-learning models to accurately predict the risk for critical COVID-19, *Intern. Emerg. Med.*, 2020, **15**(8), 1435–1443.
24. H. R. Rangel, *et al.*, SARS-CoV-2 host tropism: An *in silico* analysis of the main cellular factors, *Virus Res.*, 2020, **289**, 198154.

25. M. Hussain, *et al.*, Molecular docking between human TMPRSS2 and SARS-CoV-2 spike protein: conformation and intermolecular interactions, *AIMS Microbiol.*, 2020, **6**(3), 350–360.
26. C. G. K. Ziegler, *et al.*, SARS-CoV-2 Receptor ACE2 Is an Interferon-Stimulated Gene in Human Airway Epithelial Cells and Is Detected in Specific Cell Subsets across Tissues, *Cell*, 2020, **181**(5), 1016–1035 e19.
27. K. S. Doran, et al., Concepts and mechanisms: Crossing host barriers, *Cold Spring Harbor Perspect. Med.*, 2013, **3**(7), a010090.
28. N. J. Vielle, *et al.*, The human upper respiratory tract epithelium is susceptible to flaviviruses, *Front. Microbiol.*, 2019, **10**(MAR), 811.
29. H. Chu, *et al.*, Comparative tropism, replication kinetics, and cell damage profiling of SARS-CoV-2 and SARS-CoV with implications for clinical manifestations, transmissibility, and laboratory studies of COVID-19: an observational study, *Lancet Microbe*, 2020, **1**(1), e14–e23.
30. M. M. Lamers, *et al.*, SARS-CoV-2 productively infects human gut enterocytes, *Science*, 2020, **369**(6499), 50–54.
31. Y. Imura, *et al.*, A microfluidic system to evaluate intestinal absorption, *Anal. Sci.*, 2009, **25**(12), 1403–1407.
32. M. Chi, *et al.*, A microfluidic cell culture device (muFCCD) to culture epithelial cells with physiological and morphological properties that mimic those of the human intestine, *Biomed. Microdevices*, 2015, **17**(3), 9966.
33. L. Sharma and A. Riva, Intestinal barrier function in health and disease—any role of sars-cov-2? *Microorganisms*, 2020, **8**(11), 1–27.
34. H. Li, *et al.*, Transmission Routes Analysis of SARS-CoV-2: A Systematic Review and Case Report, *Front. Cell Dev. Biol.*, 2020, **8**, 618.
35. A. nvernizzi, *et al.*, Retinal findings in patients with COVID-19: Results from the SERPICO-19 study, *EClinicalMedicine*, 2020, **27**, 100550.
36. R. Hirose, H. Ikegaya, Y. Naito, N. Watanabe, T. Yoshida, R. Bandou, T. Daidoji, Y. Itoh and T. Nakaya, Survival of Severe Acute Respiratory Syndrome Coronavirus 2 (SARS-CoV-2) and Influenza Virus on Human Skin: Importance of Hand Hygiene in Coronavirus Disease 2019 (COVID-19), *Clin. Infect. Dis.*, 2021, **73**(11), e4329–e4335.
37. H. C. Helms, *et al.*, *In vitro* models of the blood–brain barrier: An overview of commonly used brain endothelial cell culture models and guidelines for their use, *J. Cereb. Blood Flow Metab.*, 2016, **36**(5), 862–890.
38. P. Prieto, *et al.*, Blood–brain barrier *in vitro* models and their application in toxicology. The report and recommendations of ECVAM Workshop 49, *Altern. Lab. Anim.*, 2004, **32**(1), 37–50.
39. T. P. Buzhdygan, et al., The SARS-CoV-2 spike protein alters barrier function in 2D static and 3D microfluidic in vitro models of the human blood–brain barrier, *Neurobiol. Dis.*, 2020, 146, 105131.
40. M. K. Wong, *et al.*, Establishment of an *in vitro* placental barrier model cultured under physiologically relevant oxygen levels, *Mol. Hum. Reprod.*, 2020, **26**(5), 353–365.

41. A. Nishiguchi, *et al.*, *In vitro* placenta barrier model using primary human trophoblasts, underlying connective tissue and vascular endothelium, *Biomaterials*, 2019, **192**, 140–148.
42. C. Blundell, A microphysiological model of the human placental barrier, *Lab Chip*, 2016, **16**(16), 3065–3073.
43. D. D. Mruk and C. Y. Cheng, An *in vitro* system to study Sertoli cell blood-testis barrier dynamics, *Methods Mol. Biol.*, 2011, **763**, 237–252.
44. S. W. Byers, *et al.*, Growth and characterization of polarized monolayers of epididymal epithelial cells and Sertoli cells in dual environment culture chambers, *J. Androl.*, 1986, **7**(1), 59–68.
45. K. M. Yamada and E. Cukierman, Modeling tissue morphogenesis and cancer in 3D, *Cell*, 2007, **130**(4), 601–610.
46. A. Steinberger, E. Steinberger and W. H. Perloff, Mammalian testes in organ culture, *Exp. Cell Res.*, 1964, **36**(1), 19–27.
47. S. Sakib, *et al.*, Three-dimensional testicular organoids as novel *in vitro* models of testicular biology and toxicology, *Environ. Epigenet.*, 2019, **5**(3), dvz011.
48. J. P. Alves-Lopes and J. B. Stukenborg, Testicular organoids: a new model to study the testicular microenvironment in vitro? *Hum. Reprod. Update*, 2018, **24**(2), 176–191.
49. M. A. Lancaster and J. A. Knoblich, Organogenesis in a dish: modeling development and disease using organoid technologies, *Science*, 2014, **345**(6194), 1247125.
50. J. K. Millet, J. A. Jaimes and G. R. Whittaker, Molecular diversity of coronavirus host cell entry receptors, *FEMS Microbiol. Rev.*, 2021, **45**(3), fuaa057.
51. A. M. Huertas, D. Montani, L. Savale, J. Pichon, Y. Tu, F. Parent, C. Guignabert and M. Humbert, Endothelial cell dysfunction: a major player in SARS-CoV-2 infection (COVID-19)? *Eur. Respir. J.*, 2020, **56**, 2001634.
52. W. Essahib, *et al.*, SARS-CoV-2 host receptors ACE2 and CD147 (BSG) are present on human oocytes and blastocysts, *J. Assist. Reprod. Genet.*, 2020, **37**(11), 2657–2660.
53. R. Amraie, et al., CD209L/L-SIGN and CD209/DC-SIGN Act as Receptors for SARS-CoV-2, *ACS Cent. Sci.*, 2021, **7**(7), 1156–1165.
54. G. Naffah-Mazzacoratti Mda, *et al.*, What have we learned about the kallikrein-kinin and renin-angiotensin systems in neurological disorders? *World J. Biol. Chem.*, 2014, **5**(2), 130–140.
55. P. R. Carvalho, P. Sirois and P. D. Fernandes, The role of kallikrein-kinin and renin-angiotensin systems in COVID-19 infection, *Peptides*, 2020, **135**, 170428.
56. N. Zamorano Cuervo and N. Grandvaux, ACE2: Evidence of role as entry receptor for SARS-CoV-2 and implications in comorbidities, eLife, 2020, **9**, e61390.
57. H. Tang, *et al.*, Thoughts on detecting tissue distribution of potential COVID-19 receptors, *Future Virol.*, 2020, **15**(8), 489–496.

58. M. Matusiak and C. M. Schurch, Expression of SARS-CoV-2 entry receptors in the respiratory tract of healthy individuals, smokers and asthmatics, *Respir. Res.*, 2020, **21**(3), 252.
59. A. Vaajanen, *et al.*, The expression of Mas-receptor of the renin-angiotensin system in the human eye, *Graefes Arch. Clin. Exp. Ophthalmol.*, 2015, **253**(7), 1053–1059.
60. L. Jackson, et al., Within the Brain: The Renin Angiotensin System, Int. J. Mol. Sci., 2018, **19**(3), 876.
61. Y. H. Tseng, R. C. Yang and T. S. Lu, Two hits to the renin-angiotensin system may play a key role in severe COVID-19, *Kaohsiung J. Med. Sci.*, 2020, **36**(6), 389–392.
62. I. T. Lee, *et al.*, ACE2 localizes to the respiratory cilia and is not increased by ACE inhibitors or ARBs, *Nat. Commun.*, 2020, **11**(1), 5453.
63. X. Yang, et al., Analysis of adaptive immune cell populations and phenotypes in the patients infected by SARS-CoV-2, medRxiv, 2020, DOI: 10.1101/2020.03.23.20040675.
64. Y. Zhang, Y. Chen and Z. Meng, Immunomodulation for Severe COVID-19 Pneumonia: The State of the Art, *Front. Immunol.*, 2020, **11**, 577442.
65. J. Wigen, *et al.*, Converging pathways in pulmonary fibrosis and Covid-19 - The fibrotic link to disease severity, *Respir. Med. X*, 2020, **2**, 100023.
66. B. Rockx, *et al.*, Early upregulation of acute respiratory distress syndrome-associated cytokines promotes lethal disease in an aged-mouse model of severe acute respiratory syndrome coronavirus infection, *J. Virol.*, 2009, **83**(14), 7062–7074.
67. A. Roberts, *et al.*, A mouse-adapted SARS-coronavirus causes disease and mortality in BALB/c mice, *PLoS Pathog.*, 2007, **3**(1), e5.
68. R. Channappanavar, *et al.*, Dysregulated Type I Interferon and Inflammatory Monocyte-Macrophage Responses Cause Lethal Pneumonia in SARS-CoV-Infected Mice, *Cell Host Microbe*, 2016, **19**(2), 181–193.
69. S. Ghosh, *et al.*, *In vitro* model of mesenchymal condensation during chondrogenic development, *Biomaterials*, 2009, **30**(33), 6530–6540.
70. V. G. Puelles, *et al.*, Multiorgan and Renal Tropism of SARS-CoV-2, *N. Engl. J. Med.*, 2020, **383**(6), 590–592.
71. G. Tavazzi, *et al.*, Myocardial localization of coronavirus in COVID-19 cardiogenic shock, *Eur. J. Heart Fail.*, 2020, **22**(5), 911–915.
72. D. Wichmann, *et al.*, Autopsy Findings and Venous Thromboembolism in Patients With COVID-19: A Prospective Cohort Study, *Ann. Intern. Med.*, 2020, **173**(4), 268–277.
73. S. Sala, *et al.*, Acute myocarditis presenting as a reverse Tako-Tsubo syndrome in a patient with SARS-CoV-2 respiratory infection, *Eur. Heart J.*, 2020, **41**(19), 1861–1862.
74. Z. Xu, *et al.*, Pathological findings of COVID-19 associated with acute respiratory distress syndrome, *Lancet Respir. Med.*, 2020, **8**(4), 420–422.

75. Z. Varga, *et al.*, Endothelial cell infection and endotheliitis in COVID-19, *Lancet*, 2020, **395**(10234), 1417–1418.
76. H. Li, *et al.*, SARS-CoV-2 and viral sepsis: observations and hypotheses, *Lancet*, 2020, **395**(10235), 1517–1520.
77. A. L. Bailey, et al., SARS-CoV-2 Infects Human Engineered Heart Tissues and Models COVID-19 Myocarditis, *JACC Basic Transl. Sci.*, 2021, **6**(4), 331–345.
78. M. M. Lamers, *et al.*, SARS-CoV-2 productively infects human gut enterocytes, *Science*, 2020, **369**(6499), 50–54.
79. R. Wolfel, *et al.*, Virological assessment of hospitalized patients with COVID-2019, *Nature*, 2020, **581**(7809), 465–469.
80. F. Xiao, *et al.*, Evidence for Gastrointestinal Infection of SARS-CoV-2, *Gastroenterology*, 2020, **158**(6), 1831–1833 e3.
81. H. Zhang, *et al.*, Digestive system is a potential route of COVID-19: an analysis of single-cell coexpression pattern of key proteins in viral entry process, *Gut*, 2020, **69**(6), 1010–1018.
82. H. Su, *et al.*, Renal histopathological analysis of 26 postmortem findings of patients with COVID-19 in China, *Kidney Int.*, 2020, **98**(1), 219–227.
83. X. W. Pan, *et al.*, Identification of a potential mechanism of acute kidney injury during the COVID-19 outbreak: a study based on single-cell transcriptome analysis, *Intensive Care Med.*, 2020, **46**(6), 1114–1116.
84. A. Iwasaki and P. S. Pillai, Innate immunity to influenza virus infection, *Nat. Rev. Immunol.*, 2014, **14**(5), 315–328.
85. C. P. Larsen, *et al.*, Collapsing Glomerulopathy in a Patient With COVID-19, *Kidney Int. Rep.*, 2020, **5**(6), 935–939.
86. S. J. Allison, SARS-CoV-2 infection of kidney organoids prevented with soluble human ACE2, *Nat. Rev. Nephrol.*, 2020, **16**(6), 316.
87. D. Lahiri, *et al.*, Neuroinvasive potential of a primary respiratory pathogen SARS- CoV2: Summarizing the evidences, *Diabetes, Metab. Syndr.*, 2020, **14**(5), 1053–1060.
88. L. Yang, Y. Han, B. E. Nilsson-Payant, V. Gupta, P. Wang, X. Duan, J. Zhu, Z. Zhao, F. Jaffré, T. Zhang, T. W. Kim, O. Harschnitz, D. Redmond, S. Houghton, C. Liu, A. Naji, G. Ciceri, S. Guttikonda, Y. Bram, D.-H. T. Nguyen, M. Cioffi, V. Chandar, D. A. Hoagland, Y. Huang, J. Xiang, H. Wang, D. Lyden, A. Borczuk, H. J. Chen, L. Studer, F. C. Pan, D. D. Ho, B. R. tenOever, T. Evans, R. E. Schwartz and S. Chen, A Human Pluripotent Stem Cell-Based Platform to Study SARS-CoV-2 Tropism and Model Virus Infection in Human Cells and Organoids, *Cell Stem Cell*, 2020, **27**, 125.
89. N. P. Disser, *et al.*, Musculoskeletal Consequences of COVID-19, *J. Bone Jt. Surg., Am. Vol.*, 2020, **102**(14), 1197–1204.
90. H. Lad, *et al.*, Intensive Care Unit-Acquired Weakness: Not just Another Muscle Atrophying Condition, *Int. J. Mol. Sci.*, 2020, **21**(21), 7840.
91. S. Recalcati, *et al.*, Acral cutaneous lesions in the time of COVID-19, *J. Eur. Acad. Dermatol. Venereol.*, 2020, **34**(8), e346–e347.

92. R. Gianotti, *et al.*, Cutaneous Clinico-Pathological Findings in three COVID-19-Positive Patients Observed in the Metropolitan Area of Milan, Italy, *Acta Derm.-Venereol.*, 2020, **100**(8), adv00124.
93. I. Colmenero, *et al.*, SARS-CoV-2 endothelial infection causes COVID-19 chilblains: histopathological, immunohistochemical and ultrastructural study of seven paediatric cases, *Br. J. Dermatol.*, 2020, **183**(4), 729–737.
94. S. Al-Benna, Gene Expression of Angiotensin-Converting Enzyme 2 Receptor in Skin and the Implications for COVID-19, *Adv. Skin Wound Care*, 2021, **34**(1), 31–35.
95. A. C. Gupta, *et al.*, Establishment of an *in vitro* organoid model of dermal papilla of human hair follicle, *J. Cell. Physiol.*, 2018, **233**(11), 9015–9030.
96. (a) GeneCards - the human gene databasewww.genecards.org; (b) G. Stelzer, R. Rosen, I. Plaschkes, S. Zimmerman, M. Twik, S. Fishilevich, T. Iny Stein, R. Nudel, I. Lieder, Y. Mazor, S. Kaplan, D. Dahary, D. Warshawsky, Y. Guan-Golan, A. Kohn, N. Rppaport, M. Safran and D. Lancet, The GeneCards Suite: From Gene Data Mining to Disease Genome Sequence Analysis, *Curr. Protoc. Bioinf.*, 2016, **54**, 1.30.1–1.30.33.
97. F. B. Bastian, J. Roux, A. Niknejad, A. Comte, S. S. Fonseca Costa, T. M. de Farias, S. Moretti, G. Parmentier, V. R. de Laval, M. Rosikiewicz, J. Wollbrett, A. Echchiki, A. Escoriza, W. H. Gharib, M. Gonzales-Porta, Y. Jarosz, B. Laurenczy, P. Moret, E. Person, P. Roelli, K. Sanjeev, M. Seppey and M. Robinson-Rechavi, The Bgee suite: integrated curated expression atlas and comparative transcriptomics in animals, *Nucleic Acids Res.*, 2021, **49**(D1), D831–D847.
98. O. T. Olaniyan, *et al.*, Testis and blood-testis barrier in Covid-19 infestation: role of angiotensin-converting enzyme 2 in male infertility, *J. Basic Clin. Physiol. Pharmacol.*, 2020, **31**(6), 20200156.
99. M. Yang, *et al.*, Pathological Findings in the Testes of COVID-19 Patients: Clinical Implications, *Eur. Urol. Focus*, 2020, **6**(5), 1124–1129.
100. Y. Yuan, *et al.*, *In vitro* testicular organogenesis from human fetal gonads produces fertilization-competent spermatids, *Cell Res.*, 2020, **30**(3), 244–255.
101. K. Zheng, *et al.*, COVID-19 and the bone: underestimated to consider, *Eur. Rev. Med. Pharmacol. Sci.*, 2020, **24**(20), 10316–10318.
102. M. D. Lundholm, *et al.*, SARS-CoV-2 (COVID-19) and the Endocrine System, *J. Endocr. Soc.*, 2020, **4**(11), bvaa144.
103. S. Destek, The Status of Parathyroid Hormone Secretion and Its Relationship With White Blood Cells, Neutrophil/Lymphocyte Ratio, c Reactive Protein, Fibrinogen, Procalcitonin, Ferritin and D-dimer in Hospitalized Covid-19 Patients, in ClinicalTrials.gov, 2020.
104. N. P. Somasundaram, *et al.*, The Impact of SARS-Cov-2 Virus Infection on the Endocrine System, *J. Endocr. Soc.*, 2020, **4**(8), bvaa082.
105. C. E. Duran, *et al.*, *In vitro* dynamics of parathyroid hormone secretion regulated by calcium and effects on the cell cycle: parathyroid hyperplasia *versus* adenoma, *Nefrologia*, 2010, **30**(4), 413–419.

106. G. Bellastella, M. I. Maiorino and K. Esposito, Endocrine complications of COVID-19: what happens to the thyroid and adrenal glands? *J. Endocrinol. Invest.*, 2020, **43**(8), 1169–1170.
107. L. Scappaticcio, et al., Impact of COVID-19 on the thyroid gland: an update, *Rev. Endocr. Metab. Disord.*, 2021, 22, 803–815.
108. L. Wei, *et al.*, Endocrine cells of the adenohypophysis in severe acute respiratory syndrome (SARS), *Biochem. Cell Biol.*, 2010, **88**(4), 723–730.
109. J. Gu, *et al.*, Multiple organ infection and the pathogenesis of SARS, *J. Exp. Med.*, 2005, **202**(3), 415–424.
110. L. Wei, *et al.*, Pathology of the thyroid in severe acute respiratory syndrome, *Hum. Pathol.*, 2007, **38**(1), 95–102.
111. Tracking System for Alternative methods towards Regulatory acceptance (TSAR), https://tsar.jrc.ec.europa.eu/.
112. J. Samanta, *et al.*, Coronavirus disease 2019 and the pancreas, *Pancreatology*, 2020, **20**(8), 1567–1575.
113. S. R. Bornstein, *et al.*, Endocrine and metabolic link to coronavirus infection, *Nat. Rev. Endocrinol.*, 2020, **16**(6), 297–298.
114. D. Harmer, *et al.*, Quantitative mRNA expression profiling of ACE 2, a novel homologue of angiotensin converting enzyme, *FEBS Lett.*, 2002, **532**(1–2), 107–110.
115. F. Liu, *et al.*, ACE2 Expression in Pancreas May Cause Pancreatic Damage After SARS-CoV-2 Infection, *Clin. Gastroenterol. Hepatol.*, 2020, **18**(9), 2128–2130 e2.
116. D. Fignani, *et al.*, SARS-CoV-2 Receptor Angiotensin I-Converting Enzyme Type 2 (ACE2) Is Expressed in Human Pancreatic beta-Cells and in the Human Pancreas Microvasculature, *Front. Endocrinol.*, 2020, **11**, 596898.
117. J. Li, *et al.*, COVID-19 infection may cause ketosis and ketoacidosis, *Diabetes, Obes. Metab.*, 2020, **22**(10), 1935–1941.
118. A. Gupta, *et al.*, Extrapulmonary manifestations of COVID-19, *Nat. Med.*, 2020, **26**(7), 1017–1032.
119. C. Huang, *et al.*, Clinical features of patients infected with 2019 novel coronavirus in Wuhan, China, *Lancet*, 2020, **395**(10223), 497–506.
120. F. Zhou, *et al.*, Clinical course and risk factors for mortality of adult inpatients with COVID-19 in Wuhan, China: a retrospective cohort study, *Lancet*, 2020, **395**(10229), 1054–1062.
121. X. Yang, *et al.*, Clinical course and outcomes of critically ill patients with SARS-CoV-2 pneumonia in Wuhan, China: a single-centered, retrospective, observational study, *Lancet Respir. Med.*, 2020, **8**(5), 475–481.
122. G. Chen, *et al.*, Clinical and immunological features of severe and moderate coronavirus disease 2019, *J. Clin. Invest.*, 2020, **130**(5), 2620–2629.
123. S. Recalcati, Cutaneous manifestations in COVID-19: a first perspective, *J. Eur. Acad. Dermatol. Venereol.*, 2020, **34**(5), e212–e213.
124. L. Mao, *et al.*, Neurologic Manifestations of Hospitalized Patients With Coronavirus Disease 2019 in Wuhan, China, *JAMA Neurol.*, 2020, **77**(6), 683–690.

125. J. Wan, *et al.*, Digestive symptoms and liver injury in patients with coronavirus disease 2019 (COVID-19): A systematic review with meta-analysis, *JGH Open*, 2020, **4**(6), 1047–1058.
126. L. F. Goncalves, *et al.*, Smell and taste alterations in COVID-19 patients: a systematic review, *Rev. Assoc. Med. Bras.*, 2020, **66**(11), 1602–1608.
127. R. da Rosa Mesquita, et al., Clinical manifestations of COVID-19 in the general population: systematic review, *Wien. Klin. Wochenschr.*, 2021, **133**(7–8), 377–382.
128. A. MadaniNeishaboori, *et al.*, Central Nervous System Complications in COVID-19 Patients; a Systematic Review and Meta-Analysis based on Current Evidence, *Arch. Acad. Emerg. Med.*, 2020, **8**(1), e62.
129. M. A. Islam, *et al.*, Prevalence of Headache in Patients With Coronavirus Disease 2019 (COVID-19): A Systematic Review and Meta-Analysis of 14,275 Patients, *Front. Neurol.*, 2020, **11**, 562634.
130. S. Kotta, *et al.*, Combating the Pandemic COVID-19: Clinical Trials, Therapies and Perspectives, *Front. Mol. Biosci.*, 2020, **7**, 606393.
131. L. Santacroce, *et al.*, The human coronaviruses (HCoVs) and the molecular mechanisms of SARS-CoV-2 infection, *J. Mol. Med.*, 2021, **99**(1), 93–106.
132. J. B. Moore and C. H. June, Cytokine release syndrome in severe COVID-19, *Science*, 2020, **368**(6490), 473–474.
133. J. Zhang, H. Wu and X. Yao, *et al.*, Pyroptotic macrophages stimulate the SARS-CoV-2-associated cytokine storm, *Cell. Mol. Immunol.*, 2021, **18**, 1305–1307.
134. J. Machhi, *et al.*, The Natural History, Pathobiology, and Clinical Manifestations of SARS-CoV-2 Infections, *J. Neuroimmune Pharmacol.*, 2020, **15**(3), 359–386.
135. D. E. Leisman, *et al.*, Cytokine elevation in severe and critical COVID-19: a rapid systematic review, meta-analysis, and comparison with other inflammatory syndromes, *Lancet Respir. Med.*, 2020, **8**(12), 1233–1244.
136. P. Mehta, *et al.*, COVID-19: consider cytokine storm syndromes and immunosuppression, *Lancet*, 2020, **395**(10229), 1033–1034.
137. D. S. Battagello, *et al.*, Unpuzzling COVID-19: tissue-related signaling pathways associated with SARS-CoV-2 infection and transmission, *Clin. Sci.*, 2020, **134**(16), 2137–2160.
138. Y. Guo, *et al.*, Modeling SARS-CoV-2 infection in vitro with a human intestine-on-chip device, *bioRxiv*, 2020, DOI: 10.1101/2020.09.01.277780.
139. Y. Wang, *et al.*, SARS-CoV-2 infection of the liver directly contributes to hepatic impairment in patients with COVID-19, *J. Hepatol.*, 2020, **73**(4), 807–816.
140. R. Mao, *et al.*, Manifestations and prognosis of gastrointestinal and liver involvement in patients with COVID-19: a systematic review and meta-analysis, *Lancet Gastroenterol. Hepatol.*, 2020, **5**(7), 667–678.
141. B. Zhao, *et al.*, Recapitulation of SARS-CoV-2 infection and cholangiocyte damage with human liver ductal organoids, *Protein Cell*, 2020, **11**(10), 771–775.

142. H. Ryan and C. S. Simmons, Potential Applications of Microfluidics to Acute Kidney Injury Associated with Viral Infection, *Cell Mol. Bioeng.*, 2020, **13**(4), 1–7.
143. A. Achar and C. Ghosh, COVID-19-Associated Neurological Disorders: The Potential Route of CNS Invasion and Blood–Brain Relevance, *Cells*, 2020, **9**(11), 2360.
144. D. B. Chou, *et al.*, On-chip recapitulation of clinical bone marrow toxicities and patient-specific pathophysiology, *Nat. Biomed. Eng.*, 2020, **4**(4), 394–406.
145. A. E. Matei, *et al.*, Vascularised human skin equivalents as a novel *in vitro* model of skin fibrosis and platform for testing of antifibrotic drugs, *Ann. Rheum. Dis.*, 2019, **78**(12), 1686–1692.
146. S. Al Heialy, *et al.*, Regulation of Angiotensin- Converting Enzyme 2 in Obesity: Implications for COVID-19, *Front. Physiol.*, 2020, **11**, 555039.
147. G. Pasquarelli-do-Nascimento, *et al.*, Hypercoagulopathy and Adipose Tissue Exacerbated Inflammation May Explain Higher Mortality in COVID-19 Patients With Obesity, *Front. Endocrinol.*, 2020, **11**, 530.
148. F. Ponten, K. Jirstrom and M. Uhlen, The Human Protein Atlas--a tool for pathology, *J. Pathol.*, 2008, **216**(4), 387–393.
149. *The Human Proteome Atlas*. Available from: https://www.proteinatlas.org/ENSG00000130234-ACE2/tissue.
150. S. Verma, S. Saksena and H. Sadri-Ardekani, ACE2 receptor expression in testes: implications in coronavirus disease 2019 pathogenesisdagger, *Biol. Reprod.*, 2020, **103**(3), 449–451.
151. K. C. Brannen, *et al.*, Alternative Models of Developmental and Reproductive Toxicity in Pharmaceutical Risk Assessment and the 3Rs, *ILAR J.*, 2016, **57**(2), 144–156.
152. M. J. Lespasio, N. Sodhi and M. A. Mont, Osteonecrosis of the Hip: A Primer, *Perm. J.*, 2019, **23**, 18–100.
153. B. L. Cuetara, *et al.*, Cloning and characterization of osteoclast precursors from the RAW264.7 cell line, *In Vitro Cell. Dev. Biol.: Anim.*, 2006, **42**(7), 182–188.
154. F. Liu, *et al.*, ACE2 Expression in Pancreas May Cause Pancreas Damage After SARS-CoV-2 Infection, *Clin. Gastroenterol. Hepatol.*, 2020, **18**(9), 2128–2130.e2.
155. M. Y. Li, *et al.*, Expression of the SARS-CoV-2 cell receptor gene ACE2 in a wide variety of human tissues, *Infect. Dis. Poverty*, 2020, **9**(1), 45.
156. C. M. Ferrario, ACE2: more of Ang-(1-7) or less Ang II? *Curr. Opin. Nephrol. Hypertens.*, 2011, **20**(1), 1–6.
157. I. Valencia, *et al.*, DPP4 and ACE2 in Diabetes and COVID-19: Therapeutic Targets for Cardiovascular Complications?*Front. Pharmacol.*, 2020, **11**, 1161.
158. L. Roncati and B. Lusenti, The moonlighting protein able to explain the Th1 immune lockdown in severe COVID-19, *Med. Hypotheses*, 2020, **143**, 110087.

159. E. Urciuoli and B. Peruzzi, Inhibiting Extracellular Vesicle Trafficking as Antiviral Approach to Corona Virus Disease 2019 Infection, *Front. Pharmacol.*, 2020, **11**, 580505.
160. H. Maaroufi, SARS-CoV-2 Encodes a PPxY Late Domain Motif that is Known to Enhance Budding and Spread in Enveloped RNA Viruses, *bioRxiv*, 2020, DOI: 10.1101/2020.04.20.052217.
161. L. Miorin, *et al.*, SARS-CoV-2 Orf6 hijacks Nup98 to block STAT nuclear import and antagonize interferon signaling, *Proc. Natl. Acad. Sci. U. S. A.*, 2020, **117**(45), 28344–28354.
162. J. Burtscher, *et al.*, Mitochondria: In the Cross Fire of SARS-CoV-2 and Immunity, *iScience*, 2020, **23**(10), 101631.
163. A. Kaser, Genetic risk of severe Covid-19, *N. Engl. J. Med.*, 2020, **383**(16), 1590–1591.
164. S. Pisanti, *et al.*, Correlation of the two most frequent HLA haplotypes in the Italian population to the differential regional incidence of Covid-19, *J. Transl. Med.*, 2020, **18**(1), 352.
165. Y. Hou, *et al.*, New insights into genetic susceptibility of COVID-19: An ACE2 and TMPRSS2 polymorphism analysis, *BMC Med.*, 2020, **18**(1), 216.
166. C. I. Van Der Made, *et al.*, Presence of Genetic Variants among Young Men with Severe COVID-19, *JAMA, J. Am. Med. Assoc.*, 2020, **324**(7), 663–673.
167. A. E. Gracia-Ramos, Is the ACE2 Overexpression a Risk Factor for COVID-19 Infection? *Arch. Med. Res.*, 2020, **51**(4), 345–346.
168. L. Wallentin, *et al.*, Angiotensin-converting enzyme 2 (ACE2) levels in relation to risk factors for COVID-19 in two large cohorts of patients with atrial fibrillation, *Eur. Heart J.*, 2020, **41**(41), 4037–4046.
169. J. D. Strope, C. H. C. PharmD and W. D. Figg, TMPRSS2: Potential Biomarker for COVID-19 Outcomes, *J. Clin. Pharmacol.*, 2020, **60**(7), 801–807.
170. W. Miesbach and M. Makris, COVID-19: Coagulopathy, Risk of Thrombosis, and the Rationale for Anticoagulation, *Clin. Appl. Thromb./Hemostasis*, 2020, **26**, 1076029620938149.
171. R. Laguna-Goya, *et al.*, IL-6–based mortality risk model for hospitalized patients with COVID-19,*J. Allergy Clin. Immunol.*, 2020, **146**(4), 799–807.e9.
172. Y. Zuo, *et al.*, Neutrophil extracellular traps in COVID-19, *JCI Insight*, 2020, **5**(11), e138999.
173. S. Chiappetta, *et al.*, COVID-19 and the role of chronic inflammation in patients with obesity, *Int. J. Obes.*, 2020, **44**(8), 1790–1792.
174. R. Cecchini and A. L. Cecchini, SARS-CoV-2 infection pathogenesis is related to oxidative stress as a response to aggression, *Med. Hypotheses*, 2020, **143**, 110102.
175. J. Burtscher, G. P. Millet and M. Burtscher, Low cardiorespiratory and mitochondrial fitness as risk factors in viral infections: implications for COVID-19, *Br. J. Sports Med.*, 2021, **55**, 413–415.

176. R. K. Mehlotra, Chemokine receptor gene polymorphisms and COVID-19: Could knowledge gained from HIV/AIDS be important? *Infect., Genet. Evol.*, 2020, **85**, 104512.
177. N. Kirtipal and S. Bharadwaj, Interleukin 6 polymorphisms as an indicator of COVID-19 severity in humans, *J. Biomol. Struct. Dyn.*, 2020, 1–3.
178. S. Karakas Celik, G. Cakmak Genc and A. Dursun, A bioinformatic approach to investigating cytokine genes and their receptor variants in relation to COVID-19 progression, *Int. J. Immunogenet.*, 2021, **48**(2), 211–218.
179. E. W. Fakhouri, *et al.*, Genetic Polymorphisms Complicate COVID-19 Therapy: Pivotal Role of HO-1 in Cytokine Storm, *Antioxidants*, 2020, **9**(7), 636.
180. T. Aoe, Pathological Aspects of COVID-19 as a Conformational Disease and the Use of Pharmacological Chaperones as a Potential Therapeutic Strategy, *Front. Pharmacol.*, 2020, **11**, 1095.
181. J. Fajnzylber, *et al.*, SARS-CoV-2 viral load is associated with increased disease severity and mortality, *Nat. Commun.*, 2020, **11**(1), 5493.
182. A. S. Ojo, *et al.*, Pulmonary Fibrosis in COVID-19 Survivors: Predictive Factors and Risk Reduction Strategies, *Pulm. Med.*, 2020, 6175964.
183. C. Yang, *et al.*, Myocardial injury and risk factors for mortality in patients with COVID-19 pneumonia, *Int. J. Cardiol.*, 2021, **326**, 230–236.
184. E. D. Siew and B. C. Birkelo, COVID-19–associated acute kidney injury an evolving picture, *Clin. J. Am. Soc. Nephrol.*, 2020, **15**(10), 1383–1385.
185. S. Schnaubelt, *et al.*, Atrial fibrillation: A risk factor for unfavourable outcome in COVID-19? A case report, *Eur. Heart J.*, 2020, **4**(FI1), 1–6.
186. D. Liu, *et al.*, Viral sepsis is a complication in patients with Novel Corona Virus Disease (COVID-19), *Med. Drug Discovery*, 2020, **8**, 100057.
187. E. Mahase, Covid-19: Why are age and obesity risk factors for serious disease? *BMJ Case Rep.*, 2020, **371**, m4130.
188. CDC, *COVID-19 Hospitalization and Death by Race/Ethnicity*, 2020.
189. F. T. Mendonça, Immunosuppressed Patients and the Risk of COVID-19: A Narrative Review, *Clin. Oncol. Res.*, 2020, **3**(8), 2–11.
190. Y. Alwarawrah, K. Kiernan and N. J. MacIver, Changes in nutritional status impact immune cell metabolism and function, *Front. Immunol.*, 2018, **9**, 1055.
191. J. E. Snyder, *et al.*, Measuring the frequency of mouse and human cytotoxic T cells by the Lysispot assay: Independent regulation of cytokine secretion and short-term killing, *Nat. Med.*, 2003, **9**(2), 231–235.
192. M. Plebanski, *et al.*, Methods to measure T-cell responses, *Expert Rev. Vaccines*, 2010, **9**(6), 595–600.
193. N. Bercovici, *et al.*, New methods for assessing T-cell responses, *Clin. Diagn. Lab. Immunol.*, 2000, **7**(6), 859–864.
194. E. Montecino-Rodriguez, B. Berent-Maoz and K. Dorshkind, Causes, consequences, and reversal of immune system aging, *J. Clin. Invest.*, 2013, **123**(3), 958–965.
195. N. Pavillon, *et al.*, Noninvasive detection of macrophage activation with single-cell resolution through machine learning, *Proc. Natl. Acad. Sci. U. S. A.*, 2018, **115**(12), E2676–E2685.

196. N. Ashary, *et al.*, Single-Cell RNA-seq Identifies Cell Subsets in Human Placenta That Highly Expresses Factors Driving Pathogenesis of SARS-CoV-2, *Front. Cell Dev. Biol.*, 2020, **8**, 783.
197. G. Mor and I. Cardenas, The Immune System in Pregnancy: A Unique Complexity, *Am. J. Reprod. Immunol.*, 2010, **63**(6), 425–433.
198. M. Irfan, *et al.*, Pulmonary functions in patients with diabetes mellitus, *Lung India*, 2011, **28**(2), 89–92.
199. S. Tsalamandris, *et al.*, The role of inflammation in diabetes: Current concepts and future perspectives, *Eur. Cardiol. Rev.*, 2019, **14**(1), 50–59.
200. D. J. Pattanshetty, *et al.*, Elevated troponin predicts long-term adverse cardiovascular outcomes in hypertensive crisis: A retrospective study, *J. Hypertens.*, 2012, **30**(12), 2410–2415.
201. S. Kulkarni, B. L. Jenner and I. Wilkinson, COVID-19 and hypertension, *JRAAS*, 2020, **21**(2), 1470320320927851.
202. R. Bing and M. R. Dweck, Myocardial fibrosis: Why image, how to image and clinical implications, *Heart*, 2019, **105**(23), 1832–1840.
203. M. Wu, *et al.*, Transcriptional and proteomic insights into the host response in fatal COVID-19 cases, *Proc. Natl. Acad. Sci. U. S. A.*, 2020, **117**(45), 28336–28343.
204. J. Veenstra, *et al.*, Antecedent immunosuppressive therapy for immune-mediated inflammatory diseases in the setting of a COVID-19 outbreak, *J. Am. Acad. Dermatol.*, 2020, **83**(6), 1696–1703.
205. J. Zhong, Q. Gong and A. Mima, Inflammatory Regulation in Diabetes and Metabolic Dysfunction, *J. Diabetes Res.*, 2017, 5165268.
206. T. Pillaiyar, *et al.*, The recent outbreaks of human coronaviruses: A medicinal chemistry perspective, *Med. Res. Rev.*, 2021, **41**(1), 72–135.
207. *Cytokine Adsorption Devices for Treating Respiratory Failure in People with COVID-19*, ed. National Institute for Health and Care Excellence, 2020.
208. F. Touret, *et al.*, In vitro screening of a FDA approved chemical library reveals potential inhibitors of SARS-CoV-2 replication, *Sci. Rep.*, 2020, **10**(1), 13093.
209. B. Shah, P. Modi and S. R. Sagar, In silico studies on therapeutic agents for COVID-19: Drug repurposing approach, *Life Sci.*, 2020, **252**, 117652.
210. S. A. Ahmed, *et al.*, Destabilizing the structural integrity of COVID-19 by caulerpin and its derivatives along with some antiviral drugs: An *in silico* approaches for a combination therapy, *Struct. Chem.*, 2020, **31**(6), 2391–2412.
211. C. Ge and Y. He, *In silico* Prediction of Molecular Targets of Astragaloside IV for Alleviation of COVID-19 Hyperinflammation by Systems Network Pharmacology and Bioinformatic Gene Expression Analysis, *Front. Pharmacol.*, 2020, **11**.
212. M. Garcia-Cremades, *et al.*, Optimizing Hydroxychloroquine Dosing for Patients With COVID-19: An Integrative Modeling Approach for Effective Drug Repurposing, *Clin. Pharmacol. Ther.*, 2020, **108**(2), 253–263.
213. S. Wang, *et al.*, Modeling the viral dynamics of SARS-CoV-2 infection, *Math. Biosci.*, 2020, **328**, 108438.

214. Y. H. Taguchi and T. Turki, A new advanced in silico drug discovery method for novel coronavirus (SARS-CoV-2) with tensor decomposition-based unsupervised feature extraction, *PLoS One*, 2020, **15**, e0238907.
215. A. Mulay, *et al.*, SARS-CoV-2 infection of primary human lung epithelium for COVID-19 modeling and drug discovery, *Cell Rep.*, 2021, **35**, 109055.
216. M. A. Canham, J. D. M. Campbell and J. C. Mountford, The use of mesenchymal stromal cells in the treatment of coronavirus disease 2019, *J. Transl. Med.*, 2020, **18**(1), 359.
217. M. Kavianpour, M. Saleh and J. Verdi, The role of mesenchymal stromal cells in immune modulation of COVID-19: focus on cytokine storm, *Stem Cell Res. Ther.*, 2020, **11**(1), 404.
218. D. F. McAuley, *et al.*, Clinical grade allogeneic human mesenchymal stem cells restore alveolar fluid clearance in human lungs rejected for transplantation, *Am. J. Physiol.*, 2014, **306**(9), L809–L815.
219. S. Al-Khawaga and E. M. Abdelalim, Potential application of mesenchymal stem cells and their exosomes in lung injury: an emerging therapeutic option for COVID-19 patients, *Stem Cell Res. Ther.*, 2020, **11**(1), 437.
220. Y. Han, *et al.*, Identification of Candidate COVID-19 Therapeutics using hPSC-derived Lung Organoids, bioRxiv, 2020, DOI: 10.1101/2020.05.05.079095.
221. Y. Han, *et al.*, Identification of SARS-CoV-2 inhibitors using lung and colonic organoids, *Nature*, 2021, **589**(7841), 270–275.
222. L. Si, H. Bai and M. Rodas, *et al.*, A human-airway-on-a-chip for the rapid identification of candidate antiviral therapeutics and prophylactics, *Nat. Biomed. Eng.*, 2021, **5**, 815–829.
223. C. A. Bassett, Beneficial effects of electromagnetic fields, *J. Cell. Biochem.*, 1993, **51**(4), 387–393.
224. M. Vigano, *et al.*, Mesenchymal stem cells as therapeutic target of biophysical stimulation for the treatment of musculoskeletal disorders, *J. Orthop. Surg. Res.*, 2016, **11**(1), 163.
225. A. K. Chan, *et al.*, Pulsed electromagnetic fields reduce acute inflammation in the injured rat-tail intervertebral disc, *JOR Spine*, 2019, **2**(4), e1069.
226. L. de Girolamo, *et al.*, *In vitro* functional response of human tendon cells to different dosages of low-frequency pulsed electromagnetic field, *Knee Surg. Sports Traumatol. Arthrosc.*, 2015, **23**(11), 3443–3453.
227. M. Saraiva and A. O'Garra, The regulation of IL-10 production by immune cells, *Nat. Rev. Immunol.*, 2010, **10**(3), 170–181.
228. T. Akahoshi, J. J. Oppenheim and K. Matsushima, Interleukin 1 stimulates its own receptor expression on human fibroblasts through the endogenous production of prostaglandin(s), *J. Clin. Invest.*, 1988, **82**(4), 1219–1224.
229. A. M. Petersen and B. K. Pedersen, The anti-inflammatory effect of exercise, *J. Appl. Physiol.*, 2005, **98**(4), 1154–1162.

230. J. E. de Vries, Immunosuppressive and anti-inflammatory properties of interleukin 10, *Ann. Med.*, 1995, **27**(5), 537–541.
231. J. B. Mumm, *et al.*, IL-10 Elicits IFNg-Dependent Tumor Immune Surveillance, *Cancer Cell*, 2011, **20**(6), 781–796.
232. C. J. Bonnans C and Z. Werb, Remodelling the extracellular matrix in development and disease, *Nat. Rev. Mol. Cell Biol.*, 2014, **15**(12), 786–801.
233. G. Pietramaggiori, P. Liu, S. S. Scherer and A. Kaipainen, Tensile forces stimulate vascular remodeling and epidermal cell proliferation in living skin, *Ann. Surg.*, 2007, **246**, 896.
234. O. Lieleg, R. M. Baumgärtel and A. R. Bausch, Selective filtering of particles by the extracellular matrix: an electrostatic bandpass, *Biophys. J.*, 2009, **97**(6), 1569–1577.
235. D. E. Ingber, Cellular mechanotransduction: putting all the pieces together again, *FASEB J.*, **20**, 811.
236. E. Bettelli, T. Korn and V. K. Kuchroo, Th17: the third member of the effector T cell trilogy, *Curr. Opin. Immunol.*, 2007, **19**, 652–657.
237. E. Yalcinkaya, M. Celik and B. Bugan, Extracellular matrix turnover: a balance between MMPs and their inhibitors, *Arq. Bras. Cardiol.*, 2014, **102**(5), 519–520.
238. D. A. Siwik and W. S. Colucci, Regulation of matrix metalloproteinases by cytokines and reactive oxygen/nitrogen species in the myocardium, *Heart Fail. Rev.*, 2004, **9**(1), 43–51.
239. A. K. Samanta, J. J. Oppenheim and K. Matsushima, Interleukin 8 (monocyte-derived neutrophil chemotactic factor) dynamically regulates its own receptor expression on human neutrophils, *J. Biol. Chem.*, 1990, **265**(1), 183–189.
240. E. Radice, V. Miranda and G. Bellone, Low-doses of sequential-kinetic-activated interferon-γ enhance the ex vivo cytotoxicity of peripheral blood natural killer cells from patients with early-stage colorectal cancer. A preliminary study, *Int. Immunopharmacol.*, 2014, **19**(1), 66–73.
241. W. Zhang, *et al.*, Protective effect of bone marrow mesenchymal stem cells in intestinal barrier permeability after heterotopic intestinal transplantation, *World J. Gastroenterol.*, 2014, **20**(23), 7442–7451.
242. V. García-Hernández, *et al.*, EGF regulates claudin-2 and -4 expression through Src and STAT3 in MDCK cells, *J. Cell. Physiol.*, 2015, **230**(1), 105–115.
243. K. L. Howe, *et al.*, Transforming growth factor-beta regulation of epithelial tight junction proteins enhances barrier function and blocks enterohemorrhagic Escherichia coli O157:H7-induced increased permeability, *Am. J. Pathol.*, 2005, **167**(6), 1587–1597.
244. J. Xiong, *et al.*, Effects of interleukin-4 or interleukin-10 gene therapy on trinitrobenzenesulfonic acid-induced murine colitis, *BMC Gastroenterol.*, 2013, **13**, 165.

Subject Index

Entries in **bold** refer to tables; those in *italic* to figures

Entries refer to COVID-19/SARS-CoV-2 except where otherwise specified

abbreviations of terms 26, 330
accessory proteins 38, 79, *80*, 208, **211**, 212–214, 244, *457*
ACE receptor *see* angiotensin-converting enzyme receptor
activated partial thromboplastin time (aPTT) 339, **339**; *see also* coagulopathies
ACTT *see* adaptive COVID-19 Treatment Trial
acute cerebral infarction 317
acute disseminated encephalomyelitis (ADEM) 312, 320, 326, *327*, 329
acute necrotizing encephalopathy (ANE) 320, 326, 329
acute respiratory distress syndrome (ARDS) 69, 73, 83–84, 203
  immunological response 253
  infection kinetics/dynamics 484–485
  pathogenesis of virus 381
  treatment options 138, 438
Adaptive COVID-19 Treatment Trial-1 (ACTT-1) trial 140, 413, **414**, 414, 415, 431
adaptive immune response 166, 257–262
ADE (antibody-dependent enhancement) theory 258–259
ADEM *see* acute disseminated encephalomyelitis
Adhanom Ghebreyesus, Dr Tedros xiii, 6, 64
adrenal glands 355–356, **360**
adrenocorticotropic hormone (ACTH) 356, 357
aerosols, exhaled xiii, 9, 25, 64, 243, 245
age factors 187
  case fatality rate 190
  severity of disease 258, 267–268
ageusia 70, 102, **178**, 178–179, 322
air transportation, role in global spread of infection 137
ajwain (*Trachyspermum ammi*) 144
Allium sativum (garlic) 144, **145**
alpha-coronaviruses 12, 13, *14*, 14, 38, 204
alveolar epithelial type II (ATII) cells 254
alveoli *see* lungs
amino acid transporter B(0)AT1, infection process 247
aminopeptidase N, infection process **213**
anakinra, interleukin inhibitor 429–430

ANE (acute necrotizing encephalopathy) 320, *326*, 329
angiotensin-converting enzyme (ACE) inhibitors 85–86, 268, 269
angiotensin-converting enzyme 2 (ACE2) receptor 18
 animal studies 459
 attachment of virus to host cell 41–42
 cellular factors in infection **213**, 218–220, *219*, **225**, 227
 CNS infection 312, 317, 319, 321
 endocrinopathies 353, 354, 355, 360
 factors affecting severity of disease 264–265, 267
 genomics 39, *41*
 immunological responses 243
 infection kinetics/dynamics 457, 463–471, **472–482**, 484–493, **495**, 496–497, **501**
 infection process 20–21, 78–82, 215, 247, 337–338, 457
 neurological manifestations 322
 pathogenesis of virus 381
 pathophysiology of virus 23, 74–77, *78*
 regional variations in clinical presentation 177–178
 reproductive system health 369, 370, 374
 sex-based differences 188, 265
angiotensin receptor blockers (ARBs) 85–86, 268, 269
animal origin of human coronaviruses *12*, 206; *see also* zoonotic transmission
animal studies xvi, 458–459
anosomia 70, 102, **178**, 178–179, 322
anterior nares specimen, sample collection 108
anterior segment barrier, barriers to infection 465
antibody-dependent enhancement (ADE) theory 258–259
antibody-mediated immunity 257–259 258–259
antibody testing 172–173, 176, 184, 345
antibody therapy *137*; *see also* monoclonal antibodies
antiepileptic medications 328
antiphospholipid antibody syndrome (APLA) 317
antiplatelet therapy 142
antithrombotic agents 138, 141–142, 439–440, **441**, 443
antiviral agents *137*, 159, 440–441, **441**
  clinical trials 148, 413–418, **414**, **417**
antroquinonol, antiviral activity 425
apilimod dimesylate, antiviral activity 425
APLA (antiphospholipid antibody syndrome) 317
app (smartphone application) technology 70–71
ARBs (angiotensin receptor blockers) 85–86, 268, 269
ARDS *see* acute respiratory distress syndrome
arrhythmias 85
arterial thrombosis *see* thrombosis
Asia, clinical epidemiology 179–181, *180*
asymptomatic cases xiii, 4, 9–10, 25, 64, 65, 382
 children 186
 clinical epidemiology 166, 168–169, 174–175, 194
 definition 174, 175
 diagnostics 100, 130
 global spread of infection 137
Athens, plague of vii, ix
athimaduram (*Glycyrrhiza glabra*) 144, **145**
atorvastatin, clinical trials **156**
attachment to host cell 41–42; *see also* infection process
 cellular factors **213**
 genomics 215
 pathogenesis 246

autoimmune encephalitis, radiology 326
autoimmunity, neuronal 314–315
autopsy 380–382
bone marrow 397, *398*
brain 392, 395
coagulopathies 341–343, *342*
electron microscopy 391, 400
FFPE cell pellets 399–400, 401–402
general protocols 382–383
heart 392, 394
immunohistochemistry 399
importance of 382
kidney 392, 395, 400
light microscopy *393*, 393–399, *396*
liver 392, *396*, 396–397
lungs 391–392, *393*, 393–394
molecular testing 389–390, 401–402
other organs 392, 397–399
pathological examination 391–392
procedures 387–388
protocols 383–386
sample collection 389–391
in situ hybridization 399–400
types 386–389
avian coronaviruses viii, 12, 13, 15, 38, 75, 204, **205**, 226, 397
AXL kinase inhibitors, antiviral activity 424
Azadirachta indica (neem) 144
azithromycin 141, **441**

B-cells, adaptive immune response 257–259
Bacillus Calmette-Guérin (BCG) vaccine 271–272
bamlanivimab (LY3819253/ LY-CoV55) 434–435
baricitinib 143, 415–416
barriers to infection 462–464, *463*; *see also* blood–brain barrier
eyes 465
intestinal 464–465
placental 467
respiratory 464
skin 466
retinal 465–466
testis 467–468
basigin (BSG) **213**, 222–223, **225**, 227, 246, 468, 469
bats
Middle East respiratory syndrome 17
origin of human coronaviruses 2, 165, 204, 206, 381
zoonotic transmission 12, 13, 17, 243
BCG (Bacillus Calmette-Guérin) vaccine 271–272
Bell's palsy 322
bemcentinib, antiviral activity 424
β coronaviruses 20, 38, 204
classification 11, 12, 13, 14, *14*
genomic structure 208
β-D-N4-hydroxycytidine, antiviral activity 424
β1 interferon 143
binding immunoglobin proteins **213**, 223
bioinformatics analysis 503
biomarkers, cytokine storm 176–177
biosafety guidelines, autopsy 385
bird flu viii, 12, 13, 15, 38, 75, 204, **205**, 226, 397
Black Death viii
blame attributions, for infection ix
BLAZE-1 study, monoclonal antibodies 434
blood–brain barrier 312, 314, 466
CNS infection 312–313, 314, 325
symptoms of infection *87*
blood groups
severity of disease 266
susceptibility to COVID-19 117, 118
blood pressure, CNS infection 314; *see also* hypertension
blood–retinal barrier, barriers to infection 465–466

blood–testis barrier, barriers to infection 467–468
blood tests 115–117, 120, 127, 128–130
blood vessel formation 338
body temperature, progression of illness 249; *see also* fever
bone marrow, autopsy examination 397, *398*
brain, autopsy examination 392, 395
brain damage, hypoxic 314; *see also* central nervous system disorders
Brazil xv, 190, **191**, 322, 423, 436
breastfeeding 371–372
bronchoalveolar lavage, sample collection 109
BSG (basigin) **213**, 222–223, **225**, 227, 246, 468, 469
butyrate 271

C-reactive protein (CRP), diagnostics 112, 116
C-terminal domain (CTD), spike protein 246, 247
C-X-C motif chemokine receptor 6 (CXCR6) 117
C-X-C motif chemokine ligand 10 (CXCL10) 381, 382
CAC *see* COVID-19-associated coagulopathy
Caco-2 cell lines 505
California xv
camels xii, *12*, 15, 37, 206
Camus, Albert ix
canakinumab 429–430, 442
CAN-COVID trial 430
cardiovascular system
adverse drug effects 141, 273–274
autopsy 392, 394
infection kinetics/dynamics 485–486
symptoms of infection 83, 85–86, 317
COVID-19-induced ARDS (CARDS) 138; *see also* acute respiratory distress syndrome
CARDS see COVID-19-induced ARDS
carrimycin, clinical trials **150, 153**
case fatality rate, clinical epidemiology 190–193
by age *192*
by country **191–192, 193**
definition 190
cash-less society, China's response to virus 53
cathepsin, infection process 42, **213**, 215, 223, **225**
CC motif chemokine receptor 9 (CCR9) 117
CD4 (cluster of differentiation 4) T-cells 76–77, 253, 255, 259, 260–262, 342–343, 382, **499**, 500
CD8 cytotoxic T-cells 76, 117, 186, 255, 256, 259–262, 270, 341, 382, 397
CD26 (dipeptidyl peptidase) **213**, 221–222
CD147 (basigin/EMMPRIN) **213**, 222–223, **225**, 227, 246, 468, 469
cell-based factors influencing infection
genomics **213**, 217–226, **225**
infection kinetics/dynamics 493–494, 497
risk factors for infection 497–498
viral attachment and entry **213**, 217–226, **225**
cell entry by virus *see* infection process
cell kinetics/dynamics *see* infection kinetics/dynamics
cell lines, therapeutics based on 505
cell-mediated immunity 257
cell signaling
infection kinetics/dynamics 492
therapeutics based on **501**
cell therapy, clinical trials 278, 280
Centers for Disease Control and Prevention (CDC) guidelines 138

central nervous system (CNS)
disorders 312, 316–321
acute inflammatory/infectious syndromes 326, *327*
case vignettes 318, 319, 320, 321
demyelination syndromes 320–321
encephalitis 319–320
encephalopathy 320
imaging/radiology 324–328, *325*, *326*, *327*
intracerebral haemorrhage 318
ischemic stroke 317–318
movement disorders 321
myelitis 321
seizures 318–319, 321
vascular manifestations of infection 316–318
cerebellitis, imaging/radiology 326, *327*
cerebral infarction 317
chemiluminescence, diagnostics 121, 129
chemosensory dysfunction 70, 102, **178**, 178–179, 322
chest X-rays 112, 126
children
clinical epidemiology 184–186
diagnostic criteria 173–174
infection rate/prevalence of infection 170, **171**
symptoms of infection 101
China 50–51, 58–59
case fatality rate **191**, *193*
chronology of the outbreak 2–6
comparison with SARS-CoV outbreak 56–58
contact tracing 55–56
control and prevention measures 54–56
epidemiological perspectives 53–54, 177, 179–181, *180*
initial outbreak 2, 25
mortality rates **191**
new year travel rush/spring movement 53
lessons learned 51–52, 54
Chinese Center for Disease Control and Prevention 4
chloroquine 23, 138, **139**, 141, 159, 416, 442
clinical trials **150–154**
immunological response 272–274
ineffectiveness **441**
infection kinetics/dynamics 503, 506
chronic obstructive pulmonary disease (COPD) 269–270
chymotrypsin-like protease (3CLpro) 38, *39*, 80
cigarette smoking 313, 372
civets 206
civilization, plagues as inevitable consequence vii–x
classification, coronaviruses 11–15, *14*, 37–38, 244, 337
genomics 204, **205**
symptom types 70–71, 173–174
clinical autopsy *see* autopsy
clinical epidemiology 166–168, 193–194
antibody testing 172–173
asymptomatic cases 166, 168, 168–169, 174–175, 194
case fatality rate 190–193, **191–192**
chemosensory dysfunction 178–179
classification 173–174
clinical and immunological response 176–177
clinical spectrum of disease 173–177, 194
cruise ship outbreak 167, 170, **171**, 174, 181, 185, 190, *192*
definition 166
infection 168–169
infection rate 167, 170–172, **171**, 194

infectivity 194
occupational differences 188, **189**
older adults 187
paediatric patients 184–186
population distribution 183–188, **189**
population-based infection rate 170–172, **171**
sex-based differences 187–188
testing 169–170
variations around the world 177–182, **178**, *180*, *182*, *183*
clinical presentation 63–65, 89; *see also* cytokine storm; pathophysiology; symptoms of infection
ARDS 73
clinical epidemiology 176–177
clinical trajectory and disease spectrum 65–69, **66**
comorbidities in severe illness 71–72, *73*, 89
histopathology 83
regional variations 177–178, **178**, *180*, *182*, *183*, 245
sepsis and septic shock 68–69, 73–74
smartphone application (app) technology 70–71
clinical spectrum of disease 37, 64, 65–71, **66**, 243, 456
cell kinetics/dynamics *456*
clinical epidemiology 168–169, 173–177, 194
diagnostics 99–100, 101
factors affecting 264–272
viraemia 340
clinical trials 272–280
antiviral agents 413–418, **414**, **417**
current 149
early 138, **139–140**, 140, **150–158**
endocrinopathies 356
immunomodulators 426–434
nutritional therapies 436–437
ongoing 148–149
repurposed drugs showing antiviral activity 418–426
CNS *see* central nervous system disorders
coagulopathies 336–337, 347–348; *see also* thrombosis
clinical autopsy findings 341–343, *342*
clinical features 343–344
CNS infection 313
comparative guidelines **346**
diagnostics 116, 344–345
disseminated intravascular coagulation **339**, 339–340
lymphopenia 263
management 345–347, **346**
neutrophilic extracellular traps 340–341
pathogenesis 337–341, *338*
pharmacotherapy 439–440
role of cytokines 341
severity of disease 340
colchicine, antiviral activity 426
COLCORONA trial, colchicine 426
College of American Pathologists (CAP), autopsy protocols 382–383
colloidal gold immunochromatographic assay (GICA) **110**, 121–122, 129
colonic organoids, therapeutics 506
comorbidities 166
clinical epidemiology 175–176, 192
clinical presentation 65, 66, 71–72, *73*, 89
factors affecting severity of disease 268–270
risk factors for infection 500
complement pathways, endothelial 338
computational modelling, infection kinetics/dynamics 492

computed tomography (CT)
scanning
adrenal glands 356
diagnostics 101, **110**, 112–115, 125–126
neurological infections *325*, *326*, *327*, *328*
concentration-depth ratio (PC), pharmaceutical distributions *509*
conspiracy theories ix
contact tracing 18, 55–56, 101
contaminated surfaces 9, 64, 245
control measures xiii, 7, 54–56
convalescent plasma therapy *137*, 146, 159, **441**, 442
clinical trials 278, **279**, 431–434
COPD (chronic obstructive pulmonary disease) 269–270
'coronavirus disease 2019' 243
coronaviruses 1–2; *see also* classification; Middle East respiratory syndrome; SARS-CoV-2; severe acute respiratory syndrome
animal origin *12*
epidemiology 15–19, 68
human 206
zoonotic transmission 12, 165
corticosteroids 147–148, 274–277, 356, 427–428; *see also* dexamethasone; hydrocortisone
cough 70, 101, 102, *103*
COVACTA trial, tocilizumab 429
COVID-19 *see* SARS-CoV-2/ COVID-19
COVID-19 Host Genetics Initiative 118
COVID A to Z trial, nutritional therapies 438–439
COVID Symptom Study 70–71
COVID-19-associated coagulopathy (CAC) 337, **339**; *see also* coagulopathies
COVID-19-induced ARDS (CARDS) 138; *see also* acute respiratory distress syndrome
COVID-19 Treatment Guidelines Panel 138
cranial neuropathies 322
crazy paving pattern, imaging 113
C-reactive protein (CRP), diagnostics 112, 116
Crisanti, Andrea 11
CRISPR (clustered regularly interspaced short palindromic repeats), diagnostics 128
critical illness 174, 177, 186; *see also* clinical spectrum of disease
cross-reactivity, demyelination syndromes 315
Curcuma longa (turmeric) 144
C-terminal domain (CTD), spike protein 246, 247
C-X-C motif chemokine receptor 6 (CXCR6) 117
C-X-C motif chemokine ligand 10 (CXCL10) 381, 382
cytokine biomarkers, cellular assays 504
cytokine release syndrome-like (CRSL) toxicity 263
cytokine storm xii–xiii, 8, 203, 262–264, *264*
adaptive immune response 258
biomarkers 176–177
clinical trajectory and disease spectrum 67
clinical trials 149, 275–277
CNS infection 313, 314
comorbidities 270
diagnostics 112
endocrinopathies 357
Indian traditional medicine 144
infection kinetics/dynamics 485, 490–492, *491*, 495
pathogenesis of virus 381–382
pathophysiology of virus 24, 75, *77*
pharmacotherapy 419, 431
progression of illness 248

role in coagulopathies 341
symptoms/manifestations of disease *84*
cytotoxic T-cells 76, 117, 186, 255–256, 259–262, 270, 341, 382, 397

D-dimer levels 85, 88
coagulopathies 343, 344–345, 346, 347
diagnostics **110**, 112, 116
infection kinetics/dynamics **473**, **495**, 497, 498
neurological manifestations 313–314, 317
severity of disease 88, 225
therapeutics 421, 439
DC-SIGN receptor, viral attachment and entry 224, **225**
De Quervain thyroiditis 358, 359
death rates *see* mortality/death rates
delta-coronaviruses *14*, 15, 204
demyelination syndromes 314–315, 320–321
dendritic cells, innate immune system 255–256
developed world xv
dexamethasone 25, 26, 147–148, 159, 442
clinical trials 274–275, 427–428
diabetes 269, 357, 500
diagnostics 99–101, 123, 130–131
advantages of different approaches 124–130
chemiluminescence 121, 129
coagulopathies 344–345
colloidal gold immunochromatographics 121–122, 129
common and rare symptoms *103*
CRISPR-based 128
differential diagnosis 103, **104**, 105, 106–107, 112
enzyme-linked immunosorbent assay 120–121, 128–129
false negatives/positives 123–124, 130
genetic susceptibility 117–119
guidelines/evaluation criteria 103–107
history-taking 102, 124–125
imaging 112–115, 125–126
laboratory tests 112, 115–117, 120, 127, 128–130
lateral flow tests 122, 129
molecular 111–112
neutralization assay 122–123, 130
priorities for testing 103
progression of illness over 10 day period *102*
properties of different methods **110**
rapid antigen test 122
recent tools 109, 111
RT-LAMP 127–128
RT-PCR 119–120, 127
sample collection 100, 107–109
symptom-based 99–103, *103*, 125
transcription-mediated amplification 120, 128
transmission and spread of virus **106**
diamagnetism 507–510, *509*
Diamond Princess cruise ship outbreak 167, 170, **171**, 174, 181, 185, 190, *192*
diarrhoea xii, *103*, 372
dietary factors *see* nutrition
differential diagnoses 103, **104**, 105, 106–107, 112
digestive system *see* gastrointestinal system
dipeptidyl peptidase 4 (DPP4) **213**, 221–222
disease progression *102*, 248–249
disinfection of surfaces xiii, 52
disseminated intravascular coagulation (DIC) 142, **339**, 339–340

docking studies 503
domestic animals, coronavirus infections 217
Donne, John ix
double-membrane vesicles (DMVs) 22, 42, **209**, 212, 400
doxycycline, ineffectiveness **441**
drug repurposing xvi, 507; *see also* therapeutics
dysglycaemia 357
dyspnea 66, 68, 69, 75, 268, 311, 314, **472**
    diagnostics 100, 101, 102, 103
    infection kinetics/dynamics 484, 507
    regional variations 177, 179, 181, 188

E protein *see* envelope protein
East Indian globe thistle (*Sphaeranthus indicus*) 144
Ebola viii, 140, 412–413
eculizumab 142, **155**
education of the public 52
elderly patients *see* age factors
electron microscopy, autopsy specimens *342*, 391, 400
emetine hydrochloride, antiviral activity 425
EMMPRIN (extracellular matrix metalloproteinase inducer) **213**, 222–223, **225**, 227, 246, 468, 469
EMPACTA trial, tocilizumab 429
EMRA (terminally differentiated effector memory cells) 260, 261
encephalitis 319–320
encephalopathy 320
endocrine system/endocrinopathies 353–354, 361
    adrenal glands 355–356
    clinical features and possible mechanisms **360–361**
    gonads 359–360
    hypothalamic-pituitary axis 354–355
    infection kinetics/dynamics 492–493
    multi-level interactions with virus *354*
    pancreas 356–357
    thyroid gland 355, 357–359
endoplasmic reticulum–Golgi intermediate compartment (ERGIC) 42–43, 80
endothelium, vascular 313, 338
entry of virus to host cell *see* infection process
envelope (E) protein 4, *207*, **211**
    genomics 20
    infection process 43
    pathophysiology 22
    structure 38, 40, *41*
    vaccine candidates 212
    virology 244, 245
enzyme-linked immune absorbent spot (ELISpot) assays 504–505
enzyme-linked immunosorbent assay (ELISA) **110**, 120–121, 128–129
epidemiological perspectives 165–166, 168, 193–194; *see also* clinical epidemiology; prevalence of infection
    China's response to COVID-19 53–54, 55–56
    coronaviruses 15–19, 68
    definitions and terminology 166, 167
    genetic 101
    pregnancy 370
epithelial cells, therapeutics 505
etiology of disease 11–13, 37, 194, 471
eucalyptus (*Eucalyptus globulus*) 144
Europe, clinical epidemiology 181–182, *183*
evolutionary models 12, 13
exposure-tracking, China 55–56; *see also* contact tracing
extracellular matrix metalloproteinase inducer (EMMPRIN) **213**, 222–223, **225**, 227, 246, 468, 469
extracellular vesicles, infection kinetics/dynamics 493

extracorporeal membrane oxygenation 143
eyes, barriers to infection 465

face masks xiii, 8, 52, 54–55, 64, 58
false negatives/positives
 diagnostics 123–124, 130
 viral nucleic acid testing 170
favipiravir 159, 440
 clinical trials **153**, **154**, **156**, 416–417, **417**
FDG PET (2-fluoro-2-deoxy-D-glucose positron emission tomography) 115, 126
 fear of infection 50–51, 52, 54, 55, 58, 59
females, reproductive system health 369–370
fetal infection, neurodevelopmental outcomes 329
fever 70, *84*, *103*
 diagnostics 100, 101, 102
 progression of illness 249
fingolimod, clinical trials 149, **150**
fit-for-purpose cellular assays 504–505
5G (fifth-generation wireless transmissions) ix
Flu SC2 Multiplex Assay 100
formalin-fixed paraffin-embedded (FFPE) cell pellets, autopsy specimens 399–400, 401–402
furin cleavage site
 cellular assays 504
 infection process 82, **213**, 215, 221, 227, 247
furin enzyme 23, 39
fusion peptide (FP) 247

gamma coronaviruses 12, 13, 14, *14*, 15, 204
garlic (*Allium sativum*) 144, **145**
gastrointestinal system
 autopsy examination 397
 barriers to infection 464–465
 diagnostics 101
 infection kinetics/dynamics 486
 symptoms/manifestations of disease 86
gender *see* sex-based differences
gene ontology (GO) enrichment analysis 503
genetic epidemiology 101
genetic susceptibility to COVID-19
 diagnostics 117–119
 factors affecting severity of disease 264–266
 infection kinetics/dynamics 496
genomic sequencing 4, 56, 65, 168, 186, 203
genomics 4, 12, 13, 20–22, 37–38, *80*, 203–204, 226–227
 accessory proteins 213–214
 classification of coronaviruses 204, **205**
 genomic structure 38–41, *39*, *80*, *207*, 207–208
 host cellular factors **213**, 217–226, **225**
 human coronaviruses and diseases 206
 immunological response 244
 mutations 214–215
 proteins 38, **40**, 208, **209–211**, 212–213
 translation and replication 215–217, *216*
 virion release/assembly 217
Germany 11, **191**
GICA *see* colloidal gold immunochromatographic assay
GIS-based sequencing platform xv
global initiatives ix, xvi, 52, 58
Global Initiative on Sharing All Influenza Data (GISAID) 401
globalization, role in spread of COVID-19 137
glucocorticoid therapy 356
glucose-regulated protein 78 (GRP78) 223

Glycyrrhiza glabra (liquorice) 144, **145**
gonads, endocrinopathies 359–360, **361**
granulocyte colony stimulating factor (G-CSF) 177, 263
Graves' disease 358
Great Influenza (Spanish flu) ix, 412
ground-glass opacities, imaging 4, 23, 73, 75, 83, 391, 459
    clinical epidemiology 184, 185
    clinical presentation 66
    diagnostics 101, **110**, 113, 114
    Guillain-Barré syndrome 323
    during pregnancy 184, 373
growth hormones, endocrinopathies 355
growth-regulated oncogene-α (GRO-α) 177
Guangdong Province, China xi, 4
guidelines
    autopsy 382–386
    coagulopathies **346**
    evaluation criteria based on 103–107
    pharmacotherapy 414
    sample collection 100, 107–108
    treatment options 138
Guillain-Barré syndrome 322–323, 326, *327*, 329
gut microbiota, factors affecting disease severity 270–271

H1N1 influenza (swine flu) 36, 413
hand washing xiii, 18, 52
hazard considerations, autopsy 385
headache *103*
health education 52
health and safety guidelines, autopsy 385
health service capacity, China 52, 56
healthcare workers, clinical epidemiology 174, 188
healthcare workers, training 52
heart autopsy 392, 394; *see also* cardiovascular system
hemaglutinin esterase 4, *207*
hematological manifestations of disease 88
hematopoietic system, infection kinetics/dynamics 488
hemorrhagic stroke, imaging/radiology 324–325, *326*
herbal medicines *137*, 144, **145**, 159, 271
histopathology 83, 389, *396*, 398, 439, **472–482**, 495
history-taking, diagnostics 102, 124–125
HIV, zoonotic transmission viii
host factors affecting severity of disease 264–272; *see also* cell-based factors influencing infection
Huanan seafood market, China 2, 4, 9, 242
human airway organ-on-a-chip systems 506–507
human pluripotent stem cells (hPSCs) 506
hyaluronanic acid 77
hydrocortisone 428, 442
hydroxychloroquine 141, 159, 442
    adverse effects 273–274
    clinical trials 148, **153**, **155**, **158**, 272–273
    immunological response 272–274
    ineffectiveness **441**
hyperactive platelets 142
hypercytokinemia 89, 263; *see also* cytokine storm
hypertension 85, 268–269, 500
hypothalamic-pituitary axis (HPA) 354–355, **360**
hypothyroidism 355, 357
hypoxemia, infection kinetics/dynamics 492

hypoxia
brain damage 314
early pregnancy 370
imaging/radiology 328, *328*
infection kinetics/dynamics 484–485

ICH (intracerebral haemorrhage) 318
IFN *see* interferon
IGF-1 (insulin-like growth factor-1) 355
IL-6 inhibitors 138, 159, 428–430, 442; *see also* interleukins
imaging 112–115, 125–126
neurological 324–328, *325*, *326*, *327*
immunoglobulins (IGs) 138, 371
immunohistochemistry 358, 399, **472**, **481**, 486, **495**, **496**
immunological perspectives xvi, 24, 26, 242–243, 280–281; *see also* cytokine storm
adaptive immune response 257–262
clinical epidemiology 176–177
and disease severity 89
evasion by virus 250–251
factors affecting severity of disease 264–272
infection kinetics/dynamics 494–495
innate immune system 249–257, *250*
multiple organ dysfunction 456
pathogenesis 246–248
in pregnancy 370, 372
progression of illness 248–249
risk factors for infection 500
sex-based differences 187
therapeutics, clinical trials 272–280
thrombogenicity 313
virology 244–246
immunotherapies *137*, 141, 146, 426–434, **441**, 442; *see also* convalescent plasma therapy
clinical trials 148–149
infection kinetics/dynamics 492
in silico methods
drug repurposing xvi; *see also volume 2*
mathematical-based methodology 457–459, 462
therapeutics based on cellular mechanisms 503–504
in situ hybridization, autopsy specimens 399–400
in vitro diagnostic (IVD) devices 111
in vitro studies
3D models *460*, 460–461
cell and tissue-based research methodology 457–461
incubation period 9–10, 243, 248, 382
India
mortality rates **191**, 381
traditional medicine *137*, 144, **145**
Indian Council of Medical Research (ICMR) 105
individual freedom 58
infants, clinical epidemiology 184–186; *see also* children
infection, definition 168
infection kinetics/dynamics *456*, 456–462, 510–511
3D *in vitro* models 460, *460*
barriers to infection 462–468, *463*
cell signaling pathways **501**
cellular level 493–494, 497
clinical spectrum of disease *456*
infection process 457, *457*
organ-level 469–471, *470*, **472–482**, *483*, 483–493, 498
organism-level 498

infection kinetics/dynamics (*continued*)
risk factors for infection **495**, 495–498
in silico studies 457–459, 462
therapeutics based on 501–510, **502**
tissue-level *460*, 468–469, 498
in vitro studies 457–461
vulnerabilities 498, **499**, 500
infection process 20–22, *21*, 41–43, 65
ACE2 receptor 20–21, 78–82, 215, 247, 337–338, 457
attachment of virus to host cell 41–42
cellular factors involved in attachment/entry **213, 225**
genomics 215, 225–226
infection kinetics/dynamics 457, *457*
invasion mechanism 78–82, *79*, *80*, *82*, 215
pathogenesis 246–247
replication of the virus *81*
spike protein 20–22, 41–42, 65, *79*, 225–227, 246, 247, 312, 457
infection rate 167, 170–172, **171**, 194
infectivity 18, 26
clinical epidemiology 166, 194
definition 167
inflammatory signaling 494; *see also* cytokine storm
influenza ix, 412
bird flu viii, 12, 13, 15, 38, 75, 204, **205**, 226, 397
differential diagnosis 105, 106–107
Spanish flu ix, 412
swine flu 36, 413
zoonotic transmission viii
influenza-like illness (ILI), differential diagnosis 106–107
innate immune system 166, 249–257, *250*
insulin-like growth factor (IGF)-1 355
integument *see* skin
intensive care, complications requiring 89
interferon (IFN)
BCG vaccine status 271
clinical trials **152, 153**, 275–276
cytokine storm 263
immunological response 249, 251–252
pathogenesis of virus 381
role in coagulopathies 341, 342
T-cells 260
transcription factors 243
interferon-gamma inducible protein (IP10) 24
interferon-stimulated gene 15 (ISG15) 208
interleukins 24, 262–263; *see also* cytokine storm
BCG vaccine status 271
biomarkers of inflammation 177
clinical trials 276–277
diagnostics 116
pathophysiology of virus 76–77, 382
progression of illness 248
role in coagulopathies 341
interleukin (IL-6) inhibitors 138, 159, 428–430, 442
internal fusion peptide (IFP) 247
international cooperation ix, xvi, 52, 58
International Reagent Resource Multiplex Assay 100
interventions *see* therapeutics
intestinal barrier to infection 464–465; *see also* gastrointestinal system
intracellular vesicles, infection kinetics/dynamics 493
intracerebral haemorrhage (ICH) 318

intraparenchymal haemorrhage (IPH) 318
invasion of host tissues *see* infection process
invasive mechanical ventilation (IMV) 137
ischemic stroke 317–318, 324, *325*
isolation measures 11
Italy 11, **191**
IVD (*in vitro* diagnostic devices) 111
IVERCOR clinical trials 420–421
ivermectin 143, 418–420, 442
  clinical trials 420–423

JAK (Janus kinase) inhibitors 430–431, 442
JAK-STAT inhibitors, clinical trials 277
JAK-STAT signaling pathway 275, 341, 355
Justinian, plague of viii

kidney
  autopsy examination 392, 395, 400
  infection kinetics/dynamics 486–487
  manifestations of disease 86
kinase inhibitors, clinical trials 430–431
Kyoto Encyclopaedia of Genes and Genomes (KEGG) 503

L-SIGN adhesion molecule 224, **225**
  laboratory tests **110**, 112, 115–117, 127, 128–130
  coagulopathies 344–345
lactate dehydrogenase (LDH) 112, 116, 117
lateral flow tests **110**, 122, 129
leucine-rich repeats (LRRs) 214
leukocytes, CNS infection 313
leukoencephalopathy, imaging *328*
leukopenia 88, 117
life cycle of virus 41–43, *81*
light microscopy, autopsy specimens 393–399, *396*
likelihood of exposure, epidemiology 166
liquorice (*Glycyrrhiza glabra*) 144, **145**
liver, autopsy examination 392, *396*, 396–397
lockdowns 5, 18, 53, 54
  lopinavir/ritonavir combined **139**, 140, 442
  clinical trials **150, 152, 154, 155**, 418
  ineffectiveness **441**
loss of smell and taste 70, 102, **178**, 178–179, 322
low molecular weight heparin (LMWH) 439–440
lung organoid model, human pluripotent stem cells 506
lungs
  autopsy examination 391–392, *393*, 393–394
  barriers to infection 464
  biopsy/sample collection 109
  clinical presentation 83–4
  diagnostics 101
  infection kinetics/dynamics 484–485, *491*
  infection process 247
  manifestations of disease 83–84, 89, 311
  pathophysiology 25, 75–77
  progression of illness 248
  ultrasound scanning 113–114
  treatment options 142
lymphatic system, infection kinetics/dynamics 488
lymphocyte antigen 6 (Ly6) family **213**, 223–224, **225**
lymphocytes, time-lymphocyte model (TLM) 257
lymphopenia 88, 116, 257, 259–260, 263

M protein *see* membrane protein
macrolides 141; *see also* azithromycin

macrophage colony-stimulating factor 177
macrophage inflammatory proteins 24, 262–263, 382
macrophages, immunological response 252–254
magnetic chemiluminescence diagnostics (MCLIA) **110**, 121, 129
magnetic therapy *see* diamagnetism
main protease (Mpro) 38, 80, 208
malaria, zoonotic transmission viii
males, reproductive health 374–375
malnutrition *see* nutrition
mas-related-G protein-coupled receptor (MrgD) *470*, 471
mask wearing xiii, 8, 52, 54–55, 64, 58
mathematical-based methods *see* infection kinetics/dynamics
MCLIA (magnetic chemiluminescence diagnostics) **110**, 121, 129
measles viii, 412
mechanical ventilation 67
    medicinal plants *137*, 144, **145**, 159, 271
membrane (M) protein 4, *207*, **211**
    genomics 20
    infection process 43
    structure 38, 39, *39*
    virology 244
meplazumab, clinical trials 149
merimepodib 423
MERS *see* Middle East respiratory syndrome
mesenchymal stem cell (MSC) therapy *137*, 146–147, 149, 159, 505–506
metalloproteinase inhibitors, traditional medicine 144
methylprednisolone, clinical trials **151**, **153**, **156**
microbiome, factors affecting disease severity 270–271
MicroCLOTS (microvascular COVID-19 lung vessels obstructive thrombo-inflammatory syndrome) 24
microhaemorrhages, white matter 325
Middle East respiratory syndrome (MERS-CoV) viii, xi, xiii, 2, 36, 38, 242
    comparative symptoms with COVID-19 103, **104**
    epidemiology 17
    immunology 258
    infectivity and morbidity **19**
    mortality rates 13
    origins of infection xi–xii
    pregnancy 371
    reproductive system health 369
    treatment options 140, 143
middle-income and developing countries, vaccines xv
mild illness, classification 173, 175; *see also* clinical spectrum of disease
minimally invasive autopsy (MIA) 388–389
mitochondria, infection kinetics/dynamics 494
mitochondrial antiviral-signaling protein (MAVS) 251
moderate illness 140, 173, 175; *see also* clinical spectrum of disease
MODS *see* multiple organ dysfunction syndromes
molecular diagnostics **110**, 111–112
molecular docking studies 503
molecular testing, autopsy specimens 389–390, 401–402
molnupiravir (MK-4482), clinical trials 417–418
monoclonal antibodies 25, 26, 434–435, **441**, 442
monocyte chemoattractant protein 1 (MCP1) 24
morbidity 13, 18
    endocrinopathies 357
    infection during pregnancy 373

mortality/death rates xi, xiii, 13, 37, 64, 66, 243
by age *192*
case fatality rate 190–193
children 186
comorbidities 357
comparison with SARS and MERS **19**
by country **191–192**, 193
SARS-CoV and MERS 13
by WHO region xv
worldwide 6, 381
movement disorders, CNS infection 321
Mpro (main protease) 38, 80, 208
MSC *see* mesenchymal stem cell (MSC) therapy
multiple organ dysfunction syndromes (MODS) 66
immunological response 456
progression of illness 249
treatment options 142, 143
multiplicity of infection (MOI), immunology 252
multisystem inflammatory syndrome 24, 26, 81
musculocutaneous system 88, 488–489; *see also* skin
mutations ix, xv–xvi, 214–215, 245
myalgia 88, 323–324
myelitis 321, 326
myositis 323–324

N protein *see* nucleocapsid protein
N-terminal domain, spike protein 246
NAK (numb-associated kinase) inhibitors 430–431, 442
naproxen, clinical trials **157**
nasal mid-turbinate swabs, sample collection 108
nasopharyngeal/nasal wash/aspirate 108, 118
National Medical Products Administration (NMPA) 55
native Americans, plagues viii
natural killer (NK) cells, innate immune system 256–257
nausea *103*
neem (*Azadirachta indica*), traditional medicine 144
neonatal infection, neurodevelopmental outcomes 329
nervous/neurological system 87–88, *87*, 311–312, 315–316; *see also* central nervous system; peripheral nervous system
abbreviations 330
diagnostics 101
future perspectives 329
imaging/radiology 324–328, *325*
infection kinetics/dynamics 487–488, *491*
mechanisms to explain complications 313–314
outcomes 329
pathogenesis of neuronal autoimmunity 314–315
pathophysiology of neurotropism 312–313, 313–314
relative frequencies of different presentations **315**
risk factors for neurotropism 313
routes of viral neuroinvasion 312–313
therapeutics 328–329
timeline of symptom development *316*
network pharmacology, cellular mechanisms 503
neurodevelopmental outcomes 329
neuromuscular junction disorders 323
neuronal autoimmunity 314–315
neurons, CNS infection 312–313
neurotropism 312–314
neutralization assay **110**, 122–123, 130
neutrophilia 88
neutrophilic extracellular traps (NETs) 340–341, 497

neutrophils, innate immune system 254–255
neutrophil-to-lymphocyte ratio (NLR), diagnostics 112
New Zealand xv
N-glycosylated ectodomain (*N*-GE) 244–245
niclosamide 423
nitazoxanide 423
NLRP3 (nucleotide-binding oligoerization domain) 214
N-methyl-D-aspartate receptor encephalitis 319
'no one is protected until everyone is protected' xiii
nod-like receptor family pyrin domain-containing-3 (NLRP3), clinical trials 277
non-invasive oxygen therapy 137, 138
non-structural proteins (nsp 1-16) 38
  genomics 38, **40**, 208, **209–211**, 212
  infection process 42, 79–80
  RNA replication machinery 43–45, *44*
  virology 245
novel coronavirus pneumonia *3*; *see also* SARS-CoV-2/COVID-19
nsp12 *see* RNA-dependent RNA polymerase; *see also* non-structural proteins
N-terminal domain, spike protein 246
nuclear factor (NF), transcription factors 243
nucleic acids 119–120, 169–170
nucleocapsid (N) protein 4, *207*, **211**
  genomics 20
  structure 38, 40–41, *41*
  virology 244, 245
nutrition
  factors affecting severity of disease 270–271
  risk factors for infection 500
  therapeutics 435–439, **441**, 443
  vitamin status 435–438, 443

occupational differences, clinical epidemiology 188, **189**
Ocimum tenuiflorum (tulsi) 144
older adults *see* age factors
olfactory bulb, imaging/radiology 326
olfactory nerve, CNS infection 312
oligodendrocytes, CNS infection 313
omeprazole, clinical trials **156**
  online lifestyle, China's response to virus 53
  open reading frame (ORF) accessory proteins 213–214
  animal studies 458
  genomics/genetics 20, 22, 38, *39*, 208
  immunological response 244
  infection process 80
  proteins 208, **210–211**
  RT-PCR 169
  signaling cascades 251
organ systems, infection kinetics/dynamics 469–471, **472–482**, *483*, 483–484
  cardiovascular system 485–486
  digestive system 486
  endocrine system 492–493
  immune system 494–495
  lymphatic/hematopoietic system 488
  musculocutaneous systems 488–489
  nervous system 487–488, *491*
  reproductive system 489–490
  respiratory system 484–485, *491*
  risk factors for infection 498
  skeletal system 490–492, *491*
  urinary system 486–487
organism level risk factors for infection 498

organoids 357, *460*, 461, 465, 467, **472–478**
organ-on-a-chip systems *460*, 506–507
oropharyngeal swabs, sample collection 108
ovaries 369
oxygen saturation 67, 69, 72, 138
oxygen therapy 137, 143, 148

pacritinib 431
pancreas 356–357, **361**, 397
pangolins 12, 17–18, 243, 381
papain-like protease (PLpro) 38, *39*, 80, 251
pathogenesis of disease 67–68, 381–382; *see also* clinical presentation
coagulopathies 337–341, *338*
immunological response 246–248
neuronal autoimmunity 314–315
pathological examination, autopsy specimens 391–392
pathophysiology of virus 22–25, 74–78
ACE2 receptor 74–77, *78*
chemosensory dysfunction 178–179
cytokine storm 24, 75, *77*
neurotropism 312–314
renin–angiotensin system *78*
pattern recognition receptors (PRR) 249, *250*, 250–251
PCR *see* polymerase chain reaction testing
PEMFs (pulsed electromagnetic fields) 507
peripheral blood mononuclear cell (PBMC) samples 253
peripheral nervous system (PNS) disorders 312
case vignettes 322, 323
cranial neuropathies 322
demyelination 314–315, 320–321
Guillain-Barré syndrome 322–323
imaging/radiology 326
loss of smell and taste 322
myalgia and myositis 323–324
neuromuscular junction disorders 323
personal protective equipment (PPE) 7, 8, 55, 383–385
persons under investigation (PUI) for COVID -19 100
pertussis, zoonotic transmission viii
PET (positron emission tomography) 115, 126
pharmaceutical distributions, concentration-depth ratio *509*
pharmacotherapies see therapeutics
phlegm, sample collection 109
pituitary gland, endocrinopathies 358
placenta, autopsy examination 398–399, 400
placental barrier to infection 467
PLACID trial, convalescent plasma therapy 432
plagues, as inevitable consequence of civilization vii–x
The Plague (Camus) ix
plant-based medicines *137*, 144, **145**, 159, 271
plasmins, viral attachment and entry 225–226
pleural fluid, sample collection 109
plitidepsin, antiviral activity 424
PLpro (papain-like protease) 38, *39*, 80, 251
PNS *see* peripheral nervous system
polymerase chain reaction (PCR) testing
autopsy specimens 389–390, 401–402
clinical epidemiology 169, 176
diagnostics 100, 105–107, **110**, 114, 119–120, 127

polyproteins, genomics 38, *39*
population-based infection rates 170–172, **171**
population-based seroprevalence surveys 172–173
population distribution, clinical epidemiology 183–188
positron emission tomography (PET) 115, 126
post-hypoxic leukoencephalopathy, imaging *328*
PPE (personal protective equipment) 7, 8, 55, 383–385
pregnancy 368, 369; *see also* reproductive system
  clinical epidemiology 183–184
  clinical presentation of infection 372–374
  early pregnancy 370–372
  as immunosuppressive state 370, 372
  risk factors for infection 500
  vertical transmission 369, 370, 371
presymptomatic cases, definition 175
prevalence of infection xiii, 108, *137*, 142, 159
  clinical epidemiology 170, 172
  by country *5*, 6
  by WHO region *xiv*
  worldwide *5*, 10–11, *19*, 26, 37, 243, 381
prevention of infection xiii, 7, 54–56, 420–421
progression of illness *102*, 248–249
pro-inflammatory cytokines 252; *see also* cytokine storm
proning (lying face down) 67
prostaglandins 249
protease inhibitors, clinical trials 418
proteases, viral attachment and entry 225–226
proteins; *see also* envelope protein; membrane protein; nucleocapsid protein; spike protein
  mechanism of invasion into host cells 79–80, 81–82
  non-structural proteins 208, **209–211**, 212
  structural proteins 212–213, 244–245
proteome, immunological response 244–245
prothrombin time (PT) **339**, 339; *see also* coagulopathies
PRR (pattern recognition receptors) 249, *250*, 250–251
  public health policies xvi, *3*, 18, 25, 51, 52, 54–56, 64
public service sectors, clinical epidemiology 188
public transport *3*
pulmonary system *see* lungs
pulsed electromagnetic field (PEMFs) 507

quarantine, China 53, 54, 56

radiology 112–115, 125–126
  neurological 324–328, *325*, *326*, *327*
randomized controlled trials *see* clinical trials
range of symptoms *see* clinical spectrum of disease
rapid antigen test **110**, 122, 130
Rapid Evidence Appraisal for COVID-19 Therapies (REACT) trial 356, 427–428
RAS *see* renin–angiotensin system
RdRP *see* RNA-dependent RNA polymerase
real-time PCR *see* polymerase chain reaction (PCR) testing
receptor-binding domain (RBD) 17–18
  genomics 39
  immunological response 245, 258

infection process 21, 41–42, *79*, 215
pathophysiology 75
recombinant human interferon **152**, **153**
RECOVERY trial 147, 274, 356, 427
regional variations in clinical presentation 177–178, 178, *180*, *182*, *183*, 245
remdesivir 25, 137, 138, 140, 159, 440
clinical trials 148, **150–155**, 413–416
renal system *see* kidney; urinary system
renin–angiotensin system (RAS)
CNS infection 318
infection kinetics/dynamics 468–469, *470*, 471
pathophysiology of virus *78*
viral attachment and entry 218–219, *219*
replicase–transcriptase complex (RTC) 38, 42, 43, 244, 247
replication, viral 42, *81*
genomics 215–217, 216, *216*
pathogenesis 247–248
reproductive number (R_0) 9
reproductive system 368–369, 375
females 369–370
infection kinetics/dynamics 489–490
males 374–375
pregnancy 370–374
respiratory barrier to infection 464; *see also* lungs
respiratory droplets 9, 25, 64
respiratory failure 101
retina, barriers to infection 465–466
retinoic acid-inducible gene I (RIG-I)-like receptors 249
reverse transcriptase polymerase chain reaction *see* polymerase chain reaction (PCR) testing
reversed halo sign, imaging 113
ribavirin 143
rintatolimod, antiviral activity 424
risk factors for infection **495**, 495–498; *see also* susceptibility to infection
ritonavir *see* lopinavir/ritonavir combined
RNA replication machinery 43–45, *44*, 216, *216*
RNA-dependent RNA polymerase (RdRP/nsp12) **40**, 42, 43, 80, 169, 244, 401, 501
genomics 208, 216
therapeutic options **139**, 140, 143, 413, 416
R_0 (reproductive number) 9
rodents 12, 206
Royal College of Pathologists, autopsy guidelines 382
RTC (replicase–transcriptase complex) 38, 42, 43, 244, 247
RT-LAMP (reverse transcription loop-mediated isothermal amplification) 127–128
RT-PCR *see* polymerase chain reaction (PCR) testing

S protein *see* spike protein
S2 subunit, spike protein 23, 39, 79, *82*, 220, 221, 247
safety/hazard considerations, autopsy 385
saliva, sample collection 109
sample collection 100, 107–109
sarilumab **151**, 428–429
SARS-CoV *see* severe acute respiratory syndrome coronavirus
SARS-CoV-2/COVID-19 (general information) viii, 1, 40–41, 337; *see also subject-specific entries*
cases *see* prevalence of infection
chronology of the outbreak 2–6, *3*
classification 13–15, *14*, 37–38, 70, 244

SARS-CoV-2/COVID-19 (general information) (*continued*)
comparison with SARS and MERS **19**, 103, **104**
conspiracy theories ix
etiology 11–13, 37, 194, 471
initial outbreak xiii, 1, 8–10, 25, 37, 51, 53, 57, 63, 64, 68, 242–243, 336, 369, 380–381
life cycle 41–43, *81*
progression of illness *102*, 248–249
severity *see* clinical spectrum of disease
timeline of symptom development *316*
structure *20*, *21*, 38–41, 207–208, 244
WHO–China Joint Mission 6–8
saturated fatty acids (SFA), dietary 270
seizures 318–319, 321, 328
sepsis and septic shock 68–69, 73–74, 249
sequencing-based testing 169
serological assays 115–117, 120
serum antibodies 172–173
serum troponin levels 85
severe acute respiratory infections (SARIs), differential diagnosis 105, 106
severe acute respiratory syndrome coronavirus (SARS-CoV) viii, xi, xiii, 2, 36, 38, 206, 242
China 56–58, 59
comparative symptoms with COVID-19 **19**, 103, **104**
endocrinopathies 355
epidemiological perspectives 53
epidemiology 16–17
immunology 258
infectivity and morbidity **19**
lessons learned 51–52
lethality 203
mortality rates 13
pregnancy 371
reproductive system health 369
transmission xii
severe illness; *see also* clinical spectrum of disease
classification 173–174
clinical epidemiology 175–176
clinical trials 140
T-cell response 261–262
sex-based differences
ACE2 receptor 265
clinical epidemiology 187–188
clinical trajectory and disease spectrum 68
endocrinopathies 360
factors affecting severity of disease 266–267
reproductive system health 369–370, 374–375
Shi Zhengli 2
shivering *103*
sialic acid **213**, 224–225, **225**
sick euthyroid syndrome 355
signal transducer and activator of transcription (STAT) signaling pathway 355
signaling cascades, innate immune system 251
sildenafil citrate, clinical trials **153**
SIMPLE trial 140, 148
simulations, infection kinetics/dynamics 503–504
skeletal system, infection kinetics/dynamics 490–492, *491*
skin 88
autopsy examination 397–398
barriers to infection 466
infection kinetics/dynamics 488–489
smallpox viii, 412
smartphone application (app) technology 70–71
smell, loss of sense of 70, 102, **178**, 178–179, 322
social distancing xiii, 18, 53, 64

SOLIDARITY trial 141, **414**, 415
source-tracking, China 55–56;
see also contact tracing
South Africa xv, **191**
South Korea 180–181, **191**
Spain 173, 188, **191**
Spanish flu ix, 412
Sphaeranthus indicus (East Indian globe thistle) 144
spheroids *460*, 461, 486, 487
spike (S) protein 4, *207*, **210**, 337
genomics 20
immunological response 249, 258, 260, 265
infection process 20–22, 41–42, 65, *79*, 225–227, 246, 247, 312, 457
pathophysiology of virus 22
response to vaccines 208
S2 subunit 23, 39, 79, *82*, 220, 221, 247
structure 38, *39*, 39
virology 244–245
spinal cord, myelitis 321
spleen, autopsy examination 397
sputum, sample collection 109
stannous protoporphyrin, antiviral activity 425
stem cell therapy *137*, 146–147, 149, 159, 505–506
Strobilanthes cusia, Indian traditional medicine 144, **145**
strokes 24, 25, 88, 311
hemorrhagic 324–325, *326*
ischemic 317–318, 324, 325
structural proteins 212–213, 244–245; *see also* envelope; membrane; nucleocapsid *and* spike protein
structure of virus *20*, *21*, 38–41, 207–208, 244
subacute thyroiditis 358, 359
super-spreaders 246
surface contamination 9, 64, 245
surveillance xvi
China 52, 55–56, 58
diagnostics 101–102
survival outside the body 64
susceptibility to infection 166; *see also* risk-factors
host factors associated 217
RAS system imbalance 220
sex-based differences 187–188
swabs, autopsy 390–391
Sweden xv, **191**
Switzerland 173, **191**
symptom-based diagnostics 99–103, *103*, 125
symptoms of infection xii, 4, 10, 23–24, 64–69, *84*, 89, 243, 382; *see also* clinical presentation
cardiovascular manifestations 85–86
chemosensory dysfunction **178**, 178–179
clinical epidemiology 168
common and rare symptoms *103*
common symptoms 179, *183*
COVID Symptom Study 70–71
diagnostics 101–102
gastrointestinal manifestations 86
hematological manifestations 88
histopathology 83
musculocutaneous manifestations 88
neurological manifestations *87*, 87–88
during pregnancy 372–374
pulmonary manifestations 83–84, 89
regional variations 177–182, **178**, *180*, *182*, *183*
renal manifestations 86
timeline of development *316*
variability/range *see* clinical spectrum of disease
systems biology 456, 458

T-cells, adaptive immune response 257, 259–262
early pregnancy 370
journey from protective to destructive 260–261
lymphopenia 257, 259–260
severity of disease 261–262
taste, loss of sense of 70, 102, **178**, 178–179, 322
Technical Recommendations from the Report of the WHO–China Joint Mission 6–8
temperature, environmental 13
tensor decomposition-based unsupervised feature extraction method 504
testes
barriers to infection 467–468
endocrinopathies 359, **361**
testing 18, 100, 111; *see also* diagnostics *and specific tests*
China 55–56
clinical epidemiology 169–170
priorities for 103
testosterone levels 359–360
Thailand 4, 9, 119, **153**, 180
thalidomide 149, **153**
therapeutic pulsed electromagnetic field (PEMFs) 507
therapeutics xvi, 10–11, 25, 412–413, 440–443, **441**; *see also* clinical trials; immunotherapies; *and drugs by name*
ACE inhibitors 85–86, 268, 269
adverse effects 141, 273–274, 416, 424, 458
antithrombotic agents 138, 141–142, 439–440, **441**, 443
antiviral agents *137*, 148, 159, 413–418, **414**, 440–441
based on cellular mechanisms **501**, 501–510, **502**, *509*
coagulopathies 345–347, **346**
common regimens 138–143
guidelines 138
initial treatment options 136–138, *137*, **139–140**, 159
medicinal plants *137*, 144, **145**, 159
mesenchymal stem cells *137*, 146–147, 149, 159, 505–506
monoclonal antibodies 434–435, **441**, 442
neurological manifestations 328–329
nutritional therapies 435–439, **441**, 443
other repurposed drugs 418–426, **441**, 442–443
oxygen therapy 137, 143, 148
pharmacokinetic–pharmacodynamic models 503
3D *in vitro* models *460*, 460–461
3D tissue/organoids 506
thromboembolism 84, 343–344
thrombogenicity, neurological complications 313–314, 317
thrombosis 24, 25; *see also* coagulopathies
clinical features 344
diagnostics 116
therapeutics 138, 141–142, 328
thyroid gland 355, 357–359, **361**
thyrotoxicosis 358
time-lymphocyte model (TLM) 257
tissue-based infection kinetics/dynamics *460*, 468–469, 498
TMPRSS2 protease *see* transmembrane protease serine
TNF *see* tumor necrosis factor
tocilizumab **140**, 142, 429, 442, **156**
toll-like receptors (TLRs) 249
trabedersen, antiviral activity 425
tracheal aspirate, sample collection 109
Trachyspermum ammi (ajwain) 144
trained immunity 271
training, healthcare workers 52
transcriptase polymerase chain reaction assays *see* polymerase chain reaction (PCR) testing

transcription, infection process 42, 247–248
transcription factors 243
transcription-mediated amplification, diagnostics 120, 128
translation, viral
genomics 216, *216*
pathogenesis 247–248
transmembrane protease serine 2 (TMPRSS2) 39
CNS infection 312
endocrinopathies 359
factors affecting severity of disease 265–266
genomics 40
infection kinetics/dynamics 492, **495**
infection process 42, **213**, 220–221, **225**, 227
mechanism of invasion into host cells 82, 215
pathogenesis of virus 381
pathophysiology of virus 23
reproductive system health 374
transmissibility, definition 167
transmission of SARS-CoV xii
transmission of SARS-CoV-2 xii, xiii, 4, 5, 25, 64, 381
definition 167
diagnostics **106**
immunological response 245–246
initial outbreak 8–10
Report of the WHO–China Joint Mission 6
speed of 166, 203–204, 208
vertical (pregnant women to fetus) 369, 370, 371
worldwide spread 10–11, 137
zoonotic viii, xii, 2, 12, 17–18, 26, 165, *206*, 243
trans-synaptic transfer, CNS infection 312
travel restrictions 18, 53
treatments *see* therapeutics
tropism, viral 312–313
troponin levels 85
trust, public 53, 54
tuberculosis, zoonotic transmission viii
tulsi (*Ocimum tenuiflorum*) 144
tumor necrosis factor (TNF) alpha 24
cytokine storm 262–263
pathogenesis of virus 382
progression of illness 248
T-cells 260
tumor necrosis factor-related apoptosis-inducing ligand (TRAIL) 177
Turkey xv, **191**
turmeric (*Curcuma longa*) 144

ultrasound scanning 113–114
umifenovir, clinical trials 418
United Kingdom
clinical trials 427
mortality rates **191**
vaccines xv
United States 11
clinical epidemiology 181, *182*
control measures xiii
mortality rates **191**
vaccines xv
urinary system 486–487; *see also* kidney

vaccines xi, xiii, xv, xvi, 137, 208, 413
candidate proteins 212
neurological effects 329
valosin-contacting protein **213**
variability of symptoms *see* clinical spectrum of disease
vascular endothelium 313, 338
vascular manifestations of infection 316–318
vasculitis 317
venous thromboembolism (VTE) 343–344
ventilation, invasive mechanical (IMV) 137
Vero-E6 cell lines 505

vertical transmission (women to foetus) 369, 370, 371
VIR-2703, antiviral activity 426
viraemia, and severity of disease 340
viral load profile 65
viral tropism 312–314
viral variants xv–xvi, 13; *see also* mutations
Virchow's triad, symptoms 337–339
virion release 217, 248; *see also* infection process
vitamin C 437–438, 443
vitamin D 435–437, 443
Vitex trifolia 144, **145**
vomiting *103*
von Willebrand factor (vWF) 338
VTE (venous thromboembolism) 343–344
vulnerabilities, infection kinetics/dynamics 498, **499**, 500; *see also* susceptibility to infection

Western diet 270; *see also* nutrition
wet markets 2, 4, 9, 26, 242
white blood cell count (WBC) 112
white matter, microhaemorrhages 325
whooping cough, zoonotic transmission viii
World Health Organization (WHO) 4
  Dashboard Situation Board 10
  global influenza surveillance standards 105
  mortality rates *xv*
  pharmacotherapy guidelines 414
  prevalence by region *xiv*
worldwide spread 10–11, 26
Wuhan Institute of Virology 2
Wuhan Municipal Health Commission 2, *3*
Wuhan, regional variations in clinical presentation 177

Yersinia pestis (Black Death) viii

Zhang, Yong-Zhen 2
Zika 54, 223, 418, 424
zinc 438–439, 443
zoonotic transmission viii, xii, 2, 12, 17–18, 26, 165, *206*, 243